THE CAT PRIMARY VISUAL CORTEX

THE CAT PRIMARY VISUAL CORTEX

Bertram R. Payne and Alan Peters

Boston University School of Medicine

ACADEMIC PRESS

San Diego London Boston New York Sydney Tokyo Toronto

This book is printed on acid-free paper.

Academic Press
A division of Harcourt, Inc.
525 B Street, Suite 1900, San Diego, California 92101-4495, USA
http://www.academicpress.com

Academic Press
Harcourt Place, 32 Jamestown Road, London NW1 7BY, UK
http://www.academicpress.com

Library of Congress Card Number: 2001093294

International Standard Book Number: 0-12-552104-9

PRINTED IN THE UNITED STATES OF AMERICA
Transferred to Digital Printing, 2012

CONTENTS

3

2-DEOXYGLUCOSE ARCHITECTURE OF CAT PRIMARY VISUAL CORTEX 167

SIEGRID LÖWEL

4

FUNCTIONAL MAPPING IN THE CAT PRIMARY VISUAL CORTEX USING HIGH MAGNETIC FIELDS 195

DAE-SHIK KIM, TIMOTHY Q. DUONG, KAMIL UGURBIL, AND SEONG-GI KIM

5

RELATIONSHIPS OF LGN AFFERENTS AND CORTICAL EFFERENTS TO CYTOCHROME OXIDASE BLOBS 221

JOANNE A. MATSUBARA AND JAMIE D. BOYD

6

INFLUENCE OF TOPOGRAPHY AND OCULAR DOMINANCE ON THE FUNCTIONAL ORGANIZATION OF CALLOSAL CONNECTIONS IN CAT STRIATE CORTEX 259

JAIME F. OLAVARRIA

7

ESSENTIAL AND SUSTAINING LGN INPUTS TO CAT PRIMARY VISUAL CORTEX 295

THEODORE G. WEYAND

8

INTEGRATION OF THALAMIC INPUTS TO CAT PRIMARY VISUAL CORTEX 319

R. CLAY REID, JOSE-MANUEL ALONSO, AND W. MARTIN USREY

9

THE EMERGENCE OF DIRECTION SELECTIVITY IN CAT PRIMARY VISUAL CORTEX 343

ALLEN L. HUMPHREY AND ALAN B. SAUL

10

LONG-RANGE INTRINSIC CONNECTIONS IN CAT PRIMARY VISUAL CORTEX 387

KERSTIN E. SCHMIDT AND SIEGRID LÖWEL

11

PHARMACOLOGICAL STUDIES ON RECEPTIVE FIELD ARCHITECTURE 427

ULF T. EYSEL

12

ORIENTATION SELECTIVITY AND ITS MODULATION BY LOCAL AND LONG-RANGE CONNECTIONS IN VISUAL CORTEX 471

DAVID SOMERS, VALENTIN DRAGOI, AND MRIGANKA SUR

13

RESPONSE SYNCHRONIZATION, GAMMA OSCILLATIONS, AND PERCEPTUAL BINDING IN CAT PRIMARY VISUAL CORTEX 521

WOLF SINGER

16

BEHAVIORAL ANALYSES OF THE CONTRIBUTIONS OF CAT PRIMARY VISUAL CORTEX TO VISION 655

DONALD E. MITCHELL

Preface

We hope that this volume will appeal to those with a specific interest in cat primary visual cortex, as well as neuroscientists with a general interest in the structure and operations of cerebral cortical systems. It is estimated by some that more is known about the cat primary visual cortex than about any other cerebral area, or even any other structure in the brain of any other species. In recent years, many new investigative tools and strategies have been used to probe the construction and workings of this cortex. Together they have enabled us to gather information of unparalleled quality. Much of this information is contained in this volume.

While the cat visual cortex is recognized to be a premier system for studying the mysteries of vision, visual processing, cerebral circuitry, and cerebral processing, the question has to be asked " What is cat primary visual cortex?" As we reason in Chapter 1, we believe that both areas 17 and 18 should be included under the single umbrella term of "Cat Primary Visual Cortex." The appropriateness of our view becomes apparent as we review the historical background and evolution in thinking that emerges from a large number of studies on neural architecture, connections, anatomical and functional aspects of circuitry, functional maps, receptive field properties, and contributions that areas 17 and 18 make to visual processing in the cat. Indeed, from numerous aspects areas 17 and 18 are very similar: basically, both areas process primary signals from LGN in parallel and both areas contribute to overall visual processing, but in distinct, yet overlapping, ways.

Subsequent chapters describe the representational architecture of areas 17 and 18 as revealed by optical imaging (Hübener and Bonhoeffer), 2-deoxyglucose (Löwel), and functional MRI techniques (Kim and colleagues). These chapters provide an overall view of the functional columnar architecture of cat primary visual cortex at the level of neural systems. This level of analysis provides a perspective into which anatomical and electrophysiological analyses of the connections and neuronal operations can be readily comprehended. The anatomical studies summarize the organization of afferent and efferent connections within one hemisphere (Matsubara and Boyd) and between the hemispheres (Olavarria), and they describe the structural bases for comprehending the fundamental visual cortex operations that have been revealed by elegant electrophysiological studies (Weyand; Reid, Alonso, and Usrey; and Humphrey and Saul). Next, the topic of long-range intrinsic circuitry is reviewed from an anatomical perspective (Schmidt and Löwel) and from the perspective of emergence of orientation and direction preferences of neurons as revealed by techniques to reversibly deactivate specific cerebral circuits (Eysel). These reviews are complemented nicely by the chapter in which Somers, Dragoi, and Sur discuss models to conceptualize the emergence of orientation selectivity and its modulation by ancillary circuits, while the chapter by Singer places the neuronal operations of cat primary visual cortex in the context of overall brain function, gamma oscillations, and perceptual binding.

The chapter by Payne and Rushmore provides a new view on the special relationship between β retinal ganglion cells and the primary visual cortex, and the volume is rounded out with the chapters of Scannell and Young and of Mitchell. The thought provoking chapter by Scannell and Young analyzes the brain connectivity in which the primary visual cortex is embedded, while Mitchell reviews the contributions areas 17 and 18 make to vision and visually guided behavior. From Mitchell's description we learn the importance of primary visual cortex for high acuity vision, hyperacuity, and stereoscopic depth perception. *In toto* these studies show that comprehensive analyses of enormous breadth and sophistication are required to probe the structure and function of brain regions. But even using modern techniques we have, to date, only made a small dent in our comprehension of the nature of neural signals, transformations, and syntheses carried out by neurons.

What emerges repeatedly in the chapters is the perspicacity and overarching role played by David Hubel and Torsten Wiesel in opening up meaningful investigations of the cat primary visual cortex. Even after 40 years their descriptions of receptive field properties, connections, and formulations on receptive field properties and their emergence still have a powerful hold over the way that we think about visual cortex. Interestingly, in the 1980s many felt that the neural operations of cat primary visual cortex were well understood, and little additional work needed to be done. However, all of the chapters in this volume argue otherwise. They show that while many advances have been made over the previous two decades, there is still so much more to learn. Particularly important areas that need to be investigated include the topics of receptive field dynamics, influence of feedback projections, and imaging of functional representations at ever-finer levels of resolution.

Some may wonder why the topics of dynamic receptive field properties and feedback connections are not included in this volume. Several leading authorities on these topics felt that insufficient information is available for comprehensive reviews to be written; one of the clearest recommendations to prompt future work! The future looks bright, but investigators still need to devise imaginative experiments to test ideas, for which they will require the cooperation of granting agencies so that more can be revealed about visual processing and the nature of seeing. We believe the contributions to this volume are highly persuasive in showing what can be achieved by carrying out careful and imaginative investigations.

We hope that this volume emerges as one of the essential reads for all those interested in cerebral cortical processing of visual signals.

Finally, we thank Aaron Johnson at Academic Press and Larry Meyer at Hermitage Publishing Services for their patience and diplomacy in bringing this volume to publication, and we thank the publication team at Academic Press for the high standard of presentation and for the high quality of the figures that are so important for conveying messages and summarizing data.

Bertram Payne and Alan Peters

Contributors

Numbers in parentheses indicate the pages on which the authors' contributions begin.

Jose-Manuel Alonso (319) Department of Psychology, University of Connecticut, Storrs-Mansfield, Connecticut, U.S.A.

Tobias Bonhoeffer (131) Max-Planck-Institute for Neurobiology, Martinsried, Germany

Jamie D. Boyd (221) Department of Biological Sciences, Simon Fraser University, Burnaby, British Columbia, Canada

Valentin Dragoi (471) Department of Brain and Cognitive Sciences, Massachusetts Institute of Technology, Cambridge, Massachusetts, U.S.A.

Timothy Q. Duong (195) Center for Magnetic Resonance Research, University of Minnesota Medical School, Minneapolis, Minnesota, U.S.A.

Ulf T. Eysel (427) Department of Neurophysiology, Ruhr-University, Bochum, Germany

Mark Hübener (131) Max-Planck-Institute for Neurobiology, Martinsried, Germany

Allen L. Humphrey (343) Department of Neurobiology, University of Pittsburgh School of Medicine, Pittsburgh, Pennsylvania, U.S.A.

Dae-Shik Kim (195) Center for Magnetic Resonance Research, University of Minnesota Medical School, Minneapolis, Minnesota, U.S.A.

Seong-Gi Kim (195) Center for Magnetic Resonance Research, University of Minnesota Medical School, Minneapolis, Minnesota, U.S.A.

Siegrid Löwel (167,387) Leibniz-Institute for Neurobiology, Magdeburg, Germany

Joanne A. Matsubara (221) Department of Ophthalmology, University of British Columbia, Vancouver, British Columbia, Canada

Donald E. Mitchell (655) Psychology Department, Dalhousie University, Halifax, Canada

Jaime F. Olavarria (259) Department of Psychology, University of Washington, Seattle, Washington, U.S.A.

Bertram R. Payne (1,561) Department of Anatomy and Neurobiology, Boston University School of Medicine, Boston, Massachusetts, U.S.A.

Alan Peters (1) Department of Anatomy and Neurobiology, Boston University School of Medicine, Boston, Massachusetts, U.S.A.

R. Clay Reid (319) Department of Neurobiology, Harvard Medical School, Boston, Massachusetts, U.S.A.

R. Jarrett Rushmore (561) Department of Anatomy and Neurobiology, Boston University School of Medicine, Boston, Massachusetts, U.S.A.

Alan B. Saul (343) Department of Neurobiology, University of Pittsburgh School of Medicine, Pittsburgh, Pennsylvania, U.S.A.

Jack W. Scannell (609) Neural Systems Group, Newcastle University, Newcastle-upon-Tyne, United Kingdom

Kerstin E. Schmidt (387) Max-Planck-Institute for Brain Research, Frankfurt-am-Main, Germany,

Wolf Singer (521) Max-Planck-Institute for Brain Research, Frankfurt-am-Main, Germany

David Somers (471) Department of Psychology, Boston University, Boston, Massachusetts, U.S.A.

Mriganka Sur (471) Department of Brain and Cognitive Sciences, Massachusetts Institute of Technology, Cambridge, Massachusetts, U.S.A.

Kamil Ugurbil (195) Center for Magnetic Resonance Research, University of Minnesota Medical School, Minneapolis, Minnesota, U.S.A.

W. Martin Usrey (319) Department of Neurobiology, Harvard Medical School, Boston, Massachusetts, U.S.A.

Theodore G. Weyand (295) Department of Cell Biology and Anatomy, LSU Health Sciences Center, New Orleans, Louisiana, U.S.A.

Malcolm P. Young (609) Neural Systems Group, Newcastle University, Newcastle-upon-Tyne, United Kingdom

1

THE CONCEPT OF CAT PRIMARY VISUAL CORTEX

BERTRAM R. PAYNE AND ALAN PETERS

Department of Anatomy and Neurobiology, Boston University School of Medicine, Boston, Massachusetts

PROLOGUE

Cat visual cortex is recognized to be a premier system for studying the mysteries of vision, visual processing, cerebral circuitry, and cerebral processing. A search of medical databases shows that during the last 30 years, more than 5000 articles have been published on the topic of cat visual cortex, with the overwhelming majority directed to cat primary visual cortical areas. This number is about 10 times the number published on monkey and rat visual cortices.

At the outset of this book it is important to ask: What is cat primary visual cortex? The answer to this question is not straightforward because the word *primary* has many meanings that depend on context. It may mean the largest, the major, the first in rank, the most important, the first in time, the first in a hierarchy, the distributor of signals, and so on. The many interpretations attached to the word *primary* have had strong influences on the scientific view of the meaning *primary visual cortex,* and the meaning of *cat primary visual cortex* has been in flux since the inception of the term. This flux has been influenced by the development of new techniques that result in the acquisition of new data and by the emphasis, interest, and perspectives of investigators. Thus, the term *primary visual cortex* as applied to the cat cerebrum is better considered a concept rather than an established, immutable descriptor of a particular region of brain.

It is practical and worthwhile to trace the evolution of thinking on the term *cat primary visual cortex* as a prologue to the subsequent chapters. Moreover, it is a practical step to take to establish a meaning that is broadly applicable to all chapters in the volume. In so doing, we distill the historical background and provide valuable information on the structural and functional organization of cat primary visual cortex. This information forms a good foundation for the reading of the subsequent chapters and provides us with the basis for understanding the features that most readily distinguish areas 17 and 18 from surrounding cortices and the two areas from each other (Fig. 1-1).

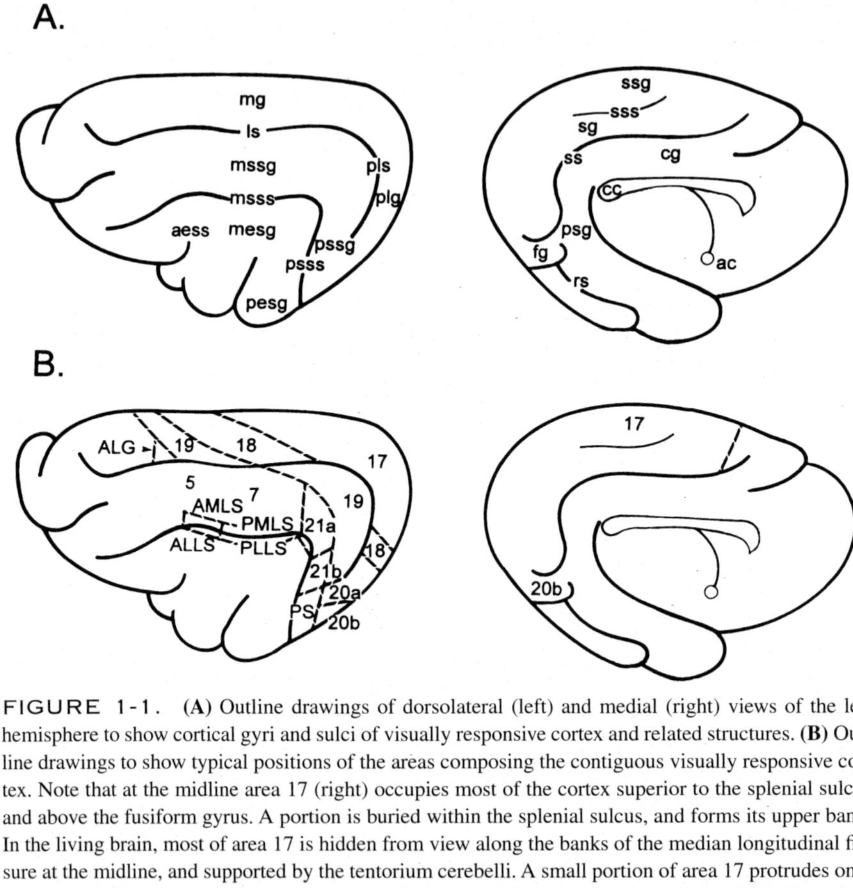

FIGURE 1-1. **(A)** Outline drawings of dorsolateral (left) and medial (right) views of the left hemisphere to show cortical gyri and sulci of visually responsive cortex and related structures. **(B)** Outline drawings to show typical positions of the areas composing the contiguous visually responsive cortex. Note that at the midline area 17 (right) occupies most of the cortex superior to the splenial sulcus and above the fusiform gyrus. A portion is buried within the splenial sulcus, and forms its upper bank. In the living brain, most of area 17 is hidden from view along the banks of the median longitudinal fissure at the midline, and supported by the tentorium cerebelli. A small portion of area 17 protrudes onto the dorsal surface of the brain at the junction of the marginal and posterolateral gyri (left). There it is visible in the exposed brain. Area 18 is a belt immediately lateral to area 17. It occupies much of the exposed portion of the marginal gyrus. However, its central portion is buried in the lateral and posterolateral sulci and the region is obscured by the junction of the middle and posterior suprasylvian gyri. Area 18 emerges, posteriorly onto the posterolateral gyrus. Area 18 is flanked laterally by area 19. At progressively more posterior positions, it spreads across the marginal gyrus, the lateral sulcus, and the middle and posterior suprasylvian gyri. Much of area 19 is buried within the lateral sulcus and forms its walls. At its caudal end, area 18, is flanked by area 20a. Abbreviations are given in Table 1-1. (Figure modified from MacNeil et al. [1996] and reproduced with the permission of Oxford University Press.)

TABLE 1-1. List of Abbreviations

ac	anterior commissure
ALG	anterior lateral gyrus area of Symonds and Rosenquist (1984a)
ALLS	anterior, lateral lateral-suprasylvian visual area of Palmer et al. (1978)
AMLS	anterior, medial lateral-suprasylvian visual area of Palmer et al. (1978)
aesg	anterior ectosylvian gyrus
C	contralateral eye innervation of LGN
cc	corpus callosum
CM	magnocellular subdivision, or layers, of LGN
CP	parvocellular subdivision, or layers, of LGN
cg	cingulate gyrus
cs	cingular sulcus
fg	fusiform gyrus
I	ipsilateral eye innervation of LGN
LGN	lateral geniculate nucleus (dorsal division)
ls	lateral sulcus
mg	marginal gyrus
mesg	middle ectosylvian gyrus
MIN	medial interlaminar nucleus subdivision of LGN
MS	middle suprasylvian visual cortex
mssg	middle suprasylvian gyrus
mass	middle suprasylvian sulcus
OR	optic radiation
OT	optic tract
pc	projection column in LGN
pesg	posterior ectosylvian gyrus
PLLS	posterior, lateral lateral-suprasylvian visual area of Palmer et al. (1978)
Pl	paralimbic cortex
plg	posterolateral gyrus
pls	posterolateral sulcus
PMLS	poterior, medial lateral-suprasylvian visual area of Palmer et al. (1978)
PS	posterior suprasylvian visual area of Updyke (1986)
psg	postsplenial gyrus
pssg	posterior suprasylvian gyrus
psss	posterior suprasylvian sulcus
rs	rhinal sulcus
sc	superior colliculus
sg	splenial gyrus
ss	splenial sulcus
ssg	suprasplenial gyrus
sss	suprasplenial sulcus
v	ventricle
1	Layer 1 of MIN
2	layer 2 of MIN
3	layer 3 of MIN
17	area 17 of Brodmann (1906)
18	area 18 of Gurewitsch and Chatschaturian (1928)
19	area 19 of Gurewitsch and Chatschaturian (1928)
20a	area 20a of Tusa and Palmer (1980)
20b	area 20b of Tusa and Palmer (1980)
21a	area 21a of Tusa and Palmer (1980)
21b	area 21b of Tusa and Palmer (1980)

INTERRELATIONS AND CONCEPT DEVELOPMENT

The concept of primary visual cortex emerged in parallel with developments in thinking on other aspects of visual processing and visual system organization. The evolution of the concept of primary visual cortex cannot be divorced from these parallel developments. The two most influential developments were: (1) the recognition that multiple and separate streams of visual signals emerge from the retina (Stone, 1983; Payne and Rushmore, this volume) and (2) the evolution of thinking on the structural, hodological, and functional organization of the lateral geniculate nucleus (LGN) (Guillery et al., 1980; Sherman, 1985). The LGN is the single structure that provides the dominant driving input to both areas 17 and 18, and in the absence of this input, neurons in areas 17 and 18 are silent (Lee et al., 1998; Malpeli, 1983; Malpeli et al., 1986; Mignard and Malpeli, 1991; Weyand, this volume). As an essential foundation for the remainder of this chapter, we briefly summarize here the visual streams and the structural organization of the lateral geniculate nucleus and the latter's afferent connections arriving from retina and its efferent projections directed to primary visual cortical areas 17 and 18.

Multiple Visual Streams

The morphologically distinct α, β, and γ retinal ganglion cells are the origin of the functionally distinct Y, X, and W signals; and each class of ganglion cell contributes to vision in its own unique way (e.g., Stone, 1983; Shapley and Perry, 1986; Wässle and Boycott, 1991). The approximate proportions of α, β, and γ ganglion cells in the retina are 5%, 55%, and 60%, respectively. Alpha retinal ganglion cells have large cell bodies and large dendritic trees (see Fig. 14-2). Their receptive fields are large, and they respond uniquely to low spatial frequencies, exhibit higher contrast sensitivity at those lower frequencies (Troy, 1987), and possess high temporal sensitivity to rapid transients in illumination and to high stimulus velocities. Alpha cells are sensitive to high spatial frequencies by virtue of their nonlinear subunits that result in increased activity but the sensitivity is without resolution (Shapley and Victor, 1979; Hochstein and Shapley, 1976). However, their large receptive fields and sparse distribution in the retina report the position of such stimuli only coarsely. They are implicated in coarse, more global aspects of spatial vision, in the detection of rapid visual transients in illumination and movement, and they are excellent at stimulus detection.

Beta cells have medium-sized cell bodies and highly compact dendritic trees (see Fig. 14-2). Their receptive fields are correspondingly smaller, and they are highly sensitive to low contrasts at mid-range spatial frequencies and to high spatial frequencies when contrast is high. Both of these sensitivities are at the expense of temporal sensitivity, which is low. Beta cells are able to report the position of small stimuli with great accuracy and minute movements of small stimuli with even greater accuracy (Shapley and Victor, 1986). Moreover, they have a relatively high density in retina (Wässle and Boycott, 1991; Stein et al., 1996), and they are implicated in local aspects of vision and high spatial acuity.

Both α and β cells give brisk responses and, under high contrast conditions, α cells generate the dominant signals up to the limit of α cell acuity. Thereafter, β cell activity dominates (Freeman, 1991; Shapley and Perry, 1986; Shapley and Victor, 1978, 1979, 1980, 1981; Troy, 1987; Wässle and Boycott, 1991). For low-contrast stimuli, both β (X) and α (Y) cells signal the contrast to the brain (Troy and Enroth-Cugell, 1993), and the X and Y cells are equally sensitive in their favored spatial and temporal frequency domains (Troy and Enroth-Cugell, 1993) as predicted by Shapley and Enroth-Cugell (1984). However, the greater numerical abundance of cells in the X-system confers a distinct advantage on it for signaling contrasts to the brain, even at low temporal and spatial frequencies (Troy, 1983).

The remaining ganglion cell types have small cell bodies and a variety of dendritic forms that have been referred to as δ, ε, γ, and η (see Fig. 14-2) (Boycott, and Wässle, 1974; Berson et al., 1998, 1999). For convenience, they are frequently combined under the single rubric of γ cells, the group from which all the types emerged. Functionally, they possess rather diverse receptive field properties, yet as a group they generate rather sluggish responses compared with the brisk reponses generated by α and β cells (Stone, 1983; Cleland and Levick, 1974). They are all classed under the banner of *W cells* (Stone, 1983). Their contributions to vision are poorly defined (Rowe and Palmer, 1995), and they bear little on the development of the 'concept of cat primary visual cortex.'

Lateral Geniculate Nucleus

Figure 1-2 summarizes both our current view of the structural and functional organization of the lateral geniculate nucleus and the relationships of its subcomponent structures to the signal streams arriving from retina and transmitted to primary visual cortex. Structly speaking, the major thalamic relay structure of visual signals to visual cortex is termed the *dorsal* division of the lateral geniculate nucleus. For brevity we refer to it in the remainder of this article as the lateral geniculate nucleus, or LGN. The ventral division of the LGN does not concern us because its connections are purely with subcortical structures.

In gross structure, the LGN has a modified sigmoid shape with an upturned tail posteriorly, an almost horizontal body in the mid-section, and a down-turned head, anteriorly (Fig. 1-2). It is composed of six laminae. The three dorsal laminae contain neurons ranging in size from small to large. Because of the presence of large neurons, these laminae are termed the magnocellular layers, and they are currently referred to as layers A, A1, and CM. Beneath CM are the three ventral laminae, which are termed the parvocellular C layers. They are highly compressed, contain small cells, and are referred to individually as layers C1, C2, and C3. Parvocellular layer C3 does not receive retinal fibers (Graham, 1977; Guillery et al., 1980). In Fig. 1-2, the three parvocellular laminae are grouped together and labeled CP. Starting with contralateral nasal input to magnocellular layer A and progressing ventrally, the retinally innervated layers receive, in alternating order, input from the contralateral nasal retina and the ipsilateral temporal retina (C and I; Fig. 1-2).

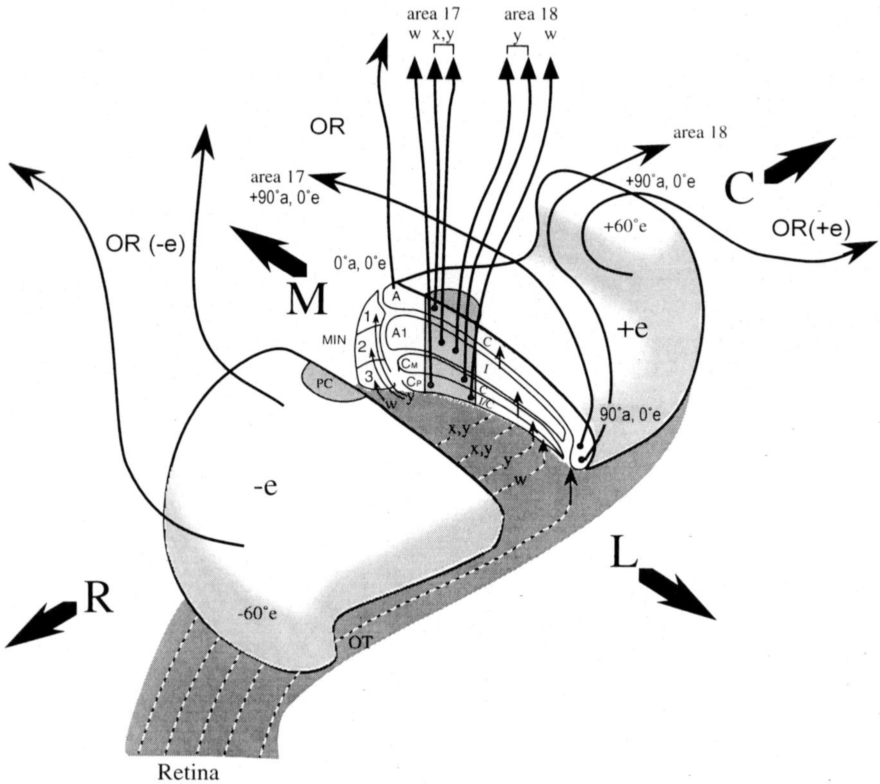

FIGURE 1-2. Anterolateral view of the left lateral geniculate nucleus. A mid-section of the nucleus has been removed to show the component laminae and compartments of the medial interlaminar component (MIN). Magnocellular layers A, A1, and CM and the group of parvocellular layers CP are indicated. Retinal afferent and cortical efferent projections are summarized by signal type. Retinal fibers penetrate through the ventral surface of the nucleus from the underlying optic tract (OT). Cortically directed fibers penetrate through the dorsal surface of the nucleus and coalesce and fan out as the optic radiation (OR). Eye of origin of retinal fibers is given as C (contralateral) and I (ipsilateral). To simplify the figure, neither axon collateral branches nor W-input to MIN are represented. The upper elevations (+e) in the visual field are represented caudally (C), and the lower elevations (–e) are represented rostrally (R). In the laminated portion of the nucleus, the vertical meridian (0° azimuth, 0°a) is represented medially (M), and the lateral periphery (90° azimuth, 90°a) of the visual field is represented laterally (L). The horizontal midline (0° elevation, 0°e) is represented approximately at the cut surface. MIN layers 1 and 2 contain a mirror-image map of the visual field representations in layers A and A1, with a line of reflection about the midline of the visual field. Thus, higher azimuths are represented medially. Upper and lower elevations are represented caudally and rostrally, respectively, as in the laminated portion of LGN. An exagerrated projection column (pc; Bishop et al., 1962) in the laminated portion of the nucleus is split by the bisection of the nucleus. All neurons in this column view the same portion of the visual field. Note the preservation of visual field elevation along the rostrocaudal axis of the optic radiation (upper field → caudal, lower field → rostral), and the increasing divergence of projections to areas 17 and 18 carrying signals from the lateral visual periphery (90°a). Those running to area 17 course medially in the optic radiation, whereas those running to area 18 course dorsally. (Diagram is based primarily on data of Garey and Powell [1967], Sanderson [1971], Lee et al. [1984]; Nelson and LeVay [1985]; Sur et al. [1987], Tamamaki et al. [1995], Bowling and Michael [1984], Guillery et al. [1980]; Holländer and Vanegas [1977], Gilbert and Kelly [1975], Niimi et al. [1981], Rosenquist [1985], and Sherman [1985] among others.)

Medial from the laminated portion of LGN is a highly compact structure, the medial interlaminar nucleus (MIN). It too is considered part of LGN (Guillery et al., 1980). The MIN occupies about half of the medial surface of the LGN and is concentrated anteriorly in the nucleus (Sanderson, 1971). It contains three numbered subdivisions. Subdivisions 1 and 2 contain the largest neurons and, in many respects, they parallel the organization of layers A and A1 in the laminated portion of the nucleus. They receive fibers from the contralateral nasal and ipsilateral temporal retinae, respectively. Layer 3 contains smaller neurons and appears in many respects to parallel the organization of the layer C1 in the parvocellular C layer complex (CP) with which it blends across an indistinct border. Layer 3 is unique because it receives retinal fibers from the contralateral *temporal* retina. (Guillery et al., 1980). The retinal recipient zone of the pulvinar nucleus (Berman and Jones, 1977) also appears to be a continuation of the parvocellular layers (Guillery et al., 1980), and it is now considered, on the basis of its retinal innervation, to be part of LGN. It is termed the *geniculate wing* by Guillery et al. (1980). Because its neurons project to lateral regions of cortex, and not to areas 17 and 18, it is not included in the summary diagram (Fig. 1-2) and is not considered further.

Retinal fibers approach from the ventral and lateral aspect in the optic tract (OT) and penetrate through the ventral surface of the nucleus to terminate in the appropriate layer according to their eye of origin and position of parent cell body in the retina. Each of the components of LGN receives a unique, signature set of fiber types from retina. Fibers carrying W-signals enter the parvocellular complex of layers (C_P), fibers carrying Y-signals penetrate through to the magnocellular layers A, A1, or C, and fibers carrying X-signals penetrate through to layers A and A1. W-fibers destined for MIN terminate in layer 3, and Y-fibers terminate in layers 1 or 2. X-innervation has been inconstantly described (Dreher and Sefton, 1979; Lee et al., 1992; Raczkowski and Sherman, 1985; Sur et al., 1987; Humphrey and Murthy, 1999). Individual W- and X-fibers usually innervate a single layer, whereas Y-fibers may innervate more than one layer by a series of collateral branches (for simplicity of the figure, these collateral branches are not shown). For example, a single Y-axon arising from the contralateral eye may innervate both magnocellular layers A and C, and MIN layer 1 (Sur et al., 1987; Tamamaki et al., 1995; Bowling and Michael, 1984). In addition to signal type, eye of origin contributes to definition of the layer of termination, as outlined here and in Figure 1-2.

The retinal fibers establish orderly point-to-point maps of considerable resolution in the laminated portion of LGN (Sanderson, 1971). Neurons at the medial end of magnocellular layer A receive fibers from ganglion cells at the nasotemporal division of the retina, and they view the vertical midline, or 0° azimuth (0°a), of the visual field (Fig. 1-2). The lateral border of the nucleus receives fibers from ganglion cells at the nasal-most extreme of the retina, and they view positions as far lateral as 90° azimuth (90°a) in the visual field. The caudal, upturned tail receives fibers from ganglion cells in inferior-most retina that view positions with elevations as far superiorly as 60° (60°e) in the visual field. The anterior, down-turned head receives fibers from ganglion cells in superior-most retina that view positions with elevations as far inferiorly as –60° (–60°e). The horizontal meridian, or 0° eleveation (0°e), is represented in the body approximately at the level of the cut surface shown in Fig. 1-2.

The neurons in the remaining layers form separate maps that are in spatial register with the map formed in magnocellular layer A. However, the lateral extent of the representations is limited to approximately 45° azimuth. Because the maps are in register, conceptual structures called projection columns (PC) are established (Bishop et al., 1962; Sanderson, 1971). A projection column is composed of a population of neurons that extends both across and perpendicular to the LGN laminar borders, and all neurons in the column view the same part of the visual field irrespective of either eye of signal origin or functional type. This arrangement brings into register all signal types from both retinae viewing the same part of the visual field before transfer of signals to visual cortex, although it also creates a single, contralateral monocular representation in the lateral component of layer A, which has no counterpart in the layers innervated by the ipsilateral eye.

The retinal fibers also establish orderly, systematic point-to-point maps in the MIN component of LGN (Lee et al., 1984). These maps are dominated by retinal fibers carrying the relatively uncommon Y-signals. As a result, compared with the laminated portion of LGN, the maps in MIN are highly compressed and the resolution is rather coarse (Lee et al., 1984). Overall, the representations of the visual field formed in MIN are dominated by the lower visual field, because MIN receives signals predominantly from ganglion cells in tapetal retina, which is located adjacent to the dorsally positioned highly reflective tapetum (Sanderson, 1971; Lee et al., 1984). In maintaining common, continuous lines of elevation with those present in the laminated portion of LGN, the MIN makes up the anterior medial surface of LGN. Layers 1 and 2 contain compressed mirror-image representations of the maps contained within layers A and A1, respectively, with the vertical meridian representation as the line of reflection. Layer 3 contains a representation of a portion of the *ipsilateral* visual field viewed by contralateral temporal ganglion cells that innervate it. It is likely that this representation is a continuation of the contralateral representation present in layer C1 (Guillery et al., 1980; Lee et al., 1984).

The axons of LGN cell bodies ascend and penetrate through the overlying layers to exit the dorsum of the nucleus (Fig. 1-2), where the fibers arising from individual projection columns coalesce into small bundles, which then fan out as the optic radiation. In the optic radiation the individual bundles are arranged parallel to each other, and together they form a thin sheet in the sagittal domain that sweeps out caudally, dorsally, and rostrally from LGN to reach all portions of visual cortex. In the long rostrocaudal dimension of the fan, the nearest neighbor relationships of fiber origins are maintained through to their terminations, thus preserving an orderly representation of visual field elevation in the optic radiation. The same cannot be written of visual field azimuth, which is highly compressed in the thin mediolateral dimension of the fan. It is not until further along the optic radiation that azimuth emerges as a systematic representation, or more properly, as two systematic representations. Those fibers destined for the medial part of area 17, where the lateral visual periphery is represented (90°a), deviate medially from partner bundles that represent equivalent visuotopic positions in lateral area 18. In this process the optic radiation establishes two broadly mirror-symmetric maps in areas 17 and 18 with an axis of reflection about the representation of the vertical midline (0°a) of the visual field.

Each of the LGN components sends a unique, signature set of projections to visual cortex. The axons arising from individual projection-columns terminate coextensively with respect to its surface domain in an area of about 1 mm^2 or more, although fibers arising from the different W, X, and Y cells in the various LGN layers terminate in largely nonoverlapping patterns in different layers (see Matsubara and Boyd in this volume). Even so, there is considerable convergence of visual signals onto cortical neurons because the dominant cortical cell type, the pyramidal cell, has a prominent apical dendrite that crosses and samples from several cortical layers. Thus, the signals carried by the different functional types of LGN fibers merge for the first time in pyramidal cells. This convergence contributes to the generation of the emergent properties of cortical neurons.

The details of the foregoing cortical innervation patterns are highly germane to the thesis of this chapter and are discussed later as we review the evolution and refinement of the concept of primary visual cortex. In the next section, we summarize the features that define cortical areas, for this too is a topic highly relevant to the emergence of the concept of cat primary visual cortex.

CEREBRAL CORTICAL AREAS

The concept of cat primary visual cortex is also bound up in the theory of the existence of distinct cerebral cortical areas. This concept too has undergone considerable evolution over the last century and especially over the last three decades. Current views of cortical areas are based on multiple complementary and reinforcing criteria. For visual cortex, these criteria include:

1. Distinct architecture
2. A unique connectional signature with other brain regions
3. Characteristic functional maps
4. A unique inventory of receptive field properties
5. A distinct catalog of contributions to visual processing and visually guided behaviors

In many senses, area 17 is considered the paradigmatic area because it fully meets all five criteria. We must ask: Are the characteristics of area 18 sufficiently clear and robust for it also to be considered an independent bona fide cortical area? On the grounds that area 18 has both a distinct architecture and a distinct representation of the visual field, the answer to the question is yes. However, as we will read, in terms of connectional signature, inventory of receptive field properties and contributions to visually guided behaviors, the answer is not so clear-cut. In many respects areas 17 and 18 have comparable features, and the functional differences that are readily identified are features that are foreshadowed by the X- and Y-afferent streams leading to the two areas. In these respects areas 17 and 18 can be considered separate areas only if they are considered to process visual signals in parallel. The appropriateness of our view will emerge as we trace the evolution of the concept of primary visual cortex. This rather complex view did not emerge easily, and it had a gestation period of over 100 years.

Area 17

Over the years, cat primary visual cortex has been equated with cytoarchitectonic area 17 of Brodmann, with the terms *striate* or *striated* cortex, and with the term *visual area VI.* In this chapter we ask whether these terms are appropriate and useful descriptors of the same entity. There was a long period when these descriptors were considered equivalent, and several of the terms remain in current usage. However, for convenience we limit ourselves to the term *Area 17* unless otherwise indicated.

Of importance until the 1960s the thinking on cat visual cortex was intimately bound with the thinking on monkey and human visual cortex. In the 1960s, the views of cat primary visual cortex started to depart from the unified view of primary visual cortex in primates. This departure was initiated by several groups of anatomists. They showed that in cats both areas 18 and 17 receive significant projections from the dorsal (LGN), which implies that in cats the concept of primary visual cortex should be extended to include the more laterally placed area 18.

Area 18

Before the 1960s the architectonically defined area 18 also acquired the term *parastriate* cortex and *visual area V2.* Are these terms equivalent? Do they describe the same entity? Are they current? Are there other features besides the connectivity that qualifies area 18 for inclusion under the umbrella term *primary visual cortex?* For example, does area 18 have a similar intrinsic circuitry to area 17, and does it process visual signals in a similar way to the processing carried out in area 17? There was a long transition period after the 1960s when considerable data, based on a number of parameters and linked to a number of criteria, revealed several substantial similarities between area 18 and area 17. As a result, the alternative terms for area 18 were used less frequently, and eventually they were dropped from the favored terminology. This fall from usage clarified terminology and thinking, and the modified concept of the components of primary visual cortex in cat finally took firm root and was accepted in the 1980s. Consequently, *Area 18* is the most frequently used term for the region and the one we adopt in this chapter.

Area 19 and More Lateral Areas

The concept of primary visual cortex has not been extended laterally beyond area 18. It does not include either area 19, which has also been termed *peristriate* cortex and *visual area V3,* or the even more lateral region composing the suprasylvian belt. There have been attempts to broaden the concept to include these more lateral areas, and there are still some recent challenges to test the present view. These challenges come from new ways of looking at cortical connections (see Scannell and Young in this volume) and to assessments of projections from LGN to suprasylvian cortex, signal timing, and linkages between the suprasylvian cortical areas and areas 17 and 18 (Dinse and Krüger, 1994; Katsuyama et al., 1996) (see page 106). The success or failure of these challenges will be decided in the future by the appropriate experiments. At present, we should consider that the concept of cat primary visual cortex is advanced and solid, but still not completely settled in all details. Our current view is that cat primary visual cortex includes both areas 17 and 18, but does not include more lateral regions of cerebral cortex. The evolution of that view is discussed next.

MILESTONES IN THE DEVELOPMENT OF THE CONCEPT OF CAT PRIMARY VISUAL CORTEX

LOCALIZATION OF CEREBRAL FUNCTIONS

The emergence of the view that both areas 17 and 18 in cat should be considered primary visual cortex can be traced back to the mid-1800s and the early stages in the vibrant era that led to the acceptance of the theory of cerebral localization of function. In efforts to establish that theory, attempts were made to reconcile experimental data derived from a variety of emerging techniques based on histological structure, electrophysiological potentials, electrical excitability, and the repercussions of surgically placed lesions. In addition, phylogenetic and ontogenetic comparative approaches were considered important and highly relevant, and they permitted transfer of knowledge among representative species spanning much of the mammalian order. Moreover, clinicopathological studies on human performance following brain lesions were also considered relevant to the discussion. This golden era of investigation culminated in the wide acceptance of the theory of cerebral localization of function, which continues to maintain a strong hold on contemporary thinking. This theory is a keystone to the acceptance of the concept of primary visual cortex.

Likewise, progress in formulating the current concept of cat primary visual cortex has rarely been uniform; and significant, rapid, and durable advances in thinking have been interspersed by long periods of intellectual and experimental consolidation. Many of the durable advances have been made in the last three decades and, by necessity, more space in this chapter is devoted to them. However, that should not diminish the significance of earlier results, which form the essential foundations on which our current views are built.

THE BEGINNINGS OF THE CONCEPT OF CAT PRIMARY VISUAL CORTEX

The initial studies in the mid-19th century tended to focus on dog cerebrum in preference to cat cerebrum. However, by the mid-20th century the bias had shifted in favor of cats because there is much less variability in the overall size and form of the multiple cat strains than in the large number of dog strains. This shift was straightforward because there are many obvious similarities in the gross and internal microscopical anatomical structure of cat and dog brain (Kreiner, 1966, 1968, 1970, 1971) that permit ready transfer of knowledge between the two species. Moreover, it was advantageous to make the shift because cats exhibit only modest variability in cerebral gyral and sulcal patterns (Kawamura, 1971), and there is great uniformity in the size and position of subcortical structures among individuals. This uniformity has permitted highly accurate and reliable stereotaxic coordinates to be developed for the cat brain (Winkler and Potter, 1914; Reinoso-Suarez, 1961; Berman, 1968; Berman and Jones, 1982; Sanderson, 1971), with consequent improvement in technique and reproducibility of results by different groups of investigators. Both have contributed considerably to progress in understanding the structural and functional organization of the cat's visual system and visual cortex.

Lesions

Panizza (1855, 1856; Polyak, 1957) appears to be the first person to test and successfully argue that bilateral posterior cerebral cortex lesions in dogs could cause blindness. He also argued that the lesions affected the faculty of vision independent of other faculties, thus helping to establish the idea that visual processing was localized to one region of the cerebrum. However, Panizza's efforts were largely ignored until Munk focused attention on cerebral cortex as the major visual analyzer (Benton, 1990). Munk (1881) described both *Seelenblindheit,* or psychic blindness, following limited lesions of posterior cortex and *Rindenblindheit,* or cortical blindness, following larger lesions that removed much of cortex that is now recognized to be visual in function. Psychic blindness is an inability to recognize the meaning or significance of objects that were familiar before the operation, although the same objects were recognized by means of other senses. True blindness was absent because animals were capable of navigating around obstacles, and they were also capable in some circumstances of relearning the significance of the objects. Psychic blindness is now usually referred to as *visual agnosia.* In contrast, cortical blindness is a more severe deficit than visual agnosia and it is much closer to a sensorial blindness that results in a failure to sense a form or stimulus. It is usually permanent. However, it is now recognized that in this condition a crude or imprecise sensibility to light may remain providing that contrast and flux are high or that transients are both large and steep (Doty, 1973).

Luciani and Tamborini (1879) and Luciani (1884) largely concurred with Munk's view, although detractors, such as Loeb, reported that the visual deficits induced by lesions were of "weakness" rather than blindness per se (Finger, 1994). The latter is closer to the modern view because it is now known that specific visual cortical lesions attenuate definable aspects of vision while leaving others partly or largely intact (see page 100).

The largely ignored Panizza (1855, 1856) also contributed to the identification of the primary optic centers and the LGN. Panizza's observations were based on enucleation experiments and gross determination of a decrease in size owing to atrophy. In this way Panizza identified the *optic thalamus,* which later came to be known as the LGN. Gudden (1869, 1874) carried out similar experiments, but on young animals. In young animals immature neurons exhibit a greater vulnerability to damage with a faster and more complete atrophy of structures (Doty, 1973; Payne and Rushmore in this volume). After both enucleations and cortical lesions in young dogs, Gudden identified a substantial atrophy of LGN. Since this result followed either enucleation or cortical lesions, it pointed to the LGN as the intermediary structure in the transfer of visual signals from retina to cerebral cortex, but the specific link to the region later termed *striate cortex* was not made.

This view for an intermediary position of LGN in the relay of visual signals to cortex was strengthened by results based on the Marchi degneration technique. Colucci (1898) showed in dogs and Jacobson (1899) showed in a variety of animals, including cats, that optic axons terminate in LGN. In addition, Probst (1900) was the first to illustrate the separate innervation of cat LGN layers by optic axons arising from the left and right eyes. However, the significance of the details in his drawing was not recognized for another 35 years (Barris, 1935).

Electrophysiology

The localization of the visual processor to occipital cortex was supported by the electrophysiological studies of Beck, who showed that the electrical conductance of dog occipital cortex changed during visual stimulation (Brazier, 1988). The observations were supported by electrophysiological studies carried out on other species by Caton (1875, 1877, 1891), and by anatomical studies of Gratiolet (1854), Meynert (1870), and Flechsig (1896) who described visual radiations linking the lateral geniculate nucleus and posterior, medial cortex. Their evidence built on the already accepted view that retinal axons project densely to the LGN.

Anatomy

The concept of cat primary visual cortex can also be traced back to the earlier identification of the prominent stripe in human brains by Gennari, Vicq d'Azyr, Sömmering, Rolando, and Baillarger in the late 18th and early 19th centuries (Finger, 1994), which led to the terms *striate, striated,* or *striped* cortex. In many instances the terms *primary visual cortex* and *striate cortex* are now used synonymously. The synonymous usage applies to primary visual cortices in a wide variety of mammals, even when the stripe is highly attenuated relative to the human and monkey pattern, or even when primary visual cortex has a different architectural character altogether (Garey et al., 1985; Garey and Leuba, 1986; Morgane et al., 1988; Revishchin and Garey, 1991). In 1914 Winkler and Potter provided the first comprehensive account of the cerebral hemispheres of cat brain. They proposed the nomenclature for the sulci and gyri of this brain and illustrated a series of coronal sections through the hemispheres. Much of this nomenclature remains in current usage.

Cats and Primates

In cats the partial extent of striate cortex was identified by Brodmann (1906, 1909; Garey, 1994) using cell body stains for Nissl substance (Fig. 1-3). Brodmann equated the highly granular cat striate cortex with both visual cortex and with area 17 (striate cortex) in the human brain, even though the internal structure

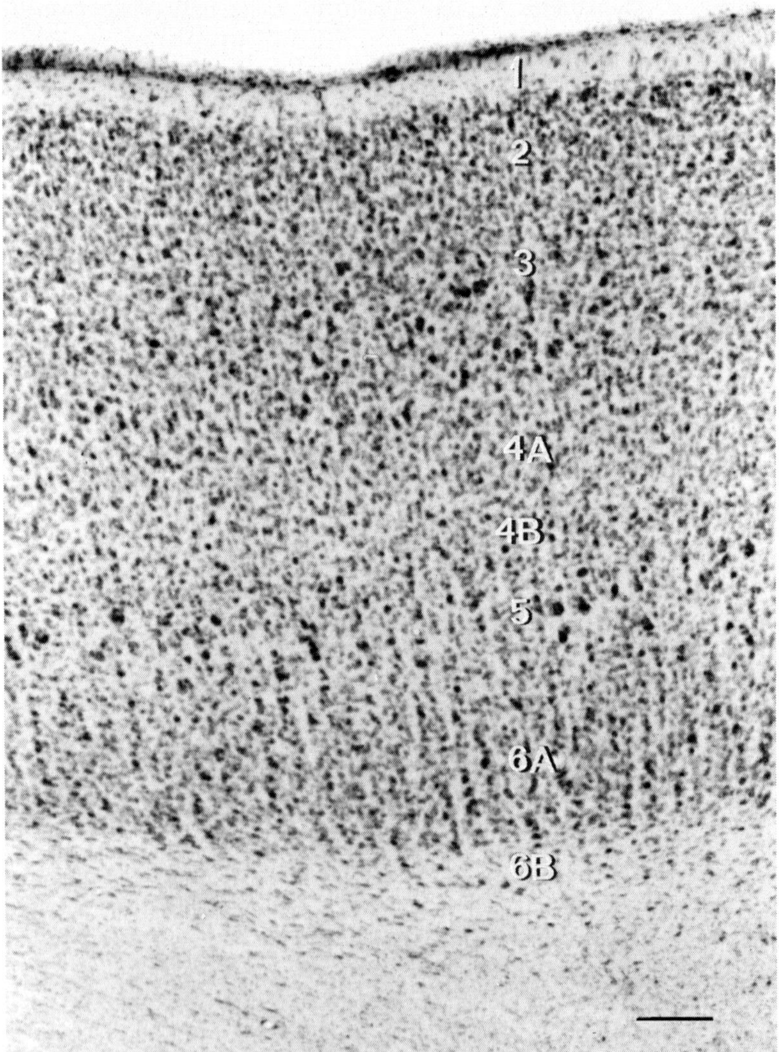

FIGURE 1-3. Nissl-stained section of area 17 of cat visual cortex showing the numbering of the layers (Calibration line, 100 μm).

of cat area 17 differed from that of area 17 in both monkey and human brains. It is important to note that from Brodmann's publications forward the term *striate* cortex usually refers to cortex containing the highly granular layer 4 in the occipital region, and it is based on cytoarchitectural appearance. Being based on Nissl stains, this view is markedly different from the myeloarchitecture, which first gave striate cortex its name.

In cytoarchitectural terms, human and monkey area 17 possesses a tristriate, Nissl architecture within layer 4, with large numbers of small, densely packed granule cells populating sublayers 4A and 4C. These sublayers are separated by layer 4B, a sublayer that contains fewer and larger cells. Many of these larger cells are dispersed by, and embedded within, a rich myelinated plexus that comprises the stria of Gennari, which corresponds to the outer line of Baillarger (Billings-Gagliardi et al., 1974; Peters, 1994). The largest of the cells correspond to the outer layer of Meynert cells in primates (Peters, 1994). In contrast, cat area 17 has neither a tristriate form nor a plexus of myelin with the same level of richness as that present in layer 4B of primates. Thus, layer 4 in cat area 17 is better classified as bistriate with two, adjacent sublayers of granule cells (Brodmann, 1909). The more superficial sublayer (4A) is comprised of larger and slightly less densely packed granule cells, and the deeper sublayer (4B) is comprised of smaller more densely packed granule cells. In this way, the terms *striate cortex* and *area 17* in cat became equated with each other.

Reconciliation

Cajal (1921a,b) gave the first comprehensive account of the lamination and neuronal composition of the cat visual cortex in an article that was written in Spanish, but to reach a wider audience, this article was subsequently translated into German and expanded by the addition of some figures (Cajal, 1922; see the annotated translation given by DeFelipe and Jones, 1988). Cajal attempted to reconcile the interpretation of the differences in the architecture of cat and human striate cortex (Cajal, 1921a,b, 1922; DeFelipe and Jones, 1988), although he was careful to stress that the structures were analogous rather than homologous. Cajal (1922) also examined visual cortex in dogs, but to our knowledge his observations on dogs were not published.

Cajal guessed correctly that for the cat, the thicker fibers forming the myelinated plexus in the granular layer had their origin in the lateral geniculate nucleus. However, he guessed incorrectly that the fibers were equivalent to the thickly myelinated fibers in the stria of Gennari in primates, which originate from other sources. Part of his error may lie in his belief that he identified large cells in cat layer 4, which he equated with the outer Meynert cells in primates located in layer 4B and embedded in the stria of Gennari. Cajal recognized the importance of this point, and the potential error in equating the cells in the two species, because he mentions that he was preparing a paper specifically designed to assess the origin of the fibers of the stria of Gennari (Cajal, 1922). Cajal identified the full extent of highly granular occipital cortex in cat and commented: "The extent of the typical configuration of the visual cortex of the cat ... coincides almost completely with that assigned by Minkowski" (DeFelipe and Jones, 1988) (see later).

Full Extent of Cat Striate Cortex

Minkowski (1911) adopted both the numerical terminology of Brodmann (1906, 1909; Garey, 1994) and the term *area striata* of Smith (1907) to identify and confirm in histological sections the location of the cortical visual processor. Moreover, based on architectural distinctions that were being refined at the time, Minkowski (1911, 1912) also established the complete extent of striate cortex, or area 17, in cat (Fig. 1-4) a decade before Cajal. Minkowski showed that area 17 encompasses the medial aspect of the marginal gyrus and posterolateral gyrus and did not extend far onto the visible, dorsal surface of the hemisphere (compare Figs. 1-1 and 1-4). However, following from von Monakow's (1882a,b) work on rabbits, Minkowski (1913) added the caveat that visual cortex may extend beyond the striate area. Finally, Minkowski (1911) also showed that cats with striate lesions are visually handicapped. These observations confirmed and refined earlier conclusions of Munk, Luciani, and Luciani and Tamburini cited previously.

Extrastriate Cortex

Beyond striate cortex, the terminology devised for human and monkey brains was also relevant to the development of terminology for cat brains. In humans, striate cortex was shown to be sequentially bounded by two architecturally distinct belts of cortex, which clinicopathological and experimental studies suggested had different functions (see Gross, 1997; Kaas, 1997; Zeki, 1993). For example, Campbell (1905) identified striate cortex as visual sensory cortex with remaining cortex of the occipital lobe composing visuopsychic cortex. Brodmann (1906, 1909) identified two areas anterior to area 17 in human and monkey brains. He numberered the areas 18 and 19, and labeled them occipital and preoccipital cortices, respectively. In 1908 Elliot Smith introduced the terms parastriate and peristriate cortices to label visuopsychic and more anterior regions of occipital cortex.

Two decades later, Gurewitsch and Bychowsky (1928) and Gurewitsch and Chatschaturian (1928), in companion studies on dog and cat cortex, concurred with Minkowski and Cajal in recognizing that striate cortex of cat includes only a small portion of the dorsal surface of the marginal gyrus immediately adjacent to the medial edge of the hemisphere, with the majority of the area lying on the medial and tentorial surfaces (Figs. 1-1B and 1-4). Gurewitsch and Bychowsky (1928) and Gurewitsch and Chatschaturian (1928) also adopted the now established terms of striate, parastriate, and peristriate cortices to describe cytoarchitectonic areas 17, 18, and 19 in the cat (Fig. 1-4). The subdivisions assigned by Gurewitsch and colleagues had significant parallels in the subdivisions of human visual cortex described 1 year later by von Economo (1929) and termed OC, OB, and OA, respectively. These parcellation schemes dividing occipital cortex into three architecturally distinct zones were largely supported by contemporary views of visual function. They suggested striate cortex as the recipient of visual

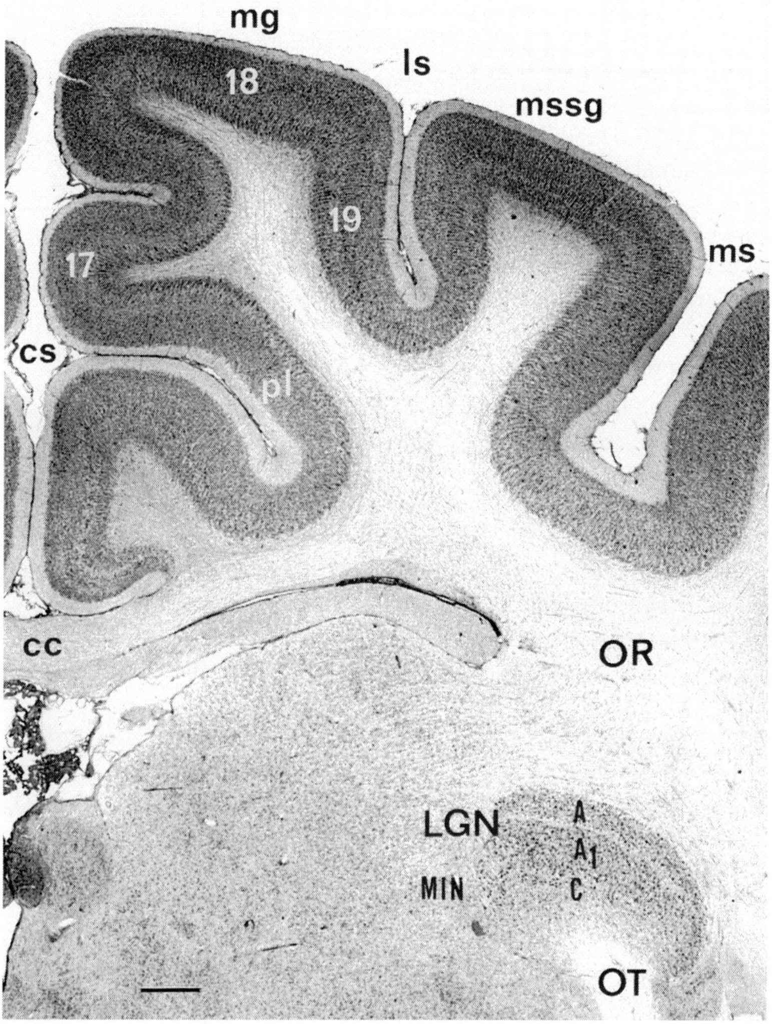

FIGURE 1-4. Low-power photomicrograph of occipital cortex in a coronal section stained for Nissl substance. At this coronal level area 17 extends from the medial margin of the hemisphere ventralwards and forms the bank of the median longitudinal fissure and the upper bank of the cingular sulcus (cs). It is flanked there by paralimbic cortex (pl). Area 18 is located on the surface of the marginal gyrus (mg). Area 19 borders area 18 laterally, forms all of the medial bank of the lateral sulcus (ls), and wraps around the fundus of the lateral sulcus onto the ventral portion of its lateral bank. In cytoarchitectonic terms the lateral border of area 19 is indistinct. Middle suprasylvian visual cortex is evident on the banks of the eponymous sulcus (ms). Major subdivisions of LGN are also shown together with the location of the optic tract (OT) and optic radiation (OR). Abbreviations are given in Table 1-1. (Scale bar = 1 mm).

signals and the site of sensation. In that sense, area 17 was equated with primary visual cortex, area 18 with parastriate cortex, and area 19 with peristriate cortex, a view maintained for more than four decades (Otsuka and Hassler, 1962; Sanides and Hoffmann, 1969).

Throughout the intervening years, both areas 18 and 19 were considered to receive sequential relays of the signals from striate cortex, and each area was considered to be involved in later stages of visual processing underlying perception and higher level abilities, respectively (Kass, 1997). This view established and perpetuated the idea of hierarchical processing of visual signals in cerebral cortex, and it was a view adopted by Hubel and Wiesel in the mid-1960s.

Cat Retinogeniculostriate Pathway

In his work on the behavioral repercussions of lesions of visual cortex, Munk (1879a,b, 1881) commented on the detailed relationship between the retina and visual cortex. He adopted the mistaken view that in both dogs and monkeys the temporal portion of the retina is represented in the lateral portion of the ipsilateral visual area and that the much larger nasal portion of the retina is represented in the larger medial portion of the contralateral visual area. He also noted, correctly, that the upper half of the retina (lower field) is represented anterior in the visual area, and that inferior retina (upper field) is represented in the posterior portion of the area (Polyak, 1957). Prophetically, Munk (1881) concluded that "... the central elements of the visual area, where the optic pathway terminates and where perception takes place, are arranged in an orderly manner and continuously, as are the receptors in the retina where the fibers arise, viz., that the adjoining retinal units remain close to the same units in the cortex" (Polyak, 1957). With this statement Munk formulated the nature of the cerebral representation of the retina.

Wilbrand (1890) accepted Munk's views on the retinotopical principle with one revision. Based on human data showing congruous scotomas following small central lesions, Wilbrand proposed in his *Hypothese der Faszikelfeldermischung* that the two homonymous retinal halves were represented in a common cortical area, instead of side by side as proposed by Munk. In this proposal Wilbrand imagined that the visual radiation was composed of multiple pairs of bundles of fibers, one bundle of the pair for each eye, that viewed identical points in the visual field and terminated in the same small region of cortex. This perspicacious concept presaged, by more than three quarters of a century, the concept of ocular dominance columns (see page 94).

The views of Munk and Wilbrand provided the basis for subsequent investigations on the organization of the visual pathways a quarter century later. A prominent investigator of this topic was Minkowski (1911, 1913, 1920), who took the analyses of Panizza, Gudden, von Monakow Munk, and Wilbrand one step further, and concentrated on pathways leading to the striate cortex, which he identified in its entirety. In his work, Minkowski established that the retinogeniculocortical pathway terminates in striate cortex, or area 17. Minkowski's work was based on detailed analyses of degeneration in the visual pathway induced by retinal or small cortical lesions. It provided direct experimental evidence of the linkage of the eye to striate cortex. Minkowski carried out parallel studies on monkeys and showed that enucleation induced an anterograde (forward) degeneration of optic axons, a transsynaptic atrophy of neurons, and a thinning of layers in LGN. He also showed that lesions confined to striate cortex induced a retrograde (backward) pallor, atrophy or death of neurons in LGN. The repercussions were more severe, and the interpretations more obvious, in monkeys than in cats. The coupling of the two experimental observations firmly established the linkages between the initial and relay limbs of the retinogeniculostriate cortex pathway and demonstrated conclusively that striate cortex was the cortical recipient of visual signals. In that sense striate cortex could be viewed as primary visual cortex. Additional experimentation to clarify the details of the organization and connections of the retinogeniculocortical pathway in cats would not take place for another half century.

Minkowski (1913, 1920) also demonstrated the topographical relationships between thalamus and cortex. He showed that the positions of the degeneration induced in LGN covaried with the loci of the lesions in striate cortex. These observations accurately established for the first time the spatial, topographic relationship of the LGN projection to striate cortex. A decade later Brouwer et al. (1923), Brouwer (1927), and Overbosch (1927) revealed the topographic representation of the cat retina on LGN. They made restricted retinal lesions and were able to show a region-to-region projection from retina to LGN. These results were confirmed in great detail several decades later by a number of investigators (Moore et al., 1966; Garey and Powell, 1968; Stone and Hansen, 1966; Guillery et al., 1980). In the meantime, the observations of Brouwer and colleagues and of Overbosch finally established, in cat, a broad retinotopic organization of the visual pathways from retina through LGN up to visual cortex. These observations confirmed the retinotopical principle of Munk (1881) formulated one-half century earlier, and they formed the foundation for the detailed electrophysiological mapping studies of visual cortex in the 1940s.

Visual Map

Evoked Potentials

The next major advance in the analyses of the visual pathways was the identification and mapping of primary visual cortex using physiological techniques. In 1940, Talbot and then Talbot and Marshall in 1941 and 1942 published the first descriptions of the arrangement of the visual field representation in cat striate cortex based on the potentials evoked by visual stimuli. They used small spots of light to selectively activate neurons in the visual pathway seeing particular parts of the visual field, and macroelectrode electrophysiological methods to record the summed potentials evoked in limited regions of cerebral cortex. In these studies Talbot and Marshall showed a correspondence between the extent of visually activated cortex and the extent of striate cortex identified with histological methods.

Moreover, Talbot and Marshall (1941) also showed that the representation of the visual field was topographically organized (Fig. 1-5): (1) the contralateral hemifield was represented in the opposite cerebral cortex, (2) the contralateral upper quadrant of the visual field was represented caudally on the posterolateral

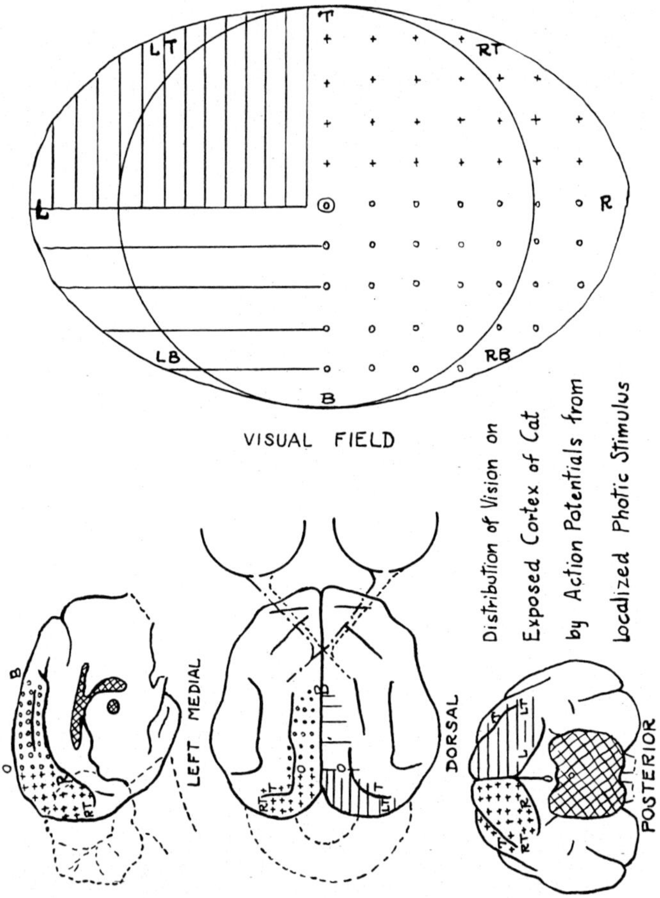

FIGURE 1-5. Projection of the visual field on the cerebral cortex of the cat. **(Top)** The upper and lower quadrants of the left and right visual hemifields of the cat. **(Bottom)** Identical coding is used to represent the location of the quadrants on the two cerebral hemispheres of the cat brain, which is shown in dorsal perspective (bottom, center) and along the banks of the medial longitudinal fissure (bottom, left and right). Interrupted lines represent the position of the brainstem and cerebellum. Compare with Fig. 1-1. Cross-hatching indicates the corpus callosum, anterior commissure, and cut brainstem. Note representation of the left visual hemifield in the right cerebral hemisphere and the right hemifield in the left hemisphere. Upper fields are represented caudally on the tentorial surface adjacent to the cerebellum, and lower fields are represented rostrally. (Figure reproduced from Talbot and Marshall [1941] with permission of Elsevier Science Ltd., modern publishers of the *American Journal of Ophthalmology.*)

gyrus and the tentorial surface of the hemisphere, (3) the lower contralateral quadrant was represented rostrally on the marginal gyrus and along the interhemispheric fissure, (4) the visual axis was represented close to the junction of the marginal and posterolateral gyri, and (5) the horizontal meridian was represented along a line extending from the junction of the marginal and posterolateral gyri in the direction of the splenium of the corpus callosum (compare Figs. 1-1 and 1-5). As part of the renewed interest in the visual system stimulated by these electrophysiological studies, O'Leary (1941) carried out detailed analyses on the structure of area 17. O'Leary's major contribution was to extend and detail the previous descriptions and brought the account of the lamination pattern more in line with the generally accepted six-layer pattern for neocortex.

In 1942 Talbot also described a more laterally positioned visual representation, "oppositely disposed, and confined anteriorly to the lateral gyrus,[1] posteriorly to the suprasylvian."

Unfortunately, the results of Talbot's study were not reported in full (Woolsey, 1971), but a drawing, showing the extent of the area, was published by Woolsey and Fairman (1946). They suggested that the visual representation in striate cortex should be termed visual area *VI* and that the more recently identified representation lateral to it be termed visual area *VII*. Thus, visual area *VI* was identified as equivalent to both striate cortex and area 17. Visual area *VII* was not correlated with a unique architecture. To avoid confusion with frequently used cortical layering schemes we have adopted the format visual areas V1, V2 and V3 rather than the format visual areas *VI*, *VII* and *VIII*.

It is important to remember that the designation visual area V2 represents second in order of discovery; it does not imply that the area constitutes a second tier in the visual processing hierarchy. Talbot (1942) noted that visual area V2 responds to photic and optic nerve stimulation independent of narcosis, cautery or convulsant drugs applied to area 17. These observations suggested that visual area V2 received visual signals independent of visual area V1.

Visual area V1 is located medially and posteriorly, and the smaller region of cortex composing visual area V2 is located more laterally. The visual representation in visual area V2 was described by Talbot (1942) as being equally as accurate as that in the striate area, but covering a narrower visual band and being concerned with the neural decussation about the vertical meridian (Woolsey, 1971). Study of rabbit visual cortex clarified understanding of the second visual representation in the cat. In the rabbit, the second visual representation encompasses a larger extent of the visual field than that in the cat. The rabbit visual area V2 has a form that is a reduced mirror image representation of the representation in striate cortex, and it has an axis of reflection centered on the vertical meridian (Thompson et al., 1950). Important for the correspondence of the cat and rabbit representations was the observation that the relative positions of visual area V1 and visual

[1] Term *lateral gyrus* is a misnomer for the more appropriate term *marginal gyrus*. According to Doty (1967) the error can be traced back to Winkler and Potter's (1914) atlas.

area V2 in the cortical sheets of the two species are similar. In addition, the line of reflection of the two visual maps in the two species is centered on the representation of the vertical meridian and the nasotemporal division of ganglion cells. In both species, this pattern ensures extension of the contralateral visual field representation to include part of the ipsilateral field. The identification of the second, independent and comprehensive visual representation in the cat outside the realm of striate cortex, and with a seemingly independent input from LGN, presaged the substantial revision in the 1960s of the concept of cat primary visual cortex, and the inclusion of area 18 under the umbrella term *primary visual cortex.*

The studies of Marshall, Talbot, and colleagues were extended (Marshall et al., 1943) to show that regions of cortex even further lateral, on the middle suprasylvian gyrus and bounding the middle suprasylvian sulcus, could also be excited by visual stimulation and electrical stimulation of the optic nerve (Figs. 1-1B and 1-4; ms). Marshall et al. (1943) labeled this more lateral region visual area V3. They also provided evidence based on lesions of afferent pathways that visual area V3 received visual input from the thalamus independent of signals transmitted to visual area V1. They even suggested that the recorded potentials might be evoked by signals transmitted directly from LGN. Clare and Bishop (1954) confirmed the visual activation of the lateral cortex yet believed, contrary to Marshall et al. (1943), that the activation was secondary to the primary activation of cortex on the marginal gyrus. However, studies first by Doty (1958) and then by Vastola (1961) provided substantial evidence that subregions within the suprasylvian belt could be activated by electrical and visual stimuli independently of area 17. For very important reasons discussed later, neither visual area V3 nor suprasylvian cortex is currently included in the concept of primary visual cortex.

Microelectrode Techniques

Microelectrode recording techniques were introduced to the study of visual cortex in the late 1950s and early part of the 1960s by Hubel and Wiesel (1959, 1962). Hubel and Wiesel studied the action potentials of individual or small clusters of neurons and plotted receptive fields for the neurons using optimally configured visual stimuli that activate the neurons strongly. This method affords a much greater level of resolution than the evoked potential method because electrode movements of less that 100 μm across the cortical surface result in distinct shifts in the positions of receptive fields within the visual field. With these methods, Hubel and Wiesel (1965a, 1969) mapped in great detail limited regions of caudal marginal gyrus, its transition to the posterolateral gyrus, and the junction of the middle and posterior suprasylvian gyri, as well as the banks of the middle suprasylvian sulcus (Fig. 1-1A). They obtained evidence to confirm that all of these regions are visually excitable and that they contain multiple, orderly, visual representations that are sequential mirror images of the representation present in area 17. Hubel and Wiesel referred to these representations as visual areas V2 and V3 (Hubel and Wiesel, 1965a) and the Clare-Bishop (Hubel and Wiesel, 1969) areas. They also observed that: "In crossing from visual areas V1, V2 and V3 from medial to lateral, the corresponding receptive field areas in the contralateral visual field moved progressively into the vertical meridian, then out into the periphery of the visual field, and then finally back to the vertical meridian" (Hubel and Wiesel, 1965a). During the same period, compatible data on visual field representations in these three areas were obtained by Bilge et al. (1967) using similar techniques.

Examples of the orderly drift in receptive field positions to and away from the vertical meridian for sequentially studied neurons in areas 18 and 17 and the intervening transition zone are shown in Fig. 1-6. This figure is taken from a more

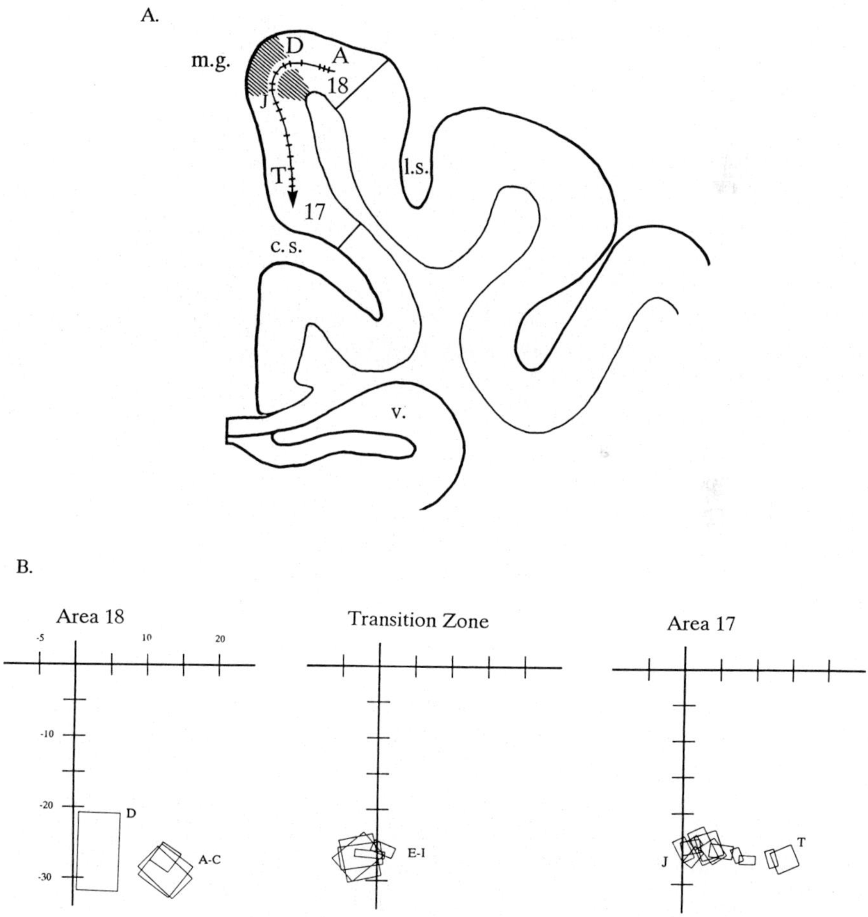

FIGURE 1-6. Mapping of the visual field in areas 17 and 18, and intervening transition zone (cross hatching). **(A)** Recording sites from two electrode tracks have been condensed onto layer 4 using the radial palisades of cells and bundles of axons as guides. **(B)** Receptive field plots have been grouped to show the orderly drift in receptive field positions for area 18 (loci A → D), transition zone (loci E → I), and area 17 (loci J → T). Intersection of thick lines represents the axis of vision, and coordinates represent azimuth and elevation in the visual field presented in degrees of visual angle. Positive azimuths represent the contralateral portion of the visual field, and negative elevations indicate the inferior portion of the visual field. Abbreviations are given in Table 1-1.

recent study by Payne (1990a). The figure summarizes the visual representation in part of area 18, part of area 17 and the intervening transition zone in a single coronal plane 9 mm anterior to the interaural plane. At this coronal level, areas 17 and 18 represent elevations at approximately –25° to –30° in the visual field. To readily appreciate the order and sequence in the visual representations, recording sites distributed across the six cortical layers have been condensed on layer 4 using the radial palisades of neurons and bundles of axons as alignment guides. It is then immediately evident that positions in area 18 sequentially closer to its medial border (loci A → D) view positions in the contralateral field (positive azimuths) progressively closer to the vertical midline. It is also obvious that positions in area 17 sequentially further from its medial border (loci J → T) view positions progressively further from the midline and into the contralateral field. Positions within the transition zone between the two areas extend the visual representation into the ipsilateral visual field (negative azimuths) (see page 73).

Hubel and Wiesel's (1965a) data on visual area V2 differed only in some details from the rough summary sketch of data on visual area V2 of Talbot and Marshall (1941) on the extent of the visual representation on the posterior suprasylvian gyrus. Yet Hubel and Wiesel's data differed considerably from the region designated visual area V2 by Woolsey (1958). In Woolsey's diagrams visual area V2 extends further laterally to encompass both areas designated visual area V2 *and* visual area V3 by Hubel and Wiesel (1965a). Of importance the electrophysiologically defined borders between visual areas V1, V2, and V3 by Hubel and Wiesel (1965a) show "a remarkably exact correspondence" with the anatomically defined borders between areas 17, 18, and 19 defined by Otsuka and Hassler (1962), and "in no case has there been any disagreement between the physiological and anatomical findings." Hubel and Wiesel used Brodmann's numerical terminology even though Brodmann did not delineate areas 18 and 19 in the cat. Regardless of terminology, here was the clinching evidence for three separate structural and functional visual areas occupying the posterior and medial aspect of the cerebral hemisphere.

Thus, the approach first used by Minkowski to correlate visual cortex with striate cortex, and by Talbot and Marshall to correlate visual area V1 with striate cortex, was extended by Hubel and Wiesel (1965a). They correlated the visual representations in visual areas V2 and V3 with areas 18 and 19, and with parastriate and peristriate cortices, respectively. Even though the terms can be used interchangeably, Hubel and Wiesel favored the use of visual areas V1, V2 and V3 when the context was physiological, and the equivalent terms areas 17, 18, and 19 when the context was anatomical. Moreover, Hubel and Wiesel recommended that the terms *parastriate* and *peristriate* of Smith (1908) and Gurewitsch and Chatschaturian (1928) should be avoided, as should the terms *occipital* and *preoccipital* of Brodmann (1906). Hubel and Wiesel considered the pairs of terms too readily confused to be useful.

Embodied in the 1965 descriptions of Hubel and Wiesel (1965a) on areas 18 and 19, there is a strong sense of cortical hierarchy. In this hierarchy, the representations in visual areas lateral to visual area V1 were considered to be secondarily derived from the primary representation in striate cortex. This view was founded on the patterns of fiber degeneration in areas 18 and 19, and beyond, induced by small lesions in area 17. These patterns suggested to Hubel and Wiesel a parallel hierarchy of projections emanating from area 17, and the parallel, sequential processing of visual signals beyond area 17. The sense of hierarchy also infused their adventurous concept of the circuitry underlying the synthesis of progressively more complicated receptive field structures as visual signals traversed cortical layers and areas (Hubel and Wiesel, 1962, 1965a, 1969).

Even with the intrinsic power of their results, Hubel and Wiesel (1965a) did not go uncriticized. Woolsey (1971) argued that in the absence of a more complete contour map, the data of Hubel and Wiesel are inconclusive because they cannot distinguish between "apparent" and "real" reversals in the visual field representation. Woolsey (1971) pointed out that this was the crux of the argument for one or two visual representations lateral to area V1. Even though Hubel and Wiesel (1965a) show one example of a reversal in the representation of the visual field at the area 18/19 border, Woolsey's perspective was not assuaged because he could not substantiate the result. The disagreement begged for more detailed mapping (see page 63.)

Regardless of the disagreements based on mapping, anatomical studies clearly showed three structurally distinct regions. These distinctions are readily identified in tissue stained for myelin. Both areas 17 and 18 are more heavily myelinated than both the more medial paralimbic cortex of the cingular sulcus or the more lateral area 19 (Fig. 1-7). However, complete maps of the visual field representations in these two areas based on microelectrode records had to wait for another decade.

CONNECTIONS

DEGENERATION METHODS: AREA 18 JOINS PRIMARY VISUAL CORTEX

In their scheme of visual processing, Hubel and Wiesel did not consider seriously the antecedent evidence for direct projections from LGN to regions or cortex lateral to striate cortex. However, very soon after Hubel and Wiesel's hierarchical scheme had been published, a reevaluation of geniculocortical connectivity using the older retrograde degeneration method, as well as evidence obtained with the newly developed and sensitive anterograde degeneration method, revealed direct projections from LGN to area 18. The anterograde degeneration technique provided then unrivalled details on the patterns of connections in the brain (Ebbesson, 1984; Nauta and Ebbesson, 1970), and it initiated a major reevaluation of the concept of primary visual cortex, which came to include both areas 17 *and* 18. It also signaled a significant departure from parallel investigations and developments on monkey visual cortex (Polyak, 1957). These developments occurred in an era when the structural organization of LGN, as the dominant relay of visual signals to visual cortex, was also being clarified (see Guillery et al., 1980 for a comprehensive description.)

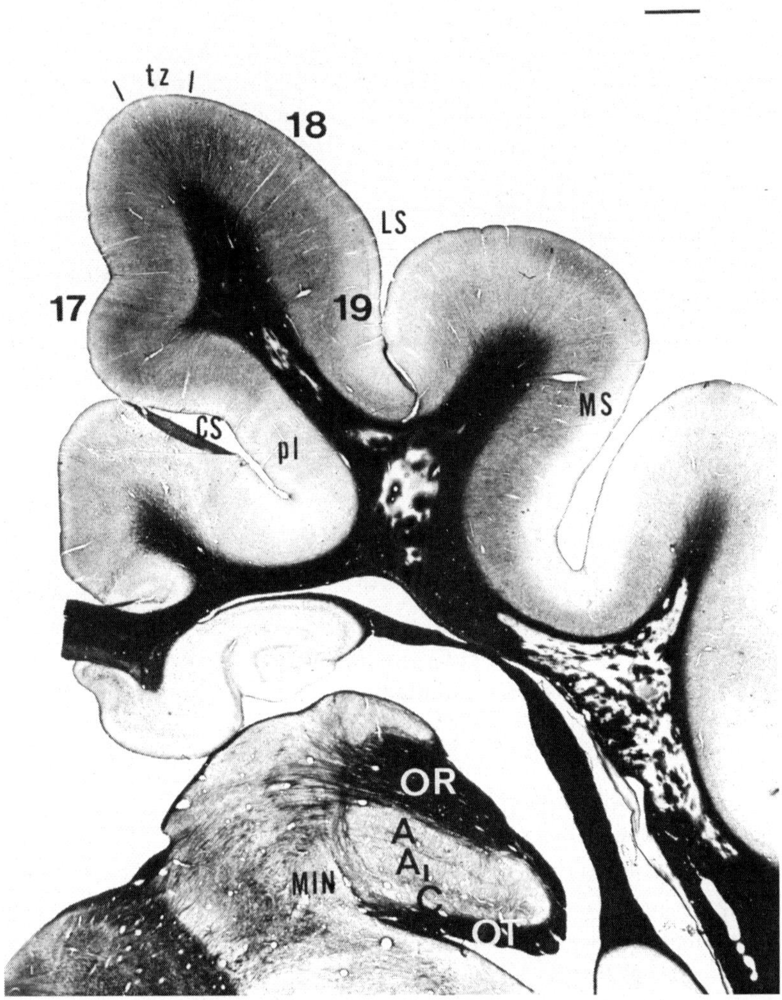

FIGURE 1-7. Low power photomicrograph of a coronal section stained for myelin. Areas 17 and 18 are obvious as more heavily stained regions on the medial and dorsal surfaces. The two areas are flanked by less intensely staining paralimbic cortex (pl) on the upper bank of the cingular sulcus (cs) and area 19 of the lateral sulcus (ls). The transitional cortex between areas 17 and 18 (tz) with its darkly staining, thicker, radiating fibers is identified. It is shown at higher power in Fig. 1-21. Major subdivisions of LGN are also shown together with the location of the optic tract (OT) and optic radiation (OR). Abbreviations are given in Table 1-1 (Scale bar = 1 mm.)

Retrograde Degeneration

The first major, clear-cut anatomical evidence for projections from LGN to regions of cortex lateral to area 17 was provided in 1967 by Garey and Powell. They were prompted to reinvestigate the connectivity of LGN to cortex for two reasons. First, the physiological evidence of Doty (1958) and Vastola (1961) supported the idea of direct, independent activation by LGN of cortex lateral to striate cortex. Second, Garey (1965) made some incidental observations on the patterns and severity of retrograde degeneration induced in LGN by cortical lesions in his study using primarily anterograde degeneration methods to examine the organization of cortical projections to the superior colliculus.

In the reanalysis of geniculocortical connections, Garey and Powell (1967) used the same retrograde degeneration technique as that used by Minkowski more than half a century earlier, but they were armed with more data on the visual cortical representations, clearer distinctions between cytoarchitectonically defined areas, and more precise questions than Minkowski had tried to answer. As a result, Garey and Powell's results were more readily interpreted, refined, and accurate. Their strategy was to make discrete lesions in area 17 or area 18 or to make larger lesions that encompassed portions of more than one area. In some instances, these lesions included more lateral regions of the marginal and suprasylvian gyri. They observed that following lesions confined to area 17, the medium and small cells in the dominant magnocellular LGN layers A and A1 degenerate, whereas lesions restricted to area 18 did not result in any localized, severe degeneration in LGN. However, combined destruction of both areas 17 and 18 caused all large, medium, and small cells in magnocellular LGN layers A, A1, and C to degenerate. Further expansion of the lesion to include area 19 resulted in additional degeneration in the medial interlaminar subdivision of LGN. With additional inclusion of suprasylvian cortex in the damage the severity of the retrograde degeneration increased still further and severe gliosis was evident. These observations showed that the LGN projects differentially on a wide region of visual cortex.

The results of Garey and Powell (1967) also provided an excellent example of the concept promulgated by Rose and Woolsey (1958) of sustaining thalamic projections on two areas of cortex, and suggested that the largest LGN neurons project to both areas 17 and 18 via collateral axon branches (Fig. 1-8). This suggestion has subsequently been confirmed in three ways:

1. Sequential electrical stimulation of areas 17 and 18 both activate antidromically a subpopulation of neurons in LGN with collateralized projections to both areas (Stone and Dreher, 1973).
2. Injections of different retrogradely transported tracers into visuotopically corresponding regions in areas 17 and 18 result in populations of neurons in LGN being double labeled (Geisert, 1980).
3. Direct intra-axonal injection of markers into geniculocortical axons reveal that some axons branch to innervate both areas 17 and 18 (Freund et al., 1985a; Humphrey et al., 1985b).

Garey and Powell (1967) took their analyses one stage further and confirmed Minkowski's observations on the topographical organization of the LGN projection onto cortex along the anterior axes of both LGN and cortex. Moreover, they showed that lateral LGN projects in mirror-image fashion to the medial part of

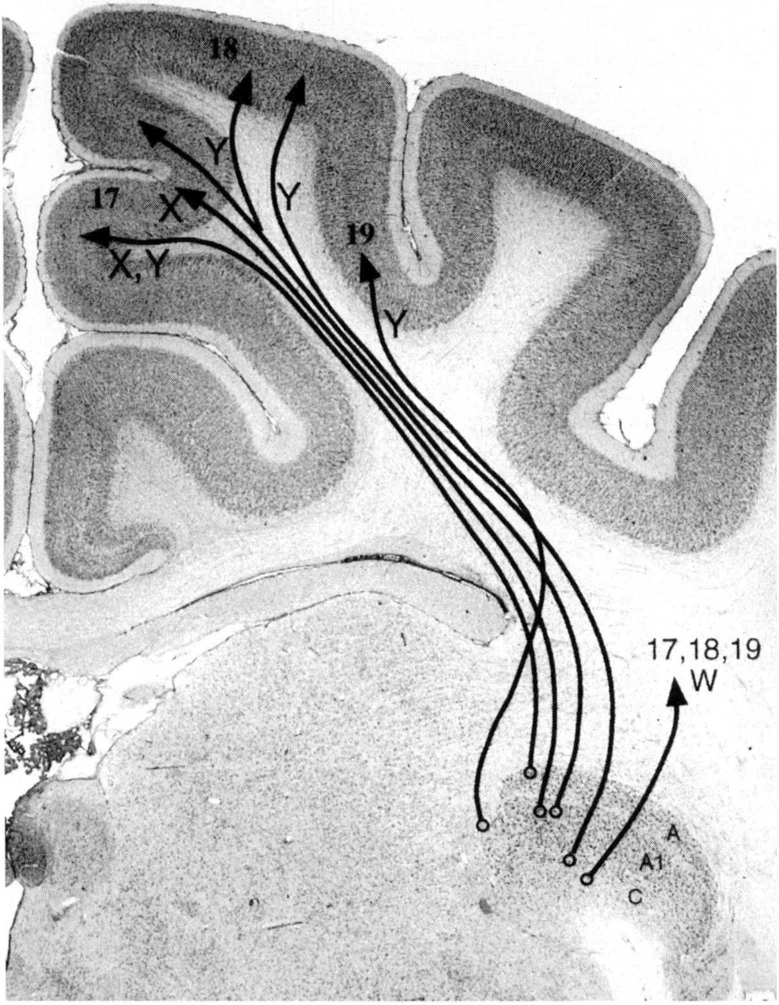

FIGURE 1-8. Summary of major patterns of innervation of visual cortical areas 17, 18, and 19 according to subcompartment origin in LGN and signal type. Note collateralized axonal projections to both areas 17 and 18 that arise in layer A1 of LGN.

area 17 and to the lateral part of area 18, thus providing independent, corroborating evidence for the mirror-image representations based on electrophysiological methods and described by Hubel and Wiesel (1965a).

Even with these gains in knowledge, it is important to recognize that ultimately the resolution of the mapping of LGN on cortex based on the retrograde degeneration method is limited because minute lesions of cortex do not produce obvious retrograde degeneration in LGN (Garey and Powell, 1967). Garey and Powell suspected that the absence of degeneration following the smallest cortical lesions revealed limited sensitivity of the method. Even so, they showed that localized lesions of sufficient size induced a column of degeneration that extended perpendicularly through the LGN layers. This observation supported the concept of the *projection column* enunciated by Bishop et al. (1962). In this concept the projection column identifies the translaminar grouping of cells in LGN that represent the same region of the visual field, regardless of whether input is derived from the ipsilateral or contralateral eye (Fig. 1-2) (Hubel and Wiesel, 1961; Bishop et al., 1962). Thus, visual signals generated in the two eyes by a single locus in the visual field are transmitted to the same representational column in LGN before relay to cortex. However, within LGN the signals from the two eyes are not intermixed as was supposed by Minkowski. As we now know, the first stage of intermixing occurs at a common locus in area 17, and again at a separate common locus in area 18. At both cortical loci the same portion of the visual field is represented as seen through the two eyes. Anatomical observations largely concordant with those made by Garey and Powell (1967) were made by Doty (1971) and Niimi and Sprague (1970) who also used the retrograde degeneration technique.

Anterograde Degeneration

Confirmative anatomical evidence for direct projections from LGN to regions of cortex lateral to striate cortex was obtained with the sensitive anterograde degeneration techniques refined in the 1960s. With these methods nerve fibers and their axon terminals were induced to degenerate by disconnecting axons and their terminals from their parent cell bodies, either by killing the neuron cell bodies directly or by damaging fibers in passage. After some days, the nerve fibers are reduced to fragments and the terminals to dust, both of which can be visualized by depositing silver precipitates on the fragments and dust in suitably prepared histological sections. Thus, this method has the distinct advantage that discrete lesions in LGN can pinpoint precisely the terminations of the projections from the damaged neurons in cortex. However, the method also has the distinct disadvantage that fibers of passage unrelated to the projection system under investigation must be avoided to eliminate false-positive results. Consequently, investigators need prior knowledge on the organization brain pathways and need to be extremely skilled at interpreting results.

Wilson and Cragg (1967) and Glickstein et al. (1967) used stereotaxic procedures to make subtotal lesions in LGN and then plotted the resulting axon and terminal degeneration in cortex. Following lesions placed in the main body of LGN, they observed separate foci of axon terminal degeneration primarily in the middle layers of areas 17, 18, and 19, with some additional degeneration in the middle suprasylvian region of cortex when the lesions included the medial interlaminar component of LGN. Moreover, they observed that the caliber of the fiber degeneration in area 18 was coarser than the degeneration induced in area 17 (Garey and Powell, 1971; Rossignol and Colonnier, 1971). This observation suggests that

LGN axons projecting to area 18 are larger than their counterparts projecting to area 17. Moreover, it is in accord with the fact that large LGN neurons undergo retrograde degeneration following combined lesions of areas 17 and 18, but not following lesions of area 17 alone (Garey and Powell, 1967). In a somewhat later study, which was combined with electron microscopy, Garey and Powell (1971) also identified degeneration in cortical layer 1.

Wilson and Cragg (1967) and Glickstein et al. (1967) also demonstrated that systematic movement of the lesion site within LGN resulted in corresponding systematic movement of the loci of terminal degeneration across the cortical surface. These observations agreed with the retrograde degeneration studies on the geniculocortical projection carried out by Garey and Powell (1967) and with the multiple representations of the visual field in LGN and cortex identified by electrophysiologists. These combined sets of observations served to cement the topographical linkages between the single visual representations in retina and the multiple visual representations in LGN and visual cortex.

Even though it was obvious that the retina mapped onto cortex in multiple orderly ways, the projections were not point-to-point. Glickstein et al. (1967) showed that "no matter how small the lesion in LGN, (they) always saw a rather large region of cortex containing dense degenerating fibers." Glickstein et al. (1967) suggested that the widespread degeneration may be either artificial and may result from unintended damage to the optic radiation, a peril of the anterograde degeneration method, or real and reveal the widespread divergence of axons from a small region of LGN to a large region of cortex. Subsequent studies using injection of markers into geniculocortical axons confirmed the latter interpretation. Moreover, they showed that some individual LGN axons have terminal arbors that can span several millimeters to innervate relatively large regions of cortex (Freund et al., 1985a; Gilbert and Wiesel, 1979, 1983; Humphrey et al., 1985a,b; Martin, 1988; Martin and Whitteridge, 1984). For these axons, parts of the termination field are dense, whereas other parts are sparse. The variation in density produces a patchy termination pattern, which in area 18 may extend to one half, or more, of the width of the area (Humphrey et al., 1985b). With these parameters it is understandable that Garey and Powell (1967) were unable to detect retrograde degeneration in LGN following minute lesions of cortex. Minute lesions eliminate neither sufficient numbers of axon branches of individual LGN neurons nor axons of a sufficient number of LGN neurons to induce detectable retrograde atrophy in LGN.

Together, these anatomical data reveal both a divergence of visual signals from a small retinal locus to large numbers of cortical neurons, and concomitantly a convergence of signals from a large number of retinal loci onto individual cortical neurons. This point was appreciated by Hubel (1959) a decade earlier when he reported that neurons in striate cortex of unanesthetized cats could be activated from large regions of the visual field. These observations demonstrated that the concept of point-to-point representation in visual pathways is true only in a statistical sense.

AXOPLASMIC TRACERS: BOTH AREAS 17 AND 18 ARE ACCEPTED AS PRIMARY VISUAL CORTICES

The introduction of axoplasmic pathway tracers in the 1970s and early 1980s ushered in an explosion in our detailed understanding of the organization of brain pathways (Cowan et al., 1972; Ebbesson, 1984; Mesulam, 1982). During this period cat area 18 was accepted as a primary visual cortical area in addition to area 17.

Anterograde and retrograde axoplasmic pathway tracers added both technical precision and sensitivity over the degeneration techniques and largely avoided the "fibers of passage problem." Anterograde tracer methods involve the injection of substances, such as tritiated amino acids, sugars, or larger molecules, into neuron rich regions. The tracers are taken up by cell bodies and transported long distances along axons to their terminals. The tracers are then visualized using autoradiographic techniques, simple biochemical reactions, or immunoreactions. In the opposite direction, retrograde tracers, substances such as horseradish peroxidase, fluorescent markers, or even colored sub-micron-sized latex spheres, are injected into the vicinity of axon terminals: the tracer is taken up by the terminals and subsequently transported backwards along axons to their cell bodies. The label is visualized in the parent cell body with the appropriate techniques, and the origin of the pathway is identified.

Anterograde Tracers

Anterograde tracers injected into LGN confirmed the results of the anterograde degeneration studies summarized previously and added a level of detail heretofore unobtained. The earliest and most significant studies were carried out by Rosenquist et al. (1975) and by LeVay and Gilbert (1976). Tritiated proline injected into the magnocellular layers A and A1 of LGN was transported to cortex along the optic radiation, and heavy amounts were identified throughout the full thickness of layer 4 in areas 17 and 18 with additional weaker labeling of layer 6 (Fig. 1-9). Concordant observations were made a decade later by Nelson and LeVay (1985). Intra-axonal injection of markers into physiologically characterized LGN axons, as they approach areas 17 and 18, reveal that axons transmitting X-type signals arborize throughout the full thickness of layer 4 in area 17 (Fig. 1-9, Area 17) (Freund et al., 1985a; Humphrey et al., 1985a,b). In contrast, those axons transmitting Y-type signals innervate the upper part of layer 4 in area 17 and the upper two thirds to full thickness of layer 4 in area 18 (Fig. 1-9, Area 18), (Freund et al., 1985a; Humphrey et al., 1985a,b). All Y axons have collateral branches in layer 6, and many have terminals dispersed through the base of layer 3 (Freund et al., 1985a; Humphrey et al., 1985a,b). However, some of the axons innervating area 18 may arise from neurons located in LGN magnocellular layer C (Fig. 1-9A; Humphrey et al., 1985b).

Autoradiographic procedures showed that the projections of parvocellular layer C neurons in LGN terminate both low and high in layer 4 of areas 17 and 18 and extend both into the base of layer 3 and into the most superficial part of layer

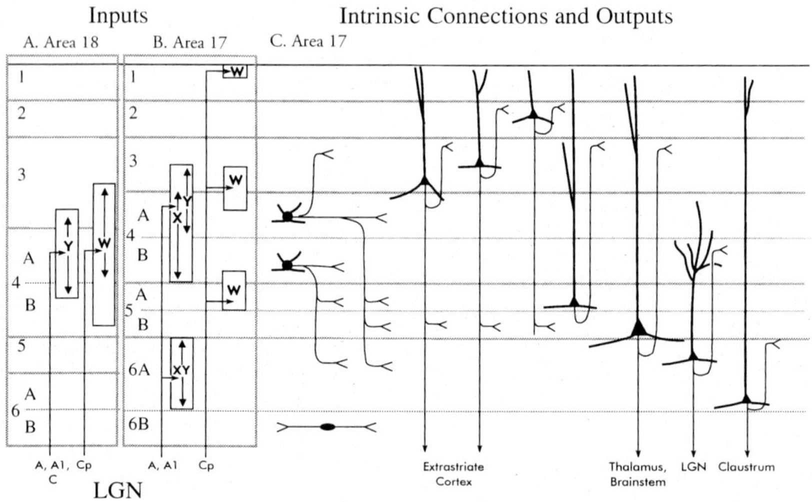

FIGURE 1-9. Diagrams of the innervation of area 18 (**A**) and area 17 (**B**) by fibers arising in LGN, and intrinsic circuitry of spiney neurons in area 17 (**C**). Spiney cells are present in all layers except layer 1. Those in layer 4 have a stellate morphology with radiating dendrites, whereas those in all other layers have a pyramidal morphology and are characterized by an ascending apical dendrite. Laminar terminations of local axon collaterals are indicated. (Figure based on data of Kawano [1998], Boyd and Matsubara [1996], Garey and Powell [1971], Humphrey et al. [1985a,b], Freund et al. [1985a], LeVay and Gilbert [1976], Rosenquist et al. [1975], Leventhal [1979], Lund et al. [1979], Katz [1987], Martin and Whitteridge [1984], and Martin [1988].) For detailed patterns of innervation in cortex relative to the cytochrome oxidase blob and interblob regions see Fig. 5-14.

5 to overlap and bracket the projections from layers A and A1 (Fig. 1-9, Area 17). The C layers also project to the superficial half of layer 1 in area 17 (Fig. 1-9, Area 18). In addition, exposure of C-layer neurons to the tracer results in transport of tracer to the middle and upper layers of areas 18 (Fig. 1-9A) and 19 and parts of the suprasylvian gyrus (not shown). These basic patterns have been substantiated recently by Boyd and Matsubara (1996) and by Kawano (1998). Thus, these results are in complete agreement with earlier anterograde degeneration studies. Moreover, they confirmed the orderly, topographic arrangement of projections from LGN to areas 17 and 18 and more lateral cortices, because they showed that systematic movement of the tracer deposit site in LGN resulted in corresponding and predictable movement of the transported tracer recipient site in cortex.

Small injections of anterograde tracer showed additional details of the form of the projections from LGN to areas 17 and 18. Tracer deposits largely confined to a single LGN layer revealed that the terminations within layer 4, base of layer 3, and layer 6 are discontinuous in areal, plan view, whereas those to the superficial half of layer 1 are continuous (Rosenquist et al., 1975; LeVay and Gilbert, 1976; Boyd and Matsubara, 1996; Kawano, 1998). These patchy terminations confirmed the absence of finite point-to-point projections between LGN and cortex. The discontinuities in upper layer 4 and lower layer 3, which are derived primarily from the LGN C layers, appear to be directed to the cytochrome oxidase rich zones that have been described in the upper half of both areas 17 and 18 (Murphy et al., 1995; Boyd and Matsubara, 1996; Matsubara and Boyd in this volume). Deeper in layer 4, the patchy terminations appear as a result of the interdigitation of fibers arising from the separate, but continuous and topographically coincident, representations in each of the LGN layers in receipt of fibers from either the right eye or the left eye. The periodicity of the patches in area 18 is greater than the periodicity of the patches in area 17.

Transsynaptic Tracing

The patchy, interdigitating nature of the projections from the two eyes to cortex was confirmed by anterograde transsynaptic transport of tritiated sugars and amino acids from one eye through LGN to cortex using the methods of Specht and Grafstein (1973) (Fig. 1-10). In LGN, the fibers from the two eyes remained segregated (Fig. 1-2), and the label was transported to layer 4 of areas 17 and 18, where they form interdigitating bands of fibers dominated by signals transmitted from either the left or the right eyes. These bands are shown in cross section in Fig. 3-5 and in tangential view of flattened cortex in Fig. 6-6. The technique also revealed that the most medial position of tracer transported from the contralateral eye coincided with the medial border of area 17 (top of Fig. 6-6A), where the number of layer 4 granule cells and myelination decrease abruptly and cortex transitions into a paralimbic form (Fig. 1-4 and 1-7). Likewise, the lateral edge of label in layer 4 coincided with the lateral border of area 18 (bottom of Fig. 6-6A). Thus, the transsynaptically transported label marked exactly the full extent of both areas 17 and 18.

The label was detected neither in layer 1 nor in layer 6 of areas 17 and 18, and none was identified in more lateral regions or cortex (Shatz et al., 1977) (see Fig. 3-5 of Löwel). The label was transported transsynaptically and apparently in sufficient quantities only by neurons in the LGN magnocellular layers A, A1, and C to layer 4 of areas 17 and 18. Subsequent studies have shown that the patches have a coarser grain in area 18 than in area 17 (Anderson et al., 1988) (Fig. 6-6). This feature corresponds to the larger and more patchy form of geniculocortical axon arbors in area 18 than in area 17 (Freund et al., 1985a; Humphrey et al., 1985a,b). In area 17 the numerically dominant X-axons usually form single continuous clumps of terminals in layer 4 with overall dimensions in the range of 0.5 to 0.8 mm^2. In area 18, the dominant and patchy Y-axons arborize over an area of 2.0 to 2.8 mm^2. This transsynaptic labeling method reemphasized the strong linkages, and even dependencies, between retinal ganglion cells, LGN neurons, and visual cortex recognized a century earlier by von Monakow (1883, 1889), and reviewed in the current volume by Payne and Rushmore. Moreover, the results confirmed Wilbrand's (1890) *Hypothese der Faszikelfeldermischung,* that fibers arising from the two retinal halves transmit signals to the same small region of cortex (see page 18).

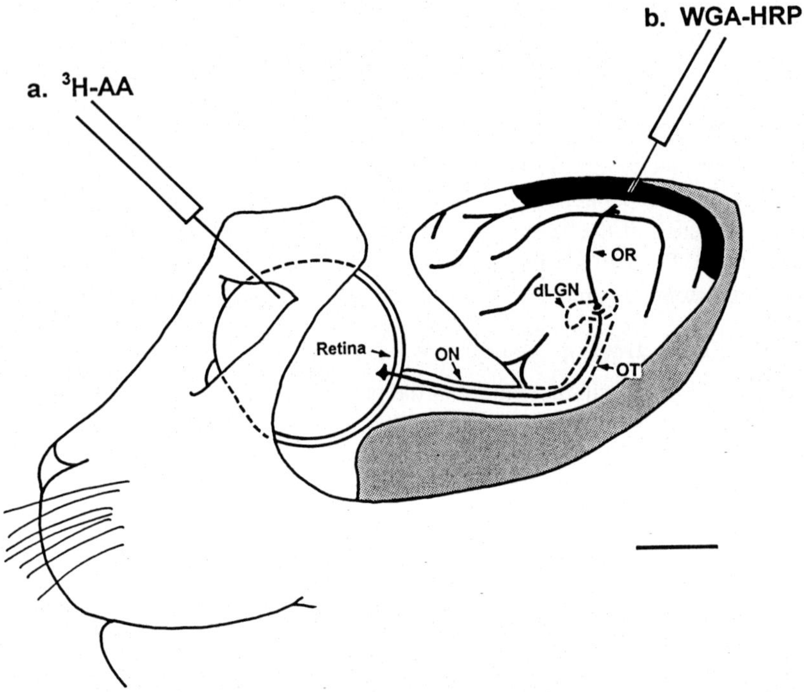

FIGURE 1-10. Diagram illustrating the transsynaptic tracing method. Either anterograde or retrograde tracers may be used. The blackened region of cortex indicates areas 17 and 18. **(A)** Tritiated amino acids (^{3}H-AA), as an anterograde tracer, are injected into the eye. The label is taken up by retinal ganglion cells and primarily transported along the optic nerve and optic tract to ganglion cell terminals in LGN. Some of the label is released from the axon terminals and is taken up by postsynaptic LGN neurons. It is secondarily transported forward along LGN fibers in the optic radiation to their terminals in areas 17 and 18. **(B)** WGA-HRP as a retrograde tracer is injected into areas 17 and 18. The label is taken up by axon terminals and it is primarily transported retrogradely to label LGN neurons. In LGN, some label is released by the neurons and is taken up by presynaptic retinal axon terminals. It is then secondarily transported backwards along the axons to label ganglion cell bodies in the retina. Anterior is left. (Illustration was prepared by Dr. Stephen Lomber and was adapted from Fig. 7-3 in Burke, 1986.) (Scale bar = 10mm.)

The interdigitation of patches in layer 4 of areas 17 and 18 in receipt of signals from the right and left eyes can be aligned in a series of sections, or viewed in flat mounts, to reveal short bands, islands, or lattices with lacunae (Fig. 6-6). With this method the demarcation between ipsilateral and contralateral eye territories is slightly sharper and more obvious for ipsilateral pathways than for contralateral pathways. This discrepancy suggests that ipsilateral pathways have a more focused innervation pattern that is confined to distinct cortical zones with little or no overlap into zones innervated most heavily by the contralateral eye. In contrast, contralateral pathways have a more diffuse and blurry pattern of innervation because small amounts of tracer are frequently detected in regions most heavily innervated by the ipsilateral eye. This feature forms the anatomical substrate for the initial convergence of signals from the two eyes (see page 94, outer dominance) and generates in areas 17 and 18 the first binocular receptive fields in the visual system (Hubel and Wiesel, 1962, 1965a). Of importance for the current topic, the bands are broader and consequently have a greater spacing in area 18 (>1 mm) than in area 17 (0.4–0.5 mm) (Shatz et al., 1977; Shatz and Stryker, 1978; Anderson et al. 1988; see page 90, Functional Columns).

The absence of detectable transsynaptically transported label from layers other than layer 4 of areas 17 and 18, and from more lateral regions of cortex, suggests that the synaptic coupling between retinal fibers and parvocellular neurons is markedly weaker than the substantial coupling between retinal axons and magnocellular layer neurons (Payne and Lomber, 1998; Payne and Rushmore, this volume). The weak coupling has been also revealed by poor labeling in parts of area 19 and the middle suprasylvian region compared with much stronger labeling in areas 17 and 18 using more sensitive transsynaptic tracers such as wheat germ agglutinin conjugated to horseradish peroxidase (Anderson et al., 1988).

Retrograde Tracers

All of the observations made in cats on projections from LGN to areas 17 and 18 and more lateral regions of cortex were amply confirmed with the use of retrograde tracers (Figs. 1-8 and 1-10) (Holländer and Vanegas, 1977; Geisert, 1980; Leventhal, 1979; Maciewicz, 1974, 1975; Niimi et al., 1981). Further, the subsequent application of more sensitive retrograde tracing techniques (Mesulam, 1982) revealed additional projections from the parvocellular C layers to a wide expanse of visually responsive cortex beyond areas 17 and 18 (Raczkowski and Rosenquist, 1980, 1983; Rosenquist, 1985; MacNeil et al., 1997a,b). Moreover, there are additional projections from the medial interlaminar and wing components of LGN to areas 17, 18, and 19, and to the middle suprasylvian region of cortex (Geisert, 1980; Gilbert and Kelly, 1975; Leventhal, 1979; Leventhal et al., 1980; Rosenquist, 1985; MacNeil et al., 1997a,b). Some of these projections are via axons that branch to innervate more than one cortical area (Birnbacher and Albus, 1987; Bullier et al., 1984; Rosenquist, 1985).

Holländer and Vanegas (1977) summarized the laminar origins of LGN projections to cortex in the following way (Fig. 1-8). Within the laminated portion of LGN most of the neurons that project to area 17 are located in the A-laminae with few in the C-laminae, and none in MIN. Most of the neurons that project to area 18 are located in layers A1 and C and in MIN. Most cells that project to area 19 are located in the C-laminae and in MIN. That is, as one goes from area 17 to 18 and then on to area 19, the source of the projection shifts within LGN from the A-laminae laminae through the C-laminae into MIN. The source of projections to area 18 covers all components of LGN, and the projecting cells are, on the whole, larger than the neurons that project to either area 17 or to area 19. As expected, there is considerable topographic order in the form of the projections, which is also visible in the optic radiation (Nelson and LeVay, 1985; Payne and Siwek, 1991a).

Transsynaptic Tracing and Neuronal Coupling in LGN

As revealed by the retrograde transsynaptic tracing method (Fig. 1-10), the coupling between primary and relay limbs of the transgeniculate pathway leading to areas 17 and 18 differs from the coupling in the transgeniculate pathway leading to area 19 and middle suprasylvian visual cortex. With optimal use and application of very sensitive procedures, localized cortical injections of retrograde tracers also result in transsynaptic transport of label through LGN to distinct foci of α and β ganglion cells in retina (LeVay and Voigt, 1990; Payne and Lomber, 1998).

LeVay and Voigt (1990) injected wheat germ agglutinin conjugated to horseradish peroxidase into areas 17 and 18. Injections into area 17 labeled neurons at visuotopically corresponding positions in LGN and, after retrograde transsynaptic transport, in retina. Labeled ganglion cells contained dense reaction product, and all cells were classifiable on the basis of soma size and initial portions of their dendritic trees as α or β types. No cells of the γ group were labeled. Similar injections into area 18 also labeled LGN neurons and both α and β retinal ganglion cells at visuotopically matching regions, although the intensity of the labeling of α cells was much greater than the intensity denser of the labeling of β cells. Given all of the information known on visual projections to areas 17 and 18, it is surprising that any β cells were labeled because it is thought that β cells do not transmit signals that reach area 18 directly from LGN. At present it is not known if the labeling of β cells reflects indiscriminate spread of tracer from LGN neurons to terminals of β optic axons unrelated to cortical target locus, or whether it is a true representation of coupling between optic axons and LGN neurons. Regardless of interpretation, there is no doubt that retinal α and β ganglion cells are strongly coupled with one or other, or both, areas 17 and 18.

In addition, these procedures confirmed the strong visuotopically, arranged linkages in the retinogeniculocortical visual system suspected from component data. They also confirmed the presence of substantial transgeniculate visual pathways to both areas 17 and 18 revealed with anterograde transsynaptically transported tracers. The success of the transsynaptic technique suggests that the synaptic linkages between primary and relay components in the magnocellular LGN layers A, A1 and C en route to areas 17 and 18 are strong, numerically large, or both. Moreover, the results have helped to explain the results of the transneuronal degeneration studies and the dependencies between visual system components revealed by Gudden (1869, 1874), von Monakow (1883, 1889), and more recently Payne et al. (1984) and made the phenomenon more understandable (see Chapter 14). However, the fundamental bases of the dependency are not known. Perhaps, it is trophic. Regardless, these features set apart the retinomagnocellular LGN pathway leading to layer 4 of areas 17 and 18 from the retinoparvocellular LGN pathway leading to the borders of layer 4 in areas 17 and 18, to area 19, and to middle suprasylvian visual cortex.

Application of the same retrograde, transsynaptic tracing procedures to area 19 and middle suprasylvian visual cortex fails to result in the presence of detectable label in retina (LeVay and Voigt, 1990; Payne and Lomber, 1998). The failure to label ganglion cells from these regions is surprising, because we know that anatomical and electrophysiological, piecemeal dissection of the retinogeniculo-cortical pathway reveals a two-stage pathway from retina, via LGN, directly to both area 19 and suprasylvian cortex. The failure of transsynaptic labeling with retrograde tracers is consistent with difficulties encountered in labeling either area 19 or middle suprasylvian visual cortex using anterograde, transsynaptic tracers injected into the eye. These difficulties suggest that the synaptic linkage between primary and relay components in the parvocellular LGN layers C1-3 en route to the borders of layer 4 in areas 17 and 18, and to area 19 and to middle suprasylvian visual cortex is weak, numerically small, or both. Moreover, there is little dependence between neurons in the parvocellular system for neuron maintenance and survival.

Corollaries and Dichotomies

Important corollary data are available to distinguish the transgeniculate projections to areas 17 and 18 from transgeniculate projections to area 19 and middle suprasylvian visual cortex. The dominant retinal input to areas 17 and 18 is derived from the functionally dominant α and β retinal ganglion cells, which are the origin of Y-and X-visual signals. Alpha and β ganglion cells are readily labeled by retrograde transsynaptic transport of tracers injected into areas 17 and 18. Both α and β ganglion cell types, as well as their magnocellular A, A1, and C LGN target neurons, have high neural activity and generate brisk responses (Stone, 1983). In contrast, area 19 and middle suprasylvian visual cortex derive visual signals from α, δ, ε, η, and γ retinal ganglion cells. These ganglion cells have not so far been labeled by retrograde transport of tracers injected into areas 17, 18, or 19 or the middle suprasylvian region in the intact adult cat. In contrast to the α and β retinal ganglion cells and magnocellular LGN neurons, the δ, ε, and γ ganglion cells, and their target neurons in parvocellular layers C1 and C2 of LGN, respond sluggishly to visual stimuli and transmit action potentials rather slowly. The α cells that contribute to the pathway have high activity and respond briskly, but they appear to be only sparsely connected to area 19 and middle suprasylvian visual cortex.

The fundamental dichotomies in activity patterns, responsiveness, and connectivity patterns described previously are buttressed by differential levels of cytochrome oxidase activity in retina, LGN, and cortex. For example, the α and β ganglion cells, their target magnocellular layer LGN neurons, and the targets of the latter neurons in layer 4 of areas 17 and 18 all exhibit high cytochrome oxidase activity (Fig. 1-11) (Kageyama and Wong-Riley, 1984, 1985, 1986; Payne and Lomber, 1996; Price, 1985; Payne, 1990a). Moreover, electron microscopy has shown that in layer 4 of area 17 much of the high cytochrome oxidase activity is located in axon terminals that form large asymmetrical synapses (Kageyama and Wong-Riley, 1986). Electron microscopy of labeled terminals has also shown that the parent cells of these axons are located in the magnocellular layers of LGN (LeVay and Gilbert, 1976; Einstein et al., 1987). It is likely that the equivalent ter-

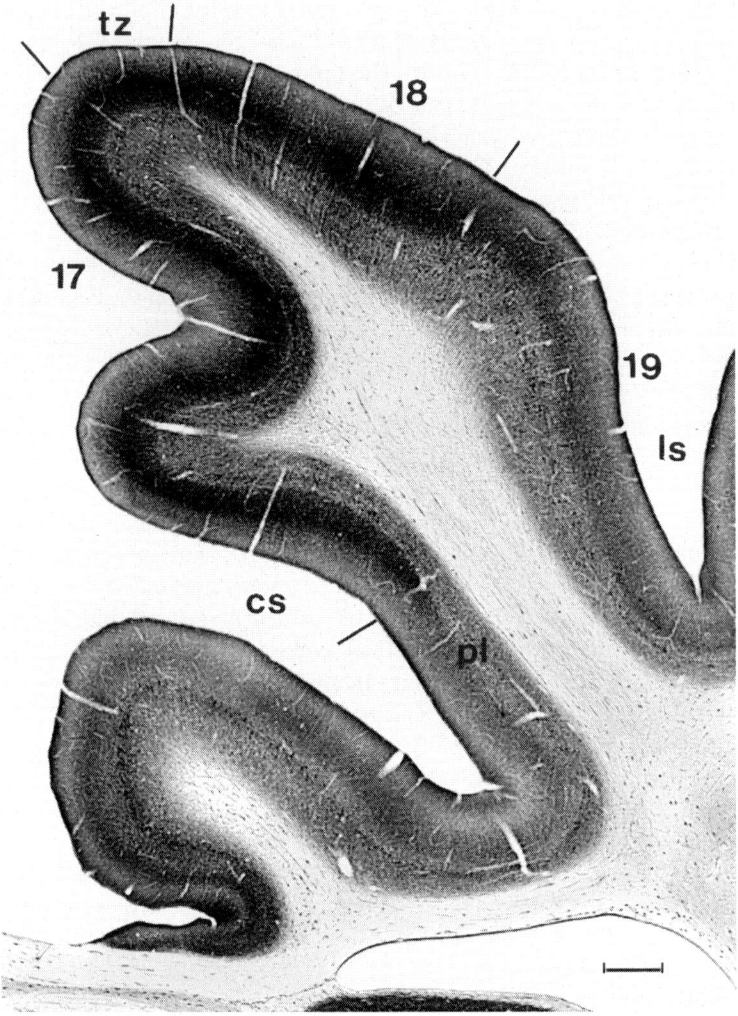

FIGURE 1-11. Low-power photomicrograph of a section reacted for the presence of cytochrome oxidase. Note high reactivity in layer 4, medium reactivity in the upper layers, and slightly lower ractivity in the deep layers of both areas 17 and 18. The particularly prominent stripe of areas 17 and 18 is absent from the middle layers of both paralimbic cortex and area 19. Moreover, density of reaction product in the upper layers of the latter two areas is lower than in the neighboring areas 17 and 18. Abbreviations are given in Table 1-1. (Scale bar = 1 mm.)

minals in area 18 also exhibit high cytochrome oxidase activity. In contrast, δ, ε, γ, and η retinal ganglion cells, their target parvocellular layer LGN neurons, and the targets of the parvocellular LGN neurons outside of layer 4 in areas 17 and 18 exhibit lower levels of cytochrome oxidase activity (Fig. 1-11) (Kageyama and Wong-Riley, 1984, 1985, 1986; Payne and Lomber, 1996; Price, 1985; Payne, 1990a). There are also lower levels of cytochrome oxidase activity in layer 4 of area 19 and the middle suprasylvian visual cortex (Price, 1985; Long et al., 1996), the additional targets of parvocellular neurons (Rosenquist, 1985; Kawano, 1998).

Multiple Dichotomies—Areas 17 and 18 Are Set Apart

These multiple dichotomies are sufficiently great that the transgeniculate pathways leading to areas 17 and 18 have been viewed differently from the transgeniculate pathways leading to area 19 and middle suprasylvian visual cortex. The former are characterized by robust activity and strong coupling, and even dependencies in Gudden's sense (Gudden 1869, 1874) between pathway components, and these two features qualify both areas 17 and 18 for inclusion under the umbrella term *primary visual cortex.* The opposite qualities of more sluggish activity, poor coupling between pathway components, and absence of dependence disqualify area 19 and middle suprasylvian visual cortex from inclusion under the same umbrella term.

The acceptance of area 18 as part of primary visual cortex in addition to area 17 merits a review of the composition of area 17 as a prelude to a comparison of the structure of area 17 with the structure of area 18. The composition of area 17 is the topic of the next section.

COMPOSITION OF AREA 17

COMPOSITION OF AREA 17

On the basis of Nissl-stained preparations of area 17, or the striate cortex, it is generally accepted that six basic layers can be discerned (Fig. 1-3), with layers 4, 5, and 6 each sometimes being further divided into two sublayers. This lamination is essentially based on the disposition of cell bodies of the various sized spiney neurons (Fig. 1-12), that is, the pyramidal cells and spiney stellate cells, which are excitatory or non-GABAergic neurons, and account for 80% of the total population of neurons in cat area 17 (Gabbott and Somogyi, 1986). They are of considerable interest because they are the sources of, and effectors of, excitatory transmission in the cortex. The remaining 20% is composed of the GABA-ergic smooth or sparsely spinous cells. They are of interest because they are the effectors of intracortical inhibition.

Many of the details of the intrinsic organization of visual cortex have been based on studies of material prepared using Golgi's technique. The most comprehensive studies were carried out by Cajal (1921a,b; 1922), O'Leary (1941), Lund

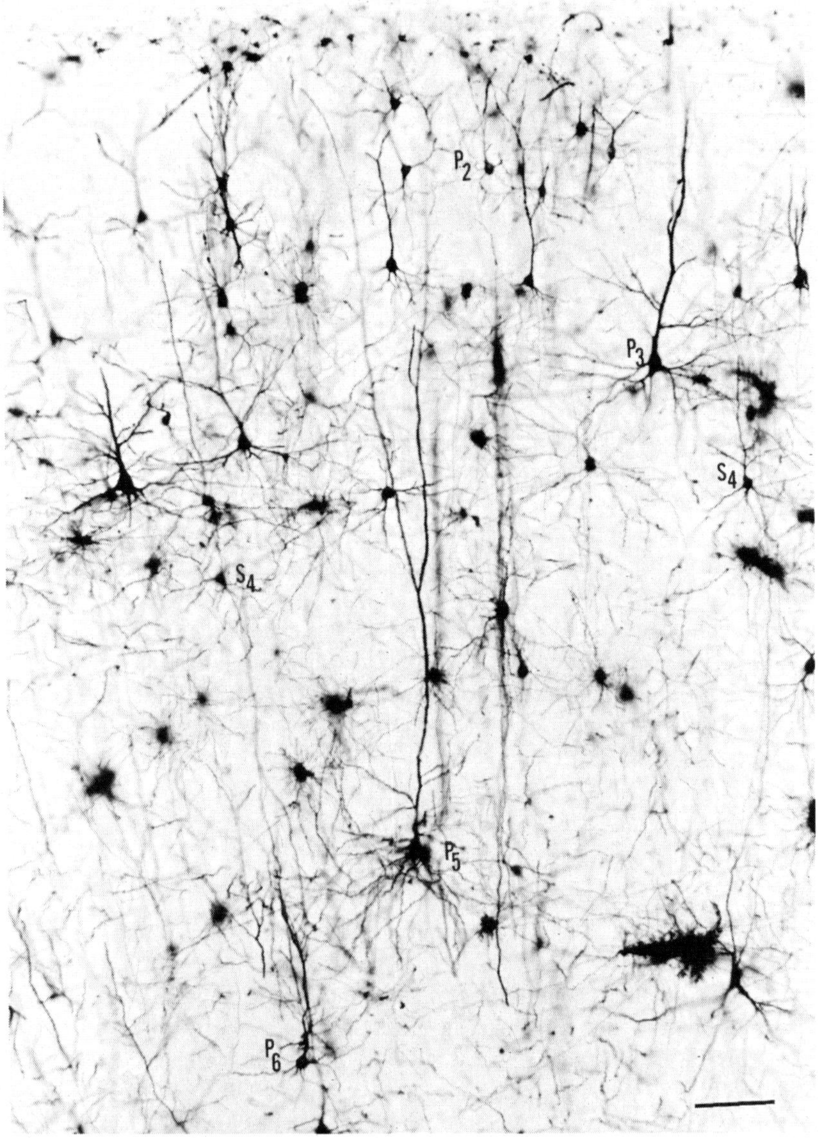

FIGURE 1-12. Golgi-Cox preparation of the supraspenial gyrus, to show the types of pyramidal cells (P_1–P_6) present in the various layers, and spiney stellate cells present in layer 4 (S_4) of area 17. (Calibration bar equals 100 μm.)

et al. (1979), and Meyer (1983). However, it is important to note that, in many Golgi impregnations, axons are rarely visible because the wrappings of myelin prevent the impregnation of the axon with precipitate. The profuseness and full complexity of axon arbors became prominently visible with intracellular injection of horseradish peroxidase and deposits of insoluble reaction products in the form of oxidized diaminobenzidine (See Fig. 10-3 and 10-4). Rather than present the complex drawings in all their glory and variety, we refer the reader to the original sources (Gilbert and Wiesel, 1979; Martin and Whitteridge, 1984; Martin, 1988), and we summarize here only the most germane features—the laminar patterns of termination of the local axon arbors (Fig. 1-9C).

Overall, the striate cortex is about 1.5 mm thick, and the most superficial, or outer layer, layer 1, has only a few neurons interspersed with other, small nuclei that belong mostly to astrocytes (Fig. 1-3). Because of the sparcity of neurons and the infiltration of this layer by the apical tufts of pyramidal cells, whose cell bodies are situated in deeper layers, layer 1 is often referred to as the plexiform layer. All of the neurons in layer 1 appear to be inhibitory (Gabbott and Somogyi, 1986), and O'Leary (1941) described two kinds of neurons in layer 1, small cells with short dendrites and locally arborizing axons, and horizontal cells. The former appears to be the type of layer 1 neuron recorded from and intracellularly filled by Martin et al. (1989). This neuron has a profuse axonal plexus that forms synapses with dendritic shafts and with spines that probably originate from the apical tufts of pyramidal cells, the prime components of the neuropil of layer 1. However, Anderson et al. (1992) subsequently injected 22 layer 1 neurons intracellularly with lucifer yellow and found that 16 of them had smooth dendrites, 2 had sparsely spined dendrites, and 4 had spiney dendrites. They also determined that despite the apparent sparcity of neurons in layer 1 compared with deeper layers, their density is sufficiently great that the dendritic trees of about six neurons cover each point in layer 1, so that there are more than sufficient neurons in layer 1 to sample signals emanating from the entire visual field multiple times.

In Nissl stained preparations (Fig. 1-3) the boundary between layers 1 and 2 is easy to define, because layer 2 is populated by small pyramidal cells that are concentrated at its upper border. With increasing depth the small pyramidal cells gradually give way to larger ones and this progressive increase in the sizes of the pyramidal cell bodies with depth continues into layer 3, for there is no distinct boundary between layers 2 and 3. Indeed, for the purposes of description, it is probably best to combine the two layers into a unified layer 2/3 that extends through about one third of the thickness of the cortex. It should be pointed out that O'Leary (1941) considered that layer 2 neurons should be differentiated from those in layer 3 on the basis that layer 2 contains pyramidal cells with short axons, whereas all of the pyramidal cells in layer 3 have axons that extend into the white matter (Figs. 1-9C and 1-13). One group of these layer 3 neurons stands out in Nissl preparations—the large pyramidal cells with irregular-shaped cell bodies that are interspersed throughout lower layer 3. O'Leary (1941) referred to these

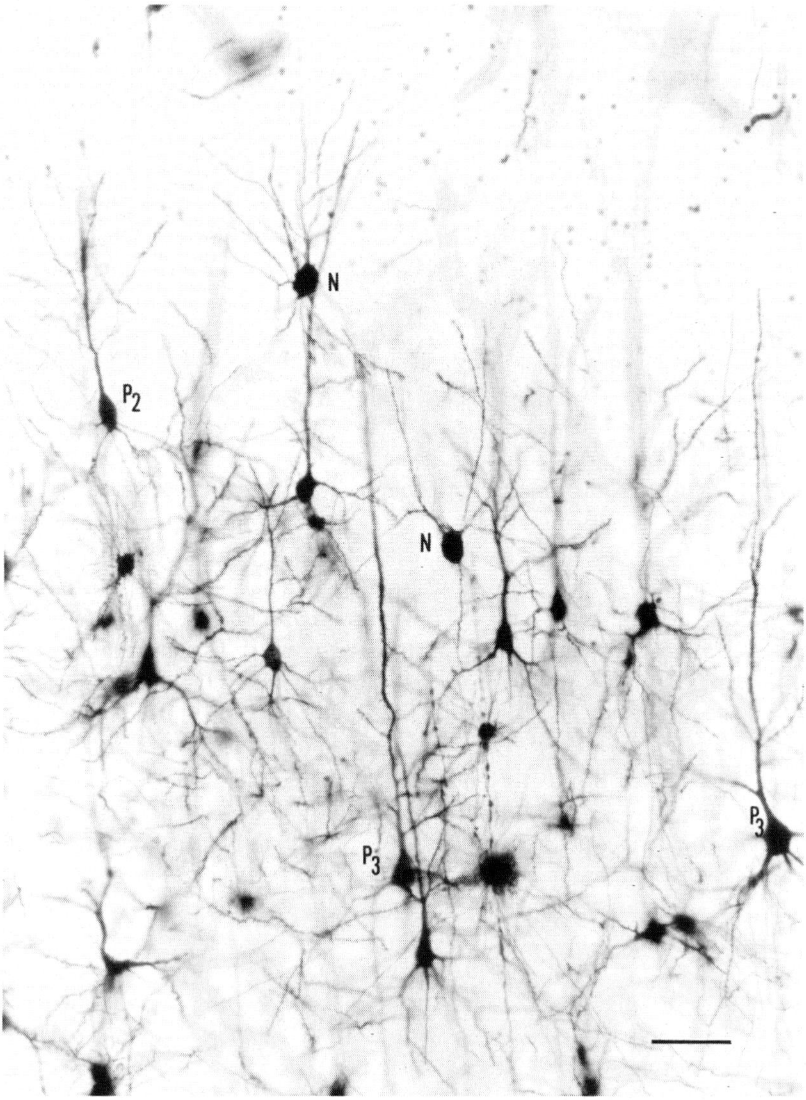

FIGURE 1-13. Golgi-Cox preparation of the outer layers of area 17 to show the small pyramidal cells (P_2) of layer 2 and the large irregular pyramidal cells (P_3) of lower layer 3. Two impregnated nonpyramidal cells (N) are also included in the field. (Calibration bar equals 50 μm.)

large pyramids as "border pyramids" since they demarcate the lower border of layer 3, where it abuts layer 4.

As is evident from studies of intracellular-filled layer 2/3 pyramidal cells by such workers as Gilbert and Wiesel (1983), Kisvárday et al. (1986), Callaway and Katz (1990), and Kisvárday and Eysel (1992), these neurons have complex projections. In addition to a branch that extends into the white matter and projects to other cortical areas (Fig. 1-9c) (Symonds and Rosenquist, 1984a; Einstein and Fitzpatrick, 1991), the axons form profuse local plexuses, as well as giving off branches that can extend horizontally in several layers (Fig. 1-14). These horizontal branches often form repeated clusters of terminals, the main targets of which are dendritic spines (84–87%), with the remainder of the terminals synapsing with dendritic shafts (Kisvárday et al., 1986). These observations suggest that the major targets of the horizontal axonal branches are other pyramidal cells, and that the prime function of the horizontal axonal spreads is to activate other pyramidal

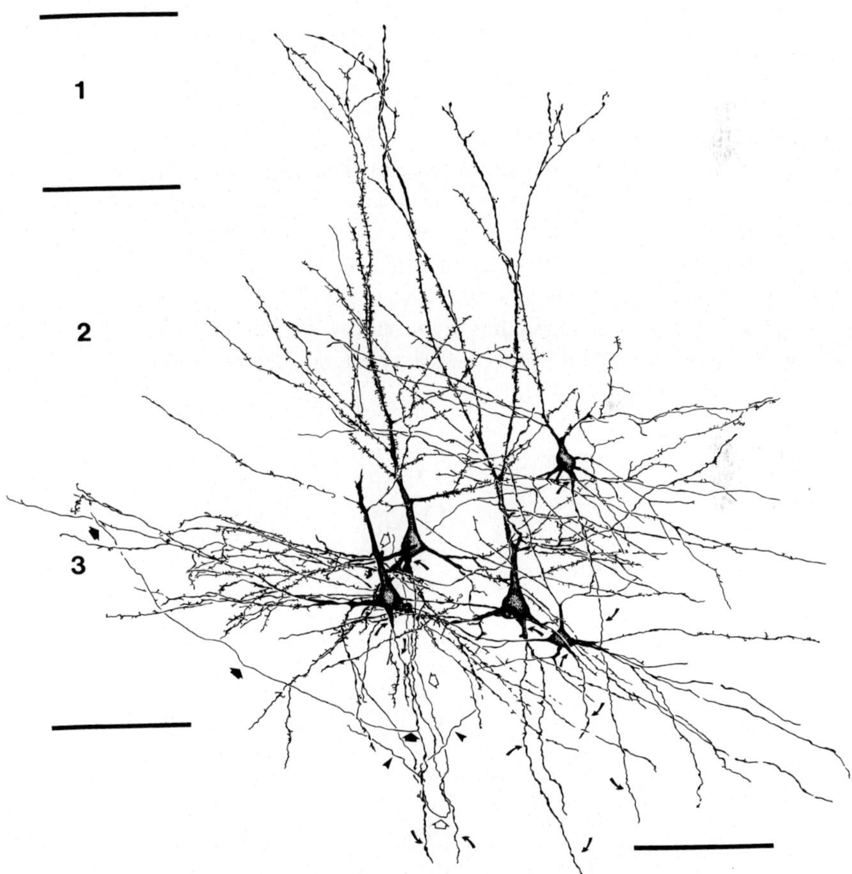

FIGURE 1-14. Camera lucida drawing of a group of layer 3 spiney pyramidal cells and two smoother cells in area 17. They were filled with oxidized diaminobenzidine following labeling with horseradish peroxidase. Note the prominent apical dendrite and substantial skirt of basal dendrites each of the three pyramidal cells possess. Curved arrows identify descending axons, and arrows and arrowheads identify collateral axon branches. (Figure reproduced from Payne [1986].)

cells. Indeed, the role of the horizontal axonal branches is most likely to connect together distant groups of pyramidal cells with similar physiological characteristics, such as similar stimulus orientation, ocular dominance, preference for movement direction, or some other receptive field property (see page 90). In terms of the extrinsic connections of the layer 2/3 pyramidal cells in area 17, their main connections are with the ipsilateral cortex (Gilbert and Kelly, 1975), although some layer 2/3 pyramidal cells close to the border with area 18 project axons through the corpus callosum to the opposite hemisphere.

Layer 4 is quite thick and occupies almost the entire middle third of the cortex. In general the neurons are smaller and more closely packed than those of layer 2/3. Most of the neurons are small spiney stellate cells (Fig. 15), but a few small pyramidal cells are scattered among them. Cajal (1921a,b) described the upper portion of layer 4 as the "layer in which stellate cells with long axons predominate." These stellate cells have spiney dendrites that extend radially away from their cell bodies and their axons may even reach the white matter and other cortical areas (Meyer and Albus, 1981; Einstein and Fitzpatrick, 1991). They resemble pyramidal cells in all of their features except that they lack an apical dendrite. Thus, their cell bodies participate only in the formation of symmetrical synapses, their dendrites bear spines on which most of the afferent axon terminals are received, and the terminals of the axons of both types of spiney neurons form asymmetrical synapses (Lund, 1984). However, O'Leary (1941) noted that the upper portion of layer 4 also contains some star pyramids, which are identified by a thin apical dendrite in addition to dendrites extending from all portions of their

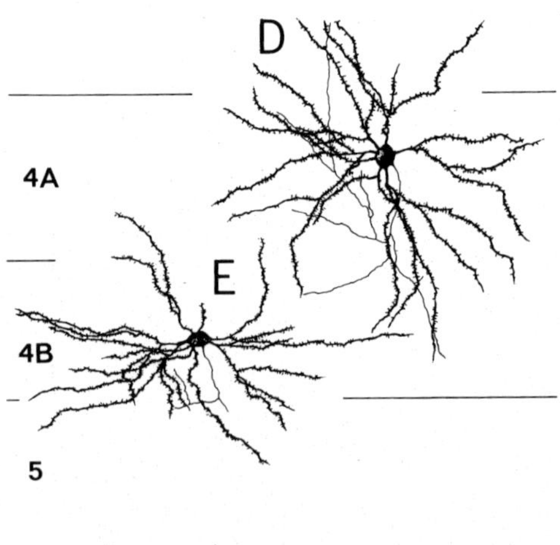

FIGURE 1-15. Camera lucida drawings of two spiney stellate cells in layer 4 (Calibration line, 100 μm). (Figure reproduced from Peters and Regidor (1981), *Journal of Comparative Neurology*, Copyright © 1981. Reprinted by permission of Wiley-Liss Inc., a subsidiary of John Wiley & Sons, Inc.)

cell bodies. The neurons in layer 4A are somewhat larger and more evenly spaced than those of layer 4B, and according to Lund et al. (1979) their axons give off two or three long trunks, which extend through layer 4A and give off primary branches that ascend into layer 2/3, although some collaterals may descend to the deeper layers (Fig. 1-9c) (Martin and Whitteridge, 1984). The stellate cells in layer 4B are somewhat smaller, and they possess fewer long dendrites than the stellate cells in layer 4A. Lund et al. (1979) point out that the 4B spiney stellate cells have axons that arise from the lower surfaces of their cell bodies and turn upwards to form a diffuse projection of fine branches that ascends into layer 2/3. However, the dominant axonal projections are directed downwards to spiney cells in layer 6 (Fig. 1-9C). On their way the descending axons give off several long collaterals that ramify throughout layer 5 and layer 6.

Apart from differing in their neurons, the separation of layer 4 into two sublayers is also justified on the basis of the inputs they receive from the lateral geniculate nucleus (Fig. 1-9B). The layer 4A input is dominated by fibers carrying Y-visual signals from LGN, whereas layer 4B receives input almost exclusively from fibers carrying X-visual signals (see pages 25 and 76). Even though layer 4 is the principal recipient layer for the geniculocortical input to area 17, only about 5% of the synapses in that layer are derived from the dLGN (Peters and Payne, 1993). Thus, the geniculocortical synapses are likely to be highly potent effectors of excitation (see Chapter 12).

Ahmed et al. (1994) attempted to analyze the complete input to layer 4 spiney stellate cells by characterizing the synapses that are derived from different sources. They did this in terms of the type of the synapse, the target of the axon terminal and the size, or area of the presynaptic bouton involved. By matching the synapses involving spiney stellate cells with their criteria for those formed by axon terminals derived from four sources, Ahmed et al. (1994) suggest that of the asymmetrical, or excitatory synapses involving spiney stellate cells, 45% could come from layer 6 pyramids, 28% from other spiney stellate cells, and 6% from thalamic afferents, a value in accord with the calculations of Peters and Payne (1993). The sources of the remaining 21% of the asymmetrical synapses could not be accounted for. Of the symmetrical or inhibitory synapses involving spiney stellate cells, they suggested that 84% of them could be formed by clutch cells (see later).

To continue the architectural description of area 17, the lower portion of layer 4B transitions into layer 5, which contains small, large- and medium-sized pyramidal cells (Fig. 1-12). O'Leary (1941) suggested that layer 5 should be divided into two substrata, with layer 5A containing irregular, discontinuous clusters of small and medium pyramidal cells with arciform axons, that is, axons that extend basally for a short distance before arching upwards into several ascending branches into layer 2/3 (Fig. 1-9C). Some of these pyramidal cells have been intracellularly filled by Martin and Whitteridge (1984), who illustrated the extensive axonal plexuses that these neurons form in upper cortical layers. Because layer 5A is of varying thickness and may even be discontinuous, its border with layer 5B is irregular, but the boundary between layer 5A and layer 5B is made obvious by the presence of the large, solitary pyramidal cells of Meynert in the upper part of layer 5B. As pointed out by Cajal (1921a,b), these large pyramidal cells form a discontinuous layer, with some being solitary, whereas others occur in pairs or triplets a short distance away. These large pyramidal cells have stout dendrites. Their apical dendrites branch first in layer 4 and then again in layer 2/3 before terminal branches extend into layer 1. As for the basal and lateral dendrites that emerge from these large neurons, Cajal (1921a,b) suggested that they give rise to a horizontal plexus of dendrites in layer 5B. However, although the solitary pyramidal cells are the most conspicuous element in layer 5B, this layer is really dominated by small pyramidal cells with apical dendrites that reach layer 1 without branching (not shown). The axons of these small pyramidal cells generally have their main projections to layer 5 and 6, as well as forming efferents that enter the white matter.

Hübener et al. (1990) examined the pyramidal cells in layer 5 in terms of their axon projections. The neurons were filled retrogradely with tracer and then stained intracellularly in brain slices to reveal their morphology. They found that all corticotectal neurons are characterized by long apical dendrites each of which forms a large terminal tuft in layer 1. These neurons have large- to medium-sized somata, and their basal dendrites form a dense and symmetrical dendritic field. This group of neurons includes the Meynert cells. In contrast, the corticocortical cells of layer 5 are small- to medium-sized cells with fewer basal dendrites than the corticotectal neurons, and they have short apical dendrites that never reach as high as layer 2/3.

In Nissl preparations layer 6A contains a preponderance of small and medium-sized pyramidal cells with rather rounded cell bodies (Fig. 1-3). These cell bodies tend to be arranged into vertical stacks and columns that are separated by clearer spaces that contain nonstaining afferent and efferent axons and dendrites of deeper lying neurons. According to O'Leary (1941), the apical dendrites of most of the pyramidal cells of layer 6A only ascend as far as layer 4 before forming their apical tufts (Fig. 1-9C). Our own examination of the layer 6A pyramids (Peters and Yilmaz, 1993) concurred with this observation, but Lund et al. (1979) suggested that although the apical dendrites of some of the pyramidal cells only attain layer 4, the majority have apical dendrites that reach layer 1 (Fig. 1-16). The axons of the layer 6A pyramids that have been intracellularly filled by Martin and Whitteridge (1984) ascend to layer 4, where they form either a patchy projection, or a more uniform spread (Fig. 1-9C). A similar result was obtained by Lübke and Albus (1989). The solution to these diverse opinions about the forms of layer 6 pyramidal neurons came from the study by Katz (1987), who labeled layer 6A neurons that project to either the dLGN or the claustrum. He used retrograde tracers to label the different populations of pyramidal cells and then filled them intracellularly. Katz found that the claustrum-projecting neurons have apical dendrites that reach layer 1 and have fine axonal collateral branches that arborized exclusively in layer 6 and in lower layer 5 (Fig. 1-9C). In contrast the corticothal-

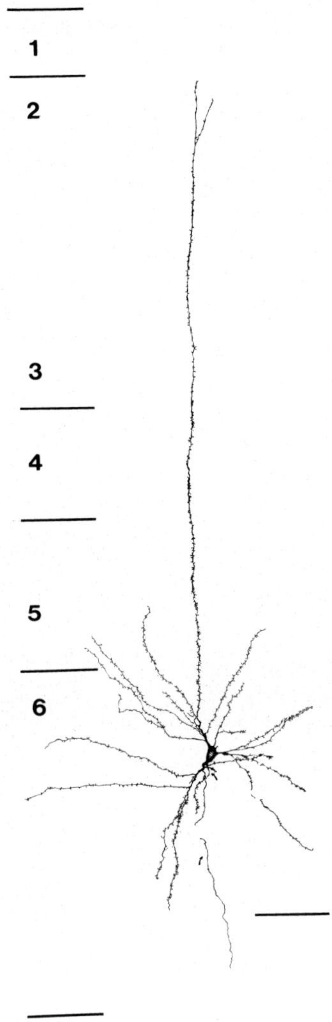

FIGURE 1-16. Camera lucid drawing of a layer 6 spiney pyramidal cell in area 17. Note apical dendrite that ascends to layer 1, and dendrites radiating around the cell body in layer 6. Neuron filled with horseradish peroxidase-catalyzed-oxidized diaminobenzidine. Curved arrow identifies descending axon.

amic neurons have apical dendrites that never reach higher than layer 3, and their axons send widespread collaterals into layer 4 (Fig. 1-9C).

Layer 6B contains the deepest neurons of the cortex, and they are embedded in large numbers of axons at the gray/white matter interface. The neurons do not have the characteristics of pyramidal cells, and the surrounding axons constrain many into multipolar forms that are horizontally aligned with the dominant trajectory of the axons (Peters and Regidor, 1981) (Fig. 1-17). To a large extent these neurons of layer 6B have been neglected, for they have been ignored in most studies of the structure of visual cortex.

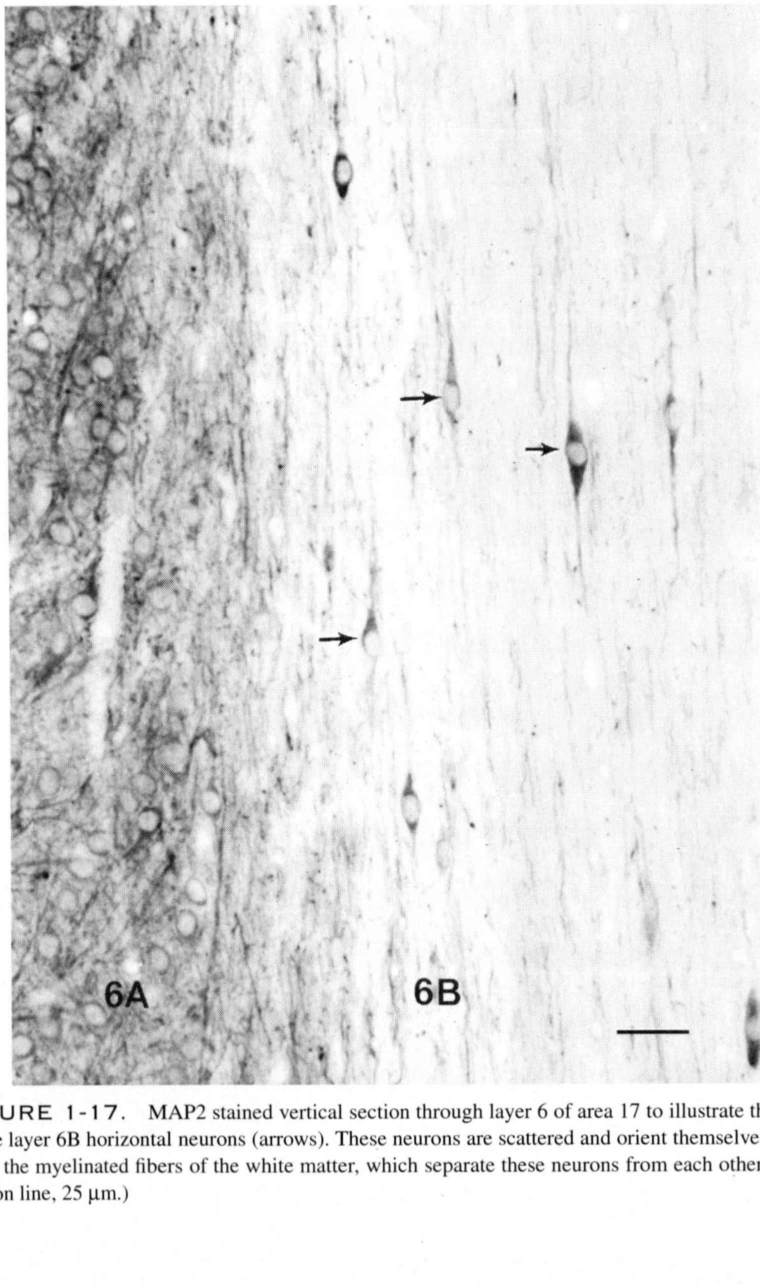

FIGURE 1-17. MAP2 stained vertical section through layer 6 of area 17 to illustrate the form of the layer 6B horizontal neurons (arrows). These neurons are scattered and orient themselves parallel to the myelinated fibers of the white matter, which separate these neurons from each other. (Calibration line, 25 μm.)

Summary of Axonal Arbors and Circuitry

The axonal arbors of the spiney cells in layers 2 through 6 in area 17 can be synthesized into a common diagram that includes the source of afferents from LGN (Figs. 9B and C). It is possible, then, in such a diagram to trace the flow of visual signals through area 17. For example, signals entering area 17 in layer 4A are transmitted elsewhere in the layer and then, in parallel, upward into the superficial layers and downward into the deep layers. From the upper layers the signals are transmitted into layer 5B to the basal dendrites of layer 5 neurons and the apical dendrites of layer 6 neurons. Signals entering cortex in layer 4B are transmitted to the deeper layers, and from there they are transmitted upward back to layer 4 or to layers 2 and 3 along axon collaterals. Some neurons, including those that project to the claustrum, maintain local axon projections within layer 6 (Katz, 1987). In addition, any neuron with an apical dendrite passing through layer 4 has the potential to receive signals directly from axons originating in LGN. This point brings us to the topic of the arrangement of apical dendrites of pyramidal cells.

ARRANGEMENT OF PYRAMIDAL CELLS

In 1993 Peters and Yilmaz made a study of the arrangement of pyramidal cells in area 17 using a antibody to MAP2 (microtubule associated protein 2), which labels microtubules in the cell bodies and dendrites of neurons. It was found that the apical dendrites of groups of the medium-and large-sized pyramidal cells in layer 5 aggregate to form clusters with a center-to-center spacing of about 56 μm as they ascend toward the outer layers. As they pass through layers 2/3, the apical dendrites of the pyramidal cells in that layer are added to the clusters. The thinner apical dendrites of the layer 6A pyramidal cells do not partake in the formation of these clusters (Fig. 1-18). Instead they form an independent grouping that we have referred to as bundles. As the Golgi data indicate, most of the apical dendrites in these layer 6A bundles reach only as far as layer 4 before terminating. It is suggested that bundles of layer 6A apical dendrites are formed to enable them to pass between the cell bodies of the layer 5 neurons as they ascend through the cortex. Thus, they contribute to the unstained clearer spaces between pallisades of layer 6 neurons as identified earlier.

It has been proposed that the clusters formed by the apical dendrites of the layer 5 and layer 2/3 pyramidal cells represent the axes of vertical modules of pyramidal cells. Moreover, it may be that these modules are recruited in various combinations to give rise to the functional columns that have been identified in visual cortex (see page 90). Of interest in this proposal are the large layer 5B pyramidal cells, because their apical dendrites can branch several times as they ascend through the cortex. This means that the apical dendrites of these large pyramidal cells may be components of several modules and that gives them the potential to integrate diverse signals processed in a number of modules.

As can be seen in Fig. 1-19, in semithick plastic sections stained with toluidine blue, and taken in the horizontal plane at the level of layer 4, the profiles of the

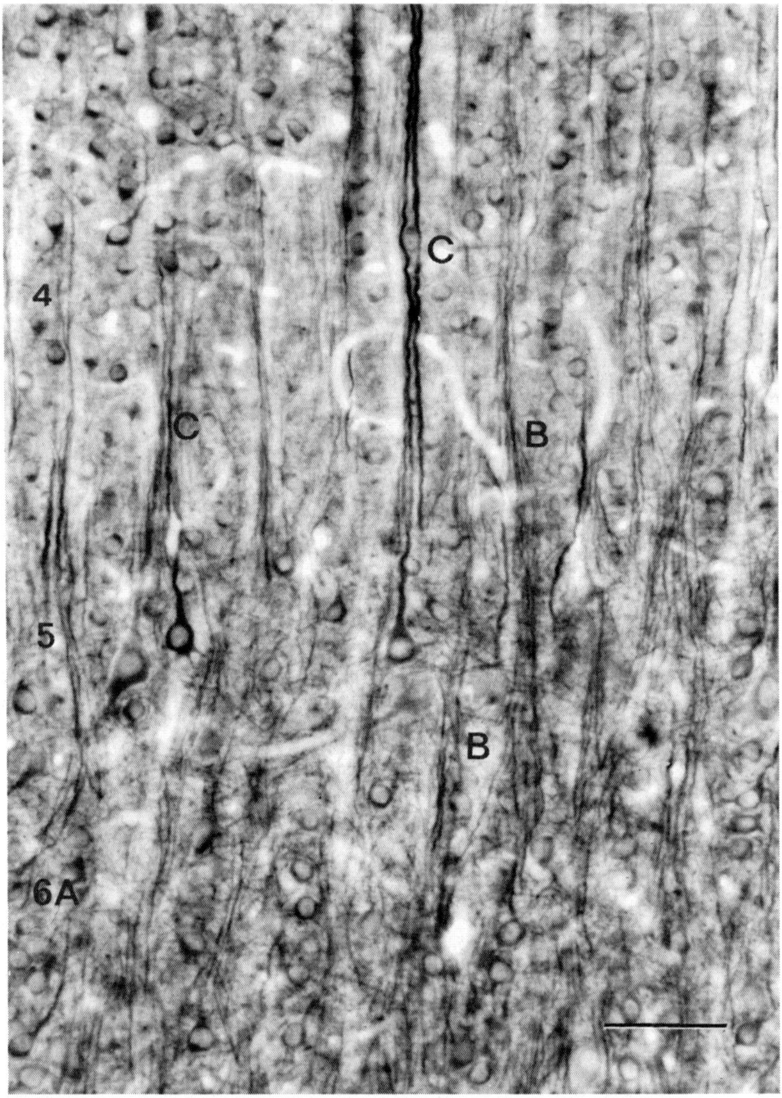

FIGURE 1-18. MAP2 stained vertical section through the lower layers of area 17. The thin apical dendrites of the layer 6A pyramidal cells aggregate to form bundles (B) that pass through layer 5 and enter layer 4, where they terminate. The apical dendrites of the layer 5 pyramids form clusters (C) that pass through layer 4 into layer 3, where the dendrites of the pyramidal cells of that layer are added to the clusters. (Calibration line, 50 μm.)

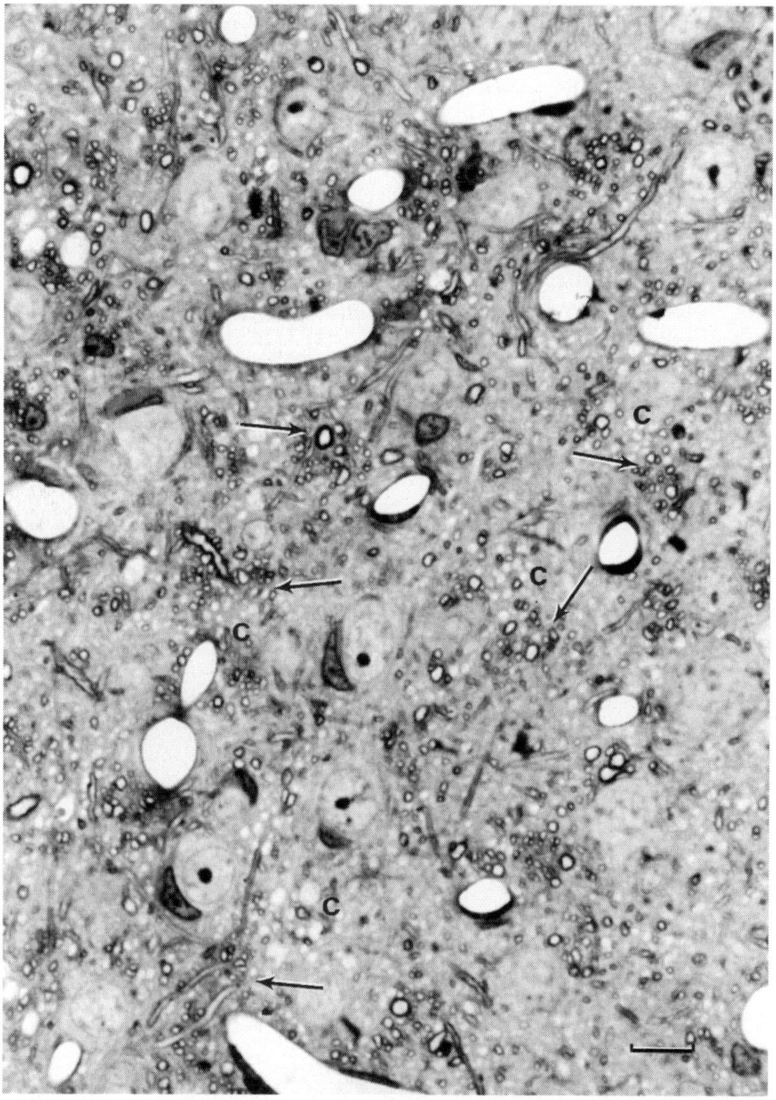

FIGURE 1-19. A 1-μm thick plastic section stained with toluidine blue, taken in the horizontal plane at the level of layer 4 in area 17. At this level the transversely sectioned bundles of vertically oriented myelinated axons (arrows) are regularly spaced, and adjacent to each one of them is a cluster of apical dendrites (c) belonging to the layer 5 pyramidal cells. (Calibration line, 10 μm.)

clusters of apical dendrites are each adjacent to a bundle of myelinated axons. This suggests, as we have described in monkey striate cortex (Peters and Sethares, 1996), that each vertical bundle of myelinated axons likely contains the efferent axons from the neurons contained within one pyramidal cell module.

Based on the average center-to-center spacing of 56 μm for the clusters of apical dendrites, if the modules are assumed to extend through the depth of the cortex, each module would contain about 200 neurons and the striate cortex would contain some 160,000 of these modules (Fig. 1-20). This number of modules is about the same as the number of X-cells in the dLGN and twice the number of β ganglion cells in the retina.

Finally, pyramidal cells in the upper and lower layer components of the modules are the source of the dominant, efferent projections from area 17 (Fig. 1-9C), yet they project axons to different classes of target structures. Most upper layer pyramidal cells project to other cortical areas and the opposite hemisphere, whereas the

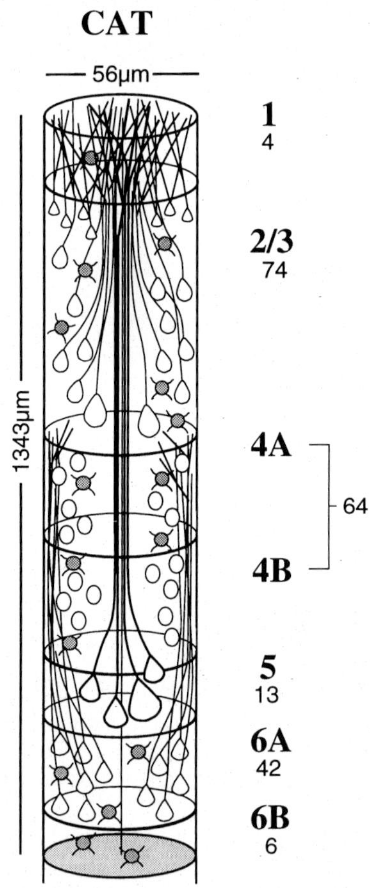

FIGURE 1-20. Diagrammatic representation of the pyramidal cell module in cat area 17. The apical dendrites of the pyramidal cells (open outlines) of layers 5 and 3 come together to form clusters that form their apical tufts in layer 1. The apical dendrites of the layer 6A pyramidal cells form bundles that ascend into layer 4 where the majority terminate. The nonpyramidal cells are stippled. The total numbers of neurons in each layer that contribute to the pyramidal cell module are given on the right, next to the number of the layer. (Adapted from Peters and Yilmaz, 1993).

majority of deep layer pyramidal cells project axons to basal forebrain, thalamus, midbrain, and hindbrain. A number of large spiney stellate cells in layer 4A also contribute to efferent projections, as they send axons to area 18 and the opposite hemisphere (Meyer and Albus, 1981; Vercelli et al., 1992; Voigt et al., 1988).

MYELIN DISTRIBUTION PATTERN

The distribution of myelinated fibers in the primary visual cortex of the cat has been described by Cajal (1921a,b) and by Sanides and Hoffmann (1969). In myelin-stained preparations of cat area 17, prominent vertically oriented bundles of myelinated axons begin to aggregate at the level of layer 3 and become more compact and prominent as they decend through the deeper layers of the cortex and pass into the white matter (Fig. 1-21). In layer 3 itself there is only a loose plexus of myelinated axons, although there are a few fine caliber vertically oriented fibers that seem to

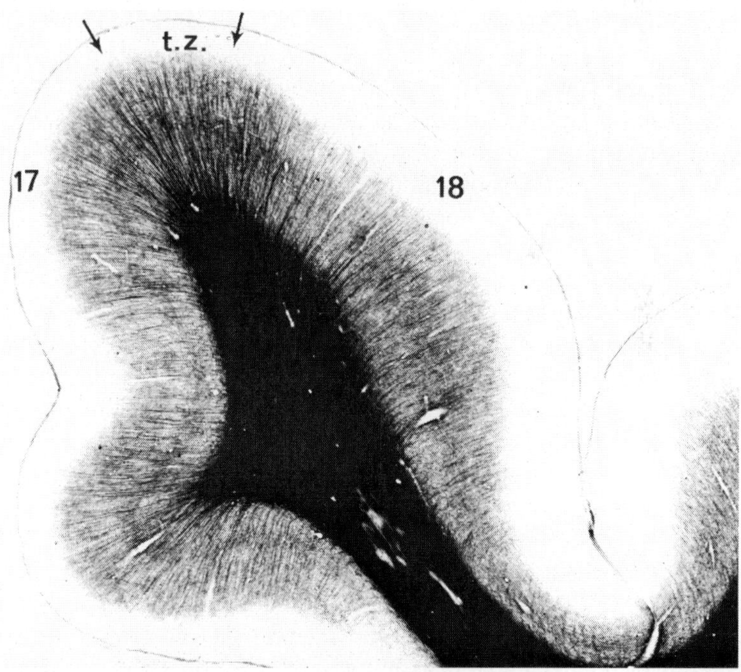

FIGURE 1-21. Medium-power photomicrograph of a coronal section through the marginal gyrus stained for myelin. Note the radiating myelinated fibers in areas 17 and 18. They are particularly prominent at the border between areas 17 and 18 where they are most obvious. They form the Grenzbuschel of Sanides and Hoffmann (1969), which demarcates the transition zone (tz) between the two areas in the marginal gyrus. At this position on the marginal gyrus the transition zone is wedge-shaped, being broader closer to the pia than at the interface with the white matter. Also note stripes of Baillarger in layers 4A and 5 that are characterized by more intense myelin staining of the horizontal plexuses in those two layers. Scale bar = 1 mm. (Figure taken from Payne [1990a] and reproduced with permission of Cambridge University Press.)

enter layer 1, where they perhaps contribute to the strong horizontal plexus of myelinated axons present in that layer. Deeper in the cortex there are other plexuses of largely horizontally oriented fibers, the inner and outer strips, or stripes, of Baillarger. The outer strip, often referred to as the line of Gennari, is present at the level of layer 4A but, as discussed earlier, it is not as prominent as the line of Gennari that typifies striate cortex in primates, and the fiber systems in the two species are not related (see page 15). The thinner, inner strip of Baillarger lies at the level of layer 5 and is probably formed by collaterals of the descending myelinated axons of layer 3 pyramidal cells innervating layer 5B (Fig. 1-9C).

NON-PYRAMIDAL CELLS

The excitatory, spiney neurons are the spiney stellate cells and pyramidal neurons. The nonpyramidal neurons are inhibitory, and they all appear to use GABA as their principal neurotransmitter. These neurons are sometimes referred to as *local circuit neurons, stellate cells,* or *interneurons.* They lack the profusion of dendritic spines that characterize the pyramidal and spiney stellate cells, their axons generally seem to be distributed locally within the cerebral cortex, and the terminals of their axons participate in the formation of symmetrical synapses.

Although the nonpyramidal cells account for only some 20% of the total neuronal population in cat visual cortex (Gabbott and Somogyi, 1986), they exhibit such a wide variety of morphologies that at present no one scheme of classification has been generally accepted. Some authors have classified the nonpyramidal cell population on the basis of the forms of their dendritic trees and the frequency with which the dendrites bear spines, whereas others have depended more on the characteristics of the axonal plexus and where their axon terminals form synapses.

Because this account is meant to serve as a background to other chapters in this volume, we will mention only nonpyramidal cell types identified in cat areas 17, 18, and 19. For more details about the varieties of nonpyramidal cells in cat visual cortex, reference should be made to the classic Golgi impregnation studies of Cajal (1921a,b; 1922) and O'Leary (1941), and to the later descriptions of Golgi impregnated nonpyramidal cells given by Colonnier (1966), Szentágothai (1969, 1973), Garey (1971), Tömböl (1978), Lund et al. (1979), Fairén and Valverde (1980), Somogyi and Cowey (1981, 1984), Peters and Regidor (1981), and Meyer (1983). Parenthetically, it should be mentioned that the account by Meyer (1983) is probably the most comprehensive one, and she described 22 different kinds of nonpyramidal cells in areas 17, 18, and 19 of cat visual cortex, each kind differing from the others in either the form of its dendritic trees or the distributions of its axonal plexus. Subsequently, much more detailed information, particularly about the distribution of the axons of these neurons and their synaptic relations emerged, as a variety of these neurons were filled with intracellular markers by a number of authors, some of whom are cited in this discussion.

There is general agreement with regard to the characteristics of some types of nonpyramidal cells. These are described in the following sections.

Chandelier Cells

These neurons have been described in cat visual cortex by Tömböl (1978), Lund et al. (1979), Fairén and Valverde (1979, 1980), Peters and Regidor (1981), Somogyi et al. (1982), and Farinas and DeFelipe (1991). Generally, these neurons are multipolar and although they have been most commonly encountered in layer 2/3, they may be also present in layer 5. The defining feature of this neuronal type is the axon. It gives rise to short, vertically oriented strings of boutons, which synapse specifically with the axon initial segments of pyramidal cells. Consequently these chandelier cells are sometime referred to as axoaxonic cells, and because of the strategic placement of the axons terminals, each chandelier cell is assumed to be a powerful inhibitor of a number of efferent pyramidal cells. The results of Farinas and DeFelipe (1991) suggest that the number of axoaxonal synapses formed on the axon initial segments of different populations of pyramidal cells is characteristic of the pyramidal cell type. Thus Farinas and DeFelipe (1991) found that the axon initial segments of layer 3 callosally projecting pyramids have 16 to 23 axoaxonal synapses formed by chandelier cells, ipsilaterally projecting pyramids in layer 3 have 22 to 28, and corticothalamic cells in layer 5 have 1 to 5 axoaxonal synapses.

Large Basket Cells (Figs. 1-22 and 1-24)

These neurons have been encountered in layers 2 through 6 (O'Leary, 1941; Szentágothai, 1973; Tömböl, 1978; Kisvárday et al., 1993; Somogyi et al., 1998; Naegele and Katz, 1990). They have rather oval, radially elongate cell bodies and vertically elongate, beaded dendritic trees, but their most distinctive feature is the axon. The main axon originates from one pole of the cell body and forms three to five branches that extend up to 1.5 mm radially, parallel to the pial surface. These branches give rise to frequent small tufts of axonal boutons, each of which form up to 5 symmetrical synapses with the soma and proximal dendrites of each of their target cells, which are principally pyramidal cells (Somogyi et al., 1983; Martin et al., 1983; Kisvárday et al., 1993). Indeed, Somogyi et al. (1983) calculated that a given pyramidal cell may receive converging inputs from as many as 10 to 25 different basket cells. However, the postsynaptic targets also include other basket cells, and Kisvárday et al. (1993) suggested that the inhibition of some of the basket cells by others allows the facilitation of particular sets of pyramidal cells by blocking the inhibition.

Other evidence suggests that the inhibition produced by large basket cells may not be entirely in the horizontal plane. For example, Kisvárday et al. (1987) found a basket cell with its cell body and dendrites in layer 5 and 6 and an axon that gave rise to both a horizontal plexus in layer 6 and ascending plexus that reached layer 2/3, where it contacted the soma and apical dendrites of pyramidal cells.

A subset of the large basket cell population, those in layer 4, are labeled by the plant lectin *Vicia villosa,* and Naegele and Katz (1990) produced illustrations of this cell type that have been filled intracellularly. They also indicated that the only other type of nonpyramidal neuron labeled by this lectin is a large neurogliaform

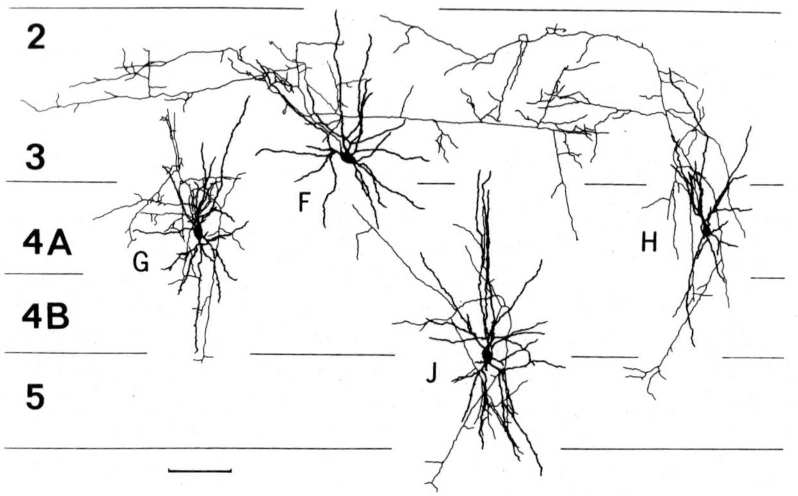

FIGURE 1-22. Camera lucida drawings of four basket cells. These neurons occur throughout layers 2 through 5. (Calibration line, 100 μm). (Figure reproduced from Peters and Regidor [1981], *Journal of Comparative Neurology,* Copyright © 1981. Reprinted by permission of Wiley-Liss Inc., a subsidiary of John Wiley & Sons, Inc.)

cell in layer 4. Other nonpyramidal cells types, such as the chandelier cells, large bitufted cells, and small basket cells do not label with the lectin.

Small Basket Cells

The clearest image of small basket cells has been obtained by intracellular filling of cells located in layer 4. Kisvárday et al. (1985) referred to these small basket cells as "clutch cells." They are multipolar cells with smooth dendrites that emerge from the upper and lower poles of the elongate cell body and their axons form local plexuses with large boutons that form clawlike clusters around somata neurons in layer 4. Kisvárday et al. (1985) indicated that the postsynaptic neurons are not immunoreactive for GABA, which suggests that they are either spiney stellate cells or star pyramids. However, the somata are not the principal postsynaptic targets of the clutch cells, as about 50% of their terminals synapse with dendritic shafts, and some 30% synapse with dendritic spines. This small basket cell has also been described in Golgi impregnated material by DeFelipe and Fairén (1982, 1988) and by Lund et al. (1979), and it is probably the identity of the neuron that Meyer (1983) described as a bitufted neuron of layer 3 and 4 with "arcade" axons. The small basket cells label with antibodies to cholecystokinin (Freund et al., 1986).

Double-Bouquet Cells

Cajal (1911) described a number of different types of neurons as being "double-bouquet cells" (Peters and Regidor, 1981), but the neurons that are now generally

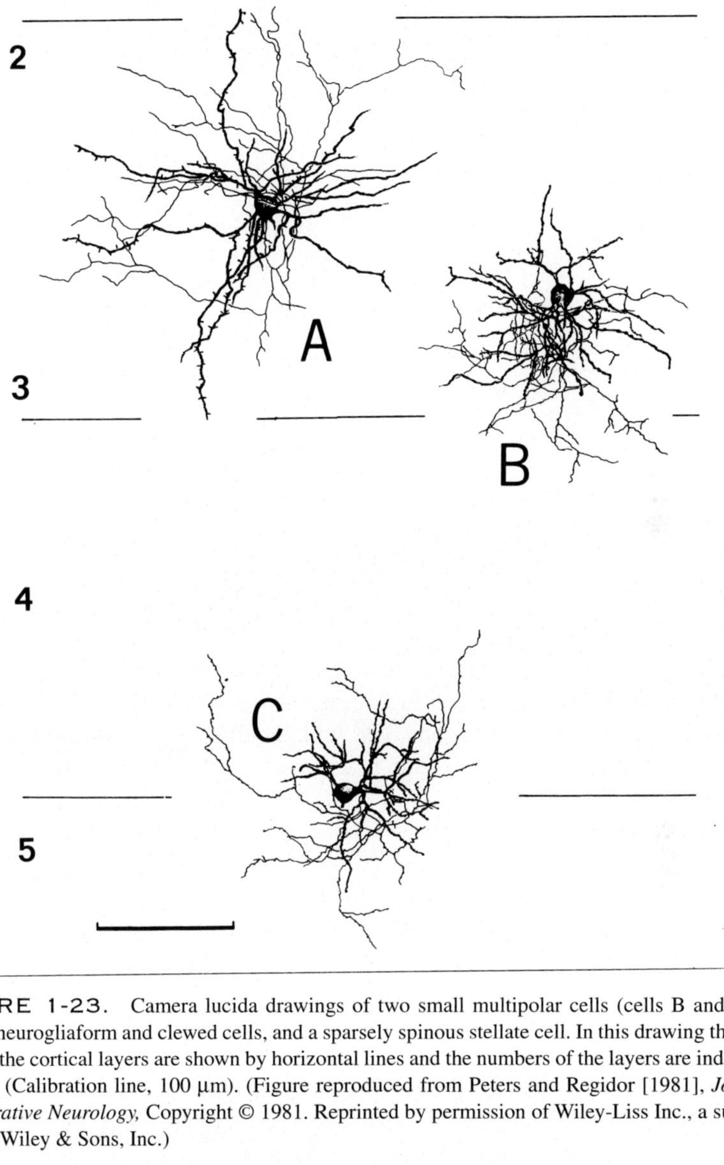

FIGURE 1-23. Camera lucida drawings of two small multipolar cells (cells B and C), also termed neurogliaform and clewed cells, and a sparsely spinous stellate cell. In this drawing the boundaries of the cortical layers are shown by horizontal lines and the numbers of the layers are indicated on the left. (Calibration line, 100 μm). (Figure reproduced from Peters and Regidor [1981], *Journal of Comparative Neurology,* Copyright © 1981. Reprinted by permission of Wiley-Liss Inc., a subsidiary of John Wiley & Sons, Inc.)

referred to as double-bouquet cells have their somata spread throughout layer 2/3. They are medium-sized multipolar cells with smooth dendrites and because of their characteristic axons, Szentágothai (1973) referred to these neurons as "horse-tailed" cells. Indeed it is the axons that makes these neurons so distinctive, because in addition to the local axonal plexus, the axon gives rise to several descending bundles of 3 to 5 closely intermingled branches (Peters and Regidor, 1981; Meyer,

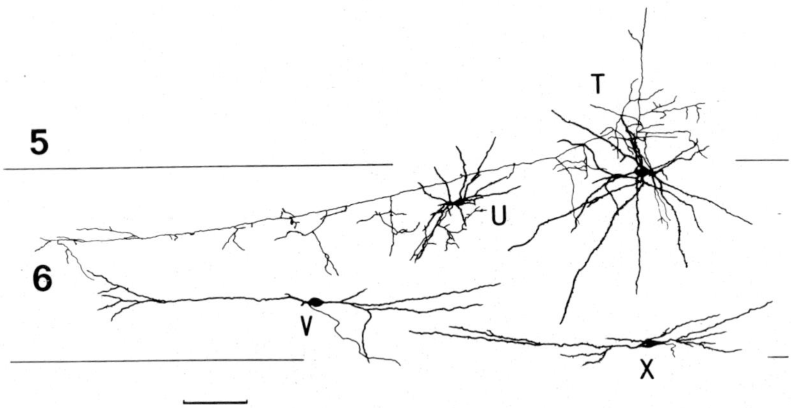

FIGURE 1-24. Camera lucida drawings of neurons in layers 5 and 6 of area 17. Cell T is a large basket cell, and Cell U is a sparsely spinous spherical multipolar neuron. Cells V and X are horizontal bipolar neurons of layer 6B. (Calibration line, 100 μm). (Figure reproduced from Peters and Regidor [1981], *Journal of Comparative Neurology,* Copyright © 1981. Reprinted by permission of Wiley-Liss Inc., a subsidiary of John Wiley & Sons, Inc.)

1983). Electron microscopic analyses (Somogyi and Cowey, 1981, 1984; Somogyi et al., 1998) have shown that this type of neuron forms symmetrical synapses, 70% of which are with dendritic spines and 30% with dendritic shafts. Because of these long axonal bundles that often traverse from layer 2 to layer 4 it is assumed that these double-bouquet cells are involved in vertical inhibition in the cortex. Some of these neurons label with antibodies to cholecystokinin (Freund et al., 1986).

Neurogliaform Cells

Neurogliaform cells are small multipolar neurons that have short and smooth dendrites and a dense local axonal plexus (Fig. 1-23). As pointed out by Jones (1984), the axon in extremely thin and rebranches many times to produce a local and entangled mass of branches. For this reason, Valverde (1971, 1978), who described such cells in monkey visual cortex, referred to them as "clewed cells": clew meaning a ball of thread or yarn. Cajal (1922) decribed such neurons in the visual cortex of the cat and called them "neurogliaform" or "arachniform" cells, and they appear to occur in all layers of cat visual cortex (Peters and Regidor, 1981). In the first Golgi-electron microscopic study of cells of this type, LeVay (1973) described one of these neurons in layer 1 of cat visual cortex and showed that the axon form symmetrical synapses. Because of the compact form of the axonal plexuses of these neurons, it is assumed that they are involved in localized inhibition of nearby cells.

Layer 6B Horizontal Neurons

Layer 6B neurons have been largely ignored by most authors including Cajal (1921a,b), O'Leary (1941), and Meyer (1983). They are difficult to impregnate by the Golgi method and are unlike the other cells in the cortex. It is clear that there are no pyramidal cells with ascending apical dendrites in layer 6B. In a general account of the neurons of layer 6B, Tömböl (1984) stated that there are horizontally oriented neurons, together with pyramidal-like neurons that are tangentially oriented (Fig. 1-24). Most of the neurons encountered by Peters and Regidor (1981) in cat visual cortex were disposed horizontally. In general they have smooth or sparsely spinous dendrites emerging from the two poles of a fusiform cell body. Thus, they have either a bipolar or a bitufted configuration, although the dendrites emerging from one pole of the cell may be much longer than those emerging from the other pole. One way to obtain a clear view of the horizontal neurons in layer 6B is through use of an antibody to MAP2 (myelin-associated protein 2), which labels the microtubules in the dendrites and cell bodies of neurons (Peters and Yilmaz, 1993). This label shows that the majority of neurons in layer 6B are horizontally oriented bipolar and bitufted neurons (Fig. 1-17). The connections of these neurons seem to be unknown.

AREA 18

Finally, it is important to recognize that even though a considerable amount is known about the composition and circuitry of area 17, similarly comprehensive data are not available on area 18. However, examination of neuron morphologies in Golgi studies and more limited studies using intracellular horseradish peroxidase/dye injection methods reveal that the same cell types are also present in area 18 in both the upper and lower layers. It is then highly likely that the basic neuronal composition and circuitry of area 18 resemble quite closely that of the detailed descriptions already given for area 17.

COMPARISON OF THE ARCHITECTONICS OF AREAS 17 AND 18 AND IDENTIFICATION OF BORDERS

In broad terms there are a great number of similarities in the architectonics of areas 17 and 18. Even so, a number of differences are obvious and should be recognized. Figure 1-25 shows the cytoarchitecture of a portion of area 17. To recapitulate in brief, in area 17 layers 2 and 3 are narrow and blended together, and the large pyramids at the base of layer 3 are located quite superficially in the cortex. Layer 4 is the thickest layer, consisting of large and medium-sized granule cells in its upper part and predominantly small granule cells in its deeper part. These two subdivisions are designated layer 4A and 4B after O'Leary (1941) and Lund et al. (1979). These two sublayers are equivalent to layer 4AB and layer 4C, respectively, of Otsuka and Hassler (1962). The lower border of layer 4B occurs at the level of the apices of the superficial pyramids in layer V. Relative to layer 4, layer 5 is thin and cell sparse, and it contains mainly medium-sized pyramids. Layer 6

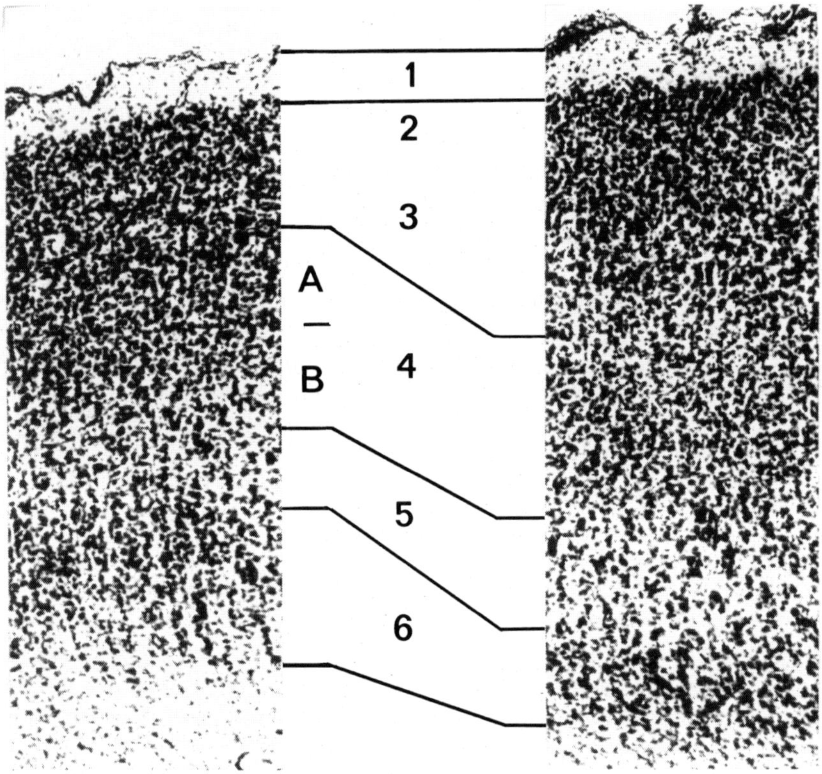

FIGURE 1-25. Sections stained for Nissl substance with thionin. Medium-power photomicrograph to compare the cytoarchitecture of areas 17 (left) with the cytoarchitecture of area 18 (right). Laminar borders are indicated. Laminae in the two areas are indicated. Note the increased thickness of layer III in area 18 and that layers 4, 5, and 6 lie deeper in the cortex. Also, note thinness of deep layers in area 18 compared with their equivalents in area 17 (Scale bar = 100 μm). Abbreviations are given in Table 1-1. (Figure taken from Payne [1990a] and reproduced with permission of Cambridge University Press.)

is conspicuous for its thickness and its large number of small neurons, which are arranged into palisades by fibers penetrating between them.

Figure 1-25 also shows the cytoarchitecture of area 18. Overall, area 18 is thicker than area 17. Layer 3 in area 18 is well developed and thicker than its counterpart in area 17, and its lower border is defined by the bases of the deepest and largest pyramidal cells. In most sections of area 18, layer 4 is about the same thickness or narrower than its counterpart in area 17. The smallest granule or spiney stellate cells in layer 4 are relatively rare in area 18, and the border between sublayer 4A and sublayer 4B is ambiguous and difficult to identify (Harvey, 1980b; Humphrey et al., 1985b). Layer 5 in area 18 lies deep in the cortex, and it contains occasional large pyramids in addition to the many medium pyramids (Otsuka and

Hassler, 1962; Sanides and Hoffmann, 1969). Except for being thinner, layer 6 in area 18 has essentially the same characteristics as its equal in area 17.

The borders of areas 17 and 18 are also readily recognizable. The border of area 17 in the splenial and retrosplenial sulci is characterized by a dramatic decrease in the number of layer 4 granule cells (Fig. 1-4). The border with paralimbic visual area 20, located at the ventral end of the posterior suprasylvian gyrus and in the parahippocampal fusiform gyrus, is also characterized by an increased prominence of layer 5 (not shown). Consonant with the decrease in layer 4 granule cell numbers at both locations, there is a dramatic decrease in the density of the cytochrome oxidase reaction product in layer 4 (Fig. 1-11). The transition from area 17 to adjacent paralimbic cortices is also indicated by a noteworthy thinning and staining intensity of myelinated fibers (Fig. 1-7) (Sanides and Hoffmann, 1969). All of these changes are instantly recognizable, and they occur over a distance of 1 mm or less.

The lateral boundary of area 18 is also recognizable by distinct decreases in both numbers of granule cells present in layer 4 and by the decreased intensity of cytochrome oxidase reaction product (Fig. 1-11). Both of these features decrease over a distance of 1 to 1.5 mm as the architecture transforms from the characteristics of area 18 into the characteristics of area 19. Contrary to the generally held view, the lateral boundary of area 18 is not always obvious in sections stained for myelin. At one or more coronal levels, and among cats, the border with area 19 is characterized by heavily myelinated radiating fibers, which Sanides and Hoffmann (1969) termed a *Grenzbuschel* (or border bundle).

TRANSITION ZONE BETWEEN AREAS 17 AND 18

Grenzbuschels also mark the transitional cortex between areas 17 and 18 (Fig. 1-21, tz). These thick, strongly radiating fibers are much more prominent and have a greater density than the radiating fibers in laterally placed area 18, and they are much more substantial than the radiating fibers in the medially placed area 17 (Payne, 1990a). Careful examination of the same regions in sections stained for cell bodies reveals a zone of cytoarchitectural transition from area 17 to area 18 (Payne, 1990a). In this zone the features characteristic of area 17 gradually transform into the features characteristic of area 18. In particularly striking examples, the zone has a width of 1 to 1.5 mm, and it is characterized by both a thickening of the upper cortical layers and a thinning of the deeper layers (Fig. 1-26, upper). There is also a concommitant downward displacement of granular layer 4 (Fig. 1-26, upper). In sections reacted for the presence of cytochrome oxidase, the downward displacement of dark, highly reactive and cytochrome oxidase-rich layer 4 is readily apparent as it translocates from a more superficial position within layer 4 of area 17 to a deeper position within layer 4 of area 18 (Fig. 1-26, lower). The zone of transition defines the transitional cortex between areas 17 and 18. The zone of transition appears slightly wider than 1 mm in the lower part of Fig. 1-26 because a coronal section cuts obliquely across the zone in the posterolateral gyrus and not perpendicular to it. On the medial bank of the posterolateral sulcus the transition zone has almost parallel borders (Fig. 1-26, upper), whereas on the crown of the marginal

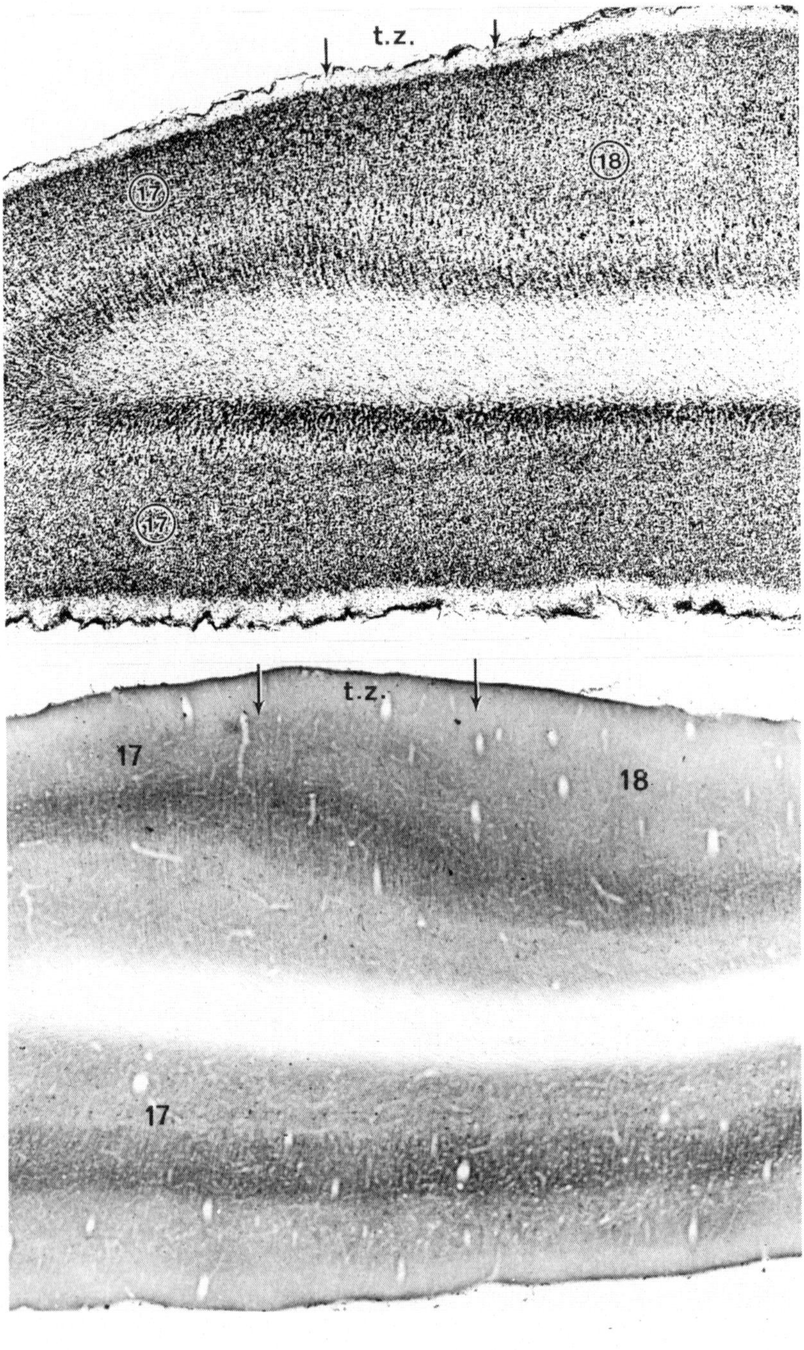
t.z.
17
18
17
t.z.
17
18
17

FIGURE 1-26. Coronal sections to compare architectures of areas 17 and 18, and to show the transition zone between the two areas. **(Upper)** Low power photomicrograph of the posterolateral gyrus. Sections stained for Nissl substance with thionin. Area 17 to the left is separated from area 18 to the right by the transition zone (tz), which is demarcated by the arrows. Note that the sides of the transition zone are approximately parallel. **(Lower)** Low power photomicrograph to show the continuity of the prominent stripe of enhanced cytochrome oxidase activity from area 17 to area 18. The drift downwards from the more superficial position in area 17 to the deeper position in area 18 also demarcates the transition zone (tz), which is again indicated by the arrows. Note that the cortex in B is thicker at the transition zone compared to the flanking cortices of areas 17 and 18. About 50% of cats show this feature. (Scale bars = 1 mm). Abbreviations are given in Table 1-2. (Figure taken from Payne [1990a] and reproduced with permission of Cambridge University Press.)

and posterolateral gyri, it is wedge-shaped, with the borders being separated by a much greater distance in layers 2 and 3 than in layer 6 (Fig. 1-21).

The Grenzbuschels marking both the medial and lateral boundaries of area 18 are more prominent in the marginal gyrus, which represents the lower visual field, than in the posterolateral gyrus, which represents the upper field (Talbot and Marshall, 1942). This feature is in accord with the dominant lower field representation in medial interlaminar nucleus (MIN) (Lee et al., 1984), which is thought to provide the dominant subcortical drive to the region (Payne, 1990a). The region also receives significant cortical input from contralateral areas 17 and 18 (Payne, 1991; Payne and Siwek, 1991b; Olavarria, this volume) via the corpus callosum.

VISUAL MAPS IN AREAS 17 AND 18

The emergence of the idea that both areas 17 and 18 should be considered components of primary visual cortex is tenable only if one assumes that the two areas deal with different aspects of visual information in parallel. A major step forward in understanding the parallel nature of the processing was taken in the late 1970s with the detailed elucidation of the organization of the visual field representations present within areas 17 and 18. These maps of the visuotopic organizations of areas 17 and 18 built on the physiological evidence provided earlier by Talbot and Marshall and by Hubel and Wiesel, as well as the anatomical studies summarized earlier in this exposition. The 1970s maps were generated using the microelectrode recording techniques pioneered by Hubel and Wiesel (1962, 1965a). With these methods the activities of single or small numbers of neurons were studied, and the position and area of the visual field most potent for activating the neurons was identified and delimited. After each plot, the electrode was then advanced 0.5 to 1.0 mm, or occasionally only 0.25 mm in critical regions, to a new site. The procedure was then repeated until a large region of cortex had been sampled. Multiple electrode penetrations were made in a limited number of coronal planes, and studies were highly systematic. Small, marking electrolytic lesions were made in cortex to permit subsequent identification and reconstruction of electrode tracks and recording sites in suitably prepared histological sections showing the cytoarchitecture and myeloarchitecture (Fig. 1-3, 1-6, 1-11, and 1-26).

The maps were drawn on representations of the cortical surface by relating the coordinates of the receptive field centers within the visual field to the individual recording sites. By this process the lines of isoazimuth and isoelevation of the visual field were plotted on the cortical surface (Fig. 1-27). This approach does not attend to depth of recording site in cortex. That parameter is usually regarded

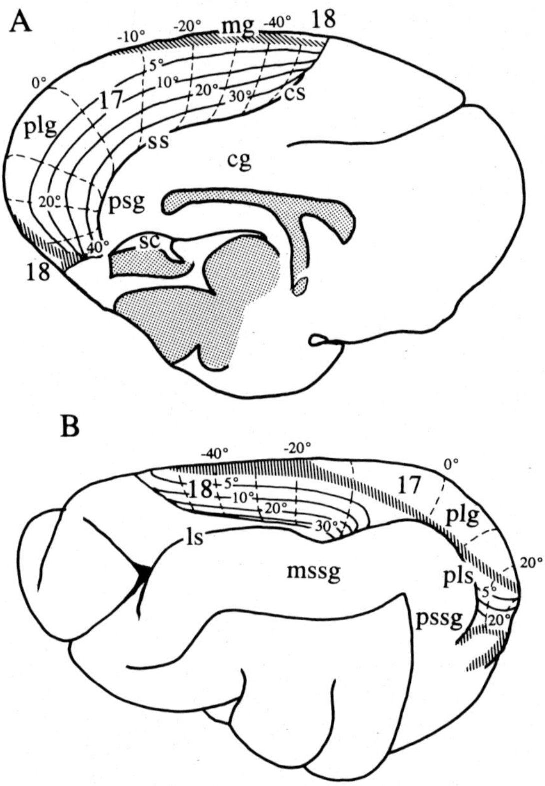

FIGURE 1-27. Outline drawing of the cat's cerebral hemispheres to show the positions of areas 17 and 18 relative to named gyri and sulci and to show the maps of the visual field the two areas contain. **(A)** Medial view. Anterior is to the right. **(B)** Dorsolateral view. Anterior in the brain is to the left. In each view, the lines of isoelevation and isoazimuth in the visual field maps of areas 17 and 18 are indicated as interrupted and continuous lines, respectively. In area 18, a type I map is shown. The hatched area in the marginal and posterolateral gyri indicates the cytoarchitectonically defined transition zone between area 17 and 18 (see page 61.). The hatched regions at the junction of the posterolateral and posterior suprasylvian gyri indicate transitional cortices between areas 17, 18, 20a, and 20b. Abbreviations are given in Table 1-1. (Figure taken from Payne [1991] and reproduced with permission of Cambridge University Press.)

to be of only limited importance because the vast majority of neurons located beneath a single point on the cortical surface have receptive fields that exhibit a great deal of spatial overlap. Even so, it is important to recognize that there is some jitter in the exact position of receptive fields along vertical electrode tracks. The basis of the jitter lies in both biological variability and the mapping method. However, for the generation of cortical maps the jitter is not of great concern, because receptive field centers are clustered about a central average location, and the dimension of the jitter is usually less than the average receptive field dimensions at that location (Albus, 1975a). The presence of receptive fields of finite size and the existence of jitter confirm that the concept of point-to-point representation in visual cortex is true, but only in a statistical sense.

Maps generated in this way encode only average receptive field center location, and they tell nothing about receptive field size of individual neurons. This limitation is of importance because receptive field area varies with neuronal depth in cortex (Payne and Berman, 1983; Berman et al., 1982; Gilbert, 1977), and it is possible to conceive that each layer within cortex contains its own map as occurs in the laminated portion of LGN. When viewed according to layer, the map in layer 4 has the greatest accuracy and is of the highest resolution, because its component neurons usually have the smallest receptive fields, and the neurons are organized with greater systematic, visuotopic precision than neurons in other layers. Conversely, the resolution is poorest in layer 5 because receptive fields are large. These large receptive fields emerge from the sampling of multiple pyramidal cell modules by branches of ascending dendrites, as mentioned earlier, and the broad arborizations of collaterals of descending layer 3 axons en route out of the cortex. Consequently, there is considerable overlap of the receptive fields, and the map in layer 5 is both coarser and fuzzier than maps in the other layers. Because recording sites may be located in any cortical layer, the published maps are composites derived from the multiple maps with coarse, intermediate, and fine resolution. In this composite form, movement of 1 to 2 mm across the cortical surface to a new locus ensures that receptive fields at the new locus do not overlap the receptive fields at the prior locus (Albus, 1975a). Thus, smaller movements are required in layer 4, and larger movements are required in layer 5. A distance of 1 mm in layer 4 also ensures that both eye domains have been crossed (see page 94). A separate measure of receptive field size or "grain" of the map is provided by magnification factor (Daniel and Whitteridge, 1961). Magnification factor is the square unit of cortex devoted to a square unit of visual field (see Rosa, 1997, for a longer discussion).

AREA 17

Area 17, or striate cortex, is the largest of the visual areas (Fig. 1-27). It encompasses approximately 300 to 400 mm^2 of cortical surface, which amounts to about one-third the total surface area of the contiguous visual areas (Anderson et al., 1988; Tusa et al., 1978; Van Essen and Maunsell, 1980).

According to Tusa et al. (1978), there is perfect agreement between the extent of cat striate cortex, or area 17, defined with architectural staining techniques and a single representation of the complete contralateral visual field. The lateral border of area 17 runs along the marginal and posterolateral gyri, and it is bounded by the architectonically defined area 18 (Fig. 1-27). The medial border is buried in a continuous sulcus that is variously named cingular, splenial, and retrosplenial sulcus; and area 17 forms much of its upper/posterior bank (Fig. 1-1, 1-6, 1-7, and 1-11). As previously described, this border is readily identified by the rapid decrease in numbers of granule cells (Fig. 1-4) and a substantial, obvious fall in the density of cytochrome oxidase reaction product in layer 4 (Fig. 1-11). There is also a concomitant decrease in myelination (Fig. 1-7 and 1-21). At this border the far periphery of the visual field is represented.

In area 17, the represented visual field extends from the midline contralaterally to an azimuth of approximately 90° on the horizon, which is buried in the postsplenial sulcus (Fig. 1-27). In the upper, contralateral visual field quadrant, the field extends gently as a curving arc from 50° elevation at the midline, to 60° at intermediate azimuths, down to 20° elevation at the 90° azimuth. It is represented primarily on the tentorial surface of the hemisphere and within the postsplenial sulcus. The representation of the lower, contralateral visual field quadrant is a mirror image, about the horizontal meridian, of the upper visual field representation. It is represented primarily on the medial surface of the hemisphere and within the cingular sulcus. Overall, the map contains a first-order transformation of the contralateral visual hemifield. This means that from a coarse viewpoint, adjacent points in the visual field are represented as adjacent points in the cortex and that the order present in the visual field is maintained without significant distortion in the same order in cortex.

Some features of the map are shown in Fig. 1-28 for multiple levels of the upper field representation. At each level, systematic movement of the assaying electrode to loci progressively further away from the lateral border of area 17 into

FIGURE 1-28. Graphic reconstruction of 5 electrode tracks (T1, 2, 4–6) and the receptive fields of the neurons recorded at each of the indicated sites. Tracks were made through the posterolateral gyrus into cortex forming the tentorial surface of area 17. The tracks were angled 12° downwards from the horizontal. Filled circles along the reconstructed tracks represent microlesions, longer cross bars along tracks indicate every fifth recording site, and the cross-hatched region indicates transitional cortex between areas 17 and 18. For the sake of simplicity, only the receptive fields plotted through the eye contralateral to the recording are shown. **(i)** Shows the receptive fields of neurons studied in area 17. Virtually all neurons represent positions exclusively in the contralateral visual field. There is virtually no overlap into the ipsilateral visual field. Note systematic drift of receptive fields away from the vertical meridian for loci progressively farther into area 17. **(ii)** Shows the receptive fields of the neurons studied in the transition zone. Note that nearly all neurons in the transition zone have receptive fields entirely within the ipsilateral hemifield. Most neurons with receptive fields that overlap the 0° meridian into the contralateral hemifield have centers in the ipsilateral hemifield. Visual field coordinates are given in degrees of azimuth and elevation. For the sake of clarity, neither track T3 nor its receptive fields are shown, for they are similar to those obtained along track T2. (Figure taken from Payne [1990a] and reproduced with permission of Cambridge University Press.)

the body of the area results in systematic movement of receptive fields away from the vertical meridian of the visual field into the contralateral visual hemifield (Fig. 1-28(i)). Movement of the sampling electrode to more ventral locations in the posterolateral gyrus (sequences T1 → T6) results in a systematic movement of receptive fields progressively higher in the visual field (Fig. 1-28(i)). Similar features characterize systematic displacements of the sampling electrode through the

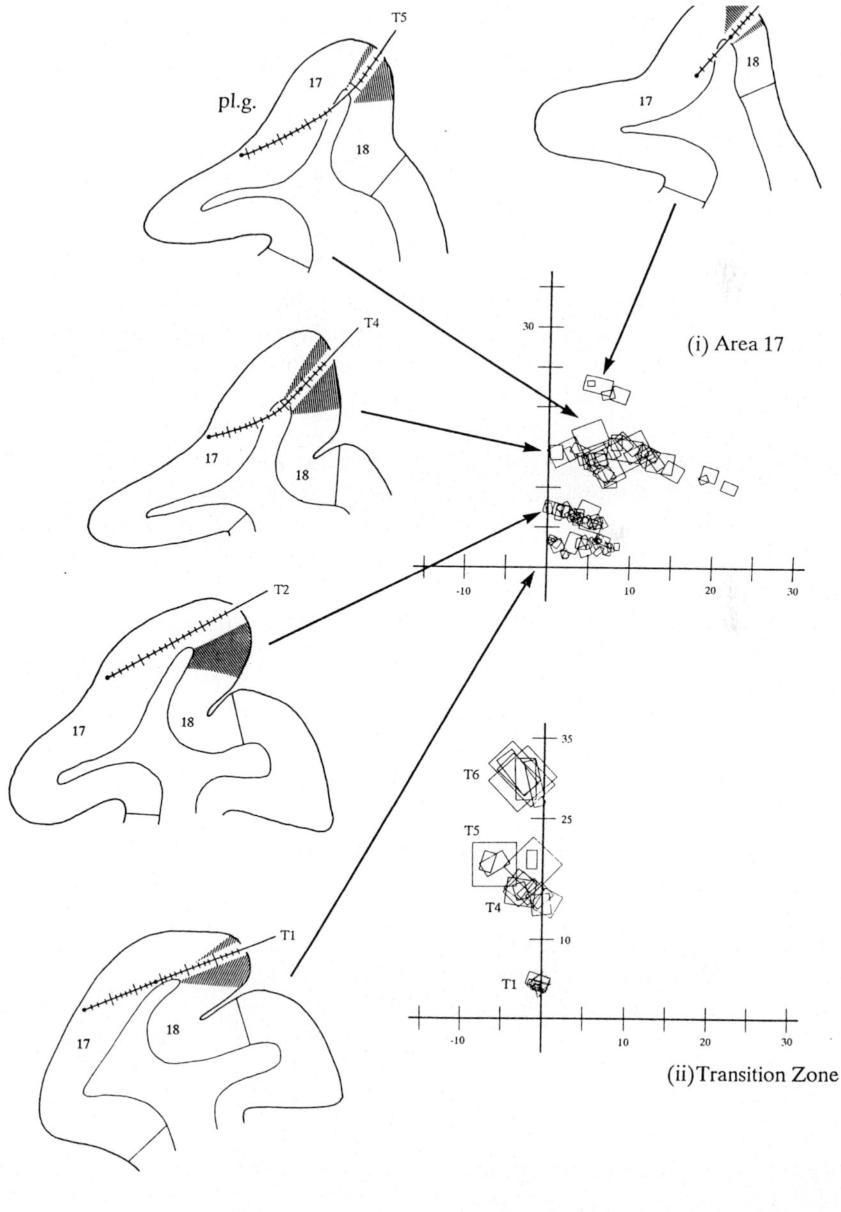

marginal gyrus, except that progressive forward displacement of the electrode samples from neurons with receptive fields sequentially lower in the visual field (not shown).

In both the posterolateral gyrus and the marginal gyrus, movement of approximately 1 mm across the cortical surface results in the attainment of a new representation of the visual field (Tusa et al., 1978). These general features of the representation are concordant both with the earlier description of the visual map in striate cortex given by Talbot and Marshall (1941, 1942) and with many aspects of the description of the visual map in the antecedent LGN given by Sanderson (1971).

Relative to LGN, the map in area 17 is inverted in the mediolaterally dimension. In LGN, the lateral visual periphery is represented laterally within the nucleus, and the midline of the visual field is represented medially (Fig. 1-2). In area 17, the lateral periphery is represented medially in the area, whereas the midline of the visual field is represented laterally (Fig. 1-27). As a result of the inversion in the maps, fibers representing the same elevation in the visual field arising from the lateral and medial components of LGN cross each other in the optic radiation as they ascend toward area 17 (Fig. 1-2) (Nelson and LeVay, 1985). A similar crossing in the anterior and posterior dimensions does not occur because the posterior pole of LGN, which represents upper visual fields, projects caudally to the posterior part of area 17, and the anterior pole of LGN, which represents lower field positions, projects dorsally to area 17.

The first-order representation of the visual field within area 17 is not completely free of distortion. Distortion is introduced because a greater surface area of cortex is devoted to central vision than to peripheral vision (Fig. 1-27). Of the 300 to 400 mm^2 surface area approximately half is devoted to the ~ 600 deg^2 represented within the central 20° radius, and an equal amount of cortex is devoted to the remaining ~ 9000 deg^2 of the visual field represented outside the central zone. This inequality in representation appears to have its basis in the unequal density of ganglion cells in the retina (Hughes, 1975; Stone, 1965, 1978, 1983). The density of ganglion cells viewing the central visual field is greater than those viewing the visual periphery. If each ganglion cell influences a similar amount of cortex, it is understandable that the representation of the central visual field is expanded relative to the periphery.

Finally, Albus (1975a), Tusa et al. (1978), Albus and Beckmann (1980), and Payne (1990a) comment on interanimal variability. Overall, variability in the position, form, and orientation of the map in area 17 is low to modest. The variation appears to be linked to overall brain size and to details of gyral and sulcal patterns. In most instances the variability is not so severe as to preclude the plotting of standard maps based on data collected from a number of cats. However, a different picture is obtained for the visual field map in area 18.

AREA 18

The architecturally defined parastriate cortex, or area 18, is a gently curved structure that is located lateral to area 17 (Fig. 1-1B and 1-27); (Gurewitsch and

Chatschaturian, 1928; Otsuka and Hassler, 1962; Tusa et al., 1979; Payne, 1990a). It is located on the exposed surface of the marginal gyrus anteriorly, and it is mostly buried in the lateral and posterolateral sulci in its middle and posterior regions.

There is some variability in the exact position of central and anterior portions of area 18. The central portion may occasionally emerge out of the junction of the lateral and posterolateral sulci onto the junction of the middle and posterior suprasylvian gyrus. In addition, all, or a large extent, of the anterior portion may or may not be visible on the dorsal surface of the marginal gyrus. The variability in visibility is linked to the presence or absence of a prominent entolateral sulcus embedded within the marginal gyrus, and to the corresponding absence or presence of a prominent suprasplenial sulcus along the medial surface of the hemisphere within area 17 (Kawamura, 1971; Tusa et al., 1979). The absence of a suprasplenial sulcus pushes the common border between areas 17 and 18 laterally, and there is a corresponding appearance of the entolateral sulcus within the marginal gyrus, which partly buries area 18 out of sight. In the opposite way, the presence of a suprasplenial sulcus is accompanied by a more medial position for the common area 17/18 boundary, and area 18 is visible as an accessible dome between the lateral sulcus and the medial margin of the hemisphere (Fig. 1-4).

Regardless of sulcal and gyral formation, the architectural boundaries of area 18 correspond exactly to the single representation of the visual field that is both lateral to the architecturally distinct area 17 and medial to an independent representation in the architecturally distinct area 19 (Hubel and Wiesel, 1965a; Tusa et al., 1979). Area 19 forms the medial bank of the lateral sulcus, extends onto the crown of the suprasylvian gyrus at the junction of its middle and posterior components, and then continues laterally on the anterior bank of the posterolateral sulcus and the adjacent exposed surface of the posterior suprasylvian gyrus (Fig. 1-1B).

Area 18 has a surface area of about 60 to 80 mm^2, so that it occupies about one fifth of the cortical area devoted to area 17 (Tusa et al., 1979; Albus and Beckmann, 1980). It represents more of the visual field than the region abutting the vertical meridian as supposed by Talbot (1942) and illustrated by Woolsey and Fairman (1946), but less than the visual field represented in area 17 (*cf* Fig. 1-27; Tusa et al., 1978, 1979). The map is also not simply a compressed mirror image of the representation in area 17 (Donaldson and Whitteridge, 1977; Tusa et al., 1979; Albus and Beckmann, 1980) as was supposed by Woolsey (1971).

The visual representation in area 18 is severely anisotropic compared with the representation in area 17 (Fig. 1-27). Area 18 is long and narrow, whereas area 17 approaches a semicircle in shape. Moreover, area 18 shares a long common boundary with area 17 that severely constrains the form of the visual representation contained within it. Further constraints are applied because of the narrowness of area 18. Thus, the representation of the vertical midline of the visual field along the anteroposterior dimension of area 18 is stretched to match the vertical midline representation in area 17. Yet the representation of the horizontal meridian in area 18 is highly compressed in the mediolateral dimension of the area. The net effect of this anisotropy is smaller receptive field dimensions in the expanded representation of elevations and larger receptive field dimensions along the compressed representation of azimuths. Overall, there is a twofold to fivefold greater magnification of vertical component of the visual field compared with the horizontal (Cynader et al., 1987).

Distortions in the visual representation in area 18 often generate a map that is more complex, and intuitively less straightforward to comprehend, than the visual map in area 17. The area 18 map contains a representation of elevations from 50° to –50° at the midline, and it has a representation extending laterally out to azimuths of about 50° in all cats (Tusa et al., 1979). The broad orientation and disposition of visual field quadrants match that of the map in LGN, and there is a more-or-less straightforward mapping of LGN onto area 18 without the inversion of projection fibers, which characterizes the visual field transformation from LGN to area 17 (Fig. 1-2) (Nelson and LeVay, 1985).

As occurs for area 17 and LGN, the lower contralateral visual quadrant is represented anteriorly in area 18, and the upper contralateral quadrant is represented caudally (Fig. 1-27). The central visual field is represented on the banks of the posterolateral sulcus and adjacent to area 17, and the common border between areas 17 and 18 represents the midline of the visual field (Fig. 1-27). Typically, movements of 1 to 2 mm across the cortical surface result in movement to a new part of the visual field and nonoverlapping receptive fields (Albus and Beckmann, 1980). This movement incorporates zones innervated by the two eyes, and it is about twice the scale of the movement necessary to effect a similar change in area 17.

The symmetry evident between the upper and lower field representations in area 17 is not evident in area 18, and internally the map may contain a second-order representation of the visual field. This type of representation was initially identified in cat by Bilge et al. (1967) and then confirmed by Donaldson and Whitteridge (1977) before becoming formalized by Tusa et al. (1979). The defining characteristic of a second-order transformation is that pairs of adjacent points in the visual field are represented by nonadjacent points in the cortex. For example, adjacent positions lying on either side of the horizontal meridian of the visual field are represented by cortical loci separated by many millimeters rather than adjacent to each other as occurs in a first-order representation. Thus, a single locus in LGN may contain neurons that project to two widely separated loci in area 18, yet project to only a single locus in the first-order representation in area 17. The second-order representation emerges because much of lateral area 18 represents positions on or close to the horizontal meridian of the visual field. Consequently, peripheral positions in the visual field may be represented internally in the map and not at its borders, as might be expected.

The map in area 18 also exhibits considerable variability from one cat to another in details of the visual field map (Fig. 1-29). For example, rostrally and laterally in area 18 neurons may view the lower periphery of the visual field (~ 50°) or positions much closer to the midline. In the former instance the representation is straightforward to comprehend, and it has been termed Type I by Tusa et al. (1979); an example is shown in Fig. 1-27. In the second instance, isolated islands

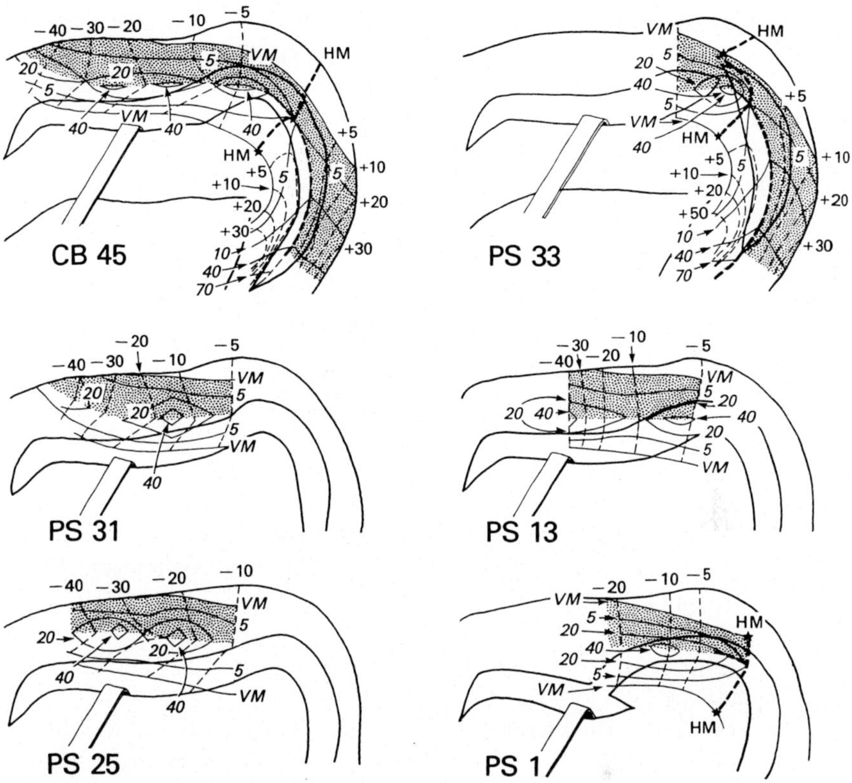

FIGURE 1-29. Six examples to show variability of the type II maps of the visual field in area18 and the confluent area 19. The figure illustrates the dorsolateral view of the brain and shows either complete or partial maps. Stippled region represents area 18. (Figure reproduced from Tusa et al. [1979], *Journal of Comparative Neurology,* Copyright © 1979. Reprinted by permission of Wiley-Liss Inc., a subsidiary of John Wiley & Sons, Inc., a subsidiary of John Wiley & Sons, Inc.)

of representation are formed, and they may be interspersed by representations of greater azimuth (Tusa et al., 1979). This type of representation has been termed Type II by Tusa et al. (1979), and six examples are shown in Fig. 1-29. When islands are present, they may be one or more in number, and it seems that the lateral visual field periphery is underrepresented. The islands can be identified in histological sections by the Grenzbuschels of strongly radiating, heavily myelinated fibers that arise from either the contralateral hemisphere or the ipsilateral medial interlaminar nucleus. The islands introduce substantial discontinuities into the visual field maps. However, the underrepresentations and discontinuities in the representations may be overcome by receptive fields of exaggerated size (Albus, 1987). For both Types I and II cats the caudal, lateral boundary representation of the juxtahorizontal meridian of region of the visual field in area 18 is much more uniform (Tusa et al., 1979; Albus and Beckmann, 1980). Even so, the presence of discontinuities and the magnitude of the variability in the lower field maps preclude the construction of a representative map of the visual field in area 18.

There is disagreement on the peak magnification factor in area 18. According to Tusa et al. (1979), the peak magnification factor, about the representation of the visual axis, is less than one-fifth the peak magnification factor present in area 17, which is in accord with the difference in size. In contrast, Albus and Beckmann (1980) reported that the peak magnification factor is only slightly less than the peak magnification factor in area 17, which is in accord with the common boundary between the two areas. Regardless of the differences, both groups report that at increasingly greater eccentricities, the magnification factor in area 18 rapidly falls to values that are well below those reported for area 17 and the differences become progressively larger with increasing eccentricities. The rate of decline varies with the visual axis used for measurements. It is less for the vertical axis, and greater for the horizontal (Albus and Beckmann, 1980), as confirmed by Cynader et al. (1987). The form of the decline resembles the form of the change in retinal ganglion density. However, it cannot be linked either specifically or numerically to the density of α retinal ganglion cells (Albus and Beckmann, 1980), which are the retinal origin of the Y signals that activate the area 18 neurons most strongly (Sherman, 1985).

MICROMAPPING

The views on the visual field representation in and adjacent to area 18 have been modified by micromapping of cortex. In earlier studies the distance between sites sampled with microelectrodes was usually between 0.5 and 1.0 mm and only occasionally 0.25 mm or less. However, in regions of highly compressed representations, a high-resolution strategy is required to generate accurate maps, and sampling distances need to be 0.1 mm or preferably much less. This point was recognized by Woolsey (1971), Donaldson and Whitteridge (1977), and Albus and Beckmann (1980). However, when a high-resolution mapping strategy is adopted, it is at the expense of the extent, or globalness, of the map that can be plotted. Even so, high-resolution mapping strategies have modified our views on the boundaries and part of the internal organization of area 18.

Large Receptive Fields

Micromapping techniques have shown that in many cats the extent of the visual representation in area 18 includes the far periphery of the visual field (Albus and Beckmann, 1980) and extends beyond the 50° azimuth (Tusa et al., 1979). Moreover, the strategy has detected neurons with extraordinarily large receptive fields that have diameters of more than 50° (Albus, 1987). These receptive fields always include the area centralis and large parts of ipsilateral and contralateral hemifields. In some instances they respond to stimulation anywhere in the visual field! These large receptive fields appear to be an adjunct to the retinotopically organized systems, which are composed of discrete, smaller receptive fields. The neurons with extraordinarily large receptive fields are frequently embedded within neuron pools that are characterized by much smaller receptive fields and high degrees of topographic order (Albus and Beckmann, 1980). Neurons with similar properties have also been detected in the common transition zone between areas 17, 18, and 20 (Payne and Siwek, 1990).

Area 17/18 Transition Zone

Micromapping of the cyto-, chemo- and myelo-architectonically distinct transition zone between areas 17 and 18 (Figs. 1-21 and 1-26) has revealed a distinct, condensed representation of the ipsilateral visual field sandwiched between the two contralateral visual hemifield representations in the flanking areas 17 and 18 (Payne, 1990a). Before 1990, there were numerous examples of neurons that had receptive fields that extended from the contralateral hemifield across the vertical meridian into the ipsilateral field (Blakemore, 1969; Blakemore et al., 1983; Leicester, 1968; Tusa et al., 1978, 1979; Albus and Beckmann, 1980), yet few clear-cut examples of neurons with receptive fields located entirely within the ipsilateral field (Whitteridge and Clarke, 1982; Harvey, 1980b). In nearly all studies, emphasis was placed on neurons representing the central portion of the visual field. The identification of numerous neurons with entire receptive fields located within the ipsilateral visual field and, on a larger scale, the identification of a distinct representation of the ipsilateral field was made possible by the application of micromapping techniques, accurate determination of the positions of the visual axes, and sampling of both upper and lower field representations (Payne, 1990a). Sampling cortical locations representing positions some distance above or below the central visual field made the representation and identification of the ipsilateral visual field more obvious because of the increasingly coarser grain of the representation (Fig. 1-28).

Application of systematic micromapping revealed a significant representation of the ipsilateral visual hemifield confined to the architecturally distinct transition zone (tz) between area 17 and area 18 (Figs. 1-6 and 1-7) (Payne, 1990a). However, as shown for the upper field representation in the posterolateral gyrus (Fig. 1-28(ii) Transition Zone), not all elevations in the visual field are represented to the same extent. A similar increasing representation of the ipsilateral visual field with increasing eccentricity is evident within the marginal gyrus (Fig. 1-30). The overall shape of the visual field represented within the area 17/18 transition zone resembles a vertically split hourglass (Fig. 1-31). It is narrowest on the 0° horizontal meridian, and it is increasingly wider at progressively more positive and negative elevations. At the 0° horizontal meridian, the representation of the ipsilateral field extends to –3.6°, and for superior and inferior elevations it extends to > –20°. Concordant results have been obtained by Diao et al. (1990). Both they and Payne (1990a) suggested that α ganglion cells in the contralateral temporal retina are the origin of many signals reaching the zone.

From Fig. 1-31 it is obvious that the extent of the ipsilateral visual field representation in the right hemisphere matches the extent of the contralateral field rep-

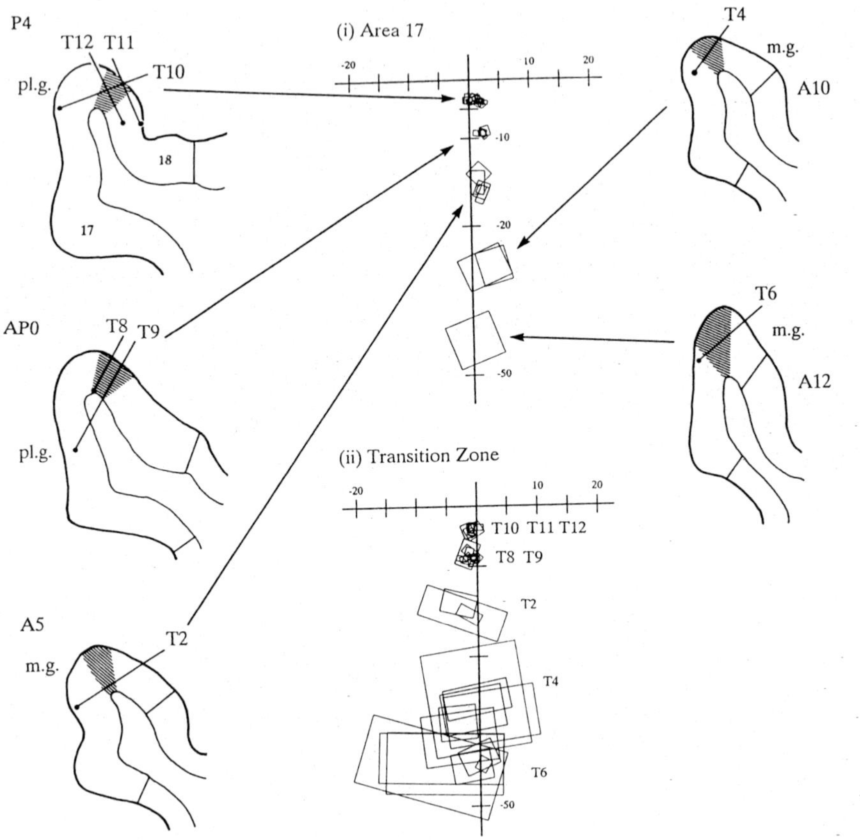

FIGURE 1-30. Reconstructions of electrode tracks made to assess the extent of the lower ipsilateral field representation. Examples of electrode tracks made at five different coronal levels through the posterolateral and marginal gyri (T2, 4, 6, 8–10). Plots of the receptive fields of neurons recorded in area 17 and the transition zone are given in **(i)** and **(ii),** respectively. For the sake of clarity, only electrode track position and trajectory are indicated, and receptive fields of neurons studied in area 18 are not shown. Conventions are given in Fig. 1-6. (Figure taken from Payne [1990a] and reproduced with permission of Cambridge University Press.)

resentation within the transcallosal sending zone in the left hemisphere (Payne, 1991, 1994). This linkage between the two representations is supported by direct demonstrations of the extent of the visual field represented by callosal fibers (Hubel and Wiesel, 1967; Payne and Siwek, 1991a) and within the callosal receiving zone in areas 17 and 18 (Payne and Siwek, 1991b). Layer 3 of the medial interlaminar component of LGN also contains a representation of the same region of the visual field (Payne, 1990a; Lee et al., 1984). It is likely that it, too, provides a substantial input to the area 17/18 transition zone. Our suggestions are supported by studies using permanent and reversible deactivation techniques

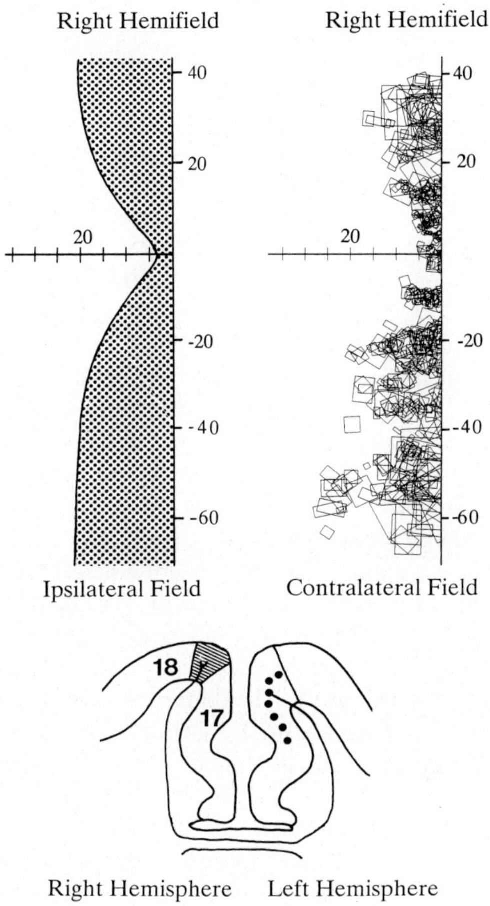

FIGURE 1-31. Diagrammatic summary of the extent of the visual field represented within the area 17/18 transition zone, and the transcallosal basis for the representation. **(Left)** Summary representation for the ipsilateral field representation is based on the receptive fields of more than 500 recording sites localized to the transition zone. **(Right)** Plot of more than 750 receptive fields of neurons localized to lie within the callosal sending zone of area 17. The callosal sending zone was identified by application of retrograde tracers to areas 17 and 18 of the right hemisphere and identification of labeled neurons in area 17 of the left hemisphere. Lower part shows the position of the transition zone in the right hemisphere (cross-hatching) and the position and extent of the callosal sending zone in area 17 of the left hemisphere (large dots). The left callosal sending zone and the right transition zone are connected via cells with axons passing through the corpus callosum. Note that the representations are narrowest on the horizontal (3 to 5° from the vertical midline), and much greater (≥ 20° from the vertical midline) at positions superior and inferior in the visual field. (Data drawn from Payne [1990a, 1991, 1994] and Payne and Siwek [1991].)

of afferent pathways (Blakemore et al., 1983; Payne, 1990b, 1994; Payne et al., 1991;). Deactivation of either the callosal or MIN pathways attenuates the extent of the representation of the ipsilateral field in the area 17/18 transition zone (Payne, 1990b, 1994; Payne et al., 1991). These comments bring us to the topics of circuitry and flow and processing of signals in areas 17 and 18.

CIRCUITRY AND SIGNAL PROCESSING IN AREAS 17 AND 18

By the mid-1970s, the search was on for evidence to support the idea of parallel processing in visual cortex or to conclusively reject it. Numerous studies had already established multiple functional streams emerge from the retina (Stone, 1983), and this evidence suggested that independent signals concerned with different aspects of the visual scene might be processed separately by neurons in areas 17 and 18. Moreover, it suggested that the processing of visual signals might be largely similar in areas 17 and 18.

Substantial effort was applied to investigate the nature of the visual signals transmitted to areas 17 and 18. Measures of similarities and differences between the two areas were obtained, and substantial attention was paid to comparisons of receptive field properties and analyses of circuitry and flow of visual signals through the two areas. The dominant visual signal type transmitted to area 17, via LGN, is the X-, brisk-sustained type. X-signals are derived from β retinal ganglion cells. The dominant visual signal type transmitted to area 18, via LGN, is the Y, brisk-transient type. Y-signals are derived from α retinal ganglion cells. In addition, there are subordinate types of signals; some Y-signals reach area 17 via LGN, and some W-signals reach both areas 17 and 18 via LGN (Fig. 1-8) (Stone, 1983). Moreover, it seems that many aspects of circuitry and signal processing in areas 17 and 18 are similar, and that the basic transformations carried out on the X- and Y-signals in the two areas are largely comparable.

CIRCUITRY AND FLOW OF SIGNALS

Measures of similarities and differences in the functional circuitry of areas 17 and 18 have been obtained by comparing receptive field properties of neurons and the consequences of electrically stimulating afferent and intrinsic circuits. The former provides information on the natural functioning of circuitry in response to naturally generated signals, whereas the latter provides measures on timing of signal transmission and processing, as well as information on functional circuitry. These measures can be acquired because of the exquisite temporal linkages that can be established between a brief electrical stimulus and its resultant evoked responses. The electrical stimulation approach has been applied to both single neurons and, via the current source density (CSD) methods, to populations of neurons. The latter method also reveals valuable information on the locations of populations of synapses. Both types of studies concentrate on the flow of excitatory signals, and data collected with the two methods are compatible. Also, the data permit the development of canonical circuits that incorporate both stages and sequences in the flow of excitatory visual signals across the cortical layers in areas 17 and 18. These circuits are valuable for comprehending the sequential transformations carried out on signals generated by natural stimuli.

Natural Stimulation and Receptive Field Properties

In their seminal studies on areas 17 and 18, Hubel and Wiesel (1959, 1962, 1965a) reported that neurons in both areas 17 and 18 could be classified in terms of the spatial organization of their receptive fields. Those fields with distinct "on" and "off" zones were referred to as *simple,* and those with coincident "on" and "off" zones were referred to as *complex.* Hubel and Wiesel (1962) considered cells with simple fields to represent an initial stage of visual processing and cells with complex fields to represent a later stage of visual processing. Neurons with simple fields were concentrated in the layers that received input from LGN, whereas many neurons with complex receptive fields were localized to other layers. Simple cells are viewed as carrying out the initial transformation on signals arriving from LGN, whereas complex cells are viewed as carrying out later and more sophisticated transformations on the visual signals (see Chapter 8). Embodied in the teminology is the concept of hierarchy, which paralleled the concept of hierarchy embodied in both the anatomical circuitry emerging from area 17 and the visual field transformations described by Hubel and Wiesel (1965a, 1969) for areas 18, 19, and Clare-Bishop.

Hubel and Wiesel (1962) reported on a number of other features about the neurons they studied. These were the emergent properties of visual cortical neurons. The neurons were sensitive to the orientation and length of lines or edges, as well as to direction and velocity of a stimulus moved through the receptive field. In addition, a large majority of neurons was activated independently by the two eyes and were thus considered to be binocularly responsive. Many of these neurons exhibit positional disparity between the left and right receptive fields (Ferster, 1981; Levick, 1977; Pettigrew and Dreher, 1987), which is a prerequisite for stereoscopic vision (Barlow et al., 1967; Nikara et al., 1968; Pettigrew et al., 1968). This catalog of receptive field properties is ubiquitous to the neuron populations composing both areas 17 and 18 (Orban, 1984), and several of the properties are represented in some sort of translaminar functional column or module (see page 90).

A plethora of subsequent studies added numerous sensitivities, such as spatial and temporal contrast sensitivity, to the catalog of receptive field properties of neurons in areas 17 and 18 (Orban, 1984). Even with the growing catalog of receptive field properties emerging from numerous laboratories, it was gradually realized that by and large, there are a great number of similarities in both the structure of receptive fields and their distribution across cortical layers in areas 17 and 18. However, there were two broad differences between the neuron populations in the two areas: size of receptive fields and responsiveness to rapid visual transients in illumination and movement. When equated for visual field position and layer, neurons in area 18 have larger receptive fields than neurons in area 17. The area 18 neurons are also more responsive to rapid visual transients than neurons in area 17 (Singer et al., 1975; Tretter et al., 1975; Orban, 1984). These differences seem not to reflect intrinsic properties of the two areas but, rather, to reflect the properties of the dominant X- and Y-subcortical afferent streams feeding into areas 17 and 18, respectively.

This view is confirmed by the results of Movshon et al. (1978), who compared the spatial frequency tuning of neurons in areas 17 and 18. For neurons in area 17, the range of preferred spatial frequencies extended from 0.3 to 3 cyc/deg, whereas neurons at the representations of equivalent eccentricities in area 18 had spatial frequency preferences that were, on average, one third of those preferred by area 17 neurons. For example, within the representation of the central 5° of the visual field, neurons in area 18 had preferred spatial frequencies between < 0.1 and 0.5 cyc/deg. Moreover, the distributions of optimum spatial frequency in the two areas were practically nonoverlapping at eccentricities as high as 15°. In contrast, the range of selectivities in the two areas was about equal with a width of 0.2 to 3.2 octaves (one octave is a factor of 2) at half amplitude. Finally, neurons in area 17 responded well to low temporal frequencies of a moving grating and less well to temporal frequencies in excess of 2 to 4 Hz. In contrast, many area 18 neurons gave optimal responses in the range of 2 to 8 Hz. Concordant values were obtained in largely equivalent studies by Berardi et al. (1982). Based on their own observations, Movshon et al. (1978) concluded that "areas 17 and 18 act in parallel to process different aspects of visual information relayed from the retina via the lateral geniculate complex. Some of these differences between the two areas may be attributable to the predominance of Y-cell input to area 18 and the predominance of X-cell input to area 17."

It is highly relevant to the subsequent discussions to acknowledge that there is a substantial amplification of the Y-system as signals are transmitted from retina to area 17. The amplification is both in absolute numerical terms and relative to the X-system (Peters and Payne, 1993). To recapitulate, the origin of Y-signals is the α cells, which comprise only about 5% of all retinal ganglion cells, and they are massively outnumbered by the 10 times more numerous β retinal ganglion cells, which are the origin of X signals. The amplification along the pathway from retina to cortex is characterized by (1) a greater field of arborization of Y-axon arbors in LGN compared to X-axons (Bowling and Michael, 1984; Sur et al., 1987; Tamamaki et al., 1995), (2) a disproportionately greater number of terminal boutons in LGN (Bowling and Michael, 1984; Sur et al., 1987; Tamamaki et al., 1995, 3) a greater proportional incidence of Y-cells than X-cells in LGN (Friedlander et al., 1981; Friedlander and Stanford, 1984), 4) broader axon arbors of Y-LGN axons in cortex (Freund et al., 1985a; Humphrey et al., 1985a), 5) a greater number of terminal boutons per Y-LGN axon than per X-LGN axon (Freund et al., 1985a; Humphrey et al., 1985a), and 6) a larger average number of synapses per Y-LGN terminal bouton than per X-LGN terminal bouton (Freund et al., 1985a). Even within LGN the 10:1 advantage of the X-system in retina is reduced to equality when all components of LGN are considered (Humphrey et al., 1985b). The Y-system continues to gain in visual cortex, where it becomes the dominant functional stream. Thus, many aspects of cat cortical visual vision is dominated by a small (~8000 cell) population of α ganglion cells.

As indicated, the X- and Y-systems extend visual processing in complementary yet overlapping ways. The X-system has great spatial acuity yet poor temporal resolution, and it extends vision in the high spatial acuity domains and low temporal domains. In contrast, the Y-system has poor spatial acuity, yet great temporal resolution, and it extends vision in the high temporal domain and in the low spatial frequency domain. Contrast levels are also signaled by both systems, with Y-cells exhibiting greater sensitivity at lower spatial frequencies, and X-cells exhibiting greater sensitivity at mid-spatial frequencies (Lee et al. 1992). Even so, it is likely that at the population level, the numerically dominant X-system dominates over the Y-system at low spatial frequencies because of signal pooling (Troy, 1983). Overall, the responsiveness of the Y-system to high temporal transients translates into sensitivity to high stimulus velocities and high flicker rates, whereas the poor temporal resolution of the X-system translates into sensitivity to slow velocities and slow temporal changes. The different characteristics of these two afferent systems imply that areas 17 and 18 are tuned to process particular types of signals derived from the visual environment. Like the ganglion cells before them, the two areas each extend visual processing capacities in specific, yet complementary and overlapping ways. This broad view based on natural stimulation is strengthened by electrical stimulation experiments.

Electrical Stimulation

Single Neurons—Extracellular Analyses

Areas 17 and 18 appear to be the sole recipients in cortex of activating X- and Y-signals arriving directly from LGN (Fig. 1-8 and 1-9). Dreher et al. (1980) obtained evidence for X- and Y-innervation of area 17 and Y-innervation of area 18, and by way of comparison, they showed that some area 17 neurons and the majority of area 19 neurons are activated by the slowly conducting W-stream. They also obtained evidence that some neurons in area 19 were activated by the much faster-conducting Y-system. However, Dreher et al. (1980) considered that these influences were mediated via other cortical areas rather than mediated directly by Y-fibers arising in LGN. Surprisingly, given the anatomically demonstrated projections by the magnocellular layer C and medial interlaminar components of LGN and their involvement in Y-signal transmission (Mitzdorf and Singer, 1977; Dreher and Sefton, 1979; Kratz et al., 1978; Raczkowski and Sherman, 1985; Lee et al., 1992) attempts, using electrical methods, have failed to activate area 19 neurons via the Y-stream. Comparable conclusions have been reached for the middle suprasylvian visual region (Berson, 1985). It is not clear at present if this absence of effect is grounded on a low number of fibers innervating area 19 and middle suprasylvian cortex, a low density of synapses, low efficacy of synaptic transmission, or failure of synaptic signal integration to generate action potentials. It may simply be that even though the pathways can be demonstrated with anatomical markers, they may be small and functionally insignificant, but other possibilities may obtain.

In more detailed extracellular analyses of area 17, Bullier and Henry (1979c) and then Henry et al. (1983) reported that neurons in layer 4 of area 17 could be activated monosynaptically by electrical stimulation of the optic radiation. Neurons in the deep layers 5 and 6 were activated either monosynaptically or polysynaptically by the same stimulus, whereas superficial layer 2 and 3 neurons were activated only polysynaptically. These patterns were obtained regardless of whether the electrical activation was of either the Y-stream alone or of the Y- and X-streams together. The one distinction that emerged about the two streams is that the monosynaptically Y-activated neurons in layer 4 dominated in the upper part of the layer, whereas monosynaptically X-activated neurons dominated in the deeper part of the layer (Martin and Whitteridge, 1984). Based on the anatomical circuitry depicted in Fig. 1-9B and 1-9C, these patterns are to be expected.

In rough correspondence, Mullikin et al. (1984) localized neurons with X-type simple receptive field characteristics to lower layer 3, the full thickness of layer 4, and layer 6. In addition, Mullikin et al. localized neurons with Y-type simple field characteristics to lower layer 3, upper part of layer 4, and layer 6. These observations are in loose agreement with the initial results of Hubel and Wiesel (1962) and later work of Gilbert (1977) Ferster (1981, 1990a) and others (Reid and Alonso, 1995, 1996; Hirsch et al., 1998) (see Chapter 8). Estimates have suggested that 50% or more of area 17 neurons are driven by the Y-system (Freund et al., 1985a), and the numerical dominance of the retinal X-system is subordinated to the numerically inferior retinal Y-system in the transfer of signals to cortex, as expected from the sequential amplification of Y signals described earlier.

In area 18, activation of neurons follows a similar laminar scheme (Neumann, 1978; Harvey, 1980a). Monosynaptic activation is most prevalent in layers 4 and 6 with additional monosynaptic activation of some neurons at the base of layer 3. Neurons activated polysynaptically by electrical stimulation of afferent pathways were identified throughout the full thickness of cortex. In area 18 all neurons are activated solely by the fast Y-stream. These observations on the laminar location of monosynaptically activated neurons in both areas 17 and 18 are in accord with the laminar positions of the densest terminations of LGN axons (Fig. 1-9A). Attempts to link afferent streams to the simple/complex dichotomy in areas 17 and 18 were not successful (Bullier and Henry, 1979a; Harvey, 1980a; Henry et al., 1979, 1983; Orban, 1984; Singer et al., 1975; Tretter et al., 1975).

Even with the gains made by combining electrical stimulation and extracellular recording, there are technical and biological limitations to the usefulness of the technique. The major limitations are that electrical stimulation of afferent fibers, especially those with the smallest diameter, may fail, and there may be insufficient integration of signals in the assayed target neuron for an action potential to be generated and thus for the activation to be detected. In the latter instance, subthreshold activation goes unseen. A remedy to this technical shortcoming is to record potentials from inside the neuron. The intracellular method permits subthreshold activation to be detected and may have a radical effect on conclusions on the types of afferent signals that influence, rather than drive, cortical neurons.

Single Neurons—Intracellular Analyses

Intracellular analyses of responses evoked by electrical stimulation of afferent pathways confirm the activation of area 18 neurons by Y afferents, and area 17 neurons by either X- or Y-afferents (Singer et al., 1975; Tretter et al., 1975). Congruent conclusions were reached for area 18 by Ferster (1990a) and Ferster and Jagadeesh (1991), but the same investigators failed to obtain evidence for Y-innervation of area 17 neurons even though they could demonstrate X-innervation. Their conclusions for absence of Y-innervation of area 17 were based on both the magnitude of the electrical current needed to activate the axons and differences in timing of axon conduction. Axons in the Y stream tend to be larger than X-axons and, thus, they require a lower activating current. Moreover, when Y-axons are activated they conduct action potentials faster than X-axons and evoke shorter latency effects on target neurons. Thus, if Y-axons innervate area 17, they should definitely be activated by X-axon activating currents, and they should be readily identified by the short latency of evoked postsynaptic potentials. Failure on both criteria suggests that the absence of Y-axon innervation of area 17 is real. However, these observations contradict the conclusions reached with the extracellular methods for the Y-axon innervation of area 17 and the intracellular analyses on area 17 by Singer et al. (1975). Moreover, they ignore data accumulated by Humphrey et al. (1985a) and Freund et al (1985a) on the two types of physiologically and morphologically characterized axons innervating area 17. Even so, the Ferster studies show monosynaptic activation by X-axons in LGN-recipient layers 4 and 6, and polysynaptic activation throughout the depth of cortex, a view that is in agreement with other studies.

Unfortunately, the discrepancy in the evidence for and against Y-innervation of area 17 has not been resolved by the Y-pathway inactivation studies of Burke et al. (1992), who blocked Y-signals by using a pressure cuff applied to the optic nerve. In those studies it is not possible to ascertain complete block of the Y-system, and it is not possible to distinguish between the effects of Y-blockade on the primary afferent pathway to area 17 from the effects mediated via association projections arising from area 18. In any event, the effects of the Y-blockade on area 17 neurons is relatively small compared with the larger deficits induced on area 18 neurons (Dreher et al., 1992). This result supports the view that the X-stream provides the dominant drive to area 17.

Neuron Populations—Current Source Density Analyses

The technique of CSD analysis provides invaluable information on the locations of synapses and target neurons, flow of primarily excitatory signals through the cortical circuits, and temporal aspects of the signal transmission. Because of its resolution and its global character, the method traces, in a rather comprehensive way, the flow of excitatory activity both within and between cortical layers.

Current source density analyses are based on field potentials evoked by electrical stimulation and synchronous activation of populations of afferent fibers. The method is applied to neuron populations, and it extracts information on the sites and polarities of major transmembrane currents (i.e., extracellular sinks and sources), which generate the evoked potentials (Freeman and Nicholson, 1975; Freeman and Stone, 1969; Mitzdorf and Singer, 1977; Nicholson and Freeman, 1975; Freeman and Singer, 1983). The CSD method is a useful supplement to single neuron methods described previously because it provides a survey of bulk activities in both space and time, rather than the activities of individual neurons. The method contrasts with single neuron studies, which require large data sets for statistically significant conclusions to be reached. These data sets may be biased by artifacts such as systematic omission of small cells from the samples (Stone, 1973; Levick and Cleland, 1974; Levick, 1975; Friedlander and Stanford, 1984). Moreover, in contrast to the single neuron studies that register extracellular action potentials, the CSD method mainly reveals subthreshold activity such as currents associated with excitatory postsynaptic potentials (EPSPs). Thus, the CSD method can trace synaptic activity even when EPSPs fail to reach threshold to activate a neuron to generate an action potential. These EPSPs are evoked by spiney stellate cells in layer 4 and pyramidal cells located above and below them in layers 2 and 3, and 5 and 6. The magnitude of the inhibitory currents evoked by nonspiney cells in application of the CSD method is unknown and of unknown significance.

Even so, because of broad similarities in circuitry within areas 17 and 18, there are some likenesses in signal processing in the two areas. These similarities are exemplified by the ability to superimpose the circuitries, as derived from CSD analyses, on each other in Fig. 1-32. However, because of differences in the characteristics of the circuits leading to the two areas, there are a number of differences in the details of signal processing that are brought out by the current source density analysis method. These differences are readily identified by following the transmission of signals through the circuits activated by Y-fibers terminating in layer 4A for both areas 17 and 18, and the circuits activated by X-fibers terminating in layer 4B of area 17, or Y-fibers terminating in layer 4B of area 18. For convenience, these signals paths are termed the 4A and 4B circuits, respectively, in Fig. 1-32.

Monosynaptic Activation According to Mitzdorf and Singer (1978) stimulation of the optic radiation activates two clearly distinct groups of afferents to area 17, and they terminate in different sublaminae of layer 4 (Fig. 1-33). The faster (Y) group terminates above the slower (X) group. Activation by both systems is also detected in layer 6 but only weakly, if at all, in layer 1. These electrophysiological data are in accord with anatomical observations on the terminations of

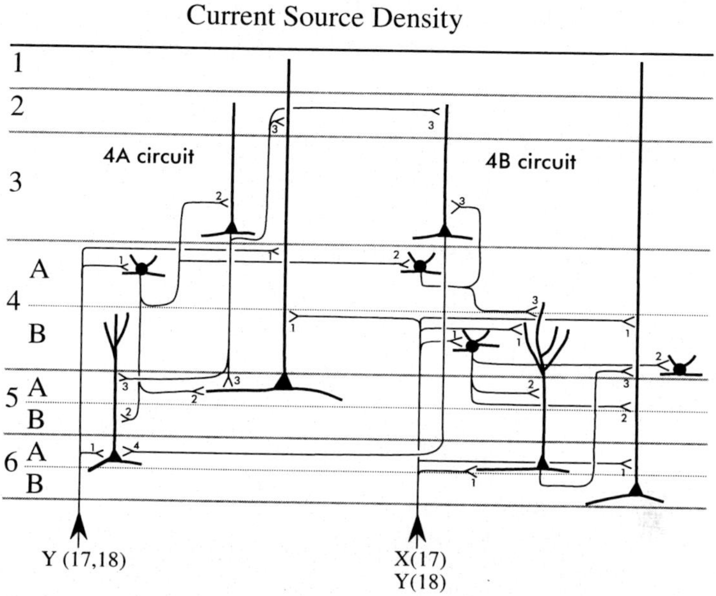

FIGURE 1-32. Composite diagram of spiney stellate and pyramidal cell architectures and circuitry elucidated with the current source density method. This method reveals the two circuits passing through, and emerging from, layers 4A and 4B. Overall, there is considerable concordance in the circuitry deduced from CSD data in areas 17 and 18. Y-fibers activate layer 4A neurons monosynaptically in both areas 17 and 18. This activation is followed by disynaptic activation of both upper and lower layers, followed by trisynaptic activation in both domains. X-fibers innervate layer 4B of area 17, and Y-fibers innervate the upper part of the equivalent layer in area 18. Activation of both afferent systems leads to monosynaptic activation of layer 4B in both areas, and then preferential disynaptic and polysynaptic activation of deeper layers. Signal types (and areas innervated) are indicated. Note broad similarity to the anatomical circuitry depicted in Fig. 1-9.

projections emanating from the magnocellular LGN layers A, A1, and C (Figs. 1-9B and 1-9C) and the single neuron studies summarized in the previous section.

In area 18, the overwhelming activation is via fast (Y) axons acting on neurons located throughout layer 4 (Mitzdorf and Singer, 1978). Again, this observation is in accord with anatomy (Fig. 1-9A). However, the fast (Y) activation of area 18 is markedly faster than the fast (Y) activation of area 17, and Mitzdorf and Singer (1978) recommend that the two fast streams originate from two separate neuronal classes. This difference in conduction speed is surprising because numerous anatomical studies have shown that the largest LGN cells, which have the thickest and fastest conducting axons, innervate both areas 17 and 18 (Birnbacher and Albus, 1987; Geisert, 1980; Freund et al., 1985a; Holländer and Vanegas, 1977; Humphrey et al., 1985b). One would expect that conduction times along the parent axon and then collateral branches would be very similar given the caliber of axons and the distances involved.

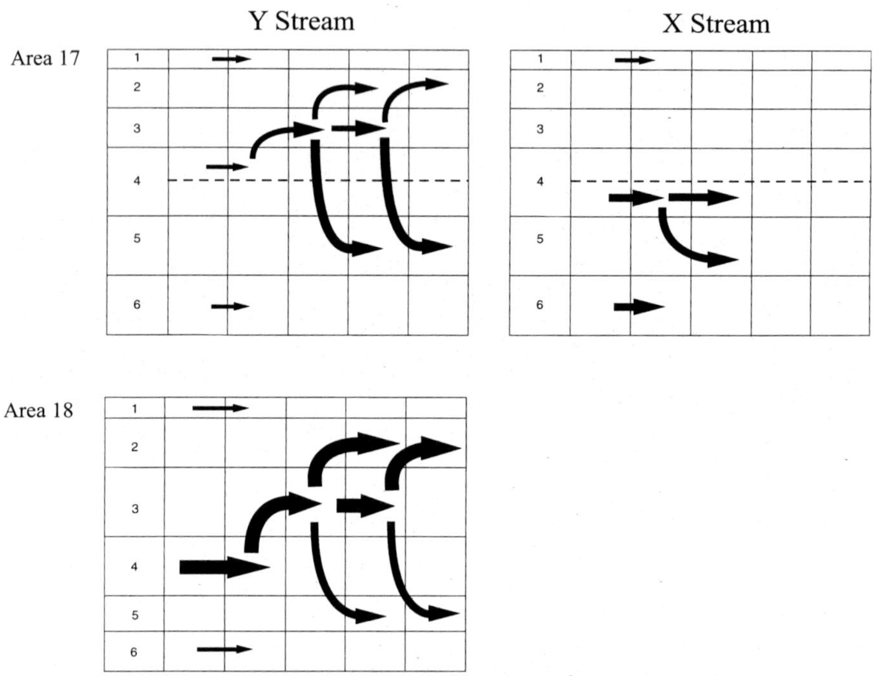

FIGURE 1-33. Summary of activation sequences by X- and Y-axons in area 17 (upper) and and by Y-axons in area 18 (lower). Arrows indicate putative causation as inferred by Mitzdorf and Singer (1978). Note: (1) area 17 Y-axon activation is primarily in upper layer 4 and spreads into both upper and lower layers; 2) there is a similar pattern of Y-axon activation in area 18; and 3) X-axon activation in area 17 is primarily in lower layer 4 and that it spreads into the deep layers. Breadth of arrows indicates relative magnitude of activation. Note selective amplification of currents. Vertical lines represent millisecond intervals. (Figure prepared by R. Jarrett Rushmore.)

In support of Mitzdorf and Singer's (1978) observations, Ferster and Lindstrom (1982), Ferster (1990b), and Ferster and Jagadeesh (1991) all reported that both the CSD and intracellular analyses show that activation of area 18 is purely by the Y-stream. Yet for area 17, the same techniques used by the Ferster group revealed activation solely by the X-stream. The basis for the failure of the Ferster group to identify innervation of area 17 is obscure, and it parallels their results using intracellular analyses. One basis may be the use of different criteria for classification of potentials into different functional types. Regardless of the explanation, the discrepancies impose untidiness on our comprehension of the innervation of area 17 by X- and Y-streams (see page 108; Epilogue).

In this regard, it is worth nothing the comments of Mitzdorf (1985). Based on CSD results, she suggested that there are three types of Y-afferents to visual cortex. One, fast-conducting class innervates and activates area 18, and, after optic nerve or visual stimulation, transgeniculate signal transmission to them is barely affected. A second independent class of slower conducting Y-type afferents innervates layer 4A in area 17 and activates that layer and the supragranular layers. Transmission through LGN modifies the signal considerably. The third class is also fast conducting, bifurcates to innervate *both* areas 17 and 18, and evokes intracortical inhibition. Studies using intra-axonal filling of axons, with physiological characterization, support Mitzdorf's view for three functional types (Humphrey et al., 1985a,b; Freund et al., 1985a) and one of the types has GABAergic, inhibitory neurons as a target (Freund et al., 1985b).

Polysynaptic Activity in Upper Layers The circuitry underlying polysynaptic activation in the upper cortical layers is more easily deciphered in area 18 than in area 17 (Figs. 1-32 and 1-33). In area 18, the initial monosynaptic activation in layer 4 is followed by disynaptic activation of layer 3 pyramidal neurons and then trisynaptic activation of layers 2 and 3 pyramidal neurons, which also may receive additional, trisynaptic inputs from layer 4 neurons (Figs. 1-32 and 1-33 lower) (Mitzdorf and Singer, 1978). There is also disynaptic and trisynaptic activation of layer 5 pyramidal neurons mediated via upper and middle layer neurons (Fig. 1-32). The flow of activity in area 17 is largely similar (Figs. 1-32 and 1-33 upper). However, in area 17 the activation in layers 4 and 6 is delayed in time, and it has a broader temporal scatter, which precludes definitive identification of long-latency monosynaptic activation by, for example, the W-stream (Dreher et al., 1980) or disynaptic activation by the X- or Y-streams. An additional difference is that activation in the upper layers of area 17 is more spatially compressed than in area 18. This observation is in accord with the anatomy, which shows that the upper layers of area 17 are only one-third to one-half the thickness of the upper layers in area 18 (Figs. 1-25 and 1-26).

The magnitude of the early activation in upper area 17 is weaker than the activation of upper area 18 (compare upper and lower components of Fig. 1-33). This weaker activation in area 17 is understandable because it is derived from a relatively small subpopulation of afferent fibers, which terminate in the upper part of layer 4 in area 17, yet functionally equivalent axons terminate throughout much of the thickness of layer 4 in area 18. As in area 18, the upper layer activity in area 17 is based on signals transmitted by spiney stellate cells located in layer 4A, which send strong projections pialwards (Figs. 1-9C and 1-31), (Gilbert and Wiesel, 1983; Martin and Whitteridge, 1984). The majority of spiny stellate neurons in layer 4A are innervated and activated by Y-LGN axons, which suggests Y-dominance of upper layer activity and further amplification of the Y-system already noted at lower levels of the visual system. The relative insignificance of the X-system to upper layer activity is supported by the absence of major pialward projections from layer 4B spiny stellate neurons (Figs. 1-9C and 1-32), (Gilbert and Wiesel, 1983; Martin and Whitteridge, 1984), which receive predominantly X-signals (Fig. 1-9B). The axons of layer 4B stellate neurons ramify densely within the same layer as the cell body of origin or in deeper layers (Martin and Whitteridge, 1984). They extend few branches into the upper layers.

This difference in the magnitude of the activation of the upper layers in areas 17 and 18 is important because the generation of the full repertoire of sophisticated receptive field properties in extrastriate regions depends on neural activity generated in the upper layers of areas 17 and 18 (Symonds and Rosenquist, 1984b; Payne and Cornwell, 1994; Payne et al., 1996a; Spear, 1991). The greater activation and the greater density of projection neurons in the upper layers of area 18 suggest that area 18 may relay more visual signals to other cortical regions than does area 17 (Mitzdorf and Singer, 1978). This view, however, needs to be tempered by the small overall size of area 18 compared with area 17. Even so, it is likely that the dominant activity flowing out of both areas 17 and 18 to other cortical areas is derived from the Y-system. However, this conclusion does not exclude the possibility that some visual signals reach extrastriate regions via other routes and that signals transmitted along them are also capable of activating neurons (Wang et al., 1997; Payne et al., 1996a).

Monosynaptic and Polysynaptic Activity in Deep Layers In both areas 17 and 18 prominent activity is also detected in the deep layers (Mitzdorf and Singer, 1978). The activity in both layers 5 and 6 is highly convergent and complex. Early signals are transmitted by both nearby and distant cells in layer 4, and later signals are carried by great numbers of layer 3 pyramidal neuron axon collaterals entering into layer 5 (Figs. 1-32 and 1-33) (Gilbert and Wiesel, 1979, 1983; Martin and Whitteridge, 1984; Martin, 1988). These signals drive the layer 5 pyramidal neurons with special-complex receptive fields, but not those with standard-complex receptive fields, which are activated via layer 4 and/or layer 6 (Schwark et al., 1986; Weyand et al., 1986b). Layer 6 neurons are also activated monosynaptically. They are activated either directly in layer 6 or via apical dendrites entering layer 4. Many of these dendrites terminate in layer 4, if arising from corticogeniculate projecting neurons, or pass through layer 4 if arising from corticoclaustral projecting neurons (Fig. 1-9C). In addition, layer 6 neurons are activated by highly convergent, polysynaptic circuits akin to those that activate layer 5 neurons (Fig. 1-32). However, layer 6 neurons are not dependent on those sources for activation (Schwark et al., 1986).

Overall, the deep activity in area 17 is greater than in area 18. In area 18, layers 5 and 6 are rather narrow (Figs. 1-25 and 1-26) and the primary input to layer 6 is rather weak, as is the activity that is relayed down from layer 4. In contrast, layers 5 and 6 in area 17 are much thicker, and they receive more substantial input from LGN. Moreover, the more slowly conducting X-axons seem to supply the deep layers in area 17 almost exclusively because few X-signals are transmitted into the upper layers (Figs. 1-32 and 1-33 upper) (Mitzdorf and Singer, 1978). However, that view does not preclude Y or other signals reaching the deep layers as anatomical circuitry and current source density analyses suggest. In accord with this latter view Henry et al. (1983) showed that some deep layer neurons are activated monosynaptically to generate action potentials by the Y-stream, as might be expected from the circuitry shown in Fig. 1-9C. The current view is that the deep pyramids activated by the Y-stream are located in layer 5, whereas the deep pyramids activated by the X-stream are located in layer 6 (Fig. 1-32) (Mitzdorf, 1985). These differences may be linked to the ascent of apical dendrites from layers 5 and 6 neurons. Layer 5 neurons possess apical dendrites that ascend through layer 4A and the field of Y-fiber terminations, whereas many layer 6 neurons possess apical dendrites that issue a profuse tuft in layer 4B and likely receive substantial X-activation from the large numbers of X-axons terminating in that layer (see page 39).

Given the targets of neurons in layers 5 and 6, we can propose that Y-signals are transmitted onto the lateral posterior nucleus (Abramson and Chalupa, 1985) and the superior colliculus (Gilbert and Kelly, 1975; Harting et al., 1992), whereas X-signals are transmitted back to LGN and possibly onto the claustrum (LeVay and Sherk, 1981). The view that all X-based activity is relayed to the infragranular layers of area 17 is supported by reversible deactivation studies. Deactivation of LGN layer A, which is the sole relay of X-signals from the contralateral eye, does not perturb receptive field properties of neurons in the superficial layers. These properties are then based on Y-signals. Because few neurons in the deep layers project to the lateral, or later, areas, it seems that virtually all the X-type visual information is processed most completely in area 17 or sent back to LGN to modify ascending signals. Mitzdorf (1985) suggested that this inference is in accord with models submitting that the high resolution X-system is not necessary for many higher brain functions mediated by more lateral, or later, areas. This suggestion has considerable significance for understanding visual processing and is worthy of additional investigation to support or refute the idea.

Natural versus Electrical Stimulation Mitzdorf (1985, 1987) compared the spatial and temporal patterns of sinks and sources evoked in areas 17 and 18 by electrical stimuli with sinks and sources evoked by visual stimuli such as strobe flash, Ganzfeld-on, Ganzfeld-off, contours, contrast, and onset of movement. She concluded that the similarities of the sinks and sources evoked by these stimuli with those evoked by electrical stimuli suggest that the visually evoked CSDs reflect the same intracortical relays, and that the same cell types are involved in all the activations regardless of stimulus type. For the present thesis, all contour-dependent activities are equally apparent in both areas 17 and 18, although the exact details and actual magnitudes may vary according to the exact nature of the stimulus. For example, moving gratings are five times more effective in activating area 17 than other natural stimuli or electrical stimuli. Alternatively, for area 18, electrical stimulation is two to three times more effective than natural stimuli.

Finally, the circuitry conceived on the basis of CSDs is in harmony with circuitry conceived on the basis of cross correlation analyses (Toyama et al., 1981a,b; Tanaka, 1983; Ts'o et al., 1986; Hata et al., 1991). However, in the main, intra-areal and translaminar transmission within area 18 is faster than within area 17 (Mitzdorf, 1985, 1987). This observation is understandable because intracortical axons of area 18 neurons are, on average, larger and faster conducting than their counterparts within area 17 (Creutzfeldt et al., 1977; Fisken et al., 1975).

Parallel Signal Processing by Areas 17 and 18

Experiments involving careful reversible deactivations of optic nerve, LGN, area 17, and area 18 have been carried out to assess the functional impact of the ascending and lateral connections on the activity in areas 17 and 18 and the shaping of neuronal responses. Overall, results confirm and extend conclusions obtained in experiments summarized in the previous sections. They also confirm that both areas 17 and 18 process signals largely autonomously and in parallel. However, deactivation studies reveal additional, important modulatory association connections between areas 17 and 18. The connections, based on the characteristics of laminar origin and termination of the pathway, are characterized as of the lateral type (Hubel and Wiesel, 1965a; Symonds and Rosenquist, 1984a,b; Price et al., 1994; Meyer and Albus, 1981; Einstein and Fitzpatrick, 1991) and not of the forward or backward types (Payne et al., 1996b). These characteristics put both areas 17 and 18 at the same level in the connection hierarchy.

Ascending LGN and Lateral Connections Between Areas 17 and 18

Experiments to assess the relative importance of activation by LGN afferent fibers versus the lateral association fibers arising in the neighboring area have been carried out. Reversible deactivation studies show that neurons located in the input layers of areas 17 and 18 depend on signals transmitted through the magnocellular layers A, A1, and C of LGN. However, there are differences in the dependencies of neurons in the two cortical areas. Neurons in layers 4 and 6 of area 17 depend on signals transmitted through layers A and A1 for activation (Fig. 1-9B), whereas neurons in layer 4 of area 18 depend on signals transmitted through layers A, A1, and C. (Fig. 1-9A) (Lee et al., 1998; Malpeli et al., 1986). However, deactivation of either layer C or MIN has little or no effect on neurons in layers 4 and 6 in area 17 (Chapter 7).

Given the elucidation of strong anatomical and functional pathways from layer 4 into the upper cortical layers, it is surprising to learn that upper layer neurons in both areas 17 and 18 remain visually active even when the neurons of the input layers 4 and 6 are silenced (Lee et al., 1998; Malpeli, 1983). Moreover, several of the properties of visual cortical neurons such as orientation and direction of motion selectivity, properties that were thought to emerge in layer 4 and be transmitted to the upper layers (Hubel and Wiesel, 1962), are preserved among the latter neurons when layer 4 is inactivated. These results show that lateral cortical association fibers between areas 17 and 18 are sufficiently strong and well organized to generate highly refined and emergent receptive field properties in upper layer neurons in the absence of visual drive from LGN to deeper layer neurons. Presumably, these receptive field properties are based on Y-signals because X-signal transmission was blocked by the silencing of layer A.

In the presence of activity in LGN, however, the association connections are not essential to drive neurons and, under normal circumstances, they only modulate the activity of target neurons. Alternatively, the connections can be viewed as extending the range of signals the target neurons process. For example, when large portions of area 17 are silenced, many neurons in area 18 remain visually active due to signals ascending directly from LGN (Donaldson and Nash, 1975; Dreher and Cottee, 1975), although their overall activity levels and sensitivity to slowly moving stimuli are reduced (Casanova et al., 1992; Dreher and Cottee, 1975; Sherk, 1978). This result is consistent with the slower speed selectivities of neurons in area 17 compared with area 18 (Orban, 1984). More subtle, yet compatible, effects were also detected with microdeactivation techniques applied to area 17 (Chabli et al., 1998; Ruan et al., 1997).

In the converse experiment on area 18, deactivation of ascending Y-signals to area 18 by pressure block of large fibers in the optic nerve leaves many area 18 neurons active (Dreher et al., 1992). However, the sensitivity of the area 18 neurons is limited to slower stimulus movements and approaches, as might be expected, the movement velocity sensitivities of area 17 neurons and the ascending afferent X-stream. These observations show that, in the absence of large numbers of ascending signals, the lateral association connections emerging from area 17 are sufficient to activate area 18 neurons and tune them more closely to the area 17 properties.

Mignard and Malpeli (1991) used double deactivation techniques to eliminate the influences of both lateral association projections emerging from area 18 and ascending signals from LGN through layer 4 from upper layer neuron activities in area 17. They blocked signal transmission first through the part of area 18 connected to the assay sites in area 17 and then through the relevant part of LGN. Finally, they deactivated both LGN and area 18 simultaneously. The result was that area 17 upper layer neurons were silenced by the dual blockade of LGN and area 18, but not by blockade of either area 18 or LGN alone. Thus, signal transmission through *either* LGN *or* area 18 is sufficient to activate neurons in the upper layers of area 17. These observations show that the ascending projections from LGN and the lateral association projections from area 18 are equally potent in their capacity to activate superficial layer neurons in area 17. The latter conclusion is confirmed by the work of Alonso et al. (1993) and Martinez-Condé et al. (1999).

Intrinsic Circuits

As revealed by electrophysiological methods and deactivation of afferent pathways, the compound afferent and intrinsic circuitries leading to corticotectal-projecting cells in areas 17 and 18 are highly similar (Fig. 1-9). These similarities provide additional evidence for parallel processing through areas 17 and 18, and the circuitry may be emblematic of other great similarities in the intrinsic circuitry of areas 17 and 18. In both areas 17 and 18, the proportions of standard- and special-complex corticotectal cells are similar (Weyand et al., 1986a,b, 1991). Moreover, they have virtually identical length-response functions, and similar orientation and direction tuning. Reversible blockade of signal transmission through one or more layers of LGN shows corticotectal cells in both areas can be subdivided into two additional types. One group, the standard complex cell, exhibits high dependence on signals transmitted via layer A of LGN; a second group, the special complex cells, receives convergent signals transmitted via *both* layers A and C *and* the upper cortical layers (Weyand et al., 1986a,b, 1991). For the special complex cells, blockade of either LGN layer A or layer C alone (for the contralateral eye signals) or the superficial cortical layers did not silence the neurons, but simultaneous blockade of both LGN layers, which also silences upper layer neurons, was effective and special complex cells could not be activated (Weyand et al., 1991). One notable difference between the corticotectal cells in the two areas is that those in area 18 conduct action potentials faster than their counterparts in area 17, but this feature is a structural and functional property of the neurons rather than a property of the afferent circuitry, and is a functional expression of the overall larger caliber of area 18 axons compared with area 17 axons (Creutzfeldt et al., 1977). Consequently, the difference between the two sets of corticotectal axons does not contradict the view that there are great similarities in the afferent and intrinsic circuitry leading to corticotectal cells in areas 17 and 18. The similarity between the two areas also appears to extend to several varieties of functional columns that have been identified in both areas 17 and 18. The one notable difference between the two areas appears to be our concept of the dimensions of the functional columns or modules.

Finally, we note that the deactivation experiments show that the silencing of layer 4 activity by blockade of transmission through specific layers of LGN bisects the cortical columns with few detrimental effects on neuron activity and receptive field properties of neurons in the upper layers. These observations bring into question the widely held belief that emergent receptive field properties of cortical neurons are constructed in layer 4 and then transmitted onto upper and lower layers for further elaboration. Clearly, alternative pathways up to visual cortical areas and lateral projections from area 18 or so-called feedback projections from higher areas both likely play a prominent role in the generation of receptive field properties in the upper and lower layers of area 17 and probably also in area 18. It will be interesting to discover how signals transmitted along very different pathways are capable of generating very similar receptive field properties to establish the cortical column.

FUNCTIONAL COLUMNS AND MODULES

Lorente de Nó (1949) realized that the functional connections between neurons in a vertical column were greater than their horizontal connections. He envisaged that the neurons forming the vertical column were linked by strong, translaminar feed-forward and feed-back pathways that operated on a discrete input in relative isolation from its neighbors. The input and output fibers of the neurons in the columns, as well as apical dendrites, arrange the cell bodies into palisades or pyramidal cell modules (see page 49; Figs 1-3, 1-18, and 1-20). These palisades are the most obvious anatomical manifestation of the columns in Nissl-stained preparations. This view of columns lay at the heart of Mountcastle's

(1957) observation of "deep" and "cutaneous" columns in the cat's primary somatosensory cortex. These columns were selectively activated by deep and cutaneous receptors, and they were intermingled on the cortical surface. Neurons below each surface domain have almost identical receptive fields, and they are responsive to the same deep or cutaneous modality.

Orientation

Hubel and Wiesel (1962) recognized a similar organization in cat visual cortex and plotted the domains on the cortical surface for the most robust and obvious columnar system: orientation selectivity (Hubel and Wiesel, 1963). They showed that, except for some small jitter about the average orientation, neurons studied along vertical electrode penetrations exhibited little change in their preference for a line stimulus of a particular orientation.

A good example of the similarity in preferred orientation in the vertical domain of cortex is shown in Fig. 1-34. This example is drawn from our own work, and data were collected from the medial part of area 18 in a cat with a flat marginal gyrus. The electrode entered the cortex normal to its surface and tracked along a radial palisade of cells except at its terminus in layer 6. In layer 6 fibers may enter the cortex obliquely and distort the palisades somewhat (Figs. 1-2 and 1-26, upper). Along the track presented, except for one neuron in layer 4 with a concentrically organized, on-center receptive field (**○**) and some slight drift in preferred orientation deep in layer 6, the majority of neurons had similar preferred orientations (●).

The first neuron encountered in upper layer 2 had a preferred orientation of 120° (left oblique along lower abscissa), and sequentially studied neurons in layers 2 through upper layer 6 had preferred orientations that deviated from this angle by $< 30°$. But the variation had no systematic trend, and the overall drift rate was negligible. As the electrode continued into layer 6, there was a gradual, sequential drift in preferred orientation through 180° (horizontal) to 15°. The small drift for the last three neurons studied likely reflects exit from the palisade of cells with common orientation preference into another column with slightly different orientation preference.

Electrode tracks that deviate significantly from vertical produce a different view of the representation of orientation preference in cortex and show systematic variation in preferred orientation. One exemplar is shown in Fig. 1-35, which is also drawn from our own work. In this example the electrode entered area 17 and tracked for a long distance obliquely through the base of layer 3, into layer 4, and onward into layers 5 and 6 for most of its distance. About 3 mm along its course, the electrode entered area 18 (interrupted vertical line). In area 17 (Fig. 1-35, left), the first neurons studied had preferred orientations between 60 and 90° (●, right ordinate), and the preferences of subsequently studied neurons drift first clockwise to 180/0°, and then counterclockwise through 0/180°, to reverse yet again and rotate clockwise to 180/0°. Of interest is the absence of any major interruption in the gradual rotations, even as the electrode crossed from area 17 into area 18. The drift rate in both areas 17 and 18 is about 120°/mm. The continuity

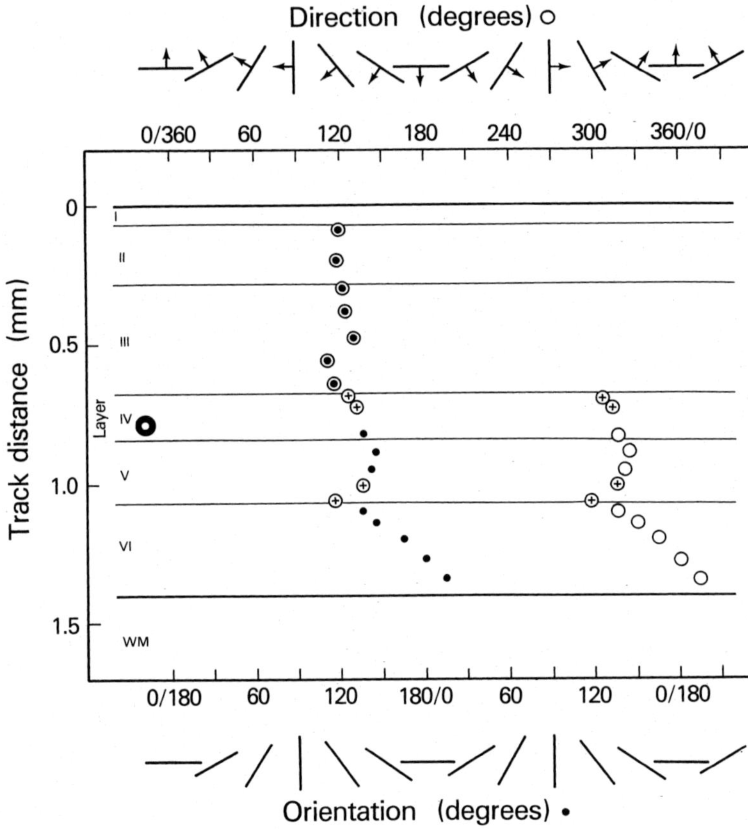

FIGURE 1-34. Diagram summarizing data on orientation and direction preferences of neurons studied along a vertical electrode track parallel to the radial palisades of neurons in the medial part of area 18. Dots (•) represent the preferred orientation of the neurons (lower abscissa). Circles (○) represent preferred directions of neurons (upper abscissa). Pluses (+) represent neurons that had no direction preference and were considered bidirectional. The annulus (**O**) offset to the left represents the concentrically organized, on-center receptive field of a layer 4 neuron. It had neither a preferred orientation nor a preferred direction. For graphical purposes, one half of the preferred direction continuum (○) studied in the upper layers has been arbitarily set to coincide with orientation continuum (•), and the • and ○ are coincident. For the second half of the two continua studied in the deep layers, the ○s of the direction continuum are displaced graphically ~180°, or one half cycle, to the right from the • of the orientation continuum. In this vertical sequence, there is only one shift of ~180° in preferred direction, and in this instance it occurs in the middle layers.

and smooth systematic rotation in orientation preference, without an abrupt, large change in angle, suggest that the basic organization of the orientation domain is both continuous and similar in the two areas.

The pooling of data from similar radial electrode tracks with little systematic variation, and from tangential tracks with considerable systematic variation led

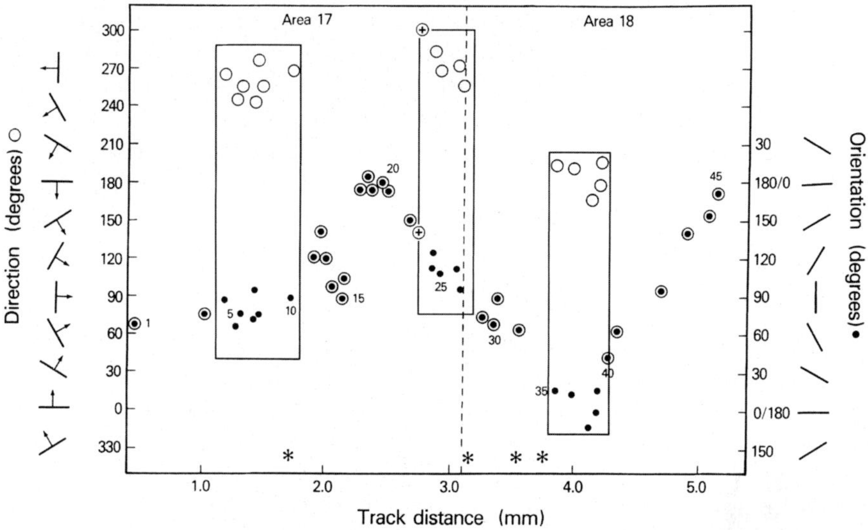

FIGURE 1-35. Diagram summarizing an oblique electrode track through visual cortex. It started in area 17 (left) and terminated in area 18 (right). The electrode crossed, and sampled from, many radial palisades of neurons and showed sequential small changes in preferred orientation (•, right ordinate) as the electrode traversed cortex. These changes rotated first in one direction and then, after a reversal at ~2.4 mm, in the opposite direction. A second reversal is evident at ~4 mm. For direction preference (○, left ordinate) the gradual drift in preference is superimposed on the gradual drift in preferred orientation (coincident ○ and •). However, in the direction continuum, the drift is punctuated by a shift in preferred direction of ~180°, and the ○ of the direction continuum and the • orientation continuum become graphically displaced from each other (rectangles). The slow drift in ○ (direction preference) continues at the same rate as the slow drift in orientation, followed by another shift of ~180° to bring the ○ and • back into coincidence. Additional punctuations and reversals characterize later portions of the sequence as it continues into area 18. Pluses (+) identify a neuron with no obvious direction preference. The asterisks identify neurons that were too weakly responsive to permit accurate identification of preferred orientation and direction.

Hubel and Wiesel (1962, 1963) to propose initially the concept of orientation columns. They later extended the concept of the larger organizational unit, the hypercolumn, which contains representations of all orientations (Hubel and Wiesel, 1977). Based on the drift rates given previously, the diameters of the hypercolumns in both areas 18 and 17 are about 1.5 mm. However, caution should be exercised in accepting this value, because it is highly probable that the assay electrode did not sample from the shortest dimension of the orientation hypercolumns, but rather passed obliquely through them. Based on available data, it is likely that the orientation hypercolumn has a dimension of 0.8 to 1.5 mm.

Although the concept of functional orientation columns and hypercolumns is appealing, it is inappropriate to view the columns as possessing rigid boundaries. For orientation columns, the boundaries cannot be determined because most neurons have broad orientation sensitivities that exceed the shifts in preference identified for sequentially studied neurons. Also of interest are the numerous point singularities that are embedded within the orientation map. At singularities orientation preference changes discontinuously, and sequentially studied neurons exhibit large jumps in preferred orientation of the order of 90° or more (Albus, 1975b). Similar observations on the characteristics and dimensions of orientation columns and hypercolumns have been made by Cynader et al. (1987), Berman et al. (1987), Swindale et al. (1987), and, of course, by Hubel and Wiesel (1965a).

Ocular Dominance

In the early work of Hubel and Wiesel (1962, 1965a), the picture for the ocular dominance columnar systems, which contribute to the integration of signals arriving from the two eyes, was not obvious. They noted that in both areas 17 and 18, they encountered a majority of neurons whose activities were dominated by one eye in preference to the other. This clarity of preference was greatest for neurons in layer 4, because a greater fraction of neurons had distinct preferences for one eye or the other. However, outside layer 4, the majority of neurons can be excited by stimuli presented to either eye, and extreme dominance by one eye or the other is rare. This factor tends to obscure the ocular dominance columns and makes them difficult to detect in the superficial and deep layers. Even so, in layer 4 of favorable preparations, the zones dominated by one or other eye may be more obvious.

A particularly striking example of ocular dominance zones in layer 4 containing sequences of neurons dominated by one eye or the other is shown in Fig. 1-36. This example is taken from our own work. In this experiment three electrode tracks were made starting at slightly different positions and passing tangentially through cortex. We assembled and juxtaposed data from neurons verified in histological sections to be located at sequential positions in layer 4. The first group of neurons was dominated by the ipsilateral eye (ocular dominance 7), including one neuron with a concentrically organized receptive field. A single neuron was encountered that was activated by both eyes, but slightly better by the contralateral eye (ocular dominance 3). Two further neurons were driven solely by the contralateral eye (ocular dominance 1). There were then additional sequences of neurons driven by, first, the ipsilateral eye, the contralateral eye, and back to the ipsilateral eye further along the track.

By fixing the borders between the ipsilateral- and contralateral-eye dominated zones at an ocular dominance rating of 4, which indicates neurons driven equally well by the two eye, we see that the widths of the ocular dominance zones vary from 0.4 to 0.7 mm. Thus, a pair of ipsilateral and contralateral zones have a combined width of 0.8 to 1.4 mm. However, as for the orientation columns, caution needs to be exercised in accepting these dimensions directly because the electrode probably did not sample the narrowest part of the ocular dominance column. With that caveat in mind, the physiological measures correspond approximately to the diameters of the ocular dominance zones defined anatomically (0.5 to 0.8 mm) and described earlier (see page 33).

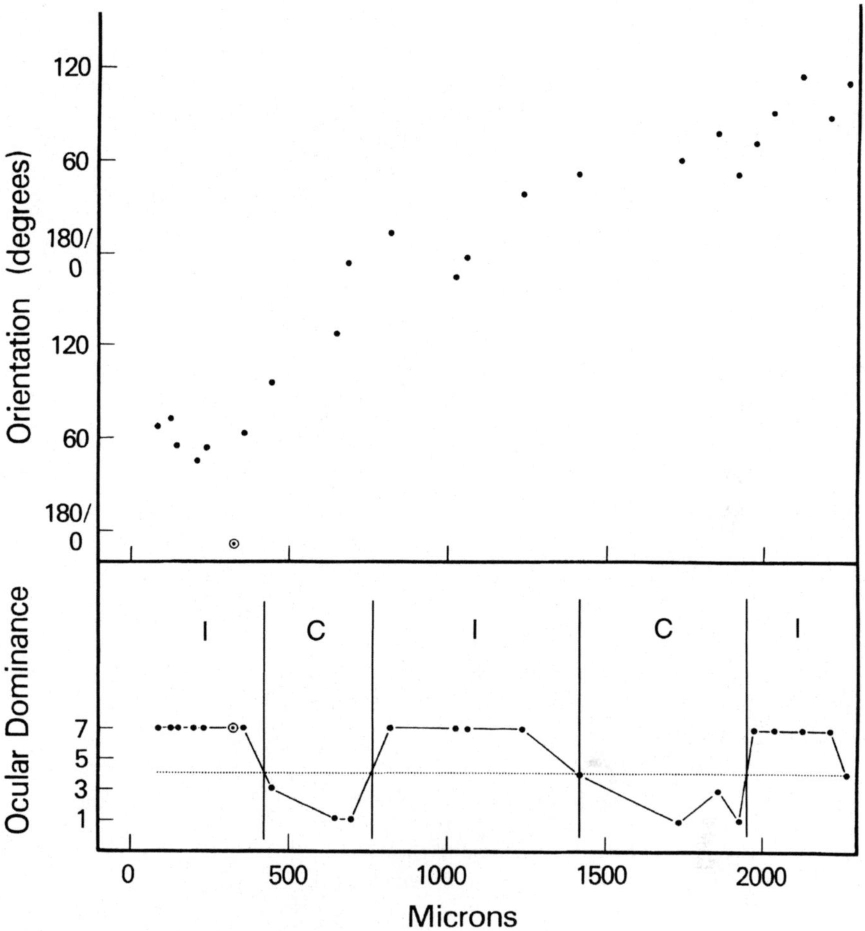

FIGURE 1-36. Diagram summarizing data drawn from sections of three electrode tracks that traversed adjacent regions in layer 4 in area 17. (Lower panel) Ocular dominance of neurons. Neurons driven by the contralateral eye (C) are designated to have an ocular dominance of 1, neurons driven equally well by both eyes have a designated ocular dominance of 4, and neurons driven solely by the ipsilateral eye (I) have a designated ocular dominance of 7. Neurons driven by both eyes but dominated to varying degrees by one or other of the eyes are designated with intermediate values. (Upper panel) Preferred orientations. Circle and dot in upper and lower panels represent a neuron with a concentrically organized, off-center receptive field.

The superimposition of the ocular dominance system on the orientation system is also evident in Fig. 1-36. Even though there were alternating zones dominated by first one eye and then the other, there was an orderly progression in orientation preferences at a rate of 90 to 120°/mm that was uninterrupted by the changes in ocular dominance. We conclude from this comparison, as did Hubel and Wiesel before us, that orientation preference is mapped on cortical neuron populations largely independent of the mapping of ocular dominance. Alternatively, if a systematic relationship exists between the two columnar systems, it continues to remain as enigmatic today as it did almost 40 years ago when Hubel and Wiesel first described similar sequences. However, important new views have been obtained with the optical imaging methods (see Fig. 2-9). In these views orientation columns seem to swirl about the irregular globular ocular dominance columns (Fig. 6-6), but nevertheless the two columnar systems intersect perpendicularly, as was presciently proposed to be advantageous by Hubel and Wiesel in 1977.

Hubel and Wiesel (1965b) obtained much more striking examples of ocular dominance zones in cats raised with artificial strabismus. In these cats the images formed on the two eyes do not meet in register in cortex, and neurons even outside of layer 4 become dominated, or even driven solely by one eye or the other. Thus, experimental sample sizes are both increased, and distinctions between ocular dominance pools are heightened to improve the visibility of the ocular dominance zones without the need to take extraordinary measures to sample solely from neurons in layer 4. The extension of the ocular dominance zones outside layer 4 in strabismic cats prompted the view that the zones should be considered columns or short slabs. This view is supported in layer 4 of normal cats by transsynaptic tracing studies after injection of tracer into one or other of the eyes (see page 33 and Fig. 1-10).

The correspondence of the anatomical ocular dominance system demonstrated with the transsynaptic tracing method and the electrophysiologically defined ocular dominance system was subsequently established by Shatz and Stryker (1978) and by LeVay et al. (1978) by combining anatomical and physiological procedures in the same cats. In these experiments four sequential steps were taken.

1. One set of "eye" columns in cortex was labeled by using transsynaptic, transgeniculate transport up to visual cortex of radiolabeled molecules injected into one eye (Fig. 1-10).
2. Physiological methods were then used to assay the ocular dominance of neurons in cortex.
3. In histological sections of the assayed cortex the electrode tracks were reconstructed, and the electrophysiologically defined ocular dominance zones in layer 4 were defined.
4. These zones were related to regions in layer 4 in receipt of large and sparse amounts of label transported transsynaptically from one eye.

As expected, most neurons in layer 4 dominated by signals generated by the labeled eye were localized to heavily labeled regions, whereas most neurons dominated by signals generated by the unlabeled eye were localized to poorly labeled regions. Neurons driven by both eyes tended to be localized to regions of intermediate label density. Thus, the link between the anatomically defined and electrophysiologically defined ocular dominance systems was proven, and Wilbrand's prophetic hypothesis of 1890 of the ocular innervation of cortex could be finally accepted almost one century later in 1978 (see page 18).

Movement Direction

The ocular dominance and orientation columnar systems have dominated thinking on the structural and functional organization of visual cortex, and they paved the way for the identification of additional columnar or modular systems in cats. The term *module* is used here to suggest a functional unit that may be shorter than a column, for it may not extend across all cortical layers. Like columns, modules are not discrete entities with rigid boundaries and independent functions because neurons do not have highly specific preferences, but they usually have some considerable tuning breadth. These broad tunings blur otherwise apparent trends in neuronal sequences and modular boundaries. Even so, the movement direction preference system is superimposed on the ocular dominance and orientation preference columns, and in intricate ways. For example, direction of movement is a property intimately linked with orientation preference in both areas 17 and 18, and many orientation selective neurons prefer movement of the stimulus in one or other of the two directions perpendicular to the preferred orientation. By convention, half the direction continuum has an angular measure that is coincident with the orientation continuum. For the second half of the direction continuum, the angular measure is 180° displaced from the orientation continuum.

As for orientation, the frequency of shifts in the direction continuum are greater in the tangential than in the vertical domains of cortex (Payne et al., 1981; Tolhurst et al., 1981; Berman et al., 1987). For example, along extended tangential electrode penetrations through areas 17 and 18 sequentially studied neurons exhibit smooth and gentle shifts in preferred orientation as noted earlier (● in Fig. 1-35). Likewise, the direction system also exhibits similar small shifts in preferred direction (**o** in Fig. 1-35) that are superimposed on the orientation continuum (coincident ● and **o**). However, the direction sequences are periodically punctuated by large changes of ~180° (Payne et al., 1981; Berman et al., 1987). These punctuations are visualized by, first, the displacement of the **o** from the ● in Fig. 1-35 (rectangles) and, then, the reversals back to coincidence and the superimposition of the **o** and ●. Six such shifts and reversals are shown in Fig. 1-35.

In the vertical domain (Fig. 1-34) the change in preferred direction from neuron to neuron is much less (Payne et al., 1981; Tolhurst et al., 1981), and there is no systematic drift in preferred direction (**o**) as an electrode tracks through the upper layers, just as there is no systematic drift in preferred orientation (●). However, in the middle layers there is a large 180° fracture in the direction continuum, which is graphically represented as a rightward displacement in **o**, and the direction continuum is displaced by 180° from the orientation continuum. Such fractures are less common in the vertical domain of cortex than in the horizontal domain (Fig. 1-34) (Berman et al., 1987).

The direction-preference modules defined in this way are also evident in surface mappings of area 18, and the preferred directions first swirl one way then the other (Swindable et al., 1987). The domains have sizes of 0.4 to 0.7 mm, or about half the size of an orientation hypercolumn. However, the domain represents only half the full compliment of directions. To our knowledge no study has shown that the direction modules penetrate through the full cortical thickness and thereby qualify as complete columns. Similar, more limited modular systems may exist for spatial frequency and spatial disparity.

Spatial Frequency

As described earlier, many neurons in areas 17 and 18 also exhibit tuning for spatial frequency (Movshon et al., 1978; Berardi et al., 1982). This parameter has an orderly and understandable representation in the laminar domain, with neurons in layer 4 preferring the highest spatial frequencies, layer 2 / 3 neurons preferring intermediate spatial frequencies, and neurons in layers 5 and 6 preferring the lowest spatial frequencies (Maffei and Fiorentini, 1977; Tolhurst and Thompson, 1981; Berardi et al., 1982). These observations are in accord with the differences in receptive field size and differences in the grain of the visual field maps in upper, middle, and lower layers described earlier (see page 63). In this tangential domain optimal spatial frequency tends to change little even though orientation preference changes sequentially from one neuron to the next (Berardi et al., 1982; Tolhurst and Thompson, 1981, 1982).

Spatial Disparity

Many neurons in areas 17 and 18 are also selective for spatial disparity (Ferster, 1980), but a search for evidence to support the idea of a columnar system linked to spatial disparity in cat areas 17 and 18 has met with only limited success. Even so, there is an ordered arrangement of disparity sensitive neurons, but they appear to be a subset of the ocular dominance system (LeVay and Voigt, 1988). A similar relationship between spatial disparity and ocular dominance has been detected in sheep visual cortex (Clarke et al., 1976). In sheep, the groupings of neurons with similar disparity selectivity are more obvious, perhaps because the disparities are greater, being based on a much greater interocular distance in sheep. The proximity of the two eyes to each other in cats generates smaller disparities, and any underlying systematic mapping of disparities in the cat may be below the resolution of present day assay methods and, thus, obscured.

Retinotopy

Finally, the map of the visual field on, and within, the cortical layers can also be considered a columnar system. As for orientation columns, neurons in a column view the same part of the visual field, with some superimposed random scatter. Moreover, neurons studied sequentially in the horizontal domain exhibit a drift in position across the visual field. The features of this system have been discussed at length previously (see page 63).

Visualization of Functional Columns

Reassuringly, most columnar systems identified with electrophysiological techniques have been visualized with the 2DG metabolic and autoradiographic technique in cross sections and flat mounts of visual cortex. The columns have been shown to extend across large portions of areas 17 and 18 and through all or most of the cortical layers (Löwel and Singer, 1987, 1993a,b; Löwel et al., 1987, 1988; Lang and Henn, 1980; Albus and Siebert, 1984; Schoppmann and Stryker, 1981; Singer, 1981; Tootell et al., 1981; Löwel, this volume). Several columnar systems have also been demonstrated with optical imaging of neural activity (Hübener and Bonhoeffer, this volume). The technique uses either voltage sensitive dyes or comparisons of relative amounts of oxyhemoglobin and deoxyhemoglobin to reveal the functional domains (Bonhoeffer and Grinvald, 1996). The technique examines global activity levels in the middle and upper layers of cortex, and numerous imaging studies have been applied to areas 17 and 18 of the cat (Hübener et al., 1997; Bonhoeffer and Grinvald, 1991, 1993; Bonhoeffer et al., 1995; Schmuel and Grinvald, 1996; Shoham et al., 1997; Kim et al., 1999; Issa et al., 2000). They have confirmed in the living system, the existence of columnar, or modular, systems for ocular dominance, and preferences for line orientation, direction of motion and high or low spatial frequencies, and show that each property is characterized by a unique functional map (Hübener and Bonhoeffer, this volume).

Theoretical studies suggest that geometrical factors do not sharply limit the ability of cortex to superimpose multiple maps providing they are of the similar periodicy, although if coverage is to be uniform for all maps, the upper numerical limit of the number of maps that can be superimposed is about 6 (Swindale, 2000). These experimental and theoretical recent studies are discussed in other Chapters (see Chapters 2-4), but it is worthy of note that experimental observations have shown that functional representations do have certain spatial relationships such as orientation columns intersecting ocular dominance columns at right angles (see page 96), and fractures in the direction map end in the immediate vicinity of orientation centers (see Fig. 2-7 and 2-9) (Hübener et al., 1997; Swindale et al., 2000). These orientation centers correspond to the singularities originally identified by Albus (1975b). We noted earlier that at singularities orientation preference changes discontinuously, and sequentially studied neurons exhibit large jumps in preferred orientation on the order of 90° or more.

As is clear from these studies, functional imaging techniques have the potential to expose the geometrical relationships between the other functional maps identified in areas 17 and 18 and also to reveal discontinuities across the transitional cortex between areas 17 and 18. Some success has been made in these directions (Hübener et al., 1997; Shoham et al., 1997; Kim et al. 1999; Ohki et al., 2000). The technique may also prove to be highly useful for assaying the functional impact of local and long-distance connections on representations (Toth et al., 1997; Galuske et al., 2000; Schmidt et al., 2000; Payne et al., 2001).

Columnar Dimensions in Areas 17 and 18

In all studies on functional columns in areas 17 and 18, the grain of the cortical column mosaic appears to be coarser in area 18 than in area 17. In area 17 the mosaic "tiles" of all modular or columnar systems so far identified have linear dimension of 0.4 to 0.7 mm, or some multiple thereof. In area 18, the dimensions of the equivalent tiles are about twice the area 17 measures at 1.0 to 1.2 mm, or some multiple thereof. These measures fit approximately with the overall with the parameters of the cytochrome oxidase blob and interblob regions (see Chapter 5), the distance between groups of neurons innervated by long-range intrinsic connections (Schmidt and Löwel, this volume), and columnar dimensions revealed by functional imaging methods (see Chapters 2-4). The basis of the difference in sizes in the two areas is obscure, but it may be linked to the smaller size of the axon arbors of the dominant X-system in area 17 compared with the larger size of terminal arbors of the Y-system in area 18. The axon arbors have spreads that approximate the area 17 (~0.6 mm) and area 18 (~1.2 mm) systems.

VISUALLY GUIDED BEHAVIOR

Attempts have been made to assess the contributions areas 17 and 18 make to visually guided behavior. Such studies have the dual challenge of completely removing one or other area without involvement of the adjoining area, and to do so bilaterally. Such attempts have rarely met with convincing success. However, attempts to remove both areas 17 and 18 in one procedure have met with greater success and have produced more convincing conclusions (Mitchell, this volume). For example, both areas 17 and 18 have been shown to be critical for a variety of fundamental and sophisticated visual attributes. These attributes include line orientation discrimination (Orban et al., 1988; Vandenbussche et al., 1991), visual acuity in the highest range and maximum acuity under conditions when contrast is low (Berkley and Sprague, 1979; Hughes and Sprague, 1986; Lehmkuhle, et al., 1982), vernier acuity (Berkley and Sprague, 1979; Mitchell, 1989), depth discrimination (Cornwell et al., 1976, 1980; Hovda et al., 1989; Kaye et al., 1981; Ptito et al., 1992; Shupert et al., 1993), forms of texture segmentation based on short line elements (De Weerd et al., 1994), and for the processing of details in patterns in the presence of visual noise (Hughes and Sprague, 1986). Some of these functions are likely to be distributed across both areas. For example, when high-contrast stimuli are used, lesions restricted to either area 17 or area 18 alone have little effect on performance, but lesions that include large parts of both areas raise thresholds significantly (Orban et al., 1990). In comparison, lesions/deactivations of lateral visual areas do not impair visual acuity or other more fundamental measures of visual processing (Sprague et al., 1977; Berkley and Sprague, 1979; Berlucchi et al., 1972), but do have a big impact on cognitive, spatial, and sensorimotor aspects of vision (Campbell, 1978; Cornwell et al., 1998; Hughes and Sprague, 1986; Lomber et al., 1996a,b; Payne et al., 1996b,c; Lomber and Payne, 1996, 2000a, b; Sprague et al., 1996).

AREA 17

Of the visual functions dependent on on areas 17 and 18, some are linked primarily to area 17. For example, high-acuity vision appears to depend on area 17. Normal cats have spatial visual acuities in the range of 6 to 8 cyc/deg when viewing high-contrast gratings under high luminance (Hall and Mitchell, 1991; Mitchell, 1990; Mitchell, this volume). These values are reduced to the range of 2 to 3 cyc/deg by ablation of areas 17 and 18 (Kaye et al., 1981; Mitchell, 1990). Compatible results have been obtained by Berkley and Sprague (1979) and others when the lower contrast and lower illumination levels are considered (Hall and Mitchell, 1991). Reassurringly, the acuity measures ascribed to area 17 resemble the highest acuities of X- (β) retinal ganglion cells under conditions of high contrast and illumination (Cleland et al., 1979), which, as we have already noted, provide the dominant visual input to area 17. Lehmkuhle et al. (1982) identified a more modest reduction in acuity following virtually complete lesion of area 17 and the immediately adjacent portion of area 18. However, in this study the initial highest acuity was lower than the measures obtained by Kaye et al. (1981) and Hall and Mitchell (1991), perhaps because of the lower level of illumination used by Lehmkuhle et al. (1982).

Other studies of discrete lesions placed *within* area 17 have also shown that X- (β) retinal ganglion cells are the origin of the dominant driving force to that area (Pasternak et al., 1995). Because the strategy used by Pasternak et al. (1995) was to make a limited lesion of part of area 17, more sophisticated procedures than usual were required to assay for the impaired visual functions. These procedures included accurate monitoring of eye position and presentation of the test visual stumuli within the cortical "scotoma" induced by the lesion. Control measures were obtained from nearby "seeing" regions of intact cortex. Using these procedures, coupled to a detection task in which cats simply indicated the presence or absence of a vertical grating, Pasternak et al. (1995) showed that contrast sensitivity loss was greatest in the middle frequencies with no loss of spatial resolution. However, when cats were required to discriminate between vertical and horizontal gratings, sensitivity loss was profound at both middle and high spatial frequencies, with an octave loss of resolution. Deficits in the ability to discriminate the direction of grating motion were also identified, but only at higher spatial and lower temporal frequencies. All of these deficits are consistent with a failure of X-signals to be processed in the ablated portion of area 17. Moreover, they are compatible with the neuronal deficits identified for area 18 neurons during deactivation of area 17 described previously (see page 78).

AREA 18

A similar approach was taken to assay the visual functions of area 18. In accord with previous views, discrete lesions contained within area 18 impair visual processing based on signals originating from Y- (α) retinal ganglion cells.

The lesions induced a loss in sensitivity to gratings of low and intermediate spatial frequency. The magnitude of the loss of sensitivity was lessened at higher spatial frequencies, and resolution limits remained intact (Pasternak and Maunsell, 1992). Moreover, deficits in discrimination of leftward and rightward motion were identified even when high contrast patterns were used (Pasternak and Maunsell, 1992). These results demonstrate that area 18 plays an important role in detecting drifting low- and intermediate-spatial-frequency targets, and it is likely to represent a critical stage in the cortical processing of motion signals. These deficits are consistent with a failure of Y-signals to be processed in the ablated portion of area 18. We are unaware of any studies that have convincingly tested acuity measures following ablation of area 18 in its entirety and alone.

CONCLUSIONS

The complementary deficits in sensitivity to low and intermediate spatial frequencies following area 18 inactivation, and middle and high spatial frequencies following area 17 inactivation, demonstrate again that areas 17 and 18 function to broaden the overall bandwidth of visual processing. A similar conclusion can be reached for motion processing. Inactivation of area 18 has a broad impact on motion discrimination, whereas the impact of area 17 deactivation is limited to targets with higher spatial and lower temporal frequencies. From these results we conclude that (1) area 18 extends the processing at the low spatial frequency end of the spectrum, (2) area 17 extends vision at the high-frequency end of the spectrum, and (3) the two areas contribute in complementary ways to visual processing of motion signals. These conclusions are concordant with the differences in receptive field properties in areas 17 and 18 as gleaned from single neuron electrophysiological studies (see page 78). Moreover, based on differences in the contrast sensitivities of their afferent neurons under a variety of adaptation levels, area 18 may be better suited to scotopic vision than area 17 because the Y-pathway exhibits greater visual sensitivity, but at the expense of acuity (Lee et al. 1992).

SYNTHESIS

Early in this chapter we identified five accepted criteria that need to be met for a cortical region to be considered a bona fide cortical area. We also asked: Are the characteristics of area 18 sufficiently clear and robust for it also to be considered a bona fide cortical area independent of area 17? On the grounds that area 18 has both a distinct architecture and a distinct representation of the visual field we answer the question with an unequivocal Yes. However, on the grounds of connectional signature, inventory of receptive field properties and contributions to visually guided behaviors, the answer is not so clear-cut, and it is much closer to an equivocal No. Much of the evidence for these views is summarized in Table 1-2, and our considered conclusion is given in the next section.

ARCHITECTURE, SUBCORTICAL INPUT, AND INTRINSIC CIRCUITRY

As can be seen in Table 1-2, areas 17 and 18 have a similar overall architecture, and their major inputs are from the same categorical subcortical structures, the lateral geniculate nucleus and the lateral posterior/pulvinar complex. In addition, the retinogeniculocortical pathways leading to both areas 17 and 18 terminate in the same layers and have many common features. Some of these features are robust and fundamentally different from the weaker retinogeniculocortical pathways leading to other visual cortical regions such as area 19 and middle suprasylvian visual cortex. These features provided the prime evidence for accepting the proposal that both areas 17 and 18 should be included under the umbrella term *primary visual cortex.* However, in terms of surface sizes and forms of the visual maps, the two areas are not exactly equivalent. Even so, the intralaminar and interlaminar flows of signals through the intrinsic circuitry of areas 17 and 18 are highly similar. These similarities generate with equal efficiency the emergent receptive field properties of orientation selectivity, movement direction selectivity, binocular interactions, and spatial disparity in the two areas.

X- AND Y-SIGNALS

The three major functional differences between the two areas are receptive field size, and spatial and temporal sensitivities of neurons (See* in Table 1-2). Neurons in area 17 tend to have smaller receptive fields, and they are sensitive to slower movements, whereas neurons in area 18 tend to have larger receptive fields, lower spatial acuities, and they have great sensitivity to temporal transients, whether flicker or fast movement (Orban, 1984). These differences do not emerge as a result of the intrinsic properties of areas 17 and 18, but seem to be properties inculcated on the two areas by the afferent X- and Y-visual streams.

EFFERENT PROJECTIONS

Other similarities between the two areas are their major efferent sets of projections to distant brain regions. These features are not dealt with in depth in this chapter because the portion of topic concerned with transcortical connections has been given extensive coverage by Rosenquist (1985). Efferent projections from both areas target the same major cortical and subcortical structures. Moreover, projections to the same target have their source in the same cortical layers (Symonds and Rosenquist, 1984b), and the termination fields of the two projection systems are in excellent, although not exact, spatial registration. This pattern is exemplified by the projections from areas 17 and 18 to, say, the superior colliculus. Layer 5 neurons in both areas 17 and 18 project heavily to the stratum gri-

TABLE 1-2. Features of Areas 17 and 18

Feature	Area 17	Area 18
Architecture		
6 layered neocortex	Yes	Yes
Layer 2/3	Thin	Thick
Layer 4	Highly granular (Bistratified)	Highly granular (Unistratified)
Layers 5 and 6	Thick	Thin
Major subcortical visual input		
LGN†		
Dominant type*	LGN X	LGN Y
Termination	Layers 4 and 6	Layers 4 (and 6)
Ocular dominance stripes	Narrower	Wider
LP/Pul†	Yes	Yes
Surface area*	300–400 mm^2	60–80 mm^2
Visual map	First order	Second order
Intrinsic circuitry		
Layer 4 ⇒ 4	Yes	Yes
Layer 4 ⇒ 3 ⇒ 2	Yes	Yes
Layer 4 ⇒ 5 and 6	Yes	Yes
Layer 3 ⇒ 5 and 6	Yes	Yes
Layer 5 ⇒ 2 and 3	Yes	Yes
Layer 6 ⇒ 4	Yes	Yes
Receptive field structure		
Simple	Yes	Yes
Complex	Yes	Yes
Hypercomplex	Yes	Yes
Binocular	Yes	Yes
Size*	Smaller	Larger
Receptive field properties		
Orientation selectivity	Yes	Yes
Movement direction selectivity	Yes	Yes
Positional disparity	Yes	Yes
Binocular interactions	Yes	Yes
Spatial frequency*	High	Low
Temporal sensitivity*:		
Movement velocity	Slower	Faster
Flicker	Low frequency	High frequency
Functional columns/modules*	Width:~0.5–0.7 mm	Width:~1–1.2 mm
Origin of major efferent projections†		
Association	Layers 3–6	Layers 3–6
Transcallosal	Layers 2–4 and 6	Layers 2–6
Transcortical	Dominated by layer 3	Dominated by layer 3
Descending to:		
Pt, SC	Layer 5	Layer 5
LP/Pul†	Layers 5 and 6	Layers 5 and 6
LGN†	Layer 6	Layer 6
Claustrum	Layer 6	Layer 6

TABLE 1-2. *(continued)*

Feature	Area 17	Area 18
Termination of major efferent projections†		
Association: 17 ⇒ 18, 18 ⇒ 17	Yes	Yes
Transcortical:		
⇒ Area 19	Yes	Yes
⇒ Middle suprasylvian sulcal cortex		
medial bank	Yes	Yes
lateral bank	Weak	Weak
⇒ Area 20	Weak	Weak
Descending:		
⇒ LGN	Yes	Yes
⇒ LP	Yes	Yes
⇒ SGS† of SC	Yes	Yes
⇒ Pontine nuclei	Yes	Yes
⇒ Claustrum	Yes	Yes

* Criteria on which areas 17 and 18 differ substantially.

† Usually reciprocal connections.

LGN, dorsal lateral geniculate nucleus; LP, lateral posterior nucleus; Pt, pretectum; Pul, pulvinar nucleus; SC, superior colliculus; SGS, stratum griseum superficiale.

seum superficiale of the superior colliculus, where they are almost in complete register except that the projections from area 17 terminate slightly closer to the collicular surface than the projections from area 18 (Harting et al., 1992). Likewise, there are great similarities in the terminations of projections from layer 6 neurons to LGN (Updyke, 1977), just as there are for projections from the upper cortical layers to other regions in cortex (Symonds and Rosenquist, 1984a).

CONCLUSIONS

When the numerous features in Table 1-2 are viewed in a composite way, the conclusion is inescapable that area 18 can be considered to be a modified duplicate of area 17, or vice-versa. The connections and properties of neurons in area 18 overlap with the connections and properties of neurons in area 17, and the two areas seem to extend each others structural and functional continua, just as the α- and β-ganglion cells extend structural and functional continua in the retina. If it were not for the distinct architecture and separate visual field representation, there is little to distinguish the features of area 18 from the details of area 17. It is, then, appropriate to view areas 17 and 18 as two complementary, parallel areas that broaden the spectrum of primary visual processing in cerebral cortex. Based on these views we become firmer in our convictions that both areas 17 and 18 qualify for inclusion under the umbrella term of *primary visual cortex.* If this is so, then we conclude that primary visual cortex occupies ~360 to 480 mm^2 and forms, by far, the largest cortical territory devoted to a single stage of visual processing. It amounts to about one half of the cortical surface devoted to vision. In those terms the combined region has the dominant rank in vision. In addition, it receives the vast majority of signals sent to cortex from LGN and virtually all signals transmitted through the magnocellular division of LGN. Following natural visual stimulation, the combined regions receive the first, activating signals from LGN. Moreover, the combined region is a major distributor of signals to numerous cortical and subcortical targets, which are activated or influenced later. Finally, the combined areas are critical for processing the broadest band of fundamental visual attributes. However, regardless of the strength of the conclusions, this view does not go without challenge.

RECENT CHALLENGES TO THE PRIMACY OF AREAS 17 AND 18

The primacy of areas 17 and 18 is settled for all except one interpretation of the word *primary.* Timing is an important component of visual processing, and it has been established that there are regions of extrastriate cortex, such as area 19 and middle suprasylvian visual cortex, that also receive inputs directly from LGN. The majority of these projections arise in the parvocellular C layers, and they can be considered part of the sluggish, slow conducting W-system. Thus, they do not bear on the topic of primacy because any activation they induce lags behind that of activity evoked by the briskly responding and rapidly conducting X- and Y-systems (Dreher et al., 1980).

However, some large neurons in LGN in the magnocellular C and medial interlaminar subdivisions have been identified to project to area 19 and to the middle suprasylvian region. Several of these neurons have a type 1 morphology (MacNeil et al., 1997a), and based on structure/function correlations established by Friedlander et al. (1981) and Raczkowski and Sherman (1985) these large cells likely transmit the brisk and fast Y-signals. Given that much of area 19 and middle suprasylvian visual cortex are closer to LGN than equivalent regions of areas 18 and 17, it is conceivable that area 19 and middle suprasylvian visual cortex might be activated before areas 17 and 18. This view is not supported because even neurons in area 19 and middle suprasylvian cortex activated by electrical stimulation of the Y-axon system respond via intermediary activity in areas 17 and 18 (Dreher et al., 1980; Berson, 1985).

TIMING

The significance of the foregoing observations has been tested in a limited way. Visual response latencies of neurons in LGN, areas 17, 18, 19, middle suprasylvian visual cortex, and area 7 have been measured by Dinse and Kruger (1994). They found an increasing lengthening of the mean response latencies across areas. These increases are consistent with the progressive advance of signals across cortex as they pass through sequential processing stages. However, they also identified a substantial, increasing scatter in the latencies that showed that neurons in several areas exhibit simultaneous activity. Overall, Dinse and Kruger noted that the earliest responses in cortex were evoked in area 18, and that they occurred some 12 msec earlier than the earliest responses in area 17. This temporal difference reflects the temporal difference in activation of the Y-and X-retinal ganglion cells (Bolz et al., 1982), which are the dominant inputs to areas 18 and 17, respectively. Both area 19 and middle suprasylvian visual cortex were activated later.

An important aspect of the work of Dinse and Kruger (1994) was that they used the same stimulus to activate neurons in the different regions. This is important because neurons are tuned to different aspects of the visual scene, and simple changes in parameters can have substantial effects on activation times. For example, both spatial frequency and contrast have significant effects on the timing of response of LGN neurons (Sestokas and Lehmkuhle, 1986). Although there is no overlap in any range of contrasts in the response latency of LGN X- and Y-cells, it may be that there are conditions when an optimally configured stimulus for neurons in one area may drive neurons earlier than a poor stimulus configured for neurons in another area.

The observations of Dinse and Kruger (1994) on the later activation of area 19 and lateral suprasylvain cortex are in accord with the substantial transcortical projections emerging from areas 17 and 18 to innervate area 19 and middle suprasylvian visual cortex. Areas 17 and 18 provide the dominant transcortical inputs to area 19 and middle suprasylvian visual cortex. Even so, it is important to recognize that large regions of cortex, including area 19 and the middle suprasylvian region, remain visually active in the absence of areas 17 and 18 (Michalski et al., 1993; Spear and Baumann, 1979) as typically occurs in regions innervated by more than one visual pathway (see page 88). However, even when residual activity remains, the levels are depressed and receptive field properties are less sophisticated in the absence of areas 17 and 18. Much of the residual activity appears to be derived from signals transmitted along the retino-collicular-lateral posterior-cortical pathway (Payne et al., 1996a). However, one study suggests that the residual activity may be based purely on signals transmitted through LGN in the absence of relays through areas 17 and 18 and superior colliculus (Gueddes et al., 1983). If validated, these data will show that even the limited inputs from dLGN directly to area 19 and middle suprasylvian visual cortex have substantial influences and are capable of activating cortical neurons. These observations demonstrate the importance of the bypass pathways in the activation of area 19 and suprasylvian cortical neurons and highlight the potential influence supplementary pathways might play in shaping neuronal properties. These same supplementary pathways may, under special circumstances, provide the dominant, driving input to their target neurons. However, under most circumstances, the drive arrives from areas 17 and 18. This discrepancy in the functional impact of LGN on neurons in area 19 and middle suprasylvian cortex is worthy of further investigation.

One special circumstance when the supplementary pathways become particularly effective is after removal of areas 17 and 18 from the immature brain. Under these circumstances, the transgeniculate pathway to the lateral visual areas, via both magnocellular and parvocellular LGN neurons, is strengthened (Lomber et al., 1993, 1995; Payne and Cornwell, 1994; Payne et al., 1996a). For example, transgeniculate projections expand and innervate layer 4 of middle suprasylvian visual cortex (Payne and Lomber, 1998). There is a corresponding increase in cytochrome oxidase activity in layer 4 (Long et al., 1996). Moreover, the coupling between residual retinal and LGN neurons increases to a level commensurate with the coupling between retinal and LGN magnocellular neurons in the intact brain that project to areas 17 and 18. This increased coupling permits the free anterograde and retrograde transsynaptic transfer of tracer molecules between primary and secondary limbs in the retinogeniculocortical pathway. The expansion of the normally small retinogeniculocortical pathway elevates it into a major conduit for the transfer of visual signals to the cerebral cortical visual system. These signals have significant impact on and form one basis for significant visual processing in lateral cortex in the absence of areas 17 and 18 (Payne et al., 1996a). These signals, as in the normal brain, also establish ordered maps of the visual field.

CROSS-CORRELATION

The second test of the temporal primacy of areas 17 and 18 involved a different strategy. Katsuyame et al. (1996) assessed temporal relations between middle suprasylvian visual cortex and area 17 with techniques designed to cross-correlate the activities of neurons in the two areas. This approach examines temporal relationships between neuron pairs in the two regions and tests for coupling between them. Katsuyame (1996) showed that a number of synaptically coupled pairs of neurons exhibited initial activation of the middle suprasylvian member of the pair and later activation of the area 17 member, a result that superficially challenges the primacy of areas 17 and 18. However, in the absence of information on the absolute timing of activation, it is not certain that middle suprasylvian visual cortex has an overall earlier activation than compared to areas 17 and 18. An equivalent conundrum pertains to its putative primate homolog (Payne, 1993), area V5, versus antecedent areas (ffytche et al., 1995; Beckers and Zeki, 1996; Raiguel et al., 1989, 1999). As a challenge to the primacy of areas 17 and 18, the data of Katsuyame et al. (1996) would be much more convincing if middle suprasylvian visual cortex was consistently activated before the faster responding area 18. Thus, even though attempts continue to challenge the primacy of areas 17 and 18, the most recent attempts have failed. We then recognize that the concept of primary visual cortex in cats should include both areas 17 and 18, but exclude all other cortical regions.

EPILOGUE

We noted a striking untidiness in our comprehension of the innervation of areas 17 and 18 by axons of the X- and Y-systems. This untidiness has a great bearing on our current and future understanding of visual processing in cat. To recapitulate briefly, there is a great deal of evidence to support the view that area 17 is innervated by both X- and Y-axons, and that area 18 is innervated by Y-axons. Moreover, based on these and other anatomical and physiological data on the circuitry of areas 17 and 18, the idea emerges that the majority of X-based signals either remain in area 17 or are transmitted to the deeper layers and back to LGN without processing in the upper layers of area 17 or further processing in the cortical visual system, a system that thus appears to be dominated by Y-based signals emerging from both area 18 and the superficial layers of area 17. This view is supported by the evidence that X-signals are not necessary for visually driven activity and receptive field properties in the upper layers of area 17 (see page 88 and Weyand, this volume). These views are disquieting because, as noted, they imply that little use is made beyond LGN and area 17 of the large numbers of signals arising from 90,000 β (X) retinal ganglion cells, and that a great deal of cortical visual processing is based on signals arising from α (Y) retinal ganglion cells that comprise a population of only about 1/10 the number β (X) cells. In our analysis, we believe that this view of X signal containment within area 17 and simply transmission back to thalamus is wrong. Our belief is supported by the observations that under certain conditions of high illumination and high contrast, high spatial-acuity and hyperacuity functions, which are based only on X-signals, are able to effectively guide visual discrimination behavior (see page 100, and Mitchell, this volume). We arrive at this conclusion because accurate reporting of these learned discriminations undoubtedly involves parts of the cerebral cortex other than area 17.

There are other dissenting views to the general scheme that the larger body of evidence supports for dominance of Y signals in area 17, and the poor representation of X signal processing. Ferster (1990a,b) presents seemingly strong evidence based on both intracellular records and current source density analyses that Y-signals contribute little to the activity in area 17, an area that is therefore dominated by X-signals. We can be secure that Ferster activated Y-fibers with his electrical stimuli because he described substantial Y-innervation of area 18 in the same work. Ferster is supported in his view by the data of Thalluri and Henry (1989). They used electrical stimulation of the superior colliculus to activate α (Y) retinal ganglion cells antidromically, and by virtue of axon collaterals, this procedure also activates the ascending Y-system, leading to LGN and primary visual cortex. In their analyses, no postsynaptic responses were evoked in neurons in the Y-terminations zone in upper layer 4 and lower layer 3 in area 17, although cells in layers 5 and 6 were activated by antidromic invasion directly from the superior colliculus.

The dissenting view is also supported by the observations of Movshon et al. (1978) and others. They compared the spatial-frequency tuning of neurons in areas 17 and 18 using drifting sinusoidal gratings, and even though they showed that the bandwidths of spatial tuning in areas 17 and 18 were nearly identical, the average preferred spatial frequencies of neurons in area 17 were three times those in area 18. This value mirrors the threefold better preferred frequencies of LGN X-cells relative to LGN Y-cells viewing the same visual field region (Derrington and Fuchs, 1979; Troy, 1983). A similar difference was not detected for neurons in layer 4B versus layer 4A in area 17 (Tolhurst and Thompson, 1981), where an equivalent difference might be expected based on the known laminar terminations of the X-and Y-systems in area 17 (Fig. 1-9). Based on such data, Movshon et al. (1978) suggest that "Y cell afferents contribute little to the properties of neurons in area 17," a view supported by Spitzer and Hochstein (1985).

If correct, the exclusion of significant Y-innervation of area 17 has a significant impact on our views of cortical visual processing. It suggests that: (1) the noted massive, sequential amplification of Y-signals in the geniculocortical portion of the retinogeniculocortical pathway is artificial, and (2) the influence of the Y-system over the cortical visual system is relegated to a relative level more commensurate with the numerical magnitude of the origins of the pathway. It also suggests a heretofore unrecognized greater processing and relevance of X-based signals in the cortical visual system, because signals that emerge from the upper layers of area 17 must be derived primarily from the X-system. Furthermore, it promotes the view that areas 17 and 18 process X- and Y-signals in parallel and, thus, each area contributes in unique yet complementary ways to overall visual processing, a view supported by behavioral studies (see page 100, and Mitchell, Chapter 16).

In conclusion, regardless of the actual bases for the untidy discrepancies in our understanding of the X- and Y-innervation patterns of areas 17, the topic is worthy of renewed investigations. Whatever the outcome of the investigations, clarification of the innervation patterns will have a significant impact on the design of models of visual processing in area 17 and the cortical visual system, and they will undoubtedly bear greatly on our understanding of the overall visual processing capabilities of cats. What is not in doubt from our discourse is that both areas 17 and 18 should be included under the umbrella term primary visual cortex.

ACKNOWLEDGMENTS

The authors acknowledge the financial support of NINDS and NIA held during the preparation of this chapter and this volume. We wish to thank Dr. Stephen Lomber for the use of Fig. 1–10, Jarrett Rushmore for preparing Fig. 1–33, and Claire Sethares for help in the preparation of many of the other figures. We also wish to thank the following publishers for permission to reproduce the following figures: Oxford University Press (Fig. 1-1); Elsevier Science (Fig. 1-5); John Wiley & Sons, Inc. (Fig. 1-14, 1-15, 1-22-1-24, 1-29); and Cambridge University Press (Fig. 1–21, 1-25-28, 1-30).

REFERENCES

Abramson, B. P., and Chalupa, L. M. (1985). The laminar distribution of cortical connections with the tecto- and cortico-recipient zones in the cat's lateral posterior nucleus. *Neuroscience* **15,** 81–95.

Albus, K. (1975a). A quantitative study of the projection area of the central and the paracentral visual field in area 17 of the cat. I. The precision of the topography. *Exp. Brain Res.* **24,** 159–179.

Albus, K. (1975b). A quantitative study of the projection area of the central and the paracentral visual field in area 17 of the cat. II. The spatial organization of the orientation domain. *Exp. Brain Res.* **24,** 181–202.

Albus, K. (1987). A neuronal subsystem in the cat's area 18 lacks retinotopy. *Brain Res.* **410,** 199–203.

Albus, K., and Beckmann, R. (1980). Second and third visual areas of the cat: interindividual variability in retinotopic arrangement and cortical location. *J. Physiol. (Lond.).* **299,** 247–276.

Albus, K. A., and Siebert, B. (1984). On the spatial arrangement of iso-orientation bands in the cat's visual cortical areas 17 and 18: A ^{14}C deoxyglucose study. *Exp. Brain Res.* **56,** 384–388.

Alonso, J. M., Cudeiro, J., Perez, R., Gonzalez, F., and Acuna, C. (1993). Orientational influences of layer V of visual area upon cells in layer V of area 17 in cat cortex. *Exp. Brain Res.* **96,** 212–220.

Ahmed, B., Anderson, J. C., Douglas, R. J., Martin, K. A. C., and Nelson, J. C. (1994). Polyneuronal innervation of spiny stellate neurons in cat visual cortex. *J. Comp. Neurol.* **341,** 39–49.

Anderson, J. C., Martin, K. A., and Picanco-Diniz, C. W. (1992). The neurons in layer 1 of cat visual cortex. *Proc. R. Soc. Lond. B. Biol. Sci.* **248,** 27–33.

Anderson, P. A., Olavarria, J., and Van Sluyters, R. C. (1988). The overall pattern of ocular dominance in cat visual cortex. *J. Neurosci.* **8,** 2183–2200.

Barlow, H. B., Blakemore, C., and Pettigrew, J. D. (1967). The neural mechanisms of binocular depth discrimination. *J. Physiol. (Lond).* **193,** 327–342.

Barris, R. (1935). Disposition of fibers of retinal origin in the lateral geniculate body. Course and termination of fibers of the optic system in the brain of the cat. *Arch. Ophthalmol.* **14,** 61–70.

Beck, A. (1891a). Die Bestimmung der Lokalisation der Gehirn- und Rückenmarkfunktionen vermittelst der elektrischen Erscheinungen. *Centralblatt für Physiologie* **4,** 473–476.

Beck, A. (1891b). Oznaczenie Loklizacyi z mózgu I Rdzeniu za Pomoca Zjawisk Elekrycznych. (Trans. Mapping of localization in brain and cortex by means of electric phenomena.) *Polska Akademica Umiejetnosci, Ser.* 2, 1, 186–232.

Beckers, G., and Zeki, S. (1995). The consequences of inactivating areas V1 and V5 on visual motion perception. *Brain* **118,** 49–60.

Benton, A. L. (1990). The fate of some neuropsychological concepts: an historical inquiry. In: *Contemporary neuropsychology and the legacy of Luria* (E. Goldberg, Ed.), pp. 171–179, Hillsdale, N. J., Lawrence Erlbaum.

Berardi, N., Bisti, S., Cattaneo, A., Fiorentini, A., and Maffei, L. (1982). Correlation between the preferred orientation and spatial frequency of neurones in visual areas 17 and 18 of the cat. *J. Physiol. (Lond.).* **323,** 603–618.

Berkley, M. A., and Sprague, J. M. (1979). Striate cortex and visual acuity functions. *J. Comp. Neurol.* **187,** 679–702.

Berlucchi, G., Sprague, J. M., Levy, J., and DiBerardino, A. C. (1972). Pretectum and superior colliculus in visually guided behavior and in flux and form discrimination in the cat. *J. Comp. Physiol. Psychol.* **78,** 123–172.

Berman, A. L. (1968). *The brain stem of the cat. A cytoarchitectonic atlas with stereotaxic coordinates.* Madison, University of Wisconsin Press,

Berman, A. L., and Jones, E. G. (1982). *The thalamus and basal telencephalon of the cat, a cytoarchitectonic atlas with stereotaxic coordinates.* Madison, University of Wisconsin Press.

Berman, N., and Jones E. G. (1977). A retino-pulvinar projection in the cat. *Brain Res.* **134,** 237–248.

Berman, N., Payne, B. R., Labar, D. R., and Murphy, E. H. (1982). Functional organization and receptive field properties of neurons in cat striate cortex, Variations in ocular dominance and receptive field types with cortical laminae and location in the visual field. *J. Neurophysiol.* **48,** 1362–1377.

Berman, N., Wilkes, M., and Payne, B. R. (1987). Organization of orientation and direction selectivity in areas 17 and 18 of cat cerebral cortex. *J. Neurophysiol.* **58,** 676–699.

Berson, D. M. (1985). Cat lateral suprasylvian cortex, Y-cell inputs and corticotectal projection. *J. Neurophysiol.* **53,** 544–556.

Berson, D., Pu, M., and Famiglietti, E. V. (1998). The zeta cell, A new ganglion cell type in cat retina. *J. Comp. Neurol.* **399,** 269–288.

Berson, D. M., Isayama, T., and Pu, M. (1999). The eta ganglion cell type of cat retina. *J. Comp. Neurol.* 408, 204–219.

Bilge, M., Bingle, A., Seneviratne, K. N., and Whitteridge, D. (1967). A map of the visual cortex in the cat. *J. Physiol. (Lond.).* **191,** 116–118P.

Billings–Gagliardi, S., Chan–Palay, V., and Palay, S. L. (1974). A review of lamination in area 17 of the visual cortex of *Macaca mulatta. J. Neurocytol.* **3,** 619–629.

Birnbacher, D., and Albus, K. (1987). Divergence of single axons in afferent projections to the cat's visual cortical areas 17, 18, and 19, A parametric study. *J. Comp. Neurol.* **261,** 543–561.

Bishop, P. O., Kozak, W., Levick, W. R., and Vakkur, G. J. (1962). The determination of the projection of the visual field on to the lateral geniculate nucleus in the cat. *J. Physiol. (Lond.).* **163,** 503–539.

Blakemore, C. (1969). Binocular depth discrimination and the nasotemporal division. *J. Physiol. (Lond.).* **205,** 471–497.

Blakemore, C., Diao, Y., Pu, M., Wang, Y, and Xiao, Y. (1983). Possible functions of the interhemispheric connexions between visual cortical areas in the cat. *J. Physiol. (Lond.).* **337,** 331–349.

Bolz, J., Rosner, G., and Wässle, H. (1982). Response latency of brisk-sustained (X) and brisk-transient (Y) cells in the cat retina. *J. Physiol. (Lond.).* **328,** 171–190.

Bonhoeffer, T., and Grinvald, A. (1991). Iso-orientation domains in cat visual cortex are arranged in pinwheel-like patterns. *Nature* **353,** 429–431.

Bonhoeffer, T., and Grinvald, A. (1993). The layout of iso-orientation domains in area 18 of cat visual cortex. Optical imaging reveals a pinwheel-like organization. *J. Neurosci.* **13,** 4157–4180.

Bonhoeffer, T., and Grinvald, A. (1996). Optical imaging based on intrinsic signals: the methodology. In: *Brain mapping: the methods* (A. W. Toga, Ed.), pp. 55–97, New York, Academic Press.

Bonhoeffer, T., Kim, D. S., Malonek, D., Shoham, D., and Grinvald, A. (1995). Optical imaging of the layout of functional domains in area 17 and across the area 17 / 18 border in cat visual cortex. *Eur. J. Neurosci.* **7,** 1973–1988.

Bowling, D. B., and Michael, C. R. (1984). Termination patterns of single physiologically characterized optic tract fibers in the cat's lateral geniculate nucleus. *J. Neurosci.* **4,** 198–216.

Boycott, B. B., and Wässle, H. (1974). The morphological types of ganglion cells of the domestic cat's retina. *J. Physiol. (Lond.).* **240,** 397–419.

Boyd, J. D., and Matsubara, J. A. (1996). Laminar and columnar patterns of geniculocortical projections in the cat: relationship to cytochrome oxidase. *J. Comp. Neurol.* **365,** 659–682.

Brazier, M. A. B. (1988). *The history of neurophysiology in the 19th* century. New York, Raven Press.

Brodmann, K. (1906). Beiträge zur histologischen Lokalisation der Grosshirnrinde. Fünfte Mitteilung, Über den allgemeinen Bauplan des Cortex pallii bei den Mammalieren und zwei homologe Rinderfelder in besonderen. Zugleich ein Beitrag zur Furchenlehre. *Journal für Psychologie und Neurologie* **6,** 275–400.

Brodmann, K. (1909). *Vergleichende Lokalisationlehre der Groβhirnrinde in ihren Prinzipien dargestellt auf Grund des Zellenbaues.* Leipzig, Johann Ambrosius Barth.

Brouwer, B., Zeeman, W. P. C., and Houwer, A. W. M. (1923). Experimentallanatomische Untersuchungen über die Projektionder Retina auf die primären Opticuszentren. *Schweiz. Arch. Neurol. Psychiat.* **13,** 118–137.

Brouwer, B. (1927). *Anatomical, phylogenetical and clinical studies on the central nervous system. The Herter lectures XVII.* Baltimore, Williams and Wilkins.

Bullier, J., and Henry, G. H. (1979a). Ordinal position of neurons in cat striate cortex. *J. Neurophysiol.* **42,** 1251–1263.

Bullier, J., and Henry, G. H. (1979b). Neural path taken by afferent streams in striate cortex of the cat. *J. Neurophysiol.* **42,** 1264–1270.

Bullier, J., and Henry, G. H. (1979c). Laminar position of first-order neurons and afferent terminals in cat striate cortex. *J. Neurophysiol.* **42,** 1271–1281.

Bullier J., Kennedy H., and Salinger, W. (1984). Bifurcation of subcortical afferents to visual areas 17, 18, and 19 in the cat cortex. *J. Comp. Neurol.* **228,** 309–328.

Burke, W., Burne, J. A., and Martin, P. (1985). Selective blockade of Y optic nerve fibers in the cat and the occurrence of inhibition in the dorsal lateral geniculate nucleus. *J. Physiol. (Lond.).* **364,** 81–92.

Burke, W., Dreher, B., Michalski, A., Cleland, B. G., and Rowe, M. H. (1992). Effects of selective pressure block of Y-type optic nerve fibers on the receptive-field properties of neurons in the striate cortex of the cat. *Vis. Neurosci.* **9,** 4–64.

Cajal, S. Ramón Y (1911). *Histologie due système nerveux de l'homme et des vertébrés.* (Trans. by L. Azoulay,) Paris, Maloine, 2 vols.

Cajal, S. Ramón y (1921a). Textura de la corteza visual del gato. *Trab. Lan. Invest. Biol. Univ. Madrid* **19,** 113–146.

Cajal, S. Ramón y (1921b). Textura de la corteza visual del gato. *Arch. Neurobiol., Madrid* **2,** 338–362.

Cajal, S. Ramón y (1922). Studien über die Sehrinde der Katze. *J. Psychol. Neurol. Leipzig* **29,** 161–181. (Translated in DeFelipe, and Jones, 1988).

Callaway, E. M., and Katz, L. C. (1990). Emergence and refinement of clustered horizontal connections in cat striate cortex. *J. Neurosci.* **10,** 1134–1153.

Campbell, A. (1978). Deficits of visual learning produced by posterior temporal lesions in cats. *J. Comp. Physiol. Psychol.* **92,** 45–57.

Campbell, A. W. (1905). *Histological studies on the localisation of cerebral function.* Cambridge, Cambridge University Press.

Casanova, C., Michaud, Y., Morin, C., McKinley, P. A., and Molotchnikoff, S. (1992). Visual responsiveness and direction selectivity in area 18 during local reversible inactivation of area 17 in cats. *Vis. Neurosci.* **9,** 581–593.

Caton, R. (1875). The electric currents of the brain. *Br. Med. J.* **2,** 278.

Caton, R. (1877). Interim report on investigation of electric currents of the brain. *Br. Med. J.* **1,** *Suppl.* L, 62.

Caton, R. (1891). Die Ströme des Centralnervensystems. *Centralblatt für Physiologie* **4,** 758–786.

Chabli, A., Ruan, D. Y., and Molotchnikoff, S. (1998). Influences of area 17 on neuronal activity of simple and complex cells of area 18 in cats. *Neuroscience* **84,** 685–698.

Clare, M. H., and Bishop, G. H. (1954). Responses from an association area secondarily activated from optic cortex. *J. Neurophysiol.* **17,** 271–277.

Clarke, P. G. H., Donaldson, I. M. L., and Whitteridge, D. (1976). Binocular visual mechanisms in cortical areas I and II of the sheep. *J. Physiol. (Lond.).* **256,** 509–526.

Cleland, B. G., Harding, T. G., and Tulunay-Keesey, U. (1979). Visual resolution and receptive field size, Examination of two kinds of cat retinal ganglion cells. *Science* **205,** 1015–1017.

Colonnier, M. L. (1966). The structural design of the neocortex. In: *Brain and conscious experience,* (J. C. Eccles, Ed.), pp. 1–23, New York, Springer.

Colucci, C. (1898). Richerche sull'anatomie e sulla fisiologia dei centri visivi cerebrali. *Acad. Med.-Chir. Napoli, Atti di Ric.* **52,** 151–241.

Cornwell, P., Overman, A., and Campbell, A. (1980). Subtotal lesions of visual cortex impair discrimination of hidden figures by cats. *J. Comp. Physiol. Psychol.* **94,** 289–304.

Cornwell, P., Overman, W. H., Levitsky, C. Shipley, J., and Lezynski, B. (1976). Performance on the visual cliff by cats with marginal gyrus lessions. *J. Comp. Physiol. Psychol.* **90,** 996–1010.

Cornwell, P., Nudo, R., Straussfogel, D., Lomber, S. G., and Payne, B. R. (1998). Dissociation of visual and auditory pattern discrimination functions within the cat's temporal cortex. *Behav. Neurosci.* **112,** 800–811.

Cowan, W. M., Gottlieb, D. I., Hendrickson, A. E., Price, J. L., Woolsey, T. A. (1972). The autoradiographic demonstration of axonal connections in the central nervous system. *Brain Res.* **37,** 21–51.

Creutzfeldt, O. D., Garey, L. J., Kuroda, R., and Wolff, J.-R. (1977). The distribution of degenerating axons after small lesions in the intact and isolated visual cortex of the cat. *Exp. Brain Res.* **27,** 419–440.

Cynader, M. S., Swindale, N. V., and Matsubara, J. A. (1987). Functional topography of area 18. *J. Neurosci.* **7,** 1401–1413.

Daniel, P. M., and Whitteridge, D. (1961). The representation of the visual field on the cerebral cortex in monkeys. *J. Physiol. (Lond.).* **159,** 203–221.

DeFelipe, J., and Fairén, A. (1982). A type of basket cell in superficial layers of the cat visual cortex. A Golgi-electron microscope study. *Brain Res.* **244,** 9–16.

DeFelipe, J., and Fairén, A. (1988). Synaptic connections of an interneuron with axonal arcades in the cat visual cortex. *J. Neurocytol.* **17,** 313–323.

DeFelipe, J., and Jones, E. G. (1988). *Cajal on the cerebral cortex, An annotated translation of the complete writings.* New York, Oxford University Press.

Derrington, A. M., and Fuchs, A. F. (1979). Spatial and temporal properties of X and Y cells in the cat lateral geniculate nucleus. *J. Physiol. (Lond.).* **293,** 347–364.

De Weerd, P., Sprague, J. M., Vandenbussche, E., and Orban, G. A. (1994). Two stages in visual texture segregation, a lesion study in the cat. *J. Neurosci.* **14,** 929–948.

Diao, Y. C., Jia, W. G., Swindale, N. V., and Cynader, M. S. (1990). Functional organization of the cortical 17/18 border region in the cat. *Exp. Brain Res.* **79,** 271–282.

Dinse, H. R., and Krüger, K. (1994). The timing of processing along the visual pathway in the cat. *Neuroreport* **14,** 893–897.

Donaldson, I. M. L., and Nash, J. R. G. (1975). The effect of chronic lesion in cortical area 17 on the visual responses of units in area 18 of the cat. *J. Physiol. (Lond.).* **245,** 325–332.

Donaldson, I. M. L., and Whitteridge, D. (1977). The nature of the boundary between cortical visual areas VII and VIII in the cat. *Proc. R. Soc. Lond. Ser. B* **199,** 445–462.

Doty, R. W. (1958). Potentials evoked in cat cerebral cortex by diffuse and by punctiform photic stimuli. *J. Neurophysiol.* **21,** 437–464.

Doty, R. W. (1967). The misnomer "lateral gyrus" in lieu of "marginal gyrus" in the cat. *Exp. Neurol.* **17,** 263–264.

Doty, R. W. (1971). Survival of pattern vision after removal of striate cortex in the adult cat. *J. Comp. Neurol.* **143,** 341–370.

Doty, R. W. (1973). Ablation of visual areas in the central nervous systm. In: *Handbook of sensory physiology, volume VII/3B, Central processing of visual information, Part B,* (R. Jung, Ed.), pp. 483–541, Berlin, Springer-Verlag,

Dreher, B., and Cottee, L. J. (1975). Visual receptive-field properties of cells in area 18 of cat's cerebral cortex before and after acute lesions in area 17. *J. Neurophysiol.* **38,** 735–750.

Dreher, B., and Sefton, A. J. (1979). Properties of neurons in cat's dorsal lateral geniculate nucleus, a comparison between medial interlaminar and laminated parts of the nucleus. *J. Comp. Neurol.* **183,** 47–64.

Dreher, B., Leventhal, A. G., and Hale, P. T. (1980). Geniculate input to cat visual cortex: a comparison of area 19 with areas 17, and 18. *J. Neurophysiol.* **44,** 804–826.

Dreher, B., Michalski, A., Cleland, B. G., and Burke, W. (1992). Effects of selective pressure block of Y-type optic nerve fibers on the receptive-field properties of neurons in area 18 of the visual cortex of the cat. *Vis. Neurosci.* **9,** 65–78.

Dreher, B., and Sefton, A. J. (1979). Properties of neurons in cat's dorsal lateral geniculate nucleus: a comparison between medial interlaminar and laminated parts of the nucleus. *J. Comp. Neurol.* **183,** 47–64.

Ebbesson, S. O. E. (1984). Evolution and ontogeny of neural circuits. *Behav. Brain Sci.* **7,** 321–366.

Einstein, G., Davis, T. L., and Sterling P. (1987). Ultrastructure of synapses from the A-laminae of the lateral geniculate nucleus in layer IV of the cat striate cortex. *J. Comp. Neurol.* **260,** 63–75.

Einstein, G., and Fitzpatrick, D. (1991). Distribution and morphology of area 17 neurons that project to the cat's extrastriate cortex. *J. Comp. Neurol.* **303,** 132–149.

Elliot Smith, G. (1907). New studies on the folding of the visual cortex and the significance of the occipital sulci in the human brain. *J. Anat.* **41,** 198–207.

Fairén, A., and Valverde, F. (1979) Specific thalamo-cortical afferents and their presumptive targets in the visual cortex. A Golgi study. *Prog. Brain. Res.* **51,** 419–438.

Fairén, A., and Valverde, F. (1980). A specialized type of neuron in the visual cortex of cat. A Golgi and electron microscope study of chandelier cells. *J. Comp. Neurol.* **179,** 761–779.

Farinas, I., and DeFelipe, J. (1991). Patterns of synaptic input on corticocortical and corticothalamic cells in the cat visual cortex. II The axon initial segment. *J. Comp. Neurol.* **304,** 70–77.

Ferster, D. (1981). A comparison of binocular depth mechanisms in areas 17 and 18 of the cat visual cortex. *J. Physiol. (Lond.).* **311,** 623–655.

Ferster, D. (1990a). X- and Y-mediated synaptic potentials in neurons of areas 17 and 18 of cat visual cortex. *Vis. Neurosci.* **4,** 115–133.

Ferster, D. (1990b). X- and Y-mediated current sources in areas 17 and 18 of cat visual cortex. *Vis. Neurosci.* **4,** 135–145.

Ferster, D., and Jagadeesh, B. (1991). Nonlinearity of spatial summation in simple cells of areas 17 and 18 of cat visual cortex. *J. Neurophysiol.* **66,** 1667–1679.

Ferster, D., and Lindstrom, S. (1982). An intracellular analysis of geniculo-cortical connectivity in area 17 of the cat. *J. Physiol. (Lond.).* **342,** 181–215.

ffytche, D. H., Guy, S., and Zeki, S. (1995). The parallel visual motion inputs into areas V1 and V5 of human cerebral cortex. *Brain* **118,** 1375–94.

Finger, S. (1994). *Origins of neuroscience,: a history of explorations into brain function.* New York, Oxford University Press.

Fisken, R. A., Garey, L. J., and Powell, T. P. S. (1975). The intrinsic, association and commissural connections of area 17 on the visual cortex. *Phil. Trans. R. Soc. Lond. Ser. B* **272,** 487–536.

Flechsig, P. E. (1896). *Gehirn und Steel.* Leipzig, Veit Press.

Freeman, A. W. (1991). Spatial characteristics of the contrast gain control in the cat's retina. *Visi. Res.* **31,** 775–785.

Freeman, B., and Singer, W. (1983). Direct and indirect visual inputs to superficial layers of cat superior colliculus, a current source-density analysis of electrically evoked potentials. *J. Neurophysiol.* **49,** 1075–1091.

Freeman, J. A., and Nicholson, C. (1975). Experimental optimization of current source-density technique for Anuran cerebellum. *J. Neurophysiol.* **38,** 369–382.

Freeman, J. A., and Stone, J. (1969). A technique for current density analysis of field potentials and its application to the frog cerebellum. In: *Neurobiology of cerebrellar evolution and development* (R. Llinás, Ed.), pp. 421–430, Chicago, American Medical Association.

Freund, T. F., Maglóczky, Z., Soltész, I, and Somogyi, P. (1986). Synaptic connections, axonal and dendritic patterns of neurons immunoreactive for cholecystokinin in the visual cortex of the cat. *Neuroscience* **19,** 1133–1159.

Freund, T. F., Martin, K. A. C., Somogyi, P., and Whitteridge, D. (1985a). Innervation of cat visual areas 17 and 18 by physiologically identified X- and Y-type thalamic afferents. I. Arborization patterns and quantitative distribution of postsynaptic elements. *J. Comp. Neurol.* **242,** 263–274.

Freund, T. F., Martin, K. A. C., Somogyi, P., and Whitteridge, D. (1985b). Innervation of cat visual areas 17 and 18 by physiologically identified X- and Y- type thalamic afferents. II. Identification of postsynaptic targets by GABA immunocytochemistry and Golgi impregnation. *J. Comp. Neurol.* **242,** 275–291.

Friedlander, M. J., Lin, C. S., Standford, L. R., and Sherman, S. M. (1981). Morphology of functionally identified neurons in lateral geniculate nucleus of the cat. *J. Neurophysiol.* **46,** 80–129.

Friedlander, M. J., and Standford, L. R. (1984). Effects of monocular deprivation on the distribution of cells types in dLGN, A sampling study with fine tipped micropipettes. *Exp. Brain Res.* **53,** 451–461.

Gabbott, P. L. A., and Somogyi, P. (1986). Quantitative distribution of GABA-immunoreactive neurons in the visual cortex (area 17). of the cat. *Exp. Brain Res.* **61,** 323–331.

Galuske, R. A. W., Schmidt, K. E., Goebel, R., Lomber, S. G., and Payne, B. R. (2000). Feedback control of orientation and direction maps in primary visual cortex. *Eur. J. Neurosci.* **12** *(Suppl.):* 73.

Ganser, S. (1882). Über die periphere und zentrale Anordnung der Sehnervenfasern and über das corpus bigeminum anterius. *Arch. Psychiat. Nervenkr.* **13,** 341–381.

Garey, L. J. (1965). Interrelationships of the visual cortex and superior colliculus in the cat. *Nature* **207,** 1410–1411.

Garey, L. J. (1971). A light and electron microscopic study of the visual cortex of the cat and monkey. *Proc. Ro. Soc. Lond. Ser. B* **179,** 21–40.

Garey, L. J. (1994). *Brodmann's Localisation in the cerebral cortex.* (Translation of Brodmann (1909), London, Smith–Gordon.)

Garey, L. J., Jones, E. G., and Powell, T. P. S. (1968). Interrelationships of striate and extrastriate cortex with the primary relay sites of the visual pathway. *J. Neurol., Neurosurg., Psychiatry* **31,** 135–157.

Garey, L. J., and Powell, T. P. S. (1967). The projection of the lateral geniculate nucleus upon the cortex in the cat. *Proc. R. Soc. Lond.* **169,** 107–126.

Garey, L. J., and Powell, T. P. S. (1968). The projection of the retina in the cat. *J. Anat.* **102,** 189–222.

Garey, L. J., and Powell, T. P. S. (1971). An experimental study of the termination of the lateral geniculo – cortical pathway in the cat and monkey. *Proc. R. Soc. Lond. B* **179,** 41–63.

Garey L. J., Winkelmann, E., and Brauer, K. (1985). Golgi and Nissl studies of the visual cortex of the bottlenose dolphin. *J. Comp. Neurol.* **240,** 305–321.

Garey, L. J., and Leuba, G. (1986). A quantitative study of neuronal and glial numerical density in the visual cortex of the bottlenose dolphin, evidence for a specialized subarea and changes with age [published erratum appears in *J. Comp. Neurol.* (1986). **250,** 263]. *J. Comp. Neurol.* **247,** 491–496.

Geisert, E. E. (1980). Cortical projections of the lateral geniculate nucleus in the cat. *J. Comp. Neurol.* **190,** 793–812.

Gilbert, C. D. (1977). Laminar differences in receptive field properties of cells in cat primary visual cortex. *J. Physiol. (Lond.).* **268,** 391–421.

Gilbert, C. D., and Kelly, J. P. (1975). The projections of cells in different layers of the cat's visual cortex. *J. Comp. Neurol.* **163,** 81–105.

Gilbert, C. D., and Wiesel, T. N. (1979). Morphology and intracortical projections of functionally characterised neurones in the cat visual cortex. *Nature* **280,** 120–125.

Gilbert, C. D., and Wiesel, T. N. (1983). Clustered intrinsic connections in cat visual cortex. *J. Neurosci.* **3,** 1116–1133.

Gilbert C. D., and Wiesel, T. N. (1989). Columnar specificity of intrinsic horizontal and corticocortical connections in cat visual cortex. *J. Neurosci.* **9,** 2432–2442.

Glickstein, M., King, R. A., Miller, J., and Berkley, M. (1967). Cortical projections from the dorsal lateral geniculate nucleus of cats. *J. Comp. Neurol.* **130,** 55–76.

Graham, J. (1977). An autoradiographic study of the efferent connections of the superior colliculus in the cat. *J. Comp. Neurol.* **173,** 629–654.

Gratiolet, P. (1854). Note sur les expansions des racines cérébrales du nerf optique et sur leur terminaison dand une region determinée de l'écorde des hémisphères. *Comptes Rendus Acad. Sci.* **39,** 274–278.

Gross, C. G. (1997). From Imhotep to Hubel and Wiesel, The story of visual cortex. In: *Cerebral cortex, volume 12, Extrastriate cortex in primates,* (K. S. Rockland, J. H. Kaas, and A. Peters, Eds.), E. G. Jones, and Peters, A. (Series Eds.), pp. 1–58, New York, Plenum Press.

Gudden, B. (1869). Experimentaluntersuchungen über das periphäre unde Zentrale nervensystem. *Arch. Psych. Nervenkr.* **2,** 693–723.

Gudden, B. (1874). Über die Kreuzung der nerven fasern im Chiasma nervorum opticorum. *Von Graefes Arch. Ophthalmol.* **20,** 249–268.

Guedes, R. S., Watanabe, S., and Creutzfeldt, O. D. (1983). Functional role of association fibers from a visual association area: the posterior suprasylvian sulcus of the cat. *Exp. Brain Res.* **49,** 13–27.

Guillery, R. W., Geisert, E. E., Polley, E. H., and Mason, C. A. (1980). An analysis of the retinal afferents to the cat's medial interlaminar nucleus and its rostral thalamic extension, the "geniculate wing". *J. Comp. Neurol.* **194,** 117–142.

Gurewitsch, M., and Bychowsky, G. (1928). Zur Architektonik der Hirnrinde (Isocortex) des Hundes. *J. Psychol. Neurol.* **35,** 283–312.

Gurewitsch, M., and Chatschaturian, A. (1928). Zur Cytoarchitektonik der Groβhirnrinde der Feliden. *Z. Anat. Entwicklg.* **87,** 100–138.

Hall, S. E., and Mitchell, D. E. (1991). Grating acuity of cats measured with detection and discrimination tasks. *Behav Brain Res.* **44,** 1–9.

Harting, J. K., Updyke, B. V., and Van Lieshout, D. P. (1992). Corticotectal projections in the cat, anterograde transport studies of twenty-five cortical areas. *J. Comp. Neurol.* **324,** 379–414.

Harvey, A. R. (1980a). The afferent connexions and laminar distribution of cells in area 18 of the cat. *J. Physiol. (Lond.)* **302,** 483–505.

Harvey, A. R. (1980b). A physiological analysis of subcortical and commissural projections of areas 17 and 18 of the cat. *J. Physiol. (Lond.)* **302,** 507–534.

Hata, Y., Tsumoto, T., and Sato, H., and Tamura, H. (1991). Horizontal interactions between visual cortical neurones studied by cross-correlation analysis in the cat. *J. Physiol. (Lond.)* **441,** 593–614.

Hayhow, W. R. (1958). The cytoarchitecture of the lateral geniculate body in the cat in relation to the distribution of the crossed and uncrossed optic fibers. *J. Comp. Neurol.* **110,** 1–64.

Henry, G. H., Harvey, A. R., and Lund, J. S. (1979). The afferent connections and laminar distribution of cells in the cat striate cortex. *J. Comp. Neurol.* **187,** 725–744.

Henry, G. H., Mustari, M. J., and Bullier, J. (1983). Different geniculate inputs to B, and C cells of cat striate cortex. *Exp. Brain Res.* **52,** 179–189.

Hirsch, J. A., Alonso, J. M., Reid, R. C., and Martinez, L. M. (1998). Synaptic integration in striate cortical simple cells. *J. Neurosci.* **18,** 9517–9528.

Hochstein, S., and Shapley, R. M. (1976). Quantitative analysis of retinal ganglion cell classifications. *J. Physiol. (Lond.)* **262,** 237–264.

Holländer, H., and Vanegas, H. (1977). The projection from the lateral geniculate nucleus onto the visual cortex in the cat. A quantitative study with horseradish peroxidase. *J. Comp. Neurol.* **173,** 519–536.

Hovda, D. A., Sutton, R. L., and Feeney, D. M. (1989). Amphetamine-induced recovery of visual cliff performance after bilateral visual cortex ablation in cats: measurement of depth perception thresholds. *Behav. Neurosci.* **103,** 574–584.

Hubel, D. H. (1959). Single unit activity in striate cortex of unrestrained cats. *J. Physiol. (Lond.)* **147,** 226–238.

Hubel, D. H., and Wiesel, T. N. (1959). Receptive fields of single neurones in the cat's striate cortex. *J. Physiol. (Lond.)* **148,** 574–591.

Hubel, D. H., and Wiesel, T. N. (1961). Integrative action in the cat's lateral geniculate bidy, *J. Physiol. (Lond.)* **155,** 385–398.

Hubel, D. H., and Wiesel, T. N. (1962). Receptive fields, binocular interaction and functional architecture in the cat's visual cortex. *J. Physiol. (Lond.)* **160,** 106–154.

Hubel, D. H., and Wiesel, T. N. (1963). Shape and arrangement of columns in the cat's striate cortex. *J. Physiol. (Lond.)* **165,** 559–568.

Hubel, D. H., and Wiesel, T. N. (1965a). Receptive fields and functional architecture in two nonstriate visual areas (18 and 19). of the cat. *J. Neurophysiol.* **28,** 229–289.

Hubel, D. H., and Wiesel, T. N. (1965b). Binocular interaction in striate cortex of kittens raised with artificial squint. *J. Neurophysiol.* **28,** 1041–1059.

Hubel, D. H., and Wiesel, T. N. (1967). Cortical and callosal connections concerned with the vertical meridian of visual fields in the cat. *J. Neurophysiol.* **30,** 1561–1573.

Hubel, D. H., and Wiesel, T. N. (1969). Visual area of the lateral suprasylvian gyrus (Clare– Bishop area). of the cat. *J. Physiol. (Lond.)* **202,** 251–260.

Hubel, D. H., and Wiesel, T. N. (1977). The Ferrier Lectrure: functional architecture of macaque monkey visual cortex. *Proc. R. Soc. Lond. B.* **198,** 1–59.

Hübener, M., Schwarz, C., and Bolz, J. (1990). Morphological types of projection neurons in layer 5 of cat visual cortex. *J. Comp. Neurol.* **301,** 655–674.

Hübener, M., Shoham, D., Grinvald, A., and Bonhoeffer, T. (1997). Spatial relationships among three columnar systems in cat area 17. *J. Neurosci.* **17,** 9270–9284.

Hughes, A. (1975). A quantitative analysis of cat retinal ganglion cell topography. *J. Comp. Neurol.* **163,** 107–128.

Hughes, H. C., and Sprague, J. M. (1986). Cortical mechanisms for local and global analysis of visual space in the cat. *Exp. Brain Res.* **61,** 332–354.

Humphrey, A. L., Sur, M., Uhlrich, D. J., Sherman, S. M. (1985a). Projection patterns of individual X- and Y- cell axons from the lateral geniculate nucleus to cortical area 17 in the cat. *J. Comp. Neurol.* **233,** 159–189.

Humphrey, A. L., Sur, M., Uhlrich, D. J., and Sherman, S. M. (1985b). Termination patterns of individual X- and Y-cell axons in the visual cortex of the cat, Projections to area 18, to the 17/18 border region, and to both areas 17 and 18. *J. Comp. Neurol.* **233,** 190–211.

Humphrey, A. L., and Murthy, A. (1999). Cell types and response timings in the medial interlaminar nucleus and C-layers of the cat lateral geniculate nucleus. *Vis. Neurosci.* **16,** 513–525.

Isayama, T., Berson, D. M., and Pu, M. (2000). Theta ganglion cell type. *J. Comp. Neurol.* **417,** 32–48.

Issa, N. P., Trepel, C., and Stryker, M. P. (2000). Spatial frequency maps in cat visual cortex. *J. Neurosci.* **20,** 8504–8514.

Jacobson, M. (1899). Cited in Probst (1900).

Jones, E. G. (1984). Neurogliaform or spiderweb cells. In: *Cellular components of the cerebral cortex. Cerebral cortex, volume 1,* (Peters, A., and Jones, E. G. Eds.), pp. 409–418, New York, Plenum Press.

Kaas, J. H. (1997). Theories of visual cortex organization in primates. In *Cerebral cortex, volume 12, Extrastriate cortex in primates,* (K. S. Rockland, J. H. Kaas, and A. Peters, Eds.), E. G. Jones, and Peters, A. (Series Eds.), pp. 91–125, New York, Plenum Press.

Kageyama, G. H., and Wong-Riley, M. T. T. (1984). The histochemical localization of cytochrome oxidase in the retina and lateral geniculate nucleus of the ferret, cat, and monkey, with particular reference to retinal mosaics and ON/OFF-center visual channels. *J. Neurosci.* **4,** 2445–2459.

Kageyama, G. H., and Wong-Riley, M. T. T. (1985). An analysis of the cellular localization of cytochrome oxidase in the lateral geniculate nucleus of the adult cat. *J. Comp. Neurol.* **242,** 338–357.

Kageyama, G. H., and Wong-Riley, M. T. T. (1986). Laminar and cellular localization of cytochrome oxidase in the cat striate cortex. *J. Comp. Neurol.* **245,** 137–159.

Kaplan, E., and Shapley, R. M. (1982). X and Y cells in the lateral geniculate nucleus of macaque monkeys. *J. Physiol. (Lond.).* **330,** 125–143.

Katsuyama, N., Tsumoto, T., Sato, H., Fukuda, M., and Hata, Y. (1996). Lateral suprasylvian visual cortex is activated earlier than or synchronously with primary visual cortex in the cat. *Neurosci. Res.* **24,** 431–435.

Katz, L. C. (1987). Local circuitry of identified projection neurons in cat visual cortex. *J. Neurosci.* **7,** 1223–1249.

Kawamura, K. (1971). Variations of the cerebral sulci in the cat. *Acta Anat.* **80,** 204–221.

Kawano, J. (1998). Cortical projections of the parvocellular laminae C of the dorsal lateral geniculate nucleus in the cat: an anterograde wheat germ agglutinin conjugate to horseradish peroxidase study. *J. Comp. Neurol.* **392,** 439–457.

Kaye, M., Mitchell D. E., and Cynader, M. (1981). Selective loss of binocular depth perception after ablation of cat visual cortex. *Nature* **293,** 60–62.

Kim, D.-S., Matsuda, Y., Ohki, Y., Ajima, A., and Tanaka, S. (1999). Geometrical and topological relationships between multiple functional maps in cat primary visual cortex. *Neuroreport* **10,** 2515–2522.

Kisvárday, Z. F., Beaulieu, C., and Eysel, U. T. (1993). Network of GABAergic large basket cells in cat visual cortex (area 18), Implication for lateral disinhibition. *J. Comp. Neurol.* **327,** 398–415.

Kisvárday, Z. F., and Eysel, U. T. (1992). Cellular organization of reciprocal patchy networks in layer III of cat visual cortex (area 17). *Neuroscience* **46,** 275–286.

Kisvárday, Z. F., Martin, K. A. C., Freund, T. F., Magloczky, Z., Whitteridge, D., and Somogyi, P. (1986). Synaptic targets of HRP-filled layer III pyramidal cells in the cat striate cortex. *Exp. Brain Res.* **64,** 541–552.

Kisvárday, Z. F., Martin, K. A. C., Friedlander, M. J., and Somogyi, P. (1987). Evidence for interlaminar inhibitory circuits in the striate cortex of the cat. *J. Comp. Neurol.* **260,** 1–19.

Kisvárday, Z. F., Martin, K. A. C., Whitteridge, D., and Somogyi, P. (1985). Synaptic connections of intracellularly filled clutch cells, a type of small basket cell in the visual cortex of the cat. *J. Comp. Neurol.* **241,** 111–137.

Kratz, K. E., Webb, S. V., and Sherman, S. M. (1978). Studies on the cat's medial interlaminar nucleus, a subdivision of the dorsal lateral geniculate nucleus. *J. Comp. Neurol.* **181,** 601–614.

Kreiner, J. (1966). Myeloarchitectonics of the occipital cortex in dog and general remarks on the myeloarchitectonics. *J. Comp. Neurol.* **127,** 531–557.

Kreiner, J. (1968). Homologies of the fissural and gyral pattern of the hemispheres of the dog and monkey. *Acta Anat.* **70,** 137–167.

Kreiner, J. (1970). Homologies of the fissural patterns of the hemispheres of dog and cat. *Acta Neurobiol. Exp.* **30,** 295–305.

Kreiner, J. (1971). The neocortex of the cat. *Acta Neurobiol. Exp.* **31,** 151–201.

Lang, W., and Henn, V. (1980). Columnar pattern in the cat's visual cortex after optokinetic nystagmus. *Brain Res.* **182,** 446–450.

Lee, D., Lee, C., and Malpeli, J. G. (1992). Acuity-sensitivity trade-offs of X and Y cells in the cat lateral geniculate complex: role of the medial interlaminar nucleus in scotopic vision. *J. Neurophysiol.* **68,** 1235–1247.

Lee, C., Malpeli, J. G., Schwark, H. D., and Weyand, T. G. (1984). Cat medial interlaminar nucleus, Retinotopy, relation to tapetum and implications for scotopic vision. *J. Neurophysiol.* **52,** 848–869.

Lee, C., Weyand, T. G., and Malpeli, J. G. (1998). Thalamic control of cat area-18 supragranular layers, simple cells, complex cells, and cells projecting to the lateral suprasylvian visual area. *Vis. Neurosci.* **15,** 27–35.

Lehmkuhle, S., Kratz, K. E., and Sherman, S. M. (1982). Spatial and temporal sensitivity of normal and amblyopic cats. *J. Neurophysiol.* **48,** 372–387.

Leicester, J. (1968). Projection of the visual vertical meridian to cerebral cortex of the cat. *J. Neurophysiol.* **31,** 371–382.

LeVay, S. (1973). Synaptic patterns in the visual cortex of the cat and monkey. Electron microscopy of Golgi preparations. *J. Comp. Neurol.* **150,** 53–86.

LeVay, S., and Gilbert, C. D. (1976). Laminar patterns of geniculocortical projection in the cat. *Brain Res.* **113,** 1–19.

LeVay S., and Sherk, H. (1981). The visual claustrum of the cat. I. Structure and connections. *J. Neurosci.* **1,** 956–980.

LeVay, S., and Voigt, T. (1988). Ocular dominance and disparity coding in cat visual cortex. *Vis. Neurosci.* **1,** 395–424.

LeVay, S., and Voigt, T. (1990). Retrograde transneuronal transport of wheat – germ agglutinin to the retina from visual cortex in the cat. *Exp. Brain Res.* **82,** 77–81.

LeVay, S., Stryker, M. P., and Shatz, C. J. (1978). Ocular dominance columns and their development in layer IV of the cat's visual cortex: a quantitative study. *J. Comp. Neurol.* **179,** 223–244.

Leventhal, A. G. (1979). Evidence that the different classes of relay cells of the cat's lateral geniculate nucleus terminate in different layers of the striate cortex. *Exp. Brain Res.* **37,** 349–72.

Leventhal, A. G., Keens, J., and Tork, I. (1980). The afferent ganglion cells and cortical projection of the retinal recipient zone (RRZ). of the cat's "pulvinar complex." *J. Comp. Neurol.* **194,** 535–554.

Levick, W. R. (1975). Form and function of cat retinal ganglion cells. *Nature* **254,** 659–662.

Levick, W. R. (1977). Participation of brisk-transient ganglion cells in binocular vision: an hypothesis. *Proc. Aust. Physiol. Pharm. Soci.* **8,** 9–16.

Levick, W. R., and Cleland, B. G. (1974). Selectivity of microelectrodes in recordings from cat retinal ganglion cells. *J. Neurophysiol.* **37,** 1387–1393.

Lomber, S. G., and Payne, B. R. (1996). Removal of the two halves restores the whole, Reversal of visual hemineglect during bilateral cortical or collicular inactivation in the cat. *Vis. Neurosci.* **13,** 1143–1156.

Lomber, S. G., and Payne, B. R. (2000a). Contributions of cat posterior parietal cortex to visuospatial discrimination. *Vis. Neurosci.* **17,** 701–709.

Lomber, S. G., and Payne, B. R. (2000b). Translaminar differentiation of visually guided behaviors revealed by restricted cerebral cooling deactivation. *Cerebral Cortex* **10,** 1066–1077.

Lomber, S. G., Payne, B. R., Cornwell, P., and Pearson, H. E. (1993). Capacity of the retinogeniculate pathway to reorganize following ablation of visual cortical areas in developing and mature cats. *J. Comp. Neurol.* **338,** 432–457.

Lomber, S. G., MacNeil, M. A., and Payne, B. R. (1995). Amplification of thalamic projections to middle suprasylvian cortex following ablation of immature primary visual cortex in the cat. *Cerebral Cortex* **5,** 166–191.

Lomber, S. G., Payne, B. R., and Cornwell, P. (1996a). Learning and recall of form discriminations during reversible cooling deactivation of ventral-posterior suprasylvian cortex in the behaving cat. *Proc. Nat. Acad. Sci. U.S.A.* **93,** 1654–1658.

Lomber, S. G., Payne, B. R., Cornwell, P., and Long, K. D. (1996b). Perceptual and cognitive visual functions of parietal and temporal cortices of the cat. *Cerebral Cortex* **6,** 673–695.

Long, K. D., Lomber, S. G., and Payne, B. R. (1996). Increased oxidative metabolism in middle suprasylvian cortex following removal of areas 17 and 18 from newborn cats. *Exp. Brain Res.* **110,** 335–346.

Lorente de Nó, R. (1949). Cerebral cortex, Architecture, intracortical connections, motor projections. In: *Physiology of the nervous system,* (J. F. Fulton, Ed.), pp. 288–312, London, Oxford University Press.

Löwel, S., and Singer, W. (1987). The pattern of ocular dominance columns in flat-mounts of the cat visual cortex. *Exp. Brain Res.* **68,** 661–666.

Löwel, S., and Singer, W. (1993a). Monocularly induced 2-deoxyglucose patterns in the visual cortex and lateral geniculate nucleus of the cat, I. Anaesthetized and paralysed animals. *Eur. J. Neurosci.* **5,** 846–856.

Löwel, S., and Singer, W. (1993b). Monocularly induced 2-deoxyglucose patterns in the visual cortex and lateral geniculate nucleus of the cat, II. Awake animals and strabismic animals. *Eur. J. Neurosci.* **5,** 857–869.

Löwel, S., Bischof, H. J., Leutenecker, B., and Singer, W. (1988). Topographic relations between ocular dominance and orientation columns in the cat striate cortex. *Exp. Brain Res.* **71,** 33–46.

Löwel, S. Freeman B., and Singer, W. (1987). Topographic organization of the orientation column system in large flat-mounts of the cat visual cortex, a 2-deoxyglucose study. *J. Comp. Neurol.* **255,** 401–415.

Lübke, J., and Albus, K. (1989). The postnatal development of layer VI neurons in the cat's striate cortex, as visualized by intracellular Lucifer yellow injections in aldehyde-fixed tissue. *Dev. Brain Res.* **45,** 29–38.

Luciani, L. (1884). On the sensorial localisations in the cerebral cortex. *Brain* **7,** 145–160.

Luciani, L., and Tamburini, A. (1879). Sulle funzioni del cervello. Regio – Emilia, S. Calderini. Abstracted by A. Rabagliati in *Brain* **2,** 234–250.

Lund, J. S. (1984). Spiny stellate neurons. In: *Cellular components of the cerebral cortex. cerebral cortex* vol. 1. A. Peters, and E. G. Jones, Eds.), pp. 255–308, New York, Plenum Press.

Lund, J. S., Henry, G. H., MacQueen, C. L., and Harvey, A. R. (1979). Anatomical organization of the primary visual cortex (area 17). of the cat. A comparison with area 17 of the macaque monkey. *J. Comp. Neurol.* **184,** 599–618.

Maciewicz, R. J. (1974). Afferents to the lateral suprasylvian gyrus of cat traced with horseradish peroxidase. *Brain Res.* **78,** 139–143.

Maciewicz, R. J. (1975). Thalamic afferents to areas 17, 18, and 19 of cat cortex traced with horseradish peroxidase. *Brain Res.* **84,** 308–312.

MacNeil, M. A., Lomber, S. G., and Payne, B. R. (1996). Rewiring of transcortical projections to middle suprasylvian cortex following early removal of cat areas 17 and 18. *Cerebral Cortex* **6,** 362–376.

MacNeil, M. A., Einstein, G. E., and Payne, B. R. (1997a). Transgeniculate signal transmission to middle suprasylvian extrastriate cortex in intact cats and following early removal of areas 17 and 18, A morphological study. *Exp. Brain Res.* **114,** 11–23.

MacNeil, M. A., Lomber, S. G., and Payne, B. R. (1997b). Thalamic and cortical projections to middle suprasylvian cortex of cats, Constancy and variation. *Exp. Brain Res.* **114,** 24–32.

Maffei, L., and Fiorentini, A. (1977). Spatial frequency rows in the striate visual cortex. *Vis. Res.* **17,** 257–264.

Malpeli, J. G. (1983). Activity of cells in area 17 of the cat in absence of input from layer A of lateral geniculate nucleus. *J. Neurophysiol.* **49,** 595–610.

Malpeli, J. G., Lee, C., Schwark, H. D., and Weyand, T. G. (1986). Cat area 17. I. Pattern of thalamic control of cortical layers. *J. Neurophysiol.* **56,** 1062–1073.

Marshall, W. H., Talbot, S. A., and Ades, W. H. (1943). Cortical responses of the anesthetized cat to gross photic and electrical afferent stimulation. *J. Neurophysiol.* **6,** 1–15.

Martin, K. A. C. (1988). The Wellcome Prize lecture. From single cells to simple circuits in the cerebral cortex. *Q. J. Exp. Physiol.* **73,** 637–702.

Martin, K. A. C., and Whitteridge, D. (1984). Form, function and intracortical projections of spiny neurones in the striate visual cortex of the cat. *J. Physiol. (Lond.)* **353,** 463–504.

Martin, K. A. C., Friedlander, M. J., and Alones, V. (1989). Physiological, morphological, and cytochemical characteristics of a layer 1 neuron in cat striate cortex. *J. Comp. Neurol.* **282,** 404–414.

Martin, K. A. C., Somogyi, P., and Whitteridge, D. (1983). Physiological and morphological properties of identified basket cells in cat's visual cortex *Exp. Brain Res.* **50,** 193–200.

Martinez-Condé, S., Cudeiro, J., Grieve, K. L., Rodriguez, R., Rivadulla, C., and Acuna, C. (1999). Effects of feedback projections fro area 18 layers 2/3 to area 17 layers 2/3 in the cat visual cortex. *J. Neurophysiol.* **82,** 2667–2675.

Mesulam, M. -M. (1982). Principles of horseradish peroxidase and their applications for tracing neural pathways—axonal transport, enzyme histochemistry and light microscopic analysis. In: *Tracing neural connections with horseradish peroxidase,* (M. -M. Mesulam, Ed.), pp. 1–151, New York, Wiley.

Meyer, G. (1983). Axonal patterns and topography of short-axon neurons in visual areas 17, 18 and 19 of the cat. *J. Comp. Neurol.* **220,** 405–438.

Meyer, G., and Albus, K. (1981). Spiny stellates as cells of origin of association fibres from area 17 to area 18 in the cat's neocortex. *Brain Res.* **210,** 3353–3341.

Meynert, T. (1870). Beiträge zur Kenntniss der centralen Projektion der Sinnesoberflächen. *Sitzungberichte der Kaiserlichten Akademie der Wissenschaften, Wien. Mathematisch—Naturwissenschaftliche Classe* **60,** 547–562.

Michalski, A., Wimborne, B. M., and Henry, G. H. (1993). The effect of reversible cooling of cat's primary visual cortex on the responses of area 21 neurons. *J. Physiol. (Lond.).* **466,** 133–156.

Mignard, M., and Malpeli, J. G. (1991). Path of information flow through visual cortex. *Science* **251,** 1249–1251.

Minkowski, M. (1911). Zur Physiologie der cortikalen Sehsphäre. *Pflügers Arch. Gesam. Physiol.* **141,** 171–327.

Minkowski, M. (1912). Experimentelle Untersuchungen über die Beziehungen des Grosshirns zum Corpus geniculatum externum. *Neurologisches Centralblatt* **31,** 1470–1472.

Minkowski, M. (1913). Experimentelle Untersuchungen über die Beziehungen des Grosshirnrinde und der Nethaut zu den primären optischen Zentren, besonders zum Corpus geniculatum externum. *Arbeiten Hirnanatomisches Institut, Zürich* **7,** 255–362.

Minkowski, M. (1920). Über den Verlauf, die Endigung und die zentrale Repräsentation von gekreuzten Sehnerven fasern bei einigen Säugetieren und bein Menschen. *Schwiezer Arch. für Neurolgie and Psychiatrie* **6,** 268–303.

Mitchell, D. E. (1989). Normal and abnormal visual development in kittens, Insights into the mechanisms that underlie visual perceptual development in humans. *Can. J. Psychol.* **43,** 141–164.

Mitchell, D. E. (1990). Sensitive periods in visual development, insights gained from studies of recovery of function in cats following early monocular deprivation or cortical lesions. In: *Vision, coding and efficiency* (C. Blakemore, Ed), pp 234–246, Cambridge, Cambridge University Press.

Mitzdorf, U. (1985). Current source density method and application in cat cerebral cortex, Investigation of evoked potentials and EEG phenomena. *Physiol. Rev.* **65,** 37–100.

Mitzdorf, U. (1987). Properties of the evoked potential generators, Current source-density analysis of visually evoked potentials in the cat cortex. *Int. J. Neurosci.* **33,** 33–59.

Mitzdorf, U., and Singer, W. (1977). Laminar segregation of afferents to lateral geniculate nucleus of the cat. An analysis of current source density. *J. Neurophysiol.* **40,** 1127–1244.

Mitzdorf, U., and Singer, W. (1978). Prominent excitatory pathways in the cat visual cortex (A17 and A18), A current source density analysis of electrically evoked potentials. *Exp. Brain Res.* **33,** 371–394.

Moore, R., Karapas, F., and Frenkel, M. (1966). Lateral geniculate projection from discrete retinal lesions in the cat. *Am. J. Ophthalmol.* **62,** 918–925.

Morgane, P. J., Glezer, I. I., and Jacobs, M. S. (1988). Visual cortex of the dolphin, an image analysis study. *J. Comp. Neurol.* **273,** 3–25.

Mountcastle, V. B. (1957). Modality and topographic properties of single neurons of cat's somatic sensory cortex. *J. Neurophysiol.* **20,** 408–434.

Movshon, J. A., Thompson, I. D., and Tolhurst, D. J. (1978). Spatial and temporal contrast sensitivity of neurones in areas 17 and 18 of the cat's visual cortex. *J. Physiol. (Lond.).* **283,** 101–120.

Mullikin, W. H., Jones, J. P., and Palmer, L. A. (1984). Receptive-field properties and laminar distribution of X-like and Y-like simple cells in cat area 17. *J. Neurophysiol.* **52,** 350–371.

Munk, H. (1879a). Physiologie der Sehsphäre der Grosshirnrinde. *Verhandlungen der Physiologischen Gesellschaft zu Berlin* **162,** 1–178.

Munk, H. (1879b). Weiteres zur Physiologie der Sehsphäre der Grosshirnrinde. *Centrallblatt für praktische Augenheilkunde* **3,** 255–266.

Munk, H. (1881). Über die Funktionene der Grosshirnrinde. Dritte Mitteilung, pp. 28–53, Berlin. In: *Some papers on the cerebral cortex.* (Translated as *'On the functions of the cortex,'* G. von Bonin, (1960), pp. 97–117, Springfield, Charles C. Thomas.

Murphy, K. M., Jones, D. G., and Van Sluyters, R. C. (1995) Cytochrome-oxidase blobs in cat primary visual cortex. *J. Neurosci.* **15,** 4196–4208.

Naegele, J. R., and Katz, L. C. (1990). Cell surface molecules containing N-acetylgalactosamine are associated with basket cells and neurogliaform cells in cat visual cortex. *J. Neurosci.* **10,** 540–557.

Nauta, W., and Ebbesson, S. O. E. (1970). *Contemporary research methods in neuroanatomy,* New York, Springer – Verlag.

Nelson, S. B., and LeVay, S. (1985). Topographic organization of the optic radiation of the cat. *J. Comp. Neurol.* **240,** 322–330.

Neumann, G. (1978). Intrinsic connectivity in area 18 of the cat. In *Developmental neurobiology of vision* (R. D. Freeman, Ed.), pp. 175–184, New York, Plenum Press.

Nicholson, C., and Freeman, J. A. (1975). Theory of current source-density analysis and determination of conductivity tensor for Anuran cerebellum. *J. Neurophysiol.* **38,** 356–368.

Niimi, K., and Sprague, J. M. (1970). Thalamo-cortical organization of the visual system in the cat. *J. Comp. Neurol.* **138,** 219–250.

Niimi, K., Matusuoka, H., Yamazaki, Y., and Matsumoto, H. (1981). Thalamic afferents to the visual cortex in the cat studied by retrograde transport of horseradish peroxidase. *Brain Behav. Evol.* **18,** 114–139.

Nikara, T., Bishop, P, O., and Pettigrew, J. D. (1968). Analysis of retinal correspondence by studying receptive fields of binocular single units in cat striate cortex. *Exp. Brain Res.* **6,** 353–372.

Ohki, K., Matsuda, Y., Ajima, A., Kim, D.-S., and Tanaka, S. (2000). Arrangement of orientation pinwheel centers around area 17/18 transition zone in cat visual cortex. *Cerebral Cortex* **10,** 593–601.

O'Leary, J. L. (1941). The structure of area striata of the cat. *J. Comp. Neurol.* **75,** 131–164.

Orban, G. A. (1984). *Neuronal operations in visual cortex* Berlin, Heidelberg, New York Tokyo, Springer–Verlag.

Orban, G. A., Vandenbussche, E., Sprague, J. M., and De Weerd, P. (1988). Stimulus contrast and visual cortical lesions. *Exp. Brain Res.* **72,** 191–194.

Orban, G. A., Vandenbussche, E., Sprague, J. M., and De Weerd, P. (1990). Orientation discrimination in the cat, A distributed function. *Proc. Natl. Acad. Sci.* **87,** 1134–1138.

Otsuka, R., and Hassler, R. (1962). Über Aufbau und Gliederung der corticalen Sehsphäre bei der Katze. *Arch. Psychiat. Zeits. Gesamm. Neurol.,* **203,** 212–234.

Overbosch J. F. A. (1927). Experimenteel-anatomische onderzoekingen over de projectie der retina het centrale zenuwstelsel. Academisch proefschrift, H. J. Paris, Amsterdam.

Palmer, L. A., Rosenquist, A. C., and Tusa, R. J. (1978). The retinotopic organization of lateral suprasylvian visual areas in the cat. *J. Comp. Neurol.* **177,** 237–256.

Panizza, B. (1855). Oservazzioni sul nervo ottico. *Istituto Lombardo di Scienze e Lettere, Milan, Gionale di Scienze e Arte* **7,** 237–252.

Panizza, B. (1856). Oservazzioni sul nervo ottico. *Memoria, Istituto Lombardo di Scienze, Lettere e Arte* **5,** 375–390.

Pasternak, T., and Maunsell, J. H. R. (1992). Spatiotemporal sensitivity following lesions of area 18 in the cat. *J. Neurosci.* **12,** 4521–4529.

Pasternak, T., Tompkins, J., and Olson, C. R. (1995). The role of striate cortex in visual function of the cat. *J. Neurosci.* **15,** 1940–1950.

Payne, B. R. (1986). The role of callosal cells in the functional organization of cat striate cortex. In: *Two hemispheres one brain functions of the corpus callosum,* (H. H. Jasper, F. Lepore, and M. Ptito, Eds.), pp. 231–254, New York, A. R. Liss.

Payne, B. R. (1990a). The representation of the ipsilateral visual field in the transition zone between areas 17 and 18 of the cat's cerebral cortex. *Vis. Neurosci.* **4,** 445–474.

Payne, B. R. (1990b). The function of the corpus callosum in the representation of the visual field in cat visual cortex. *Vis. Neurosci.* **5,** 205–211.

Payne, B. R. (1991). The visual field map in the transcallosal sending zone of area 17 in the cat. *Vis. Neurosci.* **7,** 201–219.

Payne, B. R. (1993). Evidence for visual cortical area homologs in cat and macaque monkey. *Cerebral Cortex* **3,** 1–25.

Payne, B. R. (1994). Neuronal interactions in cat visual cortex mediated by the corpus callosum. *Behav. Brain Res.* **64,** 55–64.

Payne, B. R., and Berman, N. (1983). Functional organization and receptive field properties of neurons in cat striate cortex, Variations in preferred orientation with receptive field type, ocular dominance and location in the visual field map. *J. Neurophysiol.* **49,** 1051–1072.

Payne, B. R., and Cornwell, P. (1994). System-wide repercussions of immature visual cortex damage. *Trends Neurosci.* **17,** 126–130.

Payne, B. R., and Lomber, S. G. (1996). Age dependent modification of cytochrome oxidase activity in the cat dorsal lateral geniculate nucleus following removal of primary visual cortex. *Vis. Neurosci.* **13,** 805–816.

Payne, B. R., and Lomber, S. G. (1998). Neuroplasticity in the cat's visual system: origin, termination, expansion and increased coupling in the retino-geniculo-middle suprasylvian visual pathway following early lesions of areas 17 and 18. *Exp. Brain Res.* **121,** 334–349.

Payne, B. R., and Siwek, D. F. (1990). Receptive fields of neurons at the confluence of cerebral cortical areas 17, 18, 20a and 20b in the cat. *Vis. Neurosci.* **4,** 475–479.

Payne, B. R., and Siwek, D. F. (1991a). The visual map in the corpus callosum of the cat. *Cerebral Cortex* **1,** 173–188.

Payne, B. R., and Siwek, D. F. (1991b). The visual field map in the callosal recipient zone at the border between areas 17 and 18 of the cat. *Vis. Neurosci.* **7,** 221–236.

Payne, B. R., Berman, N., and Murphy, E. H. (1981). Organization of direction preferences in cat visual cortex. *Brain Res.* **211,** 445–450.

Payne, B. R., Pearson, H. E., and Cornwell, P. (1984). Transneuronal degeneration of beta retinal ganglion cells in the cat. *Proc. R. Soc. Lond., Ser. B* **222,** 15–32.

Payne, B. R., Siwek, D. F., and Lomber, S. G. (1991). Complex transcallosal interactions in visual cortex. *Vis. Neurosci.* **3,** 283–289.

Payne, B. R., Lomber, S. G., MacNeil, M. A., and Cornwell. P. (1996a). Evidence for greater sight in blindsight following damage of primary visual cortex early in life. *Neuropsychologia* **34,** 741–774.

Payne, B. R., Lomber, S. G., Villa, A. E., and Bullier, J. (1996b). Reversible deactivation of cerebral network components. *Trends Neurosci.* **19,** 535–542.

Payne, B. R., Lomber, S. G., Geeraerts, S., Van der Gucht, E., and Vandenbussche, E. (1996c). Reversible visual hemineglect. *Proc. Nat. Acad. Sci. U.S.A.* **93,** 290–294.

Payne, B. R., Lomber, S. G., Schmidt, K. E., and Galuske, R. A. W. (2001). Feedback circuits and impact on representations in primary visual cortex. *Neurosci. Abs.* **27.**

Peters, A. (1994). The organization of the primate visual cortex in the macaque. In: *Cerebral cortex, volume 10, Primary visual cortex in primates,* (A. Peters, and K. S. Rockland, Eds.), E. G. Jones, and A. Peters (Series Eds.), pp. 1–35, New York, Plenum Press.

Peters, A., and Payne, B. R. (1993). Numerical relationships between geniculo-cortical afferents and pyramidal cell modules in cat primary visual cortex. *Cerebral Cortex* **3,** 69–78.

Peters, A., and Regidor, J. (1981). A reassesment of the forms of nonpyramidal neurons in area 17 of cat visual cortex. *J. Comp. Neurol.* **203,** 685–716.

Peters, A., and Sethares, C. (1996). Myelinated axons and the pyramidal cell modules in monkey primary visual cortex. *J. Comp. Neurol.* **365,** 232–255.

Peters, A., and Yilmaz, E. (1993). Neuronal organization of area 17 of cat visual cortex. *Cerebral Cortex* **3,** 49–68.

Pettigrew, J. D., Nikara, T., and Bishop, P. O. (1968). Binocular interaction on single units in cat striate cortex, simultaneous stimulation by single moving slit with receptive fields in correspondence. *Exp. Brain Res.* **6,** 391–410.

Pettigrew, J. D., and Dreher, B. (1987). Parallel processing of binocular disparity in the cat's retinogeniculocortical pathways. *Proc. R. Soc. Ser. B* **232,** 297–321.

Polyak, S. (1957). *The vertebrate visual system.* Chicago, University of Chicago Press.

Price, D. J. (1985). Patterns of cytochrome oxidase activity in areas 17, 18 and 19 of the visual cortex of cats and kittens. *Exp. Brain Res.* **58,** 125–133.

Price, D. J., Ferrer, J. M., Blakemore, C., and Kato, N. (1994). Functional organization of corticocortical projections from area 17 to area 18 in the cat's visual cortex. *J. Neurosci.* **14,** 2732–2746.

Probst, M. (1900). Über den verlauf der Sehnervenfasern und deren Endigungen im Zwischen-und Littelhirn. *Monatsschr. Psychiatr. Neurol.* **8,** 165–181.

Ptito, M., Lepore, F., and Guillemot, J. -P. (1992). Loss of stereopsis following lesions of cortical areas 17–18 in the cat. *Exp. Brain Res.* **89,** 521–530.

Raczkowski, D., and Sherman, S. M. (1985). Morphology and physiology of single neurons in the medial interlaminar nucleus of the cat's lateral geniculate nucleus. *J. Neurosci.* **5,** 2702–2718.

Raczkowski, D., and Rosenquist, A. C. (1980). Connections of the parvocellular C laminae of the dorsal lateral geniculate nucleus with the visual cortex in the cat. *Brain Res.* **199,** 447–451.

Raczkowski, D., and Rosenquist, A. C. (1983). Connections of the multiple visual cortical areas with the lateral posterior–pulvinar complex and adjacent thalamic nuclei in the cat. *J. Neurosci.* **3,** 1912–1942.

Raczkowski, D., and Sherman, S. M. (1985). Morphology and physiology of single neurons in the medial interlaminar nucleus of the cat's lateral geniculate nucleus. *J. Neurosci.* **5,** 2702–2718.

Raiguel, S. E., Lagae, L., Gulyas, B., and Orban, G. A. (1989). Response latencies of visual cells in macaque areas V1, V2 and V5. *Brain Res.* **493,** 155–159.

Raiguel, S. E., Xiao, D. K., Marcar, V. L., and Orban, G. A. (1999). Response latency of macaque area MT/V5 neurons and its relationship to stimulus parameters. *J. Neurophysiol.* **82,** 1944–1956.

Reid, R. C., and Alonso, J. M. (1995). Specificity of monosynaptic connections from thalamus to visual cortex. *Nature.* **378,** 281–284.

Reid, R. C., and Alonso, J. M. (1996). The processing and encoding of information in the visual cortex. *Curr. Opin. Neurobiol.* **6,** 475–480.

Reinoso-Suarez, F. (1961). Topographischer Hirnatlas der Katze für experimental-physiologische Untersuchungen (Topographical atlas of the cat brain for experimental-physiological research). Darmstadt, Merck.

Revishchin, A. V., and Garey, L. J. (1991). Laminar distribution of cytochrome oxidase staining in cetacean isocortex. *Brain Behav. Evol.* **37,** 355–367.

Rosa, M. G. P. (1997). Visuotopic organization of primate extrastriate cortex. In; *Cerebral cortex, volume 12, Extrastriate cortex in primates,* (K. S. Rockland, J. H. Kaas, and A. Peters, Eds.), E. G. Jones, and Peters, A. (Series Eds.), pp. 127–203, New York, Plenum Press.

Rose, J. E., and Woolsey, C. N. (1958). Cortical connections and functional organization of the thalamic auditory system of the cat. In: *Biological and biochemical bases of behavior,* (H. F. Harlow, and C. N. Woolsey Eds.), Madison, University of Wisconsin Press.

Rosenquist, A. C. (1985). Connections of visual cortical areas in the cat. In *Cerebral cortex, Volume 3, Visual cortex,* (A. Peters, and E. G. Jones, Eds.), pp. 81–117, New York, Plenum Press.

Rosenquist, A. C., Edwards, S. B., and Palmer. L. A. (1975). An autoradiographic study of the projections of the dorsal lateral geniculate nucleus and posterior niucleus in the cat. *Brain Res.* **80,** 71–93.

Rossignol, S., and Colonnier, M. (1971). A light microscope study of degeneration patterns in cat cortex after lesions of the lateral geniculate nucleus. *Vis. Res. Suppl.* **3,** 329–338.

Rowe, M. H., and Palmer, L. A. (1995). Spatio-temporal receptive-field structure of phasic W cells in the cat retina. *Vis. Neurosci.* **12,** 117–139.

Ruan, D. Y., Chabli, A., and Molotchnikoff, S. (1997). Spatial frequency properties in area 18 during inactivation of area 17 in cats. *Exp. Brain Res.* **113,** 431–442.

Sanderson, K. J. (1971). The projection of the visual field to the lateral geniculate and medial interlaminar nuclei in the cat. *J. Comp. Neurol.* **143,** 101–118.

Sanides, F., and Hoffmann, J. (1969). Cyto – and myeloarchitecture of the visual cortex of the cat and the surrounding integration cortices. *J. Hirnforsch.* **1,** 79–104.

Schmidt, K. E., Goebel, R., Castelo-Branco, M., Lomber, S. G. Payne, B. R., and Galuske, R. A. W. (2001). Global motion representation in cat primary visual cortex and its emergence based on feedback connections. *Eur. J. Neurosci.* **12** (Suppl.) 75.

Schmuel, A., and Grinvald, A. (1996). Functional organization for direction of motion and its relationship to orientation maps in cat area 18. *J. Neurosci.* **16,** 6945–6964.

Schoppmann, A., and Stryker, M. P. (1981). Physiological evidence that the 2-deoxyglucose method reveals orientation columns in cat visual cortex. *Nature* **293,** 574–576.

Schwark, H. D., Malpeli, J. G., Weyand, T. G., and Lee, C. (1986). Cat area 17. II. Response properties of infragranular layer cells in the absence of supragranular layer activity. *J. Neurophysiol.* **56,** 1074–1087.

Sestokas, A. K., and Lehmkuhle, S. (1986). Visual latency of X- and Y-cells in the dorsal lateral geniculate nucleus of the cat. *Vis. Res.* **26,** 1041–1054.

Shapley, R., and Perry, V. H. (1986). Cat and monkey retinal ganglion cells and their visual functional roles. *Trends Neurosci.* **9,** 229–235.

Shapley, R. M., and Enroth-Cugell, C. (1984). Visual adaptation and retinal gain controls. *Prog. Brain Res.* **3,** 263–343.

Shapley, R. M., and Victor, J. D. (1978). The effect of contrast on the transfer properties of cat retinal ganglion cells. *J. Physiol. (Lond.).* **285,** 275–298.

Shapley, R. M., and Victor, J. D. (1979). Nonlinear spatial summation and the contrast gain control of cat retinal ganglion cells. *J. Physiol. (Lond.).* **290,** 141–161.

Shapley, R. M., and Victor, J. D. (1980). The effect of contrast on the non-linear response of the Y cell. *J. Physiol. (Lond.).* **302,** 535–547.

Shapley, R. M., and Victor, J. D. (1981). How the contrast gain modifies the frequency responses of cat retinal ganglion cells. *J. Physiol. (Lond.)* **318,** 161–179.

Shapley, R. M., and Victor, J. D. (1986). Hyperacuity in cat retinal ganglion cells. *Science* **231,** 999–1002.

Shatz, C. J., and Stryker, M. P. (1978). Ocular dominance in layer IV of the cat's visual cortex and the effects of monocular deprivation. *J. Physiol. (Lond.)* **281,** 267–283.

Shatz, C. J., Lindstrom, S., and Wiesel, T. N. (1977). The distribution of afferents representing the right and left eyes in the cat's visual cortex. *Brain Res.* **131,** 103–116.

Sherk, H. (1978). Area 18 cell responses in cat during reversible inactivation of area 17. *J. Neurophysiol.* **41,** 204–215.

Sherman, S. M. (1985). Functional organization of the W-, X- and Y- cell pathways in the cat, a review and hypothesis. In: *Progress in psychobiology, and physiological psychology, volume 11,* (J. M. Sprague, and A. N. Epstein, Eds.), pp. 233–314, New York, Academic Press.

Shoham, D., Hübener, M., Schulze, S., Grinvald, A., and Bonhoeffer, T. (1997). Spatio-temporal frequency domains and their relation to cytochrome oxidase staining in cat visual cortex. *Nature* **385,** 529–533.

Shupert, C., Cornwell, P., and Payne, B. R. (1993). Differential sparing of depth perception, orienting and optokinetic nystagmus after neonatal versus adult lesions of cortical areas 17, 18 and 19 in the cat. *Behav. Neurosci.* **107,** 633–650.

Singer, W. (1981). Topographic organization of orientation columns in the cat visual cortex. *Exp. Brain Res.* **44,** 431–436.

Singer, W., Tretter, F., and Cynader, M. (1975). Organization of cat striate cortex: a correlation of receptive field properties and afferent and efferent connections. *J. Neurophysiol.* **38,** 1080–1098.

Somogyi, P., and Cowey, A. (1981). Combined Golgi and electron microscopic study on the synapses formed by double bouquet cells in the visual cortex of the cat and monkey. *J. Comp. Neurol.* **195,** 547–566.

Somogyi, P., and Cowey, A. (1984). Double bouquet cells. In: *Cellular components of the cerebral cortex, cerebral cortex, volume 1* (A. Peters, and E. G. Jones, Eds.), pp. 337–360, New York, Plenum Press.

Somogyi, P., Freund, T. F., and Cowey, A. (1982). The axo-axonic interneuron in the cerebral cortex of the rat, cat and monkey. *Neuroscience* **7,** 2577–2607.

Somogyi, P., Kisvárday, Z. F., Martin, K. A. C., and Whitteridge, D. (1983). Synaptic connections of morphologically identified and physiologically characterized large basket cells on the striate cortex of cat. *Neuroscience* **10,** 261–294.

Somogyi, P. Tamás, G., Lujan, R., and Buhl, E. H. (1998). Salient features of synaptic organisation in the cerebral cortex. *Brain Res.* **26,** 113–135.

Spear, P. D. (1995). Plasticity following neonatal visual cortex damage in cats. *Can. J. Physiol. Pharmacol.* **73,** 1389–1397.

Spear, P. D., and Baumann, T. P. (1979). Effects of visual cortex removal on the receptive field properties of neurons in lateral suprasylvian visual area of the cat. *J. Neurophysiol.* **42,** 31–56.

Specht, S., and Grafstein, B. (1973). Accumulation of radioactive protein in mouse cerebral cortex after injection of ^{3}H-fucose into the eye. *Exp. Neurol.* **41,** 705–722.

Spitzer, H., and Hochstein, S. (1985). Simple- and complex-cell response dependencies on stimulation parameters. *J. Neurophysiol.* **53,** 1244–1265.

Sprague, J. M. (1996). Neural mechanisms of visual orienting responses. *Prog. Brain Res.* **112,** 1–15.

Sprague, J. M., Levy, J., DiBerardino, A., and Berlucchi, G. (1977). Visual cortex areas mediating form discrimination in the cat. *J. Comp. Neurol.* **172,** 441–488.

Stein, J. J., Johnson, S. A., and Berson, D. M. (1996). Distribution and coverage of beta cells in the cat retina. *J. Comp. Neurol.* **372,** 597–617.

Stone, J. (1965). A quantitative analysis of the distribution of ganglion cells in the cat's retina. *J. Comp. Neurol.* **124,** 337–352.

Stone, J. (1973). Sampling properties of microelectrodes assessed in the cat's retina. *J. Neurophysiol.* **36,** 1071–1079.

Stone, J. (1978). The number and distribution of ganglion cells in the cat's retina. *J. Comp. Neurol.* **180,** 753–772.

Stone, J. (1983). *Parallel processing in the visual system. The classification of retinal ganglion cells and its impact on the neurobiology of vision. Perspectives in vision research,* (C. Blakemore, Series Editor) New York, Plenum Press.

Stone, J., and Hansen, S. (1966). The projection of the cat's retina on the lateral geniculate nucleus. *J. Comp. Neurol.* **126,** 601–624.

Stone, J., and Dreher, B. (1973). Projection of X – and Y – cells of the cat's lateral geniculate nucleus to areas 17, and 18 of visual cortex. *J. Neurophysiol.* **36,** 551–567.

Sur, M., Esguera, M., Garraghty, P. E., Kritzer, M. F., and Sherman, S. M. (1987). Morphology of physiologically identified retinogeniculate X- and Y-axons in the cat. *J. Neurophysiol.* **58,** 1–32.

Swindale, N. V. (2000). How many maps are there in visual cortex? *Cerebral Cortex* **10,** 633–643.

Swindale, N. V., Matsubara, J. A., and Cynader, M. S. (1987). Surface organization of orientation and direction selectivity in cat area 18. *J. Neurosci.* **7,** 1414–1427.

Swindale, N. V., Shoham, D., Grinvald, A., Bonhoeffer, T., and Hubener, M. (2000). Visual cortex maps are optimized for uniform coverage. *Nat. Neurosci.* **3,** 822–826.

Symonds L. L., and Rosenquist A. C. (1984a). Corticocortical connections among visual areas in the cat. *J. Comp. Neurol.* **229,** 1–38.

Symonds L. L., and Rosenquist A. C. (1984b). Laminar origins of visual corticocortical connections in the cat. *J. Comp. Neurol.* **229,** 39–47.

Szentágothai, J (1969). Architecture of the cerebral cortex. In: *Basic mechanisms of the epilepsies,* (H. H. Jasper, A. A. Ward, and A. Pope, Eds.), pp. 13–28, Boston, Little Brown and Co.

Szentágothai, J. (1973). Synaptology of the visual cortex. In: *Visual centers of the brain, handbook of sensory physiology, volume VII/3* (R. Jung, Ed.), pp. 269–324, Berlin, Springer-Verlag.

Talbot, S. A. (1940). Arrangement of visual area of cat's cortex. *Am. J. Physiol.* **129,** 477–478P.

Talbot, S. A. (1942). A lateral localization in cat's visual cortex. *Fed. Proc.* **1,** 84.

Talbot, S. A., and Marshall, W. H. (1941). Physiological studies on neural mechanisms of visual localization and discrimination. *Am. J. Ophthalmol.* **24,** 1255–1264.

Talbot, S. A., and Marshall, W. H. (1942). Physiological studies on neural mechanisms of visual localization and discrimination. *Arch. Ophthalmol.* **27,** 213–215.

Tamamaki, N., Uhlrich, D. J., and Sherman, S. M. (1995). Morphology of physiologically identified retinal X and Y axons in the cat's thalamus and midbrain as revealed by intraaxonal injection of biocytin. *J. Comp. Neurol.* **354,** 583–607.

Tanaka, K. (1983). Cross-correlation analysis of geniculostriate neuronal relationships in cats. *J. Neurophysiol.* **49,** 1303–1318.

Thalurri, J., and Henry, G. H. (1989). Neurons of the cat striate cortex driven transsynaptically by electrical stimulation of the superior colliculus. *Vis. Res.* **29,** 1319–1323.

Thompson, J. M., Woolsey, C. N., and Talbot, S. A. (1950). Visual areas I and II of cerebral cortex of the rabbit. *J. Neurophysiol.* **13,** 277–288.

Tolhurst, D. J., and Thompson, I. D. (1981). On the variety of spatial frequency selectivities shown by neurons in area 17 of the cat. *Proc. R. Soc. Ser. B* **213,** 183–199.

Tolhurst, D. J., and Thompson, I. D. (1982). Organization of neurones preferring similar spatial frequencies in cat striate cortex. *Exp. Brain Res.* **48,** 217–227.

Tolhurst, D. J., Dean, A. F., and Thompson, I. D. (1981). Preferred direction of movement as an element in the organization of cat visual cortex. *Exp. Brain Res.* **44,** 340–342.

Tömböl, T. (1978). Comparative data on the Golgi architecture of interneurons of different cortical areas in cat and rabbit. In: *Architectonics of the cerebral cortex,* (M. A. B. Brazier, and H. Petsche, Eds.), pp. 59–76, New York, Raven Press.

Tömböl, T. (1984). Layer VI cells. In: *Cellular components of the cerebral cortex. Cerebral cortex, volume 1,* (A. Peters, and E. G. Jones, Eds), pp. 479–520, New York, Plenum Press.

Tong, L., Spear, P. D., Kalil, R. E., and Callahan, E. C. (1982). Loss of retinal X-cells in cats with neonatal or adult visual cortex damage. *Science* **217,** 72–75.

Tootell, R. B. H., Silverman, M. S., and De Valois, R. L. (1981). Spatial frequency columns in primary visual cortex. *Science* **214,** 813–815.

Toth, L. J., Kim, D.-S., Rao, S. C., and Sur, M. (1997). Integration of local inputs in visual cortex. *Cerebral Cortex* **7,** 703–710.

Toyama, K., Kimura, M., and Tanaka, K. (1981a). Cross-correlation analysis of interneuronal connectivity in cat visual cortex. *J. Neurophysiol.* **46,** 191–201.

Toyama, K., Kimura, M., and Tanaka, K. (1981b). Organization of cat visual cortex as investigated by cross-correlation technique. *J. Neurophysiol.* **46,** 202–214.

Tretter, F., Cynader, M., and Singer, W. (1975). Organization of cat parastriate cortex: a primary or secondary visual area. *J. Neurophysiol.* **38,** 1099–1113.

Troy, J. B. (1983). Spatial contrast sensitivities of X, and Y type neurons in the cat's dorsal lateral geniculate nucleus. *J. Physiol. (Lond.).* **344,** 399–417.

Troy, J. B. (1987). Do Y geniculate neurons have greater contrast sensitivity than X geniculate neurons at all visual field locations? *Vis. Res.* **27,** 1733–1735.

Troy, J. B., and Enroth-Cugell, C. (1993). X, and Y ganglion cells inform the cat's brain about contrast in the retinal image. *Exp. Brain Res.* **93,** 383–390.

Ts'o, D. Y., Gilbert, C. D., and Wiesel, T. N. (1986). Relationships between horizontal interactions and functional architecture in cat striate cortex as revealed by cross-correlation analysis. *J. Neurosci.* **6,** 1160–1170.

Tusa, R. J., and Palmer, L. A. (1980). Retinotopic organization of areas 20 and 21 in the cat. *J. Comp. Neurol.* **193,** 147–164.

Tusa, R. J., Palmer, L. A., and Rosenquist, A. C. (1978). The retinotopic organization of area 17 (striate cortex). in the cat. *J. Comp. Neurol.* **177,** 213–236.

Tusa, R. J., Rosenquist, A. C., and Palmer, L. A. (1979). Retinotopic organization of areas 18, and 19 in the cat. *J. Comp. Neurol.* **185,** 657–678.

Updyke, B. V. (1977). Topographic organization of the projections from cortical areas 17, 18 and 19 onto the thalamus, pretectum and superior colliculus in the cat. *J. Comp. Neurol.* **173,** 81–122.

Updyke, B. V. (1986). Retinotopic organization within the cat's posterior suprasylvian sulcus and gyrus. *J. Comp. Neurol.* **246,** 265–280.

Valverde, F. (1971). Short axon neuronal subsystems in the visual cortex of the monkey. *Int. J. Neurosci.* **1,** 181–197.

Valverde, F. (1978). The organization of area 18 in the monkey. *Anat. Embryol.* **154,** 305–334.

Van Essen, D. C., and Maunsell, J. H. R. (1980). Two-dimensional maps of the cerebral cortex. *J. Comp. Neurol.* **191,** 255–281.

Vandenbussche, E., Sprague, J. M., De Weerd, P., and Orban, G. A. (1991). Orientation discrimination in the cat, Its cortical locus. I. Areas 17 and 18. *J. Comp. Neurol.* **305,** 632–658.

Vastola, E. F. (1961). A direct pathway from lateral geniculate body to association cortex. *J. Neurophysiol.* **24,** 469–487.

Vercelli, A., Assal, F., and Innocenti, G. M. (1992). Emergence of callosally projecting neurons with stellate morphology in the visual cortex of the kitten. *Exp. Brain Res.* **90,** 346–358.

Voigt, T., LeVay, S., and Stamnes, M. A. (1988). Morphological and immunocytochemical observations on the visual callosal projections in the cat. *J. Comp. Neurol.* **272,** 450–460.

von Economo, C. (1929). *The cytoarchitectonics of human the cortex,* Oxford, Oxford University Press.

von Monakow, C. (1882a). Über einige durch Exstirpation circumscripter Hirnrindenregionen bedingte Entwickelungenshemmungen des Kaninchengehirns. *Arch. Psych. Nervenkr.* **12,** 141–156.

von Monakow, C. (1882b). Weitere Mitheilungen über durch Extirpation circumscripter Hirnrindenregionen bedingte Entwickelungenshemmungen des Kaninchengehirns. *Arch. Psych. Nervenkr.* **12,** 535–549.

von Monakow, C. (1883). Experimentelle und pathologisch-anatomische Untersuchungen über die Beziehungen der sogenannten Sehsphäre zu den infracorticalen Opticuscentren und zum Nervus Opticus. *Arch. Psych. Nervenkr.* **14,** 699–751.

von Monakow, C. (1889). Experimentelle und pathologisch-anatomische Untersuchungen über die optischen Centren und Bahnen. *Arch. Psychiatr. Nervenkr.* **20,** 714–787.

Wang, C., Dreher, B., Huxlin, K. R., and Burke, W. (1997). Excitatory convergence of Y and non-Y information channels on single neurons in the PMLS area, a motion area of the cat visual cortex. *Eur. J. Neurosci.* **9,** 921–933.

Wässle, H., and Boycott, B. B. (1991). Functional architecture of the mammalian retina. *Physiol. Rev.* **71,** 447–480.

Weyand, T. G., Malpeli, J. G., and Lee, C. (1991). Area 18 corticotectal cells, Response properties and identification of sustaining geniculate inputs. *J. Neurophysiol.* **65,** 1078–1088.

Weyand, T. G., Malpeli, J. G., Lee, C., and Schwark, H. D. (1986a). Cat area 17. III. Response properties and orientation anisotropies of corticotectal cells. *J. Neurophysiol.* **56,** 1088–1101.

Weyand, T. G., Malpeli, J. G., Lee, C., and Schwark, H. D. (1986b). Cat area 17. IV. Two types of corticotectal cells defined by controlling geniculate inputs. *J. Neurophysiol.* **56,** 1102–1108.

Whitteridge, D., and Clarke, P. G. H. (1982). Ipsilateral visual field represented in the cat's visual cortex. *Neuroscience* **7,** 1855–1860.

Wilbrand, H. (1890). *Die hemianopischen Gesichtsfeld-Formen und das optische Wahrnehmungszentrum.* Wiesbaden, J. F. Bargman.

Wilson, M. E., and Cragg, B. G. (1967). Projections from the lateral geniculate nucleus in the cat and monkey. *J. Anat. Lond.* **101,** 677–692.

Winkler, C., and Potter, A. (1914). *An anatomical guide to experimental researches on the cat's brain.* Amsterdam, W. Versluys.

Woolsey, C. N. (1958). Organization of somatic sensory and motor areas of the cerebral cortex. In: *Biological and biochemical bases of behavior,* (H. F. Harlow, and C. N. Woolsey, Eds.), Madison, University of Wisconsin Press.

Woolsey, C. N. (1971). Comparative studies on cortical representation of vision. *Vis. Res. Suppl.* **3,** 365–382.

Woolsey, C. N., and Fairman, D. (1946). Contralateral, ipsilateral and bilateral representation of cutaneous receptors in somatic areas I and II of the cerebral cortex of pig, sheep, and other mammals. *Surgery* **19,** 684–702.

Zeki, S. (1993). *A vision of the brain* Oxford, Blackwell.

2

OPTICAL IMAGING OF FUNCTIONAL ARCHITECTURE IN CAT PRIMARY VISUAL CORTEX

MARK HÜBENER AND TOBIAS BONHOEFFER

Max-Planck-Institute for Neurobiology, Martinsried, Germany

INTRODUCTION

A hallmark of the functional architecture of neocortex is that neurons with similar response properties are clustered together in columns extending radially through the whole thickness of the cortical gray matter. This columnar organization was first described in the somatosensory cortex by Mountcastle (1) and has subsequently been found in many other primary sensory and motor areas of the neocortex, e.g. the auditory cortex (2), the motor cortex (3), but also in a number of higher cortical areas (4–7). In visual cortex, too, it was recognized early on that neurons are arranged in columns according to their key response properties, orientation preference and ocular dominance (8,9). The widespread occurrence of columnar structures in different cortical areas and in many mammalian species suggests that this specific arrangement of response properties is of fundamental importance for cortical information processing.

Much of our current knowledge concerning the columnar organization of response properties in the visual cortex is based on extracellular microelectrode recordings. Typically, such recordings revealed that neurons encountered along electrode penetrations orthogonal to the cortical surface had largely similar response properties, while a systematic shift of preferred stimulus properties was observed when the electrode was advanced parallel to the cortical layering (8–12; see Figs. 1-34–1-36). However, the systematic, high-resolution mapping of response properties across large regions of the visual cortex with electrophysiological recordings is seriously hampered by the fact that it is exceedingly time-consuming and therefore potentially susceptible to undersampling. Another technique for studying the columnar architecture of the visual cortex involves metabolic mapping with the radioactively labeled glucose-analog 2-deoxyglucose (13) (see also Chapter 3). This technique allows the visualization of functional architecture with sufficiently high resolution, but it has the disadvantage that it can only reveal the activation pattern in response to a single, or, when using two different isotopes, two visual stimuli. An alternative approach for visualizing neuronal response patterns in the brain is based on optical methods. Since its first application in the visual cortex (14,15) optical imaging has been successfully used to study the detailed layout of a number of different functional maps. Compared with the previously mentioned methods, optical imaging allows the high-resolution visualization of multiple maps in the same animal.

In this chapter, we first describe some of the theoretical and methodological aspects of optical imaging. In the following sections we review the basic findings on the structure of functional maps in the cat's primary visual cortex that have been obtained with optical imaging, and we compare these results with observations made with other methods. After demonstrating how different response properties are arranged in the cat's visual cortex, we address the issue of the spatial relationships between these different maps. The last section compares results obtained in the cat's visual cortex with those found in other species.

METHODOLOGICAL ASPECTS OF OPTICAL IMAGING

Two optical imaging methods have been used to investigate the mammalian primary visual cortex: one uses voltage-sensitive dyes to measure neuronal activity by directly assessing membrane potential changes of neurons; the second is based on intrinsic reflectance changes of active nervous tissue. Since the majority of imaging studies on functional maps in the cat's visual cortex have made use of the latter method, we concentrate here on the description of optical imaging of intrinsic signals.

A detailed description of this technique is beyond the scope of this chapter. However, to put the reader into a position to evaluate the data generated with intrinsic imaging, we briefly describe some of the essential features of the method. A comprehensive account of the technical aspects of optical imaging of intrinsic signals can be found in Bonhoeffer and Grinvald (16).

THE SOURCES OF THE INTRINSIC SIGNAL

A number of different signal sources contribute to the overall change in light reflectance associated with neural activity. The relative contributions of these sources depend on different parameters; the two most important ones are the time elapsed since the onset of the stimulus and the wavelength used to illuminate the cortex.

One important component is the so-called oxymetry signal, which is based on changes in the ratio between oxyhemoglobin and deoxyhemoglobin in the micro-capillaries upon activation of neuronal circuits. Since the absorption spectra of these two forms of hemoglobin differ significantly, these changes result in increased or decreased light absorption (depending on the wavelength) and complementary changes in reflection. Imaging spectroscopy experiments suggest that during the first 1 to 1.5 seconds after the onset of sensory stimulation, the intrinsic signal is dominated by a decrease in the concentration of oxyhemoglobin in the capillary bed owing to increased neuronal oxygen consumption (17). This early response component (initial dip) has also been observed in a number of functional magnetic resonance imaging (fMRI) studies using blood oxygenation level-dependent (BOLD) contrast, particularly when high magnetic field strengths were used (18–20). Vanzetta and Grinvald (21) have directly and elegantly demonstrated the initial decrease in the oxyhemoglobin concentration by measuring the oxygen–concentration-dependent changes in the phosphorescence decay time constants of a probe injected into the bloodstream. At wavelengths above 600 nm (those used in most optical imaging studies) this initial response leads to an increased absorption and hence a decrease in reflection in active regions of the brain.

A second component of the intrinsic optical signal related to the vascular supply of the brain is caused by changes in blood volume. Local blood flow in the brain is tightly coupled to the level of neuronal activity and blood volume increases in active regions (22,23). The corresponding increase in the total amount of hemoglobin alters the absorption properties of the tissue. Since blood flowing into active regions initially contains a high concentration of oxyhemoglobin, absorption does not necessarily increase; rather the sign of the change depends on the wavelength. In general, particularly during the later phases of the response, the component of the intrinsic signal associated with blood volume changes is less well localized to active regions of the tissue (17,24).

The third signal component is based on altered light scattering properties of active brain regions. A number of factors contribute to this signal component; movement of ions and water between different tissue compartments and accompanying cell volume changes, capillary responses, and neurotransmitter release are probably the most important ones (25). The best evidence that processes unrelated to the blood supply contribute to the intrinsic signal comes from studies using blood-free preparations such as brain slices, where strong intrinsic signals can be imaged after electrical stimulation (24,26–28). The light-scattering component has a short latency, and it makes a strong contribution to the overall optical signal at longer wavelengths (16,17).

Although the exact time courses of these three signal components differ to a certain extent, they are all in the range of hundreds of milliseconds to seconds. Thus optical imaging of intrinsic signals cannot be used to resolve fast events associated with the spiking of single neurons. Rather, its main application is the visualization of the spatial arrangement of response properties in the brain. Because components of different spatial origins contribute to the overall optical signal, it is difficult to give a definite value for the spatial resolution that can be achieved with this technique. However, based on experience from the visual cortex, it is on the order of 50 to 100 μm.

THE EXPERIMENTAL SETUP

In principle, the technique of optical imaging of intrinsic signals from the visual cortex is straightforward: the brain of an animal is exposed and illuminated with light of an appropriate wavelength (mostly between 600 and 707 nm). Images are then acquired with a CCD-camera while the animal is stimulated visually. A schematic of the experimental setup is shown in Fig. 2-1.

Despite the simplicity of the basic layout of the experimental setup, certain aspects of the technique and of the imaging procedure demand further attention. The camera system used to acquire images from the cortex must meet two important criteria: first it should have a good signal to noise ratio, and second, because the amplitude of the intrinsic signal is in the order of 0.1% or less, the system's effective digitization depth must be at least 10 bits. Most laboratories use one of two different

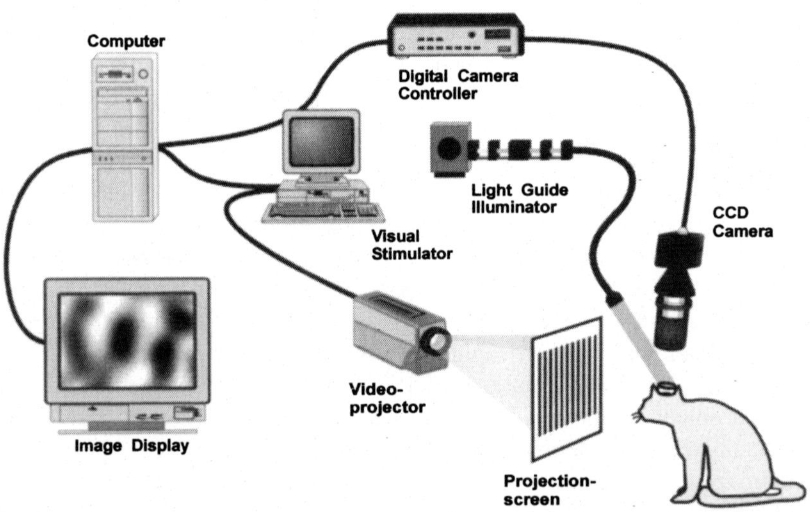

FIGURE 2-1. Experimental setup for optical imaging of intrinsic signals in the visual cortex. The exposed visual cortex is stabilized either by mounting a sealed, oil-filled chamber to the scull or by covering the cortex with warm agar and placing a glass coverslip on top of the agar. The cortex is illuminated with red light and reflectance images are acquired with a CCD-camera while the anaesthetized animal is visually stimulated. The images are digitized in the camera controller and sent to a computer, which also controls the visual stimulator.

CCD-camera systems, both of which fulfill these requirements. One is based on a slow-scan CCD camera with a digitization depth of 12 bits or more, and the other uses differential video imaging (16). The latter system utilizes analog circuitry to store a reference image and to subtract this from the images acquired with a high-quality video camera. The resulting differential image is first amplified and then digitized with an 8-bit A/D converter, giving an apparent resolution of 12 to 14 bits.

The camera's objective needs to have a high numerical aperture to collect as much light as possible. In addition the optic system should be designed such that it has a very shallow depth of field. This allows lowering the focus well below the cortical surface, thereby minimizing artifacts introduced by large blood vessels. A "macroscope" designed by Ratzlaff and Grinvald (29) combines a high numerical aperture with a shallow depth of field. In essence this lens arrangement consists of two high-performance 50-mm camera objectives mounted front to front.

COMPUTATION OF ACTIVITY MAPS

An important issue concerns the algorithms used to derive cortical maps from intrinsic imaging data. Because the amplitude of intrinsic signals is very small compared with reflectance differences caused by uneven illumination of the cortical surface, raw, unprocessed images of the cortex will usually not reveal the effect of sensory stimulation. Therefore, to visualize specific, visually evoked response patterns, images obtained under different stimulus conditions must be compared. The straightforward way to do this is to use subtraction or division of the images from stimulated and unstimulated cortex. In many instances such "blank-corrected" images result in clear maps of the functional architecture. However, quite often blank-corrected activity maps are contaminated with large blood vessel artifacts, which are partially caused by the general vascular response after activating the cortex with any visual stimulus. Consequently, these artifacts are reduced only when images of the visually stimulated cortex are compared with each other. To this end the response to a single visual stimulus is compared with a "cocktail-blank" (30), which reflects the combined response of the cortex to all visual stimuli. The disadvantage of using cocktail-blank correction is that conclusions can be drawn only about the relative effectiveness of a specific visual stimulus in activating the cortex more or less than other stimuli. In particular it is important to note that the cocktail-blank itself does not necessarily activate the cortex uniformly. Under such conditions cocktail-blank correction of an image results in a false functional map, even if the visual stimulus used to obtain this image did not activate the cortex at all.

Another beneficial effect of using the cocktail-blank as a reference image is simply based on the large number of individual images that contribute to it. Thus, biological noise present in each single reflectance image tends to cancel out. Similarly, when using images of the unstimulated cortex as a reference, the signal to noise ratio of the maps can be improved by acquiring many individual blank images and averaging over them. Ideally, blank images acquired immediately before the presentation of each visual stimulus are subtracted from the images obtained with this stimulus before any other image processing is applied. This so-called first frame analysis (16) is particularly useful to remove low-frequency components of the biological noise such as the prominent 0.1 Hz reflectance fluctuations associated with vasomotion (31).

OPTICAL IMAGING OF FUNCTIONAL MAPS IN CAT VISUAL CORTEX

Although the spatial resolution of optical imaging of intrinsic signals is clearly lower than that of single cell recordings, it easily reveals individual functional domains, which in most cases have a size of a few hundred micrometers in the cat's visual cortex. In the next sections we describe the layout of functional maps for different response properties, as they have been determined with optical imaging.

THE STRUCTURE OF THE ORIENTATION MAP

The majority of cells in the cat's primary visual cortex respond selectively to the orientation of a visual contour. In their initial studies Hubel and Wiesel (8,9) noted that neurons with similar orientation preferences were arranged in column-like structures extending radially through the cortex from the pial surface to the white matter (e.g., Fig. 1-34). Based on multiple electrode penetrations it was concluded that individual orientation domains had various shapes and often formed bands meandering across the cortex (9–11). Basically similar conclusions were reached in studies using the 2-deoxyglucose (2-DG) technique to map responses to a single orientation in cat visual cortex (32–35). In these studies the activated regions had the appearance of bands, or "beaded bands" (35), running across the cortex over a few millimeters (see Chapter 3).

In line with this observation, activity maps obtained in response to a single orientation with optical imaging were found to show a pattern of round or elongated patches with a spacing of roughly a millimeter (Fig. 2A–D) (30,36–38). In many instances the individual orientation domains are connected to each other by bridges of less activated regions, quite similar to the beaded bands seen in some of the 2-DG studies (Fig. 3-3). Optical imaging, however, offers the advantage that, for a particular region of cortex, data for four or eight different stimulus orientations can be easily obtained and combined to produce an overall picture of how orientation is represented on the cortical surface. To this end it is instructive to use a color scheme that codes the preferred orientation for each point in the cortex, a technique first applied by Blasdel and Salama (14) in their optical imaging study in monkey visual cortex. In this procedure, the responses to all stimulus orientations are vectorially summed on a pixel-by-pixel basis, and the angle of the resulting vector is indicated by the color of the pixel. When analyzing such color-coded orientation preference maps, it becomes evident, that the cortex is parceled into many patchlike regions with different orientation preferences (Fig. 2-3A).

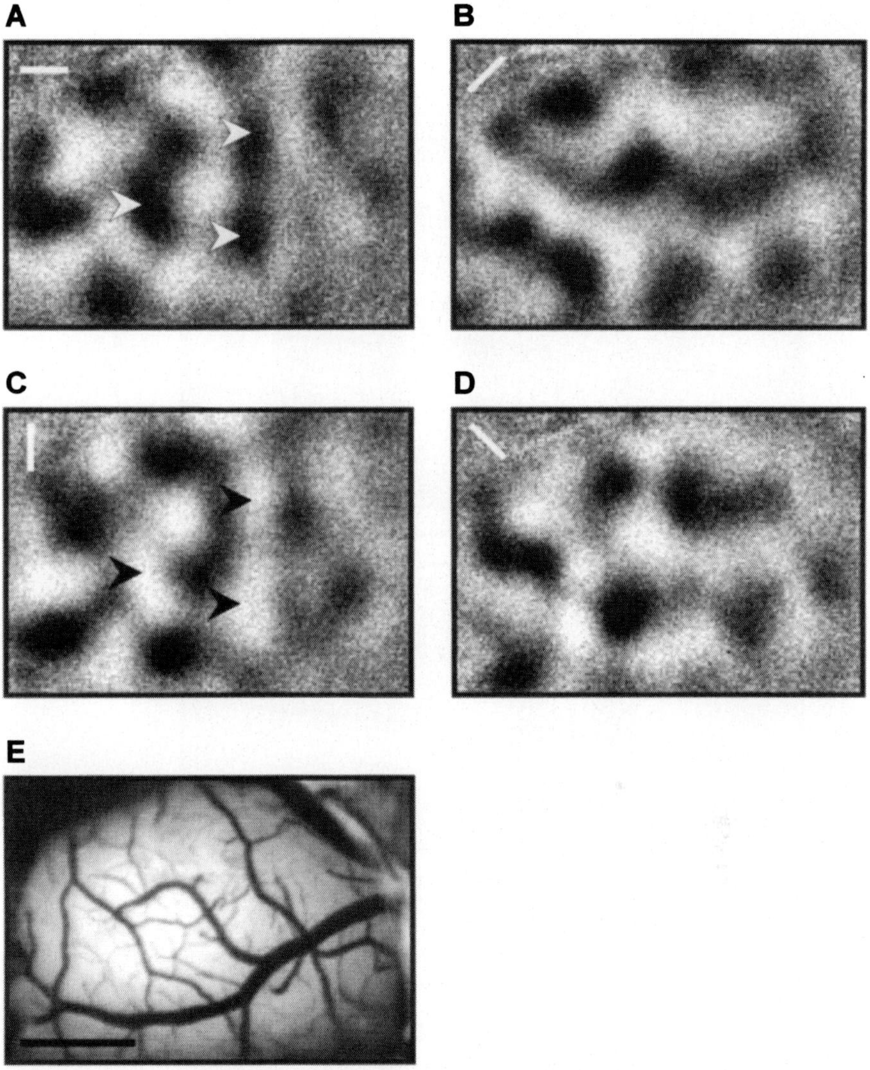

FIGURE 2-2. Activity maps in cat area 17 visualized with optical imaging of intrinsic signals **(A-D).** The animal was binocularly stimulated with gratings of four different orientations, moving back and forth during image acquisition. In each map dark regions were activated by the stimulus orientation indicated by the white bar in the upper left corner. All maps were computed by dividing the responses to a single orientation by the sum of the responses to all four orientations (cocktail-blank). The arrowheads in the maps for horizontal **(A)** and vertical **(C)** orientations are in the same positions and show that both maps are nearly complementary to each other. Thus, as expected, cortical regions activated strongly by a horizontal stimulus are least active with a vertical stimulus, and vice versa. This is also true for the two maps obtained with oblique orientations. The image in **(E)** shows the blood vessel pattern on the cortical surface of the imaged region. Note that even relatively large blood vessels are not visible in the activity maps (Scale bar: 1 mm).

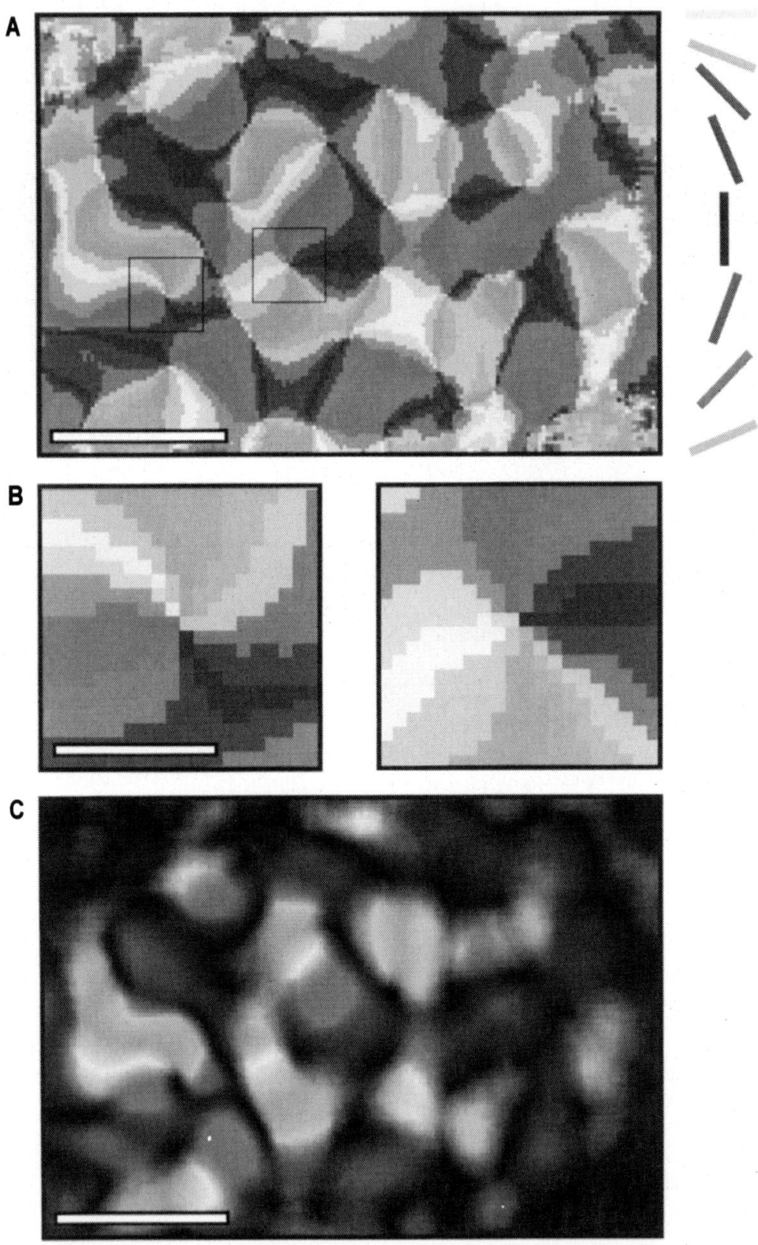
A
B
C

Only very few cases, usually in area 18 or along the border between areas 17 and 18, show regions forming bands running through the cortex (39); the much more widespread and characteristic feature of these orientation preference maps in cat visual cortex is a radial arrangement of orientation domains in a pinwheel-like fashion around orientation centers (36). Each pinwheel is composed of a complete set of orientation domains, with each orientation represented once. The radial arrangement of orientation domains implies that there are two possible types of pinwheels: one with a clockwise and one with a counterclockwise sequence of orientation domains (Fig. 2-3B). In fact, both types are found in the cat's visual cortex at about equal numbers, with the nearest neighbor of a pinwheel normally being one of the opposite sign (40). A systematic arrangement of clockwise and counterclockwise pinwheels was observed in the border region between areas 17 and 18. Pinwheels of the same type tend to form rows running parallel to the border and alternating with rows of the other type (41).

The density of the orientation centers can vary considerably between animals, with average values for area 17 of adult cats ranging between 2.0 and 2.6 per mm^2 (38,42–44). A number of factors probably contribute to the observed variation in the densities of the orientation centers: different localization algorithms were used in different studies, the signal to noise ratio of the imaged orientation maps is likely to influence the number of detected orientation centers, and the position of the imaged region within area 17 plays a role (see later). Most important, however, as with other columnar structures in the visual cortex, e.g. ocular dominance columns (37,45), there seems to be a large interindividual variation with respect to the spacing of functional domains. A considerably lower density of orientation centers (1.2/mm^2) was observed in area 18 (36). Apart from this, the overall structure of the orientation map is similar between area 17 and area 18, and the border between both areas does not induce any apparent breaks in the map (42).

Interestingly, one study using closely spaced electrode penetrations in area 18 had predicted a circular arrangement of orientation domains based on interpolation algorithms (46). A number of modeling studies also suggested a pinwheel-like organization of the orientation map, albeit with slightly different properties (47–50).

FIGURE 2-3. The arrangement of orientation preference in cat primary visual cortex. **(A)** Color-coded orientation preference map computed from the activity maps shown in Fig. 2-2. The preferred orientation of each pixel is indicated by the color code on the right of the figure. In many instances preferred orientations are arranged in a pinwheel-like fashion around orientation centers. In this example the orientations are displayed in 16 discrete steps to facilitate visualization of the essential features of the map. **(B)** Close-up view of two orientation centers outlined with black squares in **(A).** One orientation center shows a clockwise and the other a counterclockwise sequence of orientation domains. **(C)** Polar map from the same cortical region as the map shown in **(A).** The magnitude of the orientation vector is indicated by the intensity of the color; thus dark regions have a very broad orientation tuning. Note that these dark regions often coincide with orientation centers. In contrast to the map shown in **(A)** a continuous color scale was used here to demonstrate that the preferred orientation is mapped smoothly across most regions of the cortex (Scale bar: 1 mm in (**A** and **C**); 250 μm in (**B**)). See color insert for color reproduction of this figure.

Color-coded orientation preference maps like the one shown in Fig. 2-3A denote the preferred orientation for each point in the visual cortex, but they do not contain information about the selectivity of a cortical location for a particular orientation, or, in other words, its orientation tuning curve. This information can be displayed in so-called "polar maps" (30). In addition to color-coding the angle of the vector obtained by summing the responses to all orientations for each pixel, the magnitude of the vector is indicated by the intensity of the color. Thus cortical regions with broad orientation tuning appear dark in polar maps (it should be noted, though, that a vector with a small magnitude can be the result of either strong but equal responses to all orientations or a general lack of responsiveness). Figure 2-3C shows such a polar map, obtained from the same cortical region as the map depicted in Figure 3A.

The general appearance of the polar map is similar to the orientation preference map shown in Figure 3A, and it is evident that the dark regions of the map usually coincide with the orientation centers, indicating that orientation selectivity is low in these regions. However, the intrinsic optical signal from a given point in the visual cortex represents the summed response of a large number of neurons beneath this point. Therefore optical imaging alone cannot answer the question whether the broad orientation tuning of the orientation centers is caused by the collective signal from cells with sharp but very different tuning curves or by cells that are in fact less well tuned to a specific orientation. This question was addressed by combining optical imaging with electrophysiological recordings (51). In this study the electrical recordings were carried out with "tetrodes" (52,53) because these allowed, in combination with spike-sorting algorithms, the simultaneous recording of up to eight cells. These results show that cells located near orientation centers were as sharply tuned for stimulus orientation as cells anywhere else in the visual cortex. On the other hand cells located close to orientation centers exhibited a larger scatter in their preferred orientations compared with cells at other cortical locations. Moreover, the number of unresponsive cells was not higher near orientation centers. Thus, the reduced orientation tuning seen with optical imaging at these cortical locations is most likely caused by the summed response from sharply tuned cells with different preferred orientations.

In a recent study also using tetrodes to record from cat area 17, Hetherington and Swindale (54) found that single units at cortical sites with a large scatter in preferred orientation also tended to have broader tuning curves. However, without knowing the location of the less well-tuned cells with respect to the orientation map, it is difficult to judge whether this observation is at odds with the study by Maldonado et al. (51).

When Hubel and Wiesel first described the columnar organization of orientation selective neurons, they assumed that these columns were discrete entities with sudden steps in the preferred orientation occurring between neighboring columns (8–10). Maps of orientation preference obtained with optical imaging do not show any obvious step changes in the preferred orientation, apart from the orientation centers. As the spatial resolution of optical imaging is between 50 and 100 µm, one might argue that discrete columns of very small size might not be resolved with this method. However, quantitative electrophysiological studies using long horizontal penetrations also failed to find evidence for discrete columns, but rather suggested that orientation preference changed smoothly in most regions of the cortex (11,12 see Chapter 1). In their initial study using voltage sensitive dyes to image the functional architecture of monkey visual cortex, Blasdel and Salama (14) stressed the presence of "fractures" in the maps, lines of a very rapid change of the preferred orientation. Orientation maps in cat visual cortex do not contain distinct fractures, although orientation gradient maps reveal that neighboring orientation centers are sometimes connected by regions with an increased change in preferred orientation (30). However, the rate of change in these elongated regions is still much lower than in the orientation centers itself. The picture that emerges then is that in the cat's visual cortex, orientation preference appears to be mapped in a smooth and continuous fashion across the cortical surface, with sudden changes in the preferred orientation only at the orientation centers.

Diverging results have been obtained with regard to the question of whether all orientations are represented equally in the cat's visual cortex. Although a number of studies found that more neurons respond to the cardinal axis (55–57), others failed to find such an anisotropy (58,59). One reason for these discrepancies is that microelectrode recordings are potentially susceptible to sampling biases, particularly when the property under study is nonuniformly distributed, as is the case with cortical orientation preference. Optical imaging avoids this sampling problem and is therefore well suited to address this issue. The analysis of orientation maps from area 18 revealed that all orientations were equally present in the cortex (30), and only a very weak, nonsignificant overrepresentation of horizontal and vertical orientations was observed in area 17 (40,60). However, these observations hold true only for cortical regions at the surface of the brain, which are accessible to optical imaging: in area 18 regions corresponding to eccentricities between 0° and 40° from the vertical meridian and elevations of –5 to –30° and in area 17 parts near the representation of the *area centralis.* It is also important to note that some of the electrophysiological studies found the strongest overrepresentation of horizontal and vertical orientations among simple cells (55,56,61, but see 11). Simple cells are found primarily in layer 4 and thus contribute little to the optical signal, which is mainly derived from the upper cortical layers.

The orientation maps previously described are based on stimulation with luminance contrasts. In addition, orientation maps for subjective contours have been reported in cat visual cortex (62). In area 18 cortical regions activated by subjective contours were partially overlapping with regions responsive to a luminance contrast of the same orientation, whereas the maps were dissimilar in area 17. In both areas, the responsiveness to subjective contours was mapped systematically across the cortical surface.

OCULAR DOMINANCE COLUMNS

The cerebral cortex is the first station in the visual pathway where neuronal signals from both retinae converge onto single neurons. Hubel and Wiesel (8,63) coined the term *ocular dominance* to indicate that most binocular neurons do not respond equally well to stimulation of either eye but that usually one eye dominates the response. The spatial clustering of cells dominated by either one or the other eye was again initially observed in electrophysiological studies of cat area 17 (8,64,65). These observations were strengthened by studies that used transneuronal labeling of thalamocortical terminals to clearly demonstrate an anatomical segregation of afferents carrying signals from the left and the right eye (66–68). In these and subsequent studies using transneuronal transport (69–74; Fig. 1-10 this volume) or 2-DG mapping (73–76), ocular dominance columns were found to have the shape of elongated patches or bands of varying width (see Fig. 3-6 this volume). Although these bands occasionally run in parallel for a few millimeters, the general impression from these studies is that the arrangement of the ocular dominance columns in cats is rather irregular, especially compared with the orderly arrangement of ocular dominance bands in macaque primary visual cortex (77–79).

Conceptually, given the essentially binary nature of the stimulus space and the obvious segregation of the inputs from both eyes demonstrated in many anatomical studies, the visualization of ocular dominance columns with optical imaging of intrinsic signals seemed to be straightforward. In fact, clear ocular dominance maps can be readily imaged in primate striate cortex (24,80). However, initial results of experiments using optical imaging to demonstrate ocular dominance columns in the visual cortex of adult cats turned out to be rather disappointing. Bonhoeffer et al. (42) visualized ocular dominance columns in adult cat visual cortex by focusing the image-plane relatively deep (750 μm) below the cortical surface and by using near infrared light. They could thus obtain ocular dominance maps, whose amplitude was only about a third of that of the corresponding orientation maps. Most likely, the difference in the quality of ocular dominance maps obtained from monkeys and cats is due to differences between both species in the degree to which information coming from the two eyes is kept separate in the visual cortex. In macaque area V1 nearly all neurons in the input layer (IVc) are strictly monocular (81). In contrast, in the visual cortex of the cat, there is already a comparatively large number of binocular cells in layer IV, and many cells in the upper cortical layers are driven equally well by both eyes (68,82,83; see also Chapter 1). Unlike in adult cats, ocular dominance maps of relatively high amplitude can be imaged in the visual cortex of kittens or young cats (37,84,85). One possible explanation for this age dependence is that in young animals neuronal activity in layer IV makes a larger contribution to the overall optical signal than in adult cats. Differences in the tissue composition are one possible reason; it is known, for example, that myelination in the visual cortex is much less in kittens or young cats than in adult animals (86).

Figure 2-4 depicts two sets of orientation maps obtained by stimulating the two eyes of a cat separately with oriented, moving gratings. While at each orien-

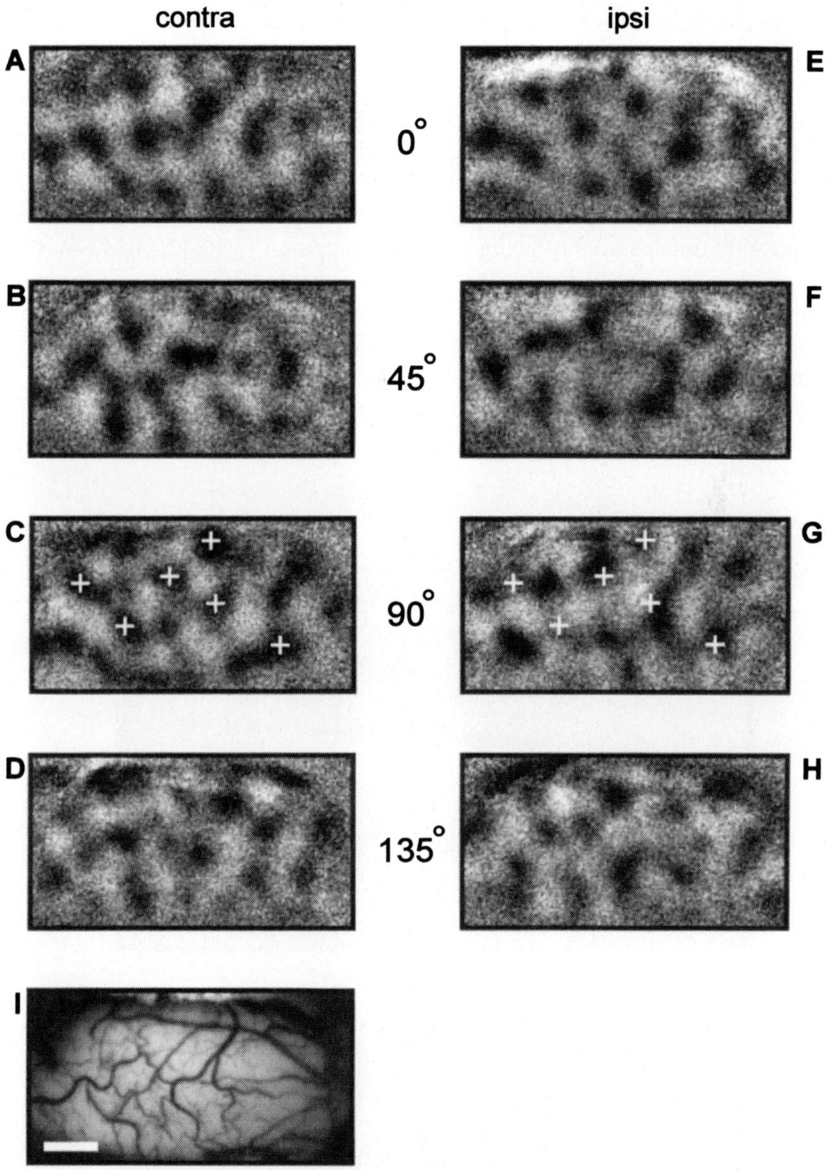

FIGURE 2-4. Orientation maps obtained with monocular stimulation of the contralateral (left) and the ipsilateral (right) eye. To facilitate comparison, active regions in one of the contralateral orientation maps **(C)** have been marked with white crosses, which were then transferred to the corresponding ipsilateral map **(G)**. **(I)** Blood vessel pattern of the imaged region in area 17. (Scale bar: 1 mm.)

tation the two maps of a pair resemble each other, clear differences between the maps are evident. From these maps the ocular dominance map can be computed by summing the maps obtained during stimulation of one eye and dividing the result by the sum of the maps obtained through the other eye. Figure 2-5 shows two examples of ocular dominance maps obtained by this algorithm. It results in a clear pattern of alternating dark and light regions, corresponding to the ocular dominance columns. Maps obtained in this manner generally confirm the results obtained with transneuronal tracing or 2-DG mapping (37,40,42,44,84,85). Ocular dominance domains have the form of elongated, sometimes irregularly shaped patches, which occasionally coalesce into bands that run across the cortex for a few millimeters. The width of the patches or bands is typically about 0.5 mm, but there is large variation within each map as well as between animals.

Most of the aforementioned anatomical and 2-DG studies reported that ocular dominance bands were preferentially oriented orthogonal to the border between

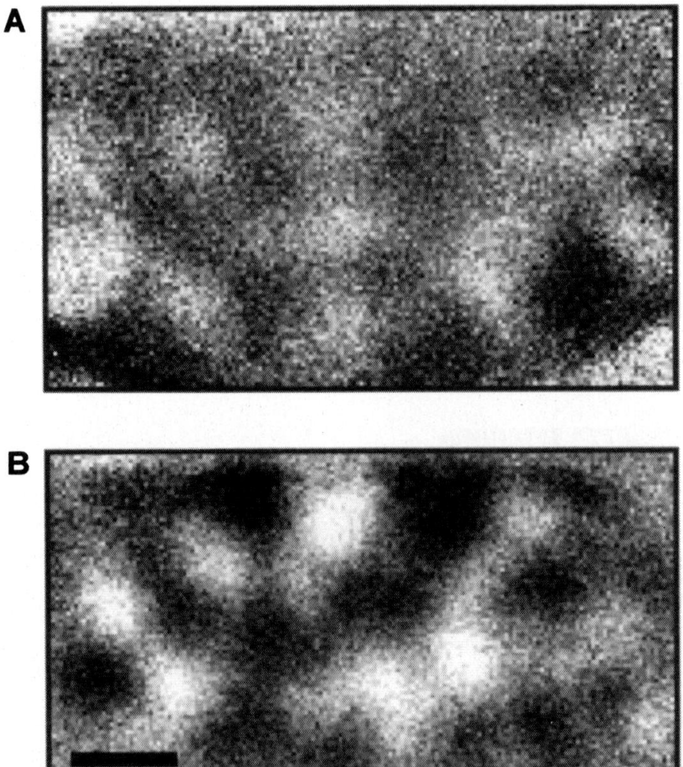

FIGURE 2-5. Ocular dominance maps in cat area 17. In both maps dark regions were activated stronger by the contralateral eye (scale bar: 1 mm). **(B)** (From Hübener et al. [37].)

areas 17 and 18. We found that very few ocular dominance maps obtained with optical imaging showed such a preference; in general the bands or patches do not have a specific orientation (37). One might attribute this to the fact that the part of area 17 accessible to optical imaging is too small to detect this preferential arrangement. However, using a quantitative analysis, Anderson et al. (70) found that the bands were oriented orthogonal to the 17/18 border in only two of six hemispheres. A random arrangement of ocular dominance columns in cat area 17 is also predicted by the model of Jones et al. (87). Thus, considerable variation seems to be present between animals, not only with respect to the width and form of individual ocular dominance columns but also regarding the overall layout of the ocular dominance map.

In primary visual cortex the contralateral eye generally exerts a stronger influence than the ipsilateral eye. This is reflected in the ocular dominance distribution, which is slightly skewed toward the contralateral eye (8). Anatomically, anterograde tracing experiments showed that columns of the ipsilateral eye are often smaller and more sharply delineated, whereas contralateral eye columns, owing to a higher level of interband labeling, were less distinct (66). However, there is electrophysiological (83) as well as anatomical (70) evidence that the overrepresentation of the contralateral eye varies with eccentricity and that both eyes are represented equally in cortical regions near the representation of the *area centralis.* Accordingly, optical imaging in cat area 17, which is confined to the representation of the central part of the visual field, did not reveal apparent imbalances between both eyes (37). In fact, quantitative analysis of optical imaging data from area 17 proved that the average cortical areas activated by the two eyes are almost equal in size (48% and 52% for the ipsilateral and contralateral eye, respectively; Sengpiel, personal communication).

DIRECTION PREFERENCE MAPS

Most neurons in the cat's visual cortex prefer moving stimuli over stationary ones. Moreover, these neurons are often direction selective (i.e., they prefer movement in one direction over movement in the other). Such cells were already described in the pioneering studies by Hubel and Wiesel (8,63,88) and, analogous to orientation preference, it was later reported that cells preferring stimulus movement in the same direction cluster in the visual cortex (46,89–92).

Shmuel and Grinvald (43) conducted a very detailed analysis of the functional organization of direction preference in cat area 18 using optical imaging. Figures 2-6A and B show two isoorientation maps from their study. Both maps were obtained with a grating pattern of the same orientation, but moving in opposite directions. A comparison of both maps reveals that the activation patterns are rather similar, but not identical. The differential direction map (Fig. 2-6C) was computed by subtracting one map from the other. Typically such direction maps obtained at a single orientation consist of dark and light patches embedded in a gray background. The dark and light patches correspond to cortical regions that

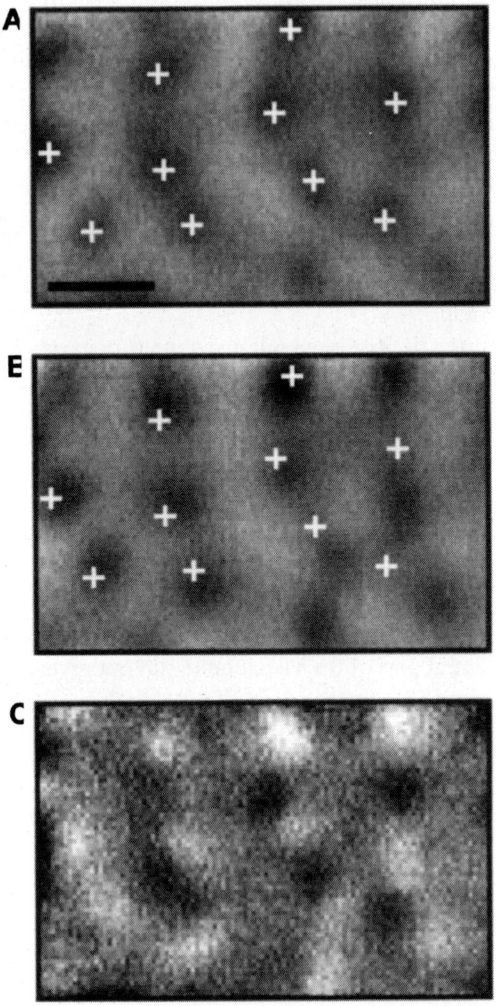

FIGURE 2-6. Clustering according to stimulus direction in area 18. **(A and B)** Orientation maps obtained with a horizontal grating moving upward (A) or downward (B). Both maps are very similar, but they are clearly not identical, as indicated by the white crosses, which were placed in identical positions in both maps. **(C)** Differential direction map computed by subtracting the map in **(B)** from that in **(A).** Dark regions prefer upward movement and lighter regions prefer downward movement (Scale bar: 1 mm) (From Shmuel and Grinvald [43].)

were activated by opposite directions of stimulus movement, whereas most of the gray regions were not activated at all by the orientation used for stimulation. Evidently, some of the patches lie in isolation, but it is more common that a light and dark patch directly bordering onto each other form a pair. Thus, in many, although not all, cases, individual orientation domains consist of two (or sometimes more) subregions, which prefer movement in opposite directions.

The overall organization of direction preference in the visual cortex can be visualized best by a color-coded direction preference map based on a pixel-wise vectorial addition of the responses to all directions, analogous to the computation of the orientation preference map described on page 136. In addition to the color-coding, the map shown in Fig. 2-7A contains arrows indicating the preferred

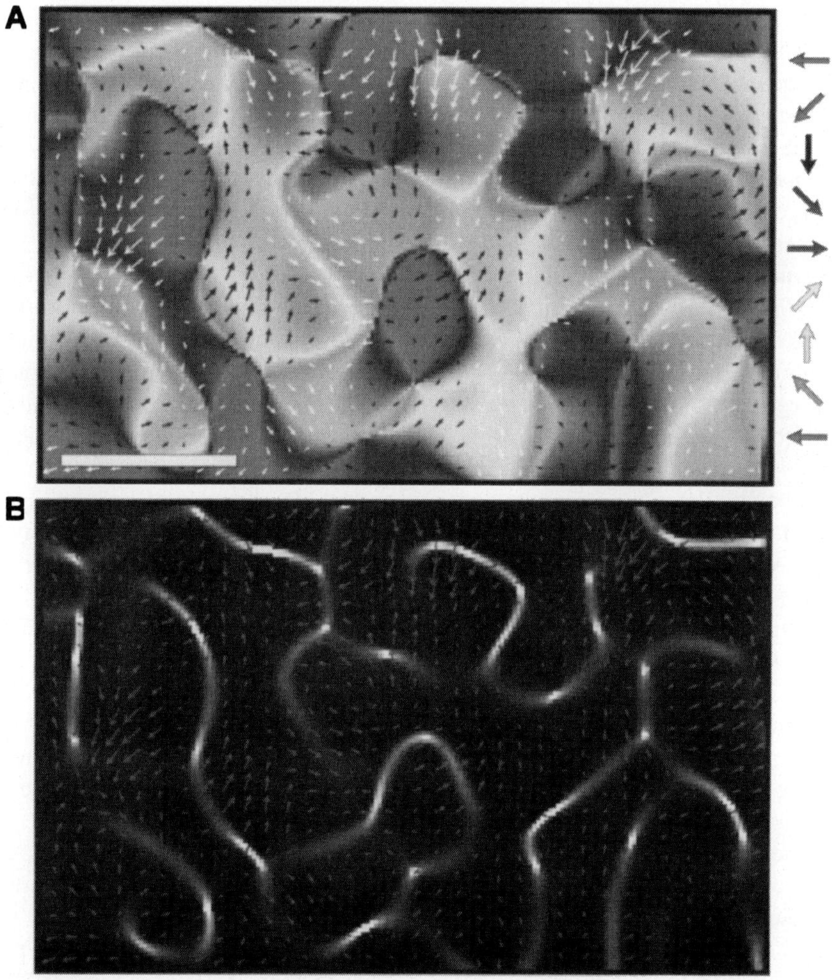

FIGURE 2-7. The layout of direction preference across the cortical surface in area 18. In **(A)** the preferred direction is color-coded according to the key shown on the right. In addition the arrows denote the local preferred direction and, by their length, the magnitude of the preference. In most regions direction preference is mapped continuously across the cortex. **(B)** Rate of change of the preferred direction. The white lines correspond to regions where the preferred direction changes rapidly. In many instances these lines separate regions with a preference for opposite directions of movement (Scale bar: 1 mm) (From Shmuel and Grinvald [43].) See color insert for color reproduction of this figure.

direction as well as the selectivity for direction of the respective cortical region. Superficially, the structure of the direction preference map resembles that of the orientation preference map. In fact, both maps have a number of features in common: the direction map is composed of patchlike regions preferring the same direction of movement, and the periodicity of the direction map is similar to that of the orientation map from the same cortical region (about 1.1 mm in area 18). Like orientation preference, direction preference is by and large mapped smoothly across the cortex (i.e., neighboring cells in the cortex tend to respond best to similar directions of motion). The major difference between both maps is that while regions of rapid change in preferred orientation are essentially point-like (the orientation centers), regions of rapid change in direction most often take the form of lines (Fig. 2-7B). These lines run across the cortical surface along curved trajectories and, in many instances, demarcate cortical regions activated by opposite directions of movement. Overlaying these lines of discontinuity in the direction map onto the corresponding orientation map reveals that the lines often end in the immediate vicinity of orientation centers.

A question that is difficult to answer with optical imaging alone is whether not only direction *preference* but also the *selectivity* for direction (i.e., the tuning) changes in a systematic fashion across the cortical surface. Shmuel and Grinvald (43) computed a map displaying the magnitude of direction selectivity and found that regions of low direction selectivity coincided with regions of rapid changes of direction preference. This is to be expected, because the optical signal pools the activity of many neurons. Thus, closely neighboring regions with a preference for opposite directions appear as zones of overall low direction selectivity. Apart from these regions most parts of the cortex appeared to be more or less equally selective for stimulus direction. In an electrophysiological study employing the reverse correlation technique, DeAngelis et al. (92) compared the receptive field properties of nearby pairs of simple cells. Although neighboring cells tended to respond to the same direction of movement, the authors found no indication of a clustering of cells according to their direction selectivity index. Similarly, Berman et al. (91) did not observe any clustering of bidirectional cells. These findings support the view that the selectivity for direction is not mapped systematically in cat visual cortex. To definitely resolve this issue, optical imaging needs to be combined with single-unit recordings.

In general, the picture that emerges for the layout of direction preference maps in cat visual cortex is in good agreement with previous findings based on electrical recordings. These studies found a columnar organization for direction in area 17 as well as area 18 (46,89–91). In most of these studies, sudden changes from one preferred direction to the opposite one were also observed (46,89,91; see Fig. 1-35 this volume). In particular, the study by Swindale et al. (46) predicted a direction map based on interpolation of closely spaced electrode recordings, which is very similar to the type of map actually observed with optical imaging, including the close spatial association between the lines of discontinuity in the direction map and the pinwheel centers in the orientation map. One unsettled issue that is also difficult to address with optical imaging is whether the direction columns extend through all layers of the visual cortex. Berman et al. (91) observed that the preferred direction frequently reversed at least once in vertical penetrations, indicating that the direction domains do not form true columns in a strict sense (also Fig. 1-34 this volume). Although such reversals were occasionally observed by Shmuel and Grinvald (43), direction preference remained constant in the majority of their vertical tracks.

SPATIAL FREQUENCY COLUMNS

When stimulated with grating patterns of different spatial frequencies, many neurons in the cat's visual cortex act as spatial band pass filters; that is, they are optimally activated by a certain spatial frequency but respond less well to finer or coarser gratings (93,94). Overall, neurons in area 17 prefer much higher spatial frequencies than cells in area 18, but in both areas, at a given eccentricity, neurons can differ considerably in their optimum spatial frequency, the best values covering a range of more than three octaves (95,96). Similarly, while cells preferring lower spatial frequencies generally have larger bandwidths, the selectivity also varies between cells. Electrophysiological studies investigating the possibility of an ordered arrangement of spatial frequency responses in cat visual cortex came to conflicting results. Maffei and colleagues found that neurons preferring similar spatial frequencies were clustered in the cortex and suggested a laminar organization of this response property (97,98). Tolhurst and Thompson (99) confirmed the clustering, but could neither obtain evidence for a layered nor a columnar organization. The issue of a columnar versus laminar organization was eventually settled when Tootell et al. (100) used the 2-DG technique to map spatial frequency preference in cat area 17. The experiments revealed that stimulation with a single spatial frequency activated columns extending through all cortical layers.

A number of recent studies have used optical imaging to visualize spatial frequency columns in cat visual cortex. In many of our experiments, we used two sets of oriented, moving gratings of two different spatial frequencies to stimulate area 17 (37,101). The spatial frequency map was then computed by dividing the sum of the orientation maps obtained with one spatial frequency by the summed maps of the other spatial frequency. Figure 2-8A displays one example of a spatial frequency map computed this way. It shows a patchy pattern, with the dark regions preferring low (0.3 cyc/deg) and the light regions high spatial frequencies (0.9 cyc/deg). The layout of this map is typical of many spatial frequency maps imaged in our labs: Isolated patches with a preference for low spatial frequencies are surrounded by a contiguous region of high spatial frequency preference. Only a few cases showed spatial frequency domains arranged in a more stripelike fashion. By and large the pattern in area 18 is very similar, with the range of spatial frequencies activating this area being much lower than in area 17. Targeted electrical recordings from the centers of the spatial frequency domains confirmed that neurons in these domains responded as predicted by their location within the map (Fig. 2-8B).

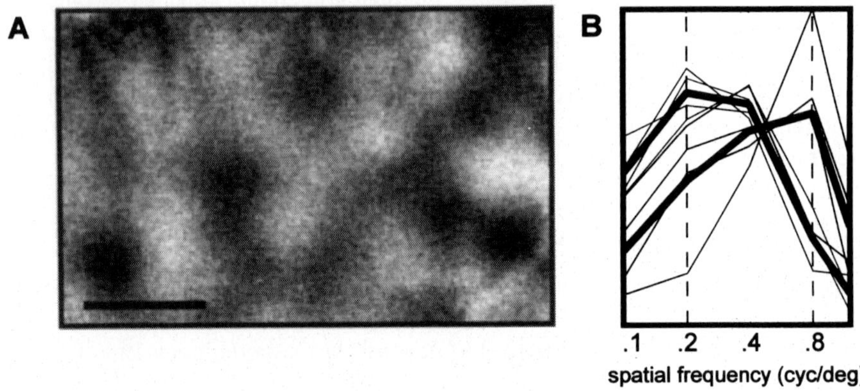

FIGURE 2-8. Spatial frequency map in cat area 17. **(A)** Dark regions were activated by low spatial frequencies (0.3 cyc/p deg), while the lighter regions responded better to high spatial frequencies (0.9 cyc/deg). In this and many other spatial frequency maps, regions preferring low spatial frequencies form islands surrounded by a contiguous region of high spatial frequency preference. **(B)** Electrophysiological confirmation of spatial frequency maps obtained with optical imaging. The thin green lines represent average spatial frequency tuning curves from four penetrations aimed at regions, which, based on optical imaging, were preferentially activated by low spatial frequencies; the thick green curve is the average tuning curve from all these penetrations. Similarly, the red curves are from recordings in regions preferring higher spatial frequencies (Scale bar: 1 mm). **(B)** from Shoham et al. [101].) See color insert for color reproduction of this figure.

Although other optical imaging studies have since confirmed the presence of spatial frequency columns in cat visual cortex (102–104), there is disagreement on how exactly different spatial frequencies are mapped across the cortical surface. Shoham et al. (101), comparing the cortical activation patterns evoked by four spatial frequencies, concluded that the visual cortex is divided into two sets of domains, one activated best by a range of low spatial frequencies and the other by higher spatial frequencies. This view was questioned by Everson et al. (102), who used principal component analysis of the optical imaging data and suggested a continuous mapping of spatial frequency in area 17. According to this study, cortical regions responding to different spatial frequencies form pinwheel-like structures resembling those found in the orientation preference maps. Issa et al. (103) also found evidence for a continuous representation, but could not confirm the pinwheel-like organization of the spatial frequency map. Resolving this issue is hampered by the fact that in general the amplitude of spatial frequency maps obtained with optical imaging is only about a third that of orientation maps from the same cortical region. Thus spatial frequency maps are relatively noisy, and averaging over a large number of stimulus repetitions is needed to produce reasonable maps. A combination of optical imaging with extensive single unit recordings is required to determine which of the proposed schemes for the mapping of spatial frequency best describes the layout of this response property in the cat's visual cortex.

The spatial frequency map has a counterpart in the anatomical organization of the cortex. As in primates, the visual cortex of the cat contains patchy regions of enhanced cytochrome oxidase activity, the so-called blobs (105–107). Combining optical imaging with staining for cytochrome oxidase revealed that cortical regions preferring low spatial frequencies coincide with the blobs, whereas the high spatial frequency domains correspond to the interblob region (101) (see Chapter 5). On the one hand, this unraveled the functional properties of neurons in the cytochrome oxidase compartments in the visual cortex of the cat; on the other hand, it also made it possible to visualize the blobs in vivo, using spatial frequency preference as the differentiating parameter. This allowed the direct addressing of the question of whether the blob and interblob regions in cat visual cortex receive different types of input from the lateral geniculate nucleus (LGN), as is the case in primates (for a review see 108). Injection of retrograde tracers into the cortical blob and interblob regions and analysis of the labeled projection neurons in the LGN indicated that both compartments receive a mixed input from different LGN cell classes. The soma-size distributions of labeled cells suggested that LGN Y-cells preferentially project into the blobs, whereas the interblobs receive a relatively stronger X-cell input (109). This observation is in line with experiments by Boyd and Matsubara (110) who came to similar conclusions based on injections of anterograde tracers into different layers of the LGN. Since Y-cells in the LGN are tuned to lower and X-cells to higher spatial frequencies (111), these results indicate that the different spatial frequency preferences of neurons in the visual cortex are at least in part based on the type of input they receive from the LGN.

REPRESENTATIONS OF OTHER STIMULUS FEATURES

In the preceding sections we have described those features of the functional architecture that have been elaborated in detail with optical imaging. In addition, optical imaging studies have pointed to the existence of other functional maps in cat visual cortex.

The experiments investigating the spatial frequency map in area 17 also allowed a comparison of the cortical responses to grating stimuli that had the same spatial frequency, but differed in their temporal frequencies. The resulting temporal frequency maps were of much lower amplitude than the spatial frequency maps from the same piece of cortex, but nevertheless a clear variation in the preferred temporal frequency across the cortical surface was noticed by Shoham et al. (101). A clustering of neurons with similar temporal frequency preferences was also found with electrophysiological methods by DeAngelis et al. (92); in line with the results from optical imaging, these authors also noted that clustering according to temporal frequency was much weaker than that according to spatial frequency. In the latter study nearby neurons tended to be similar with respect to other temporal characteristics such as response latency or response duration. Unfortunately, mapping of these properties with optical imaging of intrinsic signals is almost impossible, given the slow time course of the signal.

Perhaps expectedly, the layout of the temporal frequency map was similar to that of the spatial frequency map (and thus the cytochrome oxidase pattern): cortical regions preferring low temporal frequencies were responding better to high spatial frequencies and vice versa. Given the previously described partial segregation of X- and Y-fibers in the visual cortex (109,110), this association is easily explained because X-cells in the LGN are tuned to lower temporal frequencies than Y-cells (111). The relatively weak mapping of temporal frequency, therefore, might reflect the relatively small difference in the temporal frequency tuning curves between geniculate Y- and X-cells observed by Derrington and Fuchs (111). Another potential reason for the rather low amplitude of temporal frequency maps in the imaging experiments is that they were performed on young cats (about 8 weeks old). Although the spatial frequency tuning characteristics of cortical cells are almost mature at this age (112,113), the study by DeAngelis et al. (113) showed that the temporal frequency tuning characteristics are not yet mature and that the inverse relationship between preferred spatial and temporal frequency of single cells seen in the adult cortex is not present in young animals. Thus, in cats of this age, clustering of spatial frequency is present, but cells preferring either low or high spatial frequencies have not yet fully developed their distinct temporal tuning characteristics.

The realization that the functional properties of neurons in the blob and interblob regions of cat visual cortex might be based on the LGN cell types that target these domains prompted a search for additional functional differences between neurons in the cytochrome oxidase compartments. After visualizing the blob and interblob regions by means of their spatial frequency preferences with optical imaging, single unit recordings were used to characterize neurons in both compartments in detail. In addition to the differences expected on the basis of optical imaging data, it was found that cells in the blobs had higher contrast sensitivity than cell in the interblobs (114). Again, this matches the dichotomy observed between X- and Y-cells in the LGN, because when stimulated with their respective optimum spatial frequencies, Y-cells have a higher contrast sensitivity than X-cells (111,115; but see 116). So far, attempts to visualize a contrast map with optical imaging have not been successful (Schulze, personal communication). One reason is that in order to activate both compartments differentially, very low contrasts (around 10%) must be used. Because the visual cortex is generally not activated strongly at such low contrasts, the signal-to-noise ratio of functional maps obtained with optical imaging tends to be rather low.

RELATIONSHIPS BETWEEN COLUMNAR SYSTEMS

All of the maps described in the preceding sections coexist in the same piece of cortex. It becomes important therefore to address how the different maps are geometrically arranged with respect to each other. The first suggestion that there might be some specificity in map relationships came from Hubel and Wiesel

(117) who, mainly based on their electrophysiological and anatomical work in monkeys, suggested the so-called ice-cube model for the functional architecture of the primary visual cortex. According to this model, orientation and ocular dominance columns are arranged as parallel slabs that cross each other at right angles. The authors themselves noted that the suggestion of perpendicular crossings between the two types of columns was highly speculative, because at the time, the techniques were not available to map multiple cortical representations with sufficient resolution. Optical imaging some decades later finally offered the possibility to visualize different maps in the same cortical region simultaneously. For cat primary visual cortex, the geometric relationships among three different types of columns were analyzed by Hübener and colleagues (37; see also 104).

Figure 2-9 shows the overlay of an ocular dominance and an orientation map from cat area 17 obtained with optical imaging. It is immediately apparent that this image does not much resemble the ice-cube model, the main difference being the radial, rather than the originally suggested slablike organization of orientation columns (the different pattern of the ocular dominance columns is due mostly to species differences; ocular dominance columns in monkeys are almost as regular as those in the ice-cube model). Although the overall organization of orientation preference is markedly different from the ice-cube model, close inspection of the image reveals that one prediction of the model is fulfilled remarkably well: There

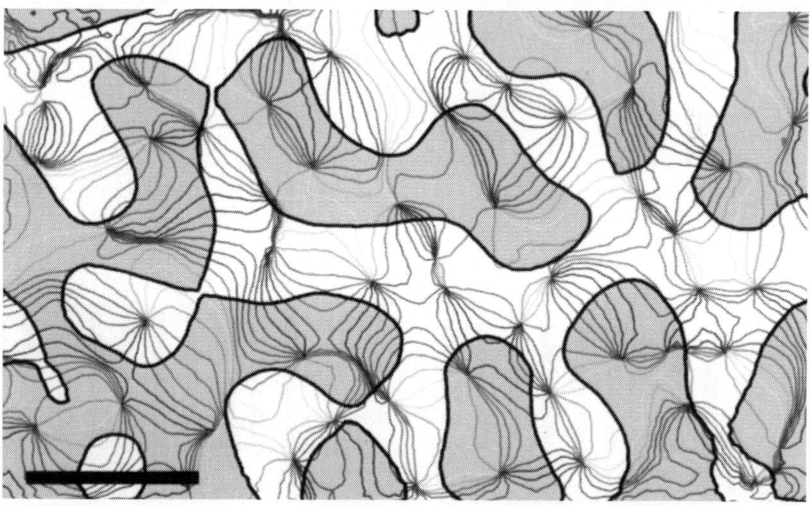

FIGURE 2-9. Overlay of an orientation and an ocular dominance map form cat area 17. The colored lines are iso-orientation lines, with each point on a line of a given color preferring the same orientation. The contours of the ocular dominance columns were obtained from the ocular dominance map of the same cortical region; gray denotes contralateral eye dominance. Many iso-orientation lines cross the borders between ocular dominance columns close to right angles, and the pinwheel centers are preferentially located in the middle of the ocular dominance columns. (Scale bar: 1 mm.) (From Hübener et al. [37].) See color insert for color reproduction of this figure.

are many instances of perpendicular crossings between orientation and ocular dominance columns. Quantitative evaluation of these intersection angles confirms the visual impression (Fig. 2-10A,B). Moreover, the overlay in Fig. 2-9 also shows that orientation centers are not randomly distributed on the ocular dominance map: rather, they are preferentially located in the middle of ocular dominance columns. This was also reported by Crair et al. (84), who additionally pointed out that orientation centers were not only lying on the center lines of ocular dominance bands but were most often located near ocular dominance peaks, i.e. regions with strong monocular responses. In cats made strabismic early in development, where the segregation into alternating ocular dominance columns is

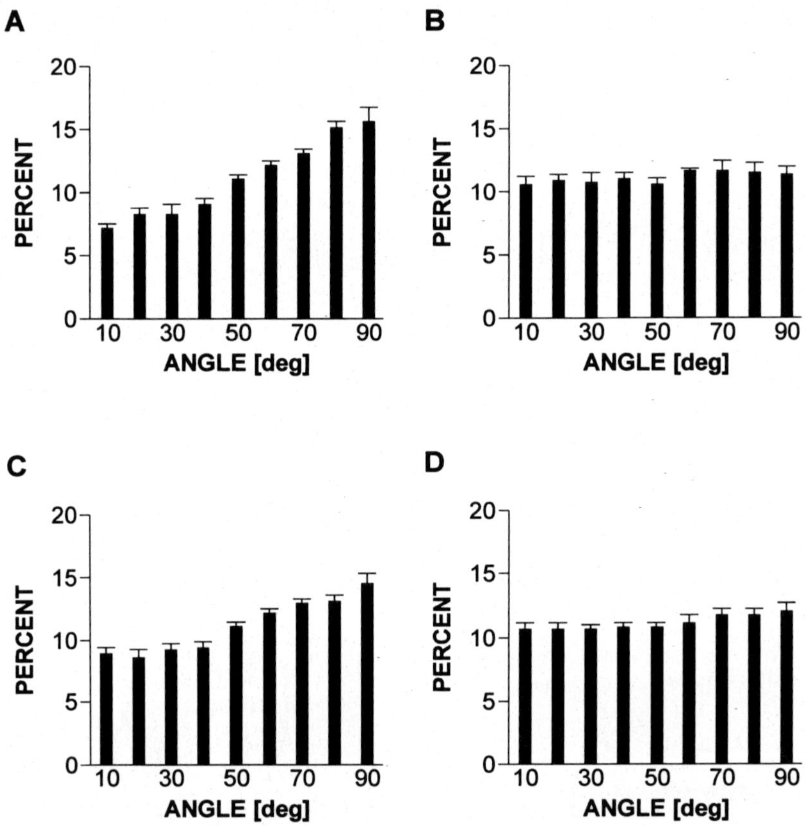

FIGURE 2-10. Quantitative analysis of intersection angles between columnar systems. **(A)** Averaged distribution (from 6 maps, error bars are given as SEM) of intersection angles between iso-orientation lines and ocular dominance borders. The histogram reveals a clear predominance of large intersection angles. **(B)** As a control the orientation preference map from one cat was overlaid with the ocular dominance map from a different cat. The averaged distribution from six such controls is essentially flat. **(C,D)** Similar analysis for the intersection angles between iso-orientation lines and borders between spatial frequency domains (n = 13 maps).

much more pronounced, similar geometric relationships between orientation and ocular dominance maps are present (44). Thus, these specific relationships are robust and are not affected by decorrelating neuronal activity between both eyes.

The geometrical relationship between the orientation and spatial frequency maps is again such that there are more perpendicular intersections than would be expected by chance (Fig. 2-10C,D). Again there is a tendency for orientation centers to lie near the centers of either high or low spatial frequency domains. Yet, both effects are clearly less pronounced.

Finally, it is also interesting to consider a systematic relationship between ocular dominance columns and spatial frequency domains. Because the spatial frequency map coincides with the cytochrome oxidase staining pattern (see page 151), this question is equivalent to asking whether the blobs are aligned on the ocular dominance columns. The latter problem has been addressed in anatomical studies of cat visual cortex, with conflicting results. Dyck and Cynader (106) did not detect any particular relationship between both structures, but Murphy et al. (107) did report a functional relationship. The results from optical imaging confirm the latter finding, as we found a weak tendency for the low spatial frequency domains and (thus the blobs) to avoid the border regions of ocular dominance columns (37). Clearly, however, the association between both systems is a loose one.

What is the possible functional significance of these specific relationships for visual information processing? It is easy to see that orthogonal relationships between columns have the advantage of allowing all combinations of stimulus properties to be represented in the smallest possible cortical area. Such an arrangement of columns in the visual cortex would help ensure that each point in the visual field is analyzed with respect to all parameter combinations or, formally, that coverage of the stimulus space is maximized.

Swindale (118) formalized the problem of cortical coverage and devised a quantitative measure for coverage uniformity. Using this measure it can be tested whether maps for orientation, ocular dominance, and spatial frequency are arranged to maximize cortical coverage (119). To this end, coverage uniformity was first calculated for the three maps in their original positions. Next, to assess the effect of disturbing the relationships between maps, one map was shifted by increasing distances relative to the others, and coverage uniformity was determined for each offset. Figure 2-11 clearly shows that shifting the maps deteriorates coverage uniformity. Obviously, this does not imply that the arrangement of different maps is solely governed by the need to optimize coverage. But this finding indicates that coverage uniformity is likely to be one of the underlying principles of the functional architecture in cat visual cortex.

The direction preference map is also spatially related to other functional maps in cat visual cortex (43,46,104). In particular, the observation that in many instances individual orientation domains are split into two subregions with preferences for opposite directions (43,46) (see Fig. 1-35 this volume) is compatible with the view that the direction preference map is arranged such that cortical coverage is optimized.

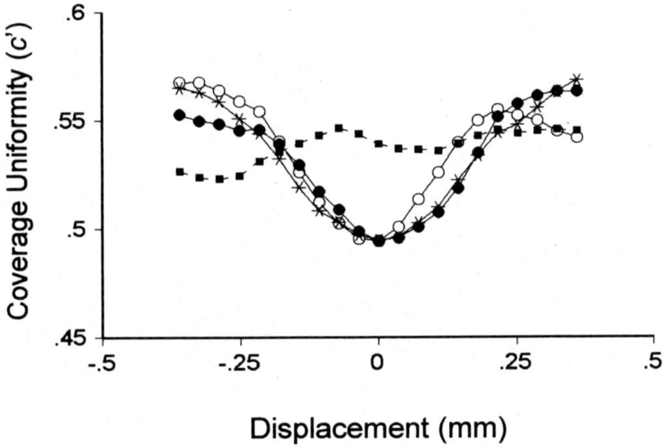

FIGURE 2-11. Effect of relative displacements between maps on coverage uniformity. Coverage uniformity *(c′)* was computed as described in Swindale (118), with small values of *c′* meaning that coverage is good. For each map, coverage always gets worse when it is shifted relative to the two other maps (closed circles: displacement of the ocular dominance map; open circles: displacement of the spatial frequency map; stars: displacement of the orientation map). As a control, one of the maps was first rotated by 180° and then displaced (squares) (From Swindale et al. [119].)

For the sake of simplicity, the computations of cortical coverage carried out by Swindale et al. (119) were performed under the assumption of a uniform retinotopic map. This assumption has been confirmed with optical imaging in tree shrews (120), but maps of retinotopy have not yet been imaged in cat visual cortex. Das and Gilbert (121) used closely spaced electrical recordings to assess the fine structure of the retinotopic map in cats and its relation to the orientation map, which had been visualized with optical imaging in these experiments. They concluded that the map of visual space contained local discontinuities, which were systematically related to the orientation map, such that large jumps in receptive field position coincided with fast changes in the preferred orientation (e.g., at orientation centers). Although this observation seems to be at odds with coverage optimization as an organization principle for the arrangement of maps in the visual cortex, the actual quantitative effects of such discontinuities in the retinotopic map on coverage uniformity have yet to be tested. It would be important therefore to image maps of visual space in cat visual cortex and to analyze these with the methods described previously.

COMPARISON WITH THE FUNCTIONAL ARCHITECTURE IN OTHER SPECIES

Optical imaging has been used to visualize the functional architecture of the primary visual cortex in a number of other mammalian species. In general, these experiments revealed that many of the characteristic features of the functional organization described in the preceding sections for the cat are also present in the visual cortex of other species. Nonetheless, there are also some interesting deviations from this common scheme.

With some variations a predominantly radial organization of orientation domains has been observed in all other species investigated to date. However, the density of orientation centers can vary quite considerably among different species. Whereas average values between 2.1 and 2.6 per mm^2 have been reported for cat area 17 (38,42–44), the density in area 17 of ferrets is notably higher (5.4–6.9/mm^2 (38,40)). Still higher values have been found in the striate cortex of various primates (macaque: around 8/mm^2 (122,123); squirrel monkey: 11/mm^2 (123); galago: around 7/mm^2 (124)). Area 17 of tree shrews also contains a very high density of orientation centers (9.4/mm^2; Bosking and Fitzpatrick, personal communication). In the latter species, optical imaging also revealed that the structure of the orientation map changes systematically across the cortical surface. Whereas the central portion of the exposed dorsal part of the striate cortex shows a patchy pattern of orientation domains, other regions, such as those close to the border between area 17 and area 18, contain a predominantly parallel arrangement of elongated orientation domains, with only relatively few orientation centers (125).

A pronounced variation between species is also present with respect to the relative representation of the different orientations in the visual cortex. As described previously, in cat visual cortex each orientation occupies similar fractions of the visual cortex (30,40,60), and similar results have been obtained in macaque monkeys (126). In contrast, optical imaging revealed a marked overrepresentation of horizontal and vertical orientations in ferret visual cortex (40,127,128). Although the developmental mechanisms that result in the formation of cortical orientation maps are unclear, one might speculate that the imbalance in the representation of different orientations in ferret visual cortex reflects the distribution of ganglion cells in its retina. In contrast to the cat, the ferret retina has a relatively well-developed visual streak (i.e., a horizontal stripe of an increased density of retinal ganglion cells (129)).

As in cats, optical imaging of ocular dominance columns in primates basically confirmed previous observations obtained with anterograde tracing or 2-DG mapping. Unlike these species, the ferret has a rather complex ocular dominance map, which has been studied in detail by White et al. (130). Whereas ocular dominance domains in the more peripheral parts of the binocular region are of the familiar patch or stripe type, the central region of area 17 often consists of a single elongated region of contralateral eye dominance. This region contains an ipsilateral visual field representation and borders in a complex fashion onto area 18, which in turn is entirely dominated by the ipsilateral eye.

So far maps of direction preference have been imaged in the primary visual cortex of one other species, the ferret (131). The structure of the direction preference map in the ferret is similar to the one in cat visual cortex, and it also contains lines of reversal of the preferred direction, which often end in orientation centers.

The ferret visual cortex also contains a spatial frequency map, which can be readily visualized with optical imaging (Schwartz and Bonhoeffer, unpublished observations). In some ferrets these maps are mostly patchy and thus resemble spatial frequency maps in the cat; in other cases, individual spatial frequency domains were rather large and irregularly shaped. While cytochrome oxidase blobs are present in ferret visual cortex (132), it has not yet been tested whether the blobs coincide with low spatial frequency domains, as is the case in cats. So far optical imaging has not been used to visualize spatial frequency domains in primate visual cortex. However, electrophysiological and 2-DG studies in macaque striate cortex have shown that cells in the cytochrome oxidase blobs are tuned to lower spatial frequencies than cells in the interblob region (133–136). The association between cytochrome oxidase-rich blobs and low spatial frequency preference is not a universal rule, though in the galago, blob cells are tuned to higher spatial frequencies than interblob cells (137), and in owl monkeys neurons in both compartments do not differ in their spatial frequency tuning (138).

The issue of spatial relationships between different columnar systems in the visual cortex has been analyzed in detail in macaque striate cortex (122,126,139). As in the cat, orientation and ocular dominance columns often cross each other at right angles and orientation centers are usually centered on the ocular dominance columns. Thus, in both species orientation and ocular dominance maps are arranged to optimize cortical coverage. In contrast, in ferret visual cortex orientation centers do not seem to be associated with the central regions of ocular dominance columns (140), and only a weak orthogonal relationship was observed between both columnar systems (Fitzpatrick, personal communication). Thus, the specific geometric relationships found in macaque and cat may not be present in the ferret. It remains to be seen whether the lack of apparent geometric relationships between orientation and ocular dominance maps in this species indicates that coverage is not optimized in ferret visual cortex, or whether other mechanisms ensure optimal coverage of the stimulus space.

CONCLUDING REMARKS

One of the great strengths of optical imaging is that it allows the visualization of different aspects of the functional architecture in the same region of the visual cortex. How do the insights gained with this new technique relate to the concept of the modular organization of the visual cortex? According to this concept, the visual cortex is composed of many identical elementary processing units, with each module containing the neuronal circuitry for the analysis of a small patch of the visual world with respect to all possible stimulus combinations. Although optical imaging has shown that each map in the cortex exhibits a certain amount of regularity, it is hard to find evidence for modules in a strict sense, which would form the building blocks of the visual cortex. It seems as if the visual cortex should rather be considered as containing mosaics of functional domains for the different properties that are arranged in a nonrandom manner. Optical imaging has therefore helped to further elucidate the fundamental question of how information about the visual world is represented in the primary visual cortex.

ACKNOWLEDGMENTS

We thank David Fitzpatrick and Frank Sengpiel for critically reading the manuscript. Work in the authors' laboratory is supported by a grant from the EC Biotech Program and the Max-Planck Gesellschaft.

REFERENCES

1. Mountcastle, V. B. (1957). Modality and topographic properties of single neurons of cat's somatic sensory cortex. *J. Neurophysiol.* **20,** 408–434.
2. Merzenich, M. M., Knight, P. L., and Roth GL (1975). Representation of cochlea within primary auditory cortex in the cat. *J. Neurophysiol.* **38,** 231–249.
3. Georgopoulos, A. P., Caminiti, R., and Kalaska, J. F. (1984). Static spatial effects in motor cortex and area 5: quantitative relations in a two-dimensional space. *Exp. Brain Res.* **54,** 446–454.
4. Albright, T. D., Desimone, R., and Gross, C. G. (1984). Columnar organization of directionally selective cells in visual area MT of the macaque. *J. Neurophysiol.* **51,** 16–31.
5. Fujita, I., Tanaka, K., Ito, M., and Cheng, K. (1992). Columns for visual features of objects in monkey inferotemporal cortex. *Nature* **360,** 343–346.
6. Malonek, D., Tootell, R. B. H., and Grinvald, A. (1994). Optical imaging reveals the functional architecture of neurons processing shape and motion in owl monkey area MT. *Proc. R. Soc. Lond. [Biol.]* **258,** 109–119.
7. Wang, G., Tanaka, K., and Tanifuji, M. (1996). Optical imaging of functional organization in the monkey inferotemporal cortex. *Science* **272,** 1665–1668.
8. Hubel, D. H. and Wiesel, T. N. (1962). Receptive fields, binocular interaction and functional architecture in the cat's visual cortex. *J. Physiol.* **160,** 106–154.
9. Hubel, D. H. and Wiesel, T. N. (1963). Shape and arrangement of columns in cat's striate cortex. *J. Physiol.* **165,** 559–568.
10. Hubel, D. H. and Wiesel, T. N. (1974). Sequence regularity and geometry of orientation columns in the monkey striate cortex. *J. Comp. Neurol.* **158,** 267–293.
11. Albus, K. (1975). A quantitative study of the projection area of the central and paracentral visual field in area 17 of the cat. II. The spatial organization of the orientation domain. *Exp. Brain Res.* **24,** 181–202.
12. Albus, K. (1985). A microelectrode study of the spatial arrangement of iso- orientation bands in the cat's striate cortex. In: Models of the visual cortex (D. Rose, and V. G. Dobson, Eds.), pp. 485–491. Chichester, Wiley.
13. Sokoloff, L., Reivich, M., Kennedy, C., Des Rosiers, M. H., Patlak, C. S., Pettigrew, K. D., Sakurada, O., and Shinohara, M. (1977). The [^{14}C] deoxyglucose method for the measurement of local cerebral glucose utilization: theory, procedure, and normal values in the conscious and anesthetized albino rat. *J. Neurochem.* **28,** 897–916.
14. Blasdel, G. G., and Salama, G. (1986). Voltage-sensitive dyes reveal a modular organization in monkey striate cortex. *Nature* **321,** 579–585.
15. Grinvald, A., Lieke, E. E., Frostig, R. D., Gilbert, C. D., and Wiesel, T. N. (1986). Functional architecture of cortex revealed by optical imaging of intrinsic signals. *Nature* **324,** 361–364.

16. Bonhoeffer, T., and Grinvald, A. (1996). Optical imaging based on intrinsic signals: The Methodology. In: Brain mapping: the methods (A. W. Toga, and J. C. Mazziotta, Eds.), pp. 55–97. San Diego, Academic Press.
17. Malonek, D., and Grinvald, A. (1996). Interactions between electrical activity and cortical microcirculation revealed by imaging spectroscopy: Implications for functional brain mapping. *Science* **272,** 551–554.
18. Menon, R. S., Ogawa, S., Hu, X., Strupp, J. P., Anderson, P., and Ugurbil, K. (1995). BOLD based functional MRI at 4 Tesla includes a capillary bed contribution: echo-planar imaging correlates with previous optical imaging using intrinsic signals. *Magn. Reson. Med.* **33,** 453–459.
19. Kim, D. S., Duong, T. Q., and Kim, S. G. (2000). High-resolution mapping of iso-orientation columns by fMRI. *Nat. Neurosci.* **3,** 164–169.
20. Logothetis, N. K., Guggenberger, H., Peled, S., and Pauls, J. (1999). Functional imaging of the monkey brain. *Nat. Neurosci.* **2,** 555–562.
21. Vanzetta, I., and Grinvald, A. (1999). Increased cortical oxidative metabolism due to sensory stimulation: implications for functional brain imaging. *Science* **286,** 1555–1558.
22. Fox, P. T., and Raichle, M. E. (1984). Stimulus rate dependence of regional cerebral blood flow in human striate cortex, demonstrated by positron emission tomography. *J. Neurophysiol.* **51,** 1109–1120.
23. Fox, P. T., Mintun, M. A., Raichle, M. E., Miezin, F. M., Allman, J. M., and Van Essen, D. C. (1986). Mapping human visual cortex with positron emission tomography. *Nature* **323,** 806–809.
24. Frostig, R. D., Lieke, E. E., Ts'o, D. Y., and Grinvald, A. (1990). Cortical functional architecture and local coupling between neuronal activity and the microcirculation revealed by *in vivo* high-resolution optical imaging of intrinsic signals. *Proc. Natl. Acad. Sci. U.S.A.* **87,** 6082–6086.
25. Cohen, L. B. (1973). Changes in neuron structure during action potential propagation and synaptic transmission. *Physiol. Rev.* **53,** 373–418.
26. Grinvald, A., Manker, A., and Segal, M. M. (1982). Visualization of the spread of electrical activity in rat hippocampal slices by voltage sensitive optical probes. *J. Physiol.* **333,** 269–291.
27. MacVicar, B. A., and Hochman, D. (1991). Imaging of synaptically evoked intrinsic optical signals in hippocampal slices. *J. Neurosci.* **11,** 1458–1469.
28. Dodt, H.-U., and Zieglgänsberger, W. (1994). Infrared videomicroscopy: a new look at neuronal structure and function. *Trends Neurosci.* **17,** 453–458.
29. Ratzlaff, E. H., and Grinvald, A. (1991). A tandem-lens epifluorescence macroscope: hundred-fold brightness advantage for wide-field imaging. *J. Neurosci. Methods* **36,** 127–137.
30. Bonhoeffer, T., and Grinvald, A. (1993). The layout of iso-orientation domains in area 18 of cat visual cortex. Optical imaging reveals a pinwheel-like organization. *J. Neurosci.* **13,** 4157–4180.
31. Mayhew, J. E. W., Askew, S., Zheng, Y., Porrill, J., Westby, G. W. M., Redgrave, P., Rector, D. M., and Harper, R. M. (1996). Cerebral vasomotion: a 0.1-Hz oscillation in reflected light imaging of neural activity. *Neuroimage* **4,** 183–193.
32. Albus, K. (1979). ^{14}C-deoxyglucose mapping of orientation subunits in the cats visual cortical areas. *Exp. Brain Res.* **37,** 609–613.
33. Singer, W. (1981). Topographic organization of orientation columns in the cat visual cortex. A deoxyglucose study. *Exp. Brain Res.* **44,** 431–436.
34. Albus, K., and Sieber, B. (1984). On the spatial arrangement of iso-orientation bands in the cat's visual cortical areas 17 and 18: A ^{14}C-deoxyglucose study. *Exp. Brain Res.* **56,** 384–388.
35. Löwel, S., Freeman, B., and Singer, W. (1987). Topographic organization of the orientation column system in large flat-mounts of the cat visual cortex. A 2-deoxyglucose study. *J. Comp. Neurol.* **255,** 401–415.
36. Bonhoeffer, T., and Grinvald, A. (1991). Iso-orientation domains in cat visual cortex are arranged in pinwheel-like patterns. *Nature* **353,** 429–431.
37. Hübener, M., Shoham, D., Grinvald, A., and Bonhoeffer, T. (1997). Spatial relationships among three columnar systems in cat area 17. *J. Neurosci.* **17,** 9270–9284.
38. Rao, S. C., Toth, L. J., and Sur, M. (1997). Optically imaged maps of orientation preference in primary visual cortex of cats and ferrets. *J. Comp. Neurol.* **387,** 358–370.

39. Shmuel, A., and Grinvald, A. (2000). Coexistence of linear zones and pinwheels within orientation maps in cat visual cortex. *Proc. Natl. Acad. Sci. U.S.A.* **97,** 5568–5573.
40. Müller, T., Stetter, M., Hübener, M., Sengpiel, F., Bonhoeffer, T., Gödecke, I., Chapman, B., Löwel, S., and Obermayer, K. (2000). An analysis of orientation and ocular dominance patterns in the visual cortex of cats and ferrets. *Neural Comput.,* **12,** 2573–2595.
41. Ohki, K., Matsuda, Y., Ajima, A., Kim, D. S., and Tanaka, S. (2000). Arrangement of orientation pinwheel centers around area 17/18 transition zone in cat visual cortex. *Cerebral Cortex* **10,** 593–601.
42. Bonhoeffer, T., Kim, D.-S., Malonek, D., Shoham, D., and Grinvald, A. (1995). Optical imaging of the layout of functional domains in Area 17 and across the Area 17/18 border in cat visual cortex. *Eur. J. Neurosci.* **7,** 1973–1988.
43. Shmuel, A., and Grinvald, A. (1996). Functional organization for direction of motion and its relationship to orientation maps in cat area 18. *J. Neurosci.* **16,** 6945–6964.
44. Löwel, S., Schmidt, K. E., Kim, D. S., Wolf, F., Hoffsümmer, F., Singer, W., and Bonhoeffer, T. (1998). The layout of orientation and ocular dominance domains in area 17 of strabismic cats. *Eur. J. Neurosci.* **10,** 2629–2643.
45. Horton, J. C., and Hocking, D. R. (1996). Intrinsic variability of ocular dominance column periodicity in normal macaque monkeys. *J. Neurosci.* **16,** 7228–7239.
46. Swindale, N. V., Matsubara, J. A., and Cynader, M. S. (1987). Surface organization of orientation and direction selectivity in cat area 18. *J. Neurosci.* **7,** 1414–1427.
47. von Seelen, W. (1970). Zur Informationsverarbeitung im visuellen System der Wirbeltiere. *Kybernetik* **7,** 89–106.
48. Braitenberg, V., and Braitenberg, C. (1979). Geometry of orientation columns in the visual cortex. *Biol. Cybern.* **33,** 179–186.
49. Swindale, N. V. (1982). A model for the formation of orientation columns. *Proc. R. Soc. Lond. [Biol.]* **215,** 211–230.
50. Obermayer, K., Ritter, H., and Schulten, K. (1990). A principle for the formation of the spatial structure of cortical feature maps. *Proc. Natl. Acad. Sci. U.S.A.* **90,** 8245–8349.
51. Maldonado, P. E., Gödecke, I., Gray, C. M., and Bonhoeffer, T. (1997). Orientation selectivity in pinwheel centers in cat striate cortex. *Science* **276,** 1551–1555.
52. Wilson, M. A., and McNaughton, B. L. (1993). Dynamics of the hippocampal ensemble code for space. *Science* **261,** 1055–1558.
53. Gray, C. M., Maldonado, P., Wilson, M., and McNaughton, B. L. (1995). Tetrodes markedly improve the reliability and yield of multiple single unit isolation from multiunit recordings in cat striate cortex. *J. Neurosci. Methods* **63,** 43–54.
54. Hetherington, P. A., and Swindale, N. V. (1999). Receptive field and orientation scatter studied by tetrode recordings in cat area 17. *Vis. Neurosci.* **16,** 637–652.
55. Pettigrew, J. D., Nikara, T., and Bishop, P. O. (1968). Responses to moving slits by single units in cat striate cortex. *Exp. Brain Res.* **6,** 373–390.
56. Leventhal, A. G., and Hirsch, H. V. B. (1977). Effects of early experience upon orientation sensitivity and binocularity of neurons in visual cortex of cats. *Proc. Natl. Acad. Sci. U.S.A.* **74,** 1272–1276.
57. Payne, B. R., and Berman, N. E. J. (1983). Functional organization of neurons in cat striate cortex: variations in preferred orientation and orientation selectivity with receptive-field type, ocular dominance, and location in visual-field map. *J. Neurophysiol.* **49,** 1051–1072.
58. Campbell, F. W., Cleland, B. G., Cooper, G. F., and Enroth-Cugell, C. (1968). The angular selectivity of visual cortical cells to moving gratings. *J. Physiol.* **198,** 237–250.
59. Rose, D., and Blakemore, C. (1974). An analysis of orientation selectivity in the cat's visual cortex. *Exp. Brain Res.* **20,** 1–17.
60. Sengpiel, F., Stawinski, P., and Bonhoeffer, T. (1999). Influence of experience on orientation maps in cat visual cortex. *Nat. Neurosci.* **2,** 727–732.
61. Mansfield, R. J., and Ronner, S. F. (1978). Orientation anistropy in monkey visual cortex. *Brain Res.* **149,** 229–234.

62. Sheth, B. R., Sharma, J., Rao, S. C., and Sur, M. (1996). Orientation maps of subjective contours in visual cortex. *Science* **274,** 2110–2115.
63. Hubel, D. H., and Wiesel, T. N. (1959). Receptive fields of single neurones in the cat's striate cortex. *J. Physiol.* **148,** 574–591.
64. Hubel, D. H., and Wiesel, T. N. (1965). Binocular interaction in striate cortex of kittens reared with artificial squint. *J. Neurophysiol.* **28,** 1041–1059.
65. Albus, K. (1975). Predominance of monocularly driven cells in the projection area of the central visual field in the cat's striate cortex. *Brain Res.* **89,** 341–347.
66. Shatz, C. J., Lindström, S., and Wiesel, T. N. (1977). The distribution of afferents representing the right and left eyes in the cat's visual cortex. *Brain Res.* **131,** 103–116.
67. LeVay, S., Stryker, M. P., and Shatz, C. J. (1978). Ocular dominance columns and their development in layer IV fo the cat's visual cortex: a quantitative study. *J. Comp. Neurol.* **179,** 223–244.
68. Shatz, C. J., and Stryker, M. P. (1978). Ocular dominance in layer IV of the cat's visual cortex and the effects of monocular deprivation. *J. Physiol.* **281,** 267–283.
69. Löwel, S., and Singer, W. (1987). The pattern of ocular dominance columns in flat-mounts of the cat visual cortex. *Exp. Brain Res.* **68,** 661–666.
70. Anderson, P. A., Olavarria, J., and Van Sluyters, R. C. (1988). The overall pattern of ocular dominance bands in cat visual cortex. *J. Neurosci.* **8,** 2183–2200.
71. Löwel, S., Bischof, H.-J., Leutenecker, B., and Singer, W. (1988). Topographic relations between ocular dominance and orientation columns in the cat striate cortex. *Exp. Brain Res.* **71,** 33–46.
72. Hata, Y., and Stryker, M. P. (1994). Control of thalamocortical afferent rearrangement by postsynaptic activity in developing visual cortex. *Science* **265,** 1732–1735.
73. Löwel, S., and Singer, W. (1993). Monocularly induced 2-deoxyglucose patterns in the visual cortex and lateral geniculate nucleus of the cat. I. Anaesthetized and paralysed animals. *Eur. J. Neurosci.* **5,** 846–856.
74. Löwel, S. (1994). Ocular dominance column development: strabismus changes the spacing of adjacent columns in cat visual cortex. *J. Neurosci.* **14,** 7451–7468.
75. Löwel, S., and Singer, W. (1992). Selection of intrinsic horizontal connections in the visual cortex by correlated neuronal activity. *Science* **255,** 209–212.
76. Löwel, S., and Singer, W. (1993). Monocularly induced 2-deoxyglucose patterns in the visual cortex and lateral geniculate nucleus of the cat. II. Awake animals and strabismic animals. *Eur. J. Neurosci.* **5,** 857–869.
77. Hubel, D. H., and Wiesel, T. N. (1972). Laminar and columnar distribution of geniculo-cortical fibers in the macaque monkey. *J. Comp. Neurol.* **146,** 421–450.
78. Wiesel, T. N., Hubel, D. H., and Lam, D. M. K. (1974). Autoradiographic demonstration of ocular-dominance columns in the monkey striate cortex by means of transneuronal transport. *Brain Res.* **79,** 273–279.
79. LeVay, S., Hubel, D. H., and Wiesel, T. N. (1975). The pattern of ocular dominance columns in macaque visual cortex revealed by a reduced silver stain. *J. Comp. Neurol.* **159,** 559–576.
80. Ts'o, D. Y., Frostig, R. D., Lieke, E. E., and Grinvald, A. (1990). Functional organization of primate visual cortex revealed by high resolution optical imaging. *Science* **249,** 417–420.
81. Hubel, D. H., and Wiesel, T. N. (1968). Receptive fields and functional architecture of monkey striate cortex. *J. Physiol.* **195,** 215–243.
82. Gilbert, C. D. (1977). Laminar differences in receptive field properties of cells in cat primary visual cortex. *J. Physiol.* **268,** 391–421.
83. Berman, N., Payne, B. R., Labar, D. R., and Murphy, E. H. (1982). Functional organization of neurons in cat striate cortex: variations in ocular dominance and receptive-field type with cortical laminae and location in visual field. *J. Neurophysiol.* **48,** 1362–1377.
84. Crair, M. C., Ruthazer, E. S., Gillespie, D. C., and Stryker, M. P. (1997). Ocular dominance peaks at pinwheel center singularities of the orientation map in cat visual cortex. *J. Neurophysiol.* **77,** 3381–3385.
85. Crair, M. C., Gillespie, D. C., and Stryker, M. P. (1998). The role of visual experience in the development of columns in cat visual cortex. *Science* **279,** 566–570.

86. Haug, H., Kolln, M., and Rast, A. (1976). The postnatal development of myelinated nerve fibres in the visual cortex of the cat: a stereological and electron microscopical investigation. *Cell Tissue Res.* **167,** 265–288.
87. Jones, D. G., Sluyters, V. R. C., and Murphy, K. M. (1991). A computational model for the overall pattern of ocular dominance. *J. Neurosci.* **11,** 3794–3808.
88. Hubel, D. H., and Wiesel, T. N. (1965). Receptive fields and functional architecture in two nonstriate visual areas (18 and 19) of the cat. *J. Neurophysiol.* **28,** 229–289.
89. Payne, B. R., Berman, N. E. J., and Murphy, E. H. (1981). Organization of direction preferences in cat visual cortex. *Brain Res.* **211,** 445–450.
90. Tolhurst, D. J., Dean, A. F., and Thompson, I. D. (1981). Preferred direction of movement as an element in the organization of cat visual cortex. *Exp. Brain Res.* **44,** 340–342.
91. Berman, N. E. J., Wilkes, M. E., and Payne, B. R. (1987). Organization of orientation and direction selectivity in areas 17 and 18 of cat cerebral cortex. *J. Neurophysiol.* **58,** 676–699.
92. DeAngelis, G. C., Ghose, G. M., Ohzawa, I., and Freeman, R. D. (1999). Functional micro-organization of primary visual cortex: receptive field analysis of nearby neurons. *J. Neurosci.* **19,** 4046–4064.
93. Campbell, F. W., Cooper, G. F., and Enroth-Cygell, C. (1969). The spatial selectivity of the visual cells of the cat. *J. Physiol.* **203,** 223–235.
94. Maffei, L., and Fiorentini, A. (1973). The visual cortex as a spatial frequency analyser. *Vis. Res.* **13,** 1255–1267.
95. Movshon, J. A., Thompson, I. D., and Tolhurst, D. J. (1978). Spatial and temporal contrast sensitivity of neurones in areas 17 and 18 of the cat's visual cortex. *J. Physiol.* **283,** 101–120.
96. Tolhurst, D. J., and Thompson, I. D. (1981). On the variety of spatial frequency selectivities shown by neurons in area 17 of the cat. *Proc. R. Soc. Lond. [Biol.]* **213,** 183–199.
97. Maffei, L., and Fiorentini, A. (1977). Spatial frequency rows in the striate visual cortex. *Vis. Res.* **17,** 257–264.
98. Berardi, N., Bisti, S., Cattaneo, A., Fiorentini, A., and Maffei, L. (1982). Correlation between the preferred orientation and spatial frequency of neurones in visual areas 17 and 18 of the cat. *J. Physiol.* **323,** 603–618.
99. Tolhurst, D. J., and Thompson, I. D. (1982). Organization of neurones preferring similar spatial frequencies in cat striate cortex. *Exp. Brain Res.* **48,** 217–227.
100. Tootell, R. B. H., Silverman, M. S., and De Valois, R. L. (1981). Spatial frequency columns in primary visual cortex. *Science* **214,** 813–815.
101. Shoham, D., Hübener, M., Schulze, S., Grinvald, A., and Bonhoeffer, T. (1997). Spatio-temporal frequency domains and their relation to cytochrome oxidase staining in cat visual cortex. *Nature* **385,** 529–533.
102. Everson, R. M., Prashanth, A. K., Gabbay, M., Knight, B. W., Sirovich, L., and Kaplan, E. (1998). Representation of spatial frequency and orientation in the visual cortex. *Proc. Natl. Acad. Sci. U.S.A.* **95,** 8334–8338.
103. Issa, N. P., Trepel, C., and Stryker, M. P. (2000). Spatial frequency maps in cat visual cortex. *J. Neurosci.* **20,** 8504–8514.
104. Kim, D. S., Matsuda, Y., Ohki, K., Ajima, A., and Tanaka, S. (1999). Geometrical and topological relationships between multiple functional maps in cat primary visual cortex. *NeuroReport* **10,** 2515–2522.
105. Murphy, K. M., Van Sluyters, R. C., and Jones, D. G. (1990). Cytochrome-oxidase activity in cat visual cortex: is it periodic? *Soc. Neurosci. Abstr.* **16,** 292.
106. Dyck, R. H., and Cynader, M. S. (1993). An interdigitated columnar mosaic of cytochrome oxidase, zinc, and neurotransmitter-related molecules in cat and monkey visual cortex. *Proc. Natl. Acad. Sci. U.S.A.* **90,** 9066–9069.
107. Murphy, K. M., Jones, D. G., and Van Sluyters, R. C. (1995). Cytochrome-oxidase blobs in cat primary visual cortex. *J. Neurosci.* **15,** 4196–4208.
108. Casagrande, V. A. (1994). A third parallel visual pathway to primate area V1. *Trends Neurosci.* **17,** 305–309.

109. Hübener, M., Schulze, S., and Bonhoeffer, T. (1996). Cytochrome-oxidase blobs in cat visual cortex coincide with low spatial frequency columns. *Soc. Neurosci. Abstr.* **22,** 951.
110. Boyd, J. D., and Matsubara, J. A. (1996). Laminar and columnar patterns of geniculocortical projections in the cat: relationship to cytochrome oxidase. *J. Comp. Neurol.* **365,** 659–682.
111. Derrington, A. M., and Fuchs, A. F. (1979). Spatial and temporal properties of X and Y cells in the cat lateral geniculate nucleus. *J. Physiol.* **293,** 347–364.
112. Derrington, A. M., and Fuchs, A. F. (1981). The development of spatial-frequency selectivity in kitten striate cortex. *J. Physiol.* **316,** 1–10.
113. DeAngelis, G. C., Ohzawa, I., and Freeman, R. D. (1993). The spatiotemporal organization of simple cell receptive fields in the cat's striate cortex. I. General characteristics and postnatal development. *J. Neurophysiol.* **69,** 1091–1117.
114. Schulze, S., Bonhoeffer, T., and Hübener, M. (1999). Response properties of neurons in the cytochrome oxidase compartments of cat visual cortex. *Soc. Neurosci. Abstr.* **25,** 1932.
115. Lehmkuhle, S., Kratz, K. E., Mangel, S. C., and Sherman, S. M. (1980). Spatial and temporal sensitivity of X- and Y-cells in the dorsal lateral geniculate nucleus of the cat. *J. Neurophysiol.* **43,** 520–541.
116. Hartveit, E., and Heggelund, P. (1992). The effect of contrast on the visual response of lagged and nonlagged cells in the cat lateral geniculate nucleus. *Vis. Neurosci.* **9,** 515–525.
117. Hubel, D. H., and Wiesel, T. N. (1977). Functional architecture of macaque monkey visual cortex (Ferrier lecture). *Proc. R. Soc. Lond. [Biol.]* **198,** 1–59.
118. Swindale, N. V. (1991). Coverage and the design of striate cortex. *Biol. Cybern.* **65,** 415–424.
119. Swindale, N. V., Shoham, D., Grinvald, A., Bonhoeffer, T., and Hübener, M. (2000). Visual cortex maps are optimized for uniform coverage. *Nat. Neurosci.* **3,** 822–826.
120. Bosking, W. H., Crowley, J. C., and Fitzpatrick, D. (1997). Fine structure of the map of visual space in tree shrew striate cortex revealed by optical imaging. *Soc. Neurosci. Abstr.* **23,** 1945.
121. Das, A., and Gilbert, C. D. (1997). Distortions of visuotopic map match orientation singularities in primary visual cortex. *Nature* **387,** 594–598.
122. Obermayer, K., and Blasdel, G. G. (1993). Geometry of orientation and ocular dominance columns in monkey striate cortex. *J. Neurosci.* **13,** 4114–4129.
123. Obermayer, K., and Blasdel, G. G. (1997). Singularities in primate orientation maps. *Neural Comput.* **9,** 555–575.
124. Bosking, W. H., White, L. E., Casagrande, V. A., and Fitzpatrick, D. (1996). Functional organization of visual areas V1 and V2 in the prosimian primate galago revealed by optical imging. *Soc. Neurosci. Abstr.* **22,** 1610.
125. Bosking, W. H., Zhang, Y., Schofield, B., and Fitzpatrick, D. (1997). Orientation selectivity and the arrangement of horizontal connections in tree shrew striate cortex. *J. Neurosci.* **17,** 2112–2127.
126. Bartfeld, E., and Grinvald, A. (1992). Relationships between orientation preference pinwheels, cytochrome oxidase blobs and ocular dominance columns in primate striate cortex. *Proc. Natl. Acad. Sci. U.S.A.* **89,** 11905–11909.
127. Chapman, B., and Bonhoeffer, T. (1998). Overrepresentation of horizontal and vertical orientation preferences in developing ferret area 17. *Proc. Natl. Acad. Sci. U.S.A.* **95,** 2609–2614.
128. Coppola, D. M., Purves, H. R., McCoy, A. N., and Purves, D. (1998). The distribution of oriented contours in the real word. *Proc. Natl. Acad. Sci. U.S.A.* **95,** 4002–4006.
129. Vitek, D. J., Schall, J. D., and Leventhal, A. G. (1985). Morphology, central projections, and dendritic field orientation of retinal ganglion cells in the ferret. *J. Comp Neurol.* **241,** 1–11.
130. White, L. E., Bosking, W. H., Williams, S. M., and Fitzpatrick, D. (1999). Maps of central visual space in ferret V1 and V2 lack matching inputs from the two eyes. *J. Neurosci.* **19,** 7089–7099.
131. Weliky, M., Bosking, W. H., and Fitzpatrick, D. (1996). A systematic map of direction preference in primary visual cortex. *Nature* **379,** 725–728.
132. Cresho, H. S., Rasco, L. M., Rose, G. H., and Condo, G. J. (1992). Blob-like pattern of cytochrome oxidase staining in ferret visual cortex. *Soc. Neurosci. Abstr.* **18,** 298.

133. Tootell, R. B. H., Silverman, M. S., Hamilton, S. L., Switkes, E., and De Valois, R. L. (1988). Functional anatomy of macaque striate cortex. V. Spatial frequency. *J. Neurosci.* **8,** 1610–1624.
134. Silverman, M. S., Grosof, D. H., De Valois, R. L., and Elfar, S. D. (1989). Spatial-frequency organization in primate striate cortex. *Proc. Natl. Acad. Sci. U.S.A.* **86,** 711–715.
135. Born, R. T., and Tootell, R. B. H. (1991). Spatial frequency tuning of single units in macaque supragranular striate cortex. *Proc. Natl. Acad. Sci. U.S.A.* **88,** 7066–7070.
136. Edwards, D. P., Purpura, K. P., and Kaplan, E. (1995). Contrast sensitivity and spatial frequency response of primate cortical neurons in and around the cytochrome oxidase blobs. *Vis. Res.* **35,** 1501–1523.
137. Debruyn, E. J., Casagrande, V. A., Beck, P. D., and Bonds, A. B. (1993). Visual resolution and sensitivity of single cells in the primary visual cortex (V1) of a nocturnal primate (bush baby): correlations with cortical layers and cytochrome oxidase patterns. *J. Neurophysiol.* **69,** 3–18.
138. O'Keefe, L. P., Levitt, J. B., Kiper, D. C., Shapley, R. M., and Movshon, J. A. (1998). Functional organization of owl monkey lateral geniculate nucleus and visual cortex. *J. Neurophysiol.* **80,** 594–609.
139. Blasdel, G. G., Obermayer, K., and Kiorpes, L. (1995). Organization of ocular dominance and orientation columns in the striate cortex of neonatal macaque monkeys. *Visu. Neurosci.* **12,** 589–603.
140. Issa, N. P., Trachtenberg, J. T., Chapman, B., Zahs, K. R., and Stryker, M. P. (1999). The critical period for ocular dominance plasticity in the Ferret's visual cortex. *J. Neurosci.* **19,** 6965–6978.

3

2-Deoxyglucose Architecture of Cat Primary Visual Cortex

Siegrid Löwel

Leibniz-Institute for Neurobiology, Magdeburg, Germany

INTRODUCTION

More than 20 years ago, Sokoloff and co-workers were the first to use a radioactively labeled sugar analog, 2-deoxyglucose (2-DG), to visualize large-scale activity patterns in the monkey visual cortex (Fig. 3-1) (Kennedy et al., 1976). Compared to single-unit recordings, which were widely used in the 1960s and 1970s and which gave detailed but only very local information about neuronal response properties, this new technique provided information about activity patterns in large brain regions and therefore represented a major advancement in our understanding of the functional architecture of the mammalian brain.

The 2-DG technique exploits the fact that the brain (in contrast to most other tissues) almost exclusively uses sugar (glucose) as a source of energy and that cellular processes involved in brain activation require increased energy metabolism (for reviews see Hand, 1981; Sokoloff, 1985, 1992; Redies and Gjedde, 1989; Duncan and Stumpf, 1991). Therefore, assessment of brain glucose utilization can provide a measure of functional activity. When 2-DG (a slightly modified version of glucose) is administered intravenously, it crosses the blood-brain barrier and, like glucose, is taken up by neurons in an activity-dependent manner. Once taken up, it is phosphorylated but cannot be metabolized further. 2-DG-6-phosphate is trapped and thus accumulates in active cells, and its distribution in large brain regions can later be visualized by autoradiography (Sokoloff et al., 1977).

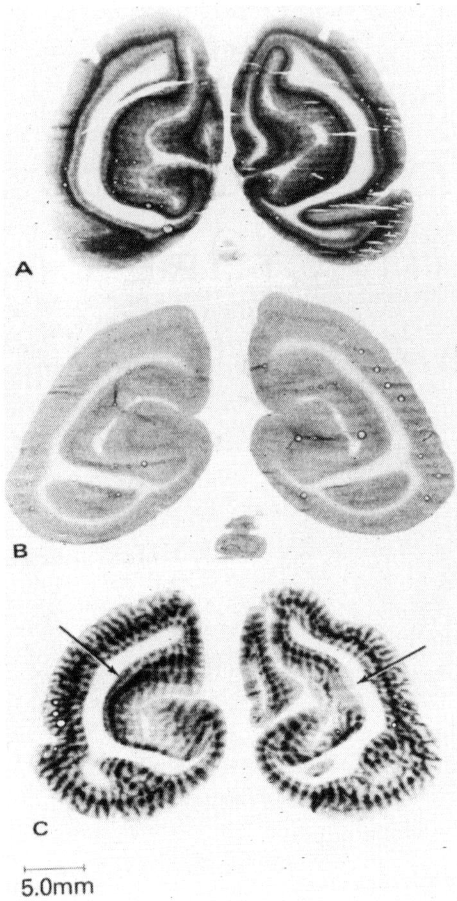

FIGURE 3-1. Visualization of ocular dominance columns in the rhesus monkey brain with 2-deoxyglucose (2-DG) autoradiography. Autoradiographs from coronal sections at the level of the primary visual cortex. **(A)** Animal with normal binocular vision. Note the laminar distribution of increased 2-DG uptake. **(B)** Animal with binocular visual deprivation. Note the almost uniform and reduced 2-DG labeling. **(C)** Animal with right eye occluded. Note the alternate dark and light striations corresponding to ocular dominance columns. These columns are most apparent in the dark band corresponding to layer IV, but extend throughout the entire thickness of the cortex. Arrows point to regions of bilateral asymmetry believed to be the loci of the cortical representations of the blind spots (densely labeled on the left, contralateral hemisphere and nearly label-free on the right, ipsilateral hemisphere). (Reproduced with permission from Kennedy C. et al. (1976). *Proc. Natl. Acad. Sci. U.S.A.* **73:** 4230–4234.

Inasmuch as local energy metabolism in the brain, as in other tissues, is closely coupled to local functional activity, increased 2-DG-6-phosphate accumulation reflects increased neuronal activity. Thus, the technique provides a "radioisotopic stain for functional activity" (Plum et al., 1976). Autoradiographic films exposed to 2-DG labeled brain sections for several weeks to months become black after photographic development where ever neurons have been active during the 2-DG experiment. The immense advantage of this technique, therefore, lies in the fact that the activity patterns of large brain regions (in fact the entire brain) can be mapped simultaneously using a relatively simple procedure that comprises a typically 45-minute period of tracer accumulation during the actual experiment, followed by histological processing and sectioning of the tissue, the exposure of brain sections to autoradiographic film, and the development of the films. It should be emphasized, however, that the 2-DG technique allows the visualization of sites where neurons respond to a particular stimulus and not necessarily where they prefer that stimulus.

After the initial description of the protocol in the monkey visual cortex in 1976, the 2-DG technique was successfully applied to analyze the functional organization of a variety of brain regions including sensory and motor cortices in a variety of animals (Hand, 1981; Sokoloff, 1985).

This chapter focuses on the 2-DG architecture of cat area 17 following the nomenclature of the anatomist Korbinian Brodmann (1909) who divided the human cortex according to cytoarchitectonic criteria and numbered consecutively all identified areas (see also Tusa et al., 1978). The classical electrophysiological studies of Hubel and Wiesel established that, in cat area 17, neurons that respond to similar visual stimuli (e.g., to lines of a particular orientation presented at a particular location in the visual field) (within their receptive field) are not distributed randomly across the cortex but are arranged in columns extending from layer I to layer VI (so-called orientation columns; Hubel and Wiesel, 1962). In addition, neurons encountered in a vertical penetration through cortical layers respond preferentially to stimulation of either the right or left eye (forming so-called ocular dominance columns; Hubel and Wiesel, 1962). However visual cortical neurons not only prefer lines of a particular orientation presented to one eye, but are also sensitive to a variety of other stimulus parameters such as the direction of motion of contours, their spatial frequency, or binocular disparity. After the initial discovery of these various response properties of visual cortical neurons, investigations using the 2-DG technique contributed much to our understanding of both the topographical arrangement of neurons with similar functional properties and spatial relationship between the various functional domains in area 17.

The following section I describe examples of all functional systems that have been mapped with the 2-DG technique such as orientation, ocular dominance, and spatial frequency domains. I then briefly discuss results on the development of these domains and their experience-dependent changes as revealed by 2-DG autoradiography. Finally, I summarize the advantages and disadvantages of the 2-DG technique in particular compared with minimally or noninvasive imaging techniques such as optical imaging of intrinsic signals and functional magnetic resonance imaging (fMRI).

ORIENTATION DOMAINS

In their pioneering studies, Hubel and Wiesel (1959, 1962) established that unlike cells in the retina and lateral geniculate nucleus, visual cortical neurons are selective for the orientation of contours presented within their receptive field. In addition, cells recorded during electrode penetrations perpendicular to cortical layers were found to prefer the same stimulus orientation, whereas a gradual shift of orientation preference was observed along tangential penetrations (Hubel and Wiesel, 1974; Albus, 1975a, 1975b, 1975c; Murphy and Sillito, 1986) (see Chapter 1). Following the terminology of Mountcastle (Mountcastle, 1957), Hubel and Wiesel referred to the cortical volume containing neurons with similar orientation preference as an orientation column. The idea that orientation columns are laid out systematically across the cortex was first suggested by Hubel and Wiesel (1963a) on the basis of surface maps of the cat cortex made from multiple, closely spaced vertical electrode penetrations. However, it was another 15 years before the development of the 2-deoxyglucose technique by Sokoloff et al. (1977) allowed direct visualization of the layout of orientation preference. The 2-DG analyses not only confirmed the previous assumptions but also allowed the topographical arrangement of orientation columns to be mapped more comprehensively than is possible with electrophysiological recordings.

The initial 2-DG studies of orientation subunits in cat area 17 revealed that 2-DG labeling is continuous through all cortical layers (Fig. 3-2A) (Albus, 1979; Lang and Henn, 1980; Singer, 1981; Schoppmann and Stryker, 1981) and that orientation domains frequently had the shape of slabs and bands rather than of isolated columns (Fig. 3-2B) (Albus and Sieber, 1984; Singer, 1981). In one study, single-unit recording and 2-DG autoradiography were combined in the same hemisphere. The results showed good correlation between the physiologically and anatomically (with 2-DG) defined orientation columns (Schoppmann and Stryker, 1981).

However, since the cat cerebral cortex is highly convoluted, autoradiographs of single (mostly horizontal) sections displayed only a small portion of area 17, so that the detailed geometry and arrangement of orientation columns could not be investigated. By combining the 2-DG technique with cortical flat-mounting (Freeman et al., 1987; Olavarria and Van Sluyters, 1985; Tootell and Silverman, 1985), these limitations could be overcome and activity maps from the entire area 17 and adjacent visual cortical regions could be visualized in single sections cut tangential to the cortical surface (Freeman et al., 1987; Löwel et al., 1987). These flat-mount studies demonstrated that 2-DG-labeled orientation domains (or ocular dominance domains, see Fig. 3-6) form highly organized periodic patterns in a plane parallel to the cortical surface; 2-DG autoradiography after stimulation with a single orientation revealed a pattern of orientation domains consisting of beaded bands and isolated patches with a periodicity of approximately 1 mm (Fig. 3-3) (Löwel et al., 1987). A similar pattern was observed after monocular visual stimulation with contours of one orientation (Löwel and Singer, 1993a). The bands

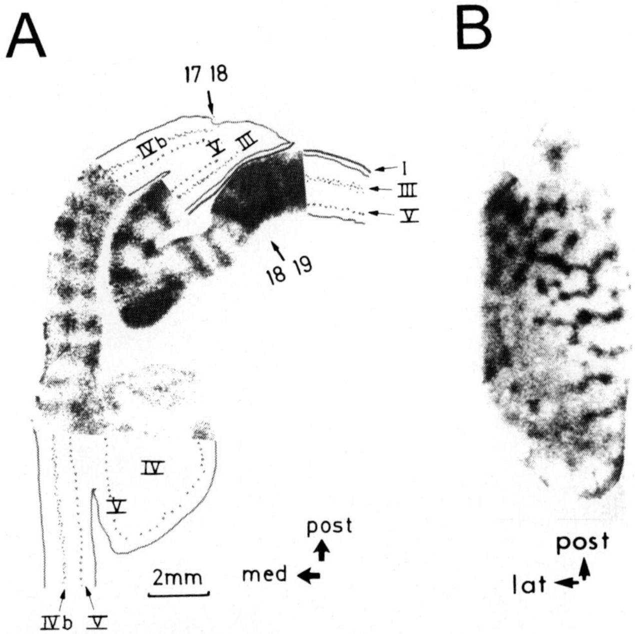

FIGURE 3-2. 2-DG labeling of orientation domains in the visual cortex of cats. **(A)** Autoradiograph of a horizontal section through a left hemisphere. The medial part of the posterior brain, including the upper bank of the splenial sulcus and the medial and lateral bank of the postlateral sulcus, are shown. Cortical layers are indicated by Roman numerals. Note that 2-DG-labeled orientation domains are continuous throughout all cortical layers. (Modified from Albus, K. (1979). ^{14}C-Deoxyglucose mapping of orientation subunits in the cat's visual cortical areas. *Exp. Brain Res.* **37:** 609–613) (Figure 3-1A, left; reproduced with permission from Springer-Verlag.) **(B)** Topographic arrangement of orientation domains visualized in a horizontal section through the right marginal gyrus. Note that orientation domains are slablike and run parallel to each other over a limited cortical distance. (Modified from Albus K., and Sieber B. [1984]. On the spatial arrangement of iso-orientation bands in the cats visual cortical areas 17 and 18: a ^{14}C-deoxyglucose study. *Exp. Brain Res.* **56:** 384–388) (Figure 3-2A, left; reproduced with permission from Springer-Verlag.)

tended to run perpendicular to the borders of area 17 but were irregular and had frequent fusions and blind endings (Löwel et al., 1987; Albus and Sieber, 1984; Singer, 1981) (for review see LeVay and Nelson, 1991). When a stimulus containing two alternating orthogonal orientations was presented, the banded pattern was similar but had a periodicity of 0.5 mm (Löwel et al., 1987). These results were inconsistent with the geometrical model of Braitenberg and Braitenberg (1979) for the layout of iso-orientation domains in which orientation domains were assumed to have a radial arrangement with every orientation appearing twice per orientation center (a conclusion also supported by a later optical imaging experiment by Bonhoeffer and Grinvald (1991)).

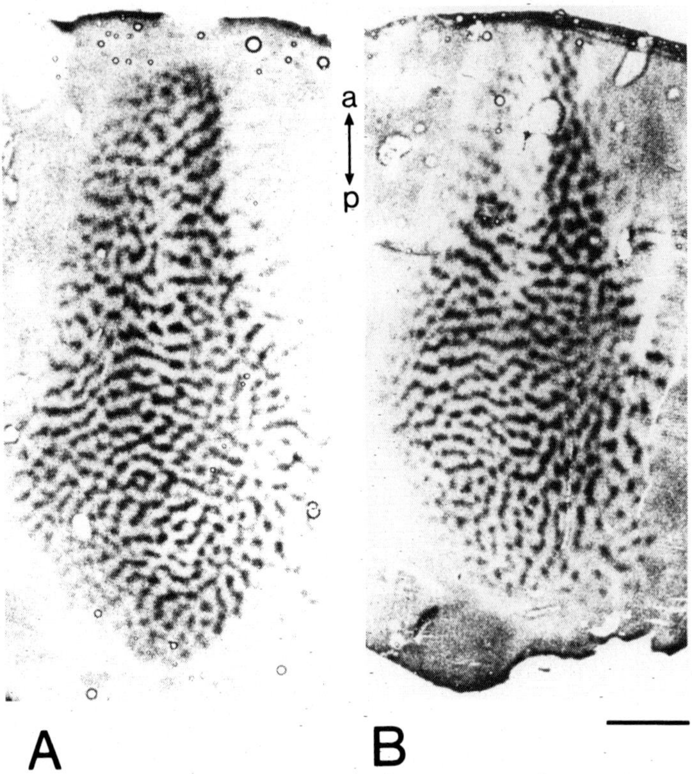

FIGURE 3-3. The complete pattern of orientation domains on flat-mount sections of the primary visual cortex of cats. 2-DG autoradiographs following binocular visual stimulation with moving contours of a single orientation: 90° **(A)** and 135° **(B).** Stimulus velocity varied between 2.5 and 28 deg/sec. Note that the 2-DG-labeled orientation domains form highly organized and periodic patterns, consisting of beaded bands that run parallel to each other over limited distances. (Abbreviations: a, anterior; p, posterior. Scale bar, 5 mm.) (Reprinted in modified form from Löwel S., and Singer W. [1990]. Tangential intracortical pathways and the development of iso-orientation bands in cat striate cortex, *Dev. Brain Res.* **56:** 99–116, Copyright 1990, with permission from Elsevier Science **(A).** Modified from Löwel S., and Singer W. [1993a]. *Eur. J. Neurosci.* **5:** 846–856, reproduced with permission from Blackwell Science [B].)

The advent of optical imaging techniques has allowed the most comprehensive visualization of orientation maps to date (see Chapter 2). Because multiple recordings can be made in a living animal, it is possible to compare, in the same cortical region, the neuronal responses to a variety of visual stimuli. These analyses have revealed that iso-orientation domains are more patchlike and less banded than one might have expected on the basis of electrophysiological and 2-DG mapping experiments. It seems that the columns can be viewed as parallel bands only on a very local scale (< 1 mm) and that iso-orientation domains are arranged radially around singularities, with each orientation appearing once per singularity (Blasdel and Salama, 1986; Bonhoeffer and Grinvald, 1991). One has to bear in mind, however, that optical imaging in cat area 17 can record data only from foveal representations close to the vertical meridian of the visual field; therefore these data cannot be considered representative of the overall organization or primary visual cortex. In the first 2-DG study in cat area 17, it was noted that "the spatial pattern of the orientation subunits seems to be more regular in the visual field periphery than in regions representing the vertical meridian" (Albus, 1979, p. 609).

OCULAR DOMINANCE DOMAINS

Historically, ocular dominance columns are famous in the context of the 2-DG technique because they were the first functional system to be mapped in 1976 in the monkey (Kennedy et al., 1976) (Fig. 3-1). Sokoloff and colleagues used the binocular visual system of the monkey to test the capacity of their newly developed technique by comparing 2-DG patterns in three groups of animals with different types of visual input: (1) intact binocular vision, (2) bilateral visual occlusion, and (3) monocular visual occlusion. In their study, visual occlusions were achieved either by enucleation or by insertion of opaque plastic discs in the eyes. Animals with intact vision in one or both eyes were either exposed to the ambient light of the laboratory or surrounded by a black-and-white complex geometric pattern that was illuminated and rotated during the experimental procedure. Qualitatively similar results were obtained with either method of visual occlusion and with or without the rotating geometric pattern. Binocular visual stimulation led to metabolic activity in the primary visual cortex, with the most intense labeling observed in the thalamocortical input layer IV. Bilateral visual occlusion lowered the rates of glucose consumption in the striate cortex and reduced the metabolic differentiation of the various layers. Unilateral visual deprivation led to columnar labeling in the primary visual cortex and thus visualization of the ocular dominance columns (Fig. 3-1) (Kennedy et al., 1976).

In 1983, ocular dominance columns were also visualized in area 17 of cats (Tieman and Tumosa, 1983) (see also Wagner et al., 1981). In these experiments, monocular visual stimulation was achieved by enucleation (three animals) or TTX-injection into one eye (one animal). During 2-DG accumulation, the animals were awake and encouraged to explore the lighted laboratory. In area 17 of all monocularly stimulated animals, there were alternating regions of light and dark label (Fig. 3-4) (Tieman and Tumosa, 1983). Near the representation of the area centralis, especially in the hemisphere ipsilateral to the stimulated eye, the labeled regions formed columns that extended from the pial surface to the white matter.

That these columns correspond to ocular dominance columns was inferred from the fact that columnar labeling was absent in control cats in which both eyes were stimulated or neither eye was stimulated. Direct proof of this correspon-

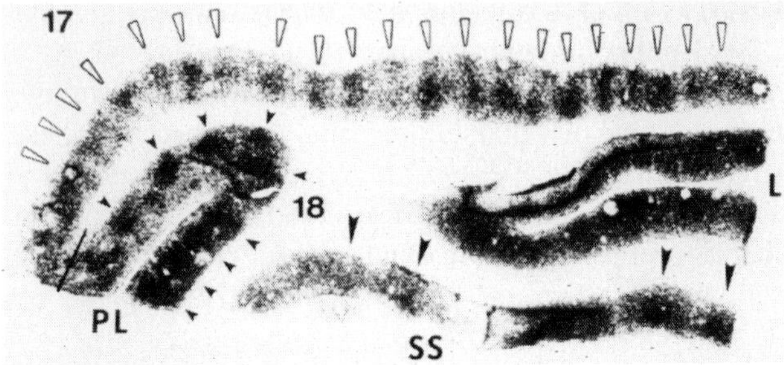

FIGURE 3-4. 2-DG labeled ocular dominance columns in the cat. 2-DG autoradiograph of a horizontal section through the occipital pole of the brain ipsilateral to the stimulated eye. The locations of the lateral (L), posterolateral (PL), and suprasylvian (SS) sulci are shown. Labeled regions in areas 17 and 18 and in the suprasylvian sulcus are indicated by open, small, and large filled arrowheads, respectively. (Reprinted in modified form from Tieman S. B., and Tumosa, N. [1983]. [^{14}C]2-deoxyglucose demonstration of the organization of ocular dominance in areas 17 and 18 of the normal cat. *Brain Res* **267:** 35–46, Copyright 1983, with permission from Elsevier Science.)

dence came from a study by Löwel and Singer (1993b). Using a double-label protocol, monocular visual stimulation in awake cats was combined with transneuronal labeling of ocular dominance domains using the classical [^{3}H]proline technique (Grafstein, 1971; Wiesel et al., 1974). In both normally raised and strabismic animals, 2-DG patterns were in precise register with the proline-labeled ocular dominance columns in layer IV (Fig. 3-5) (Löwel and Singer, 1993b). Again, regions of increased 2-DG uptake extended in a columnar fashion through all cortical layers. In strabismic cats, analyses of cortical flat-mount sections revealed that the columnar 2-DG patterns displayed all the characteristic features of ocular dominance domains (Fig. 3-6) (Löwel and Singer, 1993b; Shatz et al., 1977; Löwel and Singer, 1987; Anderson et al., 1988). (i) The optic disc representations of the stimulated and nonstimulated eye were identifiable in the posterior third of both hemispheres as oval regions that were solidly labeled ipsilateral to the stimulated eye and unlabeled contralateral to it. (ii) The monocular segment was identified by uniform labeling at the medial border of area 17 contralateral to the stimulated eye and by the absence of labeling at comparable eccentricity on the ipsilateral side. (iii) The territories of the open eye tended to be larger in the contralateral than in the ipsilateral hemisphere (see also Shatz and Stryker, 1978). These features of the ocular dominance system were expressed equally well in all cortical layers (Löwel and Singer, 1993b).

This study and a companion study (Löwel and Singer, 1993a) revealed differences in these monocularly induced 2-DG patterns between strabismic and normally raised animals depending on whether the animals were awake or anaesthetized and paralysed. First, in awake normally raised cats, in contrast to awake strabismic animals, the 2-DG patterns had a more patchy appearance in both

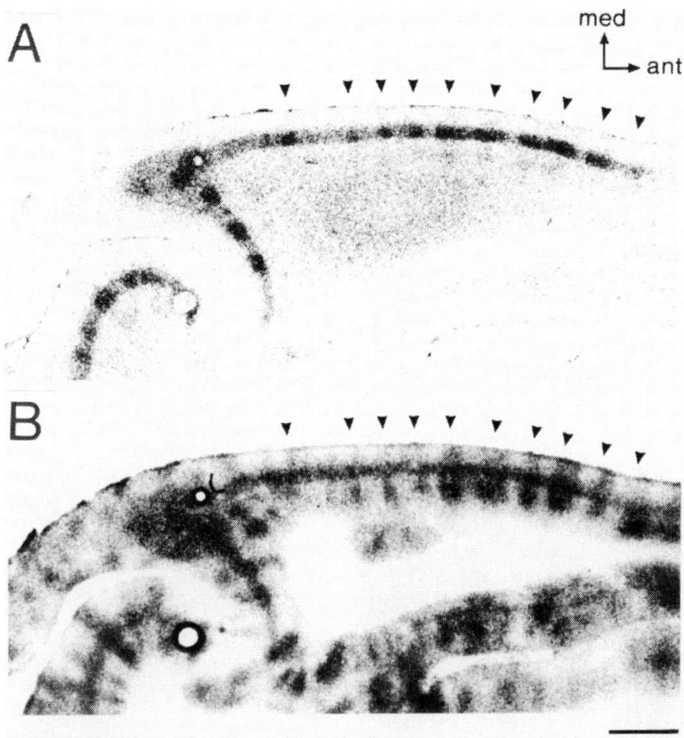

FIGURE 3-5. Ocular dominance columns in cat area 17, visualized with [^{3}H]proline and 2-[^{14}C]DG autoradiography. Comparison of ocular dominance columns in a horizontal section through the medial bank of the right area 17 as revealed by transneuronally transported [^{3}H]proline injected into the right (ipsilateral) eye **(A)** and 2-[^{14}C]DG labelling after right-eye stimulation **(B).** The two autoradiographs are from the same tissue section. Note that the columns of increased 2-DG uptake are in register with the right-eye territories in layer IV (arrowheads). (ant, anterior; med, medial. Scale bar, 2 mm.) (Modified from Löwel S., and Singer W. [1993b]. *Eur. J. Neurosci.* 5: 857–869, reproduced with permission from Blackwell Science.)

hemispheres, so that labeled patches were smaller and occupied less territory than the afferents of the stimulated eye in layer IV (Fig. 3-7) (Löwel and Singer, 1993b). Second, under anaesthesia, monocular stimulation with contours of many different orientations induced columnar 2-DG labeling only in strabismic and not in normally raised animals. These results suggested (1) the existence of a mechanism in normally raised cats that restricts cortical activation after monocular stimulation to territories that are in register with the afferents from the stimulated eye, and (2) this mechanism appears to be effective only when the animals are awake and actively exploring their environment.

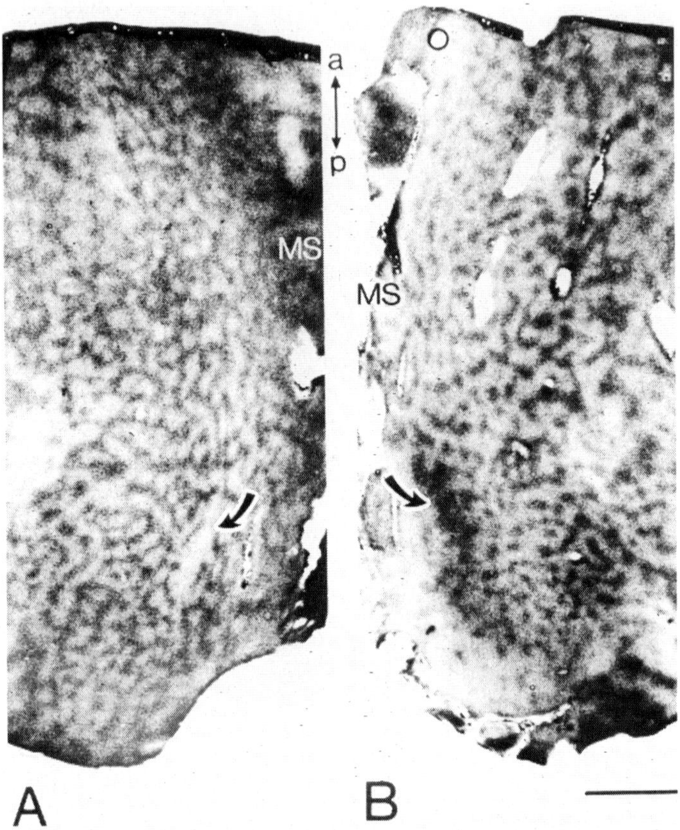

FIGURE 3-6. Ocular dominance columns in flat-mount sections of the unfolded primary visual cortex of cats. 2-DG autoradiographs of supragranular flat-mount sections from the unfolded left **(A)** and right **(B)** hemisphere of a strabismic cat that had been stimulated through the right eye. Note the sharp delineation of active (dark gray) and inactive (light gray) territories. The optic disc representations are indicated by arrows. Note that the territories of the open eye tend to be larger in the contralateral **(A)** than in the ipsilateral hemisphere **(B)**. (a, anterior; p, posterior; MS, monocular segment. Scale bar, 5 mm.) (From Löwel S., and Singer W. [1993b]. *Eur. J. Neurosci.* **5:** 857–869, reproduced with permission from Blackwell Science.)

Taken together, monocular visual stimulation does not necessarily induce 2-DG accumulation in register with and corresponding to columns above and below the sites of active thalamocortical afferents in layer IV. Monocular 2-DG patterns are sensitive to both anesthesia (and probably also attention) and to the binocularity of the experimental animals.

In contrast, visualization of orientation columns does not seem to be dramatically impaired in anesthetized cats. Using the 2-DG method, Livingstone and Hubel (1981) compared the patterns of orientation domains induced in two

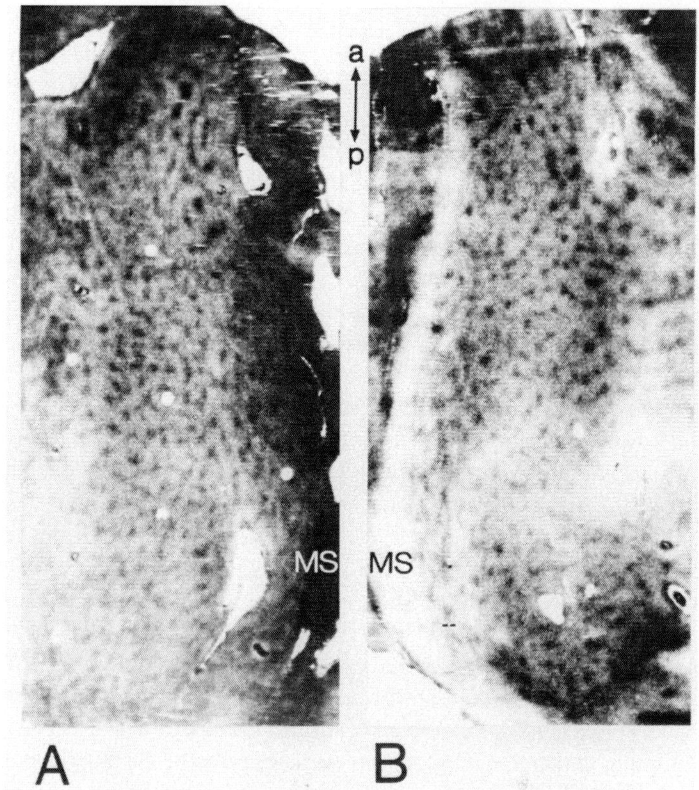

FIGURE 3-7. 2-DG-labeled ocular dominance columns in the visual cortex of awake, normally raised cats. Autoradiographs of flat-mount sections from supragranular layers of both the contralateral **(A)** and the ipsilateral **(B)** hemisphere. Note that the 2-DG patterns appear very patchy in both hemispheres. Compared with the strabismic animals (for example Fig. 3-6), characteristic features of ocular dominance columns are less well discernible: (1) the open-eye territories (dark 2-DG patches) are not larger in the contralateral than in the ipsilateral hemisphere, (2) the blind spots are not detectable at all, and (3) only the monocular segments (MS) are visible as indicated by uniform labeling at the medial border of the contralateral area 17 **(A)** and by the absence of labeling at comparable eccentricity on the ipsilateral side **(B)** (Abbreviations as in Fig. 3-6.) (From Löwel S., and Singer W. [1993b]. *Eur. J. Neurosci.* **5:** 857–869, reproduced with permission from Blackwell Science.)

paralyzed cats, one awake and the other in slow-wave sleep. Under both conditions, columns of increased 2-DG uptake were visible, the only difference being a somewhat weaker labeling in infragranular layers in sleeping cats. Thus, while it is known that anesthesia diminishes cortical responsiveness and [^{14}C]2-DG uptake (Sokoloff et al., 1977), this does not seem to impede the visualization of columnar systems in general.

SPATIAL FREQUENCY DOMAINS

Most cortical cells are selective for the spatial frequency of appropriately oriented sinusoidal gratings (Cooper and Robson, 1968; Campbell et al., 1969; Maffei and Fiorentini, 1973) (for review see Orban, 1991; LeVay and Nelson, 1991). Compared with cells at earlier stages of the visual system, cortical neurons are more sharply tuned for spatial frequency. While the bandwidth (full-width at half-height of the spatial frequency tuning curve measured on a logarithmic scale) of neurons in the lateral geniculate nucleus is typically between three to five octaves, in the cortex the average bandwidth is about 1.5 octaves (DeValois et al., 1977; Movshon et al., 1978; Tolhurst and Thompson, 1981).

On the basis of single-unit recordings in cat visual cortex, Maffei and Fiorentini (1977) reported that neurons preferring different spatial frequencies were grouped into rows corresponding to cortical layers. Tolhurst and Thompson (1981, 1982), using similar techniques, found that neurons with similar spatial frequency preferences were only loosely organized in clusters and that these were neither strictly laminar nor columnar. Again, the use of the 2-DG technique, with the advantage that the activity "of millions of cells, rather than just a few in one region" (Tootell et al., 1981) can be visualized simultaneously, allowed a direct test of these competing hypotheses about the grouping of cells preferring similar spatial frequencies. When cats were stimulated with sinusoidal gratings containing a single spatial frequency presented at all orientations, columns of increased 2-DG uptake were labeled in cat primary visual cortex (Fig. 3-8), (Tootell et al., 1981; Thompson and Tolhurst, 1979). A control stimulus containing all spatial frequencies presented at all orientations produced no columnar density differences within the striate cortex. Thus 2-DG autoradiography demonstrated the existence of spatial frequency columns (of a columnar organization of spatial frequency-specific sensitivity). The columns had a periodicity of about 1 mm and extended through all layers. High spatial frequencies labeled columns only within the central 5° of the area centralis projection region, whereas low spatial frequencies induced columnar patterns extending peripherally across the entire stimulated visual field representation (Tootell et al., 1981). These results confirmed previous single unit results in cat visual cortex showing that cells sensitive to higher spatial frequencies are confined within about 5° of the area centralis region in area 17, whereas lower frequency units are much more widely distributed (Movshon et al., 1978). Stimulation with a single spatial frequency presented at a single orientation led to a breakdown of the slablike 2-DG pattern, presumably caused by the intersection with the orientation column system (Tootell et al., 1981).

In line with the 2-DG experiments, electrophysiological recordings indicated that there is probably not such a fine-grained columnar representation of spatial frequency as has been demonstrated for orientation. Tootell et al. (1988) noted that their data were consistent with the existence of only two types of spatial frequency columns in the cat: high and low. In addition, the ranges of the two sets of columns might overlap considerably, producing an overall distribution of preferred frequencies that peaks in the center. Furthermore, Tootell et al. (1988) reported that both in the cat and the monkey, intermediate spatial frequencies

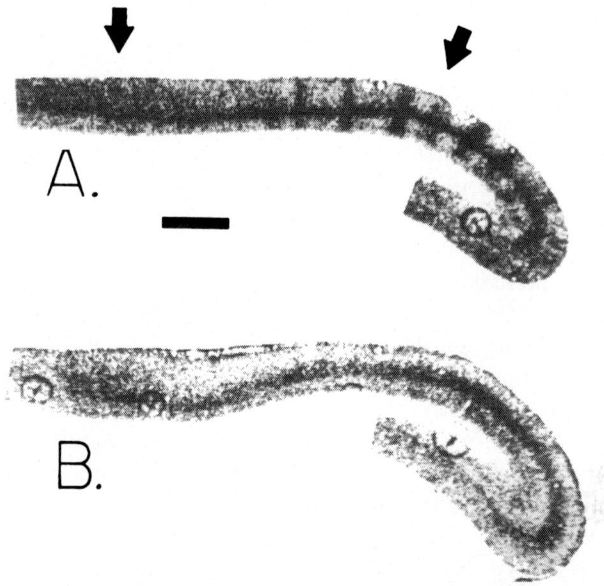

FIGURE 3-8. Spatial frequency columns in cat primary visual cortex. Comparison of autoradiographs from area 17 of two cats. One animal **(A)** viewed a pattern containing a single spatial frequency (2.0 cycldeg) and the other **(B)** a pattern containing multiple spatial frequencies. Both patterns were presented binocularly and at all orientations. Discrete dark columns are seen in the single spatial frequency case **(A)** but not in the multiple frequency animal **(B).** In animals that viewed a high-frequency pattern, the columns were restricted to the center of area 17 (arrow on the right), even though the visual pattern extended in visual space as far as indicated by the arrow on the left. The autoradiographs were taken from horizontal sections cut at the same depth. (Scale bar, 2 mm.) (Reprinted with permission from Tootell R.B.H. et al. [1981]. Spatial frequency columns in primary visual cortex. *Science* **214:** 813–815, Copyright 1981 American Association for the Advancement of Science.)

were less effective in labeling columns than were high or low frequencies. A recent optical imaging study in the cat confirmed the existence of high and low spatial frequency domains (Hübener et al., 1997). Stimulation of 2- to 3-month-old kittens with oriented gratings at a spatial frequency of 0.2 and 0.6 cyc/deg revealed two sets of activated domains: orientation maps obtained after "high" and "low" spatial frequency stimulation were similar, but not identical so that a spatial frequency map could be computed (Hübener et al., 1997).

DEVELOPMENT AND EXPERIENCE-DEPENDENT CHANGES OF CORTICAL MAPS

NORMAL DEVELOPMENT AND THE EFFECTS OF BINOCULAR DEPRIVATION

In the primary visual cortex of adult monkeys and cats, both electrophysiology and 2-DG autoradiography demonstrated that neurons are arranged in functional columns that extend through the full thickness of the cortex. The questions of how this orderly arrangement develops in very young animals and whether neuronal activity is a necessary prerequisite are still a matter of debate. The first 2-DG study, investigating the development of orientation columns in cat striate cortex, dates back to 1983. In normal animals, periodic metabolic labeling around layer IV was first clearly observed at 21 days, and by 35 days the pattern had become truly columnar (Fig. 3-9) (Thompson et al., 1983). In animals deprived of normal pattern vision (by bilateral lid suture), no differential labeling was observed except for weak periodicity in a single 35-day-old animal (Thompson et al., 1983). The authors suggested from these results that striate cortex is immature at the time of eye opening and that visual experience is crucial for maintaining the normal development of orientation columns.

An optical imaging study by Crair et al. (1998) settled the long-standing debate about the role of visual experience in the development of columns in cat visual cortex. In this study, cortical maps for both orientation and ocular dominance were observed already in 2-week-old kittens (Crair et al., 1998). Because the cortical maps were identical until nearly 3 weeks of age in both normally raised and binocularly deprived (bilaterally lid-sutured) animals (Crair et al., 1998), early pattern vision appears to be unimportant for the initial establishment of the positions of orientation and ocular dominance columns in the visual cortex. The naive maps were dominated by the contralateral eye. With continued binocular deprivation ipsilateral eye responses never became strong or very selective, so that experience was essential for responses of the other eye to become strong (Crair et al., 1998). In line with interpretations of 2-DG studies, experience is necessary for the maintenance of visual cortical responsiveness and selectivity, which, in its absence, degrade over the course of the well-documented critical period (Crair et al., 1998) (see also Frégnac and Imbert, 1984; Henry et al., 1994).

Why have orientation domains not been visualized with 2-DG in 2-week-old animals. Why have they first been detected in the cortical imput layer IV at 3 weeks? That optical orientation maps are seen in very young kittens does not necessarily indicate that the optical signals derive from all cortical layers. It is possible that the first optically visualized maps derive predominantly from cortical layer IV and that columnar activation develops over 2 to 3 weeks. Nevertheless the question remains why clear periodic labeling with 2-DG was not seen before the age of 5 weeks.

Electrophysiological studies have shown that a population of orientation-selective neurons exists in the striate cortex of very young cats, even before the age of 2 weeks (Blakemore and Van Sluyters, 1975; Hubel and Wiesel, 1963b) and that their proportion reaches adult levels by 4 to 6 weeks (Blakemore and Van Sluyters, 1975; Frégnac and Imbert, 1978; Henry et al., 1994). However, neurons in very young kittens have a tendency to habituate or fatigue (Hubel and Wiesel, 1963b; Beckmann and Albus, 1982). Given the standard protocol of 2-DG experiments, which consist of 45 minutes of constant stimulation, habituation or fatigue of visual cortical neurons is a likely explanation for the absence of peri-

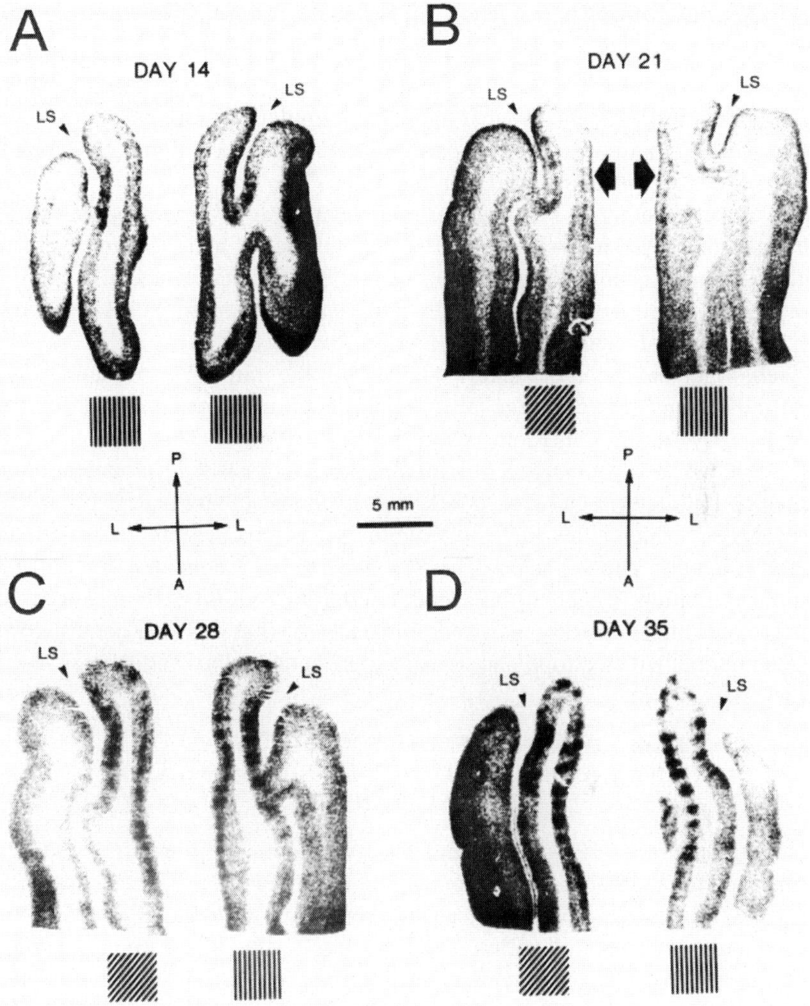

FIGURE 3-9. Development of orientation columns in cat striate cortex revealed by 2-DG autoradiography. Montages of autoradiographs in normal cats at PND 14 **(A),** 21 **(B),** 28 **(C),** and 35 **(D).** The autoradiographs show horizontal sections through the lateral and suprasylvian gyri, separated by the lateral sulcus (LS), of both hemispheres. Note that columnar label is apparent only in the posterior lateral gyrus. Note in addition that regular periodicities in label first appear at 21 days in layer IV (**B,** large arrows) and are fully columnar at 35 days of age **(D).** (a, anterior; p, posterior; l, lateral). (Modified from Thompson I.D. et al. [1983]. *Nature* **301:** 712–715, reproduced with permission from Nature © Macmillan Publishers Ltd.)

odic labeling in the youngest animals. In contrast, standard optical imaging protocols consist of only a few seconds (4.8 in Crair et al., 1998) of constant stimulation with a particular orientation. The orientation of the visual stimulus is then changed in random sequence so that the entire stimulation time (between 150 and 230 seconds in Crair et al., 1998) is partitioned into 4.8-second blocks, which are highly unlikely to cause response habituation.

That 2-DG labeling is first visible in cortical layer IV might be due to less habituation or fatigue in that layer and/or the presence of a higher number of orientation-selective neurons. Indeed, a large number of innately specified neurons in very young animals, as well as in binocularly deprived cats, is located around layer IV (Blakemore and Van Sluyters, 1975).

Taken together, the most likely reason for the failure to visualize 2-DG domains in kittens before the age of 3 weeks is habituation of visual cortical neurons caused by the stimulus protocol of the 2-DG technique. Thus optical imaging seems to be the more suitable technique for visualizing early development of cortical maps. However, 2-DG studies allow analysis of activity patterns in distinct cortical layers, which is not (yet) possible in optical experiments. In this context, the 2-DG study by Thompson et al. (1983) visualized a change in laminar distribution of orientation-dependent differential activity (from layer IV in 3-week-old animals to columnar labeling in 5-week-old animals) that was later confirmed and extended in detailed electrophysiological experiments (Albus and Wolf, 1984).

EXPERIENCE-DEPENDENT MODIFICATION OF ORIENTATION COLUMNS: STRIPE-REARING AND STRABISMUS

Although the initial layout of orientation maps seems to be determined largely by intrinsic factors, their organization can be modified by visual experience (see Chapter 10). In a study by Singer et al. (1981) (see also Flood and Coleman, 1979), which was the first detailed 2-DG study about the influence of experience on orientation columns, the visual experience of kittens was restricted to a single orientation (vertical or horizontal) using cylindrical lenses. In the 2-DG experiments, kittens were stimulated with vertical contours in one visual hemifield and with horizontal contours in the other hemifield. Interhemispheric comparisons revealed the following changes in the arrangement of orientation columns

1. Within layer IV, columns whose preference corresponded to the experienced orientation were wider and more active than those encoding the orthogonal orientation, but the columnar grid remained basically unaltered.
2. Outside layer IV, the columnar system was maintained only for columns encoding the experienced orientations, whereas deprived columns frequently failed to extend into nongranular layers (Fig. 3-10) (Singer et al., 1981).

The authors concluded from these results that within layer IV, the blueprint for the system of orientation columns is determined by genetic instruction: first-order cells in layer IV develop orientation selectivity irrespective of experience. On the other hand, the expansion of the columnar system from layer IV into nongranular layers is dependent on experience, because the malleability of the columnar system is much more pronounced in nongranular than in granular layers (Singer et al., 1981). A recent optical imaging experiment confirmed the major conclusions of this 2-DG study by showing that experienced orientations occupied significantly more cortical surface area than orthogonal ones (Sengpiel et al., 1999). Visual experience is therefore thought to play an instructive role for the developmental specification of orientation preference.

Raising animals with a divergent squint angle also has distinct effects on the layout of 2-DG-labeled orientation domains. In contrast to normally raised cats (Löwel and Singer, 1993a), monocular visual stimulation of strabismic animals with gratings of a single orientation produced isolated patches of increased 2-DG uptake (Schmidt et al., 1997). These patches closely resembled the monocular activation patterns revealed in a recent optical imaging study in strabismic animals (Löwel et al., 1998). In addition, the 2-DG maps provided direct evidence that squint had caused a disruption of cortical binocularity, because there was a consistent lack of labeling at the representation of the optic disc in the hemisphere contralateral to the stimulated eye (Schmidt et al., 1997). No obvious differences were observed between the 2-DG patterns evoked by stimulation of either the deviated or the nondeviated eye. This was consistent with electrophysiological evidence from strabismic cats, which suggested that neurons driven from the normal and deviated eyes have similar and normal orientation tuning (Hubel and Wiesel, 1965; Freeman and Tsumoto, 1983; Kalil et al., 1984; Sengpiel et al., 1994).

EXPERIENCE-DEPENDENT CHANGES OF OCULAR DOMINANCE COLUMNS: MONOCULAR DEPRIVATION, ALTERNATING MONOCULAR EXPOSURE, AND STRABISMUS

Experience-dependent changes in the arrangement of ocular dominance columns have also been investigated with 2-DG autoradiography. Monocular visual stimulation of 5-week-old kittens revealed a columnar labeling of 2-DG. After 3 days of monocular deprivation (by lid suture), stimulation of the deprived eye induced patches of higher 2-DG uptake in layer IV and in immediately adjacent parts of layers III and V, but columns were only occasionally observed extending through all cortical layers. After 7 days of monocular deprivation, even the patches of enhanced labeling in layer IV were much fainter and did not extend beyond that layer (Fig. 3-11) (Kossut et al., 1983) (see also Bonds et al., 1980). This observation was consistent with the electrophysiological findings of Hubel and Wiesel (1970), who reported a marked domination of the visual cortex by the open eye after 6 days of lid suture, but only a partial effect after 3 days. In addition, the 2-DG study indicated that the first visible changes in activation of cortical tissue occur outside layer IV, and it was suggested that intracortical synapses, and not the geniculocortical synapses terminating in layer IV, may be the first to

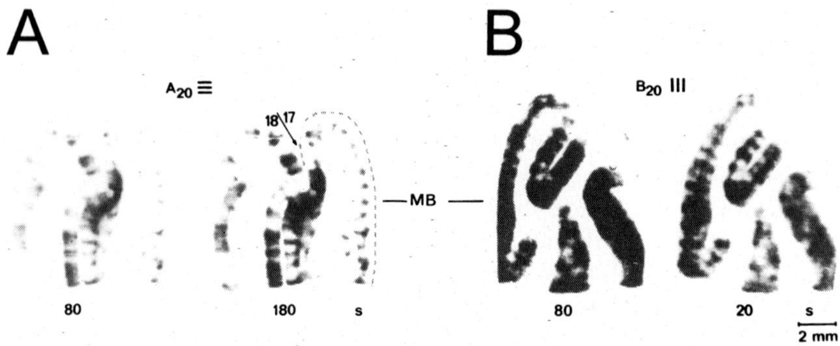

FIGURE 3-10. Experience-dependent changes of orientation columns. 2-DG autoradiographs from horizontal sections through the medial bank of a 7-week-old kitten. The animal was dark-reared from the time of eye opening to 4 weeks of age. Subsequently, it was exposed selectively to vertical contours by cylindrical lenses that had – 25 dptr in one axis and 0 dptr refractive power in the orthogonal axis. During the 2-DG experiment, the left visual hemifield was stimulated with vertical and the right visual hemifield with horizontal contours. Thus, interhemispheric comparison allowed changes in columnar systems corresponding to experienced and inexperienced orientations to be determined. Section A_{20} **(A)** is from the left, inadequately stimulated (exposed to the inexperienced orientation) and section B_{20} **(B)** from the right, adequately stimulated hemisphere. Because the activity levels in the two hemispheres were rather different, contact prints were made at varied exposure times (indicated below the sections). The beaded bands of increased activity in A correspond to layer IV. Note the incomplete translaminar extent of columns in area 17 of the left hemisphere **(A)** compared with the right hemisphere **(B).** Note in addition, that the adequately stimulated right occipital cortex **(B)** had much more total radioactivity than the left side **(A).** (Modified from Singer W. et al. [1981]. Restriction of visual experience to a single orientation affects the organization of orientation columns in cat visual cortex. *Exp. Brain Res.* **41:** 199–215, Figure 3-4, middle row; reproduced with permission from Springer-Verlag.)

be functionally altered by the asymmetry of the visual input (Kossut et al., 1983; see also Singer et al., 1981).

To test the effects of binocular competition on the development of domains activated by visual stimulation of one eye ("ocular activation columns"), cats were raised with alternating monocular exposure (Tumosa et al., 1989). Three different exposure regimes were applied: (1) equal alternating monocular exposure (AME) (the left and right eyes received 8 hours of visual stimulation every other day), (2) unequal AME (one eye received 8 hours, the other 1 hour of stimulation every other day), and (3) monocular deprivation (by lid-suture). By using the [^{14}C]2-DG technique, the average size and laminar layout of the ocular activation columns of the eye stimulated during the 2-DG experiment were compared in the three animal groups. Throughout all cortical layers, the 2-DG-labeled ocular activation columns were smaller for the disadvantaged eye than for the advantaged eye so that the size of the columns of the stimulated eye was a function of the relative competitive advantage it received during rearing: In order of increasing precentage of visual cortex activated, the eyes were (1) deprived eye of monocularly

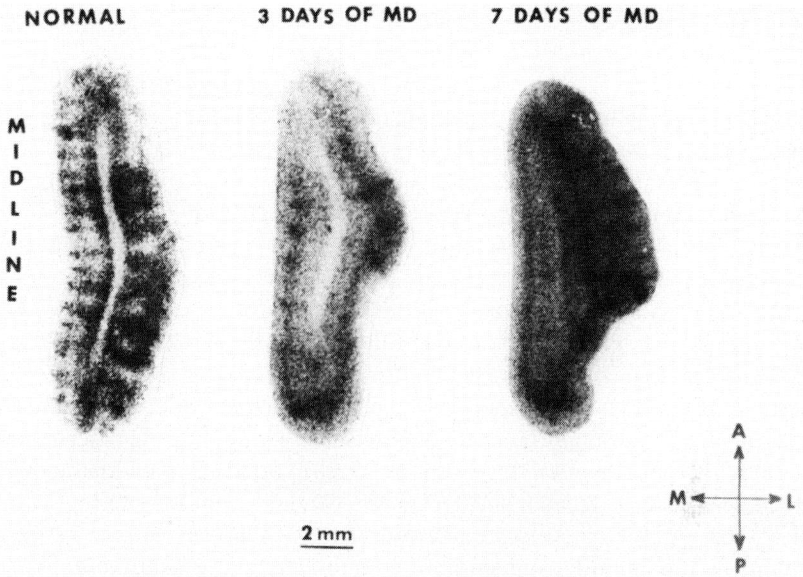

FIGURE 3-11. Effects of monocular deprivation (MD) on 2-DG labeled ocular dominance columns. 2-DG autoradiographs of horizontal sections through area 17 of a normally raised kitten (left), after 3 days of MD (middle) and after 7 days of MD (right). Sections taken from the hemisphere contralateral to the deprived eye. Note that the columnar 2-DG labeling visible in the normally raised cat (left) is already disrupted after 3 days of MD (middle): labeling is virtually absent in the lowest part of the cortex and patches of higher 2-DG uptake are more or less confined to layer IV and immediately adjacent parts of layers III and V. After 7 days of MD (right), even the patches of enhanced labeling in layer IV are much fainter and do not extend beyond that layer. (Modified from Kossut M. et al. [1983]. *Acta Neurobiol. Exp.* **43:** 273–282, reproduced with permission from the Polish Neuroscience Society.)

deprived cats, (2) less experienced eye of unequal AME, (3) either eye of equal AME, (4) more experienced eye of unequal AME, and (5) experienced eye of monocularly deprived cats. Second, the laminar layout of the ocular activation columns was differentially affected by the relative experience of the eye. Consistent with the study by Kossut et al. (1983), deprived eye columns in monocularly deprived cats were widest and darkest in layer IV, and narrower in the extragranular layers, sometimes disappearing altogether. In contrast, the ocular activation columns of either eye in AME cats were about the same width in all layers. Even the columns of the less experienced eye of AME cats, while being similar in width to the deprived eye domains of monocularly deprived cats in layer IV, extended through all layers. Thus, disruption of patterned visual input—as for the deprived eye in monocularly deprived cats—seemed to affect both the geniculocortical and intracortical connections (see also Singer et al., 1981), whereas a smaller competitive disadvantage—as for the less experienced eye in unequal

AME cats—seemed to modify intracortical connections less dramatically than inputs from the lateral geniculate nucleus.

The effect of early onset strabismus on the layout of ocular dominance columns was investigated in a combined [^{14}C]2-DG and [^{3}H]proline study (Löwel, 1994). It was known previously that elimination of correlated activity between the eyes, as it occurs in strabismus, enhances the segregation of ocular dominance columns (Hubel and Wiesel, 1965; Shatz et al., 1977; Löwel and Singer, 1993b). Whether similar mechanisms are also responsible for the final expression of the columnar pattern—that is, for the spacing of adjacent columns, their width, and location—is still a matter of debate. In the study by Löwel (1994), the spacing of adjacent columns in strabismic cats was significantly larger than in normally raised controls, thus providing direct evidence for an activity-dependent development of the pattern of ocular dominance domains. In agreement with an important role of correlated activity for the determination of columnar spacing, similar observations were reported in cats with alternating monocular exposure (Tieman and Tumosa, 1997). A recent longitudinal optical imaging study that monitored the development of columnar patterns in area 17 of kittens before and after induction of a squint angle also supported this conclusion by showing an expansion of ocular dominance column spacing of 20% between the fourth and eighth week (Sengpiel et al., 1998). However, in two conference reports (Jones et al., 1996; Rathjen et al., 1999), columnar spacing was similar in strabismic and normally raised cats. These data are not easy to reconcile with the previous observations. It is possible that the interindividual variability is much greater than previously supposed, a conclusion supported by experiments in macaque monkeys (Horton and Hocking, 1996), and that genetic differences might have an influence on columnar spacing or on the susceptibilty for activity related factors. Further experiments are necessary to clarify these issues.

ADVANTAGES AND DISADVANTAGES OF THE 2-DG TECHNIQUE

ADVANTAGES

1. The most obvious advantage of the 2-DG technique is that it allows activity patterns in the entire brain to be mapped simultaneously. This includes structures that lie deep in the brain and are largely inaccessible with intrinsic signal imaging (or even with the noninvasive magnetencephalography). But even for mapping activity patterns in cortical regions, the 2-DG technique has distinct advantages. Cat visual cortex, for example, is a highly convoluted area, with most of area 17 either hidden in sulci or buried in the medial bank. Intrinsic signal imaging is able to access only the superficial layers of the most central parts of area 17, and it is not yet possible to dif-

ferentiate between activity patterns in different layers. Although recently orientation maps have also been recorded using fMRI (Kim et al., 2000, see Chapter 4), the visualization of activity maps in deep brain structures has yet to be established.

2. Combining 2-DG with the flat-mount technique (in which the highly convoluted cat cortex is unfolded) has proven particularly successful in this respect. It allows visualization of activity patterns throughout areas 17 and 18 (and even some surrounding regions such as PMLS and area 21a) in single sections (without the need, and without the danger of artifacts, of serial reconstructioning) and in different cortical layers. That activity patterns of the entire primary visual cortex are contained in single autoradiographs allows the layout of functional domains in central and peripheral visual field representations to be compared directly.
3. 2-DG-labeled activity maps can be obtained from awake and freely moving animals. Although progress has been made in generating intrinsic signal imaging in awake primates (Vnek et al., 1999; Grinvald et al., 1991), the application of the technique to freely moving animals has yet to be established.
4. Since the 2-DG technique is noninvasive as far as the brain is concerned (one needs only to inject the sugar solution), it is applicable to very young animals, which would be difficult to prepare for more invasive intrinsic signal imaging.
5. The 2-DG technique provides single-cell resolution (Durham et al., 1981; Duncan et al., 1987). Although not routinely used for standard applications, single cell resolution might be necessary for certain questions.
6. The 2-DG technique can be combined with neuronal tract tracing (e.g., with fluorescent latex microspheres, "beads") (Katz et al., 1984; Katz and Iarovici, 1990), so that both the functional architecture and intracortical connectivity patterns can be visualized in the same brain regions (Gilbert and Wiesel, 1989; Löwel and Singer, 1992; Schmidt et al., 1997).

DISADVANTAGES

1. The most obvious disadvantage of the 2-DG technique is that usually only one stimulus can be tested before the brain has to be analyzed for its radioactive distribution. In elaborate and difficult double-label protocols, a maximum of two different stimuli can be tested (e.g., Livingstone and Hubel, 1981; Olds et al., 1985; John et al., 1986; Friedman et al., 1987; Redies et al., 1987; Geesaman et al., 1997) (for a review see Redies and Gjedde, 1989). Thus, it is impossible to conduct longitudinal (developmental) studies or to test a variety of different visual stimuli to analyze the topographic relationship among various functional modules.
2. The technical protocol that requires a 45-minute stimulation period during which the radioactively labeled 2-DG is accumulated in active neurons

makes it impossible to follow quick and adaptive activity changes. In addition, it introduces the problem of response habituation.

3. The 2-DG technique uses radioactive isotopes with a very long half-life ([^{14}C]: 5730 years, [^{3}H]: 12.4 years) so that specific laboratory equipment is needed (e.g., Redies and Gjedde, 1989).

OUTLOOK

For the analysis of cat primary visual cortex, the 2-DG method and intrinsic signal imaging or fMRI are complementary techniques. For high-resolution mapping of functional properties in superficial cortical regions, intrinsic signal imaging is the method of choice. In the future, fMRI may catch up in spatial resolution with optical imaging (see Chapter 4) and, of course, has the major advantage of being a noninvasive technique that is applicable to humans. However, to date, analysis of deep brain structures or of larger activity patterns in the entire area 17 or in awake and freely moving animals is better performed with 2-DG. Thus, although the 2-DG technique was introduced more than 20 years ago, it will continue to play an important role in the analysis of the functional architecture of the brain.

ACKNOWLEDGMENTS

I thank Wolf Singer for his cooperation in a number of the reviewed experiments. In addition, I wish to thank Steffi Bachmann for assistance with the figures and John M. Crook and Stefan Rathjen for critical reading of the manuscript. Special thanks are due to Louis Sokoloff, Suzannah Bliss Tieman, Klaus Albus, Ian D. Thompson, Malgorzata Kossut, Roger Tootell, and Wolf Singer for providing original photographs from their papers. The support of the WGL is gratefully acknowledged.

REFERENCES

Albus, K. (1975a). Predominance of monocularly driven cells in the projection area of the central visual field in cat's striate cortex. *Brain Res.* **89,** 341–347.

Albus, K. (1975b). A quantitative study of the projection area of the central and the paracentral visual field in area 17 of the cat. I. The precision of topography. *Exp. Brain Res.* **24,** 159–179.

Albus, K. (1975c). A quantitative study of the projection area of the central and the paracentral visual field in area 17 of the cat. II. The spatial organization of the orientation domain. *Exp. Brain Res.* **24,** 181–202.

Albus, K. (1979). ^{14}C-deoxyglucose mapping of orientation subunits in the cats visual cortical areas. *Exp. Brain Res.* **37,** 609–613.

Albus, K., and Sieber, B. (1984). On the spatial arrangement of iso-orientation bands in the cats visual cortical areas 17 and 18: a ^{14}C-deoxyglucose study. *Exp. Brain Res.* **56,** 384–388.

Albus, K., and Wolf, W. (1984). Early post-natal development of neuronal function in the kitten's visual cortex: a laminar analysis. *J. Physiol. (London)* **348,** 153–185.

Anderson, P. A., Olavarria, J., and Van Sluyters, R. C. (1988). The overall pattern of ocular dominance bands in cat visual cortex. *J. Neurosci.* **8,** 2183–2200.

Beckmann, R., and Albus, K. (1982). The geniculocortical system in the early postnatal kitten: an electrophysiological investigation. *Exp. Brain Res.* **47,** 49–56.

Blakemore, C., and Van Sluyters, R. C. (1975). Innate and environmental factors in the development of the kitten's visual cortex. *J. Physiol. (London)* **248,** 663–716.

Blasdel, G. G., and Salama, G. (1986). Voltage-sensitive dyes reveal a modular organization in monkey striate cortex. *Nature* **321,** 579–585.

Bonds, A. B., Silverman, M. S., Sclar, G., and Tootell, R. B. H. (1980). Visually evoked potentials and deoxyglucose studies of monocularly deprived cats. *Invest. Ophthalmol. Vis. Sci. Suppl.* **19,** 225–226.

Bonhoeffer, T., and Grinvald, A. (1991). Iso-orientation domains in cat visual cortex are arranged in pinwheel-like patterns. *Nature* **353,** 429–431.

Braitenberg, V., and Braitenberg, C. (1979). Geometry of orientation columns in the visual cortex. *Biol. Cybern.* **33,** 179–186.

Brodmann, K. (1909). *Vergleichende Lokalisationslehre der Groβhirnrinde in ihren Prinzipien dargestellt auf Grund des Zellenbaues.* Leipzig, Joh. Ambr. Barth.

Campbell, F. W., Cooper, G. F., and Enroth-Cugell, C. (1969). The spatial selectivity of the visual cells of the cat. *J. Physiol. (London)* **203,** 223–235.

Cooper, G. F., and Robson, J. G. (1968). Successive transformations of spatial information in the visual system. *IEE Conference Publication* **47,** 134–143.

Crair, M. C., Gillespie, D. C., and Stryker, M. P. (1998). The role of visual experience in the development of columns in cat visual cortex. *Science* **279,** 566–570.

De Valois, R. L., Albrecht, D. G., and Thorell, L. G. (1977). Spatial tuning of LGN and cortical cells in monkey visual system. In: *Spatial contrast* (H. Spekreijse, and L. H. van der Tweel, Eds.), pp. 60–63. North Holland, Amsterdam.

Duncan, G. E., and Stumpf, W. E. (1991). Brain activity patterns: assessment by high resolution autoradiographic imaging of radiolabeled 2-deoxyglucose and glucose uptake. *Prog. Neurobiol.* **37,** 365–382.

Duncan, G. E., Stumpf, W. E., and Pilgrim, C. (1987). Cerebral metabolic mapping at the cellular level with dry-mount autoradiography of [^{3}H]2-deoxyglucose. *Brain Res.* **401,** 43–49.

Durham, D., Woolsey, T. A., and Kruger, L. (1981). Cellular localization of 2-[^{3}H]deoxy-D-glucose from paraffin-embedded brain. *J. Neurosci.* **1,** 519–526.

Flood, D. G., and Coleman, P. D. (1979). Demonstration of orientation columns with [^{14}C]2-deoxyglucose in a cat reared in a striped environment. *Brain Res.* **173,** 538–542.

Freeman, B., Löwel, S., and Singer, W. (1987). Deoxyglucose mapping in the cat visual cortex following carotid artery injection and cortical flat-mounting. *J. Neurosci. Methods* **20,** 115–129.

Freeman, R. D., and Tsumoto, T. (1983). An electrophysiological comparison of convergent and divergent strabismus in the cat: electrical and visual activation of single cortical cells. *J. Neurophysiol.* **49,** 238–253.

Frégnac, Y., and Imbert, M. (1978). Early development of visual cortical cells in normal and dark-reared kittens: Relationship between orientation selectivity and ocular dominance. *J. Physiol. (London)* **278,** 27–44.

Frégnac, Y., and Imbert, M. (1984). Development of neuronal selectivity in the primary visual cortex of the cat. *Physiol. Rev.* **64,** 325–434.

Friedman, H. R., Bruce, C. J., and Goldman-Rakic, P. S. (1987). A sequential double-label ^{14}C-and ^{3}H-2-DG technique: validation by double-dissociation of functional states. *Exp. Brain Res.* **66,** 543–554.

Geesaman, B. J., Born, R. T., Andersen, R. A., and Tootell, R. B. H. (1997). Maps of complex motion selectivity in the superior temporal cortex of the alert macaque monkey: a double-label 2-deoxyglucose study. *Cerebral Cortex* **7,** 749–757.

Gilbert, C. D., and Wiesel, T. N. (1989). Columnar specificity of intrinsic horizontal and corticocortical connections in cat visual cortex. *J. Neurosci.* **9,** 2432–2442.

Grafstein, B. (1971). Transneuronal transfer of radioactivity in the central nervous system. *Science* **172,** 177–179.

Grinvald, A., Frostig, R. D., Siegel, R. M., and Bartfeld, E. (1991). High-resolution optical imaging of functional brain architecture in the awake monkey. *Proc. Natl. Acad. Sci. U.S.A.* **88,** 11559–11563.

Hand, P. J. (1981). The 2-deoxyglucose method. In: *Neuroanatomical tract tracing methods* (L. Heimer, and M. J. Robards, Eds.), pp. 511–538, New York, Plenum Press.

Henry, G. H., Michalski, A., Wimborne, B. M., and McCart, R. J. (1994). The nature and origin of orientation specificity in neurons of the visual pathways. *Prog. Neurobiol.* **43,** 381–437.

Horton, J. C., and Hocking, D. R. (1996). Intrinsic variability of ocular dominance column periodicity in normal macaque monkeys. *J. Neurosci.* **16,** 7228–7239.

Hubel, D. H., and Wiesel, T. N. (1959) Receptive fields of single neurones in the cat's striate cortex. *J. Physiol. (London)* **148,** 574–591.

Hubel, D. H., and Wiesel, T. N. (1962). Receptive fields, binocular interaction and functional architecture in the cat's visual cortex. *J. Physiol. (London)* **160,** 106–154.

Hubel, D. H., and Wiesel, T. N. (1963a). Shape and arrangement of columns in cat's striate cortex. *J. Physiol. (London)* **160,** 106–154.

Hubel, D. H., and Wiesel, T. N. (1963b). Receptive fields of cells in striate cortex of very young, visually inexperienced kittens. *J. Neurophysiol.* **26,** 994–1002.

Hubel, D. H., and Wiesel, T. N. (1965). Binocular interaction in striate cortex of kittens reared with artificial squint. *J. Neurophysiol.* **28,** 1041–1059.

Hubel, D. H., and Wiesel, T. N. (1970). The period of susceptibility to the physiological effects of unilateral eye closure in kittens. *J. Physiol. (London)* **206,** 419–436.

Hubel, D. H., and Wiesel, T. N. (1974). Sequence regularity and geometry of orientation columns in the monkey striate cortex. *J. Comp. Neurol.* **158,** 267–293.

Hübener, M., Shoham, D., Grinvald, A., and Bonhoeffer, T. (1997). Spatial relationships among three columnar systems in cat area 17. *J. Neurosci.* **17,** 9270–9284.

John, E. R., Tang, Y., Brill, A. B., Young, R., and Ono, K. (1986). Double-labeled metabolic maps of memory. *Science* **233,** 1167–1175.

Jones, D. G., Murphy, K. M., and Van Sluyters, R. C. (1996). Spacing of ocular dominance columns is not changed by monocular deprivation or strabismus. *Invest. Ophthalmol. Vis. Sci. Suppl.* **37,** 1964.

Kalil, R. E., Spear, P. D., and Langsetmo, A. (1984). Response properties of striate cortex neurons in cats raised with divergent or convergent strabismus. *J. Neurophysiol.* **52,** 514–537.

Katz, L. C., and Iarovici, D. M. (1990). Green fluorescent latex microspheres: a new retrograde tracer. *Neuroscience* **34,** 511–520.

Katz, L. C., Burkhalter, A., and Dreyer, W. J. (1984). Fluorescent latex microspheres as a retrograde neuronal marker for in vivo and in vitro studies of visual cortex. *Nature* **310,** 498–500.

Kennedy, C., DesRosiers, M. H., Sakurada, O., Shinohara, M., Reivich, M., Jehle, H. W., and Sokoloff, L. (1976). Metabolic mapping of the primary visual system of the monkey by means of the autoradiographic [^{14}C]deoxyglucose technique. *Proc. Natl. Acad. Sci. U.S.A.* **73,** 4230–4234.

Kim, D.-S., Duong, T. Q., and Kim, S. G. (2000). High-resolution mapping of iso-orientation columns by fMRI. *Nat. Neurosci.* **3,** 164–169.

Kossut, M., Thompson, I. D., and Blakemore, C. (1983). Ocular dominance columns in cat striate cortex and effects of monocular deprivation: a 2-deoxyglucose study. *Acta Neurobiol. Exp.* **43,** 273–282.

Lang, W., and Henn, V. (1980). Columnar pattern in the cat's visual cortex after optokinetic stimulation. *Brain Res.* **182,** 446–450.

LeVay, S., and Nelson, S. B. (1991). Columnar organization of the visual cortex. In: *Vision and visual dysfunction* (J. R. Cronly-Dillon, Ed.), pp. 266–315, Houndmills, England: Macmillan.

Livingstone, M. S., and Hubel, D. H. (1981). Effects of sleep and arousal on the processing of visual information in the cat. *Nature* **291,** 554–561.

Löwel, S. (1994). Ocular dominance column development: strabismus changes the spacing of adjacent columns in cat visual cortex. *J. Neurosci.* **14,** 7451–7468.

Löwel, S., and Singer, W. (1987). The pattern of ocular dominance columns in flat-mounts of the cat visual cortex. *Exp. Brain Res.* **68,** 661–666.

Löwel, S., and Singer, W. (1990). Tangential intracortical pathways and the development of iso-orientation bands in cat striate cortex. *Dev. Brain Res.* **56,** 99–116.

Löwel, S., and Singer, W. (1992). Selection of intrinsic horizontal connections in the visual cortex by correlated neuronal activity. *Science* **255,** 209–212.

Löwel, S., and Singer, W. (1993a). Monocularly induced 2-deoxyglucose patterns in the visual cortex and lateral geniculate nucleus of the cat. I. Anaesthetized and paralyzed animals. *Eur. J. Neurosci.* **5,** 846–856.

Löwel, S., and Singer, W. (1993b). Monocularly induced 2-deoxyglucose patterns in the visual cortex and lateral geniculate nucleus of the cat. II. Awake animals and strabismic animals. *Eur. J. Neurosci.* **5,** 857–869.

Löwel, S., Freeman, B., and Singer, W. (1987). Topographic organization of the orientation column system in large flat-mounts of the cat visual cortex: a 2-deoxyglucose study. *J. Comp. Neurol.* **255,** 401–415.

Löwel, S., Schmidt, K. E., Kim, D. -S., Wolf, F., Hoffsümmer, F., and Singer, W. (1998). The layout of orientation and ocular dominance domains in area 17 of strabismic cats. *Eur. J. Neurosci.* **10,** 2629–2643.

Maffei, L., and Fiorentini, A. (1973). The visual cortex as a spatial frequency analyser. *Vision Res.* **13,** 1255–1267.

Maffei, L., and Fiorentini, A. (1977). Spatial frequency rows in the striate visual cortex. *Vision Res.* **17,** 257–264.

Mountcastle, V. B. (1957). Modality and topographic properties of single neurons of cat's somatic sensory cortex. *J. Neurophysiol.* **20,** 408–434.

Movshon, J. A., Thompson, I. D., and Tolhurst, D. J. (1978). Spatial summation in the receptive fields of simple cells in the cat's striate cortex. *J. Physiol. (London)* **283,** 53–77.

Murphy, P. C., and Sillito, A. M. (1986). Continuity of orientation columns between superficial and deep laminae of the cat primary visual cortex. *J. Physiol. (London)* **381,** 95–110.

Olavarria, J., and Van Sluyters, R. C. (1985). Unfolding and flattening the cortex of gyrencephalic brains. *J. Neurosci. Methods* **15,** 191–202.

Olds, J. L., Frey, K. A., Ehrenkaufer, R. L., and Agranoff, B. W. (1985). A sequential double-lable autoradiographic method that quantifies altered rates of regional glucose metabolism. *Brain Res.* **361,** 217–224.

Orban, G. A. (1991). Quantitative electrophysiology of visual cortical neurones. In: *Vision and visual dysfunction* (J. R. Cronly-Dillon, Ed.), pp. 173–222, Houndmills, England: Macmillan.

Plum, F., Gjedde, A., and Samson, F. E. (Eds.) (1976). Neuroanatomical functional mapping by the radioactive 2-deoxy-D-glucose method. *Neurosci. Res. Prog. Bull.* **14,** 457–518.

Rathjen, S., Schmidt, K. E., and Löwel, S. (1999). The development of ocular dominance columns in primary visual cortex of normally raised and strabismic kittens. In: *From molecular neurobiology to clinical neuroscience, Proceedings of the 1st* Göttingen Conference of the German Neuroscience Society (27th Göttingen Neurobilogy Conference) (N. Elsner, and U. Eysel, Eds.), p. 478, Stuttgart-New York, Thieme.

Redies, C., and Gjedde, A. (1989). Double-label and conventional deoxyglucose methods: a practical guide for the user. *Cerebrovasc. Brain Metab. Rev.* **1,** 319–367.

Redies, C., Diksic, M., Evans, A. C., Gjedde, A., and Yamamoto, Y. L. (1987). Double-label autoradiographic deoxyglucose method for sequential measurement of regional cerebral glucose utilization. *Neuroscience* **22,** 601–619.

Schmidt, K. E., Kim, D. -S., Singer, W., Bonhoeffer, T., and Löwel, S. (1997). Functional specificity of long-range intrinsic and interhemispheric connections in the visual cortex of strabismic cats. *J. Neurosci.* **17,** 5480–5492.

Schoppmann, A., and Stryker, M. P. (1981). Physiological evidence that the 2-deoxyglucose method reveals orientation columns in cat visual cortex. *Nature* **293,** 574–576.

Sengpiel, F., Blakemore, C., Kind, P. C., and Harrad, R. (1994). Interocular suppression in visual cortex of strabismic cats. *J. Neurosci.* **14,** 6855–6871.

Sengpiel, F., Gödecke, I., Stawinski, P., Hübener, M., Löwel, S., and Bonhoeffer, T. (1998). Intrinsic and environmental factors in the development of functional maps in cat visual cortex. *Neuropharmacology* **37,** 607–621.

Sengpiel, F., Stawinski, P., and Bonhoeffer, T. (1999). Influence of experience on orientation maps in cat visual cortex. *Nat. Neurosci.* **2,** 727–732.

Shatz, C. J., and Stryker, M. P. (1978). Ocular dominance in layer IV of the cat's visual cortex and the effects of monocular deprivation. *J. Physiol. (London)* **281,** 267–283.

Shatz, C. J., Lindström, S., and Wiesel, T. N. (1977). The distribution of afferents representing the right and left eyes in the cat's visual cortex. *Brain Res.* **131,** 103–116.

Singer, W. (1981). Topographic organization of orientation columns in the cat visual cortex. A deoxyglucose study. *Exp. Brain Res.* **44,** 431–436.

Singer, W., Freeman, B., and Rauschecker, J. (1981). Restriction of visual experience to a single orientation affects the organization of orientation columns in cat visual cortex. *Exp. Brain Res.* **41,** 199–215.

Sokoloff, L. (1985). Mapping local functional activity by measurement of local cerebral glucose utilization in the central nervous system of animals and man. *Harvey Lect. 1983–84* **79,** 77–143.

Sokoloff, L. (1992). The brain as a chemical machine. *Prog. Brain Res.* **94,** 19–33.

Sokoloff, L., Reivich, M., Kennedy, C., DesRosiers, M. H., Patlak, C. S., Pettigrew, K. D., Sakurada, O., and Shinohara, M. (1977). The [^{14}C]deoxyglucose method for the measurement of local cerebral glucose utilization: theory, procedure, and normal values in the conscious and anesthetized albino rat. *J. Neurochem.* **28,** 897–916.

Thompson, I. D., and Tolhurst, D. J. (1979). The representation of spatial frequency in cat visual cortex: a ^{14}C-2-deoxyglucose study. *J. Physiol. (London)* **300:** 58–59P.

Thompson, I. D., Kossut, M., and Blakemore, C. (1983). Development of orientation columns in cat striate cortex revelaed by 2-deoxyglucose autoradiography. *Nature* **301,** 712–715.

Tieman, S. B., and Tumosa, N. (1983). [^{14}C]2-deoxyglucose demonstration of the organization of ocular dominance in areas 17 and 18 of the normal cat. *Brain Res.* **267,** 35–46.

Tieman, S. B., and Tumosa, N. (1997). Alternating monocular exposure increases the spacing of ocularity domains in area 17 of cats. *Vis. Neurosci.* **14,** 929–938.

Tolhurst, D. J., and Thompson, I. D. (1981). On the variety of spatial frequency selectivities shown by neurones in area 17 of the cat. *Proc. R. Soc. Lond. B Biol. Sci.* **213,** 183–199.

Tolhurst, D. J., and Thompson, I. D. (1982). Organization of neurones preferring similar spatial frequencies in cat striate cortex. *Exp. Brain Res.* **48,** 217–227.

Tootell, R. B. H., and Silverman, M. S. (1985). Two methods for flat-mounting cortical tissue. *J. Neurosci. Methods* **15,** 177–190.

Tootell, R. B. H., Silverman, M. S., and De Valois, R. L. (1981). Spatial frequency columns in primary visual cortex. *Science* **214,** 813–815.

Tootell, R. B. H., Silverman, M. S., Hamilton, S. L., Switkes, E., and De Valois, R. L. (1988). Functional anatomy of macaque striate cortex. V. Spatial frequency. *J. Neurosci.* **8,** 1610–1624.

Tumosa, N., Tieman, S. B., and Tieman, D. G. (1989). Binocular competition affects the pattern and intensity of ocular activation columns in the visual cortex of cats. *Vis. Neurosci.* **2,** 391–407.

Tusa, R. J., Palmer, L. A., and Rosenquist, A. C. (1978). The retinotopic organization of cat area 17 (striate cortex) in the cat. *J. Comp. Neurol.* **177,** 213–236.

Vnek, N., Ramsden, B. M., Hung, C. P., Goldman-Rakic, P. S., and Roe, A. W. (1999). Optical imaging of functional domains in the cortex of the awake and behaving monkey. *Proc. Natl. Acad. Sci. U.S.A.* **96,** 4057–4060.

Wagner, H.-J., Hoffmann, K.-P., and Zwerger, H. (1981). Layer-specific labelling of cat visual cortex after stimulation with visual noise: a [^{3}H]2-deoxy-D-glucose study. *Brain Res.* **224,** 31–43.

Wiesel, T. N., Hubel, D. H., and Lam, D. M. K. (1974). Autoradiographic demonstration of ocular-dominance columns in the monkey striate cortex by means of transneuronal transport. *Brain Res.* **79,** 273–279.

4

Functional Mapping in the Cat Primary Visual Cortex Using High Magnetic Fields

Dae-Shik Kim, Timothy Q. Duong, Kamil Ugurbil, and Seong-Gi Kim

Center for Magnetic Resonance Research, University of Minnesota Medical School, Minneapolis, Minnesota

Introduction

In the mammalian brain cytoarchitectonically distinct areas form the basis for functional specialization (1). Such parcellation of the cortical tissue into functional subunits is especially prominent in the primary visual cortex of cats. Here, neurons with similar response properties, such as ocular dominance (2, 3), orientation (4, 5), and direction (6, 7) preferences, are clustered into "columns," spanning the entire cortical plate from the pia to the white matter.

Since the pioneering works of Hubel and Wiesel (2), the temporal and spatial properties of cortical columns in cat primary visual cortex have been studied extensively using a variety of techniques. Understanding of the temporal properties of the cortical columns promises to reveal the "neural code" of the brain's information processing, and elucidation of its spatial properties will shed light on how such codes are physically implemented across the cortex.

To this end, however, the applicability and fidelity of the traditional techniques to achieve this goal differ markedly. The use of single and multiunit recordings provides a good tool for assessing the temporal properties of cortical columns; finding a technique that can yield the spatial properties of cortical columns with comparable accuracy, convenience, and applicability proved to be a much more difficult task.

In this chapter, we describe recent advances in functional magnetic resonance imaging (fMRI) we have made that will enable the functional organization of the cat visual cortex to be studied at "columnar" level, thereby approximating the spatial resolution of multielectrode and optical imaging techniques, but without their limitations. These advances should lay the foundation for future noninvasive exploration of animal and human brains at the fundamental level of its columnar architecture noninvasively. In Chapter 4, we describe the imaging principles and techniques of this new method and present the first MRI studies of the functional activity in cat primary visual cortex at columnar resolution (see also Kim et al. [8]).

LIMITATIONS OF CURRENT TECHNIQUES, OR WHY WE NEED ANOTHER MAPPING TECHNIQUE

The spatial layout of the functional architecture in cat primary visual cortex has been studied extensively using a variety of different techniques (see LeVay and Nelson [9] and Chapter 1 for review). Although these techniques have provided pivotal insights, each of the traditional mapping techniques suffers from a fundamental limitation. For example, the multielectrode mapping method lacks sufficient field of view and spatial resolution (7), while the 2-deoxyglucose method is not viable for in vivo mapping (10). The more recently developed "optical imaging of intrinsic signals" allows the simultaneous recording of neuronal activity over large areas of cortex (11,12). For example, in cat primary visual cortex, this technique has been utilized extensively to examine the tangential organization of multiple cortical maps and their topological relationships to each other at a columnar resolution 13, 14) (see Chapter 2). However, optical imaging at this spatial resolution is an invasive technique, because a craniotomy has to be made to expose the cortical surface. Furthermore, the applicability of optical imaging is limited to the superficial layers of the cortical surface only (15–17), thus failing to reveal activation signals originating from deeper cortical layers or from subcortical structures.

FUNCTIONAL MAGNETIC RESONANCE IMAGING

To further facilitate our understanding of the cortical information processing in adult and developing animals, it is imperative to devise a novel method that can visualize the functional architecture of the living brain without the aforementioned limitations. The rapid progress of blood oxygenation level-dependent (BOLD) functional magnetic resonance imaging (fMRI) in recent years has raised the hope that this could be achieved. Using the paramagnetic deoxyhemoglobin as an endogenous contrast agent (18, 19), BOLD-based functional images can be obtained in vivo (in contrast to the 2-deoxyglucose technique) without the use of extrinsic contrast agents (in contrast to the positron emission technique), and from the entire brain (in contrast to the optical imaging of intrinsic signals). Most importantly, the noninvasiveness of MRI suits it well for studying the functional architecture of both the human and animal brains, thus enabling direct comparisons of putative homolog functions between humans and animals.

Despite these advantages, crucial questions must be answered before functional MRI can be utilized for addressing neurophysiological questions. For example, numerous BOLD studies during cognitive (20), motor (21), and perceptual (22–25) tasks indicate a good spatial correlation between neuronal and hemodynamic responses at a scale of several millimeters to centimeters. With this coarse scale, however, it is impossible to discriminate between the cortical columns representing individual receptive field properties, as the average periodicity between two adjacent isofunctional columns is about 1mm in most mammalian species (26). Furthermore, the ultimate functional specificity of fMRI remains highly controversial because the BOLD signal represents a complex convolution of metabolic and hemodynamic processes that is poorly understood (27).

Recent results from our laboratory and other laboratories now suggest that there is an intriguing relationship between the details of cortical hemodynamic responses and the ultimate spatial resolution of functional MRI. If utilized properly, this information will enable us to improve the functional specificity of fMRI far beyond what is achievable today.

NUCLEAR MAGNETIC RESONANCE AND MAGNETIC RESONANCE IMAGING

As the majority of the readers of this volume may not have a firsthand experience with high-field fMRI, it is worthwhile to start our discussion with a brief description of the theoretical and technical foundations of fMRI.

The phenomenon of nuclear magnetic resonance (NMR) was described in landmark articles more than 50 years ago (28, 29). The proton nucleus of the hydrogen atom, which is used for most imaging studies, possesses a small magnetic moment. When placed in a strong and homogenous magnetic field ("static field", B_0), the atoms align themselves with this field and reach a thermal equilibrium. The nuclei precess about the static field at a characteristic frequency (Larmor frequency), but at a random orientation with respect to each other. Application of a brief radio frequency excitation pulse (RF pulse) with Lamor frequency induces a rotating magnetic field orthogonal to the static field, thus causing the nuclei to precess away from the equilibrium. Using a "receiver coil," the resulting rotating magnetic moment of the combined nuclei can then be observed as a time-dependent electromagnetic signal. Three of the parameters that can be observed in NMR are (1) the longitudinal relaxation time constant ("spin-lattice relaxation"); T1, is the constant that characterizes the nuclei reapproaching the thermal equilibrium; (2) the transverse relaxation time constant ("spin-spin relaxation"), T_2, is the rate at which the MR signal decays *irreversibly* after a RF pulse in a uniform magnetic field owing to spin-spin interaction and/or due to rapid motion in the presence of magnetic field gradients; and (3) the time constant T_2*, which includes T_2 relaxation, but in addition, describes also the *reversible* signal loss owing to magnetic field inhomogeneities.

In 1973, a novel concept of using NMR of the hydrogen atoms as an imaging modality was introduced by Lauterbur (30). Although all NMR applications at the time were singularly concerned with eliminating inhomogeneities in static magnetic field magnitude over the sample, this new concept embraced them and proposed to use them to extract *spatial information.* The idea behind NMR imaging (called magnetic resonance imaging, MRI) is the fact that the precessional frequency of the nuclei depends on the local magnetic field. Thus by varying the local strength of the static magnetic field, it would be possible to separate the NMR signals from different locations according to their precessional frequency. Current MRI scanners use three mutually orthogonal sets of electromagnetic gradient coils (termed x, y, z gradients) to encode the three spatial coordinates of the MR signals.

IMAGING BRAIN ACTIVITY USING MRI

Traditionally, MRI of the brain has been used mainly to study the anatomical structures of the brain. Recent developments in MR now allow visualization of the "activity" of the living brain. The first task-induced functional imaging in humans was performed using an exogenous paramagnetic contrast agent, gadolinium diethylenetriamine pentaacetic acid (Gd-DTPA), which remains intravascular in the brain in the presence of an intact blood-brain barrier (31). Following a bolus administration of Gd-DTPA, MR signal from vessels and surrounding tissue areas decreases during the first passage. By integrating over the first passage of the contrast agent, cerebral blood volume (CBV) can be determined (32). Using this technique, an increase of CBV in the human visual cortex was observed successfully during visual stimulation (31), showing that MRI can indeed detect small signal changes induced by neuronal activity in the human brain.

In 1992, based on earlier works by Ogawa et al. (18), investigators in three laboratories (19, 33, 34) reported the first functional images of the human brain using BOLD contrast. This contrast mechanism is used for most of the functional MRI studies today. The basis of the BOLD technique is that deoxyhemoglobin acts as nature's own intravascular paramagnetic contrast agent (18, 35, 36). When placed in a magnetic field, deoxyhemoglobin alters the magnetic field in its vicinity, particularly when it is compartmentalized, as it is within red blood cells and vasculature. The effect increases as the concentration of deoxyhemoglobin increases. At concentrations found in venous blood vessels, a detectable local distortion of the magnetic field surrounding the red blood cells and surrounding blood vessel is produced. This affects the magnetic resonance behavior of the water proton nuclei within and surrounding the vessels, which in turn results in *decreases* in the transverse relaxation times (T_2 and T_2*) (18, 36). During the activation of the brain, this process is reduced: an increase in neuronal and metabolic activity results in a *reduction* of the relative deoxyhemoglobin concentration owing to an increase of blood flow (and hence increased supply of fresh oxyhemoglobin) that follows. Consequently, in conventional BOLD fMRI, brain "activity" can be measured as an *increase* in T_2 or T_2* weighted MR signals (19, 33, 34).

VASCULATURE ISSUES

BOLD fMRI is founded on the pioneering discovery of Roy and Sherrington (37), that changes in electrical activity are coupled to hemodynamic responses in the vasculature system. Consequently, for a proper understanding of the BOLD signals, the size, geometry, and location of the cortical vasculature surrounding the neural tissue must be considered. For the sake of further discussion, we define a microvascular/tissue component as an area of rapid exchange between blood water and tissue water (i.e., capillaries and adjacent tissues). In the parenchymal tissue of the cat cerebral cortex, such capillaries form a dense network with an average spacing of about 20 μm (38). By contrast, the macrovascular component is composed of arteries, small arterioles, small venules, and veins, where such an exchange does not exist.

It is well known that, with typical fMRI acquisition parameters, the BOLD response is particularly sensitive in and around large "draining veins," which are usually distant from the sites of elevated neuronal activity (27, 39, 40). Thus it is reasonable to assume that conventional BOLD-based fMRI may overrepresent the "non-functional" large-vessel contribution, whereas the functional signals from small capillaries may be masked. Furthermore, recent optical spectroscopy data (41, 42) also suggest that there is an intimate relationship between the source of the hemodynamic signal and its spatial specificity. In particular, signals from large draining vessels were found to be spread out up to 3 to 7 mm from the actual site of electrical activity (43).

In any kind of brain mapping method based on hemodynamic activity, it is therefore imperative to minimize the pertubative effect of draining vessels. To this end, our attempt to visualize cortical columns with functional MRI is aided by the fact that the vascular origin of the BOLD effect depends heavily on *magnetic field strength.* BOLD response is expected to behave according to the following equations (27, 44):

$$R_2{}^* = \alpha\{\Delta\chi_o\omega_o(1\text{-}Y)\}b_{vl} \quad \text{(large blood vessels)} \tag{1}$$

$$R_2{}^* = \eta\{\Delta\chi_o\omega_o(1\text{-}Y)\}^2\, b_{vs}p \quad \text{(small blood vessels)} \tag{2}$$

where $R_2{}^* = 1/T_2{}^*$, α and η are constants, ω_o is the external magnetic field in frequency units (rad/sec), $\Delta\chi_o$ is the maximum susceptibility difference expected in the presence of fully deoxygenated blood, Y is the fraction of oxygenated blood present, $\{\Delta\chi_o\omega_o(1\text{-}Y)\}$ is the frequency shift due to the susceptibility difference between the deoxyhemoglobin containing blood vessel, b_{vl} is the blood volume for *large* blood and b_{vs} is the *small* vessel blood volume, and p is the fraction of active small vessels (i.e., filled with deoxyhemoglobin containing red blood cells).

An important feature of equation 2 is that for small blood vessels, the equation varies as the square of the external magnetic field where the effect is dominated by dynamic averaging of magnetic field inhomogeneities surrounding deoxyhemoglobin containing blood vessels. In contrast, the dependence on the external magnetic field is linear for large blood vessels (equation 1). A stronger magnetic field therefore will not only increase the fMRI signal *per se* but will specifically enhance the signal components originating from parenchymal capillary tissue. On the other hand, conventional, low-field magnets can be expected to overrepresent macrovascular signals. In fact, experimental data suggest that most of the "functional" signals obtained with clinical 1.5 Tesla scanners may represent signals from large draining vessels (45, 46).

FUNCTIONAL MRI OF THE CAT PRIMARY VISUAL CORTEX

Despite the pioneering works of Roy and Sherrington, the functional specificity of the cortical hemodynamic response remains highly controversial. The crucial question here is whether cortical hemodynamic response after sensory stimulation is confined to domains of increased neuronal activity. If it is, then ultimately BOLD fMRI should be capable of discriminating signals at individual columnar level.

THE TECHNIQUES

To resolve the question whether and to what extent the columnar architecture of the brain can be labeled using the BOLD-fMRI technique, we used ultra-high-field magnets to obtain MR signals originating from individual orientation columns in cat primary visual cortex (area 18). To this end, we took advantage of the fact that the use of ultra-high-field magnets reduces the corruptive effect of large draining vessels per se (46, 47). Furthermore, our studies were guided by the hypothesis that the strongest hemodynamic responses may not necessarily represent the most important signals in terms of their functional significance.

Figure 4-1 depicts our experimental setup. Juvenile cats were initially anesthetized and maintained under isoflurane anesthesia following standard procedures (5, 13, 14, 16). The animals were positioned in a cradle and restrained in normal postural position using a custom-designed stereotaxic frame made of nonmagnetic plexiglass. Visual stimuli consisted of high-contrast square-wave moving gratings (0.15 cyc/deg, 2 cyc/sec) of four different orientations (0°, 45°, 90°,

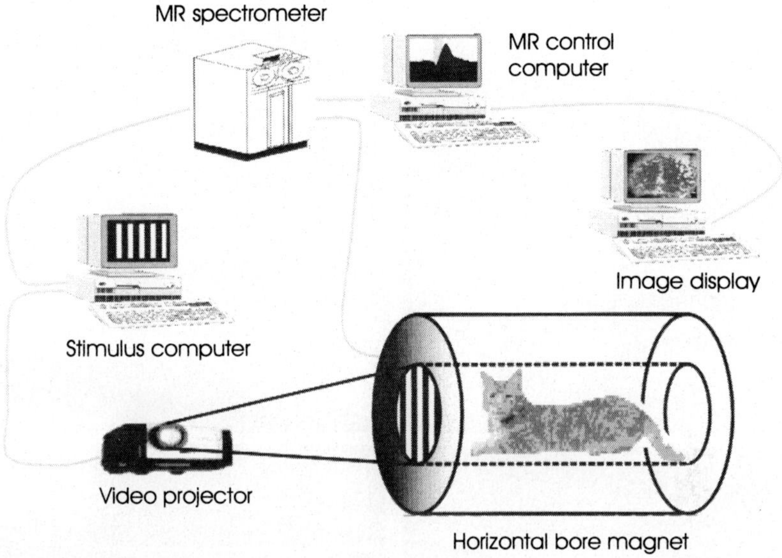

FIGURE 4-1. High-field cat MR experimental setup. Anesthetized cat was placed in a custom designed, MR-compatible stereotaxic holder and inserted into the isocenter of a 4.7 or 9.4 Tesla horizontal bore magnet. A video projector was used to stimulate the animal with moving gratings of four different orientations. Stimulus presentation was synchronized with MR data acquisition. See text for details.

135°) optimized to activate neurons in cat area 18 (48). We have used area 18 for our studies because the distance between two iso-orientation columns in this area is wider than in area 17, thus maximizing our chance of resolving individual columns. Furthermore, the cat area 18 on the lateral gyrus is essentially flat and can be covered by a single imaging slice. Area 17 is curved around the occipital pole and causes more susceptibility artifacts in MR images.

A small surface coil, 1.2 cm in diameter, was placed on top of the animal's head corresponding to the Horsley-Clark A3. All MR experiments were performed on a 4.7-T/40 cm (Oxford, UK) or 9.4-T/31 cm (Magnex, UK) horizontal MRI scanners equipped with a 15-G/cm or 30-G/cm magnetic field gradients, respectively. Figure 4-2 shows mid-sagittal (Fig. 4-2A), coronal (Fig. 4-2B), and axial (Fig. 4-2C) anatomical images of the cat brain obtained with a 9.4 Tesla magnet. The *sulcus cruciatus, sulcus splenialis,* and *sulcus suprasplenialis* are readily detectable on the mid-sagittal image. The white and gray matters in coronal and axial images are visible as bright and dark areas, respectively. Panel d displays the surface image of the left and right lateral gyri (left and right half of the image). The center of this surface image was at the anteroposterior 3 of the Horsley-Clark coordinate system, corresponding to the central visual field of the area

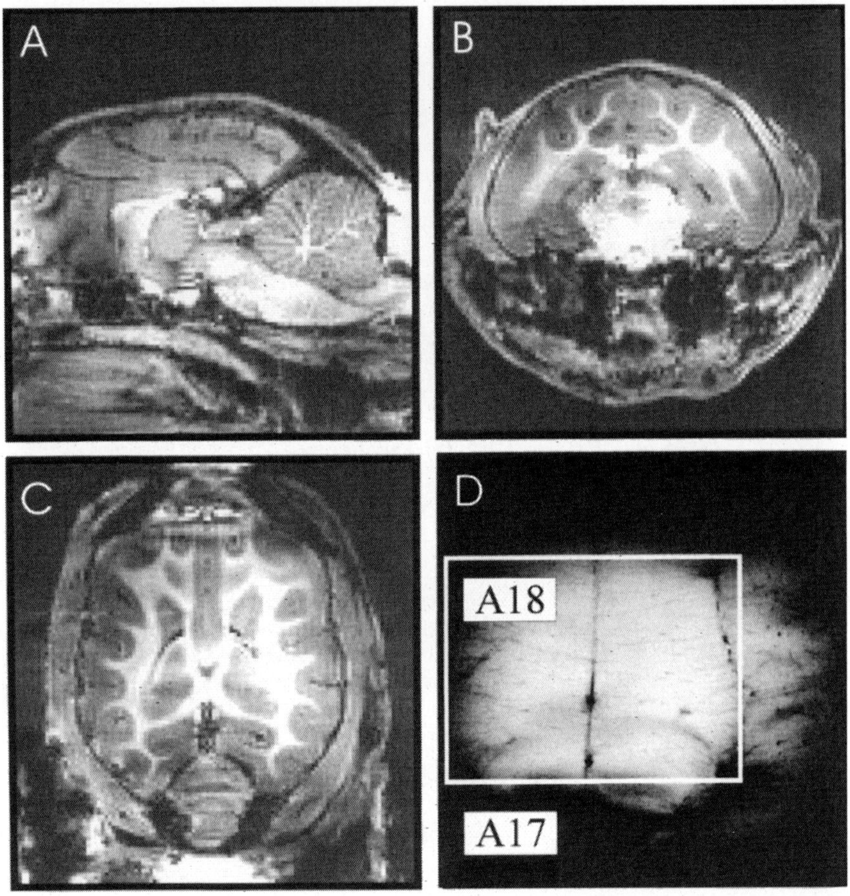

FIGURE 4-2. Mid-sagittal (**A**), coronal (**B**), and axial (**C**) anatomical images of the cat brain obtained with high-field magnet. The *sulcus cruciatus, sulcus splenialis,* and *sulcus suprasplenialis* are readily detectable on the mid-sagittal image. The white and gray matters in coronal and axial images are visible as bright and dark areas, respectively. Panel D displays the surface image of the left and right lateral gyri (left and right half of the image). The white box indicates the region of interest that has been used for the functional MR studies as described in this chapter. A17, area 17; A18, area 18.

18. The curved line at the right side of the panel is the lateral sulcus of the right hemisphere. Based on the pattern of sulci on the cortical surface, this particular cat was classified as to be a Otsuka-Hassler Type I cat (49).

All functional images described in this chapter were obtained from a region of interest (ROI) that corresponded to the white rectangle in Fig. 4-2D. BOLD measurements on a single image slice were made using gradient-echo, echo-planar imaging (EPI) technique (50,51). The imaging slice was positioned ~500 μm below pia to avoid superficial draining vessels. The MRI parameters were: data matrix = 64 × 64; single-shot (4.7T) or 2-shot (9.4T) EPI, FOV = 2 × 2 cm, slice thickness = 2 mm, TE = 31 (4.7T) or 12 (9.4T) ms, and TR = 0.5 s. A total of 160 images were acquired during each epoch: 60 images before stimulation, 20 images during stimulation, and 80 images after stimulation. Images with 64×64 matrix were zero-filled to 128×128 matrix, resulting in nominal in-plane resolution of 156×156 μm^2/pixel. Images obtained for same orientations were averaged for signal-to-noise ratio improvement (usually 5 to 10 epochs).

CONVENTIONAL BOLD FMRI IN CAT PRIMARY VISUAL CORTEX

The four panels in Fig. 4-3 show fMRI in cat primary visual cortex during stimulation of the animal with moving gratings of four different orientations. The regions of *increased* BOLD signal change are gray scale coded (bright = maximum BOLD signal). Method of image construction and statistical analysis were kept analogous to those that have been established for use in conventional human fMRI studies (52).

As indicated in the four panels, robust and homogenous activities were observed in the lateral gyri of both hemispheres. The region of activity extended several millimeters in antero-posterior and mediolateral directions. The pattern of activity was relatively uniform in all four images. This is surprising, as an overwhelming number of electrophysiological (53), 2-deoxyglucose (2-DG) (27,54), and optical imaging (5,55) studies have demonstrated that the neurons with similar orientation preferences are spatially clustered into patchy domains (see Chapters by 2 and 3). Furthermore, columnar orientation maps obtained during stimulation with "orthogonal orientations" (e.g., 0° vs. 90°, and 45° vs. 135°) are known to be spatially complementary with respect to each other. As evident in Fig. 4-3, such a "columnar" and/or "complementary" layour was not obtained in our "conventional" BOLD fMRI studies in cat primary visual cortex. In fact, all four activation maps obtained in response to moving gratings of four different orientations (0° [Fig. 4-3A], 45° [Fig. 4-3B], 90° [Fig. 4-3C], and 135° [Fig. 4-3D]) yielded homogeneous spatial distributions that were hardly distinguishable from each other (see also Jezzard et al. [56]).

Most of today's fMRI in humans and animals are performed at millimeter to centimeter resolution, naturally failing to detect the columnar organization of the brain (23,24). In these cases, the limited "camera resolution" of most available MR scanners is too coarse for columnar studies. Such a "camera resolution" argument, however, is clearly not applicable for our data from cat primary visual cortex (Fig. 4-3). As is well known, in cat area 18, the average spacing between two neighboring iso-orientation columns is about 1.2 to 1.4mm (27 (also see Chapter 1)). The nominal in-plane resolution of our "camera" (our high-field MR scanner), on the other hand, was 156×156 μm^2 per pixel. The Nyquist sampling theorem (see, for example, Gonzalez and Woods [57]) tells us, therefore, that, if the nominal camera resolution were the only problem, we should have been able to resolve individual orientation columns. Obviously, there is something more fundamental to be understood here.

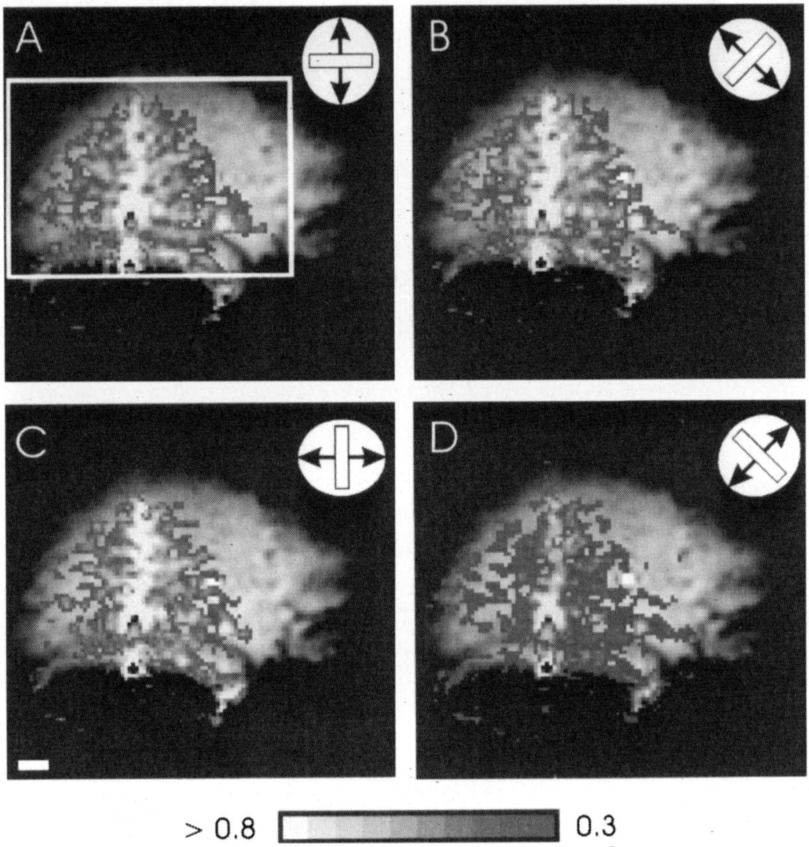

FIGURE 4-3. Functional MR images in cat primary visual cortex during stimulation of the animal with moving gratings of four different orientations. The regions of *increased* BOLD signal change are gray scale coded (bright = maximum BOLD signal). The numbers at the gray-scale key represent cross correlation coefficients. Robust and homogenous activities were observed in the lateral gyri of both hemispheres. The region of activity extended several millimeters in anterorposterior and mediolateral directions. All four activation maps obtained in response to moving gratings of four different orientations (0°, 45°, 90°, 135°) yielded homogeneous spatial distributions that were hardly distinguishable from each other. Scale bar: 1 mm.

TEMPORAL DYNAMICS OF BOLD

The most parsimonious explanation for these results can be found in the nature of the hemodynamic events during BOLD signal acquisition. Figure 4-4 summarizes, in a cartoon, the events following a hypothetical sensory stimulation.

Stimulation of the animal with moving gratings of a particular orientation, in this case 90°, will increase both the subthreshold and suprathreshold electrical activity among neurons located in the respective iso-orientation domains (see the left panel of Fig. 4-4). For the sake of simplicity, we assume the level of electrical activity in

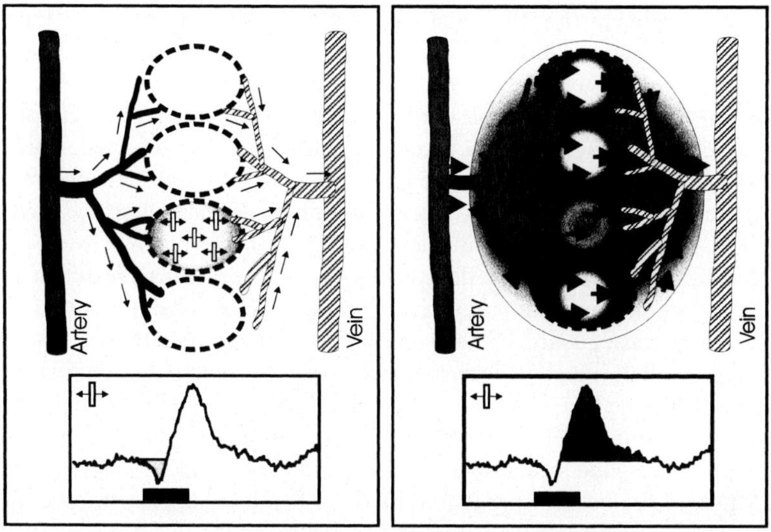

FIGURE 4-4. A "cartoon" summarizing the hemodynamic events following sensory stimulation. (Left panel) The hemodynamic events during the initial deoxygenation period. (Right panel) Hemodynamic events during the delayed oxygenation period. See text for details.

other columns—disregarding the lateral spread of activity through the plexus of horizontal connections—to remain more or less unchanged. The proposed hypothesis here is that the first event following increased neuronal activity will be a prolonged increase in oxygen consumption, caused by an elevation in oxidative metabolism of active neurons (58). Based on 2-DG data (26), we can assume the increase in oxidative metabolism in cat primary visual cortex to be colocalized with the site of electrical activity (see Chapter 3). The increase in oxidative metabolism will naturally elevate the local deoxyhemoglobin content in the parenchyma of active neurons, assuming there is no immediate commensurate change in cerebral blood flow (59). In T_2 or T_2*-weighted BOLD fMRI images, such increase in paramagnetic deoxyhemoglobin should be detectable as a transient *decrease* in observable MR signals (see the hypothetical time course in the left panel of Fig. 4-4).

Such an initial deoxygenation of the local cortical tissue (left panel) lasts only for a short time, as fresh blood (fresh oxyhemoglobin) rushes into capillaries in response to the increased metabolism (i.e., Roy and Sherrington's microcirculation response; see the right panel). Such increase in "fresh" oxyhemoglobin can be assumed to reverse the local ratio of hemoglobin in favor of oxyhemoglobin, thus resulting in an *increase* in observable MR signals. Obviously, this delayed oxygenation of the cortical tissue represents one of the bases for conventional BOLD signals (19).

Several lines of recent optical spectroscopy studies (41,42,60) suggest that, after sensory stimulation, such a "biphasic" response does indeed take place.

Their results indicate that the local deoxyhemoglobin concentration increases immediately after stimulation, yielding the maximum concentration about 2 to 3 seconds after stimulus onset. Oxygen consumption increase appears to be, at least in part, if not entirely, responsible for this early deoxyhemoglobin increase (61). The local deoxyhemoglobin concentration is subsequently reversed, ultimately resulting in a relatively large decrease in overall deoxyhemoglobin content.

The crucial question here is the "where" of the biphasic hemodynamic processes. To this end, it has been proposed that the two distinct hemodynamic events (initial deoxygenation followed by the delayed oxygenation) ought to result in fundamentally different functional specificity (41). It is assumed that the initial deoxygenation, as a consequence of an increase in oxidative metabolism, should be coregistered with the site of electrical activity up to the level of individual cortical columns. In fact, the well-established optical imaging of intrinsic signals (11,12), which has been cross-validated with single unit techniques (62–64), is similarly based on measuring the local transient increase of deoxyhemoglobin. The delayed oxygenation of the cortical tissue, on the other hand, is suggested to be far less specific. The assumption here is that the "fidelity" of the cortical vasculature is not sufficient to provide fresh oxyhemoglobin selectively to the active columns only. Instead, the entire local vicinity of the active column is flooded with fresh blood. Consequently, such delayed oxygenation signals, as detected with conventional BOLD technique, may surpass the spatial extent of the actual activated area by several millimeters (41,42).

Both the existence of "biphasic" BOLD response per se and the suggested differences in functional specificity continue to be debated. Although the initial deoxygenation signal in fMRI (termed *initial dip*) has been reported in humans (65–67) and monkeys (68), studies in rodents failed to detect any significant initial decrease in BOLD signal after sensory stimulation (69,70). Likewise, the spatial specificity of this dip remains unknown because most fMRI studies examining this phenomenon so far have been conducted in humans (65–67), thereby, by necessity using relatively coarse spatial resolution. In this context we need to address two crucial questions: (1) Are BOLD responses in cat primary visual cortex biphasic? (2) If so, does the initial deoxygenation signal label the layout of the cortical columns accurately, while the delayed oxygenation signals fail to discriminate individual cortical columns?

CAT PRIMARY VISUAL CORTEX—BIPHASIC RESPONSES?

Figure 4-5 demonstrates that, in cat primary visual cortex, the time evolution of MR signals is clearly biphasic. The temporal dynamics of the BOLD responses in cat primary visual cortex was in excellent agreement with optical spectroscopy data. After visual stimulation, the MR signal decreases, reaching a minimum of about –0.2 to –0.4% approximately 3 seconds after the stimulus onset. The signal then reverses, yielding a maximum positive signal change of 1.0 to 2.0% approximately 8 to 10 seconds after stimulus onset. Such biphasic BOLD responses after

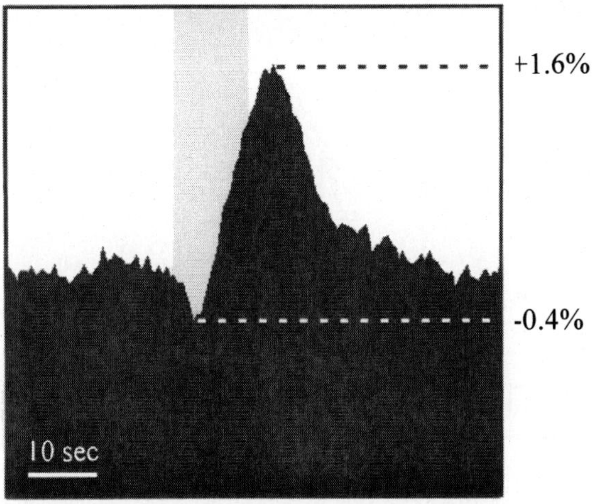

FIGURE 4-5. Biphasic time evolution of MR signals obtained in cat primary visual cortex. After visual stimulation, the MR signal decreases, reaching a minimum of about –0.2~ –0.4% approximately 3 seconds after the stimulus onset. The signal then reverses, yielding a maximum positive signal change of 1.0–2.0% approximately 8 to 10 seconds after stimulus onset. Scale bar: 10 seconds. (Adapted from Kim et al. [8].)

the visual stimulation were observed in all 10 experiments during the stimulation of the animal with moving gratings of different orientations (see Fig. 4-6). Furthermore, similar results were obtained at two different magnetic fields used for our studies (4.7 and 9.4T).

As the early negative signal changes are assumed to reflect the transient increase of local deoxyhemoglobin in parenchymal tissue, their use for functional map generation can be expected to greatly enhance the functional specificity of fMRI. To this end, to determine the time-dependent spatial layout of BOLD signals, the percent change maps within the negative and positive portions of BOLD time course were averaged into respective time-binned maps (positive and negative map, respectively). For the map of the negative response, individual pixels displaying negative percent change with at least 0.5 to 1.0 standard deviation away from the mean baseline percent change were taken to be statistically significant. For the positive response, the threshold was raised in proportion to contrast-to-noise ratio (CNR) of the maximum positive to the minimum negative BOLD response for each animal (i.e., about 5 times).

Figure 4-7 shows that the initial deoxygenation and the delayed oxygenation signals yield a fundamentally different pattern of activity in response to identical visual stimulus. Panel A depicts the spatial pattern of increased BOLD activity in response to moving gratings of 45° orientation. The positive BOLD percent changes are coded with gray scales as indicated by the respective key below. Similar to activation maps

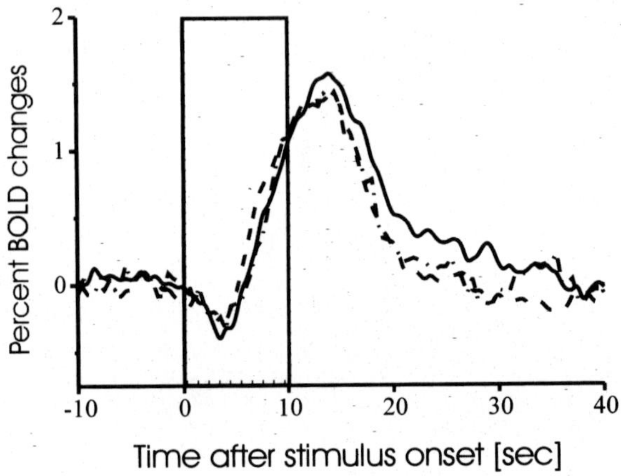

FIGURE 4-6. Biphasic BOLD responses following the visual stimulation were observed in all 10 experiments at two different magnetic fields. The depicted MR time courses represent the biphasic responses obtained during the stimulation of the animal with moving gratings of different orientations. The stimulus duration is marked by the inset rectangle. (Adapted from Duong et al. [73].)

based on conventional cross-correlation analysis (Fig. 4-3), robust and homogenous activities were observed from the entire lateral gyrus of both hemispheres. The region of activity extended several millimeters in anteriorposterior and mediolateral directions. The pattern of activity based negative BOLD percent changes (Fig. 4-7B), on the other hand, yielded strikingly different spatial layout. Here, the gray-scale coded pixels represent regions of significant decrease in MR signals after visual stimulation. In sharp contrast to the conventional fMRI maps based on positive BOLD signal changes, the foci of the negative signal changes are confined to patchy clusters. In line with the iso-orientation columns in cat primary visual cortex observed using 2-DG (26) and optical imaging (5,55) techniques, such semiellipsoidal and irregularly shaped clusters are distributed all over the approximated area 18, with an average periodicity of about 1.34 ± 0.23 mm. Note also how the areas defined by negative signal changes are located largely in tissue areas only, avoiding regions of large blood vessels including the region around the sagittal sinus.

Although the data in Fig. 4-7 indicate that the initial deoxygenation signals yield a far more column-like pattern of activity compared with those based on conventional positive BOLD signals, important questions remain to be addressed. For example, given the large difference in absolute amplitude between the negative and positive signals (Fig. 4-5), it is conceivable that the differences in spatial pattern between the positive (Fig. 4-7A) and negative (Fig. 4-7B) maps could be due simply to the difficulty in equating the signal threshold levels. We have addressed this possibility by raising the threshold for the positive response to match the number of "activated" pixels in both maps. The resulting positive map (Fig. 4-8B) largely reflects the surface vasculature pattern of the visual cortex, with the strongest signals originating from the sagittal sinus.

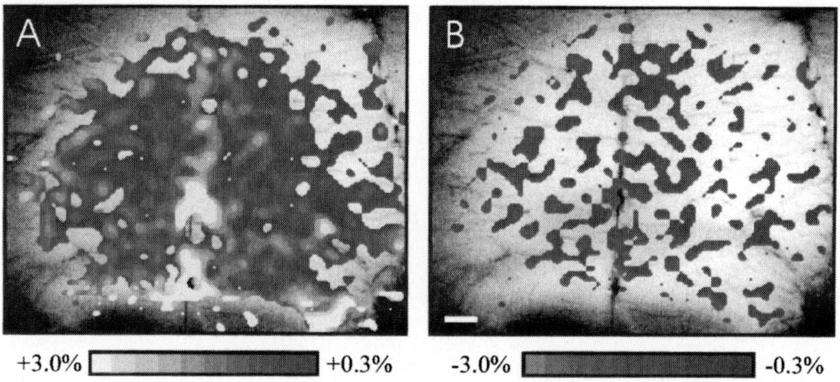

FIGURE 4-7. Improvement of spatial specificity of BOLD using the initial negative signal changes. **(A)** The pattern of increased BOLD activity in response to moving gratings of 45° orientation. The positive BOLD percent changes are marked with gray scales as indicated by the gray scale key below. For the functional map displayed in panel **(B)**, only negative BOLD percent changes occurring within the first 2 seconds after stimulus onset were utilized. The gray-scale key below indicates the negative percent changes. See text for details. Scale bar for panels: 1 mm. (Adapted from Kim et al. [8].)

These results clearly demonstrate that the differences in spatial pattern between the map of negative and positive responses are genuine. Furthermore, results in Figs. 4-7 and 4-8 show that there is no method of linear thresholding that will yield columnar pattern of activity based on positive BOLD responses. If a low threshold is used, the resulting map will be homogenous, with no differences between functional maps obtained for different receptive field properties (Fig. 4-3). Choosing a

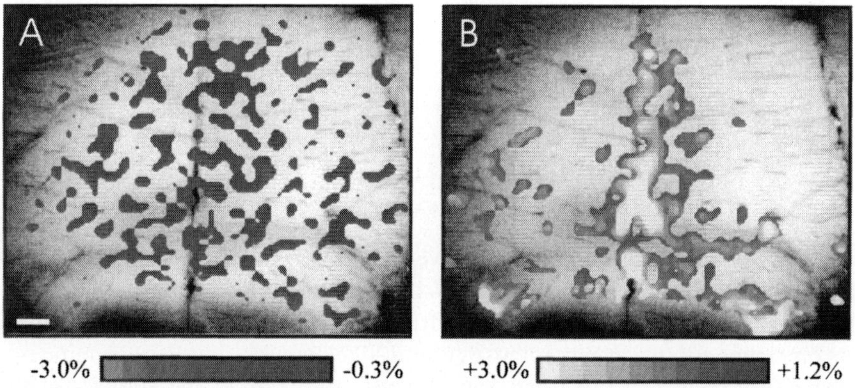

FIGURE 4-8. Differences in spatial layout of activity based on negative **(A)** and positive (B) BOLD responses. For this comparison, the threshold for the positive response was raised to match the number of activated pixels in the negative map (A). The resulting positive map **(B)** reflects largely the surface vasculature pattern of the visual cortex. Scale bar: 1 mm. (Adapted from Kim et al. [8].)

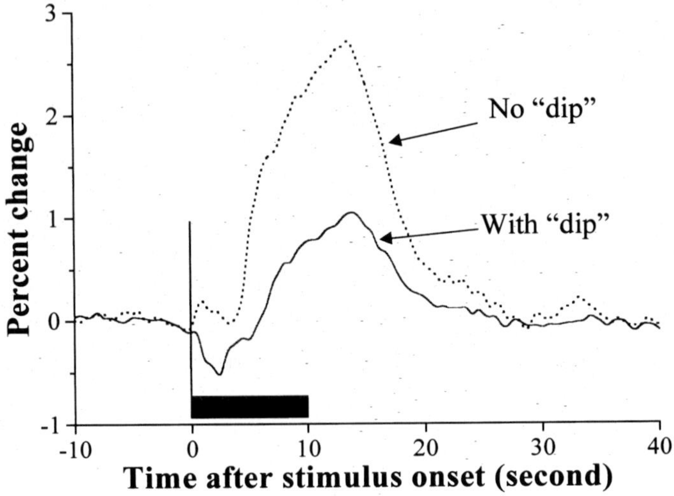

FIGURE 4-9. Expanded time courses for the pixels that exhibit early negative (solid line) and no early negative (dash line) responses, respectively. Stimulus duration is marked with the black box. Note that the pixels, which did not exhibit an early negative response, produced a much larger delayed positive response than those with an early negative response. See text for more details. (Adapted from Duong et al. [73].)

high threshold on the other hand results in activated pixels that represent predominantly nonfunctional draining vessels (Fig. 8B). This notion is further corroborated in Fig. 4-9. Here, the image pixels were classified into two disjoint sets: pixels exhibiting an initial dip response (dip-pixels) and those with no such responses (nodip-pixels). If we plot the entire signal time courses for these two groups of pixels (Fig. 4-9), it is clear that the nodip-pixels exhibit a much larger positive BOLD response than those from dip-pixels. If we assume the nodip-pixels to largely represent draining vessel signals, then choosing a high threshold in conventional BOLD fMRI, for the sake of improved statistics, may do just the opposite of what was intended, namely, selecting non-functional signals only.

FUNCTIONAL PROPERTIES OF "NEGATIVE BOLD" SIGNALS IN CAT PRIMARY VISUAL CORTEX

We have found that the initial negative BOLD signals yield, in sharp contrast to positive signals, a column-like pattern of activity in which activated pixels are clustered into isofunctional domains with a periodicity of about 1 mm (Fig. 4-7). The crucial question is whether those column-like domains represent genuine iso-orientation domains. To this end, the ultimate validation of the veracity of the fMRI-based iso-orientation columns requires simultaneous recording of single unit activity. Such simultaneous recording requires the recording setup to be placed around the isocenter of the magnetic bore, resulting in devastating magnetic field interferences. Furthermore, in such a design, it is almost impossible to change the electrode track or introduce new electrodes without repositioning the animal.

Although many groups (including our own group) are vigorously trying to overcome the technical difficulties to ensure a direct electrophysiological validation of the functional MR data, we reasoned it worthwhile to attempt a more "indirect" validation of our MR-derived columns. We took advantage of the fact that the spatial layout of the orientation columns in cat primary visual cortex displays a set of characteristic features that is well established (13,14). For example, it is well known that the orientation domains for orthogonal orientations are complementary. In fact we have used this complementary criteria in our assessment of the positive BOLD maps (Fig. 4-3).

Figure 4-10 depicts the results of such a complementarity test for negative BOLD data. Panels A through D display patterns of negative activity obtained from the same patch of cortex during the stimulation of the animal with the four different orientation stimuli. Analogous to 2-DG (26) and optical imaging (16), regions of high activity (i.e., large negative signals) are displayed as dark pixels.

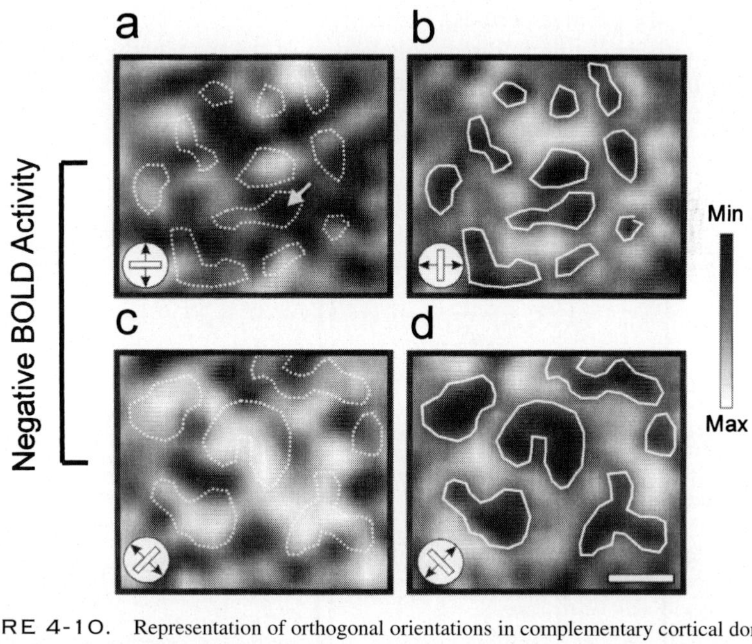

FIGURE 4-10. Representation of orthogonal orientations in complementary cortical domains. **(A-D)** Patterns of "negative" activity in response to four orientations (0°, 45°, 90°, 135°). Regions of negative signal changes during the first 2 seconds after visual stimulation are displayed as dark pixels. For visual inspection of the complementarity between the orthogonal orientation maps, the patches in panels **(B)** and **(D)** are outlined and overlaid on the maps in panels **(A)** and **(C)**, respectively. Other than a few exceptions (marked with a arrowhead), the maps for orthogonal orientations are complementary to each other. Scale bar: 1 mm. (Adapted from Kim et al. [8].)

For removing the effects of uneven MR signals across the cortical surface, the conventional "cocktail blank" method was used (16). It is evident that each pattern of negative activity is highly specific to the respective stimulus orientation. To corroborate this notion, the outlines of the orientation patches for 90° and 135° maps are overlaid on the maps obtained for the respective "orthogonal" orientations (Figure 4-10). It is clear from these panels that the patches for orthogonal orientations occupy cortical territories that are mostly complementary.

The ability of the negative BOLD signals to reflect the stimulus selective responses at columnar scale is further elucidated in Fig. 4-11. Here, the left and right panels represent the signal time courses obtained from 45° (dashed time courses) and 135° (solid time courses) orientation columns, respectively. It is clear that MR signals in pixels representing 45° columns decreased transiently during the 45° stimulation (left panel), while little or no such decrease was observed during the 135° stimulation (right panel). Likewise, MR signals from 135° columns decreased transiently during the 135° stimulation, but not during the complementary 45° stimulation.

The positive BOLD signals, on the other hand, were much less suited to discriminate between 45° and 135° columns, as they yielded largely overlapping times courses for both stimulus conditions (see the overlapping signals in either panels during the positive periods).

Finally, to permit a more direct comparison of the map quality, we have also calculated the 45° and 135° maps obtained during the positive BOLD period (Fig. 4-12). All functional maps depicted in Figs. 4-10 and 4-12 were acquired in the

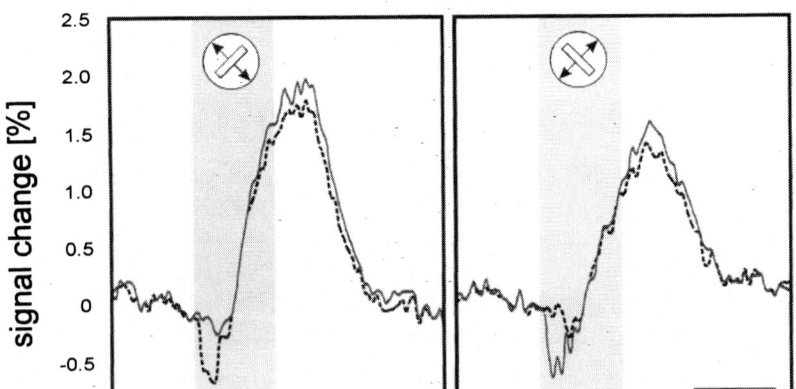

FIGURE 4-11. The signal time courses for 45° (dashed time course) and 135° (solid time course) columns during the stimulation with 45° (left panel) and 135° (right panel) orientation stimuli. The stimulation duration of 10 seconds is marked in both panels by the gray boxes behind the time courses. See text for more details. Scale bar: 10 seconds. (Adapted from Kim et al. [8].)

same fMRI studies, and analogous methods of image processing have been applied. In contrast to the patchy and interdigitized columns in negative maps, the domains of the positive BOLD responses are larger with no apparent periodicity. The outlines of the 135° map from Fig. 4-3B were then overlaid on the 45° map (Fig. 4-12A). The regions of positive activity for the orthogonal orientations were mostly overlapping.

The results shown in Figs. 4-10 through 4-12 are interesting for two reasons. First, the complementarity of the negative BOLD maps obtained during the stimulation with orthogonal orientation further corroborates the notion that the MR-based "columns" may indeed represent genuine iso-orientation columns. Furthermore, the results in Figs. 4-10 through 4-12 suggest that, although the nominal amplitude of the negative signals is far smaller than that of the positive signals (Fig. 4-5-6), the *stimulus specific contrast* of the negative signals was found to be far superior to that of the positive signals. Based on Figs. 4-10 through 4-12, we estimate the stimulus-specific contrast of the negative signals to be about 6.5 times larger compared with that of the positive signals.

FMRI-BASED PINWHEELS

Finally, as a further way of indirect validation, we attempted to derive the distinct topology of the composite orientation map in cat primary visual cortex using

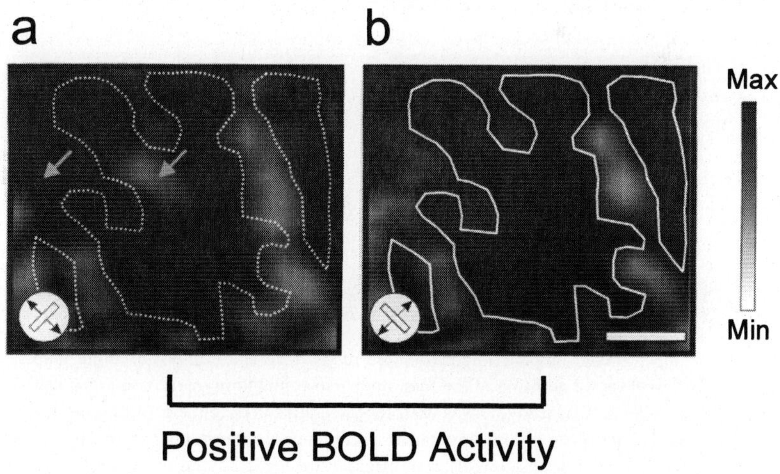

FIGURE 4-12. Representation of orthogonal orientations in positive BOLD maps. **(A,B)** Functional maps based on positive signal changes obtained during the stimulation of the animal with 45° and 135°, respectively. High positive signals are represented by dark pixels. The domains of high activity in 135° map **(B)** are outlined and overlaid on the 45° map (A). Most of the active domains in both maps are overlapping extensively. However, a few prominent exceptions exist (marked with arrowheads). Scale bar: 1 mm. (Adapted from Kim et al. [8].)

our negative BOLD maps. A characteristic and invariant feature of the mammalian orientation system is the existence of topological singularities (16) that were observed across many mammalian species using both multielectrode (7,71) and optical imaging (16) techniques.

With the use of our negative BOLD fMRI data, we obtained composite angle maps for the negative BOLD changes through a pixel-by-pixel vector addition of the four iso-orientation maps with the negative percent changes as vector amplitudes and the respective stimulus orientations as vector angles. The resulting angle at each pixel was then coded using a gray-scale code (Fig. 4-13).

In the composite map based on the early negative BOLD signals only (Fig. 4-13A), the preferred orientations change smoothly, forming a map of orientation

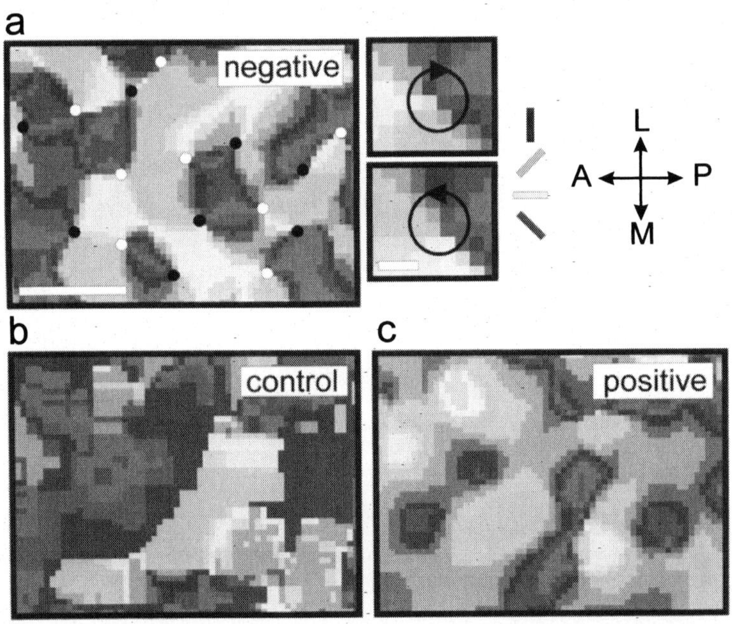

FIGURE 4-13. fMRI-based composite-angle maps. **(A)** The composite-angle map obtained through pixel-by-pixel vector addition of the four single iso-orientation maps based on negative signal changes. The gray scale key next to (A) was used for gray-scale coding the resulting orientation preferences. The overall continuity of the orientation preferences is interrupted at the orientation pinwheels where the cortical columns for different orientations are arranged in a circular manner. The white and black circles in panel (A) depict such clockwise and counterclockwise pinwheels, respectively. Two of such pinwheels are displayed, enlarged right to panel (A). Scale bar for the enlarged pinwheels: 200 μm. As a control, the composite maps based on MR signals obtained before stimulus onset **(B)** and during positive BOLD signals **(C)** are also displayed. The maps in (B) and (C) were obtained from the identical cortical region as displayed in panel (A). The control maps are devoid of topological structures that are characteristic for genuine composite angle maps. Scale bar for (A-C): 1 mm. A, anterior; P, posterior; M, medial; L, lateral. (Adapted from Kim et al. [8].)

selectivity. The continuity of orientation preferences is interrupted only at the singular points (termed orientation pinwheels) where the domains for all orientations converge, in a way analogous to that obtained using multielectrode and optical imaging techniques. Each orientation is represented only once around such a pinwheel, thereby forming two types of topological singularities according to their rotational chiralities. Two of such pinwheels are depicted enlarged to the right of Fig. 4-13A. The pinwheel density and the ratio between clockwise and counterclockwise singularities found in negative BOLD composite maps were in excellent agreement with those obtained with optical imaging techniques. In our study, a pinwheel density of 1.46 ± 0.17 pinwheels/mm^2 was found, and optical imaging studies (5,62) yielded average pinwheel densities between 1.2 and 1.95 pinwheels/mm^2. Likewise, the ratio between clockwise and counterclockwise pinwheels was found to be roughly 1:1 in both negative-BOLD and optical imaging data (see Chapter 2).

Although the preceding data agree with data from multielectrode and optical imaging studies, it is theoretically possible, that topological singularities could arise from a randomly distributed pattern of activity (72). Therefore, as a control, composite maps based on signals obtained before stimulus onset (Fig. 4-13B), and those during the delayed positive BOLD changes (Fig. 4-13C) were also calculated. All steps for composite map construction were performed identical to that for Fig. 4-13A. Unlike the composite map based on negative BOLD signals (Fig. 4-13A), the maps based on control (Fig. 4-13B) or delayed positive BOLD signals (Fig. 4-13C) displayed none of the characteristic features of the mammalian angle maps.

CONCLUSIONS

Our results from cat primary visual cortex clearly indicate that the functional specificity of BOLD at columnar scale depends highly on the temporal dynamics of the underlying signals (73). If the early negative signals are used, functional maps at columnar resolution can be obtained without the need of differential imaging (see later). The delayed positive BOLD changes, on the other hand, were clearly indicative for the pattern of the overall activation per se in the visual cortex, but were less suited to discriminate between the active and inactive columns, as they were more diffused and were less specific for the individual stimulus properties.

It is conceivable that, in principle, even such diffused positive BOLD signals could yield columnar pattern of activity, if maps of orthogonal conditions are subtracted from each other. Such differential imaging had been suggested for analyzing optical imaging (74), and conventional (i.e., positive) BOLD data (75). Although the use of differential method for optical imaging data might be defendable given the well-established verification of intrinsic signals with extensive single unit studies (62,64), the use of differential method for BOLD fMRI data, however, poses severe difficulties. In differential imaging, one activation map (e.g., to the left eye) is subtracted by the activity map that is assumed to give the complementary activation pattern (e.g., to the right eye). If this assumption is true—such as between left- and right-eye domains—then then the result of the subtraction is tautological, as it was known in advance. If the assumption is not true, however, or simply not known, as for most receptive field properties, then the subtraction method may give the wrong answer. Lacking the direct validation of BOLD using simultaneous single unit recordings to date, the applicability of the differential method for BOLD data must remain questionable.

As the T_2 and T_2*-weighted BOLD contrast predominantly originates from the regional changes in paramagnetic deoxyhemoglobin concentration (18,27), the early decrease of BOLD signals can most likely be attributed to the regional increase of deoxyhemoglobin after neuronal stimulation. Although alternative explanations exist (76), the most parsimonious explanation for such a transient deoxygenation is a transient increase in oxygen gradients from the capillaries to the tissues caused by an increase in oxygen consumption within the active orientation column without a commensurate increase in blood flow (41,42,61).

Early, transient decrease in BOLD signals has also been observed in the cortices of awake human (65–67,77) and anesthetized nonhuman primates (68) during perceptual tasks, indicating that, in principle, the capability of BOLD-fMRI to label functional columns in vivo should be applicable also in humans and monkeys. The upcoming challenge in human studies will be to attain stable and sufficient signal-to-noise-ratio from subcolumnar sized voxels, which is more difficult to achieve in awake humans than from anesthetized small animals.

Future studies in cat visual cortex will take full advantage of this novel MRI method (e.g., noninvasivity and depth resolution) to elucidate the layer-specific development and plasticity of the functional organization in multiple visual areas within the same animal. Furthermore, recent advances in diffusion tensor MRI (78) promises that high-resolution mapping of the axonal connectivity in cat visual cortex can be achieved in a completely noninvasive manner (79). Combined with columnar-resolution fMRI (8,73), this would pose a powerful new paradigm that would make it possible for the pattern of functional activity to be directly correlated with that of the underlying neuronal circuitry in a *living* animal.

As with any new technique, this novel method for noninvasive visualization of cortical columns by fMRI will naturally need further improvements and cross-validations (see Kim et al. [80] for initial attempts). We can conclude from our study, however, that noninvasive fMRI of brain functions can now be performed at columnar levels, thus bridging the gap between neurophysiology and cognitive sciences.

ACKNOWLEDGMENTS

We thank Dr. H. Merkle and J. Strupp for providing support in hard- and software. We would also like thank Drs. R. Galuske, R. Goebel, and I. Ronen helpful discussions. This work was supported by the NIH (NS38295, MH57180, NS10930, RR08079), the Minnesota Medical Foundation, and the Keck Foundation.

REFERENCES

1. Brodmann, K. (1909). *Vergleichende Lokalisationslehre der Grosshirnrinde.* Leipzig, Verlag von Johann Ambrosius Barth.
2. Hubel, D. H., and Wiesel, T. N. (1962). Receptve field, binocular interactions and functional architecture in the cat's visual cortex. *J. Physiol. (Lond)* **160,** 106–154.
3. LeVay, S., Stryker, M. P., and Shatz, C. J. (1978). Ocular dominance columns and their development in layer IV of cat's visual cortex. *J. Comp. Neurol.* **179,** 223–244.
4. Hubel, D. H., Wiesel, T. N., and Stryker, M. P. (1978). Anatomical demonstration of orientation columns in macaque monkey. *J. Comp. Neurol.* **177,** 361–380.
5. Bonhoeffer, T., and Grinvald, A. (1991). Iso-orientation domains in cat visual cortex are arranged in pinwheel-like pattern. *Nature* **353,** 429–432.
6. Payne, B. R., Berman, N. E. J., and Murphy, E. H. (1981). Organization of direction preference in cat visual cortex. *Brain Res.* **211,** 445–450.
7. Swindale, N. W., Matsubara, J. A., and Cynader, M. S. (1987). Surface organization of orientation and direction selectivity in cat area 18. *J. Neurosci.* **7,** 1414–1427.
8. Kim, D.-S., Duong, T. Q., and Kim, S.-G. (2000). Mapping iso-functional columns using magnetic resonance imaging. *Nat. Neurosci.* **3,** 164–169.
9. LeVay, S., and Nelson, S. B. (1991). Columnar organization of the visual cortex. In: *The neurological basis of visual function and visual dysfunction* (J. Cronly-Dillon, and A. G. Leventhal, Eds.), pp. 266–315, Boca Raton, CRC Press.
10. Sokoloff, L., Reivich, M., Kennedy, C., DesRosiers, M. H., Patlak, C. S., Pettigrew, K. D., Sakurada, O., and Shinohara, M. (1977). The 14C-deoxyglucose method for the measurement of local cerebral glucose utilization: theory, procedure, and normal values in the conscious and anesthetized albino rat. *J. Neurochem.* **28,** 897–916.
11. Grinvald, A., Lieke, E., Frostig, R. D., Gilbert, C. D., and Wiesel, T. N. (1986). Functional architecture of cortex revealed by optical imaging of intrinsic signals. *Nature* **324,** 361–364.
12. Frostig, R. D., Lieke, E. E., Ts'o, D. Y., and Grinvald, A. (1990). Cortical functional architecture and local coupling between neuronal activity and the microcirculation revealed by in vivo high-resolution optical imaging of intrinsic signals. *Proc. Natl. Acad. Sci. U.S.A.* **87,** 6082–6086.
13. Huebener, M., Shoham, D., Grinvald, A., and Bonhoeffer, T. (1997). Spatial relationships among three columnar systems in cat area 17. *J. Neurosci.* **17,** 9270–9284.
14. Kim, D.-S., Matsuda, Y., Ohki, K., Ajima, A., and Tanaka, S. (1999). Geometrical and topological relationships between multiple functional maps in cat primary visual cortex. *Neuroreport* **10,** 2515–2522.
15. Yae H., Elias, S. A., and Ebner, T. J. (1992). Deblurring of 3-dimensional patterns of evoked rat cerebellar cortical activity: a study using voltage-sensitive dyes and optical sectioning. *J. Neurosci. Methods* **42,** 195–209.
16. Bonhoeffer, T., and Grinvald, A. (1993). The layout of iso-orientation domains in area 18 of cat visual cortex: optical imaging reveals a pinwheel-like organization. *J. Neurosci.* **13,** 4157–4180.
17. Stetter, M., and Obermayer K. (1999). Simulation of scanning laser technique for optical imaging of blood-related intrinsic signals. *J. Opt. Soc. Am. A. Opt. Image. Sci. Vis.* **16,** 58–70.
18. Ogawa, S., Lee, T. M., Nayak, A. S., and Glynn, P. (1990). Oxygenation-sensitive contrast in magnetic resonance image of rodent brain at high magnetic fields. *Magn. Reson. Med.* **14,** 68–78.
19. Ogawa, S., Tank, D., Menon, R., Ellermann, J. M., Kim, S.-G., Merkle, K., and Ugurbil, K. (1992). Intrinsic signal changes accompanying sensory stimulation: functional brain mapping using MRI. *Proc. Natl. Acad. Sci. U.S.A.* **89,** 5951–5955.
20. Wagner, A. D., Schacter, D. L., Rotte, M., Koutstaal, W., Maril, A., Dale, A. M., Rosen, B. R., and Buckner, R. L. (1998). Building memories: remembering and forgetting of verbal experiences as predicted by brain activity. *Science* **281,** 1188–1191.
21. Kim, S.-G., Ashe, J., Hendrich, K., Ellermann, J. M., Merkle, H., Ugurbil, K., and Georgopoulos, P. (1993). Functional magnetic resonance imaging of motor cortex: hemispheric asymmetry and handedness. *Science* **262,** 615–617.

22. DeYoe, E. A., Carman, G. J., Bandettini, P., Glickman, S., Wieser, J., Coy, R., Miller, D., and Neitz, J. (1996). Mapping striate and extrastriate visual areas in human cerebral cortex. *Proc. Natl. Acad. Sci. U.S.A.* **93,** 2382–2386.
23. Engel, A. S., Glover, G. H., and Wendell, B. A. (1997). Retinotopic organization in human visual cortex and the spatial precision of functional MRI. *Cereb. Cortex* **7,** 181–192.
24. Tootell, R. B., Mendola, J. D., Hadjikhami, N. K., Ledden, P. J., Liu, A. K., Reppas, J. B., Sereno, M. I., and Dale, A. M. (1997). Functional analysis of V3A and related areas in human visual cortex. *J. Neurosci.* **17,** 7070–7078.
25. Wandell, B. (1999). Computational neuroimaging of human visual cortex. *Annu. Rev. Neurosci.* **22,** 145–173.
26. Loewel, S., Freeman, B., and Singer, W. (1987). Topographic organization of the orientation column system in large, flat-mounts of the cat visual cortex: a 2-deoxyglucose study. *J. Comp. Neurol.* **255,** 401–415.
27. Ogawa, S., Menon, R. S., Kim, S.-G., and Ugurbil, K. (1998). On the characterizatics of functional magnetic resonance imaging of the brain. *Annu. Rev. Biophys. Biomol. Struct.* **27,** 447–474.
28. Rabi, I. I., Zacharias, J. R., Millman, S., and Kusch, P. (1938). *A new method of measuring nuclear magnetic moment. Phys. Rev.* **53,** 318.
29. Bloch, F., Hansen, W. W., and Packard, M. (1946). *The nuclear induction experiment. Phys. Rev.* **70,** 474–485.
30. Lauterbur, P. C. (1973). Image formation by induced local interaction: examples employing nuclear magnetic resonance imaging. *Nature* **241,** 190–191.
31. Belliveau, J. W., Kennedy, D. N., McKinstry, R. C., Buchbinder, B. R., Weisskoff, R. M., Cohen, M. S., Vevea, J. M., Brady, T. J., and Rosen, B. R. (1991). Functional mapping of the human visual cortex by magnetic resonance imaging. *Science* **254,** 716–719.
32. Rosen, B. R., Belliveau, J. W., Arnon, H. J., Kennedy, D., Buchbinder, B. R., Fischman, A., Gruber, M., Glas, J., Weisskoff, R. M., Cohen, M. S., Hochberg, F. H., and Brady, T. J. (1991). Susceptibility contrast imaging of cerebral blood volume: human experience. *Magn. Reson. Med.* **22,** 293–299.
33. Bandettini, P. A., Wong, E. C., Hinks, R. S., Tikofsky, R. S., and Hyde, J. S. (1992). Time course EPI of human brain function during task activation. *Magn. Reson. Med.* **25,** 390–398.
34. Kwong, K. K., Belliveau, J. W., Chesler, D. A., Goldberg, I. E., Weisskoff, R. M., Poncelet, B. P., Kennedy, D. N., Hoppel, B. E., Cohen, M. S., Turner, R., Cheng, H. M., Brady, T. J., and Rosen, B. R. (1992). Dynamic magnetic resonance imaging of human brain acrivity during primary sensory stimulation. *Proc. Natl. Acad. Sci. U.S.A.* **89,** 5675–5679.
35. Pauling, L., and Coryell, C. D. (1936). The magnetic properties and structure of hemoglobin, oxyhemoglobin, and carbonmonoxyhemoglobin. *Proc. Natl. Acad. Sci. U.S.A.* **22,** 210–216.
36. Thulborn, K. R., Waterton, J. C., Mattews, P. M., and Radda, G. K. (1982). Dependence of the transverse relaxation time of water protons in whole blood at high field. *Biochem. Biophys. Acta* **714,** 265–272.
37. Roy, C. S., and Sherrington, C. S. (1890). On the regulation of blood supply of the brain. *J. Physiol.* **11,** 85–108.
38. Pawlik, G. Rackle, A., and Bing, R. J. (1981). Quantitative capillary topography and blood flow in the cerebral cortex of cat: an in vivo microscopic study. *Brain Res.* **208,** 35–58.
39. Lai, S., Hopkins, A. L., Haacke, E. M., Li, D., Wasserman, B. A., Buckley, P., Friedman, L., Meltzer, H., Hedera, P., and Friedland, R. (1993). Identification of vascular structures as a major source of signal contrast in high resolution 2D and 3D functional activation imaging of the motor cortex at 1.5T: preliminary results. *Magn. Reson. Med.* **30,** 387–392.
40. Menon, R. S., Hu, X., Adriany, G., Petersen, P., Ogawa, S., and Ugurbil, K. (1994). Comparison of spin-echo EPI, asymmetric spin-echo EPI and conventional EPI applied to functional neuroimaging: the effect of flow crushing gradients on the BOLD signal. *Proc. 2nd Soc. Magn. Reson.,* 622.
41. Malonek, D., and Grinvald, A. (1996). Interactions between electrical activity and cortical microcirculation revealed by imaging spectroscopy: Implication for functional brain mapping. *Science* **272,** 551–554.

42. Malonek, D., Dirnagl, U., Lindauer, U., Yamada, K., Kanno, I., and Grinvald, A. (1997). Vascular imprints of neuronal activity: relationships between the dynamics of cortical blood flow, oxygenation, and volume changes following sensory stimulation. *Proc. Natl. Acad. Sci. U.S.A.* **94,** 14826–14831.
43. Shmuel, A., Hu, X., Ugurbil, K., and Grinvald, A. (2000). Spread of hemo-dynamic signals in draining veins beyond the regions of electrical activation. *Proc. 8th Soc. Magn. Reson.,* 979.
44. Weisskoff, R. M., Zuo, C. S., Boxerman, J. L., and Rosen, B. R. (1994). Microscopic susceptibility variation and transverse relaxation: theory and experiment. *Magn. Reson. Med.* **31,** 601–610.
45. Song, A. W., Wong, E. C., Tan, S. G., and Hyde, J. S. (1996). Diffusion weighted fMRI at 1.5 T. *Magn. Reson. Med.* **35,** 155–158.
46. Ugurbil, K., Hu, X., Wei, C., Zhu, X.-H., Kim, S.-G., and Georgopoulos, A. (1999). Functional mapping in the human brain using high magnetic fields. *Philos. Trans. R. Soc. Lond. Biol. Sci.* **354,** 1195–1213.
47. Lee, S.-P., Silva, A. C., Ugurbil, K., and Kim, S.-G. (1999). Diffusion-weighted spin-echo fMRI at 9.4T: microvascular/tissue contribution to BOLD signal changes. *Magn. Reson. Med.* **42,** 919–928.
48. Movshon, J. A., Thompson, I. D., and Tolhurst, D. J. (1978). Spatial and temporal contrast sensitivity of neurons in areas 17 and 18 of the cat's visual cortex. *J. Physiol.* **283,** 101–120.
49. Otsuka, R., and Hassler, R. (1962). Ueber Aufbau und Gliederung der corticalen Sehsphaere bei der Katze. *Archiv fuer Psychiatrie und Zeitschrift f. d. Ges. Neurologie* **203,** 212–234.
50. Mansfield, P. (1977). Multi-planar image formation using NMR spin echoes. *J. Physiol. C* **10,** L55–L58.
51. Turner, R., LeBihan, D., Moonen, C. T., Despres, D., and Frank, J. (1991). Echo-planar time course MRI of cat brain oxygenation changes. *Magn. Reson. Med.* **22,** 159–166.
52. Bandettini, P. A., Jesmanowicz, A., Wong, E. C., and Hyde, J. S. (1993). Processing strategies for time-course data sets in functional MRI of the human brain. *Magn. Reson. Med.* **30,** 161–173.
53. Albus, K. (1975). A quantitative study of the projection area of the central and the paracentral visual field in area 17 of the cat II. The spatial organization of the orientation domains. *Exp. Brain Res.* **24,** 181–202.
54. Loewel, S., and Singer, W. (1990). Tangential intracortical pathways and the development of iso-orientation bands in cat striate cortex. *Dev. Brain Res.* **56,** 99–115.
55. Kim, D.-S., and Bonhoeffer, T. (1994). Reverse occlusion leads to a precise restoration of orientation preference maps in visual cortex. *Nature* **370,** 370–372.
56. Jezzard, P., Rauschecker, J. P., and Malonek, D. (1997). An in vivo model for functional MRI in cat visual cortex. *Magn. Reson. Med.* **38,** 699–705.
57. Gonzalez, R. C., and Woods, R. E. (1993). *Digital image processing.* New York, Addison-Wesley.
58. Roland, P. E., Eriksson, L., Stone-Elander, S., and Widen, L. (1987). Does mental activity change the oxidative metabolism of the brain? *J. Neursoci.* **7,** 2373–2389.
59. Fox, P. T., and Raichle, M. E. (1986). Focal physiological uncoupling of cerebral blood flow and oxidative metabolism during somatosensory stimulation in human subjects. *Proc. Natl. Acad. Sci. U.S.A.* **83,** 1140–1144.
60. Turner, R., and Grinvald, A. (1994). Direct visualization of patterns of deoxyhemoglobin and reoxygenation in monkey cortical vasculature during functional brain activation. *Proc. 2nd Soc. Magn. Reson. Med.,* 430.
61. Vanzetta, I., and Grinvald, A. (1999). Increased cortical oxidative metabolism due to sensory stimulation: implications for functional brain imaging. *Science* **286,** 1555–1558.
62. Shmuel, A., and Grinvald, A. (1996). Functional organization for direction of motion and its relationship to orientation maps in cat area 18. *J. Neurosci.* **16,** 6945–6964.
63. Maldonado, P. E., Goedecke, I., Gray, C. M., and Bonhoeffer, T. (1997). Orientation selectivity in pinwheel centers in cat visual cortex. *Science* **276,** 1551–1555.
64. Crair, M. C., Gillespie, D. C., and Stryker, M. P. (1998). The role of visual experience in the development of columns in cat visual cortex. *Science* **279,** 566–570.

65. Menon, R. S., Ogawa, S., Hu, X., Strupp, J. P., Anderson, P., and Ugurbil, K. (1995). BOLD based functional MRI at 4 Tesla includes a capillary bed contribution: Echo Planar Imaging correlates with previous optical imaging using intrinsic signals. *Magn. Reson. Med.* **33,** 453–459.
66. Hu, X., Le, T. H., and Ugurbil, K. (1997). Evaluation of the early response in fMRI in individual subjects using short stimulus duration. *Magn. Reson. Med.* **37,** 877–884.
67. Yacoub, E., and Hu, X. (1999). Detection of the early negaive response in fMRI at 1.5 Tesla. *Magn. Reson. Med.* **41,** 1088–1092.
68. Logothetis, N. K., Guggenberger, H., Peled, S., and Pauls, J. (1999). Functional imaging of the monkey brain. *Nat. Neurosci.* **2,** 555–562.
69. Marota, J. J. A., Ayata, C., Moskowitz, M. A., Weisskopf, R. M., Rosen, B. R., and Mandeville, J. B. (1999). Investigation of the early response to rat forepaw stimulation. *Magn. Res. Med.* **41,** 247–252.
70. Silva, A. C., Lee, S.-P., Iadecola, C., and Kim, S.-G. (2000). Early characteristics of CBF and deoxyhemoglobin changes during somatosensory stimulation. *J. Cereb. Blood Flow Metab.* **20,** 201–205.
71. Diao, Y. C., Jia, W. G., Swindale, N. V., and Cynader, M. S. (1990). Functional organization of the cortical 17/18 border region in the cat. *Exp. Brain Res.* **79,** 271–282.
72. Rojer, A. S., and Schwartz, E. L. (1990). Cat and monkey cortical columnar patterns modeled by bandpass-filtered 2D white noise. *Biol. Cybern.* **62,** 381–391.
73. Duong, T. Q., Kim, D.-S., Ugurbil, K., and Kim, S.-G. (2000). Spatio-temporal dynamics of BOLD fMRI signals: towards mapping submillimeter cortical columns using the early negative response. *Magn. Reson. Med.* **44,** 231–242.
74. Blasdel, G. G. (1992). Differential imaging of ocular dominance and orientation selectivity in monkey striate cortex. *J. Neurosci.* **12,** 3115–3138.
75. Menon, R. S., Ogawa, S. Strupp, J. P., and Ugurbil, K. (1997). Ocular dominance in human V1 demonstrated by functional magnetic resonance imaging. *J. Neurophysiol.* **77,** 2780–2787.
76. Buxton, R. B., and Frank, L. R. (1997). A model for the coupling between cerebral blood flow and oxygen metabolism during neural stimulation. *J. Cereb. Blood Flow Metab.* **17,** 64–72.
77. Ernst, T., and Henning, J. (1994). Observation of a fast response in functional MR. *Magn. Reson. Med.* **32,** 146–149.
78. Xue, R., van Zijl, P. C. M., Crain, B. J., Solaiyappan, M., and Mori, S. (1999). In vivo three-dimensional reconstruction of rat brain axonal projections by diffusion tensor-imaging. *Magn. Reson. Med.* **42,** 1123–1127.
79. Mori, S., Ugurbil, K., van Zijl P. C. M., and Kim D.-S. (2000). *In vivo* mapping of the axonal connectivity in cat visual cortex using diffusion tensor MRI. *Soc. Neurosi. Abstr.* 550.1.
80. Kim, D.-S., Duong, T. Q., Ronen, I., Ugurbil, K., and Kim, S.-G. (2000). Neural correlate of blood oxygenation level dependent functional MRI. *Soc. Neurosci. Abstr.* 309.6.

5

RELATIONSHIP OF LGN AFFERENTS AND CORTICAL EFFERENTS TO CYTOCHROME OXIDASE BLOBS

JOANNE A. MATSUBARA

Department of Ophthalmology University of British Columbia, Vancouver, British Columbia, Canada

JAMIE D. BOYD

Department of Biological Sciences, Simon Fraser University, Burnaby, British Columbia, Canada

PARALLEL PROCESSING IN THE MAMMALIAN VISUAL SYSTEM

In mammals, visual information is processed in several consecutive stages by a hierarchically organized group of interconnected structures: the retina, which performs the actual transformation of light energy into neural activity; the lateral geniculate nucleus (LGN), which is the main thalamic relay between the retina and the visual cortex; primary visual cortex, which receives the bulk of the inputs from the LGN; and a host of extrastriate areas, some of which receive direct inputs from primary visual cortex and all of which are richly interconnected among themselves.

SERIAL VERSUS PARALLEL PROCESSING

Hubel and Wiesel showed that the ascension through the hierarchy, from retinal ganglion cells through the LGN to striate and extrastriate cortex, was marked by increasingly complex receptive field properties and increasing specificity of neuronal responses for stimulus parameters (1–3). Although this implies a serial, hierarchical organization of visual pathways, each local portion of visual space is not analyzed by a single chain of neurons extending from the retina, through the LGN and striate cortex, and then through each extrastriate area in turn. Rather, multiple pathways originate at each level of the hierarchy. In cats and primates, three separate, functionally distinct pathways can be traced from the retina to the LGN, and thence to the cortex (for review, see references 4–6 and Chapter 1). Although information is not as complete for other species, it has been suggested that "...three pathways from the retina to striate cortex appear ubiquitous in mammals." (7). At the level of the cortex, each cortical area has connections with several other areas, so a single hierarchical sequence of visual processing cannot be identified here either (8,9).

ZEKI'S HYPOTHESIS

This evidence for parallel organization of visual processing led Semir Zeki (10) to suggest that one of the major functions of striate cortex might be to "...segregate out the information coming over the retinogeniculate cortical pathways and parcel this information out to different cortical areas for further analysis".

Although Zeki's hypothesis arose from his work on primates, the conditions that led him to this proposal have also been shown in the visual system of the cat. Indeed, the first evidence for parallel pathways in the mammalian visual system was from electrophysiological recordings in the retina of the cat (11), and the first demonstration of visual responses in an area outside of the geniculorecipient primary visual cortex was in the lateral suprasylvian cortex of the cat (12,13); (also see Chapter 1). Divergence of corticocortical connections between visual areas was also demonstrated early in the cat visual system (14), indicating the possibility of columnar, as well as laminar, segregation of pathways. In light of these facts, it is somewhat surprising that relatively little is known about the organization of parallel pathways in the primary visual cortex and the extrastriate visual areas of this species. In contrast, the complex and precisely detailed spatial organization of parallel pathways through the primary visual cortex and extrastriate areas has been exhaustively documented in several species of primates (7).

THE UTILITY OF CYTOCHROME OXIDASE BLOBS

The main reason for the disparity between cats and primates in the study of parallel pathway organization was the discovery of the cytochrome oxidase (CO) blobs and stripes in primates (15–17), and the key lack of several early studies to find CO blobs in cats (18,19). It quickly became apparent that the organization of the different inputs and outputs of visual cortex was intimately related to the CO architecture (20,21). More important, CO staining in primate V1 and V2 provided a needed system of cortical landmarks with which to compare the different distributions of labeling and also electrophysiological response properties. Thus, the blobs in V1 colocalize with the patchy input from the koniocellular layers of the LGN (22–24), while the thin stripes and interstripes of V2 are interconnected with the blobs and interblobs, respectively (25,26). The thick stripes receive input from layer 4B of V1 (27), and also project to MT, while the thin stripes and interstripes project to V4 (28–30). In the cat, in contrast, the difficult process of divining the relationships among the multiple inputs and outputs of visual cortical areas was severely hampered by the lack of cortical landmarks provided by CO staining. One of the few studies examining these issues used multiple retrograde labeling to examine clusters of cells projecting from V1 to different extrastriate areas (31), but no significant relationship was found between clusters projecting to different areas.

Eventually, CO blobs were demonstrated in cat visual cortex (32,33). Figure 5-1, from Boyd (34), shows an example of CO staining in cat visual cortex. The contrast between blobs and interblobs in cat visual cortex is not as great as in primates, and the blobs are visible only in a restricted region of upper layer 4 and lower layer 3 (see page 226). These problems can be overcome by flattening the cortex and sectioning it tangentially (35), as shown here. Even so, blobs are distinct only in that part of the section that grazes lower layer 3 and upper layer 4. Note that CO blobs are present in both areas 17 and 18. As discussed later, these two areas have many similarities in terms of inputs and outputs, and should perhaps be thought of as both comprosing "primary" visual cortex in the cat (see Chapter 1).

Demonstration of CO blobs in the cat finally allowed investigators to address the organization of processing streams through the visual system of the cat, and whether the CO architecture was as important to cortical organization in this species as it is in primates. Confirmation of this issue would be an important step toward generalizing the presence of segregated processing streams through visual cortex and into extrastriate areas—an organization once thought to be specific to primates (36)—to other mammalian species. This review focuses on work examining CO blobs in the cat visual cortex, and the way in which they are related to processing streams linking LGN inputs and corticocortical outputs.

ORGANIZATION OF CO STAINING IN CAT PRIMARY VISUAL CORTEX

LAMINAR ORGANIZATION OF CO STAINING

The laminar pattern of CO staining in cat visual cortex was first described by Wong-Riley (18) and later by Price (37) and Boyd and Matsubara (38). Figure 5-2, from Boyd and Matsubara (38), shows a comparison of CO staining and cytoarchitecture in area 17 and 18 (also see Chapter 1). In both areas 17 and 18, a dense band of CO staining is coextensive with layer 4 as determined by cytoarchitecture

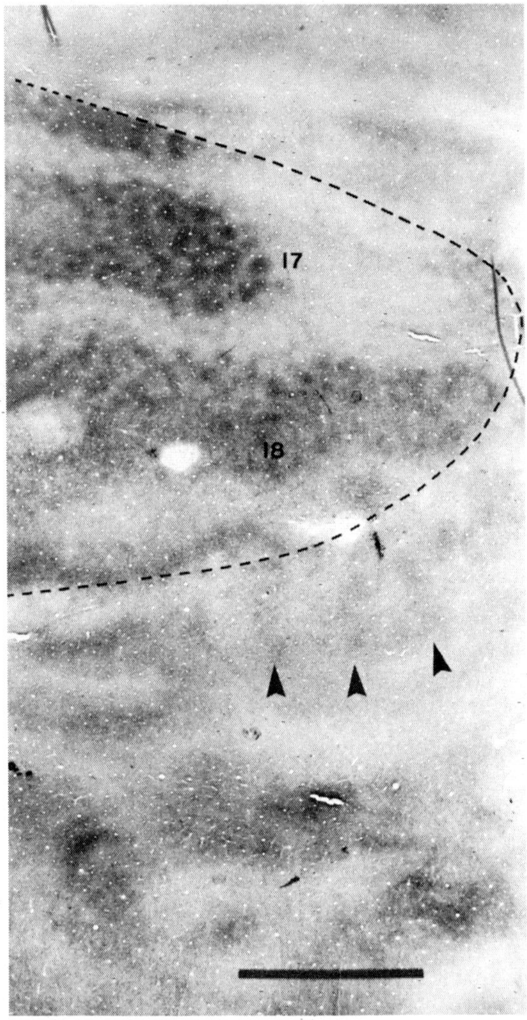

FIGURE 5-1. A section through the visual cortex of a cat, stained to visualize cytochrome oxidase (CO) activity. This section was cut tangentially after removing the cortex from the brainstem and unfolding and flattening it. Anterior is to the right and medial is to the top of the figure. The dashed line indicates the extent of areas 17 and 18, which, together, form a large oval region of increased staining density. Owing to the residual curvature in the cortical tissue present after flattening, the plane of section passes in an irregular fashion through the cortical layers. The most darkly stained parts of the section pass through the upper part of layer 4. In these regions, the CO blobs are most clearly visible. CO blobs are present in both area 17 and 18. Scalebar = 5 mm. (From Boyd [34].)

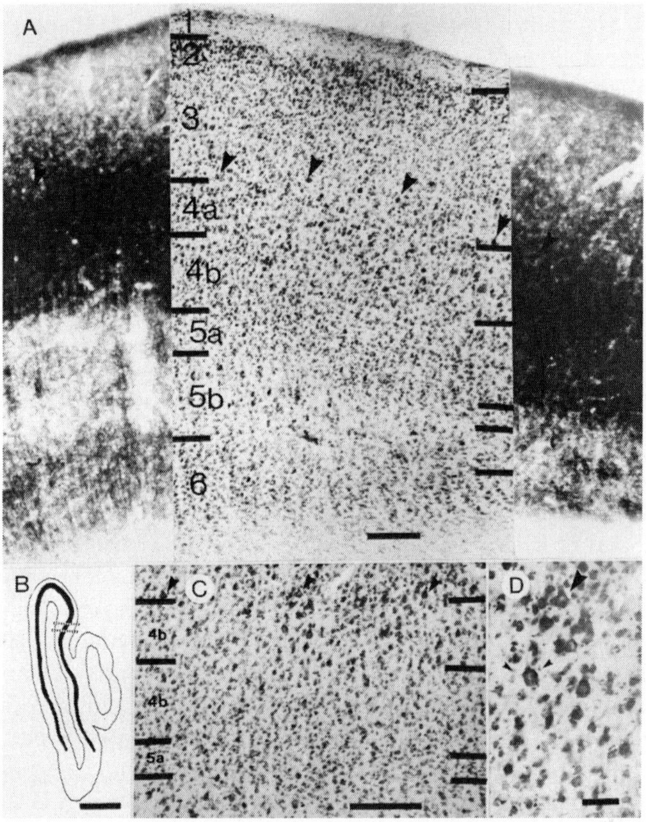

FIGURE 5-2. Laminar pattern of CO staining in areas 17 and 18. **(A)** CO staining in coronal sections through areas 17 (left panel) and 18 (right panel), respectively. These photographs were matched to an adjacent section stained for Nissl substance with cresyl violet, in the middle panel. This section includes the 17/18 border, area 17 to the left, area 18 to the right. Arrowheads mark examples of large pyramidal cells at the layer 3/4 border, stained for Nissl or CO. Scale bar = 200 μm. These sections were taken from the area between the dashed lines in B. In both area 17 and 18, the dense band of CO staining extends upwards to the level of these border pyramids. **(B)** A low power drawing of the CO stained section shown in A (A-P -2). Layer 4 is shown as a thick black line. Scale bar = 2 mm. **(C)** A high-power photomicrograph of layer 4 at the 17/18 border, area 17 to the left, area 18 to the right. Arrowheads mark three examples of large pyramidal cells at the layer 3/4 border Scale bar = 200 μm. The middle of these cells is shown at higher power in **(D)** (large arrowhead). Below this cell, in layer 4a, is a large cell, with dendrites (small arrowheads) radiating from the soma in the manner of a star-pyramid. Thus, area 18 (unlike area 17) has a population of large cells in layer 4a, but these can be distinguished from the typical pyramidal cells of layer 3. Scale bar = 50 μm. (From Boyd and Matsubara [38], with permission of Wiley-Liss, Inc.) (see also Chapter 1.)

(39–41). (The report by Price (37), that the dense band of CO staining in area 18 extends partly into lower layer 3, is probably due to a different assignation of the layer 3/4 border, rather than true differences in staining patterns.) The dense band of CO staining in layer 4 extends to just below the level of the largest pyramidal cells in the upper layers, the so-called border pyramids. Indeed, many of these border pyramids stain densely for CO (38). There is also moderately dense CO staining in layer 3, with the border between the moderate level of staining in layer 3 and the dense band of layer 4 being usually quite sharp. Layer 5a, a thin layer of very small pyramidal cells (40), does not stain for CO, providing a sharp border in the CO staining at the base of layer 4. The largest pyramidal cells in area 17 and 18, the Meynert cells, are found at the 5a/5b border, and these cells stain densely for CO (37,38,42). Layer 6 stains moderately for CO.

ORGANIZATION OF THE CO BLOBS

CO blobs in the cat's visual cortex are visible in the lower part of layer 3 and in the upper part of layer 4 (38,43). Figure 5-3, from Boyd and Matsubara (38), shows a series of nine, alternate 50 μm, tangential sections through a flattened cortex, with the appearance of the blobs in the different layers. The CO blobs appear, albeit faintly, in the middle part of layer 3, becoming more visible toward the bottom of layer 3. At the bottom of layer 3, layer 3/4 border pyramids stained for CO are more numerous within the blobs; whether only a subset of these cells are being stained or whether layer 3/4 border cells truly are more numerous within the blobs is not clear.

The blobs are most clearly visible in layer 4a. The blobs are not visible in layer 4b, and the CO staining has a fine meshwork pattern to it, which differentiates this sublayer from layer 4a. This pattern may result from the tight packing of cells in this layer, which might constrain the positions of reactive dendrites and axon terminals. Layer 6 stains moderately for CO, and the CO blobs are not visible in this layer. A similar pattern of CO staining is seen in area 18, although the contrast between the blobs and the interblobs is not as striking as in area 17 (38). We now examine how some of the different inputs and outputs of primary visual cortex in the cat are organized with respect to the columnar organization shown by the CO blobs.

RELATIONSHIP OF CO BLOBS TO OCULAR DOMINANCE COLUMNS

In Old World primates such as macaque monkeys, the most striking aspect of the CO blobs is their alignment with the centers of ocular dominance (OD) columns (15,19). In cats, however, the relationship of CO blobs with OD columns is less clear. OD columns are less regular in cats than in macaque monkeys, appearing more branched and beaded than as parallel slabs (44). Moreover, the periodicity of the CO blobs in cat visual cortex, near 1 mm, is roughly double what it would need to be to easily provide CO blobs over both ipsilateral and con-

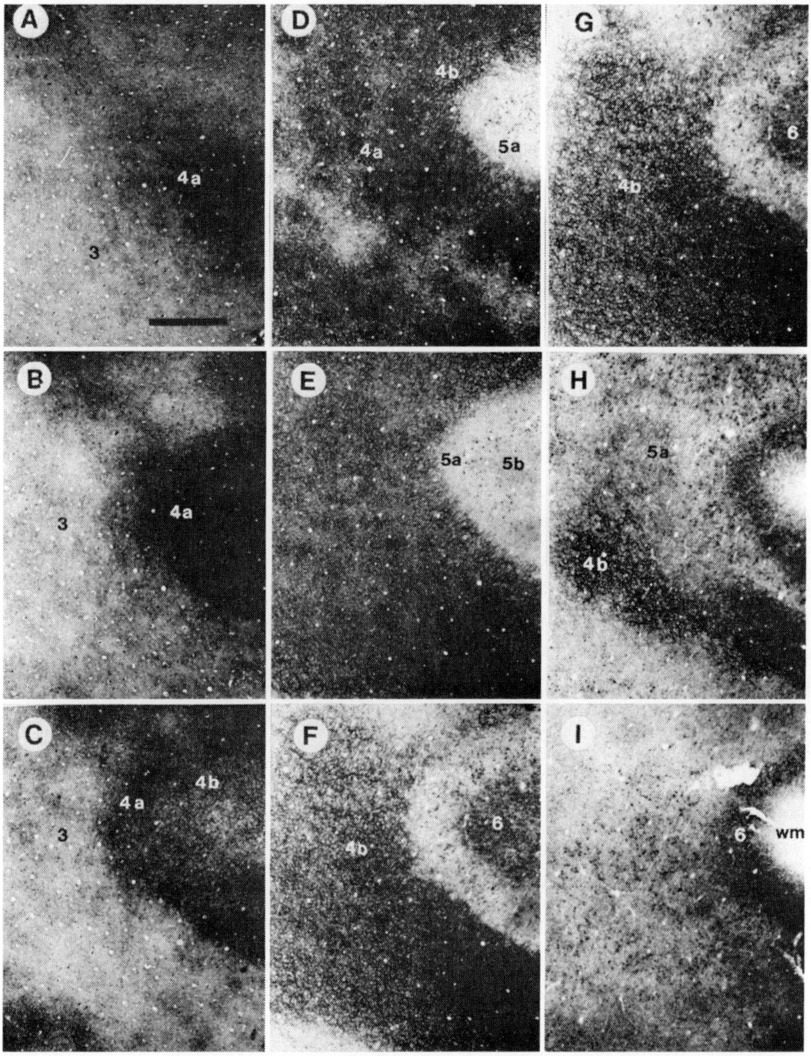

FIGURE 5-3. A series of 50-μm thick tangential sections through area 17 stained for CO. Every other section, starting from the middle of layer 3 **(5A)** and continuing to layer 6 **(5I)** is shown. Cortical layers are numbered; "wn": white matter. Blobs are first visible in the bottom part of layer 3. In layer 4, blobs are visible in 4a, but not 4b. Scale bar = 1 mm, and applies to **A-I.** (From Boyd and Matsubara [38], with permission of Wiley-Liss, Inc.)

tralateral OD columns. However, Murphy et al. (33), using transneuronal transport to label OD columns, suggested that cat CO blobs, as in primates, did tend to be more numerous near the centers of OD columns. Another study using similar methodology (45) concluded that the OD and CO blob systems were independent, with CO blobs found roughly equally over the centers and edges of OD columns. A third study (46) used optical imaging methods to examine OD columns and spatial frequency columns (which are known to correlate with CO staining (47)). This study found no obvious correlation between the two systems, although quantitative analysis showed a weak relationship between OD and spatial frequency columns/CO blobs. The relationship between CO blobs and OD columns is weak or lacking in New World primates (48,49), which also show less strong OD segregation compared with their Old World counterparts.

GENICULATE INPUTS TO THE CO BLOBS

In the cat, three main classes of LGN cells, X, Y, and W, have been distinguished on both physiological and anatomical criteria (see reference 6 and Chapter 1 for reviews). These separate populations of LGN relay cells, which are partially segregated into different layers of the LGN, each receive input from corresponding classes of retinal ganglion cells (50,51). All of the layers of the LGN project to visual cortex. The projections from the A laminae (A and A1), which include the X- and Y-cell projection, are at least partially segregated into different cortical layers than the Y- and W-cells from the C laminae (C, C1, and C2) of the LGN (51,52). The existence of three separate pathways and their segregation through the early stages of the visual system suggest theories of parallel processing in the cat's visual system (5,53). Moreover, the input from the C laminae is not continuous but is organized into patches (52), which suggests that parallel organizations might exist between different cortical columns as well as between different layers. As discussed later, both the laminar and columnar organization of the geniculostriate projections are correlated with cytochrome oxidase staining.

LAMINAR ORGANIZATION OF GENICULOCORTICAL PROJECTIONS

The input from the A laminae, containing X and Y cells, is nearly completely confined to layers 4 and 6 in areas 17 and 18 (38,52). Thus, the geniculate inputs from the A layers match perfectly the dense band of CO staining, and both match the extent of layer 4 determined by the cytoarchitecture. Some studies have suggested that Y cell projections might extend several hundred microns into layer 3, especially in area 18 (54,55). As for the similar situation with CO staining in area 18, this discrepancy is probably due to the use of different laminar schemes and to the inconsistent usage of the same scheme between different studies. Indeed, cells in layer 4a of area 18 are larger on average than cells in layer 4a of area 17, but can still be distinguished from the border pyramids above them by size and morphology. Regardless of whether it is termed layer 4a or layer 3b, in area 18 this sublayer is marked by cytology, geniculate A laminae inputs, and dense CO staining similar to layer 4a in area 17.

The C laminae contain no X cells (5,56,57); layer C proper contains mostly Y cells; and the ventral C layers, C1 and C2, contain only W cells (58–60). Label transported from the C laminae of the LGN to areas 17 and 18 shows that W cells terminate in different layers from X and Y cells. Figure 5-4, from Boyd and Matsubara (38), shows the pattern of labeling following an injection of horseradish peroxidase-conjugated wheatgerm agglutinin (WGA-HRP) into the C laminae in area 17. This injection was done in such a way that different LGN laminae were

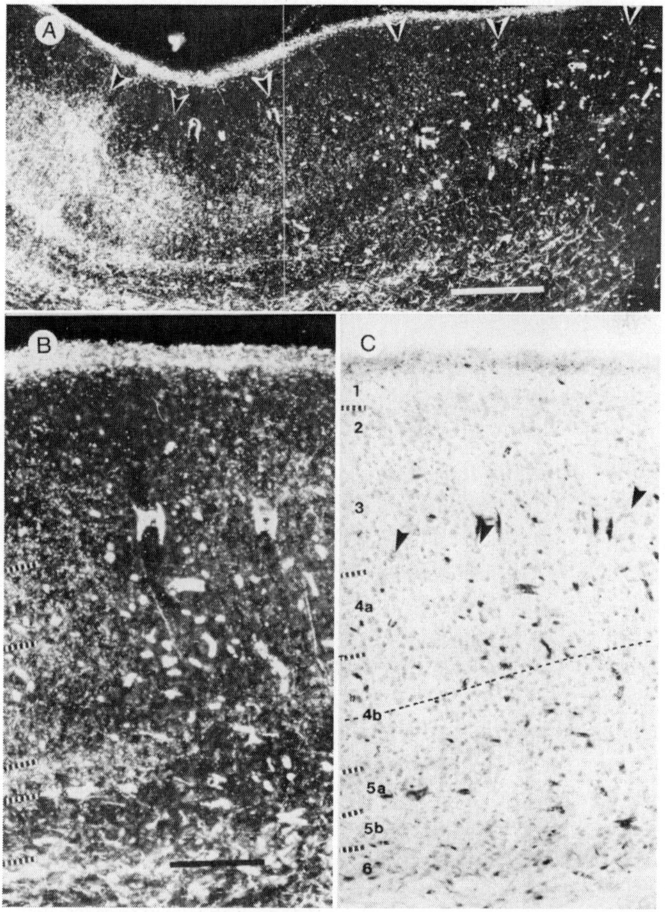

FIGURE 5-4. Labeling from the C-laminae in area 17. **(A)** A low-magnification dark-field photomicrograph of layer C labeling in area 17. The white arrows mark columns of anterograde labeling. Note the transition between two distinct patterns of labeling; the patches on the left include part of layer 4, and originates from both divisions of the C laminae; the patches on the right exclude layer 4 and originate from only the parvicellular division of the C laminae. Scale bar = 500 μm. **(B)** Another example of the transition to parvicellular C laminae labeling. This darkfield photomicrograph shows, at higher magnification, two patches of labeling, one of which involves both layer 3 and layer 4, and the other (arrowhead) involves only layer 3. **(C)** The same field of view shown in **(A)** photographed in brightfield to show the neutral red counterstain. Arrowheads show examples of layer 3/4 border pyramids. Comparison with the cytoarchitecture shows that in the patch of labeling on the right, the dense labeling in layer 4a is absent, although labeling in layer 3, 1, and 5a is present. Scale bar = 200 μm. (From Boyd and Matsubara [38], with permission of Wiley-Liss, Inc.)

injected at different isoelevation representations. Figure 5-4A shows the transition between two distinct patterns of labeling, that on the left including part of layer 4 and arising from both divisions of the C laminae, and that on the right excluding layer 4, and arising from only the parvicellular divisions of the C laminae. Figure 5-4B,C are higher power photomicrographs of another example of the transition to parvicellular C laminae labeling in area 17. Note that the patch on the left in 5-4B has dense terminal labeling in layers 4a, and less dense labeling extending upward several hundred microns from the top of layer 4 into the superficial layers. The patch of labeling on the right does not contain the dense labeling in layer 4a, although the patchy labeling in layers 3 and 5 is still present. The tier of labeling straddling the layer 3/4 border is thus actually composed of two components, since labeling from the parvocellular C layers (C1 and C2) projected to layer 3 (and also layers 5a and 1), but not to layer 4. The small W cells of layers C1 and C2 project to patches in layer 3, while the large Y cells in the top part of the C layers project to patches in layer 4a, precisely in register with the projection to layer 3.

TANGENTIAL ORGANIZATION OF GENICULOCORTICAL PROJECTIONS AND THEIR RELATIONSHIP TO THE CO BLOBS

Note that the labeling in Fig. 5-4A is patchy (arrowheads), with a spacing similar to that of the CO blobs. The relationship between the patches of geniculate labeling from the C-laminae and the CO blob system is shown in Fig. 5-5, from Boyd and Matsubara (38). Figure 5-5A shows labeling from the C-layers of the LGN in tangential sections through area 17. The labeling is arranged in patches that often align to form rows, with bridges of labeling connecting patches in a row. A similar pattern of patches, or blobs, with connecting bridges is seen in the section stained for CO in Fig. 5-5B. To facilitate comparison between the two patterns, each patch of anterograde labeling in Fig. 5-5A has been marked with a number, and these numbers have been transferred to the CO stained section using the blood vessels as landmarks. Except for the top right part of Fig. 5-5B, where the plane of section passes out of layer 4 into layer 5, there is good correspondence between the patches of anterograde labeling and the CO blobs. Similar correspondence between CO blobs and patches of inputs are seen in area 18 (38).

It has been suggested that different levels of activity and, hence, CO staining might be present in populations of cortical cells postsynaptic to different classes of geniculate inputs (61). Thus, density of CO staining in a cortical layer may be a marker for the number and type of geniculate afferents received. The close correspondences between the dense band of CO staining in layer 4 and the terminations of the more active X- and Y-cells from the LGN and between the lesser CO staining in layers 3 and 5a and the terminations of the less active W-cells are consistent with this. Similarly, the difference in CO staining between blob and interblob is probably due to the presence of a geniculate input in the blobs that is not present in the interblobs. The differences in CO staining between blobs and interblobs can be

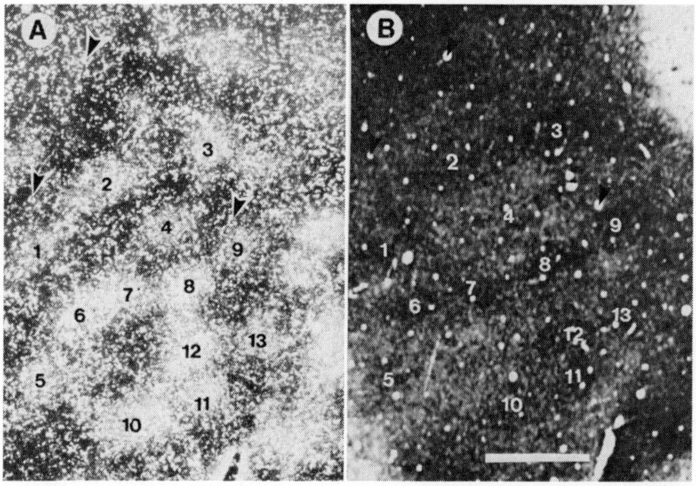

FIGURE 5-5. Comparison of cytochrome oxidase blobs and C-laminae labeling in layer 3 of area 17. **(A)** A dark-field photomicrograph of a tangential section through the lower part of layer 3 in area 17 containing anterograde labeling arising from the lateral geniculate C laminae. **(B)** A section from deeper in the cortex reacted for cytochrome oxidase. When patches of labeled terminals are compared to CO blobs, the two are most often found to align. Some examples of correspondence between the two patterns are numbered. Arrowheads show matching blood vessels used for alignment between the two sections. Scale bar = 1 mm. (From Boyd and Matsubara [38], with permission of Wiley-Liss, Inc.)

seen in layer 3, where W-cell terminations are made, and also in layer 4a, where the Y-cells from magnocellular layer C terminate. In layer 3, the correlation between LGN input and CO staining is readily apparent, as there is no geniculate input to this layer in the interblobs. In layer 4, although there is obviously geniculate input to and dense CO staining in both blobs and interblobs, Y-cells from the C layers terminate only in the blobs (38). As Y-cells in the LGN have greater spontaneous and evoked activity than X-cells, and also stain more darkly for CO than X-cells, it is suggested that Y-cell input is responsible for the more dense CO staining in layer 4a blobs versus interblobs. In area 18, the dense CO staining in layer 4a blobs is more difficult to explain, as X-cell input is lacking in this area (55,62–64; Chapter 1) so Y cell input must be made to both blobs and interblobs. Perhaps the difference between blobs and interblobs reflects differential projections from A laminae versus C laminae Y-cells, or represents a difference in the overall density of terminals (i.e., a difference in degree and not in kind).

At present, it is not clear if Y-cells from the A layers also terminate specifically in the CO blobs, as this was not possible to answer from the bulk injections used by Boyd and Matsubara (38). Recent data examining retrograde labeling from small injections targeted to blobs or interblobs suggest that A laminae cells projecting to blobs tend to be larger (65), consistent with A laminae Y-cell projections to the blobs. Y-cells in both A and C layers have similar anatomy and physiology (54,55,66) and can receive synapses from branching collaterals of the same retinal ganglion cells (67–69). Yet, C laminae Y-cells can be distinguished from those in the A laminae by their lower preferred spatial frequencies and higher contrast sensitivities (70), suggesting they may be a distinct population with possibly distinct cortical projections. Thus, there are two possibilities: both Y-cells from the A- and C-layers may be terminate selectively in the blobs, or Y-cells from the C layers may represent a distinct population that terminates selectively within the blobs, leaving Y-cells from the A-layers to terminate less selectively within both blobs and interblobs. Experiments combining intracellular staining of identified Y-cells from the A layers with CO histochemistry are needed to answer this question.

MOLECULAR MARKERS FOR OTHER BLOB/INTERBLOB INPUTS

CAT-301

The Cat-301 antibody stains a cell-surface molecule on select populations of cells in the brain (71). In the visual thalamus of the cat, Cat-301 stains Y-cells, but not X- or W-cells, both in the A and C laminae of the LGN, and in the medial interlaminar nucleus (MIN) (72). Cat-301 stained cells in the cat visual cortex are patchy (73), and unpublished experiments from this laboratory have shown that the patches of Cat-301 labeling in the cat visual cortex, just as they are in the primate, are in the CO blobs. Again, these data are consistent with the idea that the blobs in the cat preferentially receive Y-cell input. Moreover, they suggest a similarity between the organization of the magnocellular system in primates and the Y-cell system in cats.

RECEPTORS

In addition to the tracing experiments described previously, there is evidence from a variety of selective staining techniques that show different inputs to blobs and interblobs in the cat's visual cortex. Perhaps the most striking anatomically is the receptor autoradiography for certain serotonin receptor subtypes (45,74). Figure 5-6A, B, from Dyck and Cynader (45), show binding patterns for the $5HT_{1c}$ receptor ligand mesulergine in the coronal and tangential plane, respectively. Binding for mesulergine forms a continuous band in layer 4b, with evenly spaced columns extending into layer 4a. Comparison with CO staining in tangential sections shows that these columns are precisely interdigitated with the CO blobs (45). Mesulergine binding disappears after lesions of the LGN. This suggests that these receptors, like nicotinic acetylcholine receptors (75), are located on LGN terminals. Unlike the nicotine receptors, which are present throughout the thickness of layer 4 in both areas 17 and 18, the $5HT_{1c}$ receptors are restricted to area 17, mirroring the distribution of X-cell terminations, leading Dyck and Cynader

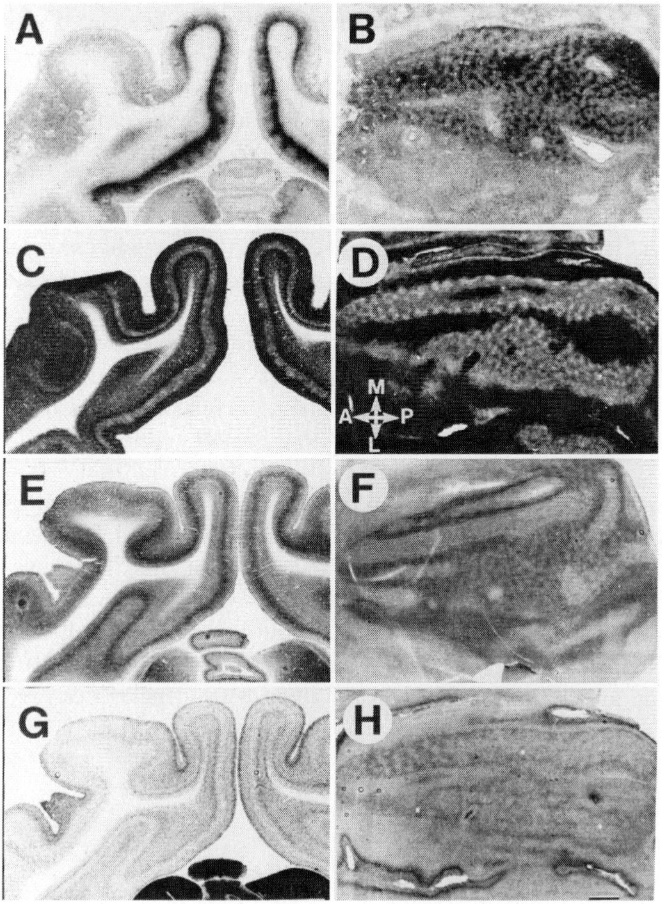

FIGURE 5-6. Multiple markers of cytochrome oxidase blobs/interblobs in cat visual cortex. Staining for each marker is shown in the coronal plane on the left column, and in the tangential plane in the right column. **A/B:** Serotonin 1C receptors. **C/D:** Synaptic zinc. **E/F:** cytochrome oxidase. **G/H:** Acetylcholine esterase. All of these markers show organization into patches of increased density in area 17, with the same center-to-center spacing of about 1 mm. As described in the text, all of these patterns are either coaligned or interdigitated with each other. Directions in D: A, anterior; P, posterior; M, medial; L, lateral. Bars = 3 mm. (From Dyck and Cynader [45], with permission of National Academy of Sciences, U.S.A.)

(74) to suggest that these receptors may be a selective marker for X-cell terminals. If this were so, then X-cell and Y-cell terminations in area 17 would be completely segregated and they are not (see Chapter 1).

Curiously, binding for another serotonin receptor, the $5HT_2$ receptor, also exhibited a patchy pattern in interblob columns in the upper part of layer 4, but binding for the $5HT_2$ receptor was mostly absent in the lower part of layer 4 (74).

Lesion experiments suggest that these receptors are also associated with LGN terminals. That these different receptors are marking previously unrecognized subclasses among LGN afferent is intriguing.

SYNAPTIC ZINC

Ionic zinc is enriched in a subset of glutamatergic axon terminals (76) and can be visualized with physical development using silver. In visual cortex, the laminar and columnar pattern of zinc staining is precisely complementary to the pattern of CO staining (45,77). Figure 5-6C,D, from Dyck and Cynader (45) shows staining for synaptic zinc in coronal and tangential sections, respectively. In the coronal section, zinc staining is relatively weak in layers 4 and 6, where the main geniculate inputs terminate. In the tangential plane, domains of slightly denser staining are present in layer 4, which correspond to the CO interblobs (77). In rat visual cortex (78), zinc-containing terminals arise from cortical, not subcortical systems. If this is also true in cat visual cortex, then the lightly staining layers and columns may merely reflect zinc-negative geniculate terminals taking up synaptic spaces that would otherwise be occupied by zinc-containing terminals of cortical origin. Alternatively, the zinc-rich terminals in the interblobs in layer 4 could represent a specific population of projection neurons that preferentially target CO interblobs. One possibility for the source of this projection is the claustrum. This region of the forebrain projects to layer 4 in primary visual cortex (79,80) and, in rodents at least, these projections are known to contain zinc (78,81). Whether there is any tangential patterning to the claustral projection to cat primary visual cortex has not yet been determined.

OTHER ENZYMES

Cytochrome oxidase (Fig. 5-6E,F) is not the only enzyme that demarcates columns in cat visual cortex. Figure 5-6 G,H, from Dyck and Cynader (45), shows patchy acetylcohlinesterase staining in coronal and tangential sections of cat visual cortex. Acetylcholinesterase patches are present in a thin band at the layer 3/4 border, and coalign with the CO blobs. Another enzyme that shows a columnar pattern of staining during development is the adenosine-producing ectoenzyme 5′-nucleotidase (82). During the period of development when this enzyme shows patchy staining in layer 4, it is associated with synaptic terminals; later in development, the enzyme increasingly becomes associated with glial membranes and the patchy axon staining is lost (82). As for the zinc terminals, the question remains whether these staining patterns represent specific inputs to blob or interblob columns, or whether the uneven staining patterns for these markers can be explained by different ratios of geniculate to cortical synapses in blobs versus interblobs, or by differences in average activity levels between blobs and interblobs. It should be noted that CO blobs in primates can also be stained with a variety of markers, the functional significance of many of which is not clear (19,83).

OUTPUTS OF THE CO BLOBS

SOFT-PATTERNING VERSUS HARD-PATTERNING

The primary visual cortex of the cat has several main projections to extrastriate areas, as shown in the exhaustive studies of Symonds and Rosenquist (8,9). Areas 17 and 18 are strongly interconnected, and both make strong projections to area 19, a belt of cortex which borders area 18 laterally. Areas 17 and 18 also project strongly to an area located on the medial bank of the lateral suprasylvian sulcus (LS area), and to area 21a, a small area located on the crown of the suprasylvian gyrus (84,85). There are also minor projections from primary visual cortex to other areas of the suprasylvian sulcus, and to area 20 on the posterior suprasylvian gyrus (86). The patchiness of most or all of these projections was noted in earlier studies (9,85,87–89).

Several later studies (31,90,91) examined how the patchiness of the projections from striate to extrastriate cortex might be organized. One possibility is that the patchiness results from the projections of adjacent columns of primary visual cortex diverging to innervate separate columns in the target area. In such a case, small extrastriate injections would preferentially label neurons projecting to a small number of columns, giving rise to patchy labeling in primary visual cortex. The intervening, unlabeled columns would have projections to uninjected columns of the labeled area, presumably columns of a different functional type. Extrastriate injections that were more extensive, involving more columns in the target area, would be more likely to include columns of all functional types and would give rise to continuous labeling in striate cortex, at least at the center of the labeling.

Theoretically, the injection site in the extrastriate area must be larger than the divergence of the striate-extrastriate projection to be sure of labeling all types of striate projections neurons (90). If an injection site is much larger than this divergence and unlabeled columns are still present in primary visual cortex, then the unlabeled columns can be concluded to genuinely not project to the target area. In such a case, the target area would receive only a subsample of the information from the efferent area. Shipp and Grant (91) called the configuration in which all the cortical columns are labeled after the injection site is made substantially larger than the divergence in the projection "soft-patterned," as distinguished from the "hard-patterned" configuration in which gaps in the labeling still remain even after large injections.

Several projections from areas 17 and 18 have been examined, with injections made purposefully large to distinguish hard-patterning from soft-patterning. The first of these was the projection from area 17 to area 18, which was shown to be hard-patterned (90,92). Another study of the same projection showed that the projection from area 17 to area 18 was orientation specific (93). That is, the patches of labeling in area 17 following a small injection into a restricted range of orientation preferences in area 18 labeled columns of the same orientation preference in area 17. As more columns in area 17 could be expected to be labeled with injections including a wider range of orientation columns in area 18, it would appear that elements of soft-patterning and hard-patterning can be present in the same projection. Other projections that have been shown to be hard-patterned include the projections from areas 17 and 18 to LS (31,91,94) and to 21a (95).

An example of hard-patterned labeling is shown in Fig. 5-7, from Conway et al. (95). This figure shows strikingly patchy labeling in a tangential section

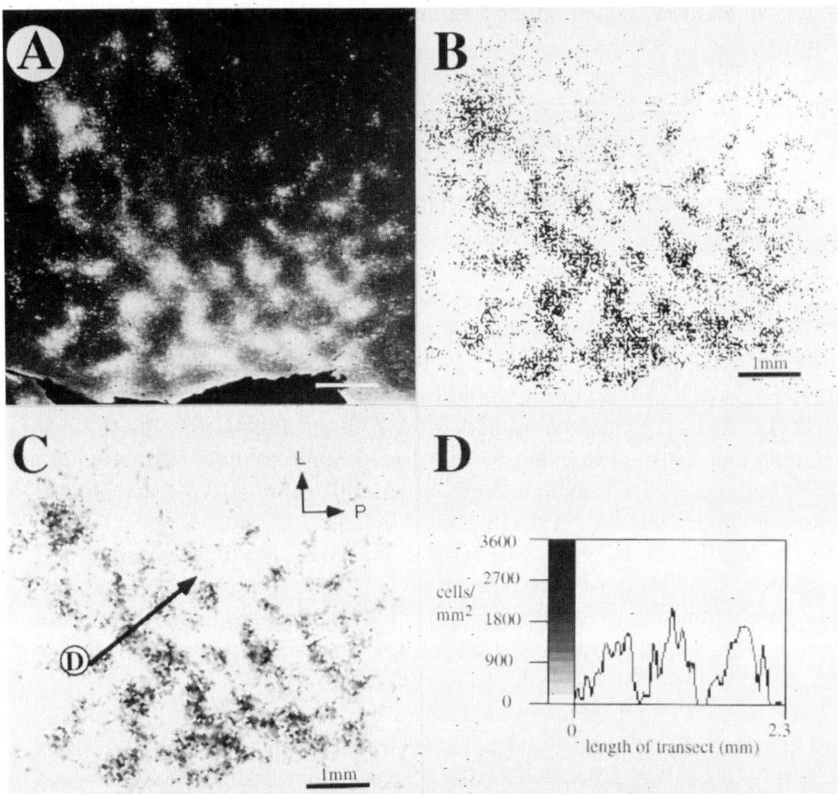

FIGURE 5-7. Retrograde labeling in areas 17 and 18 after an injection of wheatgerm agglutinin conjugated to horseradish peroxidase (HRP) in area 21a. The injection of tracer in this experiment was large enough to include the full width of area 21a. **(A)** A dark-field photomicrograph of a tangential section reacted for HRP. The labeled cells appear light on a dark background. Note how strikingly the cells are clustered into a regular array of patches. The slightly larger patches in the upper left corner are in area 18; the rest are in area 17. **(B)** A computer-aided reconstruction of 21a-projecting cells from multiple tangential sections. Even when the labeling from a full series of tangential sections is plotted and aligned, collapsing the labeling onto a single plane, the patchy pattern is still robust. **(C)** A density chart of the retrogradely labeled cells in the supragranular layers. The darkness of each (50 × 50 μm) pixel in the image is proportional to the number of cells at the corresponding location (see scale in **D**). **(D)** A plot of the cell density along the transect line shown by the arrow in **C.** This transect reveals that the center-to-center spacing of the patches is 0.7 to 0.8 mm. Also, note the magnitude of the difference in density of cells in the centers of the clusters compared to the interclusters. (From Conway et al. [95], with permission of Oxford University Press)

through areas 17 and 18 after a large injection involving the entire width of area 21a. The dark-field micrograph of a single section in Fig. 5-7A can be compared to the computer-aided reconstruction of the same area in 7B, where every labeled cell in the entire set of sections is displayed, and the image plot in 7C, where the density of cells corresponding to each pixel is represented by shading. Note the robustness of the clustering. The projections from area 17 to LS and to 18 differ from those to area 21a in that the clustering of cells is not quite so pronounced, with more cells being present between the clusters (95), but otherwise have a similar organization. All the different projection neurons have the same periodicity (slightly less than 1 mm), which matches that of the CO blobs. In area 18, the clustering of projection neurons is less robust than in area 17, and the spacing of the clusters is somewhat larger than in area 17. This compares well with the area 18 CO blobs themselves, which are more widely spaced and of less contrast than their counterparts in area 17. The finding of the hard-patterning of projection columns in primary visual cortex set the stage for an examination of how these projection columns might relate to the columns marked by CO staining.

SOME HARD-PATTERNED BLOB-EFFERENT PROJECTIONS

The first set of efferent columns to be compared to CO staining was the projection to the lateral suprasylvian area, which had previously been shown to be hard-patterned (31,91). Figure 7-8, from Boyd and Matsubara (94), shows two pairs of adjacent tangential sections stained for LS-projecting cells retrogradely labeled with wheatgerm agglutinin conjugated horseradish peroxidase (A,C) or stained for CO (B,D). In each pair, the CO- and tracer-reacted sections show corresponding areas; some of the same radially penetrating blood vessels can be seen in both sections, allowing for close alignment. Note the patchy pattern of the LS-projecting cells, although the tracer injections in these experiments were large. Some of the patches of labeled cells are marked with numbers, and the corresponding positions in the CO-stained sections are also numbered. When the positions of the patches of cells are compared to the positions of the CO blobs this way, the clusters of LS-projecting cells and the CO blobs mark the same cortical columns. CO blobs and patches of LS-projecting cells are also colocalized in area 18 (94).

The other projections from primary visual cortex that had previously been shown to be hard-patterned have also been compared to CO staining. A detailed quantitative analysis of the projections to area 21a, another of the main targets of areas 17 and 18, is shown in Fig. 5-9, from Conway et al. (95). Figure 5-9A is a bright-field photomicrograph of a tangential section through area 17; the typically patchy labeling pattern of 21a-projecting cells is seen. Figure 5-9B shows a slightly deeper section from the same experiment, stained for CO and showing the CO blobs and interblobs. The two sections were aligned using radially penetrating blood vessels (some of which are marked by arrows). Using the arrows for alignment, the positions of cell clusters and CO staining can be compared

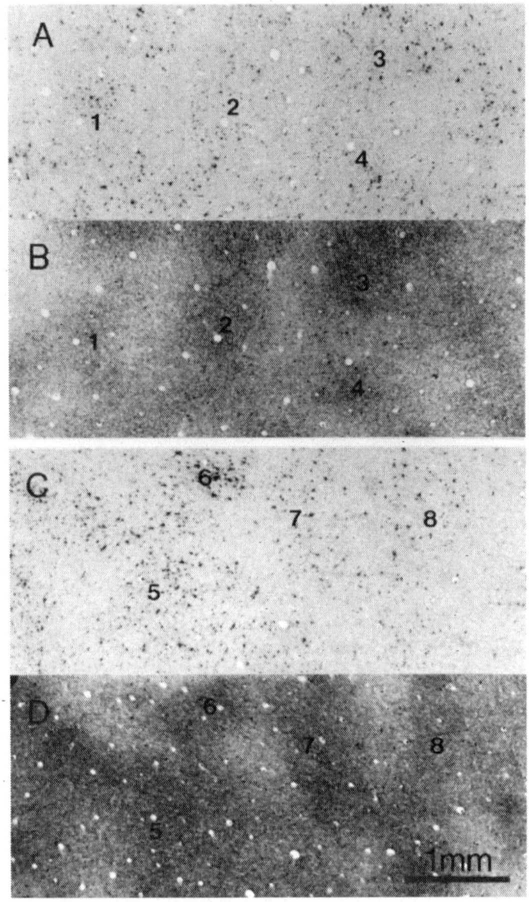

FIGURE 5-8. Comparison of retrograde labeling from LS and cytochrome oxidase (CO) staining. **(A)** A bright-field micrograph of a tangential section through area 17 containing cells retrogradely labeled from an injection in the lateral suprasylvian area (LS). Although the injection was large, the labeled cells are patchy. Several patches are indicated by numbers. **(B)** A photomicrograph of a CO-stained section from the same experiment shown in **A**, from slightly deeper in the cortex. Note the CO blobs present in this section. The two photographs in **A** and **B** have been aligned using radially penetrating blood vessels present in both sections as landmarks, allowing for appreciation of how well the patches of LS-projecting neurons align with the CO blobs. **(C** and **D)** Another aligned pair of sections processed for LS-projecting cells or CO activity, respectivly. Again, these sections show the good correspondence between the patches of LS-projecting cells and the CO blobs. Scalebar = 1 mm and applies to A-D. (From Boyd and Matsubara [94], with permission of Cambridge University Press.)

between Fig. 5-9A and 5-9B. Note that most of the patches of labeled cells align with CO-dense blob in the CO stained section. A three way chi-square analysis was performed on the data in Fig. 5-9 by dividing the cortex into three approximately equal sized compartments, blob, blob borders, and interblobs, following bandpass smoothing of the image. In Fig. 5-9C, the blobs are shown outlined and

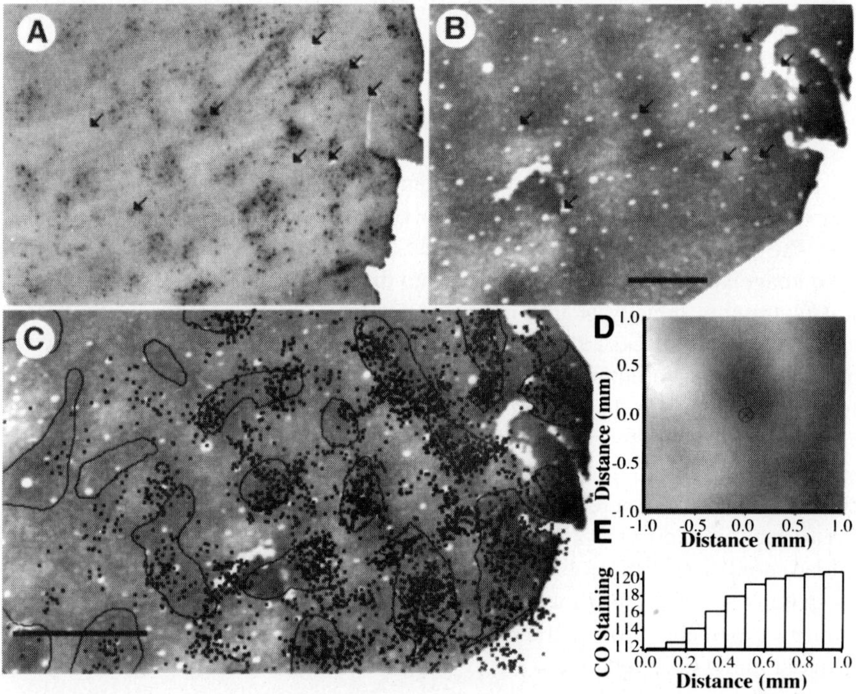

FIGURE 5-9. Quantitative analysis of the correspondence between CO staining and retrograde labeling from area 21a. **(A)** A bright-field photomicrograph of 21a-projecting cells in a tangential section area 17. Patches of labeled cells can be seen extending from the top to the bottom of the figure. **(B)** An image from a CO-stained section aligned with the data from **A** using blood vessels (arrows) as landmarks. Note that as the patches of projection cells form bands in this case, so also are the CO blobs elongated, and with the same orientation. Most of the patches of labeled cells are aligned with CO blobs. **(C)** An enlargement of the same CO image as in **B,** with the positions of the labeled neurons charted from several sections represented by black dots. The borders of the CO blobs are indicated by black lines. These borders were generated by designating the darkest third of the pixels, following bandpass smoothing of the CO images, to correspond to CO blobs. The density of labeled cells is more than twice as great in blobs compared to interblobs, significantly different ($p < .0001$) in a chi-square analysis. **(D)** The 2-dimensional spatial cross-correlation of the data shown in **C.** This image shows CO staining at different distances from the labeled cell, averaged over the entire population of labeled cells. The darkness of the image at each X-Y offset is proportional to the density of CO staining at that offset averaged over all the labeled cells. The dark spot near the origin of the plot (marked with cross-hairs) shows that CO staining was darker near labeled cells. The size, spacing, and orientation of this dark area, and the dark areas along the edges of the image (caused by the regular periodicity of the CO blobs) are similar to those of the CO blobs, suggesting that the pattern of 21a-projecting cells shows a relationship to the CO architecture. Scalebar = 1 mm. **(E)** The 1-dimensional spatial cross-correlation of the data shown in **C.** This was made by collapsing the data from **D** radially onto a single dimension and dividing it into 0.1-mm bins. Here, the average CO staining value is plotted as a function of distance from the reference position. The CO staining value is lowest (darker staining) closest to the reference position and gradually increases, leveling off at about 0.3 mm away from the reference position. (From Conway et al. [95], with permission of Oxford University Press.)

the retrogradely labeled cells (black dots) are superimposed. For clarity only the outlines of blobs are shown, so the remaining cortical area includes both interblobs and blob borders. For this experiment, the density of 21a-projecting cells in the blobs was more than twice as great as the density in the interblobs. The *p*-value of the chi-square analysis for this data was less than 0.0001, showing that more labeled cells were found in the blob columns than would be expected by chance. Fig. 5-9D shows the correlation between CO staining and 21a-projecting cells. The reference position, (0,0) is in the center of the figure, and the darkness of the image at different X,Y-distances from the reference position corresponds to the CO staining density at that distance averaged for all of the 21a-projecting cells. The dark area near the center of the image indicates that 21a-projecting cells are found preferentially in areas of darker CO staining. As the size and spacing of the dark area in the center of the image, and the dark areas along the edges of the image (caused by the regular periodicity of the CO blobs) are the same as the size and spacing of the CO blobs, it can be concluded that the pattern of 21a-projecting cells shows a relationship to the CO architecture. The two-dimensional cross-correlation in Fig. 5-9D was collapsed radially onto a single dimension and divided into 0.1-mm bins, giving the plot shown in Fig. 5-9E, where the average CO staining value is plotted as a function of distance from the reference position. Similar experiments show that the cells projecting from area 17 to area 18 are also clustered in the CO blobs (92).

SUBLAMINAR ORGANIZATION OF BLOB-EFFERENT PROJECTIONS

Thus, of the four main corticocortical outputs of area 17 (area 18, area 19, LS, 21a), three of them (area 18, 21a, LS) are clustered in the CO blobs. Although these results do show that the CO blobs in cats, like those in primates, are important in organizing corticocortical outputs, they raise certain issues. For instance, area 21a is specialized for the detection of the orientation of long, continuous elements such as lines and edges, and thus appears to be important for the processing of form vision (96–98), whereas LS is specialized for encoding the motion and direction of moving textures and background, responding only weakly to stimulus orientation (99–106). Area 18 is also functionally distinct (62,107–111). It is perhaps hard to reconcile the idea that area 17 acts to funnel different sets of information to differently specialized extrastriate areas with the finding that the projection columns of some very different areas can all align with the CO blobs. However, as reviewed below, laminar origin of projection neurons could also act to segregate different information from area 17.

Figure 5-10, from Conway et al. (95), shows a comparison of the sublaminar organization of LS- and 21a-projecting cells within layer 3 of area 17 using fluorescent double-retrograde labeling techniques and confocal microscopy. Figure 5-10A,B shows images of cells in the same cortical column of a 250-μm thick section projecting to either 21a or LS, respectively. The sublaminar organization is

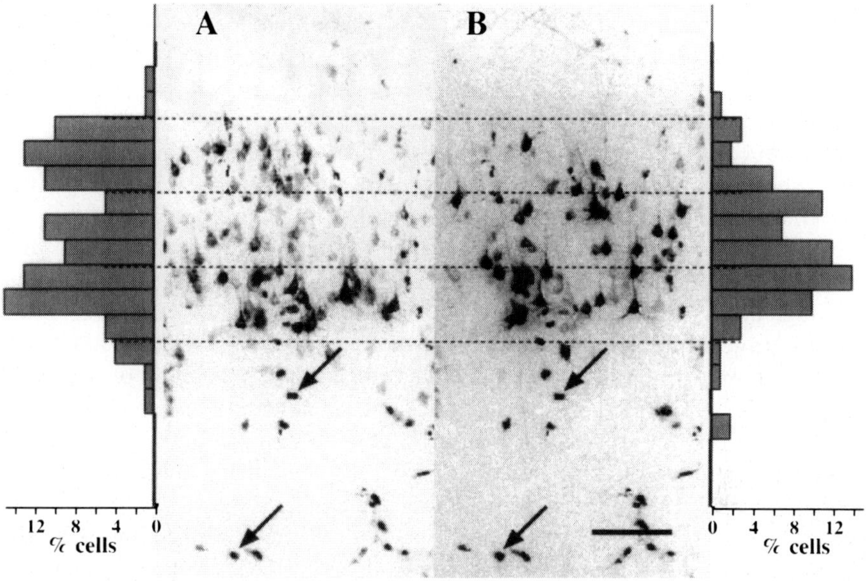

FIGURE 5-10. Laminar organization of LS- and 21a-projecting neurons in area 17. This figure shows dual channel images of double labeled tissue. The percentage of labeled cells at varying cortical depth is shown in histogram form for the 21a (left) and LS (right) projecting neurons seen in this double labeled tissue section. The dashed lines divide layer 3 into three equal sublayers. **(A)** Neurons labeled from an injection of FITC-dextran in 21a; 21a-projecting cells are most numerous in the top and bottom thirds of layer 3, with the middle third of layer 3 being relatively cell-sparse. **(B)** Neurons labeled from an injection of rhodamine-dextran into LS. Compared to 21a-projecting cells, there are relatively few LS-projecting cells in the upper third of layer 3. Instead, LS-projecting cells are numerous in the bottom two thirds of the layer. It is important to note that, even in the bottom part of layer 3 where projections neurons of both kinds are numerous, no cells are double labeled. The distinct, yet overlapping laminar distribution of these two classes of projection neurons may result in different information being sent from these two blob-efferent projections. Scalebar = 100 μm. (From Conway et al. [95], with permission of Oxford University Press.)

best seen with the help of the reference lines in Fig. 5-10, which divide layer 3 into three equal sublayers. Histograms of the percentages of labeled cells as a function of cortical depth were generated for 21a- (Fig. 5-10A, left) and LS- (Fig. 5-10B, right) projecting neurons shown in the double-labeled image. LS-projecting cells were more abundant in the middle and lower portions, with fewer cells in the upper portion of layer 3, whereas area 21A projecting cells dominate in the upper and lower thirds of layer 3. Thus, the LS-projecting neurons have a different, although overlapping, distribution with neurons that project to area 21A. Cells projecting to area 18 appear to overlap somewhat in laminar distribution with those projecting to LS (112,113), although here, too, there are laminar differences, as the cells projecting to area 18 include some stellate cells from layer 4 that do not make projections to the other areas. Although the lower part of layer 3 gives rise to projections to both 21a and LS, it is important to note that none of the cells in Fig. 5-10 projected to both areas. Even when different classes of projection neurons overlap in laminar distribution, it seems to be that few cells project to more than one area (95,114,115).

Thus, the different output channels appear to be composed of discrete populations of neurons, with unique laminar and columnar distributions. How might the differences in laminar distributions of LS- and 21a- projecting cells in the CO blobs contribute to functional differences in the output to these two areas? As discussed previously, the CO blob columns in cat visual cortex mark the termination sites of geniculocortical afferents from the C-laminae of the LGN. The tier of LS- and 21a- projecting cells in the bottom third of layer 3 are both placed to receive Y-cell input on their basal dendrites, whereas the second population of 21a- projecting cells in area 17 appear to be located superficial to the sites of geniculocortical termination. These cells may differ from the LS population by not receiving direct geniculate inputs. If this is true, it would suggest that a subclass of information going to 21a might traverse a further stage of processing in area 17, compared to that going to LS. Perhaps the intrinsic interlaminar connections that provide the input to the upper tier of 21a-projecting neurons endow these cells with a different set of functional properties than the lower tier of 21a-projecting cells.

Unfortunately, we know of no study suggesting a correlation of functional parameters such as direction/orientation selectivity or spatial/temporal tuning with cortical depth within the supragranular layers. Nor have any studies investigated differences in the interlaminar connections of the sublayers of layer 3 in cat area 17, as has been done in several primate species (116–118). Differential projections onto the various sublayers of layer 3 from the different sublayers of layer 4 suggest functional differences in the layer 3 subdivisions based on the known functional differences of the LGN inputs to the different layer 4 sublayers. Ultimately, studies combining electrophysiological examination with identification of different classes of projection neurons (119–121) will be required to examine functional differences in the anatomically distinct output pathways like the ones shown in this study.

FUNCTIONAL CONSIDERATIONS OF DIFFERENT PATHWAYS THROUGH THE BLOBS

Because LS has a high percentage of cells selective for direction of stimulus motion (98–101,103,122), patches of LS-projecting cells had been previously suggested to correspond to segregated cortical columns specialized for some aspect of visual motion processing within area 17 (91). Shipp and Grant (91) speculated that the patches of LS-projection neurons might correspond to columns of Y-cell input from the LGN. As reviewed in the previous section, at least some Y-cells (those in LGN layer C), as well as W-cells from the ventral C laminae of the LGN, project preferentially to the CO blobs.

Consistent with the notion of a preferential Y-cell projection to the CO blobs is the fact that CO blob columns respond preferentially to lower spatial frequencies than interblob columns. Figure 5-11C, from Shoham et al. (47), shows optical

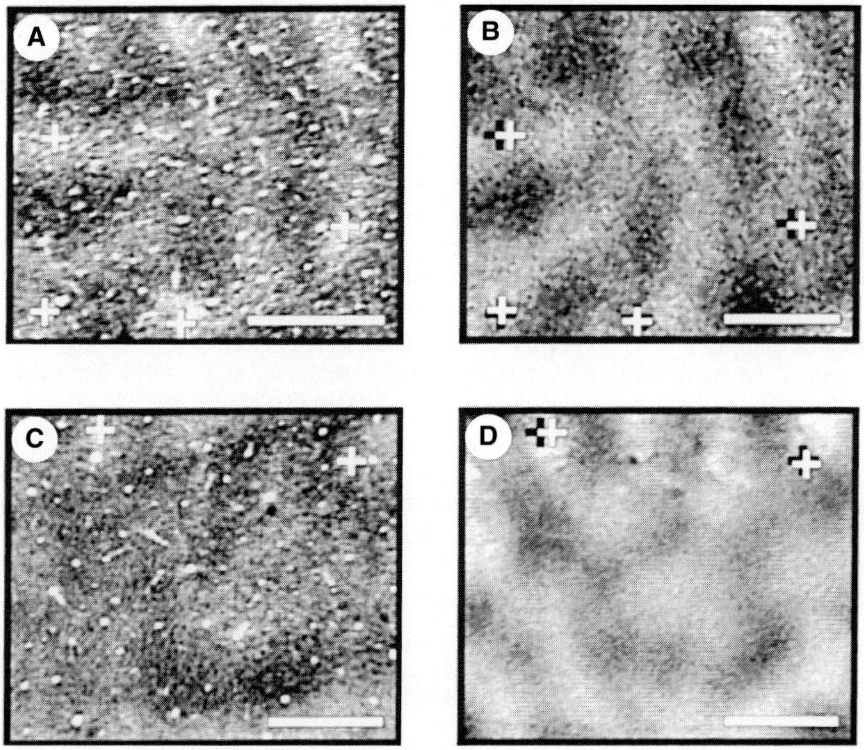

FIGURE 5-11. Comparison of spatial frequency columns and cytochrome oxidase blobs in cat area 17. **(A)** Image of CO staining in a horizontal section through upper layer 4 of cat area 17. Blobs are visible as elongated patches or short bands of darker staining. Crosses mark the sites of dye injections used for alignment with the optically recorded images. **(B)** A spatial frequency map. The dark patches correpond to regions of cortex preferentially activated by low spatial frequencies (0.2 cyc/deg). Light patches show regions responding better to higher spatial frequency gratings (0.6 cyc/deg). Crosses show the position of alignment dye injections judged at the time of making the map, compared to the actual injections recovered histologically (copied from **B**). Comparison of the two panels shows that, in most instances, there is precise correspondence between the low spatial frequency domains and the CO blobs. **(C and D)** Another example of the correlation between the spatial frequency domains (D) and the cytochrome oxidase staining **(D).** Again, there is good correspondence between CO blobs and domains preferring low spatial frequencies. Scalebars = 1 mm. (From Shoham et al. [47], with permission from *Nature.*)

imaging of spatial frequency columns in area 17 compared to CO staining. After visual stimulation with sine wave gratings of 0.2 and 0.6 cyc/deg, maps of response strength from the two stimulus conditions were subtracted from each other to give this image. Darker regions of the image correspond to areas of the cortex responding more strongly to the low spatial frequency, whereas lighter regions of the image preferred the higher spatial frequency. Comparison with the CO image of Fig. 5-11B shows that the patches of cortex preferring low spatial frequency correspond the CO blobs, and the high spatial frequency patches to the interblobs. Thus, the response properties of area 17 projection neurons targeting LS should strongly reflect (direct or indirect) input from the Y-cell and W-cell streams that might be expected to be lacking in projections from interblob columns, whereas projection neurons from both blobs and interblobs should reflect the influence of the LGN X-cell projection (115,123).

LS-projecting cells in area 18 are also preferentially found in CO blob columns (94), which are also selectively targeted by the C-layers of the LGN in this area (38). Columns preferring different spatial frequencies are present in area 18 (124), but it is not yet known if these columns correlate with the CO pattern, as they do in area 17. Other experiments are needed to address the similarities and differences in the columnar organization of areas 17 and 18.

PROJECTIONS TO AREA 19

A SOFT-PATTERNED PROJECTION

Much of the preceding discussion has been taken up with the projections of the CO blobs. However, as blob columns only make up half the cortex, it is natural to wonder about the extrastriate projections of the interblobs. The other main target of striate cortex is area 19. Cells in area 17 projecting to area 19 have been reported to have a patchy distribution (14,87,89,114,125), but these studies used rather small injections. Large injections in area 19 give rise to fairly uniform labeling in areas 17 and 18, labeling both blob and interblob columns (126). Thus, area 19 is the only extrastriate area so far examined that receives strong projections from interblob columns, as well as from blob columns, and the striate cortex to area 19 pathway is the only one shown to display soft-patterning.

Comparing the patchy results of the small injections with the continuous results of the larger injections suggests that the projections from different columns in striate cortex are segregated within different columns in area 19. To further examine the mechanism by which patchy labeling might be obtained with small injections, but not large injections, paired injections of two different, distinguishable tracers (CTB and CTB-Au) were made side by side in area 19 (126). In some cases, patches of labeling from the two tracers overlapped and in some cases they interdigitated. An example where the two tracers interdigitated is shown in Fig. 5-12, from Boyd (34). When labeling from each tracer is considered in isolation, a patchy pattern is obvious. When considered together, labeling

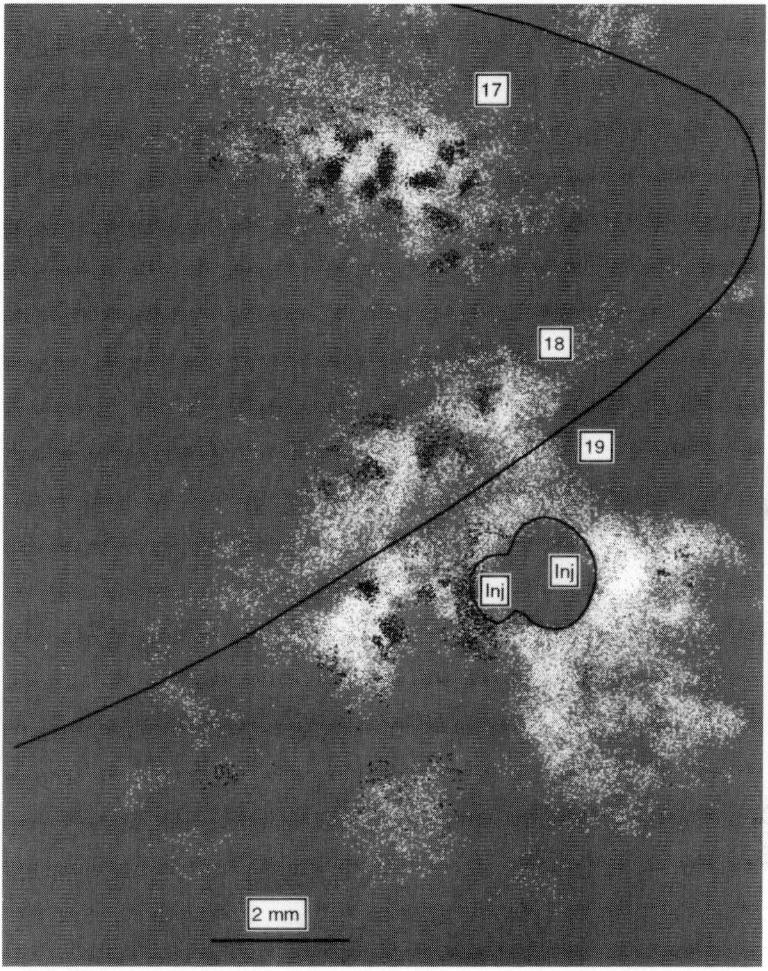

FIGURE 5-12. Organization of projections to area 19. This computer-generated plot shows the results of paired injections of unconjugated (CTB) and gold conjugated (CTB-Au) cholera toxin B subunit in area 19. A one-in-two series of tangential sections was charted and each dot represents a single labeled cell. White cells contained CTB-Au, which were labeled retrogradely from the larger injection to the right of the figure. Black cells contained unconjugated CTB, labeled from the smaller injection site. The extent of areas 17 and 18, determined from CO staining, is shown by a solid black line. In this case, labeling in areas 17 and 18 from the two injections formed an interdigitating pattern. Note the large, sometimes elongated patches of local connectivity within area 19. (From Boyd [34].)

from the two tracers fills in all of the columns of cortex, giving a continuous pattern. Note that the size and spacing of the patches are similar to the CO blobs. Compared with CO staining in adjacent sections, the projection columns to area 19 are correlated with the CO architecture (34). It is possible that, in addition to

CO blobs, another columnar system, such as orientation or ocular dominance, is also involved in the organization of the projections from areas 17 and 18 to area 19. Clustering of orientation selectivity has been noted in area 19 (2), but no evidence for ocular dominance columns has been found. Thus, columns of like orientation could be connected, as in the projection from area 17 to area 18 (93).

COLUMNAR ORGANIZATION IN AREA 19

At least two types of columns can thus be demarcated in area 19, one that receives inputs from blobs, and one that receives inputs from interblobs. Large injections would more likely include both columns of area 19 receiving inputs from blobs, and columns receiving inputs from interblobs, whereas small injections would have a greater chance of targeting a single type of column. Small injections that label cells in both blobs and interblobs could be explained by the injection straddling the borders of two such columns. (A less parsimonious but possible explanation would be the existence of a third type of column in area 19, one that receives input from both blobs and interblobs of areas 17 and 18.). A lower limit on the size of the columns of terminations from area 17 and 18 can be inferred from cases where two injections with a combined extent of about 2 mm both gave patchy labeling of blobs in area 17. Thus, axon terminals from blobs and interblobs in area 17 must segregate in area 19 into columns that are at least 2 mm wide.

Some information is available on the columnar organization of area 19. Note that Fig. 5-12 also shows local labeling in area 19. Local connections of area 19, with a spacing of about 2 mm, are organized with a periodicity larger than that of local connections within area 17. Sometimes, these patches are noticeably elongated. This suggests that the internal organization of areas 17 and 19 may be substantially different. This difference in internal organization is also suggested by arrangement of the projections from area 19 to 21a and to LS. Figure 5-13, from Boyd and Matsubara (94), shows projection columns in a tangential section through area 19 after a large injection of retrograde tracer in LS. Like the projections from area 17, the projections from area 19 are also arranged in a discontinuous fashion. However, unlike the projection columns in area 17 or 18, which have a spacing of about 1 mm, the projection columns in area 19 have a much wider spacing, of at least 2 mm (85,94). In addition, projection columns in area 19 have a tendency to group into elongated bands, often oriented perpendicular to the long axis of area 19. This larger scale of cortical organization in area 19 is apparently unique among cortical areas in the cat. Patches of labeling in other cortical areas from axonal transport studies, including areas 17,18,21a, and LS, all appear to have a spacing of about 1 mm (9,83).

Another interesting manifestation of columnar organization in area 19 comes from the study of Anderson *et al.* (44). After injections of tracer into the eye, transneuronal labeling in area 19 was discontinuous, with a similar organization to the projection columns labeled from 21a and LS. Although only one eye was

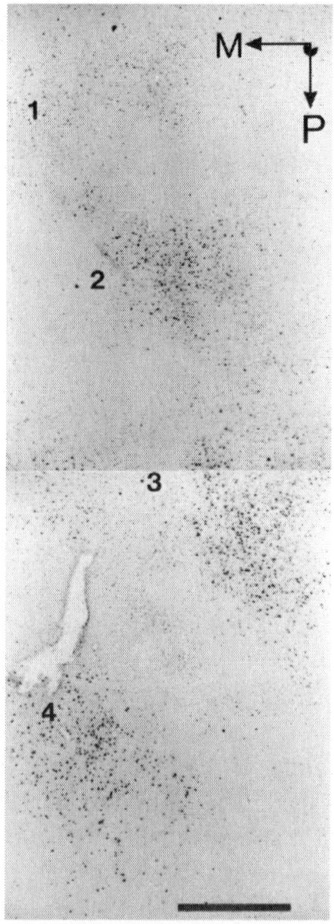

FIGURE 5-13. Organization of LS-projecting cells in area 19. This bright-field photomicrograph of a tangential section through area 19 shows cells labeled retrogradely from a large injection of gold conjugated cholera toxin in area LS. Note the four large clusters of labeling, numbered 1 to 4. The spacing between these clusters is approximately 2 mm. Connectivity within area 19 thus reveals an organization with a periodicity double that of areas 17 and 18. The mediolateral (M) and antero-posterior (P) axes are labeled. Scalebar = 1 mm. (From Boyd and Matsubara [94], with permission of Cambridge University Press.)

injected, these patches do not appear to be related to ocular dominance because this property is not organized in a columnar fashion in area 19 (127).

It is not yet known if a relationship exists between the columns containing PMLS-projecting cells and the columns receiving either blob or interblob input from areas 17 and 18. However, the fact that the input-defined columns and the output-defined columns appear to have a similar spacing is consistent with the suggestion that they are related. As PMLS receives inputs only from the blobs of areas 17 and 18, it is suggested that it may, analogously, receive input only from the columns of area 19 that also receive inputs from the blobs.

COMPARISONS WITH PRIMATES

From the findings discussed above, it appears that the CO architecture in the visual system of the cat, as in the primate visual system, is associated with the segregation of different processing streams. Below are discussed some of the similarities in how CO blobs in cats and primates are related to the organization of the visual pathways.

LGN INPUTS

W-Cells

The similarity between geniculocortical pathways in cats and primates is most striking for the comparison between the W-cell pathway of the cat and the koniocellular pathway in primates. At the level of the LGN, both the K layers of primates and the parvicellular C layers of cats receive input from the superior colliculus, but the rest of the LGN does not (128). Also, the parvicellular C laminae resemble the K layers of primate LGN in that they stain densely for one calcium binding protein (calbindin) and weakly for another (parvalbumin), whereas the opposite pattern is found in the main layers of the LGN in both species (129,130). Physiologically, the K layers contain cells with W-type properties (131). Some cells in the koniocellular layers are also color selective, receiving inputs from a particular class of blue-on ganglion cells (132,133). Interestingly, the relatively rare color-selective cells in cat LGN are found in the ventral C-layers, and are also blue-on (134,135).

Like the parvicellular C-layers of the cat, the koniocellular layers terminate in layer 3 selectively within the CO blobs (22–24,136–138). It is interesting to note that in tree shrews, the small-cell layers of the LGN also receive projections from the superior colliculus (139), stain densely for calbindin (140), and terminate specifically in patches in layer 3 (141). Although CO blobs have yet to be observed in this species, a patchy pattern can be seen in V1 using a myelin stain (142). One intriguing difference in the projections of the parvicellular C layers of the cat LGN and the small-cell layers in other species is the projection to layer 5a, which has been described only in the cat. Whether this pathway has analogs in other species is unknown.

Y-Cells

In addition to W-cell input, the Y-cell input from layer C is also confined to the CO blobs, thus accounting for the fact that the blobs are visible in layer 4a, but not in layer 4b. Although not a matter of universal agreement (143), the Y-cell pathway in the cat has traditionally been, on the basis of similarities in relative cell size and physiological properties, likened to the magnocellular pathway in primates (6). In the owl monkey (19,144) and in the prosimian galago (145), CO blobs are visible in the top part of layer 4, where the magnocellular LGN afferents terminate, similar to the arrangement of CO blobs in the cat. (This does not appear to be the case in other, diurnal new world primates such as the squirrel monkey, or in any old world primates examined [19]). It is suggested that, in these primates, the dense CO staining in blobs in layer 4Cα is a reflection of the concentration of magnocellular terminations. Indeed, some attributes of M-cells, such as lower spatial frequencies (146) and higher contrast sensitivities (147), have been reported for blob cells in primates. Moreover, the Cat-301 antibody, which stains the magnocellular LGN layers most densely (148,149), stains patches in layer 4Cα and the bottom part of layer 4B in V1, which align with the CO blobs (149,150). The presence of Cat-301 in layers of the LGN and cortex known to receive magnocellular input, as well as its presence in extrastriate areas such as V3 and MT thought to be dominated by magnocellular inputs (150), suggests that the abundance of Cat-301 in the CO blobs versus the interblobs reflects some differential magnocellular input to the blobs.

Lund (151) reviewed evidence for two classes of magnocellular cells differing in spatial and temporal responses and suggested a laminar segregation of these two types within layer $4C_{\alpha}$, although a tangential segregation into blob and interblob columns might also be worth consideration. That is, the magnocellular cells with higher contrast sensitivities and lower preferred spatial frequencies might terminate selectively in the CO blobs, whereas the other class of magnocellular axons is not blob specific, as is suggested for the C and A laminae Y-cells in the cat.

OUTPUTS

As has been shown in the cat, several extrastriate areas in primate visual cortex receive a hard-patterned projection from the blobs in V1. These include the dorsomedial visual area (152,153) and the middle temporal area (MT) (154). Like LS in the cat, MT in the primate is concerned with motion processing (155–158), and the two areas have been suggested to be homologous (159). The input from primary visual cortex to MT originates mostly from a layer of large cells at the base of layer 4B (160,161). Lund *et al.* (40) have argued that these cells might be analogs of the large pyramidal cells at the layer 3/4 border in the cat, which project to LS. In both cases, projection neurons efferent to an extrastriate area concerned with motion are clustered in the blobs, at a laminar level directly above the termination zone of the largest LGN cells, possibly receiving direct LGN contacts onto their basal dendrites. This arrangement seems ideal for segregating a fast, motion-sensitive pathway through the blobs and suggests that CO blobs in cats and primates are similarly organized.

Like the connections between striate cortex and area 19 in the cat, several connections in primate visual cortex display soft-patterning. These include the projections from V1 to V2 (25,26) and the projections from V2 to V4 (162,163). In V2, cytochrome oxidase stripes and interstripes correspond with the segregated inputs from blobs and interblobs in V1 and interblobs in V1 and tie these together with the segregated outputs to V4 and MT (28,162). Although histochemical

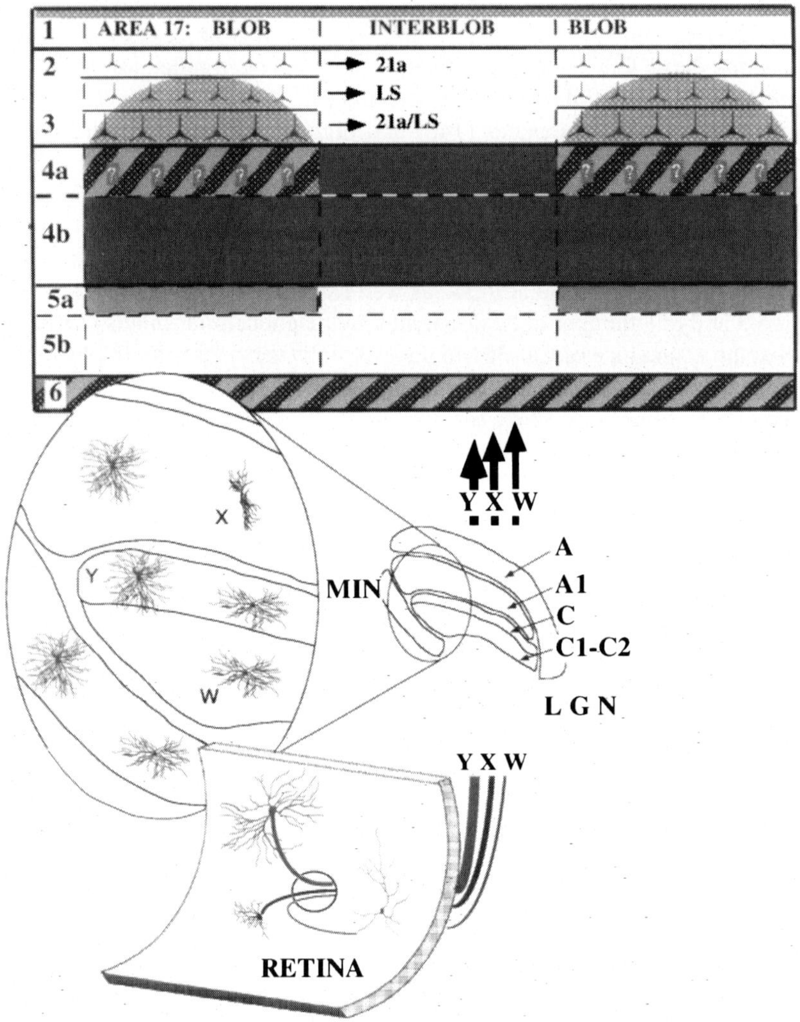

FIGURE 5-14. Summary diagram of the segregated inputs and outputs of the cat's primary visual cortex. The diagram shows the three different cell classes (X,Y,W) in the retina (bottom), and the lateral geniculate (middle). The top part of the diagram shows the termination in area 17 of W-, X-, and Y- geniculocortical fibers, shown by three different densities of shading, W being lightest and Y the darkest. The cross-hatched pattern in layer 6 and layer 4a shows mixing of X-cell and Y-cell inputs. The X-cell input to layer 4a of the cytochrome oxidase blobs is marked with question marks because it is not yet known whether X-cell terminations are found in or avoid the 4a blobs. Outputs to extrastriate areas LS and 21are shown as triangular profiles of three different sizes. Both LS- and 21a-projecting cells are clustered in the CO blobs. These outputs divide layer 2/3 within the blobs into three sublayers. The deepest sublayer projects to both 21a and LS, the middle sublayer projects to LS, and the upper sublayer projects to 21a. (Modified from Boyd and Matsubara [38], with permission of Wiley-Liss, Inc.)

markers for columnar organization are not available for area 19 in the cat, it is important to note that the close tying between input-defined and output-defined columns has been shown to exist in the absence of columnar CO staining in extrastriate areas in primates (163).

CONCLUSIONS

Segregated processing streams exist in the visual cortex of the cat, and these are intimately associated with the CO blobs in areas 17 and 18. Figure 5-14, modified from Boyd and Matsubara (38), shows a summary diagram of some of the segregated inputs and outputs of the CO blobs in the cat's primary visual cortex. The X, Y, and W pathways originating from different classes of retinal ganglion cells are shown segregated in the LGN and terminating in different layers and columns of primary visual cortex. W cells terminate in layers 3 and 5a within the CO blobs. Y cells from the C-laminae, and perhaps from the A-laminae as well, terminate in layer 4a of the blobs. X cells terminate in both blob and interblob compartments. Projections to extrastriate areas LS and 21a are concentrated in the CO blobs, with separate, but overlapping laminar distributions. Not shown are the projections to area 19, which originate from both blob and interblob columns.

CO blobs in cats and primates appear to be manifestations of similar principles of visual cortical organization. It is suggested that further study will show more similarities between blobs and interblobs in cats and primates. Moreover, given that taxa as distantly related as cats and primates have similar columnar organizations of inputs and outputs, it is likely that similar organizations are present in some other mammals.

REFERENCES

1. Hubel, D. H., and Wiesel, T. N. (1962). Receptive fields, binocular interaction and functional architecture in the cat's visual cortex. *J. Physiol.* **160,** 106–154.
2. Hubel, D. H., and Wiesel, T. N. (1965). Receptive fields and functional architecture in two nonstriate visual areas (18 and 19) of the cat. *J. Neurophysiol.* **28,** 229–289.
3. Hubel, D. H. (1988). *Eye, brain, and vision.* New York, W. H. Freeman and Company.
4. Stone, J. (1983). Parallel processing in the visual system: *The classifcation of retinal ganglion cells and its impact on the neurobiology of vision.* New York, Plenum Press.
5. Sherman, S. M. (1985). Functional organization of the W-, X-, and Y-cell pathway in the cat: a review and hypothesis. *Prog. Psychol. Physiol. Psychol.* **2,** 233–314.
6. Casagrande, V. A., and Norton, T. T. (1991). Lateral geniculate nucleus: a review of its physiology and function. In: *The neural basis of visual function* (A.G. Leventhal, Ed.), Vol. 4, pp. 41–84, London, The Macmillan Press.
7. Casagrande, V. A., and Kaas, J. H. (1994). The afferent, intrinsic, and efferent connections of primary visual cortex in primates. In: *Cerebral cortex* (A. Peters and K.S. Rockland, Eds.), Vol. 10, pp. 201–259, New York, Plenum Press.
8. Symonds, L. L., and Rosenquist, A. C. (1984). Laminar origins of visual corticocortical connections in the cat. *J. Comp. Neurol.* **229,** 39–47.

9. Symonds, L. L., and Rosenquist, A. C. (1984). Corticocortical connections among visual areas in the cat. *J. Comp. Neurol.* **229,** 1–38.
10. Zeki, S. M. (1975). The functional organization of projections from striate to prestriate visual cortex in the rhesus monkey. *Cold Spring Harb. Symp. Quant. Biol.* **40,** 591–600.
11. Enroth-Cugell, C., and Robson, J. G. (1966). The contrast sensitivity of retinal ganglion cells of the cat. *J. Physiol.* **187,** 517–552.
12. Marshall, W. H., Talbot, S. A., and Ades, H. W. (1943). Cortical response of the anesthetized cat to gross photic and electrical afferent stimulation. *J. Neurophysiol.* **6,** 1–15.
13. Clare, M. H., and Bishop, G. H. (1954). Responses from an association area secondarily activated from optic cortex. *J. Neurophysiol.* **17,** 271–277.
14. Gilbert, C. D., and Kelly, J. P. (1975). The projection of cells in different layers of the cats visual cortex. *J. Comp. Neurol.* **163,** 81–106.
15. Horton, J. C., and Hubel, D. H. (1981). Regular patchy distribution of cytochrome oxidase staining in primary visual cortex of macaque monkey. *Nature* **292,** 762–764.
16. Hendrickson, A. E. (1985). Dots, stripes and columns in monkey visual cortex. *Trends Neurosci.* **8,** 406–410.
17. Wong-Riley, M.T.T. (1989). Cytochrome oxidase: an endogenous metabolic marker for neuronal activity. *Trends Neurosci.* **12,** 94–101.
18. Wong-Riley, M. (1979). Changes in the visual system of monocularly sutured or enucleated cats demonstrable with cytochrome oxidase histochemistry. *Brain Res.* **171,** 11–28.
19. Horton, J. C. (1984). Cytochrome oxidase patches: a new cytoarchitectonic feature of monkey visual cortex. *Phil. Trans. R. Soc. Lond. B* **304,** 199–253.
20. De Yoe, E. A., and Van Essen, D. C. (1988). Concurrent processing streams in monkey visual cortex. *Trends Neurosci.* **11,** 219–226.
21. Martin, K.A.C. (1988). From enzymes to visual perception: a bridge too far? *Trends Neurosci.* **11,** 380–387.
22. Fitzpatrick, D., and Diamond, I. T. (1979). The laminar organization of the lateral geniculate body in *Galago senegalensis:* a pair of layers identified by acetylcholinesterase activity. *Brain Res.* **170,** 538–542.
23. Livingstone, M. S., and Hubel, D. H. (1982). Thalamic inputs to cytochrome oxidase rich regions in monkey visual cortex. *Proc. Natl. Acad. Sci. USA.* **79,** 6098–6101.
24. Diamond, I. T., Conley, M., Itoh, K., and Fitzpatrick, D. (1985). Laminar organization of geniculocortical projections in *Galago senegalensis* and *Aotus trivirgatus. J. Comp. Neurol.* **242,** 584–610.
25. Livingstone, M. S., and Hubel, D. H. (1983). Specificity of cortico-cortical connections in monkey visual system. *Nature* **304,** 531–534.
26. Livingstone, M. S., and Hubel, D. H. (1984). Anatomy and physiology of a color system in the primate visual cortex. *J. Neurosci.* **4,** 309–356.
27. Livingstone, M. S., and Hubel, D. H. (1987). Connections between layer 4B of area 17 and the thick cytochrome oxidase stripes of area 18 in the squirrel monkey. *J. Neurosci.* **7,** 3371–3377.
28. De Yoe, E. A., and Van Essen, D. C. (1985). Segregation of efferent connections and receptive field properties in visual area V2 of the macaque. *Nature* **317,** 5861.
29. Shipp, S., and Zeki, S. (1985). Segregation of pathways leading from area V2 to areas V4 and V5 of Macaque monkey visual cortex. *Nature* **315,** 322–325.
30. Shipp, S., and Zeki, S. (1989). The organization of connections between areas V5 and V2 in macaque monkey visual cortex. *Eur. J. Neurosci.* **1,** 333–353.
31. Ferrer, J.M.R., Kato, N., and Price, D. J. (1992). Organization of association projections from area 17 to areas 18 and 19 and to suprasylvian areas in the cat's visual cortex. *J. Comp. Neurol.* **316,** 261–278.
32. Murphy, K. M., Van Sluyters, R. C., and Jones, D. G. (1990). Cytochrome oxidase activity in cat visual cortex: is it periodic? *Soc. Neurosci. Abstr.* **16,** 292.
33. Murphy, K. M., Jones, D. G., and Van Sluyters, R. C. (1995). Cytochrome-oxidase blobs in cat primary visual cortex. *J. Neurosci.* **15,** 4196–4208.

34. Boyd, J. D. (1995). Anatomical evidence for segregated processing streams in the cat's visual cortex. Ph.D.thesis, University of British Columbia.
35. Olavarria, J., and Van Sluyters, R. C. (1985). Unfolding and flattening the cortex of gyrencephalic brains. *J. Neurosci. Methods* **15,** 191–202.
36. Kaas, J. H. (1980). A comparative study of visual cortex organization in mammals. In: Comparative neurology of the telencephalon (S.O.E. Ebbesson, Ed.), pp. 483–502. New York, Plenum Press.
37. Price, D. J. (1984). Patterns of cytochrome oxidase activity in areas 17, 18 and 19 of the visual cortex of cats and kittens. *Exp. Brain Res.* **159,** 1–9.
38. Boyd, J. D., and Matsubara, J. A. (1996). Laminar and columnar patterns of geniculocortical projections in the cat: relationship to cytochrome oxidase. *J. Comp. Neurol.* **365,** 659–682.
39. O'Leary, J. L. (1941). Structure of the area striata of the cat. *J. Comp. Neurol.* **75,** 131–164.
40. Lund, J. S., Henry, G. H., Macqueen, C. L., and Harvey, A. R. (1979). Anatomical organization of the primary visual cortex (area17) of the cat. A comparison with area 17 of the Macaque monkey. *J. Comp. Neurol.* **184,** 599–618.
41. Friedlander, M. J., and Martin, K. A. C. (1989). Development of Y-axon innervation of cortical area 18 in the cat. *J. Physiol.* **416,** 183–213.
42. Kageyama, G. H., and Wong-Riley, M. (1986). Laminar and cellular localization of cytochrome oxidase in the cat striate cortex. *J. Comp. Neurol.* **245,** 137–159.
43. Murphy, K. M., Van Sluyters, R. C., and Jones, D. G. (1991). The organization of cytochrome oxidase blobs in the cat visual cortex. *Soc. Neurosci. Abst.* **17,** 1088.
44. Anderson, P. A., Olavarria, J., and Van Sluyters, R. C. (1988). The overall pattern of ocular dominance bands in cat visual cortex. *J. Neurosci.* **8,** 2183–2200.
45. Dyck, R. H., and Cynader, M. S. (1993). An interdigiated columnar mosaic of cytochrome oxidase, zinc, and neurotransmitter-related molecules in cat and monkey visual cortex. *Proc. Natl. Acad. Sci. U.S.A.* **90,** 9066–9069.
46. Hübener, M., Shoham, D., Grinvald, A., and Bonhoeffer, T. (1997). Spatial relationships among three columnar systems in cat area 17. *J. Neurosci.* **17,** 9270–9284.
47. Shoham, D., Hübener, M., Schulze, S., Grinvald, A., and Bonhoeffer, T. (1997). Spatiotemporal frequency domains and their relation to cytochrome oxidase staining in cat visual cortex. *Nature* **385,** 529–533.
48. Horton, J. C., and Hocking, D. R. (1996). Anatomical demonstration of ocular dominance columns in striate cortex of the squirrel monkey. *J. Neurosci.* **16,** 5510–5522.
49. Livingstone, M. S. (1996). Ocular dominance columns in New World monkeys. *J. Neurosci.* **16,** 2086–2096.
50. Boycott, B. B., and Wässle, H. (1974). The morphological types of ganglion cells of the domestic cat's retina. *J. Physiol.* **240,** 397–419.
51. Leventhal, A. G. (1979). Evidence that the different classes of relay cells of the cat's lateral geniculate nucleus terminate in different layers of the striate cortex. *Exp. Brain Res.* **37,** 349–372.
52. LeVay, S., and Gilbert, C. D. (1976). Laminar patterns of geniculocortical projection in the cat. *Brain Res.* **113,** 1–19.
53. Stone, J. S., Dreher, B., and Leventhal, A. (1979). Hierarchical and parallel mechanisms in the organization of visual cortex. *Brain Res. Rev.* **1,** 345–395.
54. Humphery, A. L., Sur, M., Uhlrich, D. J., and Sherman, S. M. (1985). Projection patterns of individual X- and Y-cell axons from the lateral geniculate nucleus to cortical area 17 in the cat. *J. Comp. Neurol.* **233,** 159–189.
55. Humphery, A. L., Sur, M., Uhlrich, D. J., and Sherman, S. M. (1985). Termination patterns of individual X- and Y-cell axons in the visual cortex of the cat: projections to area 18, to the 17/18 border region, and to both areas 17 and 18. *J. Comp. Neurol.* **233,** 190–212.
56. Wilson, P. D., and Stone, J. (1975). Evidence of W-cell input to the cat's visual cortex via the C-laminae of the lateral geniculate nucleus. *Brain Res.* **92,** 472–478.

57. Wilson, P. D., Rowe, M. H., and Stone, J. (1976). Properties of relay cells in cat's lateral geniculate nucleus: a comparison of W-cells with X- and Y-cells. *J. Neurophysiol.* **39,** 1193–1209.
58. Leventhal, A. G. (1982). Morphology and distribution of retinal ganglion cells projecting to different layers of the dorsal lateral geniculate nucleus in normal and siamese cats. *J. Neurosci.* **2,** 1024–1042.
59. Sur, M., and Sherman, S. M. (1982). Linear and Nonlinear w-cells in C-laminae of the cat's lateral geniculate nucleus. *J. Neurophysiol.* **47,** 869–884.
60. Spear, P. D., McCall, M. A., and Tumosa, N. (1989). W- and Y-cells in the C layers of the cat's lateral geniculate nucleus: normal properties and effects of monocular deprivation. *J. Neurophysiol.* **61,** 58–73.
61. Kageyama, G. H., and Wong-Riley, M. (1986). The localization of cytochrome oxidase in the LGN and striate cortex of postnatal kittens. *J. Comp. Neurol.* **243,** 182–194.
62. Tretter, F., Cynader, M., and Singer, W. (1975). Cat parastriate cortex: a primary or secondary visual area? *J. Neurophysiol.* **38,** 1099–1113.
63. Dreher, B., Leventhal, A. G., and Hale, P. T. (1980). Geniculate input to cat visual cortex: a comparison of area 19 with areas 17 and 18. *J. Neurophysiol.* **44,** 804–826.
64. Harvey, A. R. (1980). A physiological analysis of subcortical and commissural projections of areas 17 and 18 of the cat. *J. Physiol.* **302,** 507–534.
65. Shoham, D., Hübener, M., Bonhoeffer, T., and Grinvald, A. (1996). Spatio-temporal frequency columns in cat visual cortex: a continuous representation or two streams? *Soc. Neurosci. Abstr.* **22,** 951.
66. Dreher, B., and Sefton, A. J. (1979). Properties of neurons In cat's dorsal lateral geniculate nucleus: a comparison between medial interlaminar and laminated parts of the nucleus. *J. Comp. Neurol.* **183,** 47–64.
67. Bowling, D. B., and Michael, C. R. (1980). Projection patterns of single physiologically characterized optic tract fibres in cat. *Nature* **286,** 899–902.
68. Bowling, D. B., and Michael, C. R. (1984). Terminal patterns of single, physiologically characterized optic tract fibres in the cat's lateral geniculate nucleus. *J. Neurosci.* **4,** 198–216.
69. Sur, M., Esguerra, M., Garraghty, P. E., Kritzer, M. F., and Sherman, S. M. (1987). Morphology of physiologically identified retinogeniculate X- and Y-axons in the cat. *J. Neurophysiol.* **58,** 1–32.
70. Frascella, J., and Lehmkuhle, S. (1984). A comparison between Y-cells in A-laminae and lamina C of cat dorsal lateral geniculate nucleus. *J. Neurophysiol.* **52,** 911–920.
71. Hockfield, S., and Mckay, R. D. (1983). A surface antigen expressed by a subset of neurons in the vertebrate central nervous system. *Proc. Natl. Acad. Sci. U.S.A.* **80,** 5758–5761.
72. Sur, M., Frost, D. O., and Hockfield, S. (1988). Expression of a surface-associated antigen on Y-cells in the cat lateral geniculate nucleus is regulated by visual experience. *J. Neurosci.* **8,** 874–882.
73. Chehil, S., Murphy, K. M., and Beaver, C. C. (1992). Tangential distribution of cat-301 neurons in cat visual cortex. *Invest. Ophthalmol Visual Sci.* **34/4,** *(suppl)* 1218.
74. Dyck, R. H., and Cynader, M. S. (1993). Autoradiographic localization of serotonin receptor subtypes in cat visual cortex: transient regional, laminar, and columnar distributions during postnatal development. *J. Neurosci.* **13,** 4316–4338.
75. Prusky, G. T., Shaw, C., and Cynader, M. S. (1987). Nicotine receptors are located on lateral geniculate nucleus terminals in cat visual cortex. *Brain Res.* **412,** 131–138.
76. Beaulieu, C., Dyck, R., and Cynader, M. (1992). Enrichment of glutamate in zinc-containing terminals of the cat visual cortex. *Neuroreport* **3,** 861–864.
77. Dyck, R., Beaulieu, C., and Cynader, M. (1993). Histochemical localization of synaptic zinc in the developing cat visual cortex. *J. Comp. Neurol.* **329,** 53–67.
78. Casanovas-Aguilar, C., Reblet, C., Perez-Clausell, J., and Bueno-Lopez, J. L. (1998). Zinc-rich afferents to the rat neocortex: projections to the visual cortex traced with intracerebral selenite injections. *J. Chem. Neuroanat.* **15,** 97–109.

79. LeVay, S., and Sherk, H. (1981). The visual claustrum of the cat I. Structure and connections. *J. Neurosci.* **1,** 956–980.
80. LeVay, S. (1986). Synaptic organization of claustral and geniculate afferents to the visual cortex of the cat. *J. Neurosci.* **6,** 3564–3575.
81. Garrett, B., Sorensen, J. C., and Slomianka, L. (1992). Fluoro-gold tracing of zinc-containing afferent connections in the mouse visual cortices. *Anat. Embryol.* **185,** 451–459.
82. Schoen, S. W., Leutenecker, B., Kreutzberg, G. W., and Singer, W. (1990). Ocular dominance plasticity and developmental changes of 5′-nucleotidase distributions in the kitten visual cortex. *J. Comp. Neurol.* **296,** 379–392.
83. LeVay, S., and Nelson, S. B. (1991). Columnar organization of the visual cortex. In: The neural basis of visual function (A. G. Leventhal, Ed.), Vol. 4, pp. 266–315. London, The Macmillan Press.
84. Heath, C. S., and Jones, E. G. (1971). The anatomical organization of the suprasylvian gyrus of the cat. *Ergebnisse der Anatomie und Entwicklungsgeschinte* **45,** 1–64.
85. Sherk, H. (1986). Location and connections of visual cortical areas in the cat's suprasylvian sulcus. *J. Comp. Neurol.* **247,** 1–31.
86. Cavada, C., and Reinoso-Suárez, F. (1983). Afferent connections of area 20 in the cat studied by means of the retrograde axonal transport of horseradish peroxidase. *Brain Res.* **270,** 319–324.
87. Gilbert, C. D., and Wiesel, T. N. (1981). Projection bands in visual cortex. *Soc. Neurosci. Abstr.* **7,** 356.
88. Montero, V. M. (1981). Topography of the cortico-cortical connections from the striate cortex in the cat. *Brain Behav. Evol.* **18,** 194–218.
89. Bullier, J., Kennedy, H., and Salinger, W. (1984). Branching and laminar origin of projections between visual cortical areas in the cat. *J. Comp. Neurol.* **228,** 329–341.
90. Ferrer, J. M. R., Price, D. J., and Blakemore, C. (1988). The organization of corticocortical projections from area 17 to area 18 of the cat's visual cortex. *Proc. R. Soc. Lond. B* **223,** 77–98.
91. Shipp, S., and Grant, S. (1991). Organization of reciprocal connections between area 17 and the lateral suprasylvian area of cat visual cortex. *Vis. Neurosci.* **6,** 339–355.
92. Matsubara, J. A., and Boyd, J. D. (1994). Modular aspects of corticocortical connectivity between areas 17 and 18. *Soc. Neurosci. Abstr.* **20,** 1742.
93. Gilbert, C. D., and Wiesel, T. N. (1989). Columnar specificity of intrinsic horizontal and corticocortical connections in cat visual cortex. *J. Neurosci.* **9,** 2432–2442.
94. Boyd, J. D., and Matsubara, J. A. (1999). Projections from V1 to lateral suprasylvian cortex: an efferent pathway in the cat's visual cortex that originates preferentially from CO blob columns. *Vis. Neurosci.* **16,** 849–860.
95. Conway, B., Boyd, J. D., Stewart, T. H., and Matsubara, J. A. (2000). The projection from V1 to extrastriate area 21a: a second patchy efferent pathway that colocalizes with the CO blob columns in cat visual cortex. *Cerebral Cortex* **10,** 149–159.
96. Wimborne, B. M., and Henry, G. H. (1992). Response characteristics of the cells of cortical area-21a of the cat with special reference to orientation specificity. *J. Physiol. Lond.* **449,** 457–478.
97. Dreher, B., Michalski, A., Ho, R. H., Lee, C. W., and Burke, W. (1993). Processing of form and motion in area 21a of cat visual cortex. *Vis. Neurosci.* **10,** 93–115.
98. Toyama, K., Mizobe, K., Akase, E., and Kaihara, T. (1994). Neuronal responsiveness in areas 19 and 21a, and the posteromedial lateral suprasylvian cortex of the cat. *Exp. Brain Res.* **99,** 289–301.
99. Hubel, D. H., and Wiesel, T. N. (1969). Visual area of the lateral suprasylvian gyrus (Clare-Bishop area) of the cat. *J. Physiol.* **202,** 251–260.
100. Spear, P. D., and Baumann, T. P. (1975). Receptive-field characteristics of single neurons in lateral suprasylvian visual area of the cat. *J. Neurophysiol.* **38,** 1403–1420.
101. Camarda, R., and Rizzolatti, G. (1976). Visual receptive fields in the lateral suprasylvian area (Clare-Bishop) area of the cat. *Brain Res.* **101,** 427–443.
102. Markuszka, J. (1978). Visual properties of neurons in the posterior suprasylvian gyrus of the cat. *Exp. Neurol.* **59,** 146–161.

103. Zumbroich, T. J., and Blakemore, C. (1987). Spatial and temporal selectivity in the suprasylvian visual cortex of the cat. *J. Neurosci.* **7,** 482–500.

104. Yin, T. C. T., and Greenwood, M. (1992). Visual response properties of neurons in the middle and lateral suprasylvian cortices of the behaving cat. *Exp. Brain Res.* **88,** 1–14.

105. Kim, J. N., Mulligan, K., and Sherk, H. (1997). Simulated optic flow and extrastriate cortex. I. Optic flow versus texture. *J. Neurophysiol.* **77,** 554–561.

106. Mulligan, K., Kim, J. N., and Sherk, H. (1997). Simulated optic flow and extrastriate cortex. II. Responses to bar versus large-field stimuli. *J. Neurophysiol.* **77,** 562–570.

107. Dreher, B., and Cottee, L. J. (1975). Visual receptive-field properties of cells in area 18 of cat's cerebral cortex before and after acute lesions in area 17. *J. Neurophysiol.* **38,** 735–750.

108. Blakemore, C., and Price, D. J. (1987). The organization and post-natal development of area 18 of the cat's visual cortex. *J. Physiol.* **384,** 263–292.

109. Cynader, M. S., Swindale, N. V., and Matsubara, J. A. (1987). Functional topography in cat area 18. *J. Neurosci.* **7,** 1401–1413.

110. Crook, J. M. (1990). Directional tuning of cells in area 18 of the feline visual cortex for visual noise, bar and spot stimuli: a comparison with area 17. *Exp. Brain Res.* **80,** 545–561.

111. Ferster, D., and Jagadeesh, B. (1991). Nonlinearity of spatial summation in simple cells of areas 17 and 18 of cat visual cortex. *J. Neurophysiol.* **66,** 1667–1679.

112. Meyer, G., and Albus, K. (1981). Spiny stellates as cells of origin of association fibrers from area 17 to area 18 in the cat's neocortex. *Brain Res.* **210,** 335–341.

113. Einstein, G., and Fitzpatrick, D. (1991). Distribution and morphology of area 17 neurons that project to the cat's extrastriate cortex. *J. Comp. Neurol.* **303,** 132–149.

114. Salin, P. A. (1989). Convergence and divergence in the afferent projections to cat area 17. *J. Comp. Neurol.* **283,** 486–512.

115. Dreher, B., Wang, C., and Burke, W. (1996). Limits of parallel processing: excitatory convergence of different information channels on single neurons in striate and extrastriate visual cortices. *Clin. Exp. Pharmacol Physiol.* **23,** 913–925.

116. Lachica, E. A., Beck, P. D., and Casagrande, V. A. (1992). Parallel pathways in Macaque monkey striate cortex: anatomically defined columns in layer III. *Proc. Natl. Acad. Sci. U.S.A.* **89,** 3566–3570.

117. Lachica, E. A., Beck, P. D., and Casagrande, V. A. (1993). Intrinsic connections of Layer-III of striate cortex in squirrel monkey and bush baby—correlations with patterns of cytochrome oxidase. *J. Comp. Neurol.* **329,** 163–187.

118. Yoshioka, T., Levitt, J. B., and Lund, J. S. (1994). Independence and merger of thalamocortical channels within macaque monkey primary visual cortex: anatomy of interlaminar projections. *Vis. Neurosci.* **11,** 467–490.

119. Henry, G. H., Lund, J. S., and Harvey, A. R. (1978). Cells of the striate cortex projecting to the Clare-Bishop area of the cat. *Brain Res.* **151,** 154–158.

120. Movshon, J. A., and Newsome, W. T. (1996). Visual response properties of striate cortical neurons projecting to area MT in macaque monkeys. *J. Neurosci.* **16,** 7733–7741.

121. Lee, C., Weyand, T. G., and Malpeli, J. G. (1998). Thalamic control of cat area-18 supragranular layers: simple cells, complex cells, and cells projecting to the lateral suprasylvian visual area. *Vis. Neurosci.* **15,** 27–35.

122. von Grünau, M. W., Zumbroich, T. J., and Poulin, C. (1987). Visual receptive field properties in the posterior suprasylvian cortex of the cat: a comparison between the areas PMLS and PLLS. *Vis. Res.* **27,** 343–356.

123. Wang, C., Dreher, B., Huxlin, K. R., and Burke, W. (1997). Excitatory convergence of Y and non-Y information channels on single neurons in the PMLS area, a motion area of the cat visual cortex. *Eur. J. Neurosci.* **9,** 921–933.

124. Bonhoeffer, T., Kim, D. S., Malonek, D., Shoham, D., and Grinvald, A. (1995). Optical imaging of functional domains in area 17 and across the area 17/18 border in cat visual cortex. *Eur. J. Neurosci.* **7,** 1973–1988.

125. Salin, P. A., Girard, P., Kennedy, H., and Bullier, J. (1992). Visuotopic organization of corticocortical connections in the visual system of the cat. *J. Comp. Neurol.* **320,** 415–434.
126. Boyd, J. D., and Matsubara, J. A. (1994). Modular organization of corticocortical inputs and outputs of area 19. *Soc. Neurosci. Abstr.* **20,** 1742.
127. Tieman, S. B., and Tumosa, N. (1983). [^{14}C]2-deoxyglucose demonstration of the organization of ocular dominance in areas 17 and 18 of the normal cat. *Brain Res.* **267,** 35–46.
128. Harting, J. K., Huerta, M. F., Hashikawa, T., and van Lieshout, D. P. (1991). Projection of the mammalian superior colliculus upon the dorsal lateral geniculate nucleus: organization of tectogeniculate pathways in nineteen species. *J. Comp. Neurol.* **304,** 275–306.
129. Stitchel, C. C., Singer, W., Heizmann, C. W., and Norman, A. W. (1987). Immunohistochemical localization of calcium-binding proteins, parvalbumin and calbindin-D 28k, in the adult and developing visual cortex of cats: a light and electron microscopic study. *J. Comp. Neurol.* **262,** 563–577.
130. Casagrande, V. A. (1994). A third parallel visual pathway to primate area V1. *Trends Neurosci.* **17,** 305–310.
131. Irvin, G. E., Norton, T. T., Sesma, M. A., and Casagrande, V. A. (1986). W-like response properties of interlaminar zone cells in the lateral geniculate nucleus of a primate *(Galago crassicaudatus). Brain Res.* **362,** 254–270.
132. Martin, P. R., White, A. J. R., Goodchild, A. K., Wilder, H. D., and Sefton, A. E. (1997). Evidence that blue-on cells are part of the third geniculocortical pathway in primates. *Eur. J. Neurosci.* **9,** 1536–1541.
133. Hendry, S. H. C., and Calkins, D. J. (1998). Neuronal chemistry and functional organization in the primate visual system. *Trends Neurosci.* **21,** 344–349.
134. Pearlman, A. L., and Daw, N. W. (1970). Opponent colour cells in the cat lateral geniculate nucleus. *Science* **167,** 84–86.
135. Pearlman, A. L., and Daw, N. W. (1971). Behavioral and neurophysiological studies on cat color vision. *Int. J. Neurosci.* **1,** 357–360.
136. Weber, J. T., Heurta, M. F., Kaas, J. H., and Harting, J. K. (1983). The projections of the lateral geniculate nucleus of the squirrel monkey: studies of the interlaminar zones and the S layers. *J. Comp. Neurol.* **213,** 113–145.
137. Lachica, E. A., and Casagrande, V. A. (1992). Direct W-like geniculate projections to the cytochrome oxidase (CO) blobs in primate visual cortex—axon morphology. *J. Comp. Neurol.* **319,** 141–158.
138. Ding, Y., and Casagrande, V. A. (1997). The distribution and morphology of LGN K pathway axons within the layers and CO blobs of owl monkey V1. *Vis. Neurosci.* **14,** 691–704.
139. Fitzpatrick, D., Carey, R. C., and Diamond, I. T. (1980). The projection of the superior colliculus upon the lateral geniculate body in *Tupaia glis* and *Galago senegalensis. Brain Res.* **194,** 494–499.
140. Diamond, I. T., Fitzpatrick, D., and Schmechel, D. (1993). Calcium binding proteins distinguish large and small cells of the ventral posterior and lateral geniculate nuclei of the prosimian galago and the tree shrew *(Tupaia belangeri). Proc. Natl. Acad. Sci. U.S.A.* **90,** 1425–1429.
141. Fitzpatrick, D., and Raczkowski, D. (1990). Innervation patterns of single physiologically identified geniculocortical axons in the striate cortex of the tree shrew. *Proc. Natl. Acad. Sci. U.S.A.* **87,** 449–453.
142. Lyon, D. C., Jain, N., and Kaas, J. H. (1998). Cortical connections of striate and extrastriate visual areas in tree shrews. *J. Comp. Neurol.* **401,** 109–128.
143. Shapley, R., and Perry, V. H. (1986). Cat and monkey retinal ganglion cells and their visual functional roles. *Trends Neurosci.* **9,** 229–225.
144. Tootell, R. B. H., Hamilton, S. L., and Silverman, M. S. (1985). Topography of cytochrome oxidase activity in owl monkey cortex. *J. Neurosci.* **5,** 2786–2800.
145. Condo, G. C., and Casagrande, V. A. (1990). Organization of cytochrome oxidase staining in the visual cortex of nocturnal primates *(Galago crassicaudatus* and *Galago senegalensis).* I. Adult patterns. *J. Comp. Neurol.* **293,** 632–645.

146. Tootell, R. B. H., Silverman, M. S., Hamilton, S. L., Switkes, E., and De Valois, R. L. (1988). Functional anatomy of macaque striate cortex. V. Spatial frequency. *J. Neurosci.* **8,** 1610–1624.
147. Edwards, D. P., Purpura, K. P., and Kaplan, E. (1995). Contrast sensitivity and spatial frequency responses of primate cortical neurons in and around the cytochrome oxidase blobs. *Vis. Res.* **35,** 1501–1523.
148. Hockfield, S., Mckay, R. D., Hendry, S. H. C., and Jones, E. G. (1983). A surface antigen that identifies ocular dominance columns in the visual cortex and laminar features of the lateral geniculate nucleus. *Cold Spring Harbor Symp. Quant. Biol.* **48,** 877–889.
149. Hendry, S. H. C., Jones, E. G., Hockfield, S., and McKay, R. D. G. (1988). Neuronal populations stained with the monoclonal antibody cat-301 in the mammalian cerebral cortex and thalamus. *J. Neurosci.* **8,** 518–542.
150. De Yoe, E. A., Hockfield, S., Garren, H., and Van Essen, D. C. (1990). Antibody labeling of functional subdivisions in visual cortex: cat-301 immunoreactivity in striate and extrastriate cortex of the macaque monkey. *Vis. Neurosci.* **5,** 67–81.
151. Lund, J. S. (1990). Excitatory and inhibitory circuitry and laminar mapping strategies in the primary visual cortex of the monkey. In: 'Signal and sense. Local and global order in perceptual Maps' (G. M. Edelman, W. E. Call, and W. M. Cowan, Eds.), pp. 51–66. New York, John Wiley and Sons.
152. Krubitzer, L., and Kaas, J. (1990). Convergence of processing channels in the extrastriate cortex of monkeys. *Vis. Neurosci.* **5,** 609–613.
153. Krubitzer, L. A., and Kaas, J. H. (1993). The dorsomedial visual area of owl monkeys: connections, myeloarchitecture, and homologies in other primates. *J. Comp. Neurol.* **334,** 497–528.
154. Boyd, J. D., and Casagrande, V. A. (1999). Relationship between cytochrome oxidase (CO) blobs in primate visual cortex (V1) and the distribution of cells projecting to the middle temporal area (MT). *J. Comp. Neurol.* **409,** 573–591.
155. Zeki, S. M. (1974). Functional organization of a visual area in the posterior bank of the superior temporal sulcus of the rhesus monkey. *J. Physiol.* **236,** 549–573.
156. Albright, T. D. (1984). Direction and orientation selectivity of neurons in visual area MT of the macaque. *J. Neurophysiol.* **52,** 1106–1130.
157. Felleman, D. J., and Kaas, J. H. (1984). Receptive-field properties of neurons in middle tempoal visual area (MT) of owl monkeys. *J. Neurophysiol.* **52,** 488–513.
158. Boussaoud, D., Ungerleider, L. G., and Desimone, R. (1990). Pathways for motion analysis: cortical connections of the medial superior temporal and fundus of the superior temporal visual areas in the macaque. *J. Comp. Neurol.* **296,** 462–495.
159. Payne, B. R. (1993). Evidence for visual cortical area homologs in cat and macaque monkey. *Cerebral Cortex* **3,** 1–25.
160. Lund, J. S., Lund, R. D., Hendrickson, A. E., Bunt, A. H., and Fuchs, A. F. (1975). The origin of efferent pathways from the primary visual cortex, area 17, of the macaque monkey as shown by retrograde transport of horseradish peroxidase. *J. Comp. Neurol.* **164,** 287–304.
161. Shipp, S., and Zeki, S. (1989). The organization of connections between areas V5 and V1 in macaque monkey visual cortex. *Eur. J. Neurosci.* **1,** 310–332.
162. Zeki, S., and Shipp, S. (1989). Modular connections between areas V2 and V4 of macaque monkey visual cortex. *Eur. J. Neurosci.* **1,** 494–506.
163. Felleman, D. J., Xiao, Y., and McClendon, E. (1997). Modular organization of occipitotemporal pathways: cortical connections between visual area 4 and visual area 2 and posterior inferotemporal ventral area in macaque monkeys. *J. Neurosci.* **17,** 3185–3200.

6

Influence of Topography and Ocular Dominance on the Functional Organization of Callosal Connections in Cat Striate Cortex

Jaime F. Olavarria

Department of Psychology, University of Washington, Seattle, Washington

INTRODUCTION

In machines engineered by humans there is a direct relationship between the function of specific parts and the structural design of these parts. Explaining the functioning of machines, therefore, implies arguing from function to structure. Although biological systems such as the nervous system are of course not engineered by humans, we nevertheless tend to apply the same heuristics for explaining the functioning of these systems. Thus, once the hypothetical function of a nervous circuit becomes widely popular, we come to expect that its anatomical organization across individuals and species will suit this function, as if the function specifies the "design" of the neural circuit in question. Moreover, there is a tendency to view the development of nervous circuits as a process for implementing structural "designs" that fit the hypothetical functions of the circuits, an approach that is necessarily limited because it depends critically on the validity of the function assigned to the circuit. For example, early studies of interhemispheric connections through the corpus callosum proposed anatomical and functional attributes that made this pathway very attractive as a model for studying the function and development of corticocortical connections. Ironically, however, some of the early notions about the function of the callosal pathway in some cortical areas have not always provided fruitful perspectives from which to explore its organization and development. In the last decades, developmental studies have taught us that, unlike the manufacture of mechanical parts, the development of neuronal circuits does not seem to simply execute structural designs suited for specific functions. Instead, it appears that the organization of specific circuits can vary across species depending on the interplay of factors and conditions that directly or indirectly facilitate or constraint the course of development in each species.

This chapter reviews some recent findings about the organization of the callosal pathway in striate cortex of the cat that cannot be readily assimilated into current thinking about the structure and function of this pathway. Hypotheses about possible developmental factors and constraints that can account for these discrepancies are formulated, and experimental evidence distinguishing between these hypotheses is evaluated. This analysis leads to a novel perspective that sheds light on how developmental factors may influence the structural organization of callosal connections, thereby also influencing the functions and interspecies variability of callosal pathways.

THE VERTICAL MERIDIAN RULE

Previous descriptions of visual callosal connections in occipital area 17 have emphasized that both the cells of origin and terminations of this interhemispheric pathway were restricted to the border of this area (Van Essen et al., 1982; Wilson, 1968; Garey et al., 1968; Ebner and Myers, 1965; Hubel and Wiesel, 1965a). Moreover, physiological recordings from neurons in area 17 revealed that the vertical meridian of the visual field was represented at the border of area 17 (e.g., Thomas and Espinoza, 1987; Zeki and Sandeman, 1976; Hubel and Wiesel, 1965a), and recordings from fibers passing through the splenium of the corpus callosum showed that the receptive fields of visual callosal axons lay on or near the vertical meridian (Shatz, 1977a; Berlucchi et al., 1967; Hubel and Wiesel, 1965a; 1967). These observations led to the notion, known as the "vertical meridian rule" (e.g., Berman and Grant, 1992; Innocenti, 1986; Berlucchi, 1981), that callosally connected cortical regions represent the vertical meridian of the visual field. Callosal fibers at the border of striate cortex would thus unite mirror-symmetrical anatomical loci representing visual fields on or near the vertical meridian (Fig. 6-1A) and mediate both the perceptual fusion of the two halves of the visual field and midline stereopsis (Blakemore et al., 1983; Berlucchi and Rizzolati, 1968; Hubel and Wiesel, 1967; Choudhury et al., 1965).

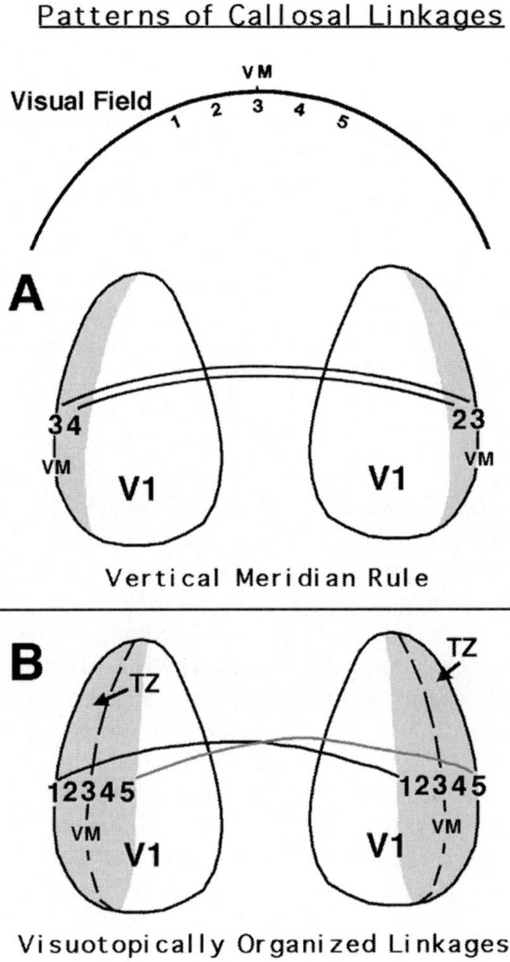

FIGURE 6-1. Diagrams relating mirror-symmetrical and non-mirror-symmetrical patterns of callosal connections to the visuotopic organization of primary visual cortex (V1). For simplicity, the visual thalamus is not represented in this diagram. In both panels, the distribution of callosal connections in lateral V1 is indicated in gray. Callosal connections in adjacent regions of area 18 are not represented. The visual field is schematically represented by the semicircular perimeter located at the top of the diagram, and the numbers 1 to 5 indicate regularly spaced loci on the visual field. **(A)** Vertical meridian rule. Callosal connections at the lateral border of V1 are thought to be restricted to a narrow cortical strip that represents the vertical meridian of the visual field (VM) and immediately adjacent regions of the visual field. Callosal fibers interlink points that are in anatomic correspondence (i.e., in the same anatomical location in each hemisphere). **(B)** Nonmirror-symmetrical callosal linkages. Callosal fibers in the border region of V1 interlink opposite points that are not in anatomic correspondence (Lewis and Olavarria, 1995; Olavarria, 1996a). Because of the representation of ipsilateral visual fields in each hemisphere (i.e., segment 3–5 of the visual field represented in the right TZ, and segment 1–3 represented in the left TZ), a central area of the visual field including the vertical meridian is represented in both hemispheres (the visual field segment 1–5). The diagram illustrates that callosal fibers, indicated by continuous lines between the hemispheres, interlink retinotopically corresponding cortical loci (i.e., both the origin and end of callosal fibers represent similar regions in the visual field). TZ, transition zone.

However, more recent studies have provided additional information about the callosal pathway that does not readily conform to the vertical meridian rule. Accumulating evidence from many species indicates that the border of area 17 does not correspond to the representation of the vertical meridian, but instead represents a strip of ipsilateral visual field. For instance, in the cat, areas 17 and 18 not only contain orderly representations of the opposite visual hemifield but also represent a substantial portion of the ipsilateral visual field (Payne, 1990; Diao et al., 1990; Pettigrew and Dreher, 1987; Blakemore et al., 1983; Whitteridge and Clarke, 1982; Harvey, 1980; Hubel and Wiesel, 1962). Payne (1990) showed that this ipsilateral representation is contained in a region between areas 17/18 which, on the basis of its architecture, has been described as a zone of transition between areas 17 and 18 (Payne, 1990; Otsuka and Hassler, 1962). Moreover, it has become evident that the extent of the callosally connected region in area 17 varies considerably among species. For example, while the distribution of callosal connections in area 17 is quite restricted in the macaque monkey (e.g., Abel et al., 2000; Olavarria and Abel, 1996; Kennedy et al., 1986) and squirrel monkey (Gould et al., 1987), it is fairly extensive in species like rats (Olavarria and Montero, 1984; Olavarria and Van Sluyters, 1983; Cusick and Lund, 1981), cats (Olavarria and Van Sluyters, 1995; Segraves and Rosenquist, 1982a; Innocenti, 1980), and galagos (Beck and Kaas, 1994; Cusick et al., 1984; Weyand and Swadlow, 1980). Finally, although the distribution of callosal cells and terminations is fairly homogeneous in the 17/18 callosal zone of species such as rodents (Olavarria and Van Sluyters, 1985a; Cusick and Lund, 1982), it is distinctly patchy in others species such as cats (Berman and Payne, 1983; Voigt et al., 1988; Payne and Siwek, 1991; Houzel et al., 1994; Boyd and Matsubara, 1994; Olavarria, 1996a; Olavarria, 2001), and galagos (Cusick et al., 1984; Beck and Kaas, 1994). These interspecies differences in the organization of visual callosal pathways may arise to fit different, albeit unknown, functional needs of the callosal pathway across species. An alternative scenario, explored later, is that these differences may be understood by considering factors and constraints that can influence the outcome of callosal development.

CALLOSAL FIBERS INTERLINK CORTICAL SITES THAT ARE IN RETINOTOPIC, RATHER THAN ANATOMICAL, CORRESPONDENCE

In the cat, the callosally interconnected region at the 17/18 border includes not only the transition zone (TZ), but also adjacent portions of areas 17 and 18 (Figs. 6-2 to 6-4). This relatively extensive distribution of callosal cells apparently violates the vertical meridian rule because callosal cells are found in cortical regions representing visual fields well away from the vertical meridian (Bourdet et al., 1996; Olavarria and Van Sluyters, 1995; Berman and Grant, 1992; Payne, 1990, 1991; Jouandet et al., 1985; Segraves and Rosenquist, 1982a). Moreover, the extent of the callosal zone raises the issue of the point-to-point organization of

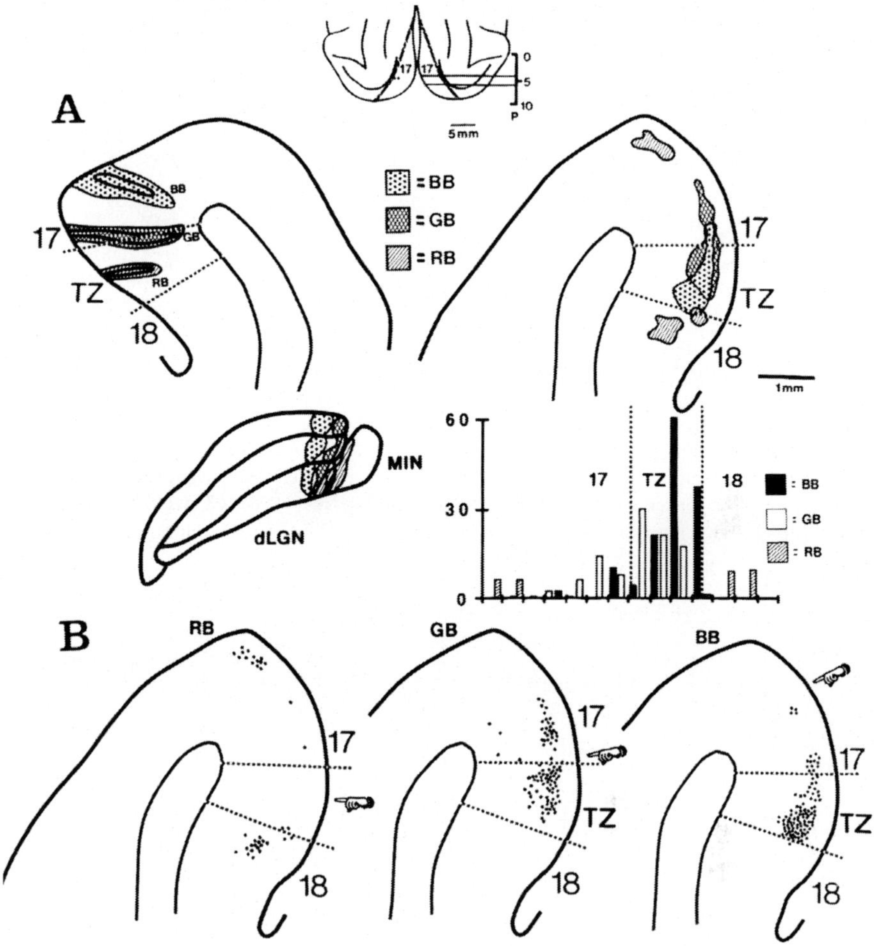

FIGURE 6-2. Nonmirror-symmetrical patterns of callosal connections in the cat. Data from brain injected with the fluorescent tracer bisbenzimide (BB) and green fluorescent latex beads (GB) into area 17, and rhodamine fluorescent latex beads (RB) into the TZ (cat C4). **(A)** Drawing of left hemisphere shows the injection of RB at the center of the TZ, and the injection of GB partly overlaps the medial border of the TZ. The injection of BB is about 1300 μm medial to the TZ. Drawing of right hemisphere shows distributions of RB-, GB-, and BB-labeled cells reconstructed from 12 sections taken from region indicated in top inset. Stippled regions indicate total areas occupied by RB-, GB-, and BB-labeled cells. Histogram shows the distribution of callosal cells labeled by each tracer in the 17/18 callosal zone (number of labeled cells/330-μm-wide bins measured along the bottom of layer 3; medial is to the left). **(B)** Separate reconstructions for RB (left), GB (center) and BB (right) labeling from the 12 sections analyzed. Each dot represents one labeled cell. Dashed lines indicate the TZ. dLGN, dorsal lateral geniculate nucleus; MIN, medial interlaminar nucleus. (From Olavarria [1996a].)

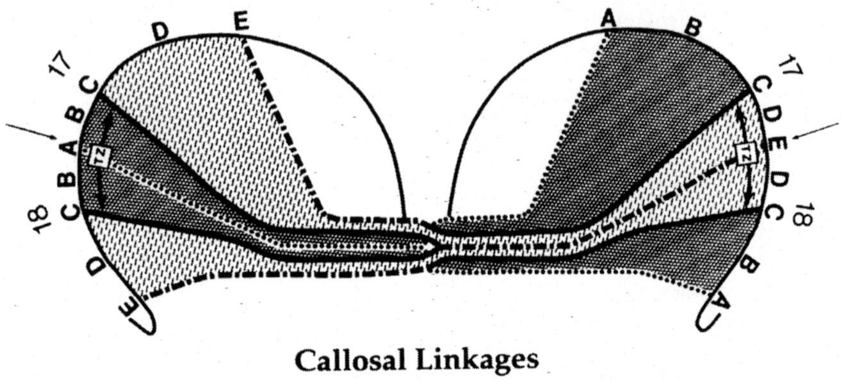

FIGURE 6-3. Schematic diagram of the mediolateral organization of callosal linkages in the 17/18 callosal zone of the cat. Cortical loci in one hemisphere marked with the letters A-E are callosally connected with loci marked with the same letters in the opposite hemisphere. Thus, loci C-E outside the TZ in the left area 17 connect with contralateral loci C-E located within the TZ, whereas loci A-C within the left TZ connect with loci A-C outside the TZ in the right area 17. A similar projection pattern is represented for area 18. Letters are closer together in the TZ than outside the TZ to indicate that the TZ receives convergent input from contralateral regions outside the TZ and, conversely, that it gives rise to divergent input to contralateral regions outside the TZ. (From Olavarria [1996a].)

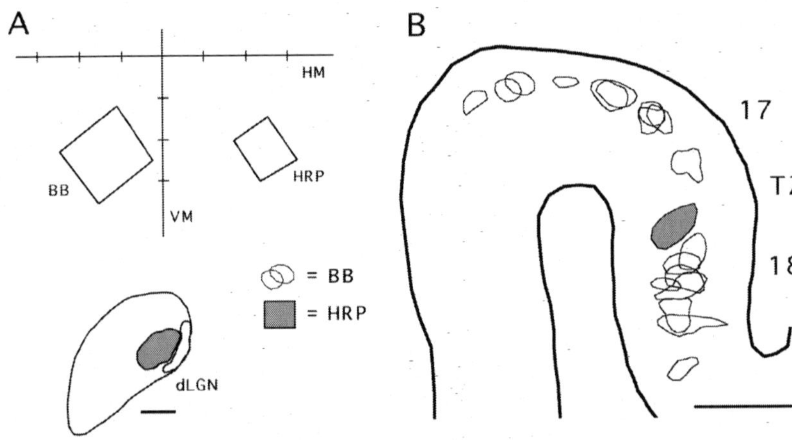

FIGURE 6-4. Mirror-reversed callosal linkages interconnect cortical loci that represent the same coordinates in the visual field. Data from experiments in which the sites injected with anatomical tracers were mapped electrophysiologically before the injections (J.F. Olavarria and H. Sherk, unpublished observations). **(A)** Receptive fields mapped through the right eye at the cortical sites injected with bisbenzimide (BB) and horseradish peroxidase (HRP) in the left hemisphere of normal cat C2. Each subdivision on the vertical meridian (VM) and horizontal meridian (HM) axes represents 1^0 of visual field. The receptive field at the injection of BB was centered at 1.5^0 into the ipsilateral hemifield (-2.5^0 elevation), whereas the receptive field at the HRP injection was centered at 2.5^0 into the contralateral hemifield (-2.3^0 elevation). The diagram at the bottom shows the location of retrogradely labeled cells in the left dLGN, medial is to the right. **(B)** Distribution of BB- and HRP-labeled cells in the right 17/18 callosal zone reconstructed from three neighboring coronal sections, medial is to the left. Patches of BB-labeled cells are distributed in areas 17 and 18 outside the TZ, and HRP-labeled cells accumulate in the TZ. Scale bars = 1 mm.

callosal linkages and its relationship to the underlying topography of visual cortex. For instance, do callosal fibers link loci that represent the same visual field coordinates, or do they, as implied by the vertical meridian rule, link anatomically corresponding loci? In the latter case, the connections may be between loci representing visual fields on either side of the vertical meridian (Fig. 6-1A).

To examine the organization of callosal linkages in the 17/18 border region of normal adult cats, small injections of different fluorescent tracers were placed into the opposite 17/18 callosal region (Olavarria, 1996a). Analysis of the distribution of retrogradely labeled callosal cell revealed that callosal fibers interconnect loci that are not mirror-symmetrical with respect to the midline. Area 17 injections placed nearly 3 mm away from the TZ produced discrete labeled areas located preferentially within the contralateral TZ. In contrast, when the injections were placed within the TZ, labeled callosal cells were found primarily outside the TZ, extending well into areas 17 and 18 (Fig. 6-2). We have described similar patterns of callosal linkages in the rat (Lewis and Olavarria, 1995).

Callosal connections originating from regions either inside or outside the TZ appear to be topographically organized. Tracer injections into cat area 17 placed away from the border of the TZ labeled cells within the contralateral TZ while injections close to or overlapping the medial border of the TZ produced largely mirror-symmetrical labeled fields straddling the medial border of the contralateral TZ (Fig. 6-2). On the other hand, large injections into the TZ produced retrograde labeling distributed in two wide regions of areas 17 and 18 adjacent to the TZ (Fig. 6-4) (Olavarria, 1996a), whereas small injections in the middle of the TZ produced distinct clusters of labeled callosal cells located well into areas 17 and 18 (Fig. 6-2). A similar pattern of connections is revealed by analyzing the distribution of labeled callosal axons following restricted injections of anterogradely transported tracers in different regions of the 17/18 callosal zone of the cat (J. F. Olavarria and F. Torrealba, unpublished observations).

These results are summarized and schematized in Fig. 6-3. Loci marked with the same letter are callosally connected: loci C-E outside the TZ in the left area 17 connect with contralateral loci C-E located within the TZ, whereas loci A-C within the left TZ connect with loci A-C outside the TZ in the right area 17. A similar projection pattern is represented for area 18.

Relating the Pattern of Callosal Connections to the Retinotopic Organization of the 17/18 Border Region

Figure 6-1B illustrates how the pattern of callosal linkages in the 17/18 border region relates to the retinotopic organization of this area. The diagram in Fig. 6-1B depicts a view of the unfolded striate cortex, and, for simplicity, the lateral portion of the TZ and area 18 are not represented. There is general agreement about the existence of an ipsilateral representation of the visual field in the 17/18 border region in the cat (i.e., segment 1 through 3 of the visual field is represented on the left TZ, and segment 3 through 5 of the visual field is represented on the right

TZ), although its extent varies among different studies (Payne, 1990; Diao et al., 1990; Pettigrew and Dreher, 1987; Blakemore et al., 1983; Whitteridge and Clarke, 1982; Albus and Beckman, 1980; Harvey, 1980; Tusa et al., 1978, 1979; Hubel and Wiesel, 1962). Payne (1990) reported that the ipsilateral representation is contained in the TZ, and that the maximum ipsilateral incursion of receptive field centers varies with elevation, being about 4^0 at the 0^0 horizontal meridian and 10^0 or more at superior or inferior elevations. Crossed projections from an area of temporal retina (Kirk et al., 1976a, 1976b; Fukuda and Stone, 1974; Stone, 1966) have been implicated in the ipsilateral visual field representation in the cat visual cortex (Fig. 6-5) (Payne, 1990; Whitteridge and Clarke, 1982). A consequence of the representations of ipsilateral visual fields in both hemispheres is that a central area of the visual field including the vertical meridian is represented in both hemispheres (the visual field segment 1 through 5).

Figure 6-1B illustrates that the mirror-reversed mapping of callosal linkages allows callosal fibers to interconnect cortical loci that represent the same coordinates in the visual field (e.g., pairs 1-1, 2-2, ..., 5-5). Data consistent with this interpretation were obtained in experiments in which the sites injected with anatomical tracers were mapped electrophysiologically prior to the injections (J. F. Olavarria and H. Sherk, unpublished observations). In one of these experiments (Fig. 6-4), the representation of the ipsilateral visual field on the 17/18 transition zone of the left hemisphere was mapped through the contralateral eye, and an injection of bisbenzimide (BB) was placed at a location whose receptive filed was centered at 1.5^0 into the ipsilateral hemifield (-2.5^0 elevation). In the right hemisphere, clusters of BB-labeled cells were found outside the TZ, including regions of area 17 and area 18 that represent the same visual field coordinates (Payne, 1990). The relatively widespread distribution of BB-labeled callosal cells probably reflects the fact that the BB injection infiltrated the entire width of the 17/18 transition zone (Olavarria, 1996a). The tracer horseradish peroxidase (HRP) was injected into a locus in the left area 17 whose receptive field center was located 2.5^0 into the contralateral hemifield (-2.3^0 elevation). In the right hemisphere, HRP labeling was restricted within the 17/18 transition zone, in accord with previous studies showing that this region of the visual field is represented in the ipsilateral TZ (Payne, 1990). The finding that callosal fibers connect cortical loci that are in retinotopic, rather than anatomical, correspondence has been recently confirmed in tree shrews by correlating callosal connections with cortical maps revealed with optical imaging techniques (Bosking et al., 2000).

Convergence/Divergence of Callosal Linkages

The schematic diagram in Fig. 6-3 also illustrates the observation that the total mediolateral extent of areas 17 and 18 labeled after a large injection in the TZ is considerably larger than the mediolateral extent of the area labeled in the TZ after injections of comparable size placed outside the TZ. This pattern of callosal linkages is consistent with previous observations in the cat indicating that callosal fibers

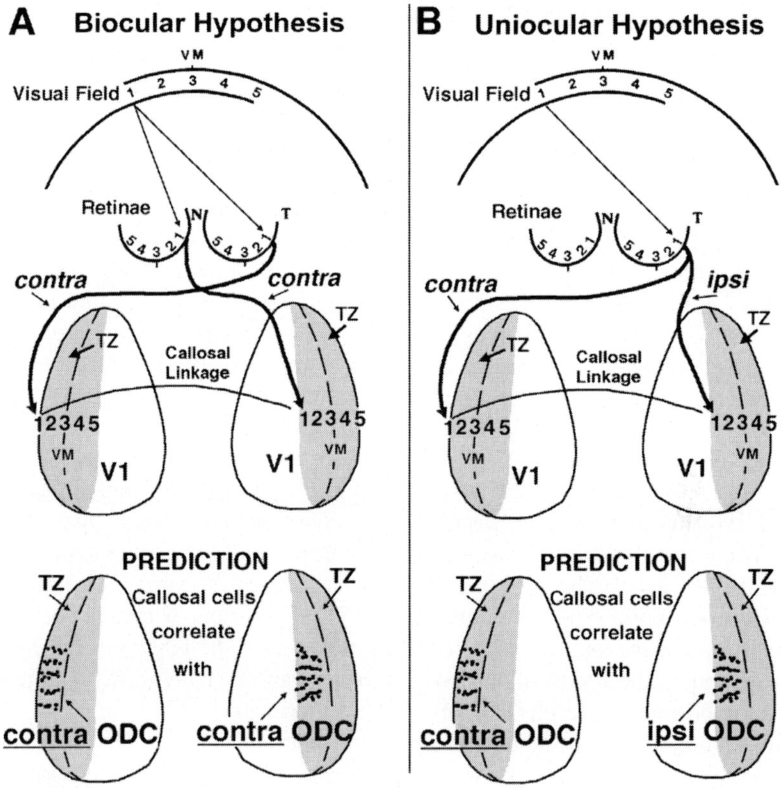

FIGURE 6-5. Two possible sources of interhemispheric correlated activity that may guide the stabilization of retinotopically matched callosal linkages. Top panels show diagrams relating the patterns of retinal projections to the pattern of callosal linkages. The visual field is schematically represented by the semicircular perimeters located at the top of the diagram. Each perimeter is completely represented in the ipsilateral retina and contralateral V1. A region of temporal retina (T) projects both ipsilaterally and contralaterally, whereas the nasal retina (N) only projects contralaterally. For simplicity, the diagrams apply only to projections from the left nasal retina and right temporal retina; the ipsilateral temporal projection has been omitted in **A,** and the contralateral nasal projection has been omitted in **B.** The panels in **A** and **B** illustrate that correlated activity between retinotopically corresponding cortical loci (loci marked 1) can potentially be elicited either from viewing the same object with both eyes (biocular hypothesis, **A**), or from activity that is relayed bilaterally from each temporal retina (uniocular hypothesis, **B**). If interhemispheric correlated activity shapes the pattern of callosal connections during development, the hypotheses in **A** and **B** lead to distinctly different predictions regarding the correlation of callosal connections with the underlying pattern of eye domains (bottom panels). Thus, in cortical regions outside the TZ, the binocularly-based hypothesis predicts that callosal connections correlate with contralateral ODCs, while the monocularly-based hypothesis predicts that callosal connections correlate with ipsilateral ODCs instead. Within the TZ, both hypotheses predict that callosal connections correlate with contralateral ODCs. ODCs, ocular dominance columns; other conventions as in Fig. 6-1. (From Olavarria [2001].)

terminate most densely in the area 17/18 transition zone, and less densely in immediately adjacent regions of areas 17 and 18 (Berman and Grant, 1992; Payne and Siwek, 1991; Voigt et al., 1988; Berman and Payne, 1983; Segraves and Rosenquist, 1982b; Innocenti, 1980; Sanides, 1978; Shatz, 1977b; Fisken et al., 1975; Garey et al., 1968; Ebner and Myers, 1965). A pattern of converging projections from regions outside the TZ into the contralateral TZ is also in accord with the difference in size between the representations of the contralateral and ipsilateral visual fields segments that are callosally interconnected: the ipsilateral field representation is quite compressed (Payne and Siwek, 1991; Payne, 1990; Diao et al., 1990; Pettigrew and Dreher, 1987; Whitteridge and Clarke, 1982). Finally, a pattern of converging callosal projections is in agreement with previous physiological data indicating that the magnification factor in the ipsilateral cortical representation is low (hamster: Tiao and Blakemore, 1976; sheep: Clarke and Whitteridge, 1976; cat: Whitteridge and Clarke, 1982), and that cells in this region tend to have large receptive fields (Payne and Siwek, 1991; Payne, 1990; Diao et al., 1990; Pettigrew and Dreher, 1987; Whitteridge and Clarke, 1982). This convergent/divergent pattern of callosal projections may be reflected in the morphology of individual callosal axons. The terminal arbors of axons projecting from the TZ to contralateral regions outside the TZ may be broader than the terminal arbors of axons projecting from regions outside the TZ into the contralateral TZ. Consistent with this possibility, Houzel et al. (1994) reported that the tangential area of the territories covered by terminal axons in the 17/18 callosal zone of the cat varied widely from no more than a few hundred square microns to several square millimeters.

Callosal linkages between noncorresponding anatomical loci in the 17/18 border region were first described in "Boston" Siamese cats (Shatz, 1977b). Because of the increased crossed retinal projection that occurs in these hypopigmented mutants (Guillery and Kaas, 1971), each cerebral hemisphere has an identical representation of the visual field extending roughly 20° to either side of the vertical midline (Berman and Grant, 1992; Shatz, 1977a). Unusually large parts of area 17 in the two hemispheres were reciprocally interconnected in such a way as to link cortical regions representing similar visual field coordinates (Shatz, 1977b). The data from normal cats (Olavarria, 1996a) suggest that the organization of callosal linkages in the 17/18 border region is similar in both normal and mutant cats, although the representation of ipsilateral visual fields is expanded in Siamese cats. Patterns of callosal linkages similar to those in the cat have been reported in lateral striate cortex of the rat (Lewis and Olavarria, 1995), and nonsymmetrical callosal linkages have also been found in tree shrews (Pritzel et al., 1988; Bosking et al., 2000), and galagos (J.F. Olavarria, H. Qi, and J.H. Kaas, unpublished observations), suggesting that this pattern of callosal linkages occurs across diverse groups of mammalian species.

INTERHEMISPHERIC CORRELATED ACTIVITY GUIDES CALLOSAL DEVELOPMENT

The finding that callosal fibers interlink cortical loci that are in retinotopic, rather than anatomical, correspondence (Fig. 6-1B) is consistent with the idea that visual input plays an important role in callosal development. In the cat, numerous studies using different experimental paradigms have shown quite conclusively that abnormal visual input leads to abnormal patterns of callosal connections in the 17/18 callosal zone (see references below). However, these studies have not identified the precise mechanisms through which visual input exerts its influence on callosal development.

Clues about to the nature of the activity-dependent mechanisms guiding the development of callosal connections came from experiments using the split chiasm preparation (Antonini et al., 1983; Lepore and Guillemot, 1982; Berlucchi and Rizzolatti 1968). These studies showed that, for some binocular cells representing fields on the vertical meridian, the contribution from the ipsilateral eye came directly through the retinothalamocortical pathway, whereas the contribution from the contralateral eye came indirectly through the corpus callosum. Moreover, both the subcortical and callosal components of the binocular receptive field were matched for size, position along the vertical meridian, orientation, and directional selectivity (Berlucchi, 1981). The similarity in the visually evoked responses mediated by the callosal and subcortical pathways led to the proposal (Berlucchi, 1981) that cells located at or near the cortical representation of the vertical meridian in both hemispheres maintain their callosal connections because they receive synchronous and congruent visual input through callosal and subcortical pathways, whereas those cells receiving mismatching inputs through both pathways lose their interhemispheric connection. This idea led to commonly accepted notion that development of callosal connections depends on interhemispheric correlated activity (i.e., cortical cells maintain their interhemispheric synaptic connection if they fire in synchrony with the postsynaptic element in the opposite hemisphere), whereas asynchronous firing leads to the loss of callosal connections. Perhaps the fact that in the split chiasm preparation callosal fibers bring together the input from both eyes contributed to the widespread belief that binocular viewing plays an important role in callosal development. In sharp contrast, the possibility that monocular mechanisms play a role in callosal development in the cat has not been considered. These two possibilities are discussed next.

Biocular or Uniocular Mechanisms can Potentially Drive Interhemispheric Correlated Neuronal Activity

Figure 6-5 illustrates two retinally driven mechanisms that can potentially stabilize retinotopically corresponding patterns of callosal linkages during development. These two mechanisms are possible because in the cat, a region of the temporal retina projects bilaterally (Kirk et al., 1976a, 1976b; Fukuda and Stone, 1974; Stone, 1966). In Fig. 6-5, those portions of temporal retina that also project contralaterally are represented by the retinal segments containing the loci 1,2,3 (right retina) and 3,4,5 (left retina). To simplify the diagram, only the projections from the left nasal retina and the right temporal retina are represented. The contralateral projection from each temporal retina relays a representation of ipsilateral visual fields on the contralateral TZ (Payne, 1990), and, in consequence, the central region of the visual field (points 1–5) is represented in both hemispheres (Olavarria, 1996a; Payne, 1990, 1994). As proposed previously (Olavarria, 1996a; Olavarria and Li, 1995), the width of the callosal zone in striate cortex reflects the extent of the crossed temporal projection. Figures 6-1B and 6-5 illustrate that the callosal zone in striate cortex of each hemisphere (indicated in gray) contains the representation of visual points 1-5.

Figure 6-5A illustrates that the contralateral projection from right temporal retina relays a representation of the loci 1,2,3 onto the left TZ (the ipsilateral projection from the right temporal retina has been omitted for clarity). In regions outside the TZ on the right hemisphere, a representation of the same loci 1,2,3 is relayed by the contralateral projection from the left nasal retina. Thus, Fig. 6-5A illustrates the hypothesis (termed the *biocular hypothesis*) that correlated activity between retinotopically corresponding loci in both hemispheres can be elicited from viewing the same object (e.g., point 1 in the visual field) with both eyes. This correlated activity would lead to the stabilization of synaptic contacts between retinotopically corresponding loci in both hemipsheres. Moreover, callosal fibers would be capable of generating binocular cells because they would bring together the inputs from both eyes. In the *Xenopus* frog, a species in which retinal projections are completely crossed, a similar biocular mechanism appears to be responsible for the development of the intertectal commissure and the generation of binocular tectal cells (Gaze et al., 1970; Keating and Feldman, 1975).

Figure 6-5B illustrates the alternative hypothesis (termed the *uniocular hypothesis*) that correlated activity between retinotopically corresponding loci in both hemispheres can be elicited by activity in each eye, by way of the bilateral projections from each temporal retina (Olavarria, 2001). As depicted in Fig. 6-5B, the contralateral projection from the right temporal retina relays a representation of the loci 1,2,3 onto the left TZ. In regions outside the TZ on the right hemisphere, a representation of the same loci 1,2,3 is relayed by the ipsilateral projection from the same right temporal retina (the contralateral projection from the left nasal retina has been omitted for clarity). Thus, Fig. 6-5B illustrates that correlated activity between retinotopically corresponding loci in both hemispheres can also be elicited by activity (either spontaneous or evoked) relayed to both hemispheres from the same temporal retina. This correlated activity would lead to the stabilization of synaptic contacts between retinotopically corresponding loci in both hemipsheres. Studies in the rat suggest that a uniocular mechanism driven by spontaneous retinal activity (Maffei and Galli-Resta, 1990; Galli and Maffei, 1988) is sufficient for specifying retinotopically corresponding callosal linkages before the onset of visual experience (Lewis and Olavarria, 1995; Olavarria and Li, 1995). In the cat, a uniocular mechanism could potentially be driven by spontaneous retinal activity before the eyes open (Meister et al., 1991), or by patterned visual experience through each eye after the eyelids open. Unlike the biocular mechanism, callosal fibers stabilized by a uniocular mechanism would not generate binocular cells because synchronous activity in both the cells of origin as well as the targets of callosal axons would be driven by the same eye. In other words, the callosal input to a cortical cell would largely duplicate the subcortical input from one of the eyes to the same cell. However, it is important to note that the *uniocular hypothesis does not require that callosal cells nor the targets of callosal axons be monocular.* In effect, the terms *biocular* and *uniocular* (instead of *binocular* and *monocular*) were chosen to emphasize that they refer to mechanisms driving interhemispheric correlated activity, not to mechanisms associated with binocular properties of cortical cells.

Although it is widely believed that binocular viewing plays an important role in callosal development in the cat, the possibility that stabilization of callosal linkages is preferentially guided by uniocular mechanisms, or by biocular and monocular mechanisms acting synergistically, has not been investigated. Support for the idea that interhemispheric correlated activity driven by either one, or both, of these mechanisms guides callosal development comes from the observation that neonatal section of the optic chiasm, a procedure that abolishes both the biocular and uniocular mechanisms while preserving visual experience (Fig. 6-5), leads to a loss of callosal connections in striate cortex. This loss is as drastic as that reported in bilaterally enucleated and dark-reared cats (Boire et al., 1995).

Previous Tests of the Role of Biocular Mechanisms

The hypothesis that binocular viewing plays an important role in callosal development has been tested by studies using various rearing paradigms designed to either manipulate or prevent binocular vision, including alternating monocular occlusion (Frost et al., 1990), monocular (Innocenti and Frost, 1979) or binocular (Innocenti et al., 1985; Innocenti and Frost, 1979, 1980) eye lid suture, monocular (Innocenti and Frost, 1979) or binocular enucleation (Olavarria and Van Sluyters, 1995; Olavarria, 1995; Innocenti and Frost, 1980), dark-rearing (Olavarria, 1995; Frost and Moy, 1989; Lund and Mitchell, 1979a), and section of the optic chiasm (Boire et al., 1995). These studies have generally reported that these rearing conditions induce changes in the number and overall distribution of callosal cells in the 17/18 callosal zone, but they have not clarified whether visual input exerts a permissive or instructive role on callosal development.

Strabismus as a Strategy for Investigating the Role of Binocular Viewing on Callosal Development

An instructive role of binocular viewing on callosal development has been proposed by studies investigating the effect of misalignment of visual axes (strabismus) on callosal development (Elberger et al., 1983; Berman and Payne, 1983; Lund and Mitchell, 1979b; Innocenti and Frost, 1979; Lund et al., 1978). These studies are based on the premise that, under conditions of normal eye alignment, synchronicity occurs at the lateral border of both striate cortices, regions that see the same fields on or near the vertical meridian. However, by shifting the cortical representation of the vertical meridian from its normal position, strabismus displaces the cortical sites in striate cortex that are activated in synchrony by stimuli located on the vertical meridian (see Fig. 3 in Berman and Payne, 1983; or Fig. 1 in Lund and Mitchell, 1979b). These shifts in cortical representation would cause the stabilization of projections between opposite sites that normally do not remain callosally connected, leading to increases in the width of the 17/18 callosal zone that are proportional to the degree of strabismus.

In apparent agreement with the hypothesis that correlated biocular input can lead to the stabilization of juvenile callosal axons, several previous experiments in strabismic cats reported moderately expanded callosal cell zones in area 17 (Elberger et al., 1983; Berman and Payne, 1983; Innocenti and Frost, 1979). Expansions of the callosal terminal territories in the same area were also observed in some animals (Berman and Payne, 1983; Lund and Mitchell, 1979b; Lund et al., 1978). These results lend support to the hypothesis that convergent or divergent strabismus leads to the stabilization of some callosal axons that normally would have been lost. However, other earlier observations were not completely in line with the predictions. For instance, expansion of callosal connections in striate cortex was reported in both hemispheres when it was predicted in only one of them, and no systematic relation was found between the degree of expansion and the angle of strabismus induced (Berman and Payne, 1983; Elberger et al., 1983). In addition, the distribution of callosal cells and terminations produced by strabismus at the 17/18 callosal zone has been described by some workers as "only slightly wider than normal" (Payne et al., 1988), or "moderately enlarged" (Innocenti, 1991). Most of the callosal cells and terminations still accumulate at the 17/18 border rather than at the displaced vertical meridian representation within striate cortex (Berman and Payne, 1983; Elberger et al., 1983; Innocenti and Frost, 1979).

The effect of strabismus on callosal development was recently reexamined by Bourdet et al. (1996). They used quantitative techniques to analyze the distribution of visual callosal cells in eight cats reared with divergent or convergent strabismus and compared with the distribution found in four normally reared cats. Although the data that Bourdet et al. (1996) obtained from strabismic cats do not differ greatly from those reported for strabismic cats in several previous studies, these authors failed to find a significant difference between normally reared and strabismic cats in both the number and overall distribution of callosal cells in striate cortex. These results, therefore, do not bear out the prediction that surgically shifting the visual axes leads to the stabilization of juvenile callosal axons in anomalous places within striate cortex.

Failure to see marked changes in the tangential distribution of callosal cells in strabismic cats may indicate that the deviation of visual axes caused by muscle surgery has not been large enough. In the study by Bourdet et al. (1996), the amount of deviation was about 12^0 on average, whereas in previous studies, unilaterally induced deviations ranged from 4^0 to 22^0 (Berman and Payne, 1983; Elberger et al., 1983), and bilaterally induced strabismus ranged from 26^0 to 45^0 (Innocenti and Frost, 1979). Although these deviations are substantial, it is possible that even larger deviations are necessary for producing marked increases in the width of the 17/18 callosal zone.

Another possibility considered by Bourdet et al. (1996) is that surgical strabismus may not meet all the requirements for adequately testing the hypothesis that synchronous activity between cortical sites in opposite hemispheres leads to the stabilization of callosal fibers. The use of strabismus for testing this hypothesis assumes that shifts in ocular alignment do indeed lead to synchronous evoked activity in cortical loci that, under normal conditions, view quite disparate regions of the visual field. More important, it is also assumed that this synchronous activity is of sufficient strength and duration to induce stabilization of immature callosal projections in regions that normally do not remain callosally connected. It is known that strabismus markedly reduces the population of binocular cells (Movshon and Van Sluyters, 1981; Hubel and Wiesel, 1965b) and can cause suppression and other functional changes that reduce the efficacy of cortical input from the deviated eye (Sengpiel et al., 1994; Chino et al., 1994). These other effects of strabismus may render the anomalous correlation in interhemispheric activity it produces too weak to stabilize juvenile callosal connections. Bourdet et al. (1996) also pointed out that, despite the presence of apparently symmetrical optokinetic and vestibular reflexes in the two eyes of the strabismic cats they used, it is possible that the deviations of the eyes were significantly noncomitant. If this is the case, then the periods of synchronous activity between anomalous regions in the two hemispheres may not have been long enough to promote stabilization of callosal connections between them. Strabismus induced optically (Van Sluyters and Levitt, 1980) offers the opportunity for studying the effects of eye misalignment on callosal development without compromising ocular motility (Elberger et al., 1983). Unfortunately, this approach has other technical drawbacks that limit its utility for these kinds of studies (Bourdet et al., 1996). In summary, previous data from strabismic cats do not offer compelling evidence that interhemispheric correlated activity driven by binocular cues plays an important role in the development of the callosal pathway, but it remains possible that induced strabismus is not a fully appropriate model for testing this hypothesis.

Callosal Connections Versus Ocular Dominance Columns: an Alternative Strategy for Investigating the Role of Biocular and Uniocular Mechanisms on Callosal Development

An alternative strategy for testing the role of interhemispheric correlated activity on the development of callosal connections, and whether correlated activity is driven by either biocular or uniocular cues, was recently presented by Olavarria (1996a, 1996b; 2001). This strategy entails correlating the distribution of callosal connections with the underlying pattern of ocular dominance columns (ODCs). An advantage of this approach over previous studies is that the observations can be performed in normal cats, thereby avoiding surgical interventions such as those required to induce strabismus.

Figure 5 illustrates that the biocular and uniocular hypotheses lead to distinctly different predictions (Olavarria, 1996a, 1996b; 2001) regarding the relation of callosal connections with the underlying patterns of ODCs. According to the biocular hypothesis, synchronous firing in both hemispheres would be elicited in territories dominated by the *contralateral* eye. Assuming that input from each eye preferentially activates callosal cells and targets of callosal axons that are located in spatial register with the territory dominated by the same eye (Toyama et al., 1974; McCourt et al., 1990), the biocular mechanism in Fig. 6-5A predicts that callosal cells associated with contralateral ODCs will maintain their callosal axons, whereas callosal cells overlying ipsilateral ODCs will tend to lose their callosal axons. This process would sculpt the initially homogeneous distribution of callosal cells (Innocenti and Clarke, 1984), leading to the segregation of callosal clusters that are preferentially correlated with *contralateral* ODCs both in the TZ as well as in regions located outside the TZ (see bottom of Fig. 6-5A; corresponding territories activated by left temporal retina and right nasal retina are not represented). On the other hand, according to the uniocular hypothesis, synchronous firing in both hemispheres would occur between territories dominated by the *contralateral* eye in the TZ (loci 1,2,3 in the left TZ, Fig. 6-5B), and territories dominated by the *ipsilateral* eye in regions outside the TZ (loci 1,2,3 in the right striate cortex, Fig. 6-5B). Thus, a uniocular mechanism would lead to the segregation of callosal clusters that are correlated with *contralateral* ODCs within the TZ, resembling the outcome from the biocular mechanism. Outside the TZ, however, callosal clusters would correlated with *ipsilateral* ODCs, in sharp contrast with the outcome predicted by the biocular hypothesis (cf. bottoms of Fig. 6-5A,B).

These predictions assume that cortical responses are segregated into eye-specific domains at the time the pattern of callosal connections is developing. This is likely because in kittens, afferents from the dorsal lateral geniculate nucleus appear segregated into ODCs by the sixth postnatal week (LeVay et al., 1978), long before maturation of the callosal pathway is complete (Innocenti and Caminiti, 1980; Innocenti et al., 1977). These hypotheses also predict that interhemispheric correlated activity driven by eye-specific sets of thalamocortical afferents sculpts the initially homogeneous distribution of callosal neurons (Innocenti and Clarke, 1984; Innocenti and Caminiti, 1980) into a patchy distribution. Consistent with this prediction, several studies have shown that callosal connections in the 17/18 callosal zone of the cat are distributed in discrete clusters rather than homogeneously (Olavarria, 1996a, 2001; Boyd and Matsubara, 1994; Houzel et al., 1994; Payne and Siwek, 1991; Voigt et al., 1988; Berman and Payne, 1983), suggesting a relation with columnar patterns.

To test the hypotheses illustrated in Fig. 6-5, retrogradely labeled callosal neurons were correlated with the patterns of ODCs revealed anatomically (Olavarria, 2001). Neurons in the 17/18 callosal zone of one hemisphere were labeled following injections of fluorescent tracers into different regions within the opposite 17/18 callosal zone (Olavarria, 1996a). The patterns of ODCs were revealed by the transneuronal transport of horseradish peroxidase conjugated with wheat germ agglutinin (WGA-HRP) after intraocular injections of this enzyme (Anderson et al., 1988).

Callosal Connections Versus Ocular Dominance Columns: Results Support the Uniocular Hypothesis

The results from a normal cat (CJ) are illustrated in Figs. 6-6 and 6-7. Figure 6-6 shows overall views of unfolded and flattened cortical tissue (Olavarria and Van Sluyters, 1985b) containing the complete reconstructions of the patterns of ODCs in areas 17 and 18 in the right (Fig. 6-6A) and left (Fig. 6-6B) hemispheres. The dark areas in Fig. 6-6A correspond to territories dominated by the contralateral eye, whereas dark areas in Fig. 6-6B form the pattern of ipsilateral ODCs. Tracer injections into the right hemisphere are indicated by colored areas, and the resulting labeled callosal cells in the left hemisphere are indicated by dots of the corresponding color. In Fig. 6-6A, the location of the border between area 17 and 18 (indicated by the black line) was recognized by the abrupt change in the pattern of ODCs that occurs on passing from one area to the other (Anderson et al., 1988). The 17/18 border is not indicated in Fig. 6-6B to avoid covering the fields of callosal labeling located along this border. The TZ, approximately 1.0 to 1.5 mm wide, straddles the 17/18 border (Olavarria, 1996a; Payne and Siwek, 1991; Payne, 1990).

Figure 6-6 illustrates once again that callosal fibers connect loci in both hemispheres that are not in anatomical correspondence. After injections of fluorescent tracers into loci of area 17 located outside the TZ (the posterior injection of bisbenzimide, BB, and the injection of green latex beads, GB), labeled callosal cells were found predominantly within the contralateral TZ. On the other hand, after fluorescent tracer injections into the TZ (the injection of rhodamine latex beads, RB, and the anterior injection of BB) labeled callosal cells were found predominantly in regions of areas 17 and 18 located outside the TZ. These results are in agreement with a previous study of callosal linkages in which areas 17, 18, and the TZ were identified cytoarchitectonically in coronal sections (Olavarria, 1996a). Figure 6-6 also illustrates that callosal cells accumulate in clusters of various sizes and shapes separated by regions containing few or no labeled cells, in agreement with previous reports (Olavarria, 1996a, 2001; Boyd and Matsubara, 1994; Houzel et al., 1994; Payne and Siwek, 1991; Voigt et al., 1988; Berman and Payne, 1983).

Data from cat CJ (Fig. 6-6B) are illustrated at higher magnification in Fig. 6-7. Figure 6-7A shows the pattern of ipsilateral, WGA-HRP labeled, ODCs from the region of the left hemisphere indicated by the box in the inset. This region contains callosal cells labeled by the injections of RB and GB into the right hemisphere. Dark areas in Fig. 6-7A are represented in black in the thresholded version shown in Fig. 6-7B, G, and in gray in Fig. 6-7D, F. Figure 6-7C shows the distribution of labeled callosal cells in regions of areas 17 and 18 located outside the TZ after the injections of RB into the TZ of the opposite hemisphere. The clusters of RB-labeled cells in the upper half of Fig. 6-7C are located in area 17 and the clusters in the lower half are located in area 18. Figure 6-7D illustrates that in both areas 17 and 18, RB-labeled cells are distributed preferentially over ipsilateral (gray) ODCs. Figure 6-7E shows the distribution of GB-labeled cells

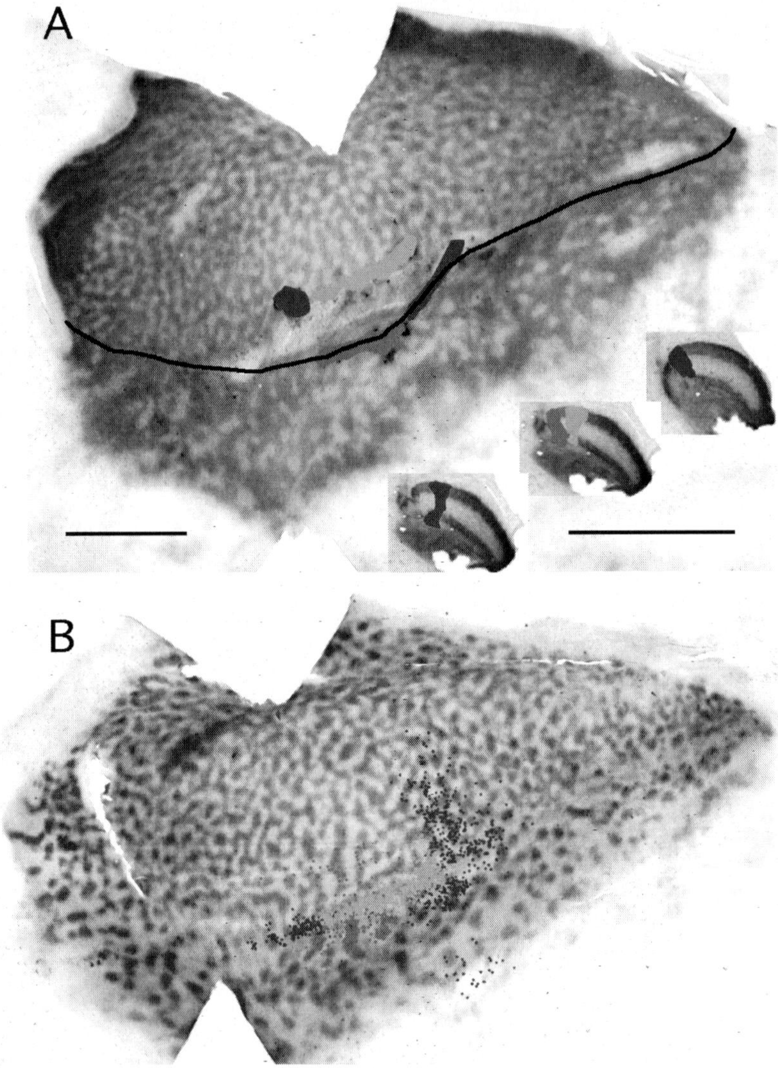

FIGURE 6-6. Correlation of labeled callosal cells and patterns of ODCs in areas 17 and 18 (normal cat CJ). Overall views of unfolded and flattened cortical tissue containing the complete reconstructions of the patterns of ODCs in areas 17 and 18 in the right **(A)** and left **(B)** hemispheres. In both panels, posterior is to the left, medial is up. The dark areas in **(A)** and **(B)** correspond to territories dominated by the contralateral and ipsilateral eye, respectively. The black line **(A)** indicates the location of the border between area 17 and 18. Tracer injections into the right hemisphere are indicated by colored areas (red, red beads; green, green beads; blue, bisbenzimide), and the resulting labeled callosal cells in the left hemisphere are indicated by dots of the corresponding color. The inset shows the fields of retrogradely labeled cells in three sections from the right dorsal lateral geniculate nucleus (dLGN), ordered from posterior (to the left) to anterior. Layers of the dLGN receiving input from the left eye (injected with WGA-HRP) appear dark. Scale bar = 5 mm. (From Olavarria [2001].) See color insert for color reproduction of this figure.

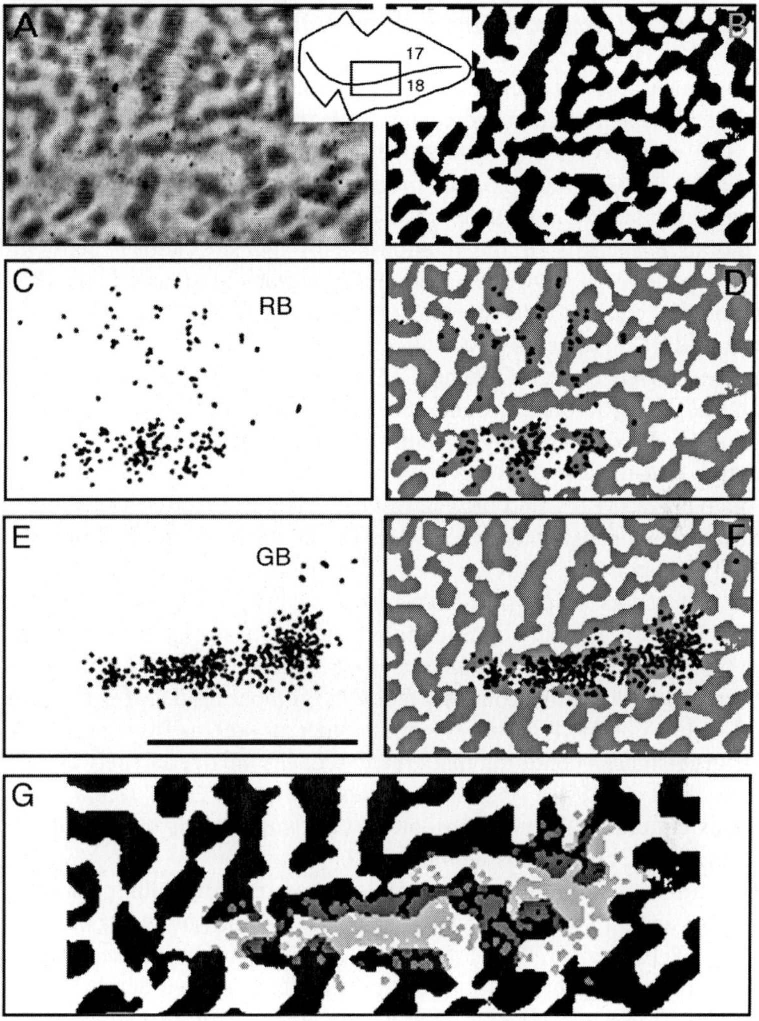

FIGURE 6-7. Correlation of labeled callosal cells and patterns of ODCs in areas 17 and 18 (normal cat CJ). Higher magnification views of the data shown in Fig. 6-6 for the injections of red beads (RB) into the TZ, and green beads (GB) into area 17 outside the TZ. Data come from portion of left hemisphere indicated in the box (inset) which includes the TZ and adjacent portions of areas 17 and 18. **(A)** pattern of WGA-HRP labeling in which dark areas correspond to ipsilateral ODCs. The dark spots correspond to blood vessels. Thresholded versions of this pattern are shown in black in **B, G** and in gray in **D, F. (C)** Distribution of RB-labeled callosal cells in regions of areas 17 and 18 located outside the TZ. The clusters of labeled cells in the upper half of panel are located in area 17, whereas the clusters in the lower half are located in area 18. **(D)** Correlation of RB-labeled cells with the underlying pattern of ODCs. **(E)** Distribution of GB-labeled cells in the TZ. **(F)** Correlation of GB-labeled cells with the underlying pattern of ODCs. **(G)** Enlarged view of **F** to better show the relationship of GB-labeled cells (gray) with both the cores (white) of contralateral ODCs and the adjacent WGA-HRP labeled (black) regions. Scale bar = 5 mm. (From Olavarria [2001].)

after injections of GB in area 17 of the opposite hemisphere (see Fig. 6-6A). The clusters of GB-labeled cells are arranged anteroposteriorly, occupying a strip of cortex that measures between 1 and 2 mm wide. Figure 6-7F, G shows that the clusters of GB-labeled cells are centered on the (white) cores of contralateral ODCs. The edges of these clusters overlap with adjacent, WGA-HRP-labeled regions (Fig. 6-7G), regions that probably correspond to zones of overlap between contralateral and ipsilateral ODCs.

Quantitative analysis of the relationship of labeled callosal cells with the underlying patterns of ODCs revealed that in regions of areas 17 and 18 located outside the TZ, callosal cells show a significant preference for ipsilateral ODCs (Olavarria, 2001). The relative densities of callosal neurons over ipsilateral and contralateral ODCs were, on average, 72.4% and 27.6%, respectively (data from cats CJ, CK, using fluorescent latex beads; Olavarria, 2001). In contrast, in the TZ, callosal cells showed a significant preference for contralateral ODCs. Here, the relative densities of callosal cells over ipsilateral and contralateral ODCs were, on average, 31.8% and 68.2%, respectively (data from cats CJ, CM, using fluorescent latex beads; Olavarria, 2001). These values are plotted in Fig. 6-8 to graphically illustrate the switch in the preference of callosal cells for specific eye domains that occurs on passing from regions located outside the TZ to regions within the TZ. A potentially confounding factor in these experiments stems from the fact that ipsilateral and contralaterall ODCs have a zone of overlap in normal cats. Olavarria (2001) ruled out this factor by comparing data from normal cats injected with WGA-HRP in one eye or the other, as well as from cats raised with surgical strabismus, a condition that greatly reduces the overlap between adjacent ODCs (Shatz et al., 1977). As in normal cats, Olavarria (2001) found that in strabismic cats callosal cells prefer ipsilateral ODCs in regions outside the TZ, and contralateral ODCs in regions within the TZ.

The finding that callosal neurons correlate preferentially with ipsilateral ODCs in regions of areas 17 and 18 located outside the TZ and with contralateral ODCs within the TZ (Olavarria, 2001) favors the hypothesis that in cats the pattern of callosal linkages is specified by interhemispheric correlated activity driven primarily by uniocular mechanisms (Fig. 6-5B). This result also suggests that callosal fibers predominantly link cortical regions that are driven by the same temporal retina. Thus, in cats, as in rats, the specification of callosal linkages in striate cortex appears to be under the guidance of uniocular mechanisms in spite of the marked difference between these species in the timing of callosal development with respect to the onset of visual experience (much of callosal development occurs after eye opening in cats, and before eye opening in rats).

The results from correlating callosal cells and patterns of ODCs further suggest that the pattern of interhemispheric activity that appears to be more effective in stabilizing callosal axons is that driven independently from each eye, rather than that driven by seeing the same object through both eyes. This suggestion is surprising in view of the prevalent notion that binocular cues plays an important role in the development and/or refinement of circuits subserving binocularity in

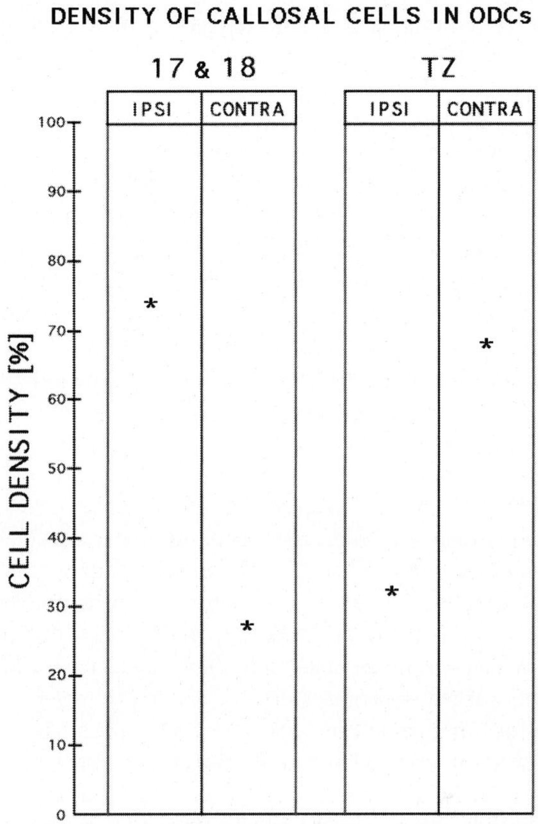

FIGURE 6-8. Relative densities of labeled callosal cells over ipsilateral and contralateral ODCs in regions located either outside the TZ (17 and 18) or within the TZ. Relative density values were calculated according to procedures described in Olavarria (2001). Values for the 17 and 18 region are averages of data from two cats (CJ, CK) that received injections of fluorescent latex beads into the TZ, whereas values for the TZ are averages of data from two cats (CJ, CM) that received injections of fluorescent latex beads into cortical regions located outside the TZ (Olavarria, 2001).

mammals (Hubel and Wiesel, 1965b) and amphibians (Gaze et al., 1970), as well as in lateral intrinsic interactions in striate cortex (Ruthazer and Stryker, 1996; Löwel and Singer, 1992; Callaway and Katz, 1991). What could be the basis for a reduced role of biocular mechanisms on callosal development? A biocular mechanism such as that in Fig. 6-5A is likely to be sensitive to instabilities of eye alignment. It is possible therefore that binocular visual experience is less effective during callosal development because, in kittens, accurate eye alignment develops during the first months of life (Sherman, 1972; Olson and Freeman, 1978). In contrast, a uniocular mechanism would not be dependent on eye alignment. Also consistent with a greater role of uniocular mechanisms on callosal development are the observations that raising cats with deviations of the visual axes does not affect the overall distribution and density of callosal cells in striate cortex (Bourdet et al., 1996), and that callosal connections do not maintain binocularity in striate cortex of strabismic cats (Hubel and Wiesel, 1965b; Van Sluyters and Levitt, 1980). Crair et al. (1998) recently showed that cortical responses are segregated into eye-specific patches at an age significantly earlier than such patches have been demonstrated anatomically, and that binocular viewing, even when present, does not play a role during early development of ODCs maps. Thus, early during development, the pattern of callosal linkages and its association to specific sets of ODCs may become specified primarily by uniocular cues, even if the period in which this occurs extends beyond the onset of visual experience. However, a biocular contribution cannot be ruled out entirely because a relatively small proportion of labeled callosal cells were in register with contralateral ODCs in regions outside the TZ (Fig. 8). It will be important to further assess the role of binocular viewing on callosal development by analyzing the correlation between callosal cells and ODCs in animals raised under conditions that completely eliminate biocular mechanisms without affecting uniocular mechanisms, such as alternating monocular occlusion. Finally, it is of interest to note that although the precise retinal and thalamic origins of ipsilateral subcortical afferents to the TZ have been speculated upon (Payne, 1990), they remain uncertain (not represented in Fig. 6-5). The present results suggest that they do not play a major role in the stabilization of callosal connections in the TZ.

In both normal and strabismic cats, Olavarria (2001) found that callosal neurons prefer ipsilateral ODCs in areas 17 and 18 in spite of the changes in the overall pattern and sizes of ODCs that occur on passing from area 17 into area 18 (e.g., Fig. 6-6; Anderson et al., 1988). Moreover, the results are similar in area 17 of normal and strabismic cats even though in strabismic cats some ODCs in regions representing central visual fields appeared to be wider than the ODCs located in corresponding regions of area 17 in normal cats (see Fig. 6 in Olavarria, 2001; Tieman and Tumosa, 1996; Löwel, 1994). These observations further support the idea that the development of periodicities in the distribution of callosal connections is influenced by neuronal activity relayed through eye-specific thalamic afferents.

PREVIOUS OBSERVATIONS IN SIAMESE AND STRABISMIC CATS ARE CONSISTENT WITH THE UNIOCULAR HYPOTHESIS

The puzzling observation that callosal connections are drastically reduced in striate cortex of "Boston" Siamese cats (Berman and Grant, 1992; Tremblay et al., 1987) cannot be explained adequately by the biocular hypothesis (Fig. 6-5A) because this mechanism is, if anything, stronger than normal owing to the increase in the contingent of crossed retinal projections that occurs in these cats (Guillery and Kaas, 1971). On the other hand, the concomitant decrease in ipsilateral projections that occurs in Siamese cats (Leventhal, 1982; Stone et al.,

1978) may significantly alter the balance of contralateral and ipsilateral connections emanating from temporal retina, thereby reducing the efficacy of temporal bilateral projections for driving interhemispheric correlated activity. A weakened uniocular mechanism in Siamese cats may thus be unable to guide the stabilization of a normal contingent of callosal connections. If this explanation is correct, an even greater reduction in the density of callosal connections in striate cortex can be expected in tyrosinase-negative albino cats because the ipsilateral projection in these cats is weaker than that in Siamese cats (Leventhal and Creel, 1985). Reduced callosal connections in striate cortex would also be expected in achiasmatic subjects because these lack uniocular mechanisms due to the reverse condition, namely, failure of retinal crossing (Hogan and Williams, 1995; Apkarian et al., 1994).

Correlating the distribution of callosal cells with the underlying patterns of ODCs in normal cats (Olavarria, 2001) yielded results consistent with the study by Bourdet et al. (1996) of the effects of strabismus on callosal development. Both approaches failed to bear out the hypothesis that binocular viewing plays an important role on callosal development. If in the cat interhemispheric correlated activity is driven primarily by a uniocular mechanism, then such a mechanism would not be affected by deviations of the visual axes. As originally formulated, the hypothesis that callosal linkages are stabilized by interhemispheric correlated activity driven by binocular viewing implies that callosal fibers both interconnect territories dominated by different eyes, and are capable of generating binocular cells (Berlucchi, 1981). The fact that in both Siamese cats (Hubel and Wiesel, 1971) and in cats raised with strabismus (Van Sluyters and Levitt, 1980) callosal connections are unable to maintain binocularity in lateral striate cortex further supports the idea that binocular viewing does not play a major role in callosal development.

ROLE OF UNIOCULAR MECHANISMS ON CALLOSAL DEVELOPMENT: ORGANIZATIONAL AND FUNCTIONAL IMPLICATIONS

The results reviewed here are consistent with the idea that the development of callosal connections in striate cortex is under the guidance of patterns of interhemispheric correlated activity driven primarily by temporal retina. This section discusses some of the implications of these findings on the functional organization of callosal connections.

Width of Callosal Zone in Striate Cortex Correlates with the Extent of Temporal Crossing

The observation that the width of the callosally connected region in striate cortex varies widely across species is not consistent with the vertical meridian rule. On the other hand, relating temporal retina to the stabilization of retinotopically corresponding callosal linkages (Fig. 6-5) implies that the width of the callosally connected region is striate cortex reflects the extent of temporal retina from which crossed projections originate (i.e., the extent of visual field represented in both hemispheres) (Fig. 6-5; Lewis and Olavarria, 1995). Thus, in animals in which all or a large portion of temporal retina contains ganglion cells projecting contralaterally, the width of the lateral callosal region should approach the width of the binocular field in striate cortex, whereas in animals with a small nasotemporal retinal overlap, callosal connections should be restricted to a narrow region along the border of striate cortex. These predictions are consistent with data on callosal connections and retinal projection patterns from a variety of species. For instance, relative to the size of striate cortex, the callosal band along the border of striate cortex is very narrow in macaque monkeys (e.g., Olavarria and Abel, 1996; Kennedy et al., 1986), a species with a relatively small naso-temporal retinal overlap (Chalupa and Lia, 1991; Fukuda et al., 1989; Leventhal et al., 1988; Bunt et al., 1977; Stone et al., 1973), whereas the callosal band is proportionately much wider in species like rats (e.g., Olavarria and Van Sluyters, 1985a), rabbits (Swadlow et al., 1978), cats (e.g., Olavarria and Van Sluyters, 1995; Segraves and Rosenquist, 1982a; Innocenti, 1980), and galagos (Beck and Kaas, 1994; Cusick et al., 1984; Weyand and Swadlow, 1980) in which either all or a substantial portion of the temporal retina has been shown to contain cells innervating the contralateral thalamus (rat: Reese and Cowey, 1987; Cowey and Perry; 1979; Lund et al., 1974; rabbit: Provis and Watson, 1981; galago: Sesma and Casagrande, 1986; cat: Rowe and Dreher, 1982; Cooper and Pettigrew, 1979; Sanderson and Sherman, 1971).

Crossed temporal projections may be necessary for the stabilization of callosal cells in lateral striate cortex. However, reports that neonatal enucleation or congenital anophthalmia in rodents (Olavarria et al., 1987, 1988; Olavarria and Van Sluyters, 1984; Rhoades and Fish, 1983; Cusick and Lund, 1982) do not prevent the development of a robust callosal zone in striate cortex indicate that, at least in rodents, retinal activity is not critical for the permanent stabilization of large numbers of callosal neurons in striate cortex. Instead, crossed projections from temporal retina, as part of the uniocular mechanism described above, may be primarily involved in guiding the development of retinotopically corresponding callosal linkages (Olavarria and Li, 1995). Development of retinotopically corresponding callosal linkages may reduce or prevent the erosion of visual topography that would result in striate cortex from non retinotopically corresponding callosal linkages. As noted above, in several species the width of the callosal zone in striate cortex correlates with the extent of the crossed temporal projection, suggesting that the extent of the crossed temporal projection is what is required to guide the development of topographically matched callosal connections in striate cortex of each species. At the low end of the spectrum, striate cortex of macaque monkeys would require little topographical guidance for ingrowing callosal axons, because in this species, callosal connections are virtually absent from most of striate cortex throughout development (Chalupa et al., 1989; Dehay et al., 1988). Consistent with this reduced need, the nasotemporal overlap in macaque retina is correspondingly small. If a major goal of the process that guides the development of retinotopically organized callosal linkages is to minimize erosion of the topographic map in striate cortex, then the solution adopted by macaque monkeys (i.e., virtually eliminating the callosal pathway in striate cortex) affords the maximum protection against erosion of topography. At the high end of the spectrum, callosal connections in lateral striate cortex of the rat occupy the entire binocular zone (Olavarria and Van Sluyters, 1985a), and, correspondingly, virtually the entire temporal crescent projects contralaterally in these species (Reese and Cowey, 1987; Cowey and Perry; 1979; Lund et al., 1974).

These considerations suggest that, during evolution, the size of the crossed temporal projection changes in parallel with the size and extent of the callosal pathway innervating lateral striate cortex, thus ensuring that callosal connections in striate cortex will be established between topographically matched loci. The possibility that changes at both retinal and cortical levels of the visual pathway are regulated by a common mechanism will remain plausible until species are discovered in which exuberant distributions of callosal connections in lateral striate cortex coexist with small crossed projections from temporal retina, or vice versa. Topographic maps in extrastriate areas are typically less precise than in striate cortex, and in some of these areas callosal connections connect loci that are not topographically matched (reviewed in Abel et al., 2000).

Interhemispheric versus Intrinsic Lateral Connections

The callosal pathway has been commonly regarded as an extension of intrinsic patterns of horizontal connections because it is thought that commissural fibers fulfill the need to complete intrinsic connectivity brought about by the separation of the two visual hemifields in each of the cortical hemispheres (Kennedy et al., 1991; Hubel and Wiesel, 1967). If the callosal pathway has the same organization and function as intrinsic connections, then one would expect that during development both systems of connections are specified by a common mechanism. It appears, however, that this is not the case. The results from Olavarria (2001) summarized previously indicate that in both normal and strabismic cats callosal neurons are aggregated into clusters that correlate with ipsilateral ODCs in regions outside the TZ, and with contralateral ODCs in regions outside the TZ. A collorary of these results is that in both normal and strabismic cats, callosal connections are sparse or absent in contralateral ODCs in regions outside the TZ, and in ipsilateral ODCs in regions within the TZ. In contrast, the distribution of intrinsic lateral connections in cat striate cortex does not appear to be biased toward specific eye domains (Löwel and Singer, 1992). Moreover, Löwel and Singer (1992) found that, in normal cats, tracer injections restricted to one ocular domain produced an unbiased distribution of retrograde labeling over neighboring left and right eye columns. On the other hand, in strabismic cats similar injections revealed patterns of intrinsic connections significantly biased toward same-eye columns, suggesting that there is selective stabilization of fibers between neurons that exhibit correlated activity owing to binocular viewing. Considered together, the results obtained by Olavarria (2001) and Löwel and Singer (1992) suggest that different sources generate the patterns of correlated activity that guide the development of callosal and intrinsic connections in striate cortex. Whereas the stabilization of callosal connections appears to rely primarily on interhemispheric correlated activity driven by uniocular mechanisms (Olavarria, 2001), the development of normal intrinsic connections depends on binocular activity cues (Löwel and Singer, 1992; Callaway and Katz, 1991; Luhmann et al., 1989). Differences in the effects of uniocular and biocular cues appear to be reflected in the distinct ways in which the emerging connections are shaped by the underlying patterns of ODCs. Callosal neurons correlate with specific sets of ODCs in different regions of the 17/18 callosal zone (Olavarria, 2001), whereas neurons in territories associated with either eye give raise to intrinsic connections in striate cortex (Löwel and Singer, 1992). These striking differences between interhemispheric and intrinsic lateral connections are not consistent with the notion that callosal connections can be simply viewed as "stretched out" intrinsic connections. These observations are important because they emphasize that rules of organization and development derived from one of these pathways may not be readily generalized to the other.

In another study, Schmidt et al. (1997) correlated the distribution of callosal connections with the pattern of ODCs following small tracer injections aimed at individual eye domains. These workers used this paradigm in one normal cat and reported that retrogradely labeled callosal neurons did not show a bias for eye-specific eye domains. A balanced distribution of labeled cells was described over territories dominated by the right and left eyes. It is difficult to determine the location of the injections of Schmidt et al. (1997) with respect to the callosal regions located either within or outside the TZ because these authors did not recognize these subdivisions. It is possible therefore that in the normal cat they studied they sampled areas that inadvertently included regions both inside and outside the TZ. In contrast, in strabismic cats Schmidt et al. (1997) found that tracer injections into a territory dominated by one eye produced contralateral labeling that was restricted to territories dominated by the same eye. Although these results are compatible with those obtained in strabismic cats by Olavarria (2001), they differ in that they do not reveal that in strabismic cats, as in normal cats, callosal connections are correlated with eye-specific domains in the different regions of the 17/18 callosal zone. Rather, the study of Schmidt et al. (1997) appears to have been designed under the assumption that callosal connections show a balanced distribution over right-and left-eye ODCs throughout the 17/18 callosal zone.

Contribution of the Callosal Pathway to the Representation of Ipsilateral Visual Field in the 17/18 Transition Zone of the Cat

Within the TZ, callosal neurons correlate preferentially with contralateral ODCs (Figs. 6-6, 6-7) (Olavarria, 2001). If one assumes that the distribution of callosal axons mimics the distribution of callosal cells (J. F. Olavarria and F. Torrealba, unpublished observations), then callosal input to ipsilateral eye domains within the TZ is, at most, sparse. This would significantly limit the transcallosal contribution of the ipsilateral eye to the representation of ipsilateral visual fields in the TZ, which is consistent with the results of previous physiological studies (reviewed in Olavarria, 1996a). On the other hand, callosal input to contralateral eye domains within the TZ is likely to be robust, which is in accord with the demonstrated transcallosal contribution of the contralateral eye to the representation of ipsilateral visual fields in the TZ (reviewed in Payne, 1990; Antonini et al., 1985; Lepore and Guillemot, 1982; Berlucchi and Rizzolatti, 1968).

Contribution of the Callosal Pathway to Binocularity in Striate Cortex of the Cat

The biocular and uniocular hypotheses also make different predictions regarding the role of callosal fibers in the generation of binocular cells in striate cortex. Experiments using the split chiasm preparation have shown that callosal fibers are capable of bringing together the inputs from both eyes at the cortical level, at least in the region representing the vertical meridian (Berlucchi and Rizzolatti, 1968; Cynader et al., 1986). However, this observation does not mean that, under normal conditions, the binocular cells found in cats with split chiasm receive input from one of the eyes solely by way of the callosal commissure. In fact, the existence of callosum-dependent binocularity in cat striate cortex is still a matter for debate. No clear picture has emerged from numerous studies that have used a variety of approaches to determine the extent to which binocularity in cat striate cortex depends on callosal input. These approaches include unilateral cooling (Blakemore et al., 1983), unilateral decortication (Cynader et al., 1986; Blakemore et al., 1983; Dreher and Cottee, 1975), and section of either the corpus callosum (Elberger and Smith, 1985; Payne et al., 1984; Minciacchi and Antonini, 1984; Elberger, 1981; Payne et al., 1980), optic chiasm (Lepore et al., 1992; Cynader et al. 1986; Lepore and Guillemot, 1982), or optic tract (Lepore et al., 1983). Some of these workers reported that these manipulations produce a substantial reduction in binocularly activated cells (Payne et al., 1984; Blakemore et al. 1983; Payne et al. 1980; Elberger, 1981; Dreher and Cottee, 1975), some reported that callosal projections make only modest contributions to the ocular dominance distribution of cortical cells (Cynader et al. 1986; Lepore et al., 1983), and still others failed to find changes in cortical binocularity after eliminating the callosal pathway in normal adult cats (Berlucchi and Antonini, 1990; Elberger and Smith, 1985; Minciacchi and Antonini, 1984).

The proposal that callosal fibers preferentially link cortical loci that are under the influence of the same eye (Figs. 6-5B) (Olavarria, 2001) may shed light on this debate because this idea implies that callosal fibers are not primarily involved in generating binocular cells. This result leads to the prediction that section of the callosal commissure should not have a significant effect on the binocularity of cat striate cortex, in agreement with some previous reports (e.g., Minciacchi and Antonini, 1984). As illustrated in Fig. 6-5B, callosal input to a cortical cell largely duplicates the subcortical input from one of the eyes to the same cell (see also Harvey, 1980). This would allow callosal connections to *sustain* the binocularity of cortical cells when subcortical input from one eye is abolished, which may explain the finding of binocular cells at the vertical meridian representation of cats with section of the optic chiasm (Lepore et al., 1992; Cynader et al. 1986; Lepore and Guillemot, 1982; Berlucchi and Rizzolatti, 1968). In cat primary auditory cortex, regions in which cells exhibit binaural summation or ipsilateral dominance contain higher concentrations of callosal connections than adjacent regions in which cells exhibit contralateral dominance (Imig and Brugge, 1978). Moreover, in spite of this clear correlation between callosal connections and binaural organization, callosal innervation is not necessary for binaural responsiveness in primary auditory cortex (Brugge and Imig, 1978). This comparison raises the possibility that mechanisms that guide the development of callosal connections in cat areas 17 and 18 share some similarities with those guiding the development of callosal connections in auditory and perhaps other, primary sensory areas.

Further evidence that callosal connections do not necessarily contribute to cortical binocularity in striate cortex comes from the ferret. In this species, the 17/18 border region and area 18 are exclusively monocular (White et al., 1999) even though these areas are richly innervated by the callosal commissure (Rockland, 1985; Grigonis et al., 1992). In ferrets, inputs from both eyes are segregated into ocular dominance patches of irregular size (Ruthazer et al., 1999; White et al., 1999; Redies et al., 1990; Law et al., 1988), and it will be interesting to determine whether callosal connections form clusters that show the same relationship with specific eye domains as they do in cats. In some species, however, binocularity in portions of lateral striate cortex does appear to depend on callosal input. In the rat, as in the sheep (Pettigrew et al., 1984), a region in lateral striate cortex that likely corresponds to the TZ receives subcortical input only from the contralateral eye (Thurlow and Cooper, 1988; J. F. Olavarria and P. L. Abel, unpublished observations). Diao et al. (1983) showed that ipsilateral eye input to lateral striate cortex in the rat was reduced or abolished by cooling the contralateral hemisphere, whereas input from the contralateral eye was largely unaffected, thus demonstrating that ipsilateral eye input to this region comes indirectly via the corpus callosum. Callosal connections have also been shown to contribute to binocularity in some extrastriate areas in Siamese cats, animals in which ipsilateral striate projections to extrastriate cortex are largely monocular (Zeki and Fries, 1980; Marzi et al., 1980). These considerations suggest that generation of binocular cells cannot be regarded as a basic, defining function of visual callosal connections in all species, and in all visual areas. Whether callosal input contributes to the generation of binocular cells appears to depend to a large extent on the specific factors and constraints that play a role in the stabilization of callosal connections during development.

SUMMARY AND CONCLUDING REMARKS

The findings from our laboratory reviewed here offer a novel perspective for analyzing the visual callosal pathway that is consistent with several, previously unexplained, features of the organization and function of this pathway. This novel perspective stems not so much from current notions about the organization and function of the callosal pathway as from the analysis of factors that may guide its development. The stabilization of callosal connections in striate cortex is thought to depend on patterns of interhemispheric correlated activity, but the sources generating these activity patterns have not been identified. It is widely believed that, in the cat, callosal connections are stabilized at the vertical meridian representations of both hemispheres by interhemispheric correlated activity driven by binocular visual experience, but the evidence supporting this hypothesis is inconclusive. Based on findings from our laboratory, we have suggested that interhemispheric correlated activity guiding callosal development is driven instead by a uniocular mechanism (i.e., by synchronous activity relayed bilaterally from the same temporal retina). We recently investigated the role of uniocular and biocular mechanisms on callosal development by using a novel strategy that entails correlating the distribution of callosal connections with the underlying patterns of ODCs. This study is consistent with the idea that the specification of callosal connections in cat striate cortex is primarily under the influence of uniocular cues. This finding is surprising given that much of the development of the callosal pathway occurs after kittens have opened their eyes. This result, as well as that from species in which callosal connections mature before eye-opening, suggests that a common mechanism operates across species in spite of marked differences in the timing of callosal development with respect to the onset of visual experience.

Features of the callosal pathway, such as the interspecies variability in the width of the callosal zone in striate cortex and the patchiness of callosal connections in striate cortex of some species, are difficult to explain from a purely functional point of view. In the context of the data reviewed here, these features may now be regarded as reflecting constraints imposed by developmental mechanisms rather than simply implementing specific functional designs. Also challenging to explain functionally is the finding from our laboratory that callosal cells prefer ipsilateral ODCs in regions outside the TZ and contralateral ODCs in regions within the TZ. This unique organization of callosal connections follows readily from the analysis of the effect of uniocular cues on callosal development in an animal such as the cat, in which afferent from both eyes are segregated into ODCs. The analysis of mechanisms that may guide the development of callosal pathways also offers clues about some of the functions of these pathways. For instance, the finding that the stabilization of callosal connections in the cat is under the influence of uniocular cues leads to the suggestion that callosal connections do not contribute importantly to the generation of binocular cells in striate cortex of the cat. That callosal connections appear to be able to generate binocular cells in striate cortex of other species and in some extrastriate visual areas further illustrates the extent to which developmental constraints may shape the functional organization of mature pathways. Finally, the finding that the stabilization of callosal connections in the cat is under the influence of uniocular cues contrasts with previous studies showing that lateral intrinsic connections in cat striate cortex develop under the influence of binocular cues, which points to fundamental and striking differences between interhemispheric and intrinsic connections. In conclusion, this chapter illustrates that identifying the source of correlated activity guiding the stabilization of neuronal connections may be important, not only for clarifying the role of activity cues on the development of specific neural connections but also for gaining insight into the functional organization and interspecies variability of mature pathways.

ACKNOWLEDGMENTS

Work in the author's laboratory was supported by NIH grant EY09343.

REFERENCES

Abel, P. L., O'Brien, B. J., and Olavarria J. F. (2000). The organization of callosal linkages in visual area V2 of macaque monkey. *J. Comp. Neurol.* **428,** 278–293.

Anderson, P. A., Olavarria J., and Van Sluyters, R. C. (1988). The overall pattern of ocular dominance bands in cat visual cortex. *J. Neurosci.* **8,** 2183–2200.

Antonini, A., Berlucchi, G., and Lepore, F. (1983). Physiological organization of callosal connections of a visual lateral suprasylvian area in the cat. *J. Neurophysiol.* **49,** 902–921.

Antonini, A., Di Stefano, M., Minciacchi, M., and Tassinari, G. (1985). Interhemispheric influences on area 19 of the cat. *Exp. Brain Res.* **59,** 171–184.

Albus, K., and Beckman, R. (1980). Second and third visual areas of the cat: interindividual variability in retinotopic arrangement and cortical location. *J. Physiol. (London)* **299,** 247–276.

Apkarian, P., Bour, L., and Barth, P. G. (1994). A unique achiasmatic anomaly detected in non-albinos with misrouted retinal-fugal projections. *Eur. J. Neurosci.* **6,** 501–507.

Beck, P. D., and Kaas, J. H. (1994). Interhemispheric connections in neonatal owl monkeys *(Aotus trivirgatus)* and galagos *(Galago crassicaudatus). Brain Res.* **651,** 57–75.

Berlucchi, G. (1981). Recent advances in the analysis of the neural substrates of interhemispheric communication. In: Brain mechanisms and perceptual awareness (O. Pompeiano, and C. A. Marsan, Eds.), pp. 133–152. New York, Raven Press.

Berlucchi, G., and Antononi, A. (1990). The role of the corpus callosum in the representation of the visual field in cortical areas. In: Brain circuits and functions of the mind: Essays in honor of R. W. Sperry (C. Trevarthen, Ed.), pp. 129–139. Cambridge, United Kingdom: Cambridge University Press.

Berlucchi, G., and Rizzolatti, G. (1968). Binocularly driven neurons in visual cortex of split chiasm cats. *Science* **159,** 308–310.

Berlucchi, G., Gazzaniga, M. S., and Rizzolatti, G. (1967). Microelectrode analysis of transfer of visual information by the corpus callosum. *Arch. Ital. Biol.* **105,** 583–596.

Berman, N. E., and Grant, S. (1992). Topographic organization, number, and laminar distribution of callosal cells connecting visual cortical areas 17 and 18 of normally pigmented and Siamese cats. *Vis. Neurosci.* **9,** 1–19.

Berman, N. E., and Payne, B. R. (1983). Alterations in connections of the corpus callosum following convergent and divergent strabismus. *Brain Res.* **274,** 201–212.

Blakemore, C., Diao, Y., Pu, M., Wang, Y., and Xiao, Y. (1983). Possible functions of the interhemispheric connections between visual cortical areas in the cat. *J. Physiol. (London)* **337,** 334–349.

Boire, D., Morris, R., Ptito, M., Lepore, F., and Frost, D. O. (1995). Effects of neonatal splitting of the optic chiasm on the development of feline visual callosal connections. *Exp. Brain Res.* **104,** 275–286.

Bosking, W. H., Kretz, R., Pucak, M. L., and Fitzpatrick, D. (2000). Functional specificity of callosal connections in tree shrew striate cortex. *J. Neurosci.* **20,** 2346–2359.

Bourdet, C., Olavarria, J. F., and Van Sluyters, R. C. (1996). The distribution of visual callosal neurons in normal and strabismic cats. *J. Comp. Neurol.* **366,** 259–269.

Boyd, J., and Matsubara, J. (1994). Tangential organization of callosal connectivity in the cat's visual cortex. *J. Comp. Neurol.* **347,** 197–210.

Brugge, J. F., and Imig, T. J. (1978). Some relationships of binaural responses patterns of single neurons to cortical columns and interhemispheric connections of auditory area AI of cat cerebral cortex. In: Evoked electrical activity in the auditory nervous system (R. F. Naunton, Ed.), pp. 478–503, New York, Academic Press.

Bunt, A. H., Minckler, D. S., and Johanson, G. W. (1977). Demonstration of bilateral projection of central retina of the monkey with horseradish peroxidase neuronography. *J. Comp. Neurol.* **171,** 619–630.

Callaway, E. M., and Katz, L. C. (1991). Effects of binocular deprivation on the development of clustered horizontal connections in cat striate cortex. *Proc. Natl. Acad. Sci.* **88,** 745–749.

Chalupa, L. M., and Lia, B. (1991). The nasotemporal division of retinal ganglion cells with crossed and uncrossed projections in the fetal rhesus monkey. *J. Neurosci.* **11,** 191–202.

Chalupa, L. M., Killackey, H., Snider, C. J., and Lia, B. (1989). Callosal projection neurons in area 17 of the fetal rhesus monkey. *Dev. Brain Res.* **46,** 303–308.

Chino, Y. M., Smith III, E. L., Yoshida, K., Cheng, H., and Hamamoto, J. (1994). Binocular interactions in striate cortical neurons of cats reared with discordant visual inputs, *J. Neurosci.* **14,** 5050–5067.

Choudhury, B. P., Whitteridge, D., and Wilson, M. E. (1965). The function of the callosal connections of the visual cortex. *Q. J. Exp. Physiol.* **50,** 214–219.

Clarke, P. G. H., and Whitteridge, D. (1976). The cortical visual areas of the sheep, *J. Physiol.* **256,** 497–508.

Cooper, M. L., and Pettigrew, J. D. (1979). The decussation of the retinothalamic pathway in the cat, with a note on the major meridians of the cat's eye. *J. Comp. Neurol.* **187,** 285–312.

Cowey, A., and Perry, V. H. (1979). The projection of the temporal retina in rats, studied by retrograde transport of horseradish peroxidase. *Exp. Brain Res.* **35,** 457–464.

Crair, M. C., Gillespie, D. C., and Stryker, M. P. (1998). The role of visual experience in the development of columns in cat visual cortex. *Science* **279,** 566–570.

Cusick, C. G., and Lund, R. D. (1981). The distribution of the callosal projection to the occipital cortex in rats and mice. *Brain Res.* **214,** 239–259.

Cusick, C. G., and Lund, R. D. (1982). Modification of visual callosal projections in rats, *J. Comp. Neurol.* **212,** 385–398.

Cusick, C. G., Gould, H. J., III, and Kaas, J. H. (1984). Interhemispheric connections of visual cortex of owl monkeys *(Aotus trivirgatus),* marmosets *(Callithrix jacchus)* and galagos *(Galago crassicaudatus), J. Comp. Neurol.* **230,** 311–336.

Cynader M., Gardner, J., Dobbins, A., Lepore, F., and Guillemot, J. P. (1986). Interhemispheric communication and binocular vision: functional and developmental aspects. In Two hemispheres-one brain: Functions of the corpus callosum (F. Lepore, M. Ptito, and H. H. Jasper, Eds.), pp. 189–209. New York, Liss.

Dehay, C., Kennedy, H., Bullier, J., and Berland, M. (1988). Absence of interhemispheric connections of area 17 during development in the monkey. *Nature* **331,** 348–350.

Diao, Y.-C., Wang, Y.-K., and Pu, M.-L. (1983). Binocular responses of cortical cells and the callosal projection in the albino rat. *Exp. Brain Res.* **49,** 410–418.

Diao, Y.-C., Jia, W.-G., Swindale, N. V., and Cynader, M. S. (1990). Functional organization of the cortical 17/18 border region in the cat. *Exp. Brain Res.* **79,** 271–282.

Dreher, B., and Cottee, L. J. (1975). Visual receptive-field properties of cells in area 18 of cat's cerebral cortex before and after acute lesions in area 17. *J. Neurophysiol.* **38,** 735–750.

Ebner, F. F., and Myers, R. E. (1965). Distribution of corpus callosum and anterior commissure in cat and raccoon. *J. Comp. Neurol.* **114,** 353–366.

Elberger, A. J. (1981). Ocular dominance in striate cortex is altered by neonatal section of the posterior corpus callosum in the cat. *Exp. Brain Res.* **41,** 280–291.

Elberger, A. J., and Smith, E. L., III (1985). The critical period for corpus callosum section to affect cortical binocularity. *Exp. Brain Res.* **57,** 213–223.

Elberger, A. J., Smith, E. L., and White, J. M. (1983). Spatial dissociation of visual inputs alters the origin of the corpus callosum. *Neurosci. Lett.* **35,** 19–24.

Fisken, R. A., Garey, L. J., and Powel, T. P. S. (1975). The intrinsic, association and commissural connections of area 17 of the visual cortex. *Phil. Trans. R. Soc. London B.* **272,** 487–536.

Frost, D. O., and Moy, Y. P. (1989). Effects of dark rearing on the development of visual callosal connections. *Exp. Brain Res.* **78,** 203–213.

Frost, D. O., Moy, Y. P., and Smith, D. C. (1990). Effects of alternating monocular occlusion on the development of visual callosal connections. *Exp. Brain Res.* **83,** 200–209.

Fukuda, Y., and Stone, J. (1974). Retinal distribution and central projections of Y, X, and W cells in the cat's retina. *J. Neurophysiol.* **37,** 749–772.

Fukuda, Y., Sawai, H., Watanabe, M., Wakakuwa, K., and Morigiwa, K. (1989). Nasotemporal overlap of crossed and uncrossed retinal ganglion cell projections in the Japanese monkey *(Macaca fuscata). J. Neurosci.* **9,** 2353–2373.

Galli, L., and Maffei, L. (1988). Spontaneous impulse activity of rat retinal ganglion cells in prenatal life. *Science* **242,** 90–91.

Garey, L. J., Jones, E. G., and Powell, T. P. S. (1968). Interrelationship of striate and extrastriate cortex with the primary relay sites of the visual pathway. *J. Neurosurg. Psychiatry* **31,** 135–157.

Gaze, R. M., Keating, M. J., Szekely, G., and Beazley, L. (1970). Binocular interactions in the formation of specific intertectal neuronal connexions. *Proc. R. Soc. London B* **175,** 107–147.

Gould, H. J., III, Weber, J. T., and Rieck, R. W. (1987). Interhemispheric connections in the visual cortex of the squirrel monkey *(Saimiri Sciureus). J. Comp. Neurol.* **256,** 14–28.

Grigonis, A. M., Rayos del Sol-Padua, R., and Murphy, E. H. (1992). Visual callosal projections in the adult ferret. *Vis. Neurosci.* **9,** 99–103.

Guillery, R. W., and Kaas, J. (1971). A study of normal and congenitally abnormal retinogeniculate projections in cats. *J. Comp. Neurol.* **143,** 73–100.

Harvey, A. R. (1980). A physiological analysis of subcortical and commissural projections of area 17 and 18 of the cat. *J. Physiol.* **302,** 507–534.

Hogan, D., and Williams, R. W. (1995). Analysis of the retinas and optic nerves of achiasmatic belgian sheepdogs. *J. Comp. Neurol.* **352,** 367–380.

Houzel, J. C., Milleret, C., and Innocenti, G. (1994). Morphology of callosal axons interconnecting areas 17 and 18 of the cat. *Eur. J. Neurosci.* **6,** 898–917.

Hubel, D. H., and Wiesel, T. N. (1962). Receptive fields, binocular interaction, and functional architecture in the cat's visual cortex. *J. Physiol. (London)* **160,** 106–154.

Hubel, D. H., and Wiesel, T. N. (1965a). Receptive fields and functional architecture in two non-striate visual areas (18 and 19) of the cat. *J. Neurophysiol.* **28,** 229–289.

Hubel, D. H., and Wiesel, T. N. (1965b). Binocular interactions in striate cortex of kittens reared with artificial squint. *J. Neurophysiol.* **28,** 1041–1051.

Hubel, D. H., and Wiesel, T. N. (1967). Cortical and callosal projections concerned with the vertical meridian of visual fields in the cat. *J. Neurophysiol.* **30,** 1561–1573.

Hubel, D. H., and Wiesel, T. N. (1971). Aberrant visual projections in the Siamese cats. *J. Physiol. (London)* **218,** 33–62.

Imig, T. J., and Brugge, J. F. (1978). Sources and terminations of callosal axons related to binaural and frequency maps in primary auditory cortex of the cat. *J. Comp. Neurol.* **182,** 637–660.

Innocenti, G. M. (1980). The primary visual pathway through the corpus callosum: morphological and functional aspects in the cat. *Arch. Ital. Biol.* **118,** 124–188.

Innocenti, G. M. (1986). General organization of callosal connections in the cerebral cortex. In: Cerebral Cortex (E. G. Jones, and A. Peters, Eds.), Vol. 5, pp. 291–353. New York, Plenum Press.

Innocenti, G. M. (1991). The development of projections from cerebral cortex. *Prog. Sensory Physiol.* **12,** 65–114.

Innocenti, G. M., and Caminiti, R. (1980). Postnatal shaping of callosal connections from sensory areas. *Exp. Brain Res.* **38,** 381–394.
Innocenti, G. M., and Clarke, S. (1984). The organization of immature callosal connections. *J. Comp. Neurol.* **230,** 387–390.
Innocenti, G. M., and Frost, D. O. (1979). Effect of visual experience on the maturation of the efferent system to the corpus callosum. *Nature* **280,** 231–233.
Innocenti, G. M., and Frost, D. O. (1980). The postnatal development of visual callosal connections in the absence of visual experience or of the eyes. *Exp. Brain Res.* **39,** 365–375.
Innocenti, G. M., Fiore, L., and Caminiti, R. (1977). Exuberant projection into the corpus callosum from the visual cortex of newborn cats. *Neurosci. Lett.* **4,** 237–242.
Innocenti, G. M., Frost, D. O., and Illes, J. (1985). Maturation of visual callosal connections in visually deprived kittens: A challenging critical period. *J. Neurosci.* **5,** 255–267.
Jouandet, M. L., Lachat, J. J., and Garey, L. J. (1985). Distribution of the neurons of origin of the great cerebral commissures in the cat. *Anat. Embryol.* **171,** 105–120.
Keating, M. J., and Feldman, J. (1975). Visual deprivation and intertectal neuronal connection in *Xenopus laevis. Proc. R. Soc. London B* **191,** 467–474.
Kennedy, H., Dehay, C., and Bullier, J. (1986). Organization of the callosal connections of visual areas V1 and V2 in the macaque monkey. *J. Comp. Neurol.* **247,** 398–415.
Kennedy, H., Meissirel, C., and Dehay, C. (1991). Callosal pathways and their compliance to general rules governing the organization of cortical connectivity. In: Neuroanatomy of the visual pathways and their development (B. Dreher, and S. Robinson, Eds.), pp. 324–359. New York, Macmillan.
Kirk, D. L., Levick, W. R., and Cleland, B. G. (1976a). The crossed or uncrossed destination of axons of sluggish-concentric and non-concentric cat retinal ganglion cells, with an overall synthesis of the visual field representation. *Vis. Res.* **16,** 233–236.
Kirk, D. L., Levick, W. R., Cleland, B. G., and Wassle, H. (1976b). Crossed and uncrossed representation of the visual field by brisk-sustained and brisk-transient cat retinal ganglion cells. *Vis. Res.* **16,** 225–231.
Law, M. I., Zahs, K. R., and Stryker, M. P. (1988). Organization of primary visual cortex (area 17) in the ferret. *J. Comp. Neurol.* **278,** 157–180.
Lepore, F., and Guillemot, J.-P. (1982). Visual receptive field properties of cells innervated through the corpus callosum in the cat. *Exp. Brain Res.* **46,** 413–424.
Lepore, F., Samson, A., and Molotchnikoff, S. (1983). Effects on binocular activation of cells in visual cortex of the cat following the transection of the optic tract. *Exp. Brain Res.* **50,** 392–396.
Lepore, F., Samson, A., Paradis, M.-C., Ptito, M., and Guillemot, J.-P. (1992). Binocular interaction and disparity coding at the 17–18 border: contribution of the corpus callosum. *Exp. Brain Res.* **90,** 129–140.
LeVay, S., Stryker, M. P., and Shatz, C. J. (1978). Ocular dominance columns and their development in layer IV of the cat's visual cortex: a quantitative study. *J. Comp. Neurol.* **179,** 223–244.
Leventhal, A. G. (1982). Morphology and distribution of retinal ganglion cells projecting to different layers of the lateral geniculate nucleus in normal and Siamese cats. *J. Neurosci.* **2,** 1024–1042.
Leventhal, A. G., and Creel, D. J. (1985). Retinal projections and functional archietecture of cortical areas 17 and 18 in the tyrosinase-negative albino cat. *J. Neurosci.* **5,** 795–807.
Leventhal, A. G., Ault, S. J., and Vitek, D. J. (1988). The nasotemporal division in primate retina: the neural basis of macular sparing and splitting. *Science* **240,** 66–67.
Lewis, J. W., and Olavarria, J. F. (1995). Two rules for callosal connectivity in striate cortex of the rat. *J. Comp. Neurol.* **361,** 119–137.
Löwel, S. (1994). Ocular dominance column development: strabismus changes the spacing of adjacent columns in cat visual cortex. *J. Neurosci.* **14,** 7451–7468.
Löwel, S., and Singer, W. (1992). Selection of intrinsic horizontal connections in the visual cortex by correlated activity. *Science* **255,** 209–212.
Luhmann, H. J., Singer, W., and Martinez-Millan, L. (1989). Horizontal interactions in cat striate cortex: anatomical substrate and postnatal development. *Eur. J. Neurosci.* **2,** 344–357.

Lund, R. D., and Mitchell, D. E. (1979a). The effect of dark-rearing on visual callosal connections of cats. *Brain Res.* **167,** 172–175.

Lund, R. D., and Mitchell, R. E. (1979b). Asymmetry in the visual callosal connections in strabismic cats. *Brain Res.* **167,** 176–179.

Lund, R. D., Lund, J. S., and Wise, R. P. (1974). The organization of the retinal projection to the dorsal lateral geniculate nucleus in pigmented and albino rats. *J. Comp. Neurol.* **158,** 383–404.

Lund, R. D., Mitchell, R. E., and Henry, G. H. (1978). Squint-induced modification of callosal connections in cats. *Brain Res.* **144,** 169–172.

Maffei, L., and Galli-Resta, L. (1990). Correlation in the discharges of neighboring rat retinal ganglion cells during prenatal life. *Proc. Natl. Acad. Sci.* **87,** 2861–2864.

Marzi, C. A., Antonini, A., Di Estefano, M., and Legg, C. R. (1980). Callosum-dependent binocular interactions in the lateral suprasylvian area of Siamese cats wich lack binocular neurons in areas 17 and 18. *Brain Res.* **197,** 230–235.

McCourt M. E., Thalluri J., and Henry G. H. (1990). Properties of area 17/18 border neurons contributing to the visual transcallosal pathway in the cat. *Vis. Neurosci.* **5,** 83–98.

Meister, M., Wong, R. O. L., Baylor, D. A., and Shatz, C. J. (1991). Synchronous bursts of action potentials in ganglion cells of the developing mammalian retina. *Science* **252,** 939–943.

Minciacchi O., and Antonini, A. (1984). Binocularity in the visual cortex of the adult cat does not depend on the integrity of the corpus callosum. *Behav. Brain Res.* **13,** 183–192.

Movshon, J. A., and Van Sluyters, R. C. (1981). Visual neural development. *Annu. Rev. Psychol.* **32,** 477–522.

Olavarria, J. F. (1995). The effect of visual deprivation on the number of callosal cells in the cat is less pronounced in extrastriate cortex than in the 17/18 border region. *Neurosci. Lett.* **195,** 147–150.

Olavarria, J. F. (1996a). Non-mirror-symmetric patterns of callosal linkages in areas 17 and 18 in cat visual cortex. *J. Comp. Neurol.* **366,** 643–655.

Olavarria, J. F. (1996b). Visual callosal fibers link cortical columns dominated by the same eye in the cat. *Soc. Neurosci. Abstr.* **22,** 491.

Olavarria, J. F. (2001). Callosal connections correlate preferentially with ipsilateral cortical domains in cat areas 17 and 18, and with contralateral domains in the 17/18 transition zone. *J. Comp. Neurol.* **433,** 441–457.

Olavarria, J. F., and Abel, P. L. (1996). The distribution of callosal connections correlates with the pattern of cytochrome oxidase stripes in visual area V2 of macaque monkeys. *Cerebral Cortex* **6,** 631–639.

Olavarria, J. F., and Li, C.-P. (1995). Effects of neonatal enucleation on the organization of callosal linkages in striate cortex of the rat. *J. Comp. Neurol.* **361,** 138–151.

Olavarria, J., and Montero, V. M. (1984). Relation of callosal and striate-extrastriate cortical connections in the rat: Morphological definition of extrastriate visual areas. *Exp. Brain Res.* **54,** 240–252.

Olavarria, J., and Van Sluyters, R. C. (1983). Widespread callosal connections in infragranular visual cortex of the rat. *Brain Res.* **279,** 233–237.

Olavarria, J., and Van Sluyters, R. C. (1984). Callosal connections of the posterior cortex in normal-eyed, congenitally anophthalmic and neonatally enucleated mice. *J. Comp. Neurol.* **230,** 249–268.

Olavarria, J., and Van Sluyters, R. C. (1985a). Organization and postnatal development of callosal connections in the visual cortex of the rat. *J. Comp. Neurol.* **239,** 1–26.

Olavarria, J., and Van Sluyters, R. C. (1985b). Unfolding and flattening the cortex of gyrencephalic brains. *J. Neurosci. Meth.* **15,** 191–202.

Olavarria, J., and Van Sluyters, R. C. (1995). Overall pattern of callosal connections in visual cortex of normal and enucleated cats. *J. Comp. Neurol.* **363,** 161–176.

Olavarria, J., Malach, R., and Van Sluyters, R. C. (1987). Development of visual callosal connections in neonatally enucleated rats. *J. Comp. Neurol.* **260,** 321–348.

Olavarria, J., Bravo, H., and Ruiz, G. (1988). The pattern of callosal connections in posterior neocortex of congenitally anophthalmic rats. *Anat. Embryol.* **178,** 155–159.

Olson, C. R., and Freeman., R. D. (1978). Eye alignment in kittens. *J. Neurophysiol.* **41,** 848–859.

Otsuka, R., and Hassler, R. (1962). Uber Aufbau and Gliederung der corticalen Sehesphare bei der Katze. *Arch. Psychiatr. Nervenkr.* **203,** 213–234.

Payne, B. R. (1990). Representation of the ipsilateral visual field in the transition zone between areas 17 and 18 of the cat's cerebral cortex. *Vis. Neurosci.* **4,** 445–474.
Payne, B. R. (1991). Visual field map in the transcallosal sending zone of area 17 in the cat. *Vis. Neurosci.* **7,** 201–219.
Payne, B. R. (1994). Neuronal interactions in cat visual cortex mediated by the corpus callosum. *Behav. Brain Res.* **64,** 55–64.
Payne, B. R., and Siwek, D. F. (1991). Visual-field map in the callosal recipient zone at the border between areas 17 and 18 in the cat. *Vis. Neurosci.* **7,** 221–236.
Payne, B. R., Elberger, A. J., Berman, N., and Murphy, E. H. (1980). Binocularity in the cat visual cortex is reduced by sectioning the corpus callosum. *Science* **207,** 1097–1099.
Payne, B. R., Pearson, H. E., and Berman, N. (1984). Role of corpus callosum in functional organization of cat striate cortex. *J. Neurophysiol.* **52,** 570–594.
Payne, B., Pearson, H., and Cornwell, P. (1988). Development of visual and auditory cortical connections in the cat. In: Cerebral cortex (A. Peters, and E. G. Jones, Eds.), Vol. 7, pp. 309–389. New York, Plenum Press.
Pettigrew, J. D., and Dreher, B. (1987). Parallel processing of binocular disparity in the cat's retinogeniculate pathways. *Proc. R. Soc. B* **232,** 297–321.
Pettigrew, J. D., Ramachandran, V. S., and Bravo, H. (1984). Some neural connections subserving binocular vision in ungulates. *Brain Behav. Evol.* **24,** 65–93.
Pritzel, M., Kretz, R., and Rager, G. (1988). Callosal projections between areas 17 in the adult tree shrew *(Tupaia belangeri). Exp. Brain Res.* **72,** 481–493.
Provis, J. M., and Watson, C. R. R. (1981). The distribution of ipsilaterally and contralaterally projecting ganglion cells in the retina of the pigmented rabbit. *Exp. Brain Res.* **44,** 82–92.
Redies, C., Diksic, M., and Rimi, H. (1990). Functional organization in the ferret visual cortex: a double-labe 2-deoxyglucose study. *J. Neurosci.* **10,** 2791–2803.
Reese, B. E., and Cowey, A. (1987). The crossed projection from the temporal retina to the dorsal lateral geniculate nucleus in the rat. *Neuroscience* **20,** 951–959.
Rhoades, R. W., and Fish, S. E. (1983). Bilateral enucleation alters visual callosal but not corticotectal or corticogeniculate projections in hamster. *Exp. Brain Res.* **51,** 451–462.
Rockland, K. S. (1985). Anatomical organization of primary visual cortex (area 17) in the ferret. *J. Comp. Neurol.* **241,** 225–236.
Rowe, M. H., and Dreher, B. (1982). Retinal W-cell projection to the medial interlaminar nucleus in the rat: implications for ganglion cell classification. *J. Comp. Neurol.* **204,** 117–133.
Ruthazer, E. S., and Stryker, M. P. (1996). The role of activity in the development of long-range horizontal connections in area 17 of the ferret. *J. Neurosci.* **16,** 7253–7269.
Ruthazer, E. S., Baker, G. E., and Stryker, M. P. (1999). Development and organization of ocular dominance bands in primary visual cortex of the sable ferret. *J. Comp. Neurol.* **407,** 151–165.
Sanderson, K. J., and Sherman, S. M. (1971). Nasotemporal overlap in visual field projected to lateral geniculate nucleus in the cat. *J. Neurophysiol.* **34,** 453–466.
Sanides, D. (1978). The retinotopic distribution of visual callosal projections in the suprasylvian visual areas compared to the classical visual areas (17, 18, 19) in the cat. *Exp. Brain Res.* **33,** 435–443.
Schmidt, K. E., Kim, D.-S., Singer, W., Bonhoeffer, T., and Lowel, S. (1997). Functional specificity of long-range intrinsic and interhemispheric connections in the visual cortex of strabismic cats. *J. Neurosci.* **17,** 5480–5492.
Segraves, M. A., and Rosenquist, A. C. (1982a). The distribution of the cells of origin of callosal projections in cat visual cortex. *J. Neurosci.* **2,** 1079–1089.
Segraves, M. A., and Rosenquist, A. C. (1982b). The afferent and efferent callosal connections of retinotopically defined areas in cat cortex. *J. Neurosci.* **2,** 1090–1107.
Sengpiel, F., Blakemore, C., Kind, P. C., and Harrad, R. (1994). Interocular suppression in the visual cortex of strabismic cats. *J. Neurosci.* **14,** 6855–6871.
Sesma, M. A., and Casagrande, V. A. (1986). The nasotemporal division of the retina in a nocturnal primate. *Invest. Ophthalmol. Vis. Sci. (Suppl.)* **27,** 222.

Shatz, C. J. (1977a). Abnormal interhemispheric connections in the visual system of Boston Siamese cats: a physiological study. *J. Comp. Neurol.* **171,** 229–246.

Shatz, C. J. (1977b). Anatomy of interhemispheric connections in the visual system of Boston Siamese and ordinary cats. *J. Comp. Neurol.* **173,** 497–518.

Shatz, C. J., Lindstrom, S., and Wiesel, T. N. (1977). The distribution of afferents representing the right and left eyes in the cat's visual cortex. *Brain Res.* **131,** 103–116.

Sherman, S. M. (1972). Development of interocular alignment in cats. *Brain Res.* **37,** 187–203.

Stone, J. (1966). The naso-temporal division of the cat's retina. *J. Comp. Neurol.* **126,** 585–600.

Stone, J., Leicester, J., and Sherman, S. M. (1973). The nasotemporal division of the monkey's retina. *J. Comp. Neurol.* **150,** 333–348.

Stone, J., Campion, J. E., and Liecester, J. (1978). The nasotemporal division of the retina in the Siamese cat. *J. Comp. Neurol.* **180,** 783–798.

Swadlow, H. A., Weyand, T. G., and Waxman, S. G. (1978). The cells of origin of the corpus callosum in rabbit visual cortex. *Brain Res.* **156,** 129–134.

Tieman, S. B., and Tumosa, N. (1996). Alternating monocular exposure increase the spacing of ocularity domains in area 17 of cats. *Invest. Ophthalmol. Vis. Sci. (Suppl.)* **37,** 425.

Thomas, H. C., and Espinoza, S. G. (1987). Relationship between interhemispheric cortical and visual areas in hooded rat. *Brain Res.* **417,** 214–224.

Thurlow, G. A., and Cooper, R. M. (1988). Metabolic activity in striate and extrastriate cortex in the hooded rat: contralateral and ipsilateral eye input. *J. Comp. Neurol.* **274,** 595–607.

Tiao, Y-C., and Blakemore, C. (1976). Functional organization in the visual cortex of the golden hamster. *J. Comp. Neurol.* **168,** 459–482.

Toyama, K., Matsunami, K., Ohno, T., and Tokashiki, S. (1974). An intracellular study of neuronal organization in the visual cortex. *Exp. Brain Res.* **21,** 45–66.

Tremblay, F., Ptito, M., Lepore, F., Miceli, D., and Guillemot, J.-P. (1987). Distribution of visual callosal projection neurons in the Siamese cat. *J. Hirnforsch.* **28,** 491–503.

Tusa, R. J., Palmer, L. A., and Rosenquist, A. C. (1978). The retinotopic organization of area 17 (striate cortex) in the cat. *J. Comp. Neurol.* **177,** 213–236.

Tusa, R. J., Rosenquist, A. C., and Palmer, L. A. (1979). Retinotopic organization of areas 18 and 19 in the cat. *J. Comp. Neurol.* **185,** 657–678.

Van Sluyters, R. C., and Levitt, F. B. (1980). Experimental strabismus in the kitten, *J. Neurophysiol.* **43,** 686–699.

Van Essen, D. C., Newsome, W. T., and Bixby, J. L. (1982). The pattern of interhemispheric connections and its relationship to extrastriate visual areas in the macaque monkey. *J. Neurosci.* **2,** 265–283.

Voigt, T., LeVay, S., and Stammes, M. A. (1988). Morphological and immunocytochemical observations on the visual callosal projections in the cat. *J. Comp. Neurol.* **272,** 450–460.

Weyand, T. G., and Swadlow, H. A. (1980). Interhemispheric striate projection in the prosimian primate. *Galago segalensis, Brain Behav. Evol.* **17,** 473–477.

White, L. E., Bosking, W. H., Williams, S. M., and Fitzpatrick, D. (1999). Maps of central visual space in ferret V1 and V2 lack matching inputs from the two eyes. *J. Neurosci.* **19,** 7089–7099.

Whitteridge, D., and Clarke, P. B. H. (1982). Ipsilateral visual field represented in the cat's visual cortex. *Neuroscience* **7,** 1855–1860.

Wilson, M. E. (1968). Cortico-cortical connections of the cat visual areas. *J. Anat.* **102,** 375–386.

Zeki, S. M., and Sandeman, D. R. (1976). Combined anatomical and electrophysiological studies on the boundary between the second and third visual areas of rhesus monkey cortex. *Proc. R. Soc. London (Biol)* **194,** 555–662.

Zeki, S., and Fries, W. (1980). A function of the corpus callosum in the Siamese cat. *Proc. R. Soc. Lond. B* **207,** 249–258.

7

ESSENTIAL AND SUSTAINING LGN INPUTS TO CAT PRIMARY VISUAL CORTEX

THEODORE G. WEYAND

Department of Cell Biology and Anatomy LSU Health Sciences Center, New Orleans, Louisianna

INTRODUCTION

A number of observations support the tenet that vision is both a parallel and serial process (1,2). Among the key observations is that different types of ganglion cells (based on both morphological and physiological criteria) have spatially overlapping receptive fields, different patterns of axonal projections, and exhibit different characteristic axonal conduction velocities. Every layer of the cat's dorsal lateral geniculate nucleus (LGN) receives a unique set of inputs from the retina, which are then relayed to visual cortex (see Chapter 1). One approach to understanding how parsing retinal afferents into these layers contributes to visual processing is to reversibly inactivate individual LGN layers while recording from neurons in visual cortex. Cortical neuron responses that survive this manipulation could be considered as not requiring the targeted LGN layer for visually driven activity, whereas disruption could be taken as evidence that the layer is necessary. In addition, it is conceivable that some of the trigger features of cortical neurons, such as selectivity for stimulus orientation, direction of motion, or length, might be a product of integrating inputs across LGN layers. The initial results obtained by Malpeli (3) could not have been anticipated. Inactivating layer A, the major contralateral LGN layer, effectively silenced activity in cortical layers 4 and 6. However, activity in the supragranular layers (2 + 3) was largely *un*affected; and selectivity for stimulus orientation, direction of motion, and stimulus length appeared intact among those cells that remained active.

This chapter describes results obtained in these and similar experiments that involved either reversibly inactivating different layers in the LGN complex or inactivating portions of visual cortex. Then, what these results tell us about the cortical circuitry that supports visual perception is summarized.

THE REVERSIBLE INACTIVATION TECHNIQUE

These experiments were done on paralyzed, anesthetized cats. The general procedures have been described (3,4). Because some of the results are contrary to common views of the functional organization of cortex, the experimental procedures and arguments for the adequacy and selectivity of the inactivations are reviewed.

PROTOCOL AND AGENTS

Cellular activity in visual cortex was recorded using either lacquer- or glass-coated tungsten microelectrodes and conventional methods. Visual receptive fields were initially mapped and characterized using hand-held stimuli, but quantitative measures were always obtained from visual stimuli (light bars) under computer control. Once a cortical site (single-cell or multiunit cluster) was selected, the retinotopically corresponding site in the LGN complex was identified electrophysiologically using specially manufactured metal-plated pipettes that facilitate the recording of cellular activity at the pipette tip (5). Geniculate inactivations were usually achieved by pressure injection of 115 nanoliters (nl) of 4 mM cobaltous chloride or 2% lidocaine chloride delivered through the pipettes. Cobalt blocks neurotransmission by interfering with calcium entry at the presynaptic terminal (6), whereas lidocaine is a potent blocker of sodium channels. Cobalt has the advantage that it does not disrupt fibers-of-passage, but the disadvantage that it fails to be reversible after four to six injections and its effects last only approximately 1.5 minutes. Lidocaine has the advantages that it is reversible over many (> 10) applications and its effects last 10 to 20 minutes, but the disadvantage of blocking fibers-of-passage. Because retinal fibers that terminate in the A layers course through the C layers and because axons of C layer cells project through layer A (see Fig. 1-2) inactivation of both layer A and the C layers were accomplished using cobalt.

Effects on visually-driven activity were assessed by comparing response magnitude during the blockade, with the average response prior to and following recovery from inactivation. The quantitative measure was called the blocked/normal (B/N) ratio and was simply the response during blockade divided by averaged response before and after recovery from inactivation (Further details can be found in Malpeli [3]). Thus, inactivations that were completely ineffective would have a B/N ratio of 1.0, whereas silencing cortical activity would yield a B/N ratio of 0.0. With the exception of the corticotectal cells that were assumed to be in layer 5 (7–10), accurate localization of cortical recording sites was achieved by making microlesions no further than 150 μm from any tested site.

All of the results were obtained by blocking the contralateral eye layers only. This was largely a pragmatic decision. The architecture of the cat's LGN complex makes it easier to selectively inactivate and evaluate the effects on the contralateral eye over the ipsilateral eye (c.f., Malpeli [3]). In fact, the major ipsilateral eye layer, A1, served as a buffer between layer A and the magnocellular portion of the C layers. Because of the parallels in circuitry, the effects observed from blockade of layer A would likely also apply to layer A1.

ADEQUACY AND SELECTIVITY OF THE INACTIVATIONS

The results testify that the injections were adequate. One potential objection to the adequacy of the injections would have occurred if the effects were more pronounced among neurons with small receptive fields than with large receptive fields. Fortunately, this was not the case. Activity among layer 6 cells, which contain many of the largest receptive fields in area 17, was consistently abolished, whereas supragranular layer neurons, many with receptive fields $< 1°$ across, were usually unaffected. The selectivity of injections is supported by two observations. First, injections were effective only when the cortical receptive field was in retinotopic alignment (usually within 1°) with the geniculate activity. Second, cobalt injections into layer A that abolished visually driven activity in cortex through the contralateral eye were ineffective in blocking activity through the ipsilateral eye (e.g., see Figs. 1 and 2 in Malpeli [3]). Because projection columns pass uninterrupted across layers A and A1 in the LGN (11), this result would indicate that both the cells and axons of A1 were unaffected. The adequacy and selectivity of blockade of the C layers are supported by similar logic. In addition, specific experiments in which we placed a second recording electrode close to the injection pipette in layer C testifies to the adequacy of using cobalt to block activity in the C layers (e.g., Malpeli et al. [4], Fig. 4). Because of the proximity of the parvocellular C layers (i.e., C_1, C_2, C_3; [12, 13]), it is assumed that injecting 115 nl of $CoCl_2$ also silenced activity in these layers as well. Blocking fibers-of-passage to other geniculate layers was not a concern with the medial interlaminar nucleus (MIN), and consequently lidocaine was used to inactivate all three layers of that structure. Inactivations of the MIN are ineffective in blocking activity in area 17, but can be highly effective in the lateral suprasylvian visual area (14), indicating that these injections were also selective and effective.

TWO CIRCUITS IN AREA 17

Inactivation of individual LGN layers, superficial cortical layers, and/or area 18 reveals two circuits in area 17, each of which can operate independently. One is heavily dependent on an intact LGN layer A; the other is not. These two circuits appear to drive two different types of projection cells in layer 5, which direct their output at the superior colliculus, a structure intimately associated with eye and head movements. Figure 7-1 shows a schematic of the proposed circuits, which are the logical consequences of the following observations:

CORTICAL LAYER PREDICTS DEPENDENCE ON LGN LAYER A

Figure 7-2 shows the visually driven responses of a neuron in layer 4 (left) and layer 6 (right) before (top), during (middle), and after recovery from (bottom) reversible inactivation of layer A using 115 nl of 4 mM $CoCl_2$. For both neurons, visually driven activity during blockade is essentially abolished (B/N

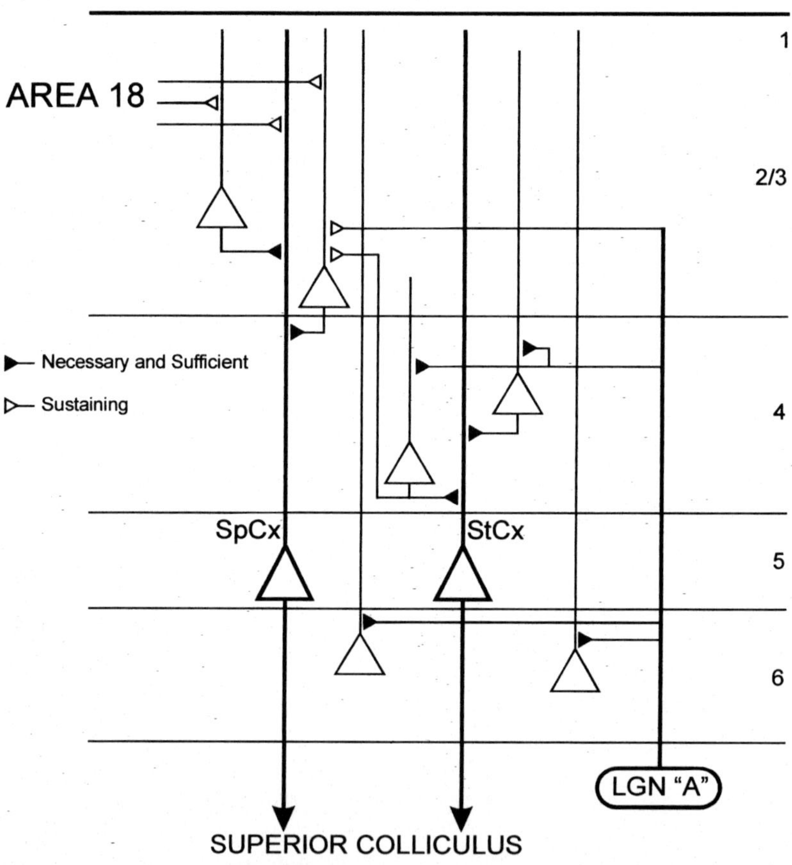

FIGURE 7-1. Proposed intracortical circuits in area 17 that result in driving either standard complex (StCx) or the special complex (SpCx) cells in layer 5, which project to the superior colliculus.

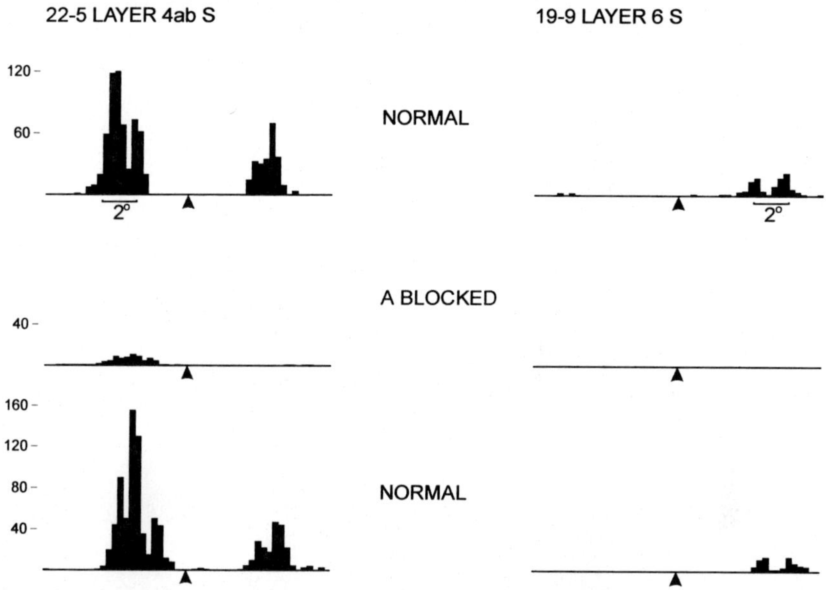

FIGURE 7-2. Reversible inactivation of layer A silenced visually driven activity in layers 4 and 6 of area 17. (Left) Peristimulus histogram (summed response of 30 stimulus sweeps) of a simple cell in layer 4ab before (top), during (middle), and after recovery from (bottom) inactivation of layer A using 115 nl of $CoCl_2$. Activity is substantially reduced, yielding a B/N ratio of 0.10. The arrow marks the point at which the stimulus reversed direction. Notice that the relative directionality of the response did not shift. For all subsequent response histograms presented, the arrow indicates the point at which the stimulus reversed direction, and the histograms are the summed responses of 30 trials. (Right) Peristimulus histogram of a simple cell in layer 6 before (top), during (middle), and after recovery from (bottom) reversible inactivation of layer A. Activity during blockade is totally abolished (B/N ratio = 0). (From Malpeli [3]. Used by permission of the American Physiological Society.)

ratio = 0.10 for the layer 4 neuron, B/N ratio = 0.00 for layer 6). In contrast to these profound effects, Figure 7-3 shows the visually driven response of a layer 3 neuron before (top), during (middle), and after recovery from (bottom) reversible inactivation of LGN layer A using 125 nl of 4 mM $CoCl_2$. Visually driven activity is not compromised (B/N ratio = 1.12). The ratio of stimulus size to receptive field size was similar in both figures. In fact, this ratio was smaller in the layer 6 cell than in the layer 3 cell. Figure 7-4 shows the results of a single penetration through area 17 in which visual activity was assessed at 17 sites in layers 3 and 4 before, during, and after recovery from inactivation of layer A of the LGN using $CoCl_2$. Visually driven activity in cortical layer 4 is virtually silenced by injecting 4 mM cobalt into layer A, whereas visually driven activity in layer 3 is largely unaffected. Because nearly all simple cells are found in layers 4 and 6 (15,16), simple cells in area 17 are dependent on LGN layer A.

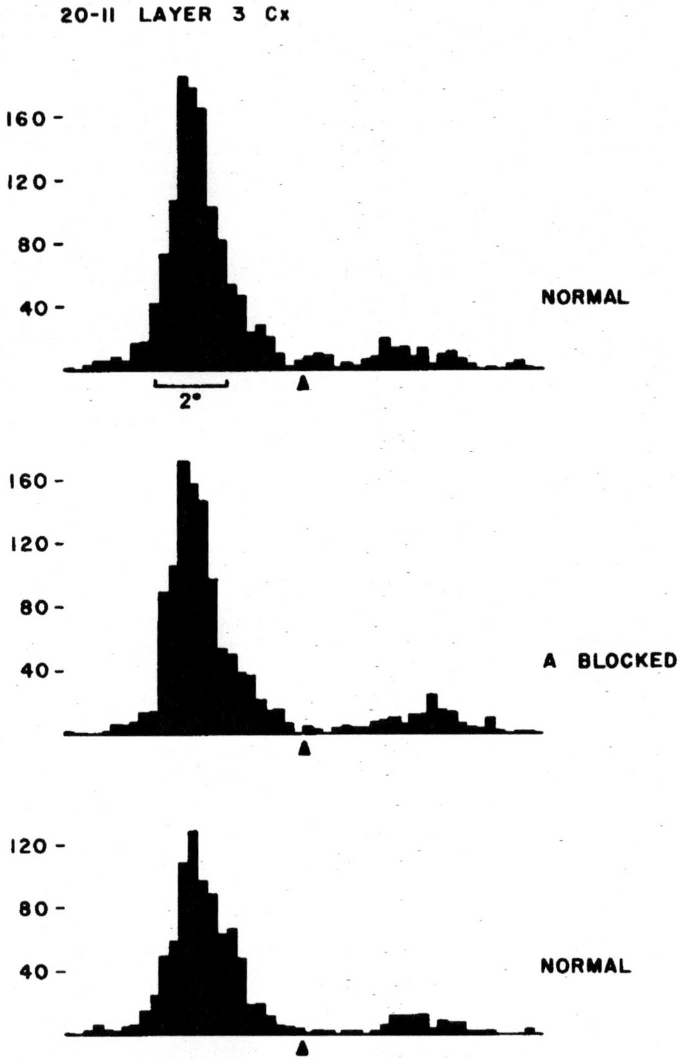

FIGURE 7-3. Reversible inactivation of layer A had little effect on visually driven activity among neurons in layers 2 and 3 in area 17. Shown is a peristimulus histogram of a layer 3 complex cell before (top), during (middle), and following recovery from (bottom) inactivation of layer A. The observed B/N ratio was 1.12. (From Malpeli [3]. Used by permission of the American Physiological Society.)

However, complex cells, when encountered in layers 4 or 6, were just as dependent on layer A as the simple cells found in these layers. With the exception of layer 5 (described later), cortical layer, rather than receptive field type, was the major predictor of dependence on LGN layer A. Figure 7-5A summarizes the results of inactivation of layer A for activity in area 17.

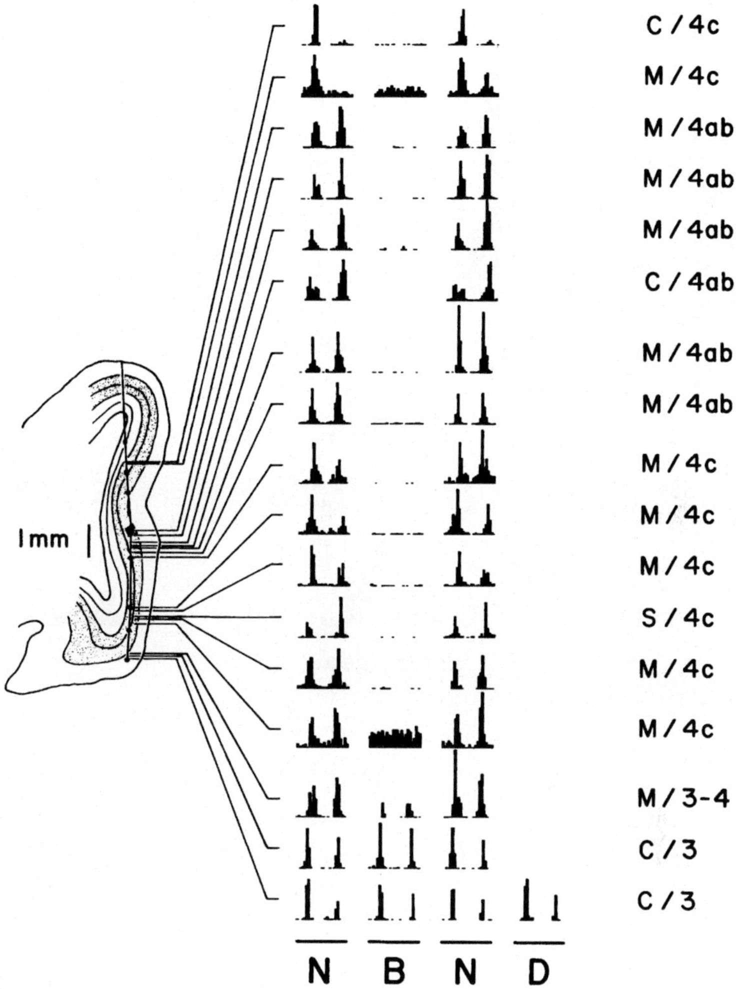

FIGURE 7-4. Reconstruction of a single electrode penetration through area 17 showing the effectiveness of inactivating LGN layer A varied with cortical layer. On the left is a drawing showing the trajectory of the electrode through the cortical layers, and indicating the recording sites. The middle of the figure shows single orientation peristimulus histograms of the activity before (left), during (middle), and after recovery from (right) reversible inactivation at each site. The "D" at the bottom indicates that at the last site tested, double the normal amount of $CoCl_2$ (230 nl) was injected. On the right are labels for the particular type of activity evaluated and cortical layer. C, complex cell; M, multiunit site; S, simple cell. (From Malpeli [3]. Used by permission of the American Physiological Society.)

ACTIVITY REMAINING DURING BLOCKADE RETAINS SPECIFICITY

The activity remaining during reversible blockade not only continues to be visually driven but also retains its selectivity for stimulus trigger features. Figure 7-6 shows the activity of 2 simultaneously recorded layer 3 complex cells during

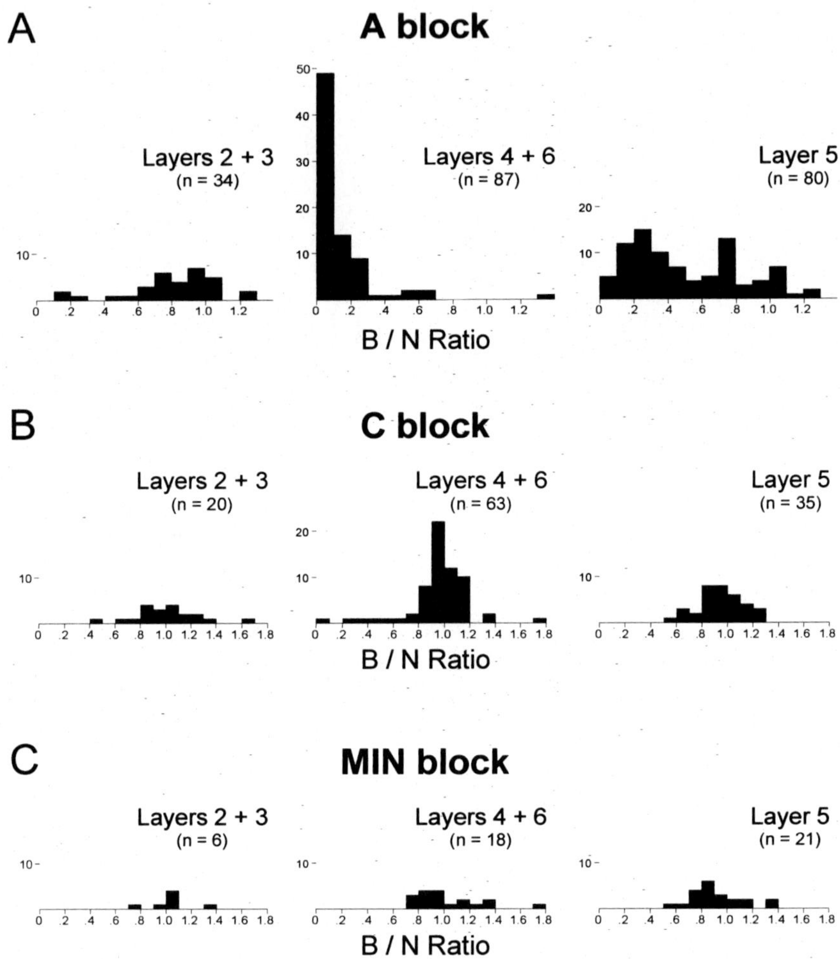

FIGURE 7-5. Summary of the effects of blocking individual LGN layers on the visually driven activity in the different layers of cortex in area 17. (Adapted from Malpeli et al. [4]. Used by permission of the American Physiological Society.)

inactivations of both layer A *and* the C layers. The top panels show the single orientation peristimulus response histogram before (left), during (middle), and after (right) recovery from geniculate inactivation. Directional selectivity, evident by the asymmetry in response profiles, was only slightly affected. Similarly, selectivity for stimulus length (Fig. 7-6, middle panels) and stimulus orientation (Fig. 7-6, bottom panels) appeared unaffected.

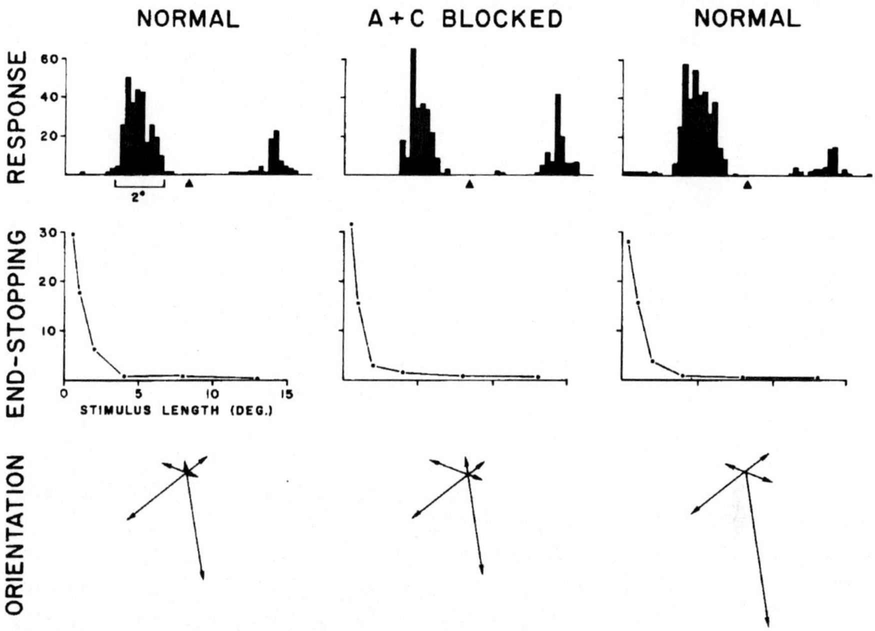

FIGURE 7-6. Simultaneous inactivation of layer A and C layers was ineffective in altering response properties of neurons in layers 2 and 3. Shown are the responses of two concurrently recorded complex cells in layer 3. For each row of results, each panel shows response characteristics before (left), during (middle), and after recovery from (right) simultaneous inactivation of the retinotopically corresponding regions of layer A and C. **(Top).** Peristimulus single orientation histograms generated by passing a 0.7 × 0.3° bar through the receptive fields. The B/N ratio for this manipulation was 0.93 **(Middle).** Length response curves for a 0.2° wide bar. **(Bottom)** Direction preferences indicated by polar plots of response magnitude. These plots were derived from three single orientation peristimulus histograms in which each orientation was separated by 60°. The stimulus used was a bar 0.2 × 1.0°. For this cortical site, we used larger volumes of cobalt to ensure complete inactivations (230 nl in layer A, 460 nl in the C layers). (From Malpeli et al. [4]. Used by permission of the American Physiological Society.)

THE C LAYERS AND THE MIN ARE NOT NECESSARY FOR VISUALLY DRIVEN ACTIVITY

Although layer A provides the major contralateral eye inputs to area 17 (17–20), the C layers and the MIN also provide either direct (17,19,20) or indirect inputs via corticocortical connections (10,21–23). In evaluating more than 160 sites (summarized in Figs. 7-5B,C), we found that reversible inactivation of the C layers or layer 1 of the MIN was largely ineffective at compromising visually driven activity in area 17.

BLOCKING LGN LAYER A REVEALS TWO TYPES OF CORTICOTECTAL CELLS

For cortical neurons in layer 5, reversible inactivation of LGN layer A yielded mixed results; some depended heavily on layer A, whereas others were largely unaffected. Layer 5 is the origin of the corticotectal pathway, a major cortical efferent. Based on responses to varying stimulus lengths, we found this output is composed of two types of complex cells (24). Figure 7-7A shows the length response characteristics of 28 corticotectal cells. For responses to the right of the arrow in the figure, increasing stimulus length usually yields increased response. These are the complex neurons originally described by Hubel and Wiesel (15), and called "standard complex" by Gilbert (16). For the cells whose responses are plotted to the left of the arrow, increasing stimulus length either has no effect or substantially decreases the response. These were originally described by Palmer and Rosenquist (8) and named "special complex" by Gilbert (16). Figure 7-7B shows the consequences of reversibly blocking LGN layer A on these corticotectal neurons (25). In this figure, the responses to the shortest stimulus are plotted against the response to the longest stimulus segregating the neurons into standard (left foreground) and special (right foreground) complex corticotectal neurons. The "Z" axis indicates the observed B/N ratio and shows that although there were exceptions, standard complex cells did and special complex cells did not heavily depend on LGN layer A for visually driven activity. The average B/N ratio for the standard complex corticotectal neurons was 0.34 versus 0.72 for the special complex corticotectal neurons ($p <$ 0.005). Figure 7-7C shows representative examples of special (left) and standard (right) complex corticotectal neurons before (top), during (middle), and after recovery from (bottom) reversible inactivation of LGN layer A.

Although the special complex corticotectal cells appear not to depend on layer A, they may depend on intact supragranular layers. In a separate series of experiments (26), we either reversibly or irreversibly cooled the supragranular layers to

→

FIGURE 7-7. Standard and special complex corticotectal cells varied in their dependence on LGN layer A. **(A)** Length response characteristics for 28 corticotectal cells. Each neuron is rank-ordered along the X axis, according to the size of the response (plotted in standard deviation units) to the shortest (0.5°) stimulus used. Along the Y-axis are plotted the responses at each of 6 stimulus lengths (0.5°, 1°, 2°, 4°, 8°, and 13°) for each cell. The arrow indicates our categorical break for standard (right) and special (left) complex cells. **(B)** Three-dimensional plot showing the relative dependence of standard complex neurons on LGN layer A. On the X and Y axes are plotted the response of the shortest stimulus (Y axis) against the response to the longest stimulus (X axis), showing segregation into the standard (left foreground) and special (right background) complex corticotectal cells. On the Z-axis are plotted the B/N ratios observed by reversibly inactivating LGN layer A. **(C)** Single-orientation peristimulus histograms of representative special (left) and standard (right) complex corticotectal cells before (top), during (middle), and after recovery from (bottom) reversible inactivation of LGN layer A. The observed B/N ratios for these example were 0.73 for the special and 0.25 for the standard complex corticotectal cell. (From Weyand et al. [25]. Used by permission of the American Physiological Society.)

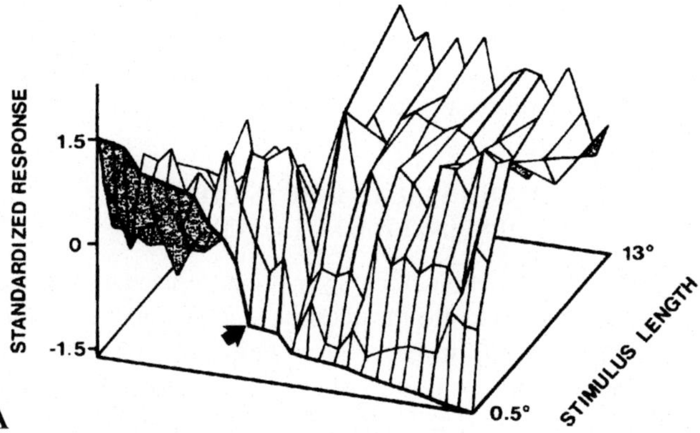
STANDARDIZED RESPONSE
1.5
0
-1.5
13°
STIMULUS LENGTH
0.5°

A

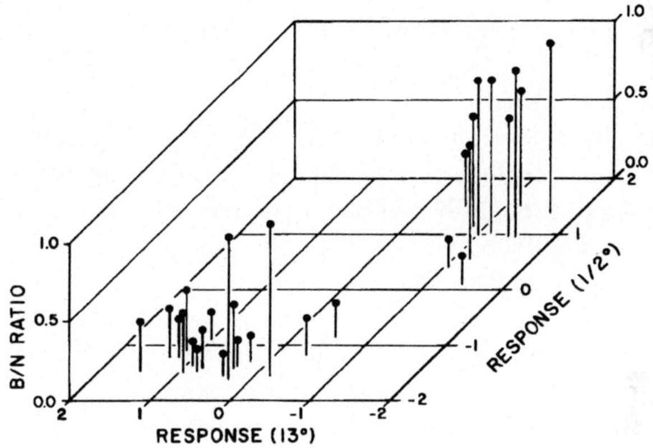
1.0
0.5
0.0
B/N RATIO
2
1
0
-1
-2
RESPONSE (1/2°)
RESPONSE (13°)

B

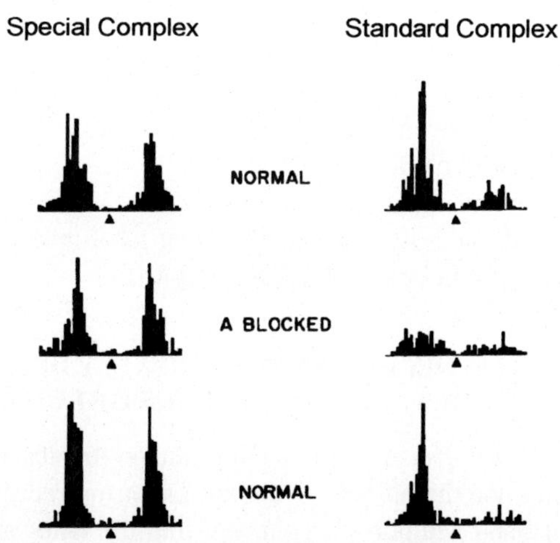
Special Complex
Standard Complex
NORMAL
A BLOCKED
NORMAL

C

determine if, and to what degree, infragranular layer cells depend on the integrity of the superficial layers. Although corticotectal neurons were not identified in these experiments, when supragranular layer input was removed, special complex neurons in layer 5 were not observed.

SUSTAINING VISUAL INPUTS COME FROM IPSILATERAL THALAMUS

Once it has been shown that blocking layer A can be used to distinguish two circuits that can operate independently, the question emerges as to the source of the sustaining inputs in the absence of layer A. Several observations indicate the source is the ipsilateral thalamus, although it is apparently not restricted to the LGN complex.

We destroyed the entire LGN complex with ibotinic acid (4). The next day we recorded from extensive regions of the ipsilateral cortex. Except for some unresolved multiunit activity near the vertical meridian, we were unable to find any visually driven activity. After approximately 20 hours of searching, we switched to the contralateral cortex, from which we immediately recorded apparently normal visually driven activity.

We destroyed the LGN complex but spared layer A by making multiple small injections of ibotenic acid (4). As depicted in Figure 7-8 (left), the lesion extended beyond the borders of the LGN complex to include portions of the medial wing and lateral posterior (LP) complex (27). The next day, we began recording in area 17 in a region in approximate retinotopic correspondence with the center of our lesion of the C layers. Activity appeared normal in all cortical layers, suggesting an intact LGN layer A is sufficient for all visually driven activity in area 17. Figure 7-8 shows that for six sites tested five supragranular and one in layer 5) visually driven activity could be abolished by now blocking layer A. Thus, in the absence of other geniculate layers, layer A appears sufficient and necessary.

Although layer A, the C layers and the MIN would seemingly exhaust all potential avenues by which area 17 neurons might receive sufficient visual drive, this did not appear to be the case. In 12 cases, we simultaneously inactivated layer A, the C layers, and the MIN. For nine of these sites (all in layers 2 and 3), we failed to silence visually driven activity. These observations indicate that sustaining visual inputs originate in either the geniculate "wing" (28,29), or emerge from tectal-recipient portions of the cat's LP complex (30–32). Table 7-1 summarizes the results of evaluating activity at sites where we simultaneously blocked more than one contralateral eye layer of the LGN complex.

AREA 18 IS NECESSARY FOR VISUALLY DRIVEN ACTIVITY IN AREA 17 IN THE ABSENCE OF LGN LAYER A

As described previously, sustaining visual inputs in the absence of layer A can reach area 17 directly, via thalamocortical neurons, or indirectly, via corticocortical circuits. Mignard and Malpeli (33) showed that the latter alternative appears

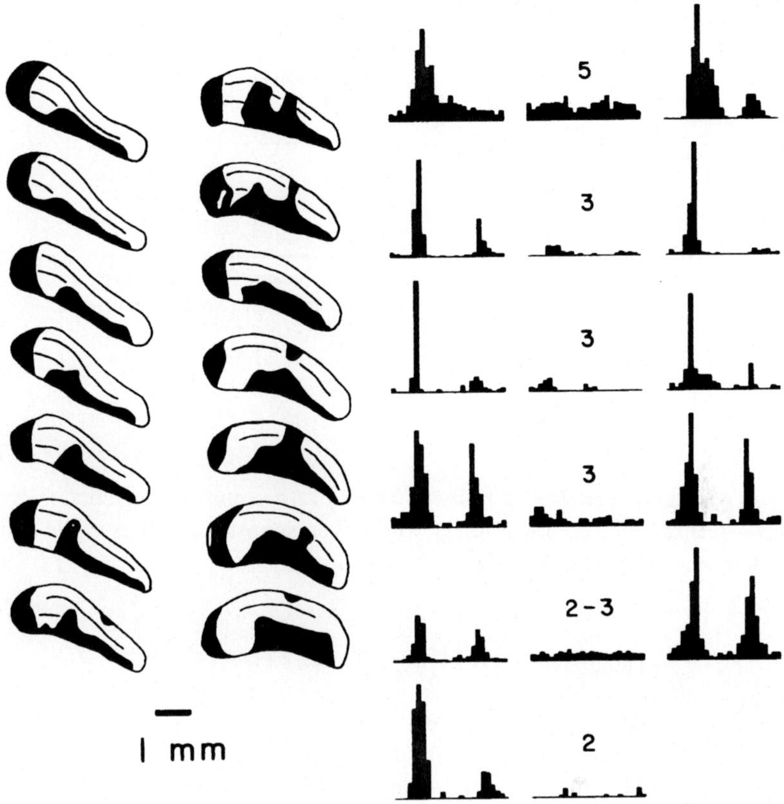

FIGURE 7-8. Layer A is sufficient to drive visual activity in area 17. **(Left)** Tracings of coronal sections through most of the LGN showing the damage (black) resulting from multiple injections of ibotenic acid delivered to the C layers and the MIN. Sections are 120 μm apart and sequenced from posterior (upper left) to anterior (lower right). The lesion extended medially through the posterior nucleus and pulvinar (not shown). **(Right)** Sets of peristimulus histograms at six different sites, showing visual responses as a stimulus is swept back and forth through the cell's receptive field before (left), during (middle), and after recovery from (right) inactivation of layer A using lidocaine (58 nl; top, third and fifth set) or cobalt (115 nl; second and sixth sets). The bottom set does not show recovery because the cell was lost before full recovery. The number above each histogram refers to the cortical layer in which the activity was recorded. The top set shows the activity of a special complex corticotectal cell, and the bottom set shows the activity of a layer 2 complex cell. The remaining sites were multipleunit clusters. (From Malpeli et al. [4]. Used by permission of the American Physiological Society.)

likely. After inactivation of area 18 by irreversible cooling, injections of $CoCl_2$ in layer A now routinely abolished visually driven activity among neurons in layers 2 and 3. Reversible inactivation experiments by Alonso and colleagues (34) indicate that inputs from area 18 alone can alter activity in area 17. Their extremely precise injections (they estimated inactivations of ~300 μm diameter) increased

TABLE 7-1. Summary of B/N Ratios for Multiple Inactivations*

Targets inactivated	Cortical layer	
	2 + 3	5
A + C	0.70 (0.89, 11)	0.57 (0.78, 11)
A + M	0.68 (0.73, 1)	0.79 (0.76, 6)
A + C + M	0.73 (0.72, 9)	0.29 (0.94, 3)

* B/N (blocked/normal) ratios are given as means. The two numbers in parentheses are the B/N ratio for layer A inactivation alone and number of sites in the sample. "M" refers to MIN.

From Malpeli et al. (3). Used by permission of the American Physiological Society.

activity in 43% and decreased activity in 21% of retinotopically matched neurons in area 17. Decreases were most likely when the response properties at the site of inactivation matched those at the site in area 17.

AREA 18: MORE INTEGRATIVE THAN AREA 17

Unlike the primate, area 18 of the cat receives a strong direct projection from the LGN (see Chapter 1). However, the pattern of connectivity from the LGN is different from area 17. Area 18 receives a more prominent projection from the C layers than area 17 (17,18,20,35). We studied the effects of inactivating individual or combinations of LGN layers on visually driven activity in supragranular layer (36) and corticotectal (37) neurons in area 18. Strong effects specific to cortical layer, such as those observed in area 17, were not as striking in area 18. However, as in area 17, proximity to LGN axons had some predictive power. Unlike area 17, both LGN layer A and the C layers innervate the ventral half of layer 3 in area 18 (17,38) (see Chapter 5). We found that the simple cells in the ventral half of layer 3 were significantly more affected by layer A inactivation than simple cells in the dorsal half (mean B/N ratios of 0.22 vs. 0.64, $p < 0.04$). This distinction did not apply to complex cells (mean B/N ratios of 0.71 vs. 0.72). Overall (including multiunit sites), blocking layer A was significantly more effective in reducing activity in the ventral half of layer 3 than the dorsal half (0.42 vs. 0.69, $p < 0.001$).

As in area 17, blocking the LGN layer A effectively identified two groups of corticotectal cells in area 18 (Fig. 7-9A). However, unlike area 17, there was no linkage between dependence on layer A and the length response characteristics of the corticotectal cell. One group of corticotectal cells (mean B/N ration ~ 0.25) appeared to depend heavily on input from layer A. The remaining cells, largely unaffected by blocking layer A, showed B/N ratios ranging from 0.60 to 1.25.

Blocking the C layers was more effective in reducing activity in area 18 than in area 17. Whereas blocking C layers alone was largely ineffective in area 17

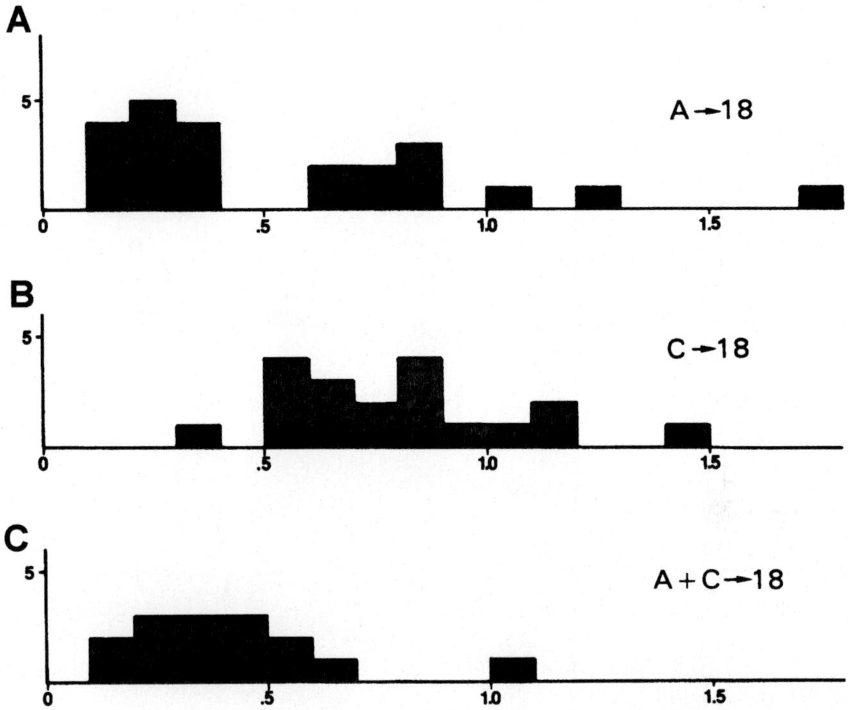

FIGURE 7-9. Area 18 corticotectal cells depend on LGN layers A and C. Shown are frequency histograms of B/N ratios for individual corticotectal cells. **(A)** Selective inactivation of layer A revealed two groups of corticotectal cells (those with B/N ratios greater than 0.60 and those with B/N ratios less than 0.40). **(B)** Selective inactivation of the C layers resulted in moderate effects (average B/N ratio 0.79) and only 1 cell dropped below 50% of its average normal response. **(C)** Combined inactivations of layer A and the C layers usually further decreased the visual responsiveness among area 18 corticotectal cells. The average B/N ratio for combined inactivations was 0.36. (From Weyand et al., [37]. Used by permission of the American Physiological Society.)

(mean B/N ratio 0.95, 120 sites), blocking the C layers commonly had some effect (for superficial layers: mean B/N ration 0.50 for 8 simple cells, 0.68 for 10 complex cells; for corticotectal cells: 0.79, Fig. 7-9B). Figure 7-10 shows a simple cell in layer 4 that was largely dependent on the C layers for visually driven activity. As shown, reversible inactivation of layer A had some effect (B/N ratio 0.66), but inactivation of the C layers alone reduced activity to near zero (B/N ratio 0.04).

Although not routinely tested, combined blockade of LGN layers A and C in area 18 was typically more effective than either blockade alone. Our most extensive use of such combined inactivations was for corticotectal cells. Of 15 corticotectal cells, only 1 was completely unaffected by such combined inactivations (Fig. 7-9C). As with other area 18 complex cells, for corticotectal cells, inactivating the C layers had an additional effect over inactivating layer A alone. Figure 7-11

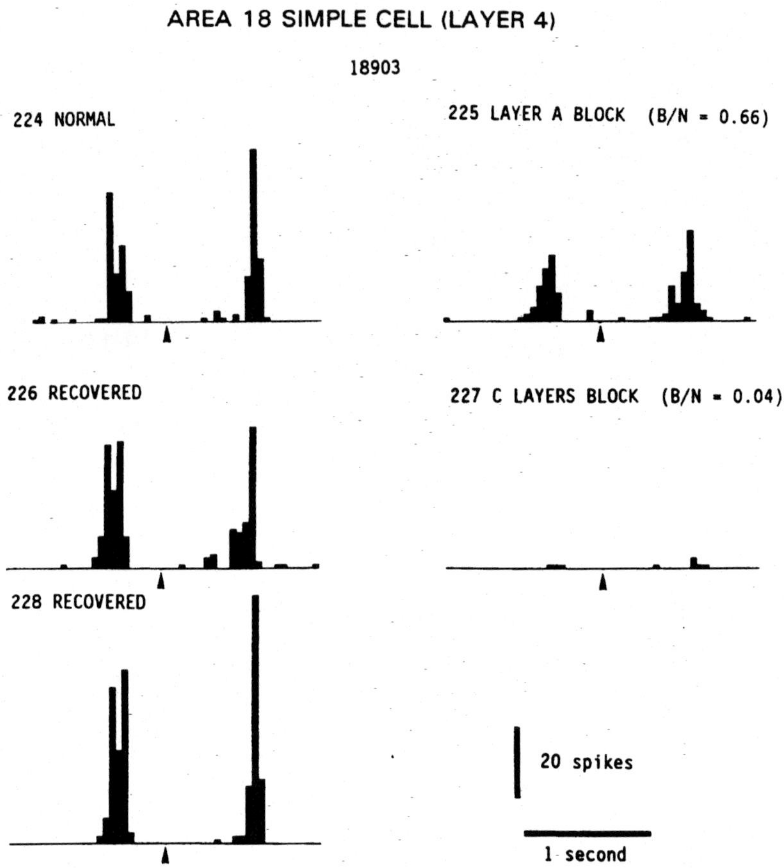

FIGURE 7-10. A layer 4 simple cell in area 18 that heavily depends on input from the C layers. Shown are five single orientation peristimulus histograms. The Arabic numerals to the upper left of each histogram indicate the sequence the data was collected. The neuron appeared mildly affected by blocking layer A with 115 nl of $CoCl_2$ (B/N ratio = 0.66), but blocking the C layers alone was devastating (B/N ratio = 0.04). (From Malpeli, Weyand and Lee, unpublished).

shows an example of such convergence of influence of both layer A and the C layers on a corticotectal cell.

FUNCTIONAL ARCHITECTURE OF VISUAL CORTEX

Hubel and Wiesel (15,39) first described the primary attributes of cortical neurons in area 17: selectivity for stimulus orientation, direction of motion, and stimulus length. Seizing on the concept of the cortical column described by Mountcastle (40), they proposed that synthesis of these properties emerged from a chain

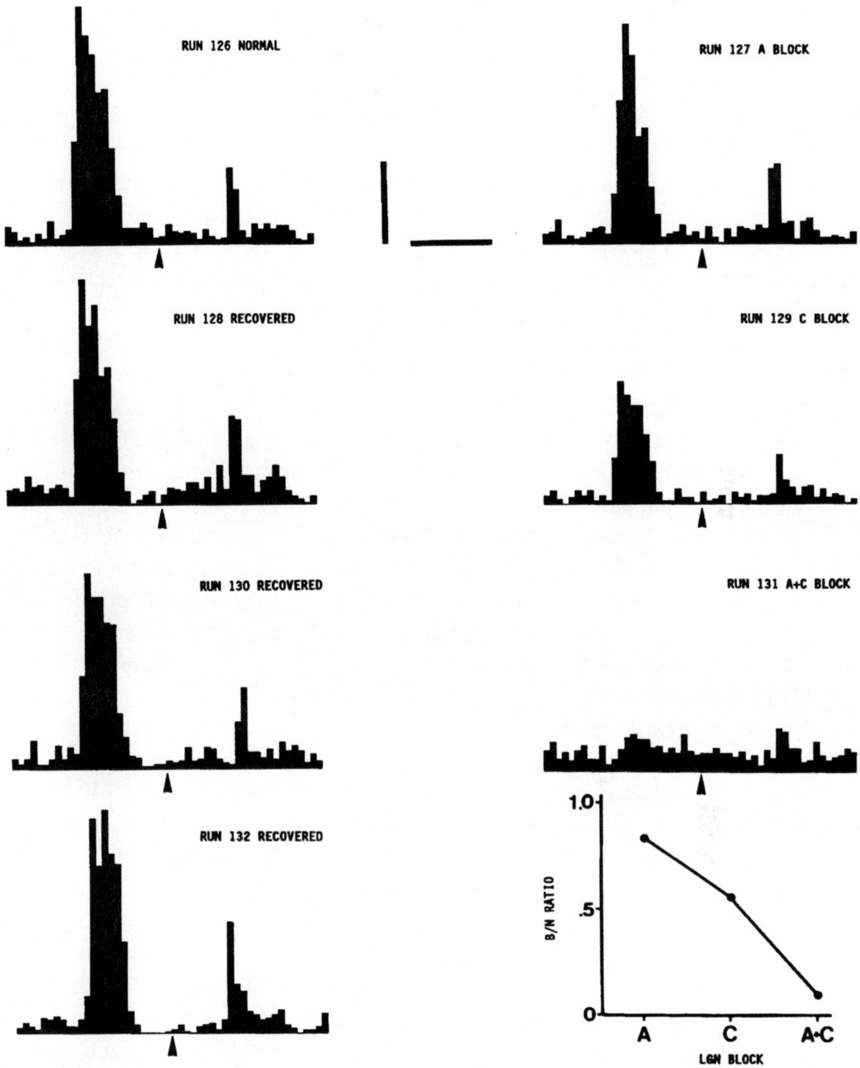

FIGURE 7-11. A standard complex corticotectal cell in area 18 that received sustaining inputs from both layer A and the C layers. Each histogram represents the cumulative response (30 trials) as a 2° stimulus passed through the receptive field, reversed direction (arrow), and passed back through the receptive field. **(Bottom right)** B/N ratio for each of the three manipulations. Calibration bars, 500 msec, 150 spikes. (From Weyand et al. [37]. Used by permission of the American Physiological Society.)

linking the concentrically organized LGN cells to the simple cells to the complex cells in a single cortical column. The hypercomplex (end-stopped) cells were believed to emerge in area 18 from combinations of complex cells. Although of heuristic value, a simple hierarchical model is limited. Reversible inactivation is a powerful tool for testing models of information flow through the brain.

The following sections describe how the results summarized above impact on our understanding of the operations of visual cortex.

THE CORTICAL COLUMN

The initial experiments by Malpeli (3) demonstrated that simple and complex cells in layer 4 are not necessary for driving complex cells of layers 2 and 3 and many cells in layer 5. Most activity remaining during the blockade was robust, and more important, these cells retained the orientation selectivity, direction, and stimulus length preferences they possessed before blockade. With the cortical column effectively transected, these experiments allowed a direct test of the hierarchical model in which complex cell receptive field properties are generated by the summation of simple cell outputs. Such observations indicate that these attributes are synthesized at least twice within the column, once in layers 4 and 6, and again in the superficial layers. Experiments in which we inactivated the superficial layers (26) indicate that layers 2 and 3 do not necessarily (exception noted later) impose orientation, direction, and length specificity on neurons in the deep layers.

The identification of two corticotectal circuits in area 17 retains some aspects of intracolumnar dependence. As depicted in Fig. 7-1, the standard complex corticotectal cells receive a necessary and sufficient input from LGN layer A via neurons in layer 4, whereas the special complex corticotectal cells are driven by the complex neurons of layers 2 and 3. In addition to the data presented, this speculation is based on several other observations. First, intracellular studies indicate that corticotectal cells do not receive monosynaptic excitation from the LGN (41,42). Because rich connections exist between neurons in layers 4, 5, and 6 (42–45), it seems reasonable that the neurons in layers 4 and 6 could easily provide a necessary and sustaining input to the standard complex corticotectal cells. There are also intralaminar connects between the neurons in layers 2 and 3 and those in layer 5 (21,42,45–47). When we inactivated the superficial layers (26), special complex cells were no longer observed in layer 5. This result suggests that superficial layer input is necessary for the activity of special complex corticotectal cells, which are largely independent of blockade of layer A. Additional evidence for this linkage comes from development. During the first 8 postnatal weeks, Tsumoto and Suda (48) found that standard complex corticotectal cells appear before special complex corticotectal cells. Because cortical neurons mature from deep layers to superficial (49), the delay in expression of special complex neurons may be related to the delayed maturation of superficial layer neurons.

The physiological distinctions between standard and special complex corticotectal neurons in area 17 may have morphological correlates. Hübener and colleagues (50) have described corticotectal neurons as "spine-free" or "spined." In consideration of biophysical consequences of these two types of morphologies, they speculated that the spine-free corticotectal cells might be special complex, whereas the spined corticotectal cells might possess the receptive field attributes of the standard complex cells.

PARALLEL AND SERIAL PROCESSING

At the level of the thalamus, each LGN layer can be considered an independent, parallel input to cortex. The identification of A-dependent cells suggests that the A-layer channel is uniquely maintained at early stages of processing in visual cortex. The longest serial chain reflecting dominance of a single LGN layer links layer A to cells of layers 4 and 6, and through these, to standard complex corticotectal cells. However, beyond this level, such relatively pure dominance by layer A does not continue. Although we identified circuits that are relatively A-dependent and A-independent in area 18 (36,37), and in lateral suprasylvian cortex (14), visually driven activity was rarely totally abolished at these levels. For example, even for area 17 standard complex corticotectal cells (a population that is relatively A-dependent), activity was reduced to only about half its normal level by inactivations of layer A. In a purely serial propagation of a parallel channel, blockade of a node early in the chain should result in complete failure at subsequent stages. This is the case only for neurons in layers 4 and 6 of area 17, whose major projections are either intracortical (21,42,44,45,51) or a return path to the LGN (10,52). At subsequent stages, dominance by layer A is only relative, and most cells integrate two or more channels. The relationship between LGN layers and cortical dependence is best conceptualized as a continuum between absolute dominance by a single LGN channel and complete integration of two or more channels. The latter can provide independence from any single LGN layer, such as appears to be the case for the special complex neurons in layer 5 of area 17, the bulk of neurons in layers 2 and 3, and many cells in area 18. Intermediate between these two ends of the continuum is integration of multiple channels for which one is relatively dominant. Such relative dominance by layer A characterizes standard complex corticotectal cells in area 17, as well as a portion of the corticotectal cells in area 18. It is also an appropriate descriptor for many simple cells in area 18 that are primarily under the control of the C layers (e.g., Fig. 7-10).

Integration does not necessarily mean response appears as the algebraic sum of its inputs. The supragranular layers receive extensive interlaminar connections from layers 4 and 6, yet as originally reported by Malpeli (3), blocking these signals seemed to have little impact on visual response or stimulus selectivity. Furthermore, inactivation of multiple LGN layers often failed to silence activity in layers 2 and 3 or layer 5 and never seemed to affect stimulus selectivities. One is left with the rather nonintuitive conclusion, that only a few residual thalamic neurons can provide sufficient drive for some cortical neurons to operate at full vigor and selectivity.

A popular descriptor for parallel circuits is the X, Y, W distinction based on response attributes and axonal conduction velocities of retinal ganglion and geniculate neurons. These groups in the cat are not parsed according to geniculate layer, so the experiments cited previously do not necessarily lend themselves to evaluate the validity of these circuits at the level of visual cortex. One exception may be X-cell inputs. To the extent that X-cells are restricted to the A-layers (reviewed in

Sherman [53]; but see [54,55]), the observation that activity in layers 2 and 3 was largely unaffected by layer A blockade would argue that X-cell input was not necessary for visually driven activity in area 17 (see Chapter 1, this volume).

DYNAMIC ANALYZERS

The demonstration that visual cortical neurons can continue to operate robustly and selectively in the face of significant loss of afferent input provides a different, albeit nonintuitive, view of the cortical neuron. An individual neuron possesses thousands of synapses. Our results indicate that for many of these neurons, the ability to achieve threshold (i.e., produce an action potential) may be relatively easy, even under conditions of sudden silence by other (perhaps normally dominating) inputs. In the context of their afferents then, cortical neurons function more like logic circuit "OR" gates rather than integrators or summators of analog signals. In dorsal column nuclei, Pettit and Schwark (56) showed the emergence of "latent" receptive fields within *minutes* after anesthesia of the cell's traditional receptive field by local application of lidocaine to the cat's paw. Similar phenomena have been observed by Pettet and Gilbert (57) in area 17. By occluding a neuron's "classic" receptive field (a state they refer to as an "artificial scotoma") but providing visual patterns nearby, the receptive field expanded within a few minutes. They argued that these distant afferents must always be there but are apparently suppressed by other, dominating afferents (whose exact circuitry remains mysterious, but originates at the level of cortex). Once the dominating afferents are themselves silenced, the erstwhile silent inputs apparently become sufficiently powerful to exceed threshold.

This theme of dynamic receptive fields was extended to cortical "modules" in interpreting the influence of geniculate subdivisions on activity in the lateral suprasylvian cortex (14). We found that dependence on layer A correlated with the "patchy" distribution of terminals from area 18. These "slabs" of cortex, defined both by dependence on layer A and recipients of input from area 18, were adjacent to other slabs of cortex, relatively independent of layer A influences and not directly linked to area 18. Curiously, these adjacent slabs tended be dependent on the MIN such that dependence on MIN and layer A were negatively correlated. This segregation was not sharp, and these modules appear to extensively overlap. We speculate that this modularization represents a solution that optimizes the utilization of cortical resources by different geniculate subdivisions. The MIN appears specialized for dim-light vision. Its map of the visual field is largely restricted to that portion of the retina containing the tapetum (58) and owing to increased convergence from retinal inputs, MIN cells show approximately one log unit advantage in detecting low-spatial frequency stimuli in very dim light (54). We speculate that in the extensive regions of overlap between modules, inputs from the MIN dominate LS cortex at low illumination levels, whereas under photopic conditions, inputs from layer A dominate. Thus, the same cortical resources are dynamically exploited by different geniculate subdivisions depending on nighttime or daytime viewing conditions.

GENERALIZABILITY AND LIMITATIONS

The observations described here are repeatable and reliable within the constraints of the preparation. However, the paralyzed, anesthetized cat is removed from the natural state. Constraints on these circuits may be radically different in an awake animal free to inspect its environment. Furthermore, other circuits, not realized in the paralyzed, anesthetized cat may have profound influence. For example, gaze angle influences the gain of visual responsiveness of at least one third of the neurons in area 17 of the awake cat (59).

Support for the validity of some of the observations obtained in the anesthetized animal comes from a recent study by Tate and Malpeli (60). They showed that reversible inactivation of layer A disrupted, but did not abolish, performance in a visual detection task. Their interpretation was that the defect was attributable to perceptual difficulties, which they speculated affected the ability to make saccades to visual targets because of disruption of corticotectal downflow from the standard complex cells in area 17.

Because reversible inactivation is a subtractive method (i.e., describing the competence of systems in the absence of inputs), it does not necessarily tell us what an LGN layer may contribute. Connections between layers 4 and 6 to the supragranular layers are extensive (21,42–45,51,61), and no doubt supply significant input to the supragranular layers. Most of our data involve evaluating the response of neurons under conditions of passing a suprathreshold bar stimulus back and forth through the receptive field. We could evaluate basic response and directional preference, and occasionally length and orientation selectivity. However, there are other properties we never tested such as spatial frequency or velocity tuning, or binocular interactions, and it is possible that some of these properties were disrupted.

Bolz and Gilbert (62) showed that inactivation of layer 6 by local injection of GABA effectively abolished length selectivity among neurons in the more superficial layers 4 and 2 and 3. Although their manipulation of the cortical microcircuitry is not identical to ours, given our results, theirs are surprising. Bolz and Gilbert (62) ventilated their animals with 100% oxygen and noted that this appears important for maintaining a high proportion of end-stopped cells observed in layers 2 and 3. Because we ventilated our animals with room air, it could be that many end-stopped cells were either not operational or not end-stopped in our preparation. Regardless of such complications, it is remarkable that cortical neurons can undergo significant functional deafferentation and yet retain full responsiveness and response selectivity previously believed to emerge from intracortical circuits.

ACKNOWLEDGMENTS

The experiments described in this chapter were funded by the National Eye Institute (EY02695) and done while the author was a postdoctoral fellow in the laboratory of Joseph Malpeli, who developed the reversible inactivation techniques used here. The author is in debt to Drs. Choongkil Lee,

Harris Schwark, and especially Joseph Malpeli for both the fruitful collaborations and the many discussions of the ideas presented here. The author acknowledges the support of the National Eye Institute (EY11144) in writing this review.

REFERENCES

1. Stone, J. (1983). *Parallel processing in the visual system.* New York, Plenum Press.
2. Stone, J., Dreher, B., and Leventhal, A. G. (1979). Hierarchical and parallel mechanisms in the organization of visual cortex. *Brain Res. Rev.* **1,** 345–394.
3. Malpeli, J. G. (1983). Activity of cells in area 17 of the cat in absence of input from layer A of the lateral geniculate nucleus. *J. Neurophysiol.* **49,** 595–610.
4. Malpeli, J. G., Lee, C., Schwark, H. D., and Weyand, T. G. (1986). Cat area 17. I. Pattern of thalamic control of cortical layers. *J. Neurophysiol.* **56,** 1062–1079.
5. Malpeli, J. G., and Schiller, P. H. (1979). A method of reversible inactivation of small regions of brain tissue. *J. Neurosci. Methods* **1,** 143–151.
6. Hagiwara, S., and Byerly, L. (1981). Calcium channel. *Annu. Rev. Neurosci.* **4,** 69–125.
7. Hollander, H. (1974). The origin of corticotectal projections in the cat. *Exp. Brain Res.* **21,** 433–440.
8. Palmer, L. A., and Rosenquist, A. C. (1974). Visual receptive fields of single striate cortical units projecting to the superior colliculus in cats. *Brain Res.* **67,** 27–42.
9. Magalhaes-Castro, H. H., Saraiva, P. E. S., and Magalhaes-Castro, B. (1975). Identification of corticotectal cells of the visual cortex by means of horseradish peroxidase. *Brain Res.* **83,** 474–479.
10. Gilbert, C. D., and Kelly, J. P. (1975). The projections of cells in different layers of the cat's visual cortex. *J. Comp. Neurol.* **163,** 81–106.
11. Sanderson, K. J. (1971). The projection of the visual field to the lateral geniculate and medial intralaminar nuclei in the cat. *J. Comp. Neurol.* **143,** 101–118.
12. Guillery, R. W. (1970). The laminar distribution of retinal fibers in the dorsal lateral geniculate nucleus of the cat: a new interpretation. *J. Comp. Neurol.* **138,** 339–368.
13. Hickey, T. L., and Guillery, R. W. (1974). An autoradiographic study of retinogeniculate pathways in the cat and the fox. *J. Comp. Neurol.* **156,** 239–254.
14. Lee, C., Weyand, T. G., and Malpeli, J. G. (1998a). Thalamic control of cat lateral suprasylvian visual area: relation to patchy association projections from area 18. *Vis. Neurosci.* **15,** 15–25.
15. Hubel, D. H., and Wiesel, T. N. (1962). Receptive fields, binocular interaction and functional architecture in the cat's striate cortex. *J. Physiol. (London)* **160,** 106–154.
16. Gilbert, C. D. (1977). Laminar differences in receptive field properties of cells in cat primary visual cortex. *J. Physiol. Lond.* **268,** 391–421.
17. LeVay, S., and Gilbert, C. D. (1976). Laminar patterns of geniculocortical projection in the cat. *Brain Res.* **113,** 1–19.
18. Hollander, H., and Vanegas, H. (1977). The projection from the lateral geniculate nucleus onto the visual cortex in the cat. A quantitative study with horseradish peroxidase. *J. Comp. Neurol.* **173,** 519–536.
19. Leventhal, A. G. (1979). Evidence that different layers of relay cells of the cat's lateral geniculate nucleus terminate in different layers of the striate cortex. *Exp. Brain Res.* **37,** 349–372.
20. Geisert, E. E. (1980). Cortical projections of the lateral geniculate nucleus in the cat. *J. Comp. Neurol.* **190,** 793–812.
21. Gilbert, C. D., and Wiesel, T. N. (1979). Morphology and intracortical projections of functionally characterized neurones in the cat visual cortex. *Nature* **280,** 120–125.
22. Symonds, L. L., and Rosenquist, A. C. (1984). Corticocortical connections among visual areas in the cat. *J. Comp. Neurol.* **229,** 1–38.

23. Henry, G. H., Salin, P. A., and Bullier, J. (1991). Projections from areas 18 and 19 to cat striate cortex: divergence and laminar specificity. *Eur. J. Neurosci.* **3,** 186–200.
24. Weyand, T. G., Malpeli, J. G., Lee, C., and Schwark, H. D. (1986a). Cat area 17. III. Response properties and orientation anisotropies of corticotectal cells. *J. Neurophysiol.* **56,** 1088–1101.
25. Weyand, T. G., Malpeli, J. G., Lee, C., and Schwark, H. D. (1986b). Cat area 17. IV. Two types of corticotectal cells defined by controlling geniculate inputs. *J. Neurophysiol.* **56,** 1102–1108.
26. Schwark, H. D., Malpeli, J. G., Weyand, T. G., and Lee, C. (1986). Cat area 17. II. Response properties of infragranular layer cells in the absence of supragranular layer activity. *J. Neurophysiol.* **56,** 1074–1087.
27. Updyke, B. V. (1983). A reevaluation of the functional organization and cytoarchitecture of the feline lateral posterior complex, with observations of adjoining cell groups. *J. Comp. Neurol.* **219,** 143–181.
28. Guillery, R. W., Geisert, E. E., Jr., Polley, E. H., and Mason, C. A. (1980). An analysis of the retinal afferents to the cat's medial interlaminar nucleus and to its rostral thalamic extension, the "geniculate wing." *J. Comp. Neurol.* **194,** 117–142.
29. Leventhal, A. G., Keens, J., and Tork, I. (1980). The afferent ganglion cells and cortical projections of the retinal recipient zone (RRZ) of the cat's "pulvinar complex." *J. Comp. Neurol.* **194,** 535–554.
30. Berson, D. M., and Graybiel, A. M. (1978). Parallel thalamic zones in the LP-pulvinar complex of the cat identified by their afferent and efferent connections. *Brain Res.* **147,** 139–148.
31. Berson, D. M., and Graybiel, A. M. (1983). Organization of the striate-recipient zone of the cat's lateralis posterior-pulvinar complex and its relations with the geniculostriate system. *Neuroscience* **9,** 337–372.
32. Raczkowski, D., and Rosenquist, A. C. (1983). Connections of the multiple visual cortical areas with the lateral posterior pulvinar complex and adjacent thalamic nuclei in the cat. *J. Neurosci.* **3,** 1912–1942.
33. Mignard, M., and Malpeli, J. G. (1991). Paths of information flow through visual cortex. *Science* **251,** 1249–1251.
34. Alonso, J. M., Cudeiro, J., Perez, R., Gonzalez, F., and Acuna, C. (1993). Orientational influences of layer V of visual area 18 upon cells in layer V of area 17 in cat cortex. *Exp. Brain Res.* **96,** 212–220.
35. Niimi, K., Matusuoka, H., Yamazaki, Y., and Matsumoto, H. (1981). Thalamic afferents to the visual cortex in the cat studied by retrograde axonal transport of horseradish peroxidase. *Brain Behav. Evol.* **18,** 114–139.
36. Lee, C., Weyand, T. G., and Malpeli, J. G. (1998b). Thalamic control of cat area-18 supragranular layers: simple cells, complex cells, and cells projecting to the lateral suprasylvian visual areas. *Vis. Neurosci.* **15,** 27–35.
37. Weyand, T. G., Malpeli, J. G., and Lee, C. (1991). Area 18 corticotectal cells: response properties and identification of sustaining geniculate inputs. *J. Neurophysiol.* **65,** 1078–1088.
38. Humphrey, A. L. Sur, M., Uhlrich, D. J., and Sherman, S. M. (1985). Termination patterns of individual X- and Y-cell axons of the cat: projections to area 18, to the 17–18 border and to both areas 17 and 18. *J. Comp. Neurol.* **233,** 190–212.
39. Hubel, D. H., and Wiesel, T. N. (1965). Receptive field and functional architecture in two nonstriate visual areas (18 and 19) of the cat. *J. Neurophysiol.* **28,** 229–289.
40. Mountcastle, V. B. (1957). Modality and topographic properties of single neurons of cat's somatic sensory cortex. *J. Neurophysiol.* **20,** 408–434.
41. Singer, W., Tretter, F., and Cynader, M. (1975). Organization of cat striate cortex: a correlation of receptive-field properties with afferent and efferent connections. *J. Neurophysiol.* **38,** 1080–1098.
42. Ferster, D. A., and Lindstrom, S. (1983). An intracellular analysis of geniculo-cortical connectivity in area 17 of the cat. *J. Physiol. Lond.* **342,** 181–215.
43. Gilbert, C. D., and Wiesel, T. N. (1983). Clustered intrinsic connections in cat visual cortex. *J. Neurosci.* **3,** 1116–1133.

44. Ferster, D., and Lindstrom, S. (1985). Synaptic excitation of neurons in area 17 of the cat by intracortical axon collaterals of corticogeniculate cells. *J. Physiol. Lond.* **367,** 233–252.
45. Martin, K. A. C., and Whitteridge, D. (1984). Form, function and intracortical projections of spiny neurones in the striate visual cortex of the cat. *J. Physiol.* **353,** 463–504.
46. Mitzdorf, U., and Singer, W. (1978). Prominent excitatory pathways in the cat visual cortex (A17 and A18): a current source density analysis of electrically evoked potentials. *Exp. Brain Res.* **33,** 371–394.
47. Kisvarday, Z. F., Martin, K. A .C., Freund, T. F., Magloczky, Zs., Whitteridge, D., and Somogyi, P. (1986). Synaptic targets of HRP-filled layer III pyramidal cells in the cat striate cortex. *Exp. Brain Res.* **64,** 541–552.
48. Tsumoto, T., and Suda, K. (1983). Postnatal development of corticotectal neurons in kitten striate cortex: an electrophysiological study. *Dev. Brain Res.* **11,** 29–38.
49. Tsumoto, T., and Suda, K. (1982). Laminar differences in development of afferent innervation to striate cortex neurons in kittens. *Exp. Brain Res.* **45,** 433–446.
50. Hübener, M., Schwarz, C., and Bolz, J. (1990). Morphological types of projection neurons in layer V of cat visual cortex. *J. Comp. Neurol.* **301,** 655–674.
51. McGuire, B. A., Hornung, J.-P., Gilbert, C. D., and Wiesel, T. N. (1984). Patterns of synaptic input to layer 4 of cat striate cortex. *J. Neurosci.* **4,** 3021–3033.
52. Tsumoto, T., and Suda, K. (1980). Three groups of cortico-geniculate neurons and their distribution in binocular and monocular segments of cat striate cortex. *J. Comp. Neurol.* **193,** 223–236.
53. Sherman, S. M. (1985). Functional organization of the W-, X-, and Y-cell pathways in the cat: a review and hypothesis. In: *Progress in psychobiology and physiological psychology* (J. M. Sprague and A. M. Epstein, Eds.), Vol. 11, pp. 233–314. New York, Academic Press.
54. Lee, D., Lee, C., and Malpeli, J. G. (1992). Acuity-sensitivity trade-offs of X and Y cells in the cat lateral geniculate complex: role of the medial interlaminar nucleus in scotopic vision. *J. Neurophysiol.* **68,** 1235–1247.
55. Humphrey, A. L., and Murthy, A. (1999). Cell types and response timings in the medial interlaminar nucleus and C-layers of the cat lateral geniculate nucleus. *Vis. Neurosci.* **16,** 513–525.
56. Pettit, M. J., and Schwark, H. D. (1993). Receptive field organization in dorsal column nuclei during temporary denervation. *Science* **262,** 2054–2056.
57. Pettet, M. W., and Gilbert, C. D. (1992). Dynamic changes in receptive field size in cat primary visual cortex. *Proc. Natl. Acad. Sci.* **89,** 8366–8370.
58. Lee, C., Malpeli, J. G., Schwark, H. D., and Weyand, T. G. (1984). Cat medial interlaminar nucleus: retinotopy, relation to tapetum and implications for scotopic vision. *J. Neurophysiol.* **52,** 848–869.
59. Weyand, T. G., and Malpeli, J. G. (1993). Responses of neurons in primary visual cortex are modulated by eye position. *J. Neurophysiol.* **69,** 2258–2260.
60. Tate, A., and Malpeli, J. G. (1998). Effects of focal inactivation of dorsal or ventral layers of the lateral geniculate nucleus on cats' ability to see and fixate small targets. *J. Neurophysiol.* **80,** 2206–2209.
61. Kisvarday, Z. F., Martin, K. A. C., Friedlander, M. J., and Somogyi, P. (1987). Evidence for interlaminar inhibitory circuits in the striate cortex. *J. Comp. Neurol.* **260,** 1–19.
62. Bolz, J., and Gilbert, C. D. (1986). Generation of end-inhibition in the visual cortex via interlaminar connections. *Nature* **320,** 362–365.

8

INTEGRATION OF THALAMIC INPUTS TO CAT PRIMARY VISUAL CORTEX

R. CLAY REID

Department of Neurobiology, Harvard Medical School, Boston, Massachusetts

JOSE-MANUEL ALONSO

Department of Psychology, University of Connecticut, Storrs-Mansfield, Connecticut

W. MARTIN USREY

Department of Neurobiology, Harvard Medical School, Boston, Massachusetts

INTRODUCTION

Primary sensory areas in the cerebral cortex have classically been defined based on the organization of their thalamic inputs. At a gross level, thalamic inputs are used to define the borders of a cortical sensory area. For example, axons from the dorsal lateral geniculate nucleus (LGN) of the thalamus project to primary visual cortex (often called V1 or area 17), whereas axons from the medial geniculate nucleus (MGN) project to primary auditory cortex (A1 or area 41). At a finer level, thalamic inputs are used to distinguish layers and modules within a cortical area (e.g., layer 2/3 blobs in primate visual cortex, centers of layer 4 barrels in rodent somatosensory cortex). Although there is little argument about the central importance of the anatomical organization of thalamic inputs to cortex, the functional roles these inputs play in the cortical circuit have been the subject of much debate.

Recent studies in the cat, however, are creating a foundation for understanding the importance of thalamic inputs in visual cortical processing.

In the cat, neurons in the principal layers of the LGN—layers A and A1—provide the majority of thalamic input to visual cortex (see Chapters 1 and 5). This input is directed most densely to neurons in cortical layer 4, predominantly to simple cells. As is clear from all of the chapters in this volume, the visual cortex of the cat has unique features that should not be confused with those of other species. In particular, the simple cell in the cat need not bear any relation to first-order neurons in the primate visual cortex. In this chapter, we discuss the functional organization of the geniculocortical pathway in the cat with respect to the response properties of layer 4 simple cells. We review experiments that have examined the specificity and strength of geniculate inputs to layer 4 neurons. With regard to specificity of geniculate inputs, it is important to know the extent to which cortical receptive fields "match" their thalamic input. With regard to strength of geniculate inputs, an important issue is the extent to which monosynaptic inputs from the LGN drive layer 4 responses. This issue can be further expressed in several ways. In the most extreme version, what would be the influence of thalamic input if all cortical spiking activity were silenced? In a less extreme version, what percentage of the excitatory synaptic input to a layer 4 neuron is derived directly from the thalamus, and what percentage from intracortical sources? Finally, what is the strength of multiple, convergent inputs from the thalamus, and what is the time course for their interactions?

SIMPLE RECEPTIVE FIELDS

In their 1962 paper on visual cortical receptive fields, Hubel and Wiesel first proposed that simple cells of cat area 17 are dominated by their thalamic input. The proposal was based primarily on the structure of simple receptive fields, which in principle could be explained by an orderly set of connections from multiple thalamic afferents. The possible importance of direct thalamic input was also suggested by their finding that the great majority of layer 4 neurons were simple cells. Although less was known about the laminar organization of inputs than is known today, it was recognized at least from the time of Cajal (DeFelipe and Jones, 1988) that most thalamic inputs terminate in layer 4.

In the original definition (Hubel and Wiesel, 1962), the receptive fields of simple cells

> were termed "simple" because like retinal and geniculate receptive fields (1) they were subdivided into distinct excitatory and inhibitory regions; (2) there was summation within the separate excitatory and inhibitory parts; (3) there was antagonism between excitatory and inhibitory regions; and (4) it was possible to predict responses to stationary or moving spots of various shapes from a map of the excitatory and inhibitory areas.

The first part of this definition, the existence of separate excitatory and inhibitory regions (also called *on* and *off* regions; Fig. 8-1) deserves emphasis, because it is

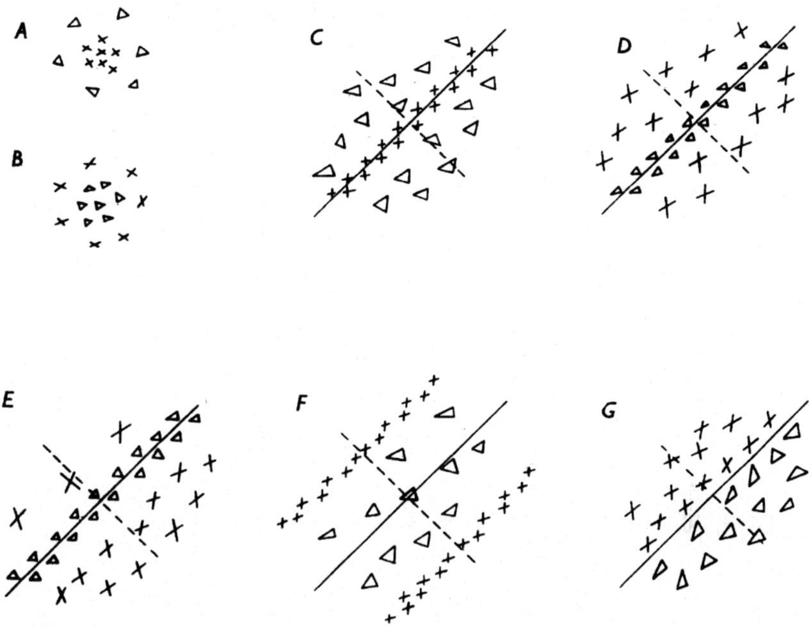

FIGURE 8-1. Receptive fields of two LGN cells **(A, B)** and five simple cells **(C-G).** Regions giving excitatory responses to light (*on* responses) are marked with Xs; regions giving inhibitory responses (*off* responses) are marked with Δs. LGN cells have center-surround receptive fields. Simple cells have receptive fields with elongated and adjacent *on* and *off* subregions. Receptive fields of simple cells are shown with an oblique orientation, but all orientations are present in area 17. (From Hubel and Wiesel, 1962, Figure 2.)

occasionally ignored. The last three criteria in this definition can be grouped together under the heading of linear spatial summation (Skottun et al., 1991). Although simple receptive fields generally have two or three subregions (sometimes four) (Movshon et al., 1978; Kulikowski and Bishop, 1981; Mullikin et al., 1984; Jones and Palmer, 1987; Reid et al., 1997), cortical neurons with only a single *on* or *off* region would not be termed simple cells by the original definition, even though they might behave linearly. These single-subunit neurons are occasionally encountered in layer 4 of the cat (although more common in layer 2/3) (Kato et al., 1978) but are much more common in other species, such as the macaque monkey. It is important, therefore, that linearity of summation not be the only criterion used to classify simple cells (however, see Skottun et al., 1991).

Given the elongated *on* and *off* regions of a simple receptive field, Hubel and Wiesel (1962) proposed that the receptive field might be built up from the convergence of multiple geniculate neurons whose receptive-field centers were aligned in a row (Fig. 8-2). In this example, an obliquely oriented simple receptive field with a central *on* region flanked by two *off* regions might be built up from multi-

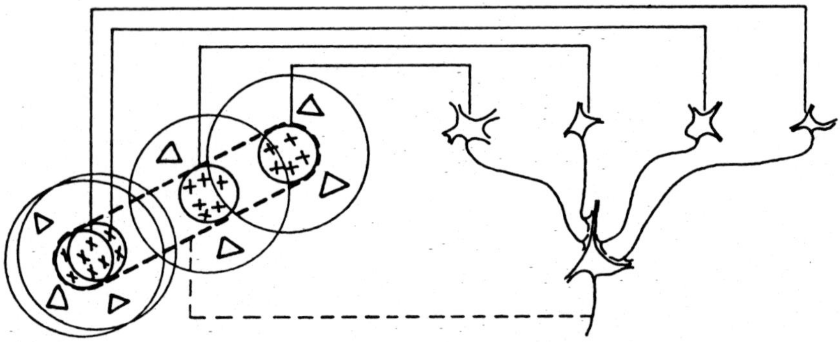

FIGURE 8-2. Hubel and Wiesel's proposed model for explaining the generation of simple-cell receptive fields. The simple cell receives convergent input from multiple LGN cells (only four shown) whose receptive fields are of the same sign (in this example, *on* center) and are located along a line of visual space. As a result, the simple cell has a receptive field with an elongated *on* subregion (indicated by the interrupted lines in the receptive-field diagram) flanked by two off subregions. (From Hubel and Wiesel, 1962, Figure 19.)

ple geniculate afferents whose *on* centers are aligned appropriately. In addition, there might be convergence of both *on* and *off* geniculate neurons, for instance if the flanks were "reinforced or enlarged by appropriately placed off-center geniculate neurons" (Hubel and Wiesel, 1962, p. 141; referring to cell **C** in Fig. 8-1).

In their 1962 paper, Hubel and Wiesel distinguished a second class of cortical cells, called complex cells. By the original definition, complex cells are orientation-tuned cells that are not simple cells. In the cat, most complex cells have both *on* and *off* responses at each position in their receptive field. Hubel and Wiesel (1962) proposed a hierarchical model to account for the complex receptive field: complex cells receive input from multiple simple cells (see Alonso and Martinez, 1998; cf. Ghose et al., 1994).

The Hubel and Wiesel model of simple receptive fields can be interpreted in several forms. In its original form, the receptive fields of geniculate afferents to a simple cell are aligned in rows that match, and generate, the *on* and *off* subregions of the simple receptive field. This need not exclude a role for intracortical processes in shaping a simple cell's behavior. An extended form of the model would be that other properties of simple cells, such as the sharpness of orientation tuning and direction selectivity, are derived from thalamic input. The most extreme form of the model, that *all* simple-cell properties are derived entirely from thalamic input, of course cannot be defended. Numerous aspects of cortical processing—contrast gain-control, push-pull organization of receptive-field regions (see later), influences beyond the classical receptive field—cannot be explained in terms of the geniculate afferents.

Since the inception of the Hubel and Wiesel model of the simple cell there has been considerable debate over the degree to which any of its forms is true. Much of the debate, however, has been based on differing conceptions of the model. There is increasing evidence supporting the role of direct geniculate afferents in constructing the *on* and *off* subregions of the simple cell. Similarly, evidence is accumulating that much of orientation and even direction selectivity might also be explained by the thalamic afferents, although these issues are still actively debated. In this chapter, we examine the evidence for different forms of the model.

NUMERICAL ASPECTS OF THE GENICULOCORTICAL PROJECTION

The first question one could ask about the geniculocortical pathway is, how many different geniculate afferents converge onto a layer 4 simple cell? Several different strategies can be used to estimate this number. Ideally, the answer would come from an individual study, but, in practice, disparate data must be combined across studies. Here, we examine the question from several different standpoints.

The anatomy of the geniculocortical system can give an upper bound on the number of geniculate afferents to a region of visual cortex. As reviewed by Peters and Payne (1993) the upper limit of potential X-cell afferents to any point in area 17 can be estimated by multiplying the total number of afferents by the size of their axonal arbors, then dividing by the total cortical area (240,000 cells $\times$ 0.6 to 0.9 mm^2/399 mm^2 = 360 to 540 cells). Similarly, the number of Y cells projecting to each point on the cortex is roughly 300 to 540 (120,000 cells $\times$ 1.0 to 1.8 mm^2/399 mm^2). Each of the numbers in these calculations is prone to considerable error, so the estimates are quite rough.

These estimates refer not to the number of afferents to a single target neuron, but to a single cortical column. Independently, one could also estimate the maximum number of geniculate synapses onto a single layer 4 spiny stellate cell. Again following Peters and Payne (1993), this number can be estimated using two different approaches: either 1) by estimating the number of synapses from each thalamic afferent, or 2) by estimating the number of synapses onto each layer 4 cell.

In the first approach, the strategy is first to estimate the total number of geniculate synapses to all of area 17 and then divide this value by the total number of potential targets. There are roughly 3300 synapses per X-cell arbor (Freund et al., 1985b; Humphrey et al., 1985; see Peters and Payne, 1993), which gives a total of 8×10^8 synapses for the afferent pool of 240,000 X-cells. There are roughly 8.4×10^8 Y-cell synapses in area 17 (7,000 synapses/cell $\times$ 120,000 cells). There are roughly 10 to 12×10^6 neurons in layer 4 (Beaulieu and Colonnier, 1987; Peters and Yilmaz, 1993). Therefore, assuming that all cells in layer 4 receive an equal number of thalamic inputs, there are roughly 150 afferent synapses per layer 4 neuron (16.4×10^8 synapses/11×10^6 neurons). Finally, this estimate of synapses per cell (150) needs to be corrected by subtracting the synapses onto the dendrites of cells from other layers. As outlined by Peters and Payne (1993), approximately 62.5% of synapses in layer 4 are onto layer 4 neurons.

Thus a spiny stellate cell in layer 4 should receive approximately 94 synapses from the LGN (150 × 62.5%).

In the second approach, the number of thalamic synapses per cortical target can be estimated from the standpoint of the cortical cell: by dividing the total number of excitatory synapses by the percentage of these synapses that come from the thalamus. As shown in two separate studies (Beaulieu and Colonnier, 1987; Anderson et al., 1988), there are roughly 2500 excitatory synapses onto a layer 4 neuron. Estimates of the percentage of thalamic synapses spans a broad range: from roughly 25% (LeVay and Gilbert, 1976; Einstein et al., 1987) down to 5% (Ahmed et al., 1994; also calculated by Peters and Payne, 1993, from the data of Humphrey et al., 1985 and Beaulieu and Colonnier, 1985). If the higher value is used, then this would yield 625 geniculate synapses per layer 4 neuron (2500 × 25%). A number more in line with the first estimate (from the standpoint of the geniculate synapses) is obtained if the lower value is used: 5% × 2500 = 125 synapses per cell.

Although these anatomical arguments provide an estimate of the number of individual geniculate synapses onto any given simple cell, it is a separate question to ask how many *different* geniculate neurons provide these synaptic contacts. One way of addressing this question would be to determine the number of synapses a single geniculate afferent provides to a single layer 4 neuron. Freund and colleagues (Freund et al., 1985a) attempted to determine this number anatomically. They labeled geniculate axons with horseradish peroxidase and then Golgi-stained the tissue to label cortical neurons. In these difficult experiments, they found that most LGN cells provided only one synapse onto any given target, although one Y cell provided eight synapses onto a layer 3 pyramidal cell. Unfortunately, from such a small sampling (two X axons and two Y axons in area 17; two Y axons in area 18), it is difficult to know if the axons studied were representative.

A second way of addressing the same question—how many geniculate afferents converge onto a single simple cell?—relies on a combination of physiological and anatomical data. First, one would need to know the relative size of geniculate and cortical receptive fields. Second, one would need to know the coverage factor of geniculate neurons, that is, the number of receptive-field centers that overlap at any given point in visual space. Finally, one would need to know the probability that an LGN cell is connected to a simple cell, given that the LGN center overlaps some part of the simple receptive field. Given these three numbers, the total number of geniculate afferents that connect to a given simple cell, N, would be given by the product:

$$N = A \cdot C \cdot p, \tag{1}$$

where A is the area of a simple receptive field (as a multiple of a typical geniculate center), C is the coverage factor of geniculate centers, and p is the probability that cells with overlapped receptive fields are monosynaptically connected. Although some of these values have been known for quite some time, recent studies have provided numbers that complete the analysis.

The relative size of simple receptive fields (A) can be estimated from the physiological literature. From several studies (Bullier et al., 1982; Reid and Alonso, 1995; Alonso et al., 2001), it is clear that the average width of a simple-cell subregion is roughly the same size as a geniculate receptive-field center at the same eccentricity (as in the diagram of the Hubel and Wiesel model, Fig. 8-2). Estimates of the length divided by the width of a simple-cell subregion, or its *aspect ratio,* show greater variability. Because the aspect ratio is closely related to the predicted orientation tuning of a simple cell, it has been the subject of many studies. Several studies of simple receptive fields have each found a broad range of aspect ratios: range 1.1 to 6, median ~3 (Watkins and Berkley, 1974); range 1.7 to 12, median ~5 (Jones and Palmer, 1987); range 2 to 12, median ~3.5 (Gardner et al., 1999). An intracellular study has given a smaller estimate: range 1 to 4, mean 1.7 (Pei et al., 1994). Finally, two extracellular studies that concentrated on layer 4 neurons have found lower values as well: ~2 in one study (Bullier et al., 1982) and a range from 1.2 to 5.5, median 2.3, in another (Alonso et al., 2001).

We use this latter value (2.3) here because our study concentrated on layer 4 simple cells; simple cells outside of layer 4, particularly in layer 6, have been found to have greater length summation (Gilbert, 1977; Bolz and Gilbert, 1989; see also Grieve and Sillito, 1995). As noted by Bullier and colleagues (1982), the anatomical spread of a geniculate arbor (Ferster and LeVay, 1978; Bullier and Henry, 1979c; Freund et al., 1985b; Humphrey et al., 1985) extends over a distance equivalent to approximately two geniculate centers. From these arguments, we can say that a typical simple cell has two to three subregions, each with an aspect ratio of 2.3, so $A \approx 6$ in equation 1.

Our estimate of the coverage factor, C, of geniculate X cells is based on the coverage factor of retinal ganglion cells and on the divergence from retina to LGN. First, the dendritic arbors of *on* and *off* retinal X cells have been shown each to have a coverage factor of roughly three (Wässle et al., 1981). This number is derived by first counting the number of *on* or *off* retinal ganglion cells in 1 mm^2 of retina at a given eccentricity, and then determining the average area occupied by each dendritic arbor. The product of these values yields the anatomical coverage factor, that is, the total dendritic area of all ganglion cells per mm^2 of retina, which is equivalent to the average number of dendritic arbors that overlap any point in the retina. Because the size of the dendritic arbor has been shown to be approximately equal to the size of the receptive-field center (as has been shown for Y cells, Wässle et al., 1983; compare also the size of X-cell arbors, Wässle et al., 1981, with X-cell receptive field centers, Linsenmeier et al., 1982), the anatomical coverage factor is equal to the physiological coverage factor—the number of times each point in visual space is represented by a given class of visual neuron. Thus each point in visual space is represented by the receptive-field centers of three *on*-center and three *off*-center retinal ganglion cells. Finally, because there are roughly 2.5 times as many LGN relay X cells as retinal X cells (reviewed in Peters and Payne, 1993; discussed previously) and their receptive-field centers are essentially the same size (Hubel and Wiesel, 1961; Cleland and

Lee, 1985; Mastronarde, 1987b; Usrey et al., 1999), the total coverage factor of LGN relay cells is: $C = 6 \times 2.5 = 15$.

Therefore we can estimate the total number of geniculate cells whose receptive-field centers overlap the receptive field of a layer 4 simple cell: $N \times C = 6 \times 15 = 90$. In our own work (discussed later, Reid and Alonso, 1995; Alonso et al., 2001; see also Tanaka, 1983, 1985), we have found that if there is such overlap between an LGN and a simple receptive field, there is roughly a one in three chance that the cells are monosynaptically connected ($p = 0.33$). Thus, approximately 30 different LGN X cells provide monosynaptic input to a layer 4 simple cell. Y cells also provide monosynaptic input to layer 4 simple cells, but the number of Y cells that provide input to individual simple cells is difficult to estimate from similar lines of reasoning.

FEEDFORWARD (THALAMIC) CONNECTIONS AND SIMPLE CELL RESPONSES

Hubel and Wiesel (1962) proposed that the basic structure of the simple cell receptive field—elongated and adjacent, *on* and *off* subregions—might result from a convergence of LGN inputs whose receptive fields were (1) located along a line of visual space and (2) were of the same sign (Fig. 8-2). Over the years, this model has been the subject of tremendous controversy. Much of the controversy stemmed from studies showing that LGN synapses onto simple cells actually represent the minority of excitatory synapses, the vast majority of excitatory synapses onto simple cells come instead from intracortical sources. The number of synapses, however, may not be a good indicator of the strength of a given class of inputs. For example, cells in the LGN receive roughly three times more synapses from the cortex than from the retina (Guillery, 1969; Wilson et al., 1984; Erisir et al., 1998), but the retinogeniculate connection is clearly the strongest (Sherman and Guillery, 1998). Indeed, the retinogeniculate connection is perhaps the strongest in the visual system; in some geniculate cells virtually all spikes are driven by a single retinal afferent (Cleland et al., 1971; Cleland and Lee, 1985; Mastronarde, 1987b; Usrey et al., 1999).

Recent studies have demonstrated that although thalamic inputs to cortex may be sparse (described previously), their strength goes beyond what might be expected from a simple comparison of bouton counts. First, thalamocortical synapses are significantly larger and contain more release sites than synapses from intracortical sources (Ahmed et al., 1994). Second, excitatory currents are larger for thalamocortical synapses than for intracortical synapses (Stratford et al., 1996; see Gil et al., 1999). Third, it has long been known that firing rates in the LGN are on average higher than those in visual cortex. Finally, precisely correlated inputs from the LGN interact synergistically in driving their cortical targets (Alonso et al., 1996; Usrey et al., 2000; see later). Taken together, these results indicate that the influence of LGN inputs to simple cells is most likely quite strong. But even if thalamic inputs to simple cells are strong, the question remains: Is the structure of a simple cell receptive field the result of an organized pattern of connections from the LGN?

EVIDENCE OF MONOSYNAPTIC INPUT TO SIMPLE CELLS: ELECTRICAL STIMULATION

Early studies of the physiology of geniculate inputs to area 17 examined the combined influence of ensembles of afferents. In numerous studies from several groups (Creutzfeldt and Ito, 1968; Stone and Dreher, 1973; Singer et al., 1975; Bullier and Henry, 1979a, 1979b, 1979c; Ferster and Lindstrom, 1983; Martin and Whitteridge, 1984), single neurons in area 17 were studied extracellularly or intracellularly while their inputs were stimulated electrically. By comparing the latencies of activation after stimulation at different sites, either retinal axons or geniculate axons, these studies could determine whether fast or slowly conducting afferents were responsible for the evoked activity, as well as whether the thalamocortical influence was monosynaptic or polysynaptic. Cortical neurons were thus classified according to their afferent streams (fast or slow) and their ordinal position in the cortical circuit (that is, monosynaptically or polysynaptically activated from the LGN).

Several important conclusions were drawn from these studies. First, they distinguished two major pathways from LGN to visual cortex, which correspond to the X and Y cells. These two pathways could be roughly separated by the conduction velocities of both the retinogeniculate and geniculocortical axons. Second, they found consistent laminar segregation of inputs. Both pathways terminated predominantly in layer 4 (although layer 3 and layer 6 neurons also received monosynaptic input). Within layer 4, the fast Y pathway terminated above the slower X pathway, although with some overlap (which is consistent with the anatomy: LeVay and Gilbert, 1976; Bullier and Henry, 1979c; Humphrey et al., 1985). Virtually all layer 4 simple cells received monosynaptic input from the LGN, which is consistent with the first stage of the Hubel and Wiesel model (1962; LGN $\rightarrow$ simple cells). In apparent contradiction to the second stage of the model (simple cells $\rightarrow$ complex cells), they also found that some complex cells received monosynaptic input from the LGN, although many complex cells received only polysynaptic input (see Alonso and Martinez, 1998).

INDIVIDUAL LGN INPUTS TO SIMPLE CELLS: STRENGTH AND RECEPTIVE-FIELD PROPERTIES

Although electrical stimulation provided strong evidence for monosynaptic input to simple cells, as a technique it was unable to demonstrate either the strength of single inputs or their receptive-field properties. These difficulties were overcome by studies that used simultaneous recordings of cortical cells and their afferents (Lee et al., 1977; Tanaka, 1983, 1985; Reid and Alonso, 1995; Alonso et al., 1996, 2001). In the first of these studies, Lee and colleagues recorded from six disynaptically connected pairs of retinal ganglion cells and cortical simple cells. Later, Tanaka recorded from a much larger sample of monosynaptically connected pairs in the LGN and cortex. Both studies demonstrated that (1) individual connections were fairly strong, often accounting for up to 10% of the spikes in a simple cell, (2) connected cells had appropriately overlapped receptive fields (e.g., retinal *on* centers over *on* subregions of a simple cell), and (3) both sustained and transient (X and Y) afferents projected to simple cells, sometimes both onto a single neuron. In addition, Tanaka also showed that simple cells can receive input from both *on* and *off* LGN afferents and confirmed that some complex cells can receive monosynaptic input from the LGN. These studies were the first of their kind, but they did not systematically explore the factors determining whether a specific afferent would be connected to any given cortical target.

Implicit in the Hubel and Wiesel model (Fig. 8-2) is that certain geniculate afferents should have a high probability of connection to a particular simple cell, while others should not. In more recent studies of simultaneously recorded LGN cells and simple cells, our laboratory (Reid and Alonso, 1995; Alonso et al., 2001) has specifically examined the relationship between receptive field overlap and probability of connection (Fig. 8-3). These studies confirmed that connections were made between pairs of cells with appropriately overlapped receptive fields. Further, they demonstrated that the probability of connection approached 100% when the center of an LGN receptive field was perfectly overlapped with the exact center of a simple subregion. When the sign of the two receptive fields were the same, but overlap was less precise—either along the length or the width of the subregion—then the probability of a connection decreased. Perhaps more important, the same studies also demonstrated that inappropriate geniculocortical connections are rarely made; that is, if the receptive field of an LGN cell is nonoverlapping or of the opposite sign *(on* versus *off)* of an overlapped simple-cell subregion, monosynaptic connections are extremely rare.

LGN AFFERENTS AND ORIENTATION SELECTIVITY

Indirect evidence of the relationship between the orientation tuning of cortical neurons and the organization of LGN afferents was demonstrated by Chapman and colleagues (Chapman et al., 1991). In ferret visual cortex, they compared the orientation preference of neurons in a column to the receptive-field locations of LGN axons that provided input to that column. Their results demonstrated that the distribution of LGN receptive fields was often elongated, and that the axis of orientation matched the orientation preference of nearby layer 4 neurons. Although this study did not examine the geniculate inputs to individual layer 4 neurons, it demonstrated that the ensemble of inputs to a column are not random, but rather match the orientation preference of potential target cells.

More direct evidence for the role of thalamic input for orientation selectivity came from a series of studies by Ferster and colleagues, who examined the contribution made by LGN inputs to simple cell responses (in the cat) in the absence of corticocortical inputs. In these studies, whole-cell recordings were made from simple cells while cortical activity was silenced or greatly diminished by either cortical cooling (Ferster et al., 1996) or by electrically evoked suppression (Chung and Ferster, 1998). In both studies, the orientation tuning of synaptic

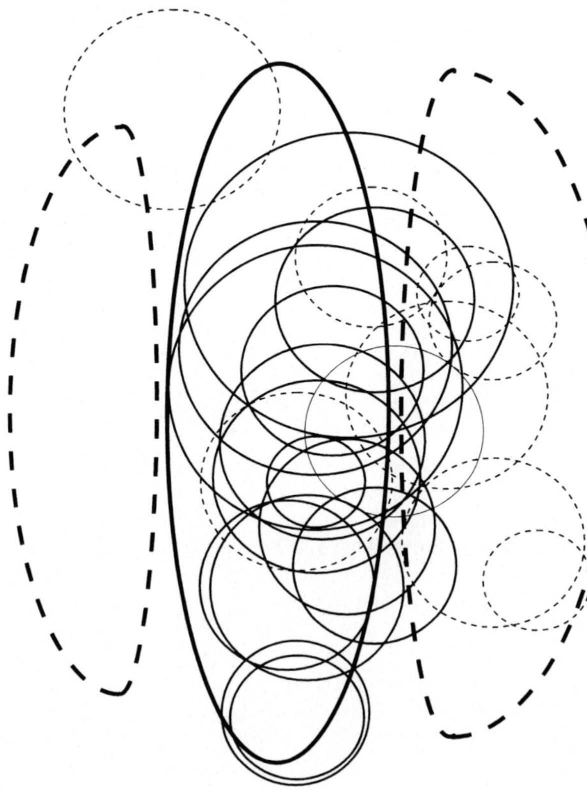

FIGURE 8-3. Summary diagram showing the relationship between the receptive fields of monosynaptically connected pairs of LGN cells and simple cells. Receptive fields were mapped with white-noise stimuli; connectivity was assessed with cross-correlation analysis. For this summary diagram, the receptive field of each simple cell was transformed into a stylized receptive field (shown in the figure with a central *on* subregion, indicated with a solid ellipse, flanked by two *off* subregions, indicated with dashed ellipses). The dashed and solid circles indicate the relative size, sign, and locations of the receptive fields of monosynaptically connected LGN cells. Of the 23 pairs of connected cells (out of a total of 74 pairs recorded), there were very few mismatches in the sign of the geniculate center and the cortical subregion. (From Reid and Alonso, 1996.)

inputs was measured either in the intact circuit or when the geniculate inputs were isolated. As shown in Fig. 8-4 (Chung and Ferster, 1998), the tuning curves under both conditions were quite similar. This result suggests that LGN inputs alone should be sufficient to generate orientation-selective responses of cortical simple cells. It should be noted, however, that in both studies the size of the currents evoked by the visual stimulus were greatly diminished in the absence of intracortical input. Thus it was estimated that the thalamus provides on average 35% (Ferster et al., 1996) to 46% (Chung and Ferster, 1998) of the total excitatory drive to simple cells.

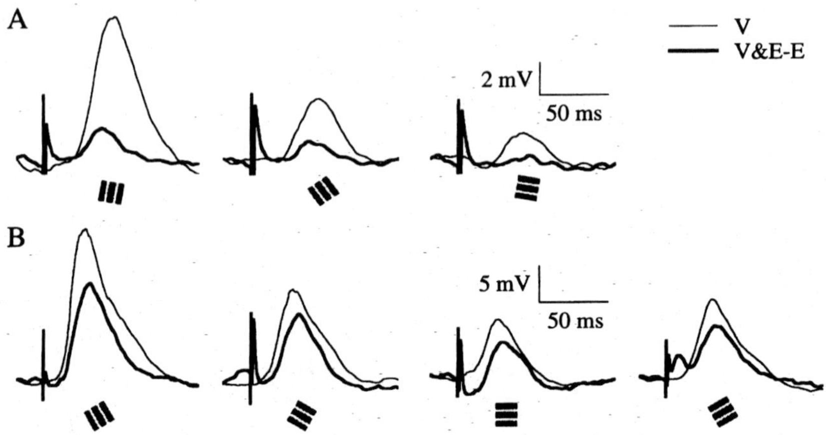

FIGURE 8-4. Intracellular responses of cortical simple cells to flashed stimuli of different orientations, either under normal conditions (thin lines) and in the absence of intracortical inputs (thick lines). In the normal condition, intracellular voltage was measured after the visual stimulus (V) alone. To silence intracortical inputs, an electrical stimulus (E) was delivered to the nearby cortex. The thick line represents the visually evoked responses following electrical stimulation, minus the effect of the electrical stimulation alone (V&E-E). For both neurons shown (**A** and **B**), the responses in the absence of cortical input were smaller, but the relative tuning for orientation remain unchanged. At all orientations, the proportion of visual responses remaining during cortical suppression was ~25% for cell **A** and ~65% for cell **B**. (From Chung and Ferster, 1998.)

Recent models of the generation of orientation selectivity (Ben-Yishai et al., 1995; Somers et al., 1995; Vidyasagar et al., 1996) have proposed a dominant role for intracortical inputs—excitatory inputs in particular. These models are based on weak input from the LGN, with slight (Vidyasagar et al., 1996) or moderate (Somers et al., 1995) orientation bias, that is both amplified—reminiscent of other amplification models (Douglas et al., 1989; Douglas and Martin, 1991; Maex and Orban, 1992; Douglas et al., 1995; Suarez et al., 1995)—and sharpened by intracortical interactions. The results of Reid and Alonso (1995) and, even more so, of Ferster and colleagues (Ferster et al., 1996; Chung and Ferster, 1998) contradict the notion that the sum of the LGN inputs is poorly oriented. Nevertheless, Ferster's finding that the strength of excitatory input decreases during cortical blockade is not inconsistent with the basic idea of cortical amplification, which in one version has the LGN providing 18% of the excitatory drive to simple cells (Douglas et al., 1995).

LGN AFFERENTS AND DIRECTION SELECTIVITY

Many simple cells in layer 4 of cat visual cortex are selective for the direction of stimulus motion. One of the most important mechanisms in setting up this receptive-field property relies on the timing of visual responses at different posi-

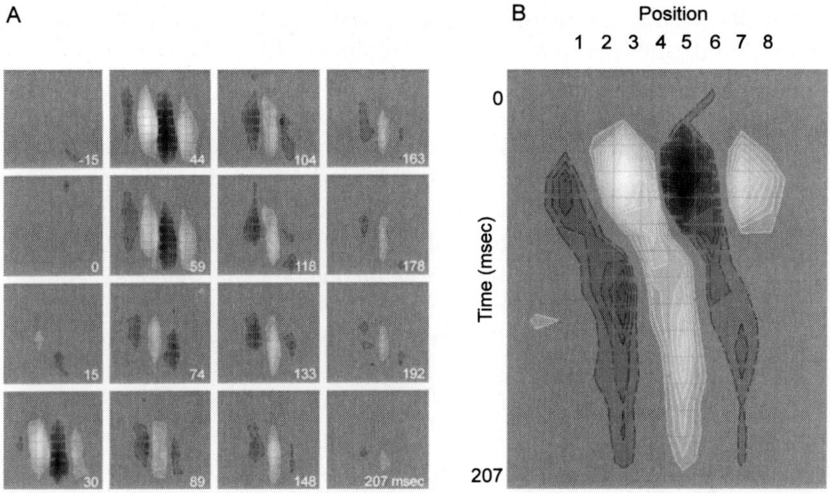

FIGURE 8-5. Spatial and temporal aspects of a simple receptive field. **(A)** Time-evolution of the spatial receptive field, as mapped with a white-noise visual stimulus. Regions that responded to a light stimulus (*on* responses) are shown lighter than the gray background; regions that responded to a dark stimulus (*off* responses) are shown darker than the gray background. Each frame represents the average spatial stimulus that preceded a spike from the layer 4 simple cell, for delays ranging from –15 to 207 msec, in steps of 14.8 msec. For instance, in the first frame of the second row (delay: 44 msec), there are four vertically oriented subregions, two *on* (shown in white) and two *off* (shown in black). The entire "movie" illustrates how these subregions shift over time, particularly in the second column. **(B)** Spatiotemporal representation of the same simple receptive field. Each row corresponds to one frame in **A** (from –15 to 207 msec); each column corresponds to the receptive field profile along the X-dimension of each frame, obtained by summing all points along the vertical (Y) axis. Note that the spatial profile of the receptive field shifts over time (that is, the "spatiotemporal receptive field" is oriented), so that the neuron was more strongly stimulated by a stimulus moving with a similar space-time orientation, or velocity. (From Reid et al., 1997)

tions within the simple receptive field. As demonstrated by a number of studies (Movshon et al., 1978; Dean and Tolhurst, 1986; Reid et al., 1987; McLean and Palmer, 1989; Reid et al., 1991; Jagadeesh et al., 1993; McLean et al., 1994), there are steady changes in response timing for stimuli located at different positions along the width of the receptive field. This property is best appreciated in a space-time plot of a receptive field (Fig. 8-5), in which the different time-courses of response to flashed stimuli are plotted as a function of spatial position. Assuming linear summation, the responses to a drifting bar should be predictable from such a space-time plot. If the bar traverses the receptive field in the preferred direction, the inputs from different positions arrive at the same time and produce a strong response. In the nonpreferred direction, the inputs arrive at different times and the peak firing rate is smaller.

Given this model, the question arises: What establishes the time course of responses at different positions within the receptive field? The simplest proposal was put forth by Saul and Humphrey (1992; see also Murthy and Humphrey, 1999 and Chapter 9), based on Mastronarde's description of two classes of geniculate relay cells with dramatically different response characteristics (Mastronarde, 1987a, 1987b). The cells, termed lagged and non-lagged cells, exhibit a range of temporal properties similar to the range found within a directional simple cell (Saul and Humphrey, 1992). Hubel and Wiesel proposed that simple cell subregions were set up by segregated input from *on* and *off* cells. Saul and Humphrey extended this idea by proposing that there was an additional segregation of lagged and non-lagged thalamic inputs. Subsequent work by Alonso et al. (2001) found some support for this model, by examining the relationship between the temporal properties of individual LGN cells and the probability of their connecting to an overlapped simple cell. Although few overtly lagged cells were studied, LGN cells with a range of temporal properties were examined. Monosynaptic connections were most likely to be found when LGN and cortical neurons responded with a similar time-course to stimuli at the same point in visual space.

Although the organization of afferents might predict the preferred direction of a simple cell, the inputs by themselves might not be sufficient to produce the degree of direction selectivity observed (see Murthy and Humphrey, 1999). The intracellular experiments of Ferster and colleagues (Ferster et al., 1996) also examined this issue. Several of the neurons that were recorded in the cortical cooling experiment were direction selective. Direction-selectivity was not diminished during cortical inactivation; in fact, some cells that were not initially direction selective developed direction selectivity in the absence of cortical input. Thus afferent input appears sufficient to create substantial direction selectivity, at least in some cells.

INTEGRATION OF THALAMIC INPUTS: SYNCHRONY AND SYNERGY

In the previous section, we reviewed a number of studies that examined the specificity of connections present in the pathway from LGN to visual cortex. These studies demonstrated that many of the response properties of layer 4 simple cells are a reflection of their thalamic input. Implicit in these studies is the notion that many thalamic inputs converge onto simple cells to help generate their visual responses. This section examines how simple cells integrate input from multiple LGN inputs. Anatomical and physiological estimates suggest that simple cells receive convergent input from approximately 30 LGN cells (see preceding). The question therefore arises: How do multiple LGN inputs interact to drive simple cells?

LGN inputs that arrive simultaneously to a simple cell have been shown to interact in a nonlinear or synergistic fashion to drive simple cell action potentials (Alonso et al., 1996). As shown in Fig. 8-6, inputs that arrive within 1 msec of each other from two presynaptic LGN axons are more likely to drive a simple-cell spike than would be expected if the inputs were to interact in a simple linear fash-

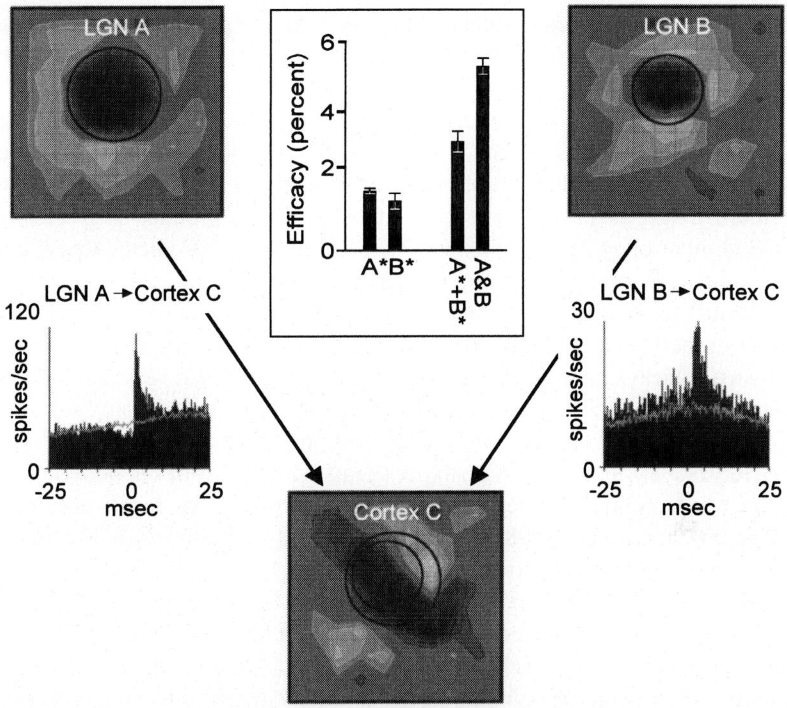

FIGURE 8-6. Action potentials that arrive simultaneously from two LGN cells interact synergistically to drive simple cells. The receptive fields of three simultaneously recorded neurons are shown: two *off*-center LGN cells (**A** and **B**) and a simple cell (**C**), all with overlapping receptive fields. Receptive fields were mapped with a white-noise stimulus, as in Fig. 8-5. The cross-correlograms at the sides of the figure represent the average firing rate of the cortical cell with respect to geniculate spikes (negative times before the geniculate spike, positive times after). The short latency and fast rise time of the peak, at approximately 2.5 msec, indicates the presence of a monosynaptic connection between the LGN cell and the simple cell. The histogram in the center of the figure shows the efficacy—the percentage of LGN spikes that evoked a simple cell spike—of simultaneous LGN spikes (A&B; defined as <1 msec apart) and nonsimultaneous LGN spikes (A*, B*; >1 msec apart). The efficacy of LGN spikes that occur simultaneously (A&B) between the two LGN cells was 70% greater than the sum of the efficacies of nonsimultaneous LGN spikes (A*+B*). Thus LGN spikes that occur simultaneously reinforce each other and are better at driving a simple-cell action potential. (From Alonso et al., 1996.)

ion. An examination of the time course for these interactions further suggests that LGN inputs must arrive within a narrow time window for reinforcement to occur. Interactions are greatest when spikes arrive within 1 msec of each other, interactions are modest when they arrive within 2 to 5 msec of each other, and spikes that arrive at times greater than approximately 6 msec of each other display little or no interaction (Usrey et al., 2000). Azouz and Gray (2000) suggested that the enhanced sensitivity of cortical neurons to synchronous synaptic input might rely, in part, on the voltage-gated currents that underlie action potential generation (see also Hirsch et al., 1995; Mainen and Sejnowski, 1995; Carandini et al., 1996; Nowak et al., 1997; Volgushev et al., 1998).

In the ensemble of LGN afferents to a given simple cell, what percentage of LGN spikes occur within the time window of maximal reinforcement? Although any pair of LGN cells will randomly produce a fraction of their spikes simultaneously, there exists a specific subset of LGN cells that produce a greater than expected number of spikes simultaneously. Cleland (1986) predicted that if a pair of LGN cells received divergent input from a common retinal ganglion cell, then spikes occurring in this retinal ganglion cell should often drive synchronous responses in the two target cells. Recordings from nearby cells in the LGN with multielectrode arrays have shown that pairs of LGN cells with very similar receptive fields—in terms of receptive-field location and size, *on* versus *off* responses, and X versus Y classification—fire a large percentage of their spikes simultaneously (Alonso et al., 1996). Further, experiments with electrodes in both the retina and LGN have shown that these synchronous spikes are the result of common retinal input (Usrey et al., 1998). As shown in Fig. 8-7, the time window for synchronous responses between two tightly correlated LGN cells is quite fast, usually within 1 msec. Further, the percentage of spikes occurring synchronously can be quite high. Up to 40% of the spikes between two tightly correlated LGN cells can occur within 1 msec of each other (Alonso et al., 1996; Usrey et al., 1998).

In addition to interacting synergistically to drive simple cells, are synchronous spikes between two tightly correlated LGN cells special in other respects? Examination of the patterns of retinal spikes that drive geniculate responses indicates that when a pair of retinal spikes occur within approximately 3 to 40 msec of each other, the second spike is much more likely than the first to drive a postsynaptic action potential in the LGN (Mastronarde, 1987b; Usrey et al., 1998). These more effective second retinal spikes are also more likely to drive synchronous spikes in two LGN targets. Thus the degree of synchronous firing in the LGN increases when there is a transition from low to high firing rates in the retina. Synchronous LGN spikes are also special with respect to information coding. Dan and colleagues (1998) have shown that synchronous spikes between pairs of tightly correlated LGN cells can carry additional information to the cortex. Thus synchronous LGN spikes not only have an increased likelihood of driving a simple-cell response, but also supply the simple cell with added information about the visual stimulus.

INTRINSIC CONNECTIONS AND SIMPLE-CELL RESPONSES

Although they are the subject of this review, afferents from the LGN cannot be viewed in isolation from other inputs to layer 4 simple cells. Other important inputs originate locally within the cortical column or distally through horizontal

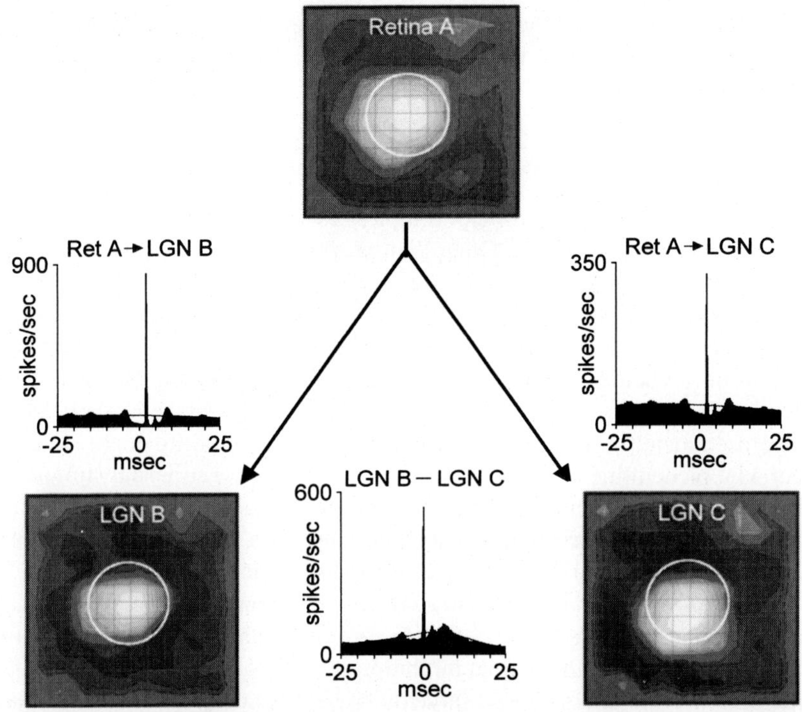

FIGURE 8-7. Retinal ganglion cells that provide divergent input to two LGN cells synchronize the firing of these LGN cells. Shown are the receptive fields of three simultaneously recorded *on*-center neurons (one retinal ganglion cell, **A,** and two LGN cells, **B** and **C**). Receptive fields were mapped with a white-noise stimulus, as in Figs. 8-5 and 8-6. The cross-correlograms at the sides of the figure each have a peak, displaced approximately 2 msec to the right of zero. These peaks indicate the presence of a monosynaptic connection between the retinal ganglion cell and each of the LGN cells. The cross-correlogram shown at the bottom of the figure (made by cross-correlating the spike trains of the two LGN cells) has a peak centered at zero. This peak is extremely narrow (full width at 1/2 max amplitude ~0.5 msec) and strong. Analysis of these synchronous LGN spikes shows that ~80% were triggered by a spike in the single retinal ganglion cell **(A).** (From Usrey et al., 1998.)

and feed-back connections (see Chapters 1,10, and 12). Layer 4 cells also receive nonthalamic subcortical inputs from the brainstem and claustrum. This section focuses only on the local intracolumnar connections, primarily inhibitory.

Most of the intracolumnar input to simple cells is from other simple cells. In cat area 17, these inputs originate primarily from neurons in layers 4 and 6. As discussed previously, most layer 4 cells are simple, as are many layer 6 cells (Hubel and Wiesel, 1962; Kelly and Van Essen, 1974; Gilbert, 1977; Kato et al., 1978; Bullier and Henry, 1979c), particularly those that project to layer 4 (Hirsch et al., 1998b). The receptive fields of layer 4 simple cells, therefore, seem to originate from the convergence of thalamic inputs and local simple cell connections.

Simple cells in layer 6 receive monosynaptic input from the LGN, but unlike layer 4 neurons, they receive a strong input from layer 5, where most cells are complex (see Figure 1-9).

INHIBITION AND ORIENTATION TUNING

In layer 4 of cat visual cortex, roughly 20% of synapses are inhibitory (Beaulieu and Colonnier, 1985). Other than preventing runaway excitation in the cortical circuit, what role does inhibition play in the generation of simple receptive fields? Soon after the availability of the GABA-A antagonist bicuculline, it became clear that many receptive-field properties are seriously affected by blocking inhibition. Whether injected intravenously (Pettigrew and Daniels, 1973; Daniels and Pettigrew, 1975), applied topically (Rose and Blakemore, 1974), or iontophoresed locally (Sillito, 1975; Tsumoto et al., 1979; Sillito et al., 1980; see Chapter 11), bicuculline made cells much more responsive to visual stimuli, but also less selective. As demonstrated by Sillito, when inhibition is blocked, simple cells lose much of their orientation selectivity, direction selectivity, ocular dominance, end-inhibition, and segregation of *on* and *off* subregions.

A generation of cross-orientation models was inspired by Sillito's findings (reviewed in Reid and Alonso, 1996). According to these models, orientation selectivity emerges from the mutual inhibition among cortical cells with orthogonal orientation preferences. These models, however, suffered a serious setback when a main prediction—that a bar moved orthogonal to the cell's preferred orientation should evoke maximal inhibition—was found to be incorrect. Data from intracellular recordings demonstrated that the strongest inhibition was evoked by stimuli at the cell's preferred orientation (Ferster, 1986; Douglas et al., 1991). A new interpretation of Sillito's finding thus seems attractive: Blocking inhibition allows otherwise weak inputs to undergo unchecked amplification by recurrent excitation (Somers et al., 1995; see Chapter 12). This view is supported by the finding that blocking inhibition intracellularly in a single neuron preserves its orientation selectivity (Nelson et al., 1994).

PUSH-PULL MODELS OF SIMPLE-CELL FUNCTION

So far, we have discussed intracortical pathways with respect to orientation tuning. Intracortical pathways, inhibition in particular, have also been suggested to help reinforce the *on* and *off* subregions in the simple receptive field. In addition to aligned excitatory LGN inputs, Hubel and Wiesel suggested that simple cells might receive precisely aligned inhibitory input of the opposite response polarity (Hubel and Wiesel, 1962, page 142):

> In the proposed scheme one should, however, consider the possibility of direct inhibitory connexions. In Text-fig. 19 [see Figure 8-2] we may replace any of the excitatory endings by inhibitory ones, provided we replace the corresponding geniculate cells by ones of the opposite type ('on'–centre instead of 'off'–centre, and conversely). Up to the present the

two mechanisms have not been distinguished, but there is no reason to think that both do not occur.

Although later studies have shown that all geniculate input is excitatory, a modified version of their proposal, later termed the push-pull model, has been put forward by a number of groups.

In most extracellular studies, the *on* and *off* subregions of simple cells have been defined in terms of the responses to light increment and decrement, or to light and dark stimuli, rather than in terms of excitatory or inhibitory currents. Even when the spatial antagonism between *on* and *off* subregions has been studied, the extracellular data cannot differentiate between active inhibition or the removal of excitation. For example, a light spot in an *off*-subregion should silence the activity of the *off*-center geniculate inputs and consequently reduce the response of the simple cell. Nevertheless, from several extracellular studies (Heggelund, 1981; Palmer and Davis, 1981; Glezer et al., 1982; Tolhurst and Dean, 1987) a push-pull mechanism was proposed in which *off*-inhibition was superimposed with *on*-excitation and vice-versa.

More recently, studies that involve intracellular recordings have explicitly examined the roles that excitation and inhibition play in the generation of simple receptive fields (Ferster, 1988; Borg-Graham et al., 1998; Hirsch et al., 1998a). These studies found strong inhibition in all simple cells, generally evoked by stimuli of the opposite sign that evoked excitation (but see Borg-Graham et al., 1998). Although the strongest push was found in the same receptive field position as the strongest pull (Hirsch et al., 1998a), push and pull were not always perfect mirror images of each other. Further, inhibitory inputs were overall stronger than excitatory inputs (Hirsch et al., 1998a). As suggested in a recent theoretical study (Troyer et al., 1998), a push-pull mechanism with dominant inhibition should allow simple cells to maintain their sharp selectivity for orientation over a wide range of stimulus contrasts, as is found experimentally (Sclar and Freeman, 1982).

CONCLUSIONS

Beginning in the 1970s, there has been slow but steady progress in our understanding of the physiology of the geniculocortical projection. In early studies involving electrical stimulation of geniculate afferents, cells that received strong thalamic input were identified, both in terms of their receptive-field physiology and their laminar location. In later work, intracellular recordings and cross-correlation analysis provided a closer look at the thalamic inputs to single neurons. In recent years, these two techniques have provided a qualitative characterization of the thalamic input to layer 4 simple cells, as well as a quantitative estimate of the strength of thalamic input.

In support Hubel and Wiesel's original model (1962), our laboratory has shown that the geniculate input to individual simple cells is precise. Geniculate cells provide synaptic input to simple cells only when their receptive fields spatially overlap a simple receptive field and match the sign *(on* or *off)* of the overlapping subregion (Reid and Alonso, 1995; Alonso et al., 1996, 2001). Ferster and colleagues (Ferster et al., 1996; Chung and Ferster, 1998) have shown that this thalamic input is sufficient to account for the orientation selectivity (and perhaps direction selectivity) of simple cells. These experiments also demonstrated, however, that thalamic input alone cannot account for all of the excitatory drive to a simple cell; approximately 50–70% (Ferster et al., 1996; Chung and Ferster, 1998) of the drive to a simple cell comes from intracortical sources. The question therefore remains: How does thalamic and intracortical input, both excitatory and inhibitory, interact to influence simple cell activity?

Intracellular studies performed in vivo have begun to provide an increasingly mechanistic understanding of the three major influences on simple-cell function: thalamocortical excitation, intracortical inhibition, and intracortical excitation. Of the three, thalamocortical excitation is certainly the best understood. New evidence for the push-pull model underscores the central role played by inhibition in the organization of *on* and *off* subregions. Finally, although there is less evidence for specific roles played by intracortical excitation in layer 4, the problem is certainly not intractable. As sophisticated methods developed in vitro become adapted to work in vivo, we should get an increasingly clear picture of the role of all three pathways and their interactions in the generation of visual responses.

ACKNOWLEDGMENTS

This work was supported by NIH grants EY06604, EY10115, EY12196, EY05253 and the Harvard Mahoney Neuroscience Institute.

REFERENCES

Ahmed, B., Anderson, J. C., Douglas, R. J., Martin, K. A., and Nelson, J. C. (1994). Polyneuronal innervation of spiny stellate neurons in cat visual cortex. *J. Comp. Neurol.* **341,** 39–49.

Alonso, J. M., and Martinez, L. M. (1998). Functional connectivity between simple cells and complex cells in cat striate cortex. *Nat. Neurosci.* **1,** 395–403.

Alonso, J. M., Usrey, W. M., and Reid, R. C. (1996). Precisely correlated firing in cells of the lateral geniculate nucleus. *Nature* **383,** 815–819.

Alonso, J. M., Usrey, W. M., and Reid, R. C. (2001). Rules of connectivity between geniculate cells and simple cells in cat primary visual cortex. *J. Neurosci.,* **21,** 4002–4015.

Anderson, P. A., Olavarria, J., and Van Sluyters, R. C. (1988). The overall pattern of ocular dominance bands in cat visual cortex. *J. Neurosci.* **8,** 2183–200.

Azouz, R., and Gray, C. M. (2000). Dynamic spike threshold reveals a mechanism for synaptic coincidence detection in cortical neurons in vivo. *Proc. Natl. Acad. Sci.*U.S.A. **97,** 8110–8115.

Beaulieu, C., and Colonnier, M. (1985). A laminar analysis of the number of round-asymmetrical and flat-symmetrical synapses on spines, dendritic trunks, and cell bodies in area 17 of the cat. *J. Comp. Neurol.* **231,** 180–189.

Beaulieu, C., and Colonnier, M. (1987). Effect of the richness of the environment on the cat visual cortex. *J. Comp. Neurol.* **266,** 478–494.

Ben-Yishai, R., Bar-Or, R. L., and Sompolinsky, H. (1995). Theory of orientation tuning in visual cortex. *Proc. Natl. Acad. Sci. U.S.A.* **92,** 3844–3848.

Bolz, J., and Gilbert, C. D. (1989). The role of horizontal connections in generating long receptive fields in the cat visual cortex. *Eur. J. Neurosci.* **1,** 263–268.

Borg-Graham, L. J., Monier, C., and Frégnac, Y. (1998). Visual input evokes transient and strong shunting inhibition in visual cortical neurons. *Nature* **393,** 369–373.

Bullier, J., and Henry, G. H. (1979a). Ordinal position of neurons in cat striate cortex. *J. Neurophysiol.* **42,** 1251–1263.

Bullier, J., and Henry, G. H. (1979b). Neural path taken by afferent streams in striate cortex of the cat. *J. Neurophysiol.* **42,** 1264–1270.

Bullier, J., and Henry, G. H. (1979c). Laminar distribution of first-order neurons and afferent terminals in cat striate cortex. *J. Neurophysiol.* **42,** 1271–1281.

Bullier, J., Mustari, M. J., and Henry, G. H. (1982). Receptive-field transformations between LGN neurons and S-cells of cat-striate cortex. *J. Neurophysiol.* **47,** 417–438.

Carandini, M., Mechler, F., Leonard, C. S., and Movshon, J. A. (1996). Spike train encoding by regular-spiking cells of the visual cortex. *J. Neurophysiol.* **76,** 3425–3441.

Chapman, B., Zahs, K. R., and Stryker, M. P. (1991). Relation of cortical cell orientation selectivity to alignment of receptive fields of the geniculocortical afferents that arborize within a single orientation column in ferret visual cortex. *J. Neurosci.* **11,** 1347–1358.

Chung, S., and Ferster, D. (1998). Strength and orientation tuning of the thalamic input to simple cells revealed by electrically evoked cortical suppression. *Neuron* **20,** 1177–1189.

Cleland, B. G. (1986). The dorsal lateral geniculate nucleus of the cat. In: *Visual neuroscience.* (J. D. Pettigrew, K. S. Sanderson, and W. R. Levick, Eds.), pp. 111–120, London, Cambridge University Press.

Cleland, B. G., Dubin, M. W., and Levick, W. R. (1971). Simultaneous recording of input and output of lateral geniculate neurones. *Nature New Biology* **231,** 191–192.

Cleland, B. G., and Lee, B. B. (1985). A comparison of visual responses of cat lateral geniculate nucleus neurones with those of ganglion cells afferent to them. *J. Physiol. (London)* **369,** 249–268.

Creutzfeldt, O., and Ito, M. (1968). Functional synaptic organization of primary visual cortex neurones in the cat. *Exp. Brain Res.* **6,** 324–352.

Dan, Y., Alonso, J. M., Usrey, W. M., and Reid, R. C. (1998). Coding of visual information by precisely correlated spikes in the lateral geniculate nucleus. *Nat. Neurosci.* **1,** 501–507.

Daniels, J. D., and Pettigrew, J. D. (1975). A study of inhibitory antagonism in cat visual cortex. *Brain Res.* **93,** 41–62.

Dean, A. F., and Tolhurst, D. J. (1986). Factors influencing the temporal phase of responses to bar and grating stimuli for simple cells in the cat striate cortex. *Exp. Brain Res.* **62,** 143–151.

DeFelipe, J., and Jones, E. G. (1988). *Cajal on the cerebral cortex. An annotated translation of the complete writings.* New York, Oxford University Press.

Douglas, R. J., Koch, C., Mahowald, M., Martin, K. A., and Suarez, H. H. (1995). Recurrent excitation in neocortical circuits. *Science* **269,** 981–985.

Douglas, R. J., and Martin, K. A. (1991). A functional microcircuit for cat visual cortex. *J. Physiol. (London)* **440,** 735–769.

Douglas, R. J., Martin, K. A., and Whitteridge, D. (1989). A canonical microcircuit for neocortex. *Neural Comp.* **1,** 480–488.

Douglas, R. J., Martin, K. A., and Whitteridge, D. (1991). An intracellular analysis of the visual responses of neurones in cat visual cortex. *J. Physiol. (London)* **440,** 659–696.

Einstein, G., Davis, T. L., and Sterling, P. (1987). Pattern of lateral geniculate synapses on neuron somata in layer IV of the cat striate cortex. *J. Comp. Neurol.* **260,** 76–86.

Erisir, A., Van Horn, S. C., and Sherman, S. M. (1998). Distribution of synapses in the lateral geniculate nucleus of the cat: differences between laminae A and A1 and between relay cells and interneurons. *J. Comp. Neurol.* **390,** 247–255.

Ferster, D. (1986). Orientation selectivity of synaptic potentials in neurons of cat primary visual cortex. *J. Neurosci.* **6,** 1284–1301.

Ferster, D. (1988). Spatially opponent excitation and inhibition in simple cells of the cat visual cortex. *J. Neurosci.* **8,** 1172–1180.

Ferster, D., Chung, S., and Wheat, H. (1996). Orientation selectivity of thalamic input to simple cells of cat visual cortex [see comments]. *Nature* **380,** 249–252.

Ferster, D., and LeVay, S. (1978). The axonal arborizations of lateral geniculate neurons in the striate cortex of the cat. *J. Comp. Neurol.* **182,** 923–944.

Ferster, D., and Lindstrom, S. (1983). An intracellular analysis of geniculo-cortical connectivity in area 17 of the cat. *J. Physiol. (London)* **342,** 181–215.

Freund, T. F., Martin, K. A., Somogyi, P., and Whitteridge, D. (1985a). Innervation of cat visual areas 17 and 18 by physiologically identified X- and Y- type thalamic afferents. II. Identification of postsynaptic targets by GABA immunocytochemistry and Golgi impregnation. *J. Comp. Neurol.* **242,** 275–291.

Freund, T. F., Martin, K. A., and Whitteridge, D. (1985b). Innervation of cat visual areas 17 and 18 by physiologically identified X- and Y- type thalamic afferents. I. Arborization patterns and quantitative distribution of postsynaptic elements. *J. Comp. Neurol.* **242,** 263–274.

Gardner, J. L., Anzai, A., Ohzawa, I., and Freeman, R. D. (1999). Linear and nonlinear contributions to orientation tuning of simple cells in the cat's striate cortex. *Vis. Neurosci.* **16,** 1115–1121.

Ghose, G. M., Freeman, R. D., and Ohzawa, I. (1994). Local intracortical connections in the cat's visual cortex: postnatal development and plasticity. *J. Neurophysiol.* **72,** 1290–1303.

Gil, Z., Connors, B. W., and Amitai, Y. (1999). Efficacy of thalamocortical and intracortical synaptic connections: quanta, innervation, and reliability. *Neuron* **23,** 385–397.

Gilbert, C. D. (1977). Laminar differences in receptive field properties of cells in cat primary visual cortex. *J. Physiol. (London)* **268,** 391–421.

Glezer, V. D., Tsherbach, T. A., Gauselman, V. E., and Bondarko, V. M. (1982). Spatio-temporal organization of receptive fields of the cat striate cortex. The receptive fields as the grating filters. *Biol. Cybern.* **43,** 35–49.

Grieve, K. L., and Sillito, A. M. (1995). Differential properties of cells in the feline primary visual cortex providing the corticofugal feedback to the lateral geniculate nucleus and visual claustrum. *J. Neurosci.* **15,** 4868–4874.

Guillery, R. W. (1969). A quantitative study of synaptic interconnections in the dorsal lateral geniculate nucleus of the cat. *Zeitschrift Fur Zellforschung Und Mikroskopische Anatomie* **96,** 39–48.

Heggelund, P. (1981). Receptive field organization of simple cells in cat striate cortex. *Exp. Brain Res.* **42,** 89–98.

Hirsch, J. A., Alonso, J. M., and Reid, R. C. (1995). Visually evoked calcium action potentials in cat striate cortex. *Nature* **378,** 612–616.

Hirsch, J. A., Alonso, J. M., Reid, R. C., and Martinez, L. M. (1998a). Synaptic integration in striate cortical simple cells. *J. Neurosci.* **18,** 9517–9528.

Hirsch, J. A., Gallagher, C. A., Alonso, J. M., and Martinez, L. M. (1998b). Ascending projections of simple and complex cells in layer 6 of the cat striate cortex. *J. Neurosci.* **18,** 8086–8094.

Hubel, D. H., and Wiesel, T. N. (1961). Integrative action in the cat's lateral geniculate body. *J. Physiol. (London)* **155,** 385–398.

Hubel, D. H., and Wiesel, T. N. (1962). Receptive fields, binocular interaction and functional architecture in the cat's visual cortex. *J. Physiol. (London)* **160,** 106–154.

Humphrey, A. L., Sur, M., Uhlrich, D. J., and Sherman, S. M. (1985). Projection patterns of individual X- and Y-cell axons from the lateral geniculate nucleus to cortical area 17 in the cat. *J. Comp. Neurol.* **233,** 159–189.

Jagadeesh, B., Wheat, H. S., and Ferster, D. (1993). Linearity of summation of synaptic potentials underlying direction selectivity in simple cells of the cat visual cortex. *Science* **262,** 1901–1904.

Jones, J. P., and Palmer, L. A. (1987). The two-dimensional spatial structure of simple receptive fields in cat striate cortex. *J. Neurophysiol.* **58,** 1187–1211.

Kato, H., Bishop, P. O., and Orban, G. A. (1978). Hypercomplex and simple/complex cell classifications in cat striate cortex. *J. Neurophysiol.* **41,** 1071–1095.

Kelly, J. P., and Van Essen, D. C. (1974). Cell structure and function in the visual cortex of the cat. *J. Physiol. (London)* **238,** 515–547.

Kulikowski, J. J., and Bishop, P. O. (1981). Linear analysis of the responses of simple cells in the cat visual cortex. *Exp. Brain Res.* **44,** 386–400.

Lee, B. B., Cleland, B. G., and Creutzfeldt, O. D. (1977). The retinal input to cells in area 17 of the cat's cortex. *Exp. Brain Res.* **30,** 527–538.

LeVay, S., and Gilbert, C. D. (1976). Laminar patterns of geniculocortical projection in the cat. *Brain Res.* **113,** 1–19.

Linsenmeier, R. A., Frishman, L. J., Jakiela, H. G., and Enroth-Cugell, C. (1982). Receptive field properties of x and y cells in the cat retina derived from contrast sensitivity measurements. *Vis. Res.* **22,** 1173–1183.

Maex, R., and Orban, G. A. (1992). A model circuit for cortical temporal low-pass filtering. *Neural Comp.* **4,** 932–945.

Mainen, Z. F., and Sejnowski, T. J. (1995). Reliability of spike timing in neocortical neurons. *Science* **268,** 1503–1506.

Martin, K. A., and Whitteridge, D. (1984). Form, function and intracortical projections of spiny neurones in the striate visual cortex of the cat. *J. Physiol. (London)* **353,** 463–504.

Mastronarde, D. N. (1987a). Two classes of single-input X-cells in cat lateral geniculate nucleus. I. Receptive-field properties and classification of cells. *J. Neurophysiol.* **57,** 357–380.

Mastronarde, D. N. (1987b). Two classes of single-input X-cells in cat lateral geniculate nucleus. II. Retinal inputs and the generation of receptive-field properties. *J. Neurophysiol.* **57,** 381–413.

McLean, J., and Palmer, L. A. (1989). Contribution of linear spatiotemporal receptive field structure to velocity selectivity of simple cells in area 17 of cat. *Vis. Res.* **29,** 675–679.

McLean, J., Raab, S., and Palmer, L. A. (1994). Contribution of linear mechanisms to the specification of local motion by simple cells in areas 17 and 18 of the cat. *Vis. Neurosci.* **11,** 271–294.

Movshon, J. A., Thompson, I. D., and Tolhurst, D. J. (1978). Spatial summation in the receptive fields of simple cells in the cat's striate cortex. *J. Physiol. (London)* **283,** 53–77.

Mullikin, W. H., Jones, J. P., and Palmer, L. A. (1984). Periodic simple cells in cat area 17. *J. Neurophysiol.* **52,** 372–387.

Murthy, A., and Humphrey, A. L. (1999). Inhibitory contributions to spatiotemporal receptive-field structure and direction selectivity in simple cells of cat area 17. *J. Neurophysiol.* **81,** 1212–1224.

Nelson, S., Toth, L., Sheth, B., and Sur, M. (1994). Orientation selectivity of cortical neurons during intracellular blockade of inhibition. *Science* **265,** 774–777.

Nowak, L. G., Sanchez-Vives, M. V., and McCormick, D. A. (1997). Influence of low and high frequency inputs on spike timing in visual cortical neurons. *Vis. Res.* **7,** 487–501.

Palmer, L. A., and Davis, T. L. (1981). Receptive-field structure in cat striate cortex. *J. Neurophysiol.* **46,** 260–276.

Pei, X., Vidyasagar, T. R., Volgushev, M., and Creutzfeldt, O. D. (1994). Receptive field analysis and orientation selectivity of postsynaptic potentials of simple cells in cat visual cortex. *J. Neurosci.* **14,** 7130–7140.

Peters, A., and Payne, B. R. (1993). Numerical relationships between geniculocortical afferents and pyramidal cell modules in cat primary visual cortex. *Cereb. Cortex* **3,** 69–78.

Peters, A., and Yilmaz, E. (1993). Neuronal organization in area 17 of cat visual cortex. *Cereb. Cortex* **3,** 49–68.

Pettigrew, J. D., and Daniels, J. D. (1973). Gamma-aminobutyric acid antagonism in visual cortex: different effects on simple, complex, and hypercomplex neurons. *Science* **182,** 81–83.

Reid, R. C., and Alonso, J. M. (1995). Specificity of monosynaptic connections from thalamus to visual cortex. *Nature* **378,** 281–284.

Reid, R. C., and Alonso, J. M. (1996). The processing and encoding of information in the visual cortex. *Curr. Opin. Neurobiol.* **6,** 475–480.

Reid, R. C., Soodak, R. E., and Shapley, R. M. (1987). Linear mechanisms of directional selectivity in simple cells of cat striate cortex. *Proc. Natl. Acad. Sci U.S.A.* **84,** 8740–8744.

Reid, R. C., Soodak, R. E., and Shapley, R. M. (1991). Directional selectivity and spatiotemporal structure of receptive fields of simple cells in cat striate cortex. *J. Neurophysiol.* **66,** 505–529.

Reid, R. C., Victor, J. D., and Shapley, R. M. (1997). The use of m-sequences in the analysis of visual neurons: linear receptive field properties. *Vis. Neurosci.* **14,** 1015–1027.

Rose, D., and Blakemore, C. (1974). Effects of bicuculline on functions of inhibition in visual cortex. *Nature* **249,** 375–377.

Saul, A. B., and Humphrey, A. L. (1992). Temporal-frequency tuning of direction selectivity in cat visual cortex. *Vis. Neurosci.* **8,** 365–372.

Sclar, G., and Freeman, R. D. (1982). Orientation selectivity in the cat's striate cortex is invariant with stimulus contrast. *Exp. Brain Res.* **46,** 457–461.

Sherman, S. M., and Guillery, R. W. (1998). On the actions that one nerve cell can have on another: distinguishing "drivers" from "modulators". *Proc. Natl. Acad. Sci. U.S.A.* **95,** 7121–7126.

Sillito, A. M. (1975). The contribution of inhibitory mechanisms to the receptive field properties of neurones in the striate cortex of the cat. *J. Physiol. (London)* **250,** 305–329.

Sillito, A. M., Kemp, J. A., Milson, J. A., and Berardi, N. (1980). A re-evaluation of the mechanisms underlying simple cell orientation selectivity. *Brain Res.* **194,** 517–520.

Singer, W., Tretter, F., and Cynader, M. (1975). Organization of cat striate cortex: a correlation of receptive-field properties with afferent and efferent connections. *J. Neurophysiol.* **38,** 1080–1098.

Skottun, B. C., De Valois, R. L., Grosof, D. H., Movshon, J. A., Albrecht, D. G., and Bonds, A. B. (1991). Classifying simple and complex cells on the basis of response modulation. *Vis. Res.* **31,** 1079–1086.

Somers, D. C., Nelson, S. B., and Sur, M. (1995). An emergent model of orientation selectivity in cat visual cortical simple cells. *J. Neurosci.* **15,** 5448–54465.

Stone, J., and Dreher, B. (1973). Projection of X- and Y-cells of the cat's lateral geniculate nucleus to areas 17 and 18 of visual cortex. *J. Neurophysiol.* **36,** 551–567.

Stratford, K. J., Tarczy-Hornoch, K., Martin, K. A., Bannister, N. J., and Jack, J. J. (1996). Excitatory synaptic inputs to spiny stellate cells in cat visual cortex. *Nature* **382,** 258–261.

Suarez, H., Koch, C., and Douglas, R. (1995). Modeling direction selectivity of simple cells in striate visual cortex within the framework of the canonical microcircuit. *J. Neurosci.* **15,** 6700–6719.

Tanaka, K. (1983). Cross-correlation analysis of geniculostriate neuronal relationships in cats. *J. Neurophysiol.* **49,** 1303–1318.

Tanaka, K. (1985). Organization of geniculate inputs to visual cortical cells in the cat. *Vis. Res.* **25,** 357–364.

Tolhurst, D. J., and Dean, A. F. (1987). Spatial summation by simple cells in the striate cortex of the cat. *Exp. Brain Res.* **66,** 607–620.

Troyer, T. W., Krukowski, A. E., Priebe, N. J., and Miller, K. D. (1998). Contrast-invariant orientation tuning in cat visual cortex: thalamocortical input tuning and correlation-based intracortical connectivity. *J. Neurosci.* **18,** 5908–5927.

Tsumoto, T., Eckart, W., and Creutzfeldt, O. D. (1979). Modification of orientation sensitivity of cat visual cortex neurons by removal of GABA-mediated inhibition. *Exp. Brain Res.* **34,** 351–363.

Usrey, W. M., Reppas, J. B., and Reid, R. C. (1998). Paired-spike interactions and synaptic efficacy of retinal inputs to the thalamus. *Nature* **395,** 384–387.

Usrey, W. M., Reppas, J. B., and Reid, R. C. (1999). Specificity and strength of retinogeniculate connections. *J. Neurophysiol.***82,** 3527–3540.

Usrey, W. M., Alonso, J. M., and Reid, R. C. (2000). Synaptic interactions between thalamic inputs to simple cells in cat visual cortex. *J. Neurosci.* **20,** 5461–5467.

Vidyasagar, T. R., Pei, X., and Volgushev (1996). Multiple mechanisms underlying the orientation selectivity of visual cortical neurones. *Trends Neurosci.* **19,** 272–277.

Volgushev, M., Chistiakova, M., and Singer, W. (1998). Modification of discharge patterns of neocortical neurons by induced oscillations of the membrane potential. *Neuroscience* **83,** 15–25.

Wässle, H., Boycott, B. B., and Illing, R. B. (1981). Morphology and mosaic of on- and off-beta cells in the cat retina and some functional considerations. *Proc. R. Soc. London Ser. B* **212,** 177–195.

Wässle, H., Peichl, L., and Boycott, B. B, (1983). A spatial analysis of on- and off-ganglion cells in the cat retina. *Vis. Res.* **23,** 1151–1160.

Watkins, D. W., and Berkley, M. A. (1974). The orientation selectivity of single neurons in cat striate cortex. *Exp. Brain Res.* **19,** 433–446.

Wilson, J. R., Friedlander, M. J., and Sherman, S. M. (1984). Fine structural morphology of identified X- and Y-cells in the cat's lateral geniculate nucleus. *Proc. R. Soc. London Ser. B* **221,** 411–436.

9

THE EMERGENCE OF DIRECTION SELECTIVITY IN CAT PRIMARY VISUAL CORTEX

ALLEN L. HUMPHREY AND ALAN B. SAUL

Department of Neurobiology, University of Pittsburgh School of Medicine, Pittsburgh, Pennsylvania

OVERVIEW

Direction selectivity (DS) is a prominent feature of most neurons in cat primary visual cortex (areas 17 and 18). As originally described by Hubel and Wiesel (1959, 1962), DS cells discharge vigorously to motion in one direction across their receptive field and weakly or not at all to motion in the opposite direction. These cells form the first stage in a multilevel analysis of image motion, which is further carried out by DS cells in extrastriate cortex (Spear, 1991). DS is important perceptually because specifically eliminating it by developmental manipulations causes deficits in psychophysical direction discrimination (Pasternak et al., 1985).

In the cat, direction selectivity emerges at the earliest synaptic stages in cortex, among simple cells in layer 4 that receive monosynaptic input from nondirectional afferents of the dorsal lateral geniculate nucleus (LGN) (Ferster, 1992; Bullier and Henry, 1979). Despite numerous studies, our understanding of the mechanisms that produce DS remains limited and under debate. However, progress has been made over the last few years in elucidating important computational principles and in identifying likely biological mechanisms. This review focuses mainly on directional mechanisms among simple cells of layer 4 in adult cat area 17, where progress has been most rewarding. For reviews on related topics see Ulinski (1999), Orban (1994), Hildreth and Koch (1987), and Nakayama (1985).

As background, we first describe key features of directional tuning as they are manifest under different stimulus conditions. We then consider the main computational principle underlying DS—a cell must receive inputs that are separated in space and time—and show how it is instantiated in terms of the spatiotemporal (S-T) receptive-field structure of layer 4 cells. We next discuss evidence that LGN cells with different response timings provide these spatiotemporally distinct signals, and we compare a model based on LGN inputs with other recent models. The additional role of excitatory and inhibitory inputs from cortical cells is reviewed, and we show how inhibition sculpts the timing structure, and hence directional tuning, of the receptive field. Having considered likely sources of the inputs, we then review evidence for how the inputs are combined via linear and nonlinear mechanisms.

DIRECTIONAL TUNING: THE BASICS

Virtually all neurons in cat area 17 are sensitive to stimulus orientation. Thus, when using elongated stimuli, the axes of motion to which cells respond are constrained by their orientation tuning, the optimal axis being orthogonal to the optimal orientation. Despite this link, separate mechanisms likely underlie each property, as evidenced by the fact that rearing kittens in stroboscopic illumination abolishes DS in virtually all cells but leaves orientation tuning intact (Humphrey and Saul, 1998; Humphrey et al., 1998; Pasternak et al. 1985; Cynader and Chernenko, 1976).

The optimal direction of motion is invariant with image contrast: a cell will prefer the same direction to a moving bar whether it is bright or dark. This invariance cannot be explained by Hubel and Wiesel's (1962) model, which posited that in simple cells DS reflects interactions between adjacent ON and OFF zones. That is, the cell should respond best to a bright bar exiting the OFF zone and entering the ON zone, owing to the synergistic effects of rebound excitation in the OFF zone and activation of the ON zone. However, a reversal in optimal direction would occur when a dark bar is used. Some cells display such reversals, but they are considered directionally asymmetrical rather than DS (Emerson and Gerstein, 1977b; Pettigrew et al., 1968).[1] Selectivity implies contrast invariance. Further, most investigators require a criterion response at least twice as great in the preferred direction (PD) as in the nonpreferred direction (NPD).

Direction selectivity generally varies with the rate of stimulus motion (Orban, 1984). Most cells are maximally selective near the velocity or temporal frequency

[1] The term *directional asymmetry* is used differently by others (Orban et al., 1981) to denote a directional bias and a PD/NPD ratio <3 that can be contrast invariant. For this review, we use a common metric of directional tuning: a directional index (DI) = (PD response − NPD response)/(PD response + NPD response) (Tolhurst and Heeger, 1997a; DeAngelis et al., 1993a; Saul and Humphrey, 1992b; Reid et al., 1991; Albrecht and Geisler, 1991). DI ranges from 0 to 1; values >0.33 correspond to PD/NPD ratios >2.

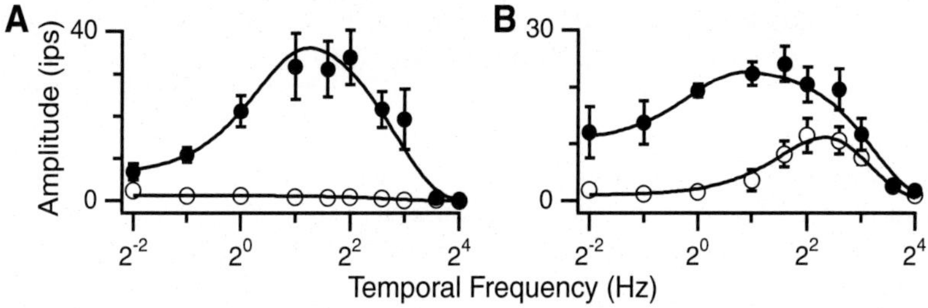

FIGURE 9-1. Directional tuning as a function of stimulus temporal frequency for two typical simple cells. Open and filled circles represent mean fundamental response amplitudes (±SE) to sinewave gratings moving in opposite directions. Responses were fit by a difference-of-gaussians function. The cell in **A** was DS across a broad range of TFs. The cell in **B** was selective only at low TFs; at ~4Hz responses in the NPD increased to roughly match that in the PD. Additional examples of tuning behaviors can be found in Saul and Humphrey (1992b).

(TF) that elicits the greatest discharge in the preferred direction. This is typically about 2–4°/sec or 2 Hz (Fig. 9-1A). Above and below this, tuning varies among cells, but many remain selective at low velocities and TFs. Interestingly, ~25% of simple and complex cells are DS only at TFs below 4 Hz (Fig. 1B) (Saul and Humphrey, 1992b). With increasing temporal frequency, the response in the non-preferred direction increases and often matches that in the preferred direction. As a result DS is lost even though the cell may continue to respond up to ~16 Hz.

The frequency of directionally tuned cells reported for area 17 varies widely across studies, ranging from about 25% to 75% for cells examined with moving bars (e.g., Orban et al., 1981; Albus, 1980; Kato et al., 1978; Goodwin and Henry, 1975; Bishop et al., 1971). Some of the variation reflects differences in the criterion PD/NPD ratio for DS, in the use of peak versus mean discharge rates, and in the range of velocities over which DS is tested. Because many cells are DS only for a subset of velocities to which they otherwise respond, averaging DS values across the full velocity range (Duysens et al., 1982; Orban et al. 1981) underestimates the prevalence of direction sensitivity. When drifting gratings of optimal temporal frequency are used, the sampling of DS cells is much higher (65–80%) (Humphrey and Saul, 1998; Casanova et al., 1992; Pasternak et al., 1985). Also, most cells prefer the same direction to gratings and bars, although the degree of DS may differ between stimuli (Casanova et al., 1992).

The minimum spatial displacement of an edge needed to produce a completely DS response is about 1 minute of arc for simple and complex cells (Goodwin and Henry, 1975; Goodwin et al., 1975), a traverse much smaller than their receptive fields. These and similar findings (Ganz and Felder, 1984; Emerson and Gerstein, 1977a, 1997b) show that retinotopically local interactions operating across limited distances in cortex are sufficient to produce DS. Most studies of directional mechanisms, and consequently this review, focus on the source and nature of the locally interacting elements.

COMPUTATIONAL REQUIREMENTS FOR DIRECTION SELECTIVITY

Here we present a theoretical and methodological framework for understanding direction selectivity, to place in perspective why certain experiments are done and how data are interpreted. The key requirement for direction selectivity is that a cell's inputs be appropriately separated in space and time.

REPRESENTING INPUT RELATIONS IN THE SPACE-TIME DOMAIN

This space-time relationship is illustrated in Fig. 9-2A, where two detectors projecting to, and exciting, a common output unit have receptive fields separated by a distance Δx and a delay Δt. A stimulus moving leftward first activates detector 2, which elicits a response that is delayed by Δt before impinging on the output unit. The stimulus traverses the distance Δx and activates detector 1, whose response is transmitted immediately to the output. If the stimulus takes a time Δt to travel to detector 1, then the two signals will arrive simultaneously at the output and optimally activate it. For rightward motion, however, the order of detector activation is reversed and signals from the two inputs arrive at different times, producing subop-

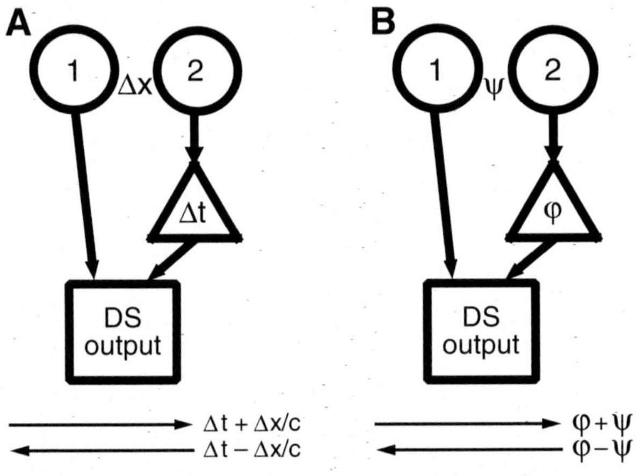

FIGURE 9-2. Conceptually, direction selectivity requires two inputs that differ spatially and temporally to provide signals whose timing varies with direction. **(A)** Space-time domain and **(B)** frequency-domain versions of the basic model. Detectors 1 and 2 are separated by a distance Δx and Δt in time. In the frequency domain, these become spatial and temporal phase differences ψ and φ. The net difference in timing for each direction of motion is given next to the arrows. In one direction (rightward), the signals reaching the DS unit (square) arrive at different times. In the opposite direction (leftward), the signals can arrive at the same time if $c = \Delta x/\Delta t$ or $\psi = \varphi$. The direction-dependent difference in timing creates different response magnitudes. The ideal case is when the signals are completely out of phase (a half-cycle) in one direction and completely in phase (zero cycles) in the other direction. This occurs when both ψ and φ are equal to a quarter-cycle, a condition known as spatiotemporal quadrature.

timal activation. The different relative timings of the input signals therefore produce directionally sensitive responses. These inputs can be combined in many ways; some combination rules are discussed later (see page 368). As a simple example, if one of the input paths is inhibitory, the preferred direction will be reversed, because the signals will cancel for leftward, but not rightward motion.

REPRESENTING INPUT RELATIONS IN THE FREQUENCY DOMAIN

The space-time formulation of this model can be more difficult to work with than the frequency-domain version (Fig. 9-2B). Here, the spatial and temporal differences (Δx and Δt) are replaced by spatial and temporal *phase* differences (ψ and φ). This leads to a simple statement of what is needed to obtain DS: The differences in timing between the inputs in the two directions are simply the sum and difference of ψ and φ. If each of the spatial and temporal phase differences is a quarter-cycle, then subtracting them gives 0 cycles (i.e., responses are completely in phase) for

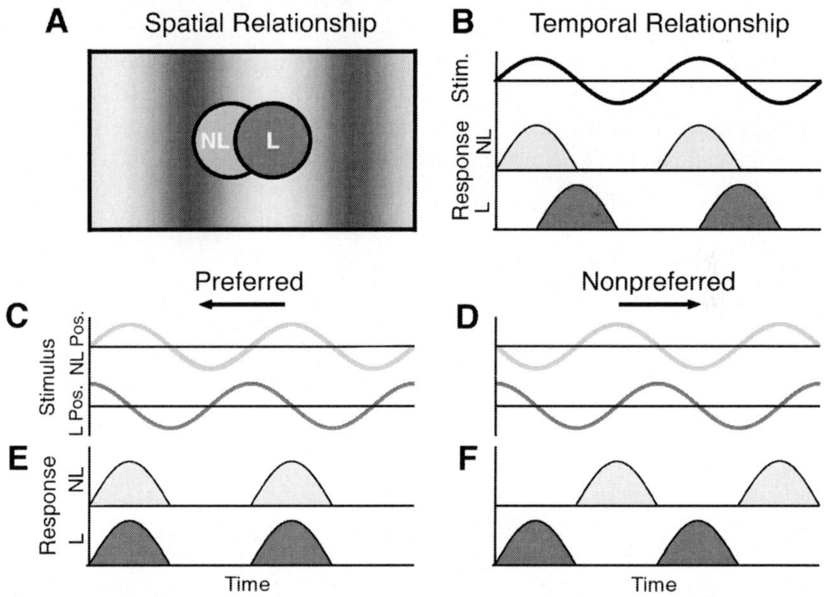

FIGURE 9-3. How spatiotemporal quadrature produces directional tuning. **(A)** The two detectors are a quarter-cycle apart in space with respect to the grating. **(B)** The quarter-cycle temporal relationship of the detectors as measured by their response to a stationary stimulus in their receptive-field centers that undergoes sinusoidal luminance modulation. In panels **C-F,** motion in the two directions is simulated. Panel A shows the starting spatial phase of the grating. (**C** and **D**) The luminance profiles to which the detectors are exposed as the grating drifts leftward and rightward, respectively. (**E** and **F**) The response of each detector after taking into account their differences in temporal phase. For the leftward direction excitation is completely in phase and overlaps in time. For the rightward direction excitation is completely out of phase.

one direction, and adding them gives a half cycle (responses are completely out of phase) for the opposite direction. This structure of quarter cycle phase differences in space and time is known as *spatiotemporal (S-T) quadrature* (Watson and Ahumada, 1985). Figure 9-3 clarifies how two detectors in S-T quadrature can establish DS. The detectors are a quarter cycle apart in space with respect to a grating (A). Both inputs project to a common output unit (not illustrated). A quarter cycle temporal relationship is illustrated in B, which shows how each detector might respond to a stationary, sinusoidally modulated spot (black line) in its receptive field. Detector 1 discharges in phase with the bright portion of the cycle; detector 2 responds a quarter cycle later. Panels C–F show the resulting directional tuning to the drifting grating. For leftward motion the detectors are completely in phase, and their combined excitatory signals drive the output unit above threshold. For rightward motion they are a half-cycle out of phase and thus less effective postsynaptically.

Stimuli and responses that are extended in time, as most are, can be analyzed easily in terms of phase. One can think of this model in terms of either its space/time or frequency domain representation, but the former can cause confusion when attempting to relate timing to DS. For example, a common misconception is that latency—the delay between stimulus onset and a criterion response—is an appropriate measure of timing. However, it predicts directional tuning poorly; latencies to flashing stimuli may be very similar across a receptive field or different but erratically organized (e.g., Barlow and Levick, 1965), or latency differences between positions may be too small to yield DS. Latency ignores the shape of the extended spike train (e.g., sustained versus transient), which can vary markedly among receptive-field positions in a cell. In contrast, spike train profiles are well captured by measuring the phase of the response. The phase difference between a completely sustained and completely transient response is a quarter-cycle, for example (Saul and Humphrey, 1990).

BIOLOGICAL INSTANTIATION OF COMPUTATIONAL PRINCIPLES

RECEPTIVE FIELDS

Based on the 2-input model, one would expect to record a progression of response timings across the receptive field of a direction-selective simple cell, and this is observed when stimulating with bars or gratings placed in different receptive-field positions (Jagadeesh et al., 1997; DeAngelis et al., 1993a; Saul and Humphrey, 1992a; McLean and Palmer, 1989; Reid et al., 1987; Dean and Tolhurst, 1986; Movshon et al., 1978). Figure 9-4A illustrates a directional cell's S-T structure assessed using a stationary sinewave grating undergoing sinusoidal luminance modulation at different positions, or spatial phases, in the receptive field. The cell's timing, or response phase, shifted gradually as a function of grating position, occurring progressively later in the stimulus cycle with increasing spatial phase. This progression is summarized quantitatively in C,

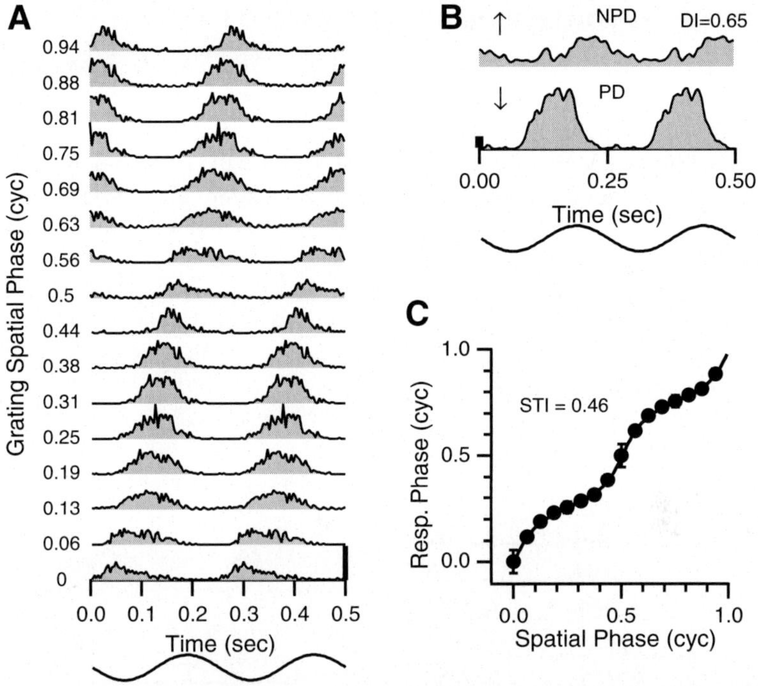

FIGURE 9-4. Average responses of an S-T oriented, direction-selective simple cell in layer 4B to sinewave luminance-modulated gratings presented at 4 Hz. Two cycles of the luminance profiles and responses are shown for clarity. **(A)** Responses to a counterphasing grating at 16 spatial phases spanning one cycle. Note the progressively delayed responses with increasing phase. **(B)** DS responses to a grating drifting upward and downward through the receptive field (DI = 0.65). **(C)** Average response phase (±SE) versus stimulus spatial phase. A linear model-based fit to the data (solid line) yielded an STI of 0.46, which is the linear prediction of DS. Calibration bars in A and B indicate 65 and 15 ips, respectively. (Data are from Murthy et al. [1998].)

which plots the Fourier-derived fundamental response phase as a function of spatial phase. The cell's S-T structure predicts that a grating moving downward through the receptive field will activate positions with progressively earlier response phases and thus will elicit stronger responses than motion in the opposite direction. That is, net excitation in the downward direction will be strong because component responses overlap in time and sum well; for upward motion they are more spread out in time and sum less. The predicted preferred direction was confirmed using a moving grating (B). For comparison, Fig. 9-5 illustrates the structure of a cell that was not DS. It displayed nearly uniform timing across the receptive field except for a half-cycle jump at about spatial phase 0.5, which merely reflects a half-cycle temporal displacement of the grating. Lacking prominent timing gradients, no directional tuning was expected and none was observed (B).

SPATIOTEMPORAL INSEPARABILITY AND DS

These DS and non-DS-receptive fields typify *S-T inseparable* and *S-T separable* structures, respectively.[2] The receptive fields are often described as S-T oriented and S-T unoriented, respectively, based on the presence or absence of a space-time tilt (e.g., McLean et al., 1994). We will use the terms S-T inseparability and S-T orientation interchangeably. S-T orientation and DS vary continuously. Murthy et al. (1998), using a quadrature-based model, quantified S-T orientation by fitting a function to the response phase versus spatial phase plots derived from counterphasing gratings. The resulting *spatiotemporal index (STI)* ranges from 0 to 1, for completely separable and inseparable receptive fields, respectively. As examples, the STIs for the DS and non-DS cells in Figs. 9-4 and 9-5 are 0.46 and 0.06, respectively.

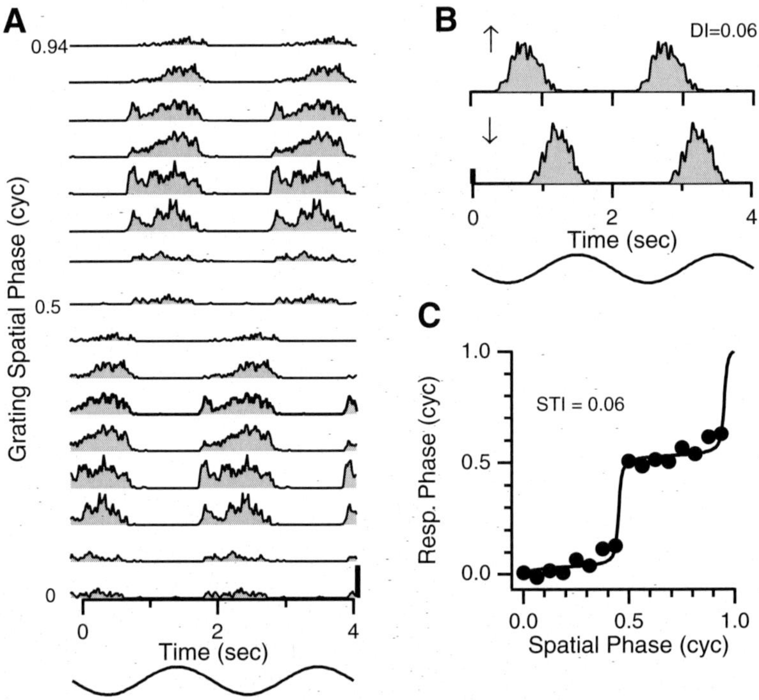

FIGURE 9-5. Responses of an S-T unoriented, directionally nonselective simple cell in layer 4A. Conventions are as in Fig. 9-4. **A** and **C** reveal that response phase was relatively constant within each half of the grating cycle (STI = 0.06). **(B)** Responses to opposite directions of motion were virtually identical (DI = 0.06). Calibration bars in A and B indicate 150 and 30 ips, respectively.

[2] An S-T separable field can be expressed as the product of a spatial and a temporal function. Response timing is the same at all positions. An S-T inseparable field cannot be so expressed; timing varies with spatial position.

The link between S-T structure and DS in layer 4 simple cells is also supported by a study in which kittens were reared in 8 Hz stroboscopic illumination (Humphrey and Saul, 1998; Humphrey et al., 1998), a paradigm that selectively abolishes DS in nearly all cortical cells (Cynader and Chernenko, 1976). Critically, strobe rearing also eliminated S-T inseparable structures in layer 4; nearly all receptive fields were S-T separable, like non-DS cells in normal cats. The few remaining DS cells continued to be S-T inseparable. Additional evidence for the link comes from a recent study (Murthy and Humphrey, 1999), which showed that blocking $GABA_A$-mediated inhibition produces highly correlated reductions in direction selectivity and S-T orientation in single layer 4 cells (see page 361 and Chapter 11).

The bulk of the evidence thus shows that directional tuning in layer 4 simple cells is linked to the presence of response timing gradients within their receptive fields. Later we examine how well S-T structure accounts for directional tuning. Now we focus on the sources of the timings that produce these structures.

ORIGINS OF CORTICAL TIMINGS: THE LAGGED/NONLAGGED CELL MODEL

LGN CELLS AND THEIR TIMINGS

Because the S-T inseparable structures first appear in area 17, it was assumed that the the timings that underlie them are all generated within cortex. The LGN was not thought to be a critical timing source because latency differences between X- and Y-cells, the major inputs to area 17, seemed too small. This perspective began to change with Mastronarde's (1987a, 1987b) reports of a previously undiscovered class of geniculate relay cells, termed lagged (X_L) X-cells. These cells are distinguished from classically studied X-cells, now termed nonlagged (X_N) cells, by their responses to flashing spots. For example, an on-center X_N-cell reacts to the onset of a bright spot in its receptive-field center with a brisk, short latency discharge (Fig. 9-6A, upper), and its response profile essentially mimics that of its X-retinal input (Mastronarde 1987a). In contrast, an on-center X_L-cell responds to spot onset with a profound dip in discharge, and its subsequent excitatory response occurs much later than that in X_N-cells (Fig. 6A, lower). Also, at spot offset the decay in X_L-cell firing is relatively prolonged and often preceded by a transient discharge. These differences between cells are robust and remain during variations in stimulus size (Humphrey and Weller, 1988a), stimulus contrast (Hartveit and Heggelund, 1992; Saul and Humphrey, 1990), inactivation of visual cortex (Mastronarde, 1987a), and activation of brainstem reticular inputs to the LGN (Humphrey and Saul, 1992), which have a net facilitatory influence on retinogeniculate transmission. Lagged and nonlagged LGN Y-cells also exist and display responses like their X-counterparts (Mastronarde et al., 1991). The mechanisms that produce lagged responses appear to involve intrageniculate inhibition (Heggelund and Hartveit, 1990; Humphrey and Weller, 1988b; Mastronarde, 1987a, 1987b) and probably a strong dependence on NMDA-mediated retinal

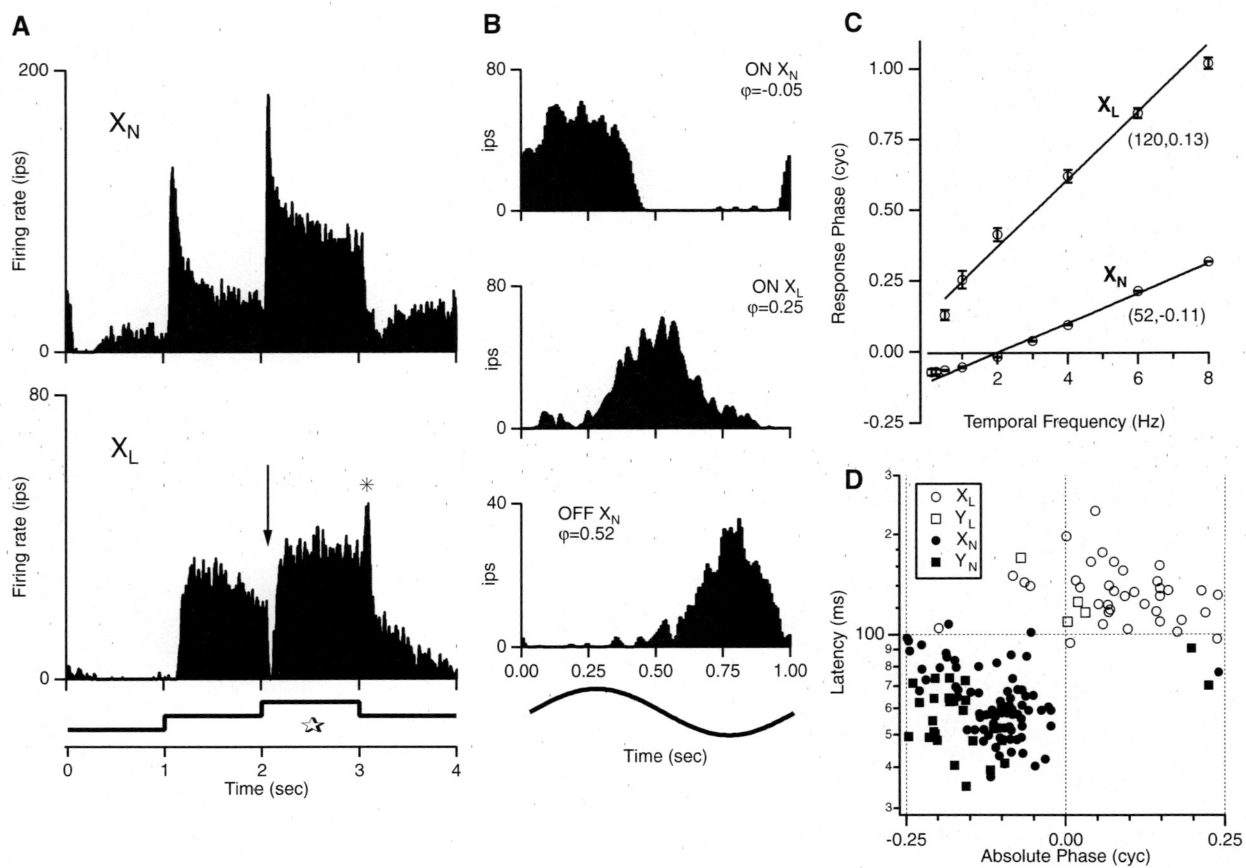

A
B
C
D
Firing rate (ips)
200
0
80
X_N
X_L
Time (sec)
ips
ON X_N
φ=-0.05
ON X_L
φ=0.25
OFF X_N
φ=0.52
40
0.00
0.25
0.50
0.75
1.00
Response Phase (cyc)
Temporal Frequency (Hz)
(120,0.13)
(52,-0.11)
Latency (ms)
Absolute Phase (cyc)
X_L
Y_L
X_N
Y_N

excitation (Kwon et al., 1991; Heggelund and Hartveit, 1990), the details of which are beyond the scope of this review.

Use of sine-modulated stimuli has provided a robust characterization of timing differences between lagged and nonlagged cells (Saul and Humphrey, 1990). Figure 9-6B illustrates responses of 3 X-cells to small stationary spots undergoing 1 Hz sinusoidal luminance modulation. As expected, the ON- and OFF-center X_N-cells discharged with increasing and decreasing luminance, respectively. The peristimulus time histograms (PSTHs) were Fourier analyzed to obtain the temporal phase of the fundamental response relative to the luminance cycle; zero stimulus phase corresponds to peak luminance. The ON-center X_N-cell peaked slightly ahead of the stimulus peak; it displayed a slight phase lead. The OFF-center cell response was simply shifted by a half-cycle. In contrast, the ON-center X_L-cell responded midway between the X_N-cells; it had a response phase lag that was shifted by one-quarter cycle relative to the other cells. This quarter-cycle timing difference between lagged and nonlagged cells is typical at low temporal frequencies (<2 Hz) (Saul and Humphrey, 1990). Thus, at least at low TFs, these cells could provide cortex with the timings necesary to produce S-T oriented, directionally tuned receptive fields, as simulated in Fig. 9-3.

The timing differences between the X_L- and X_N-cells could reflect variations in latency, phase, or both. The two factors can be distinguished by measuring response phase over a range of temporal frequencies and fitting the data with a line (Fig. 9-6C). The line intercept is *absolute phase,* which reflects where excitation arises in the stimulus cycle (increasing or decreasing luminance, or elsewhere). The slope corresponds to *latency,* which reflects delays and integrative processes between the stimulus and the cell discharge. Figure 9-6C shows that the X_N-cell had a short latency and absolute phase lead (i.e., intercept below zero), whereas the X_L-cell had a longer latency and absolute phase lag (intercept above zero). Figure 9-6D shows that these cell class differences are general; most nonlagged X- and Y-cells have latencies <100 msec and absolute phase leads, whereas most lagged cells have latencies >100 msec and absolute phase lags. Together, the two timing signatures distinguish virtually all lagged and nonlagged cells.

FIGURE 9-6. Key differences between lagged and nonlagged LGN cells. **(A)** Response histograms from an X_N-cell (upper) and an X_L-cell (lower) to a flashing spot. Star indicates luminance step most appropriate for exciting these ON-center cells. Arrow and asterisk mark an inhibitory dip and offset discharge. **(B)** PSTHs from three X-cells to a spot undergoing sinusoidal luminance modulation. Response phase (φ) is indicated for each cell. **(C)** Phase versus TF plots for 2 ON-center cells. Values in parentheses indicate latency (in msec) and absolute phase (in cycles), respectively. **(D)** Latency versus absolute phase values for 158 lagged and nonlagged X- and Y-cells. Absolute phase was normalized for OFF-center cells by subtracting one-half cycle from their measured phase. Phase leads and lags are indicated respectively by negative and positive values.

CORTICAL TIMINGS AND THEIR RELATION TO LGN INPUTS

Does cortex use the geniculate timing information? It is estimated that X_N- and X_L-cells provide ~60% and ~40% of the X-input to area 17 (Humphrey and Weller, 1988b), so each group should have a discernible influence in cortex. This was confirmed by Saul and Humphrey (1992a), who measured responses of simple cells to stationary, optimally oriented, narrow bars undergoing sinewave luminance modulation at different positions in their receptive fields. Figure 9-7A shows S-T maps obtained at 3 TFs for a DS cell in layer 5A. For each receptive-field position, absolute phase and latency were measured, as in the LGN. For six positions with reliable timing, three had short latencies (<100 ms) and absolute phase leads like nonlagged LGN cells, whereas the other three had long latencies (>100 ms) and absolute phase delays typical of lagged cells (Fig. 9-7B).

Across the simple-cell population, roughly 70% of receptive-field positions display LGN-like timing signatures (Humphrey and Saul, 1998; Saul and Humphrey, 1992a), as shown in Fig. 9-7C. Comparison with the LGN (Fig. 9-6D) reveals a sizable number of cortical positions that are nonlagged- or lagged-like. Also, as in the LGN, there is a dearth of positions with short latencies and absolute phase lags. The two regions differ mainly in the greater frequency of cortical positions with long latencies and absolute phase leads.

The geniculate-like timings also are differentially distributed among the cortical layers. Figure 9-8 summarizes the locations of simple cells that were categorized according to the proportion of timing types in their receptive fields (see legend for details). Most cells with significant nonlagged timing are recorded in layers 4, 6, and lower 3. This matches the termination sites of axons from nonlagged X- and Y-cells (Humphrey et al., 1985). In comparison, nearly all cells with lagged timing are restricted to a region from mid-layer 4 through layer 5A. Although direct anatomical confirmation of lagged-cell terminations is lacking, they appear to arborize mainly to layer 4B (see Saul and Humphrey, 1992a for details).

Further support for a geniculate contribution to cortical timings comes from observations on how DS changes with the rate of stimulus motion. As noted previously, many cortical cells are DS at low temporal frequencies but become unselective at and above ~4 Hz (Fig. 9-1B). This curious behavior is predicted by the temporal-frequency-dependent timing relationship between lagged and nonlagged cells. At low TFs (~1 Hz) the two groups respond approximately one-quarter cycle apart (Fig. 9-6C). With increasing TF, response phase increases faster in lagged than in nonlagged cells so that by ~4 Hz the cells respond about one-half cycle apart. (The changes in phase relationship reflect the different "latencies" of the two cell types, as shown in Fig. 9-6C.) In cortex, this behavior is evident in the relative timings of nearby receptive-field positions in some simple cells. In Fig. 9-7B, for example, adjacent lagged- and nonlagged-like positions are roughly in temporal quadrature at 1 to 2 Hz. They progressively deviate from quadrature at and beyond 4 Hz, and large (~ one-half cycle) jumps in response phase become progressively more evident in the field. Also, response amplitude in the lagged-like

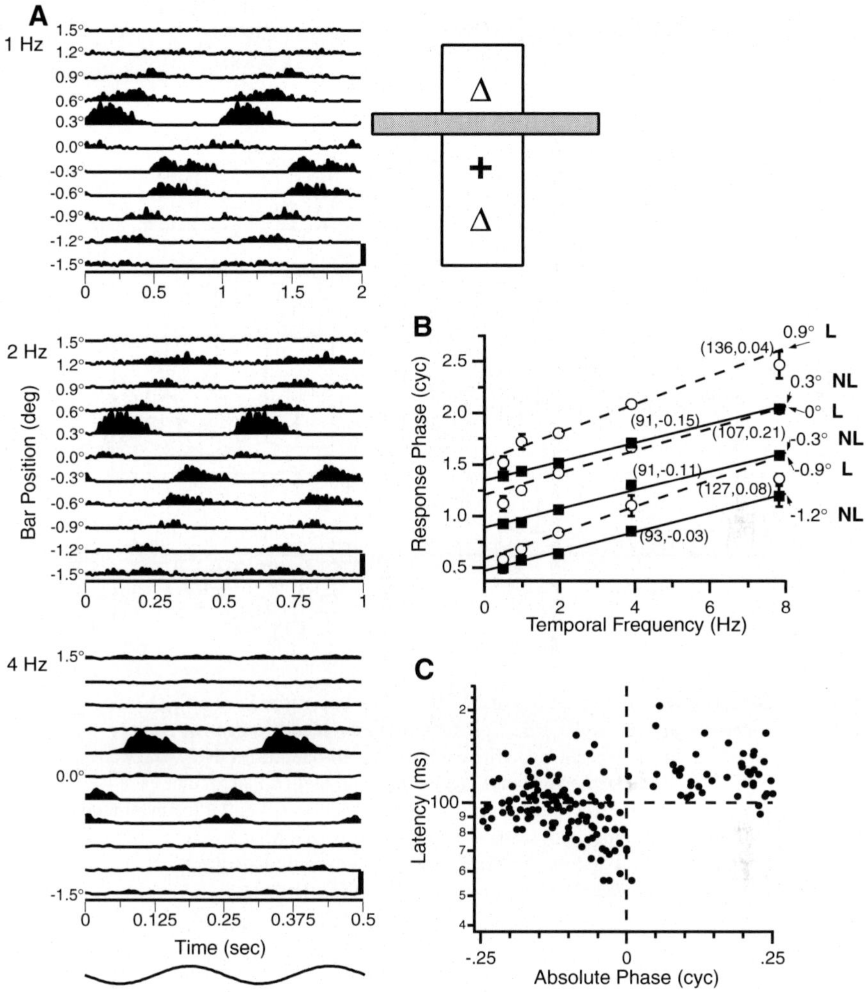

FIGURE 9-7. **(A)** Spatiotemporal maps of a DS simple cell in layer 5A obtained using a counterphasing bar with luminance modulated at three temporal frequencies. Numbers on the left indicate bar position. Cartoon on the right shows the approximate bar dimensions relative to the receptive field. Two cycles of response are shown. The receptive field was most S-T oriented below 4 Hz. **(B)** Phase versus TF plots for 6 bar positions tested in **A.** Numbers in parentheses indicate latency and normalized absolute phase, respectively. Positions with lagged- or non-lagged-like timing are spatially interdigitated in this cell. **(C)** Absolute phase and latency values for 163 receptive-field positions tested in 33 simple cells recorded in and near the borders of layer 4. (Data are from Saul and Humphrey [1992a] and Humphrey et al. [1998].)

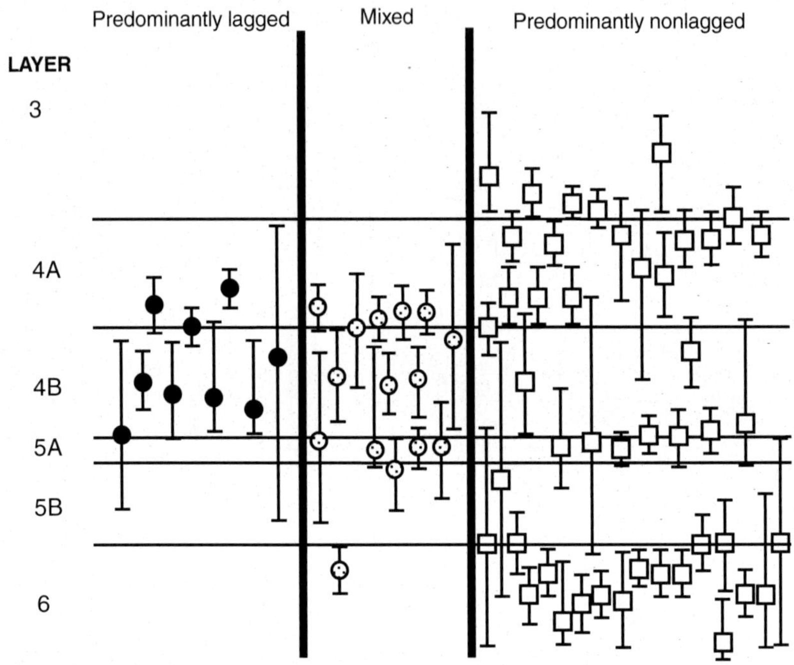

FIGURE 9-8. Laminar distributions of 69 simple cells categorized according to their response timings to counterphasing bars. Cells were classified as predominantly lagged or nonlagged if >50% of their receptive-field positions were lagged-like or non-lagged-like, respectively, and <20% were of the opposite type. Cells were deemed mixed if lagged- and non-lagged-like timings each were identified in >20% of the positions. Error bars indicate our subjective assessment of the maximum error in localizing recording sites histologically. Cells with significant lagged timing were restricted mainly to the lower two thirds of layer 4 and the 4/5 border region. Cells with nonlagged timings were found mainly in layers 4, lower 3 and 6. (Data are from Humphrey et al. [1998].)

zones weakens at the higher TFs, whereas the dominant nonlagged-like zones continue to discharge vigorously and account for most responses above 3 Hz. The timing and amplitude shifts in this cell were accompanied by changes in direction selectivity; the cell was strongly selective at 1 to 2 Hz but not at 4 Hz or higher (not shown). Thus, the loss of quadrature between lagged- and nonlagged-like positions paralleled the loss of DS. These data indicate that *the geniculate afferents contribute to DS by setting up timing gradients across the receptive field of the simple cell.*

It is important to note that convergence of lagged and nonlagged inputs is not the only way to produce timing gradients via the afferents. As revealed in Fig. 9-6D, a full quarter cycle of timing variation exists within each afferent group. Among X_N-cells, 0 cycles and –0.25 cycles correspond to sustained and transient responses, respectively. Thus, convergence of nonlagged afferents with phasic and tonic behaviors could induce S-T inseparability, and we have observed evidence for this (Humphrey et al., 1998).

Ferster and colleagues (Jagadeesh et al., 1997) provided additional evidence for a geniculate contribution to DS in layer 4 simple cells. They measured the direction and orientation selectivity of membrane potentials to drifting gratings before and during a cooling procedure designed to suppress intracortical interactions while maintaining thalamocortical activity. Although cooling markedly dampened response amplitudes, it had little effect on selectivity. Similar findings were obtained when electrical stimulation was used to suppress cortical activity during visual stimulation (Chung and Ferster, 1998). Assuming the methods suppressed all relevant intracortical interactions, the studies indicate that the spatial and temporal information conveyed by LGN afferents suffices to initiate direction and orientation tuning in many layer 4 simple cells. Finally, Jagadeesh et al. (1997) (also Kontsevich, 1995), using principal component analysis of DS-cell responses, confirmed that direction selectivity in simple cells could be generated by inputs from a minimum of two subunits whose time courses are similar to those of lagged and nonlagged cells.

It is not clear at what stage LGN timings might be combined to influence S-T structure. Direct convergence onto layer 4 cells is likely given that all such cells receive monosynaptic input from the LGN (Martin and Whitteridge, 1984; Ferster and Lindstrom, 1983; Bullier and Henry, 1979). Indirect convergence via other cortical cells undoubtedly plays a role, if for no other reason than the vast majority of inputs to cortical cells arise within cortex (Ahmed et al., 1994; Peters and Payne, 1993). An S-T inseparable cell, for instance, could be created by two separable simple cells in approximate S-T quadrature, with one input having lagged timing and the other nonlagged timing. Various combinations of excitatory and/or inhibitory interactions are possible (see pages 357 and 364). *Our studies indicate that the LGN is the wellspring of timings that underlie these interactions. We assume the timings are distributed through, and further modified by, intracortical networks, and combined in different ways to produce S-T inseparability and DS.* Before considering experimental evidence concerning intracortical interactions that influence DS, we briefly review some other current directional models that posit alternative mechanisms for producing timing delays and directional tuning.

COMPARISON WITH RECENT MODELS

Numerous models have been proposed to explain the direction selectivity of simple cells (e.g., Heeger, 1993; Worgotter and Holt, 1991; Worgotter and Koch, 1991; Ganz and Felder, 1984; Emerson and Gerstein, 1977b). Most rely exclusively on intracortical interactions and assume, implicitly or explicitly, that the geniculocortical afferents are nonlagged with identical timings (cf., Emerson, 1997; Hamada et al., 1997). Two recent models (Maex and Orban, 1996; Suarez et al., 1995) have gained attention, in part because they attempt to incorporate in some detail current data on functional and structural properties of cortical circuits. We focus mainly on the Maex and Orban (1996) model because it is the most comprehensive in attempting to account for directional behaviors and it explicitly addresses S-T structure.

MAEX AND ORBAN (1996)

The model posits that layer 4 simple cells are interconnected in a feedback network of partially overlapping receptive fields. A temporally homogeneous LGN input mediated by non-NMDA glutamate receptors would, in the absence of other inputs, produce S-T separable structure. The excitatory feedback is spatially anisotropic and is mediated by NMDA receptors whose time course of activation is determined by the synaptic weights of the cortical inputs. The weights increase in the preferred direction of motion across the receptive field. This anisotropy introduces progressively decreasing timing delays in the PD that shift S-T structure toward inseparability and induce a weak directional bias. The feedback excitation also amplifies preferred-direction responses. Excitation in the nonpreferred direction is suppressed, and DS thereby greatly enhanced, by inhibition from cells that respond best to that direction. This cross-directional inhibition essentially lowers cell membrane potential below spike threshold. Spatially anisotropic nonopponent inhibition also adds to inseparability, although its relative contribution is unclear. Other inhibitory connections are introduced to reinstate linear summation and to reinforce network stability.

The Maex and Orban model is highly constrained but appealing because it accounts for directional behaviors of cells to a variety of stimuli using mechanisms that likely exist in cortex (although whether they contribute as proposed remains to be determined). The behaviors include, among others, DS evoked by small spatial displacements, stimulus polarity (light versus dark) invariance, directional sensitivity to apparent motion (i.e., 2-bar) stimuli, and some TF-dependent directional tuning. The model differs from ours (Saul and Humphrey, 1990, 1992a, 1992b) in the source and nature of the timing delays that produce S-T inseparability. For Maex and Orban, the timings reflect latency shifts derived intracortically. These maintain DS over drift rates of 1 to 16 Hz. To produce DS at lower drift rates (e.g., 1.0 to 0.25 Hz), progressively longer latencies (250 to 1000 msec, respectively) must be produced by slowing the time-to-peak of the NMDA-mediated EPSPs or by adding more polysynaptic delays. In contrast, in the geniculocortical model, the necessary timings reflect to a large degree the phase relations of the thalamic afferents, which are near quadrature at low temporal frequencies. Whether DS in a given cell is maintained at higher TFs depends on the phase stability of its inputs. If quadrature phase is stable with changes in temporal frequency, then DS could be maintained; if quadrature deteriorates, then so should DS. These two possibilities were simulated in Saul and Humphrey (1990). We note again that, in reality, the LGN timings must be conveyed to a layer 4 cell both directly as well as indirectly via other cortical cells, probably using the types of intracortical circuits modeled in Maex and Orban (1996). Our model posits, however, that the timings underlying the interactions in the cortical circuits reflect to a large degree the geniculocortical inputs.

Maex and Orban (1996) noted that adding convergent input from lagged and nonlagged cells to the cortical network also produced DS, but the results were not unequivocally better than with nonlagged input only. However, using the geniculate timings would allow relaxation of some of the connectional and synaptic weighting constraints in the model.

SUAREZ, KOCH, AND DOUGLAS (1995)

The directional model of Suarez et al. (1995), based on the "canonical microcircuit" of Douglas and Martin (1991), differs in structural detail from Maex and Orban (1996) but shares key operations. These include (1) the introduction of temporal delays intracortically via excitatory receptors with slow dynamics and inhibitory receptors with fast time courses (they include, additionally, inhibition with slow dynamics), (2) feedback excitation that greatly amplifies geniculocortical signals in the preferred direction of motion, and (3) feedforward inhibition that is spatially offset from the excitation and suppresses responses in the nonpreferred direction. The model appears to produce some S-T receptive-field orientation, but it is unclear how it comes about or how well it relates to direction selectivity. Nevertheless, the model nicely reproduces a number of directional tuning behaviors.

GENERAL PROBLEMS IN REPRODUCING ABSOLUTE PHASE LAGS

Synaptic depression has been proposed recently as another mechanism for creating temporal phase shifts (Chance et al., 1998; Abbott et al., 1997). It reduces the steady-state synaptic efficacy of inputs firing at high rates, thus causing a cell to fire more transiently than its inputs. However, depression only generates phase advances and hence cannot account for the absolute phase lags recorded in area 17 (Fig. 9-7C).[3] Similarly, the massive recurrent excitation in the model of Suarez et al. (1995) appears to advance response phase (see Fig. 5A in Suarez et al., 1995). Such excitation would be ill suited to produce the observed absolute phase lags, particularly if the geniculocortical inputs were all nonlagged.

RELATIVE EFFICACY OF GENICULOCORTICAL VERSUS INTRACORTICAL INPUTS?

It is worth noting that recent models (Maex and Orban, 1996; Suarez et al., 1995; Douglas and Martin, 1991) assume that LGN inputs are weak and that intracortical amplification is essential to boost their signals. The assumption is based largely on

[3] Synaptic depression increases latency and could contribute to cortical receptive-field positions that display absolute phase leads and long latencies (Fig. 9-7C).

anatomical findings that (1) the vast majority (>75%) of synapses onto cortical cells come from other cortical neurons, even in layer 4 (reviewed by Peters and Payne, 1993) and (2) the few individual LGN axons that have been reconstructed appear to provide only a small number of synapses (<10) onto any given cortical cell (Freund et al., 1985). In contrast to the structural evidence, intracellular studies show that electrical stimulation of the LGN elicits strong monosynaptic excitation in layer 4 (Ferster and Lindstrom, 1983), and EPSPs generated by geniculocortical axons may be twice the size of those arising intracortically (Stratford et al., 1996; also see Gil et al., 1999, for somatosensory cortex). The disadvantage of the LGN inputs in terms of synapse number thus may be offset partially by their greater EPSP amplitudes. On another front, cross-correlation studies (Reid and Alonso, 1995, 1996; Tanaka, 1983; Toyama et al., 1981) reveal that LGN spikes are much more effective in driving cortical neurons than intracortically generated spikes. Synchronous firing of the LGN cells may enhance their synaptic efficacy (Usrey and Reid, 1999; Alonso et al., 1996). Finally, the similarities between LGN and layer 4 in response timing signatures implies a greater dependence of cortical cells on thalamic inputs than is indicated by the anatomy (Hirsch et al., 1998; Humphrey et al., 1998; Saul and Humphrey, 1992a). Thus, although intracortical amplification undoubtedly influences visual processing, there is compelling evidence for a larger contribution by the LGN than is often currently assumed. The teasing out of the relative roles of various inputs remains an important objective.

INTRACORTICAL INHIBITION: EXPERIMENTAL EVIDENCE

What do experimental studies tell us about intracortical interactions that influence direction selectivity? Although recent models have provided insights into the possible roles of recurrent excitation in influencing DS, so far there has been little experimental manipulation of excitatory circuits to critically test these ideas. Because most of our knowledge about cortical interactions comes from manipulating inhibition (which must indirectly affect excitation), this section focuses on inhibitory processes.

INHIBITORY SCULPTING OF S-T RECEPTIVE-FIELD STRUCTURE

Blocking $GABA_A$-mediated inhibition by iontophoretically applying bicuculline reduces or abolishes DS in most simple cells and some complex cells (Vidyasagar and Heide, 1986; Wolf et al., 1986; Sillito et al., 1980; Sillito, 1975; 1977, Tsumoto et al., 1979; also see chapter 11). Antagonizing $GABA_B$-mediated inhibition reportedly does not influence DS (Baumfalk and Albus, 1988). However, as noted by Suarez et al. (1995), testing has not been done using very low drift rates, where prolonged inhibitory currents might be expected to influence

DS. Returning to $GABA_A$, the bicuculline-induced reductions in DS usually reflect an increase in discharge to the NPD, although responses in the PD may also change (Murthy and Humphrey, 1999; Tsumoto et al., 1979; Sillito, 1977). Although many laboratories have confirmed these actions, it has been unclear exactly how inhibition produces its effects. One previously untested hypothesis is that inhibition creates or enhances S-T orientation. If so, then blocking inhibition should reduce S-T orientation along with DS. An alternative, though not mutually exclusive, hypothesis is that inhibition is "flat", and merely suppresses weak responses. It might act by lowering membrane potentials relative to spike threshold. This iceberg effect should enhance an initial directional bias but not significantly alter response timing. To test these possibilities, we measured S-T structure and DS in simple cells before and during iontophoretic application of bicuculline methiodide (BMI) (Murthy and Humphrey, 1999).

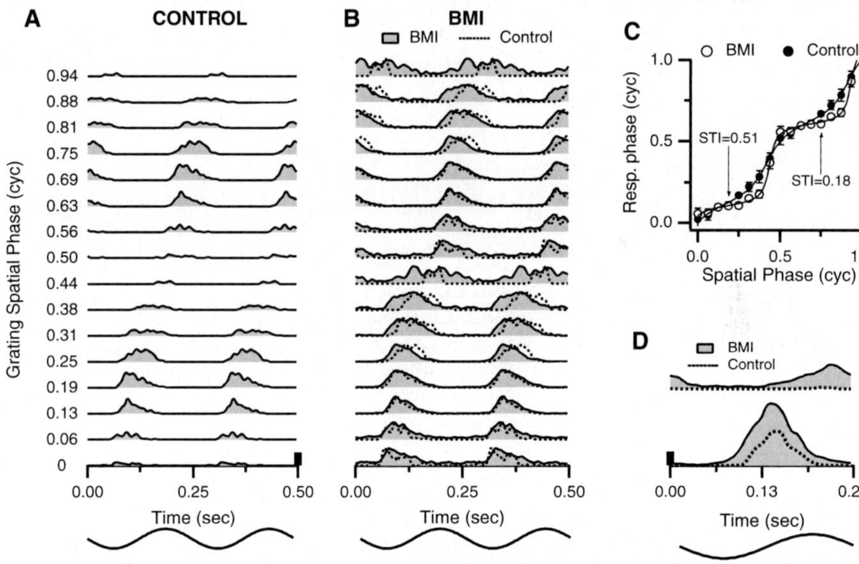

FIGURE 9-9. Action of bicuculline methiodide (BMI) on the spatiotemporal structure and directional tuning of a layer 4 simple cell. **(A)** Control responses obtained using a stationary counterphasing grating, revealing moderate S-T orientation. **(B)** Responses during iontophoresis of BMI. To clarify the changes in response timing, BMI (filled) and control (dotted) responses are superimposed, and each pair of histograms is normalized to equate peak firing rates. BMI produced a leftward shift (i.e., phase advance) in response timing at spatial phases 0.25 to 0.38 cycles. **(C)** Response phase vs. spatial phase for control and BMI conditions. BMI reduced S-T orientation. **(D)** Responses to a grating drifting in opposite directions during the two conditions. BMI reduced DS by about half (control DI = 0.93; BMI DI = 0.44). Although not illustrated, the cell's S-T orientation and directional tuning returned to control values following cessation of BMI. Calibration bars in **A** and **D** indicate 18 and 22 ips, respectively. (Data are from Murthy and Humphrey [1999].)

Figure 9-9 illustrates results from a layer 4 cell that was highly DS in control trials. Application of BMI increased responses to both directions (D), but the relative increase in the NPD was greater and the DI value declined by about half. The decline was associated with changes in S-T structure as measured with counterphasing gratings (A and B). The receptive field was moderately S-T oriented in control trials (A); response phase increased fairly smoothly with stimulus spatial phase, producing an STI of 0.51 (C, filled circles). BMI application produced an advance in response phase at certain spatial phases (0.25–0.38), which produced more uniform timing across the field and hence less S-T orientation (C, open circles; STI = 0.18). The effect seen here was typical of more than half the layer 4 sample—a reduction but not elimination of S-T orientation and DS. One of three cells with more striking effects is shown in Fig. 9-10; BMI eliminated the normally strong DS and S-T orientation (B and D). Again, the latter loss reflected systematic shifts in response phase that produced more uniform timing across the receptive field (C). Overall, BMI reduced DS and S-T orientation in 15 of 17 layer 4 cells examined. The changes in the two measures were well correlated (r = 0.8) and ~70% of the reduction in directional index was attributable to a weakening of S-T orientation.

Inhibition thus impacts direction selectivity in layer 4 simple cells by creating or enhancing S-T orientation. An additional role for flat inhibition remains

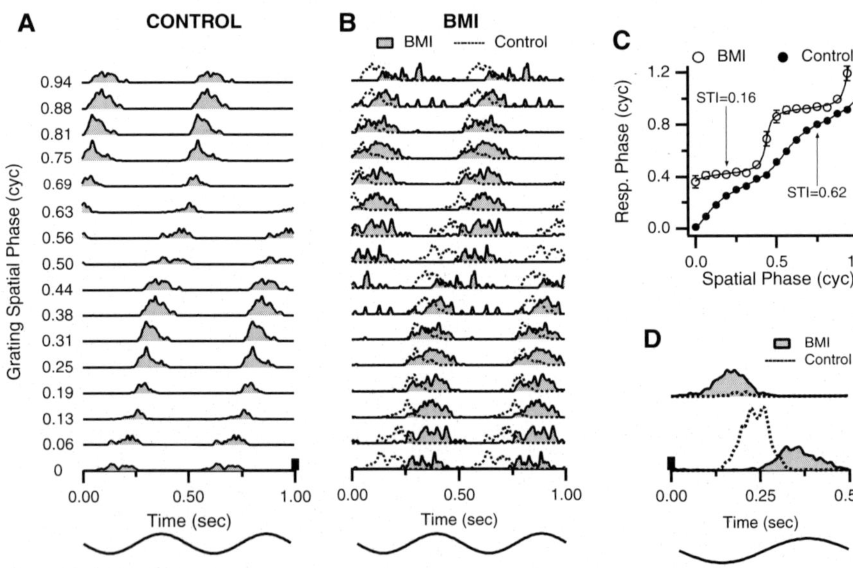

FIGURE 9-10. Action of BMI on the S-T structure and DS of another layer 4 simple cell. Conventions are as in Fig. 9-9. **(A-C):** BMI produced clear changes in timing at all spatial phases, resulting in a loss of S-T orientation. **(D)** The DI was reduced from 0.92 to 0.0 by BMI. (Data are from Murthy and Humphrey [1999].)

(Murthy and Humphrey, 1999), and requires further study, but the sculpting of S-T structure appears to be the more critical contribution. Related to this, the BMI data do not support one aspect of the Maex and Orban (1996) model, which predicts that reducing directional inhibition should decrease DS without necessarily affecting S-T orientation. In fact, in their model, S-T orientation is weak in all cells; it plays only a minor role in creating DS, by producing small directional biases that are accentuated by cross-directional inhibition. Clearly, however, S-T orientation is more strongly linked to DS, and S-T relations between the inhibitory elements and their targets are important. Finally, we note that most layer 4 cells continued to display some DS and S-T orientation during BMI, even during prolonged periods of drug application. This is consistent with other bicuculline studies (Vidyasagar and Heide, 1986; Wolf et al., 1986; Tsumoto et al., 1979 (see Chapter 11)) and with Nelson et al. (1994), who showed that intracellular blockade of $GABA_A$ and $GABA_B$ receptors weakens but generally does not eliminate DS. All of these studies are consistent with the notion that excitatory inputs from LGN and cortex, which convey different timings, generate the remaining directional tuning.

INHIBITORY CONNECTIVITY

What is the source and nature of the inhibition that alters the timing and S-T structure of layer 4 simple cells? The suppression must arise from other simple cells with receptive fields that are spatially and temporally offset from their targets. This is certainly possible. Receptive fields of adjacent simple cells simultaneously recorded have spatial phase offsets in the range needed to produce inseparability (DeAngelis et al., 1999; Pollen and Ronner, 1981), although their temporal offsets and synaptic signs have yet to be characterized. Complex cells do not likely play a role in sculpting S-T orientation because they lack spatiotemporally modulated responses that are necessary to alter the timing structure of their targets. Complex cells might contribute to setting thresholds, however (Heeger, 1993). Are the inhibitory cells themselves direction selective? Although theoretically not necessary (Suarez et al., 1995; Worgotter et al., 1991), many probably are, since most area 17 cells are DS, including some confirmed inhibitory neurons (Martin et al., 1983; Somogyi et al., 1983; cf., Azouz et al., 1997).

Does the net inhibitory input to a DS cell have the same or opposite preferred direction? The evidence on this is mixed. Comparisons of spiking responses to two bars flashed sequentially so as to induce apparent motion generally reveal suppression in the NPD (Ganz and Felder, 1984; Emerson and Gerstein, 1977a) (for more details, see page 374). Also, motion in the NPD can suppress spontaneous and elevated activity (Hammond and Kim, 1996; Bishop et al., 1980). The suppression is assumed to be cross-directional (i.e., exclusively from cells that prefer the opposite direction), as that makes sense intuitively, although seeing inhibition in the NPD does not imply its absence in the PD. In contrast to these data, intracellular studies indicate that inhibition, as evidenced by IPSP amplitudes, is more prominent in the PD than in the NPD (Douglas et al., 1991; Ferster, 1986; cf., Creutzfeldt et al., 1974), which might imply some form of isodirectional suppression (i.e., from cells with the same PD) or a general covariation of excitation and inhibition (Douglas and Martin, 1991). Shunting inhibition, which would produce large conductance changes with minimal hyperpolarization, is a potential mechanism for suppression in the NPD (Torre and Poggio, 1978), but little evidence for it has been found (Ferster and Jagadeesh, 1992; Berman et al., 1991; Pei et al., 1994). However, Borg-Graham et al. (1998) demonstrated strong shunting associated with orientation tuning. It remains to be determined whether their method also might reveal a critical contribution of shunting to directional tuning.

Adaptation paradigms have been used to infer the nature of mechanisms underlying neural processing, by selectively inactivating elements of presumed circuits (Saul, 1995; Vautin and Berkley, 1977; Maffei et al., 1973). A grating drifting across a cell's receptive field reduces subsequent responses to stimuli similar to that grating. If a DS cell's poor response to the NPD depends on cross-directional inhibition, then a stimulus moving in the NPD should adapt the inhibitory inputs, and the reduced inhibition should result in a greater response to the NPD in the DS cell. However, this does not occur; the NPD response remains low or is reduced even further by the adaptation (Giaschi et al., 1993; Saul and Cynader, 1989a, 1989b; Marlin et al., 1988). It is the PD response that has been reported to increase (Giaschi et al., 1993; Marlin et al., 1988). Assuming that inhibitory cells adapt under this paradigm, these results do not support cross-directional inhibition.

Eysel's laboratory has performed analogous experiments using a pharmacological method of "remote inactivation" to probe the direction and orientation properties of laterally interacting cortical networks (Crook et al., 1991, 1996; Worgotter and Eysel, 1991; Eysel et al., 1988; (see chapter 11)). In a study of area 17, Crook et al. (1997) used pairs of electrodes ~500 μm apart in layers 2/3 and 4 to record cells with overlapping receptive fields and similar preferred orientations (Fig. 9-11). The directional tuning of a cell at one site (RS) was then tested, and multiunit activity at the adjacent site (IS) was silenced by applying GABA. The rationale was that small (≈200–300 μm) clusters of cells tend to display similar preferred directions (Shmuel and Grinvald, 1996; Berman et al., 1987; Tolhurst et al., 1981; also see Weliky et al., 1996), so inactivating one cluster should reduce directionally congruent signals projecting to nearby sites.

Inactivation effects were seen in about two thirds of the pairings. Specifically, if cells at the two sites preferred opposite directions, then silencing one site reduced or eliminated DS at the adjacent site by increasing responses to the NPD (Fig. 9-11C versus 9-11D). Litte or no change in response occurred in the PD. Although GABA presumably affected all neurons at the inactivation site, the increased NPD response was interpreted as a silencing of cells that normally provided cross-directional inhibition. In contrast, if the inactivation and recording sites preferred the same direction, then GABA either decreased (Fig. 9-11C versus 9-11F) or

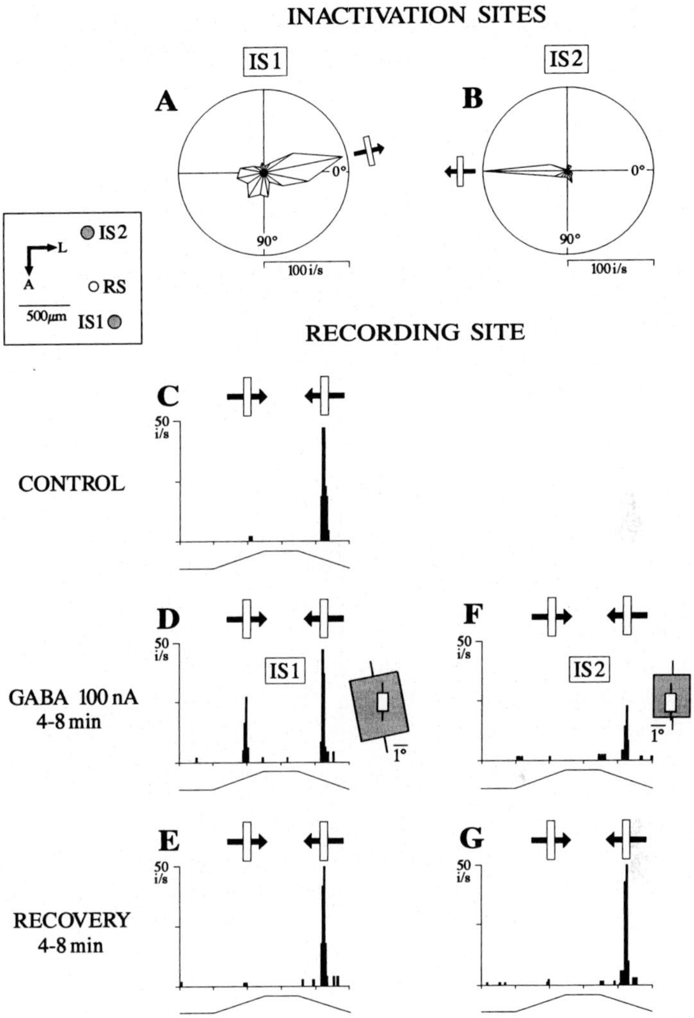

FIGURE 9-11. Differential effects on a simple cell's directional tuning by alternately inactivating two adjacent sites (IS1 and IS2) with opposite direction preferences. The DS cell (RS) was recorded at the border of layers 3 and 4. The GABA-containing pipettes at IS1 and IS2 were positioned in layers 4A and 2, respectively. (**A** and **B**) Polar plots obtained from multiunit activity showing preferred stimulus orientation and direction at each inactivation site. (**C** and **D**) The recorded cell's average responses to an optimally oriented bar drifting in opposite directions before (**C**) and during (**D**) inactivation by GABA at IS1. RS and IS1 had opposite direction preferences. Inactivation caused an increase in response to the nonpreferred direction, with little change in the preferred direction. (**F**) During iontophoresis at IS2, where the preferred direction matched that of the recorded cell, the cell's preferred-direction response decreased (cf. **C**) but there was no change in the nonpreferred direction. (**E** and **G**) After cessation of GABA, the cell's directional tuning returned to control levels. (Reprinted from Crook et al. [1997] with permission of Cambridge University Press.)

increased the PD response and had little effect on the NPD response. The reduced PD responses may reflect less excitation from cells with similar preferred directions. The increased PD responses have multiple interpretations. Overall, these results reveal a directional logic to some of the local interactions in layers 2/3 and 4 and provide compelling evidence for cross-directional inhibition. Whether the apparent contradiction between these and the adaptation data noted previously reflects the use of different methods remains to be determined.

Whatever form directional inhibition takes, basket cells are likely substrates; many have axonal arbors that collateralize locally (Crook et al., 1998; Kisvárday, 1992) as well as more broadly (Kisvárday et al., 1994; Kisvárday and Eysel, 1993), and some are known to be directionally tuned (Martin et al., 1983). However, their degree of connectional specificity is unclear. For example, Crook et al. (1998) retrogradely labeled GABAergic neurons projecting from an inactivation site to an adjacent recording site. Cells at the two sites preferred opposite directions, and inactivating one site increased responses to the NPD in the other, thus reducing DS. This suggests that the labeled inhibitory cells at the inactivation site normally played a role in suppressing responses to the NPD at the nearby site. However, the recording site received projections from scores of GABAergic cells, only a few of which were located in the inactivation region, and nothing is known about their properties. One would like to know, for instance, whether GABAergic projections from an inactivation zone that is *in*effective at reducing DS at a nearby site also avoid targeting that site. This would be compelling evidence for connectional specificity.

The remote inactivation studies provide important insights into the general types of directional interactions that exist. They leave open questions about the specific nature of the interactions. For instance, is the inhibition that is inferred from the experiments "flat" or spatiotemporally structured? The bicuculline results of Murthy and Humphrey (1999), summarized previously, suggest that directional inhibition on simple cells arises from other S-T oriented simple cells with specific spatiotemporal relations to their targets. For cross-directional inhibition, then, the S-T orientation of the aggregate inhibitory receptive field would tilt at an angle opposite to that of the target field, as suggested by Maex and Orban (1996). An open question is whether the remote inactivation effects are mediated by changes in S-T structure.

Might isodirectional inhibition play a role in some circuits? Although it may seem counterintuitive that cells with the same PD could enhance DS via inhibition, Saul (1999) showed how this may occur. The key, again, is in appreciating how cortical cells interact in space and time. Figure 9-12 models two mutually inhibitory simple cells that receive excitatory lagged or nonlagged LGN inputs. The ON- and OFF-center LGN cells have spatially offset receptive fields, producing ON and OFF zones in each simple cell that lie a half-cycle apart in space. The lagged and nonlagged LGN cells are spatially interdigitated, producing lagged and nonlagged zones in cortex that are a quarter-cycle apart in space. The net inputs to the two simple cells thus are S-T separable and lie in S-T quadrature (Fig. 9-12B, upper). Comparing the upper to the lower plots in B, we see that the

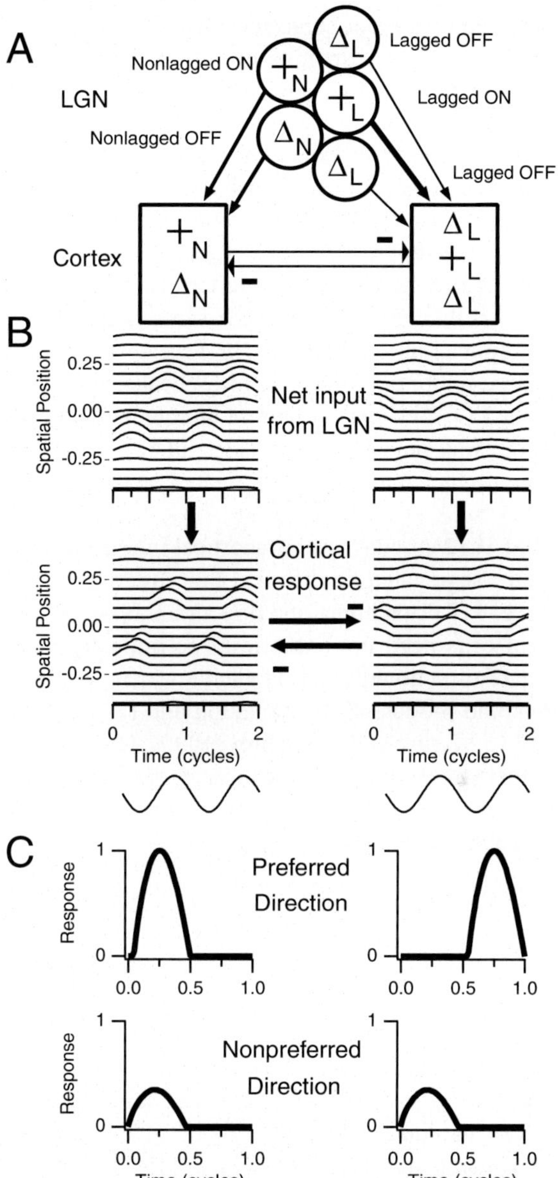

FIGURE 9-12. Model of inhibitory interactions between direction-selective simple cells that produces S-T orientation and DS. **(A)** Cortical cells receive ON- and OFF-center lagged and non-lagged inputs from the LGN. Only one spatial dimension is represented; the lagged and nonlagged receptive fields overlap but are shown horizontally offset for clarity. **(B)** Space-time receptive-field maps of the combined inputs (top) and of the simple cells (bottom). The maps can be interpreted as responses to sinusoidally modulated stationary bars. The net cortical activity shows the effects of the mutual inhibition, which reduces responses at certain positions and times. **(C)** Simulated responses of the simple cells to a grating drifting in each direction (PD is downward) across the receptive field. See text for further details. (Adapted from Saul [1999].)

afferent excitation is suppressed at some receptive-field positions and times but not at others, depending on the responses providing the inhibition. Because the excitation and inhibition are offset in space and time, their interaction creates S-T orientation (B, lower).

Figure 9-12C shows simulated responses to drifting gratings. The two model neurons respond out of phase with each other in their PD, silencing one another during alternate half-cycles. Although the strongest inhibition arrives onto each cell for the PD, it does not suppress responses because it occurs when there is little excitation. In the NPD, however, the quarter-cycle differences between the cells cancel, the two cells are activated at the same time and they inhibit each other. Note that the mutual inhibition in the NPD creates strong DS. Nonetheless, the amplitude of these inhibitory signals is low, because each presynaptic cell is not as active as in the PD. The apparent contradiction that inhibition is crucial yet weak arises from the way negative feedback regulates activity. The strength of inhibition is directly, rather than inversely, related to the strength of excitatory responses.

The model in Fig. 9-12 emphasizes intracortical interactions by having lagged and nonlagged inputs segregated on different first-order cortical cells. However, afferent convergence, S-T orientation, and directional bias could begin at the first stage. Also, cortical cells need not necessarily be mutually inhibitory. Alternatively, convergence of directionally biased excitatory and inhibitory simple cells in S-T quadrature could produce S-T orientation and DS in a cell to which they project (see Fig. 12 in Murthy and Humphrey, 1999). Overall, the message from these examples is that knowledge of the spatiotemporal receptive-field characteristics of interacting neurons is critical to understanding directional tuning, as well as other response properties.

HOW ARE INPUTS COMBINED?

We have focused so far on the sources of inputs to layer 4 simple cells that produce S-T inseparability. We now turn to how the inputs are combined. That is, are they summed or instead subjected to arithmetically more elaborate processing? It should be understood that the combination rule on inputs can usually be decomposed into series of elementary functions, such as a Taylor or Volterra series, which contain linear and nonlinear parts. A basic question then is, to what extent can cell responses be described by just the linear part? We will show that linear summation provides a reasonable approximation to DS in most layer 4 simple cells, but nonlinearities also must be invoked. We also will show that DS in layer 6 simple cells (and all complex cells) can only be understood in terms of nonlinear interactions. The questions then become, what sorts of nonlinearities operate and are they the same for all cells?

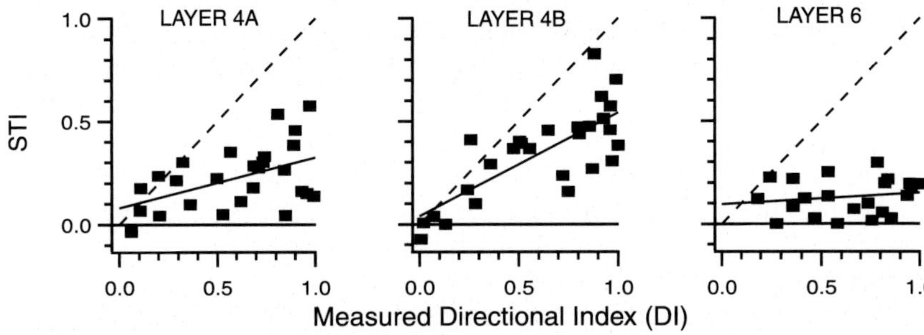

FIGURE 9-13. Laminar differences in the relationship between directional tuning and S-T orientation among simple cells. Each point represents a single cell. The dashed lines of unity slope indicate the ideal relationship assuming a strictly linear model. The solid lines are best fits to the data. The correlation was highest in layer 4B (r = 0.76; m = 0.5), lowest in layer 6 (r = 0.16; m = 0.06) and intermediate in layer 4A (r = 0.48; m = 0.24). (Data are from Murthy et al. [1998].)

EVIDENCE FOR LINEAR SUMMATION

Linearity, or properly, its failure, can be tested by predicting a cell's response to a moving stimulus from the sum of its responses to a stationary stimulus presented in different positions of the receptive field. For grating-based data, this can be done by measuring response amplitude and phase at each stimulus spatial phase, summing the responses after shifting them in time to mimic activation by a moving grating, and comparing them to responses evoked by motion (Jagadeesh et al., 1997; Murthy et al., 1998).[4] By whatever method used, the linear predictions have yielded variable results. Some groups (McLean et al., 1994; DeAngelis et al., 1993b; Albrecht and Geisler, 1991; Reid et al., 1991) reported moderate correlations (~0.5) between predicted and observed DS, with the linear predictions accounting for about one third to one half of directional tuning. However, the relationship was highly variable among cells in each study. Further, Tolhurst and Dean (1991) found no consistent relationship and concluded that DS in simple cells could only be attributed to nonlinear mechanisms.

None of the studies noted the laminar locations of the cells, which prompted us to ask whether the variable and conflicting data might reflect sampling differences. We reexamined this by distinguishing simple cells in layers 4 and 6 (Murthy et al., 1998). We relied mainly on the spatiotemporal index (STI) (page 350) as the metric of S-T structure, because the STI and DI are equivalent in a strictly linear model. Figure 9-13 summarizes their actual relationship. In layer 4B the two measures are well correlated (r = 0.76), with S-T orientation accounting for about half the directional tuning, on average. The relationship is present but weaker in layer 4A. In layer 6, however, there is virtually no correlation because most receptive fields there have little or no S-T orientation despite being

[4] For other procedures, see McLean et al. (1994), DeAngelis et al. (1993a, 1993b), Emerson and Citron (1992); Reid et al. (1991), Albrecht and Geisler (1991), and Tolhurst and Dean (1991).

as DS as those in layer 4. (Even for small STIs, however, the sign of S-T orientation nearly always predicts the preferred direction of motion.) Thus, the role of linear, or quasi-linear, summation in influencing DS varies with cell location, being most prominent in layer 4B and least important in layer 6. Recall that layer 4B appears to be the convergence zone of lagged and nonlagged afferents. Relatively simple summation of those inputs may account for the high degree of S-T orientation observed there.

Further evidence for a linear model in layer 4 comes from intracellular recordings by Jagadeesh et al. (1997). They used the modulation of simple cell membrane potentials to counterphasing gratings (Fig. 9-14A) to linearly predict modulations to drifting gratings (B). The predictions were strikingly accurate,

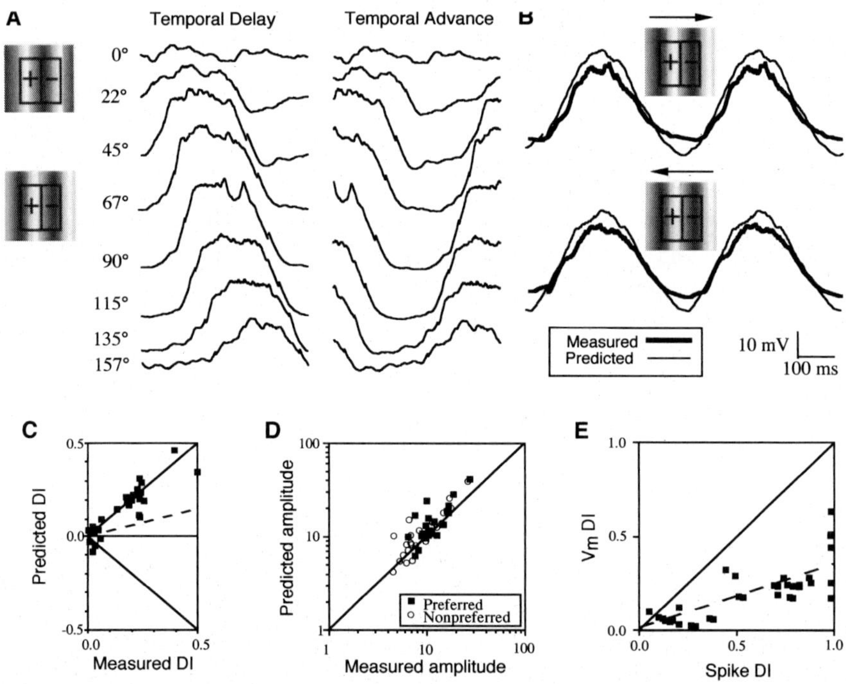

FIGURE 9-14. Directional tuning predicted from intracellular recordings of layer 4 simple cells. **(A)** Fluctuations in a non-DS cell's membrane potential evoked by a stationary grating *counterphasing* at 2 Hz and presented at eight spatial phases. Each trace was shifted in time from its original phase to mimic the response to a grating *drifting* rightward (temporal delay) or leftward (temporal advance) through the receptive field. **(B)** The summed counterphase responses were then used to predict the membrane potential actually evoked by the moving grating. For the population, the predicted versus measured directional index **(C)** and predicted versus measured amplitude of the membrane potential fluctuations **(D)** were in good agreement. However, the directional index measured from membrane potentials (V_m DI) was up to three times lower than that measured from action potentials **(E).** Data are from Jagadeesh et al. (1997), reprinted with permission of the American Physiological Society.

accounting for the amplitude, timing, and directional tuning of responses to motion (B-D). Thus, despite the many nonlinear processes that influence neurons, directional tuning seems attributable to linear summation of synaptic potentials arising from different regions of the receptive field. However, when DS was evaluated using the cells' action potentials, it was two to three times greater than predicted from membrane potentials (Fig. 9-14E). Invariably, discharge rates for the nonpreferred direction were less than expected, and responses to the preferred direction were sometimes greater than predicted. A similar, twofold to threefold underestimation of DS is routinely observed in predictions based solely on spiking data (Murthy et al., 1998; McLean et al., 1994; DeAngelis et al., 1993b; Albrecht and Geisler, 1991; Reid et al., 1991). The discrepancies reflect the action of additional, nonlinear processes.

NONLINEARITIES

The nature of the nonlinearities that affect DS has been hotly debated (Emerson, 1997a, 1997b; Baker and Boulton, 1994; Baker and Cynader, 1994; Heeger, 1993; Emerson and Citron, 1992; Albrecht and Geisler, 1991; Grzywacz and Koch, 1987; Adelson and Bergen, 1985; van Santen and Sperling, 1985; Watson and Ahumada, 1985; Reichardt, 1961). A detailed examination of this complex topic is beyond the scope of this review. We focus instead on some general aspects of nonlinearities and how they may operate. A common view is that quasilinear summation of inputs initiates directional tuning, which is then enhanced by nonlinearities. Rectification is a pervasive *intensive* nonlinearity, that is, one that specifically affects response amplitude. The spike threshold prevents the conversion of small depolarizations in the NPD into action potentials, thus suppressing weak responses. The conversion of suprathreshold depolarizations to spike firing rates is thought to reflect an additional intensive nonlinearity. Thus, thresholding of weak responses to the NPD and amplifying stronger responses to the PD could accentuate directional biases created by linear S-T summation. Albrecht and Geisler (1991) tested such a linear/nonlinear (LN; a.k.a. exponent) model on grating-based data and observed that linear summation followed by rectification and an approximate squaring amplification predicted DS in most of their simple cells. Other versions of this model that incorporate contrast normalization processes (Tolhurst and Heeger, 1997a, 1997b; Heeger, 1992, 1993) also account moderately well for DS in simple cells. The modeled amplification is biologically reasonable; for instance, the relationship between input and output in simple cells has been measured and is well fit by a power function with an exponent near 2.5 (Anzai et al., 1999; Carandini et al., 1997; Albrecht and Geisler, 1991; Albrecht and Hamilton, 1982).

Murthy et al. (1998) reevaluated the LN model in light of the previously noted laminar differences in S-T structure. The model accounted well for DS in most layer 4 simple cells, particularly in layer 4B; the required exponent was roughly 2.5, on average. However, the model failed in layer 6, and in a few layer 4A cells, because the receptive fields had little or no S-T orientation. Unrealistically large exponents would be required to produce strong DS, and the exponents would distort response amplitudes to counterphasing gratings in a manner that is not observed (see Murthy et al., 1998). How then might one describe the directional processes in these cells?

To place this question in perspective, recall that the general problem is to combine spatially and temporally distinct inputs. We may decompose these inputs into linear combinations of spatiotemporally separable quadrature pairs. This fact permits a simplification. We can assume that the inputs to be combined are in S-T quadrature and are separable, as if they were LGN cells. If x and y are the separable quadrature components, the nonlinear system would, in general, depend on higher order terms in these components, such as x^2, y^2, xy, as well as x and y. The cross-term xy provides the nonlinear interaction that contributes most to DS. The interaction can be thought of as an AND gate, where output occurs only when both the spatially and temporally distinct inputs are active (Barlow and Levick, 1965). This corresponds logically to multiplication, a process originally proposed by Reichardt (1961) to account for DS in insect eyes, and subsequently used in models of human motion perception and cortical DS (van Santen and Sperling, 1985).

How are nonlinear interactions measured in cortical cells? Typically, multiple stationary stimuli are briefly presented at different times and positions in the receptive field, and responses are plotted relative to the position and time of a single stimulus (the first-order response) and in terms of the position and time *differences* of two stimuli presented in succession (second-order response).[5] Figure 9-15A, from Emerson and Citron (1992), shows a first-order S-T map for a DS simple cell. The map is S-T unoriented and thus predicts no DS.

Although the cell's laminar location was not determined, its unoriented S-T structure is similar to that commonly seen in first-order maps of DS simple cells in layer 6 (Humphrey and Saul, 1998). In contrast, Fig. 9-15B illustrates the second-order structure derived from responses to two bars. The contours now represent nonlinear facilitation (solid lines) or suppression (dashed lines) evoked by different spatial (Δs) and temporal ($\Delta\tau$) offsets of the bars. The 2-bar interaction function, or motion kernel (Emerson et al., 1987), is obliquely oriented, revealing that the receptive field is directionally tuned. A stimulus moving upward through the field (the PD) evokes primarily strong facilitation; downward movement evokes less facilitation as well as suppression.

Similar nonlinear interactions in simple cells with first-order separable receptive fields have been observed by others (Baker and Boulton, 1994). Not surprisingly, nonlinear interactions also appear in cells that are first-order inseparable

[5] These 2-bar stimuli, which produce "second-order" maps, should not be confused with another set of stimuli referred to as "second-order" or "non-Fourier" (Chubb and Sperling, 1988). An example of the latter would be a low spatial frequency pattern (envelope) that modulates the contrast of a high spatial frequency grating (carrier). Neurons in cat cortex can respond to the envelope pattern despite the absence of Fourier components within the pattern (Zhou and Baker, 1996) to which they respond.

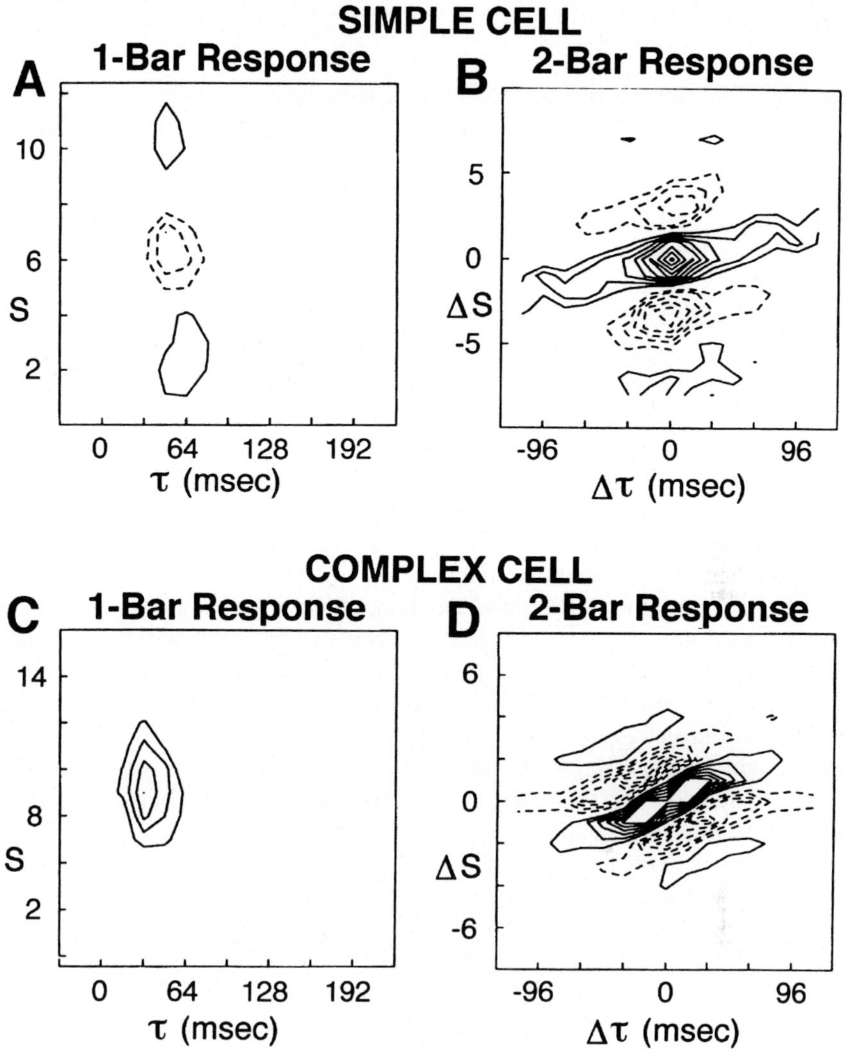

FIGURE 9-15. Space-time maps for a DS simple cell **(A,B)** and complex cell **(C,D)**. **(A)** Map of the response to single bright and dark bars flashed briefly in different positions across the simple-cell receptive field. The ordinate and abscissa plot space (S) and time (τ), respectively. Three receptive-field zones (2 ON, 1 OFF) are revealed but they are S-T unoriented and predict no DS. **(B)** Map obtained using 2 bars presented sequentially. The map is an average of numerous 2-bar interactions of bright and dark bars tested over multiple positions and time delays. It is normalized such that point $\Delta s = 0$ and $\Delta\tau = 0$ corresponds to the position and timing of one bar of the pair (the reference bar) and the contours map responses evoked by the other bar at different displacements relative to the reference bar. Negative and positive $\Delta\tau$ values correspond respectively to the second bar appearing before or after the reference bar. Negative and positive Δs values indicate, respectively, the second bar appearing below or above the reference bar. Solid and dashed contours indicate space-time displacements that evoke facilitation and suppression, respectively. The oblique orientation predicts a preferred direction of motion in the upward direction, which was confirmed using a moving bar *(not shown)*. **(C)** 1-bar responses to a dark bar revealing a uniform, S-T separable discharge region typical of complex cells. **(D)** 2-bar interactions in the complex cell are similar to those in the simple cell, indicating that DS subunits in the two cells are similar and that both cells depend on spatiotemporal nonlinearities. (Data are from Emerson and Citron [1992].)

(Emerson, 1997; Baker and Boulton, 1994), although the relative contributions of linear and nonlinear components in these cells have rarely been evaluated directly. Emerson (1997), using a model-based analysis of first- and second-order maps from a simple cell, showed that the nonlinear components provided a better match to drifting bar data, although the linear approximation was not unreasonable. The model that best fit the drifting bar data and second-order maps consisted of a pair of linear filters, spatially and temporally offset by a quarter-cycle, that were summed after each was half-squared (i.e., half-wave rectified and squared). This arrangement emphasizes the separate components (x^2, y^2) at the expense of their product. The modeled first-order plots were oriented but less than predicted from the cell's DS. The second-order plots had a peak at the origin, similar to that in Fig. 9-15B. This central peak is characteristic of an intensive nonlinearity like squaring. The model thus specifies one type of nonlinear interaction that might account for DS in simple cells of layer 6. Further, the model can be easily extended to complex cells (Emerson et al., 1987, 1992), whose directional behavior seems attributable to nonlinear interactions (Fig. 9-15C, D).

Baker and colleagues (Baker and Boulton, 1994; Baker and Cynader, 1994) have contended that, for some simple cells, receptive-field structure is best modeled by a multiplicative, rather than squaring, mechanism. For instance, separable inputs would be combined in such a way that one of the inputs is subthreshold by itself, but facilitates responses when jointly activated with the other input. This system might be described in a simplified scheme as $x(1 + y)$, where x is the suprathreshold and y the subthreshold input. This model is closely related to one that combines inputs via divisive inhibition (i.e., $x/(1 - y)$, since $1 + y \approx 1/(1 - y)$ when y is small). Such a mechanism does not necessarily show first-order inseparability; it may reflect only the separable suprathreshold input. The second-order plots in this case can have peaks away from the origin, which distinguishes them from the squaring nonlinearities emphasized by Emerson and colleagues. Baker's (1997) model is more elaborate than the simplified version presented here and permits varying degrees of first-order inseparability and a range of second-order interactions. Experimentally, both forms of second-order plots are observed in different simple cells (Baker, 1997, 1998), as might be expected from a general form that includes all terms (constant, x and y, x^2 and y^2, and xy) to varying degrees.

What neuronal mechanisms might implement these various arithmetic operations for combining inputs? Addition and subtraction can be realized by summation of excitatory and inhibitory synaptic inputs (Ferster, 1994; Ferster and Jagadeesh, 1992). Multiplication and division might occur through dendritic interactions (Mel, 1993; Langmoen and Andersen, 1983) and receptor/channel/second-messenger properties such as calcium spikes, NMDA receptors, or shunting inhibition mediated by $GABA_A$ receptors (Borg-Graham et al., 1998; van Santen and Sperling, 1985; Torre and Poggio, 1978; Rose, 1977; Blomfield, 1974; Thorson, 1966).

Multiplication could also be generated by assuming that addition of voltages is accompanied by nonlinearities that occur between spike activity and membrane

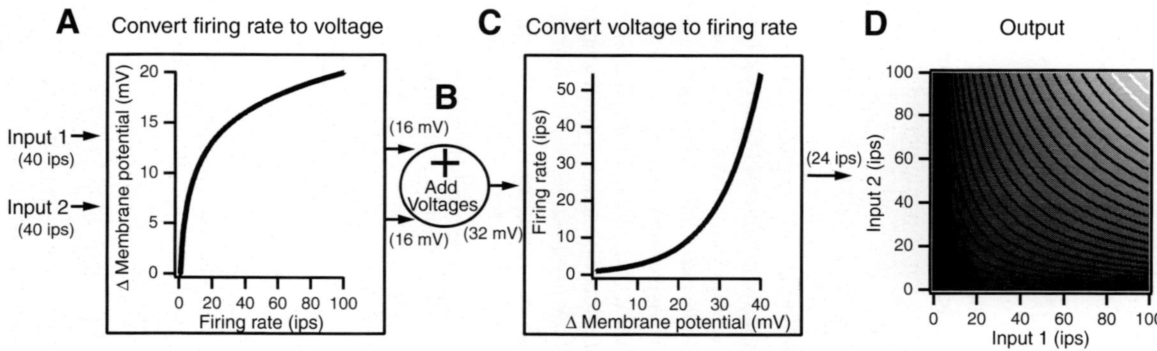

FIGURE 9-16. Multiplication can be implemented through a pair of nonlinearities surrounding addition. **(A)** A compressive nonlinearity at the synapse produces voltage from spike activity. Membrane potentials are added in the soma **(B),** and generate spikes at the axon hillock through an accelerating process **(C).** Output firing rates **(D)** are shown as grayscale, with darker grays corresponding to smaller values. If the nonlinearities were exactly logarithimic and exponential (before saturation), the output would be the product of the inputs, $xy = \exp(\ln(x) + \ln(y))$. In the example shown in parentheses, if the inputs each had firing rates of 40 ips, they would each change the membrane potential by 16 mV. Individually, a 16 mV change in membrane potential might evoke a firing rate change of about 5 ips, but together they produce 24 ips.

potential (Fig. 9-16). Given two spike-based inputs, the temporally integrated postsynaptic voltages they evoke can be roughly logarithmic as a function of input firing rate (Fig. 9-16A), owing to presynaptic failures, temporal summation, saturation, depression, additional tonic inhibitory and excitatory inputs, and other mechanisms (Gabbiani and Koch, 1998; Abbott et al., 1997; Tal and Schwartz, 1997; Bernander et al., 1991; Cleveland et al., 1981). These potentials add intracellularly (B) and are then converted to an output firing rate by a process that is roughly exponential prior to saturation (C) (Suarez et al., 1995; Carandini and Heeger, 1994; Heeger, 1992). The result of these three steps (compression, addition, and expansion) is to convert the input firing rates to an output rate that behaves like their product (D), requiring activity in both inputs and exceeding the linear expectation for strong inputs. Division is implemented by making one of the inputs inhibitory.

Networks contribute to the nonlinearities. In particular, intracortical feedback can mediate either accelerating or decelerating nonlinear functions (Suarez et al., 1995; Heeger, 1992). Suarez et al. (1995) modeled the effects of massive recurrent excitation on cortical activity, showing how it can amplify weak inputs and contribute to directional tuning. Heeger (1992) considered the effect of inhibitory feedback on regulating response amplitude, especially in regard to reducing gain at high contrast. In Heeger's model, the divisive inhibition is nonspecific, but more specific versions (Saul, 1999; Carandini et al., 1997; 1999, Borg-Graham et al., 1998) could account for some of the previously discussed nonlinear interactions.

Considering these issues from another perspective, the laminar differences in S-T structure may also reflect variations in the degree to which simple and complex cells interact. Complex cells are most common outside of layer 4 (Gilbert, 1977) and evidence exists for complex-to-simple cell interactions (Ghose et al., 1994; Hammond and MacKay, 1981). It is plausible that simple cells in layer 6 receive more input from complex cells than do most layer 4 cells. The two cell classes display similar directional tuning (Saul and Humphrey, 1992b), but complex cells lack first-order S-T inseparable receptive fields. As noted previously, their DS reflects higher-order interactions. Input from complex cells might therefore confer on layer 6 simple cells a DS that correlates only with second-order structure and that reflects complex-cell-based facilitation and inhibition, respectively, to the preferred and nonpreferred directions of motion. For most layer 4 simple cells, in contrast, evidence suggests that quasilinear interactions among simple cells play a more prominent role.

INTRALAMINAR AND INTERLAMINAR INTERACTIONS

A final comment concerns the role of interlaminar connections in producing DS. Anatomically, the predominant output from layer 4 is to the supragranular layers and from there to the infragranular layers (Gilbert, 1983; see Chapter 1). Do cells in these other layers depend on layer 4 for their directional tuning? Studies by Malpeli and colleagues suggest not. Reversible inactivation of the geniculate A-laminae was used to silence cortical activity in retinotopically corresponding regions of layer 4 (Malpeli, 1983). Surprisingly, the direction and orientation selectivity of cells in layers 2 and 3 were unaffected by the block. The selectivity of these cells could be abolished only by also inactivating the medial interlaminar nucleus (MIN), the C-laminae of the LGN, or all three regions (Malpeli et al., 1986; see Chapter 7). Similarly, direction and orientation selectivity of cells in layer 6 survived inactivation of the supragranular layers (Schwark et al., 1986). Thus it seems that direction and orientation selectivity, along with other visual response properties, can be generated independently in different layers. Layers 2 and 3 receive direct inputs from the MIN and C-laminae (Rosenquist, 1985); these inputs seem sufficient to drive the appropriate circuitry. Related to this, Humphrey and Murthy (1999) showed that lagged and nonlagged varieties of X-, Y- and non-directional W-cells reside in these two geniculate divisions. Perhaps the cells' different timings contribute to directional tuning in the supragranular layers.

Other evidence, however, suggests that interlaminar circuits do play a role in establishing selectivity. For instance, remote inactivation by GABA in layer 3 (see page 364) reduced or eliminated DS in some layer 4 cells (Crook et al., 1996, 1997 (see Chapter 11). Assuming the drug did not diffuse into layer 4, this argues for a strong laminar dependence, at least among some cells, and it is congruent with some of the intralaminar connections that are now known to exist in cat area 17 (Kisvárday, 1992; Martin and Whitteridge, 1984). Further work is needed to clarify the role of intralaminar and interlaminar processing in direction selectivity.

SUMMARY AND CONCLUSIONS

The emergence of direction selectivity in layer 4 of cat primary visual cortex depends on a number of mechanisms, both thalamocortical and intracortical. Lagged and nonlagged LGN cells provide cortex with a range of timings, or response phases, that serve as initial substrates for producing response timing gradients across receptive fields (i.e, S-T inseparability). This is particularly important at low temporal frequencies. These gradients induce directional tuning. Gradients might be established by direct convergence of afferents with spatially and temporally offset receptive fields, by indirect convergence via other simple cells with certain spatiotemporal relationships to their targets, or, most likely, by both mechanisms. Inhibitory interactions among simple cells appear to contribute to DS mainly by creating or enhancing S-T inseparable receptive-field structure. Recurrent excitatory interactions enhance DS by amplifying suprathreshold responses. Whether they also enhance S-T inseparability, as some models posit, remains to be determined experimentally. Overall, the key to elucidating how these various mechanisms contribute requires an understanding of the spatiotemporal properties of the elements and their interactions.

Direction selectivity among most layer 4 simple cells can be explained by LN models in which quasilinear summation of synaptic potentials across an S-T inseparable receptive field induces directional tuning that is then enhanced by relatively simple nonlinear processes associated with spike generation. Whether this model holds true biologically, or whether more complicated processes masquerading as simpler ones operate (Suarez et al., 1995; Maex and Orban, 1996), remains to be determined.

The LN model fails to account for DS in a few layer 4 simple cells, most simple cells in layer 6, and all complex cells, indicating a more prominent role for nonlinear mechanisms in these cells. Various nonlinearities have been modeled but experimental evidence concerning actual mechanisms remains sparse. In general, the laminar analyses of relationships between S-T structure and DS indicate that a variety of mechanisms contribute to DS. It remains to be determined whether a common set of mechanisms operates, with variations in the relative contributions of its elements, or whether qualitatively different operations obtain in different layers. At present, the heterogeneity among simple cells suggests to us that no single extant model will account for DS in all layers.

FUTURE DIRECTIONS

Solving the puzzle of DS will require new experimental approaches in addition to those now commonly used. Four main approaches can be readily identified. First, DS reflects network interactions, so analysis of ensembles of dynamically interacting neurons is needed. Cross-correlation of activity in multiple, simultaneously recorded units is possible, though technically and computationally challenging. Nevertheless, it offers one the potential to investigate the S-T properties and interactions of circuit elements in response to different types of stimuli. Second, intracellular approaches are needed to tease out the S-T properties of excitatory and inhibitory inputs and their interactions. Third, the relative contributions of geniculocortical and intracortical mechanisms need to be distinguished. The putative role of lagged inputs might be examined by determining whether their selective inactivation produces predictable changes in DS. No reliable methods for testing this possibility presently exist. New pharmacological or molecular (Tonegawa et al., 1996) approaches will likely be needed. Fourth, better and more efficient ways of analyzing circuit structures are needed, such as methods that allow chains of synaptically coupled cells to be visualized. Perhaps new tracing methods using viruses (Hoover and Strick, 1999; Card, 1998) can be improved for application to intracellular labeling to accomplish this goal.

Finally, it must be noted that most of what is known about directional mechanisms in primary visual cortex comes from work in the cat. Little is known about such mechanisms in primate V1. Differences exist between cat and monkey that may be relevant to DS. For example, DS cells are largely restricted to layers 4B and 6 in the monkey, rather than existing in all layers as in the cat, and layer 4B gets no direct LGN input (Levitt et al., 1996). These and other species variations may indicate differences in mechanisms that create DS, or in where the mechanisms come into play in the geniculocortical system. It is not yet known whether lagged cells exist in monkey LGN, although lagged-like timings can be observed in V1 (unpublished observations). DeValois and Cottaris (1999) recently suggested that magnocellular and parvocellular LGN cells might be a source of quadrature inputs to directional mechanisms, but this and other possibilities remain to be examined. S-T inseparability has been reported for some DS receptive fields in macaque V1 (Livingstone, 1998; Jacobson et al., 1993), but many important S-T aspects of these fields remain unexplored. Elucidating directional mechanisms in primates is obviously an important endeavor for the future.

ACKNOWLEDGMENTS

We wish to thank Drs. Curtis Baker and Robert Emerson for helpful discussions on nonlinear models. The studies from our lab reported herein were supported by grants NIH-EY04091 and EY06459 to A.L.H.; NIH-EY010826 and NSF-BNS9021495 to A.B.S.; and a Core Grant for Vision Research (NIH-EY08098) to the Eye and Ear Institute of Pittsburgh.

REFERENCES

Abbott, L. F., Varela, J. A., Sen, K., and Nelson, S. B. (1997). Synaptic depression and cortical gain control. *Science* **275,** 220–224.

Adelson, E. H., and Bergen, J. R. (1985). Spatiotemporal energy models for the perception of motion. *J. Opt. Soc. Am. A.* **2,** 284–299.

Ahmed, B., Anderson, J. C., Douglas, R. J., Martin, K. A. C., and Nelson, J. C. (1994). Polyneuronal innervation of spiny stellate neurons in cat visual cortex. *J. Comp. Neurol.* **341,** 39–49.

Albrecht, D. G., and Geisler, W. S. (1991). Motion selectivity and the contrast-response function of simple cells in the visual cortex. *Vis. Neurosci.* **7,** 531–546.

Albrecht, D. G., and Hamilton, D. B. (1982). Striate cortex of monkey and cat: contrast response function. *J. Neurophysiol.* **48,** 217–237.

Albus, K. (1980). The detection of movement direction and effects of contrast reversal in the cat's striate cortex. *Vis. Res.* **20,** 289–293.

Alonso, J. M., Usrey, W. M., and Reid, R. C. (1996). Precisely correlated firing in cells of the lateral geniculate nucleus. *Nature* **383,** 815–819.

Anzai, A., Ohzawa, I., and Freeman, R. D. (1999). Neural mechanisms for processing binocular information. I. Simple cells. *J. Neurophysiol.* **82,** 891–908.

Azouz, R., Gray, C. M., Nowak, L. G., and McCormick, D. A. (1997). Physiological properties of inhibitory interneurons in cat striate cortex, *Cerebral Cortex* **7,** 534–545.

Baker, C. L., Jr. (1997). A linear-gating model of direction selectivity in visual cortex neurons. *Invest. Ophthal. Vis. Sci. (Suppl.)* **38,** 624.

Baker, C. L., Jr. (1998). Testing models of direction selective visual cortex neurons: analysis with dense noise. *Soc. Neurosci. Abstr.* **24,** 144.

Baker, C. L., Jr., and Boulton, J. C. (1994). Neurobiological mechanisms of cortical direction selectivity. In: *Computational vision based on neurobiology. S.P.I.E. Proceedings.* Vol. 2054 (T. B. Lawton, Ed.), pp. 104–123. Bellingham, Washington, S.P.I.E.

Baker, C. L., Jr., and Cynader, M. S. (1994). A sustained input to the direction-selective mechanism in cat striate cortex neurons. *Vis. Neurosci.* **11,** 1083–1092.

Barlow, H. B., and Levick, W. R. (1965). The mechanism of directionally selective units in rabbit's retina. *J. Physiol. (Lond.)* **178,** 477–504.

Baumfalk, U., and Albus, K. (1988). Phacolofen antagonizes baclofen-induced suppression of visually evoked responses in the cat's striate cortex. *Brain Res.* **463,** 398–402.

Berman, N. E. J., Wilkes, N. E., and Payne, B. R. (1987). Organisation of orientation and direction selectivity in areas 17 and 18 of cat cerebral cortex. *J. Neurophysiol.* **58,** 676–699.

Berman, N. J., Douglas, R. J., Martin, K. A. C., and Whitteridge, D. (1991). Mechanisms of inhibition in cat visual cortex. *J. Physiol. (Lond.)* **440,** 697–722.

Bernander, Ö., Douglas, R. J., Martin, K. A. C., and Koch, C. (1991). Synaptic background activity influences spatiotemporal integration in single pyramidal cells. *Proc. Natl. Acad. Sci. U.S.A.* **88,** 11569–11573.

Bishop, P. O., Coombs, J. S., and Henry, G. H. (1971). Responses to visual contours: spatio-temporal aspects of excitation in the receptive fields of simple striate neurones. *J. Physiol. (Lond.)* **219,** 659–687.

Bishop, P. O., Kato, H., and Orban, G. A. (1980). Direction-selective cells in complex family in cat striate cortex. *J. Neurophysiol.* **43,** 1266–1283.

Blomfield, S. (1974). Arithmetical operations performed by nerve cells. *Brain Res.* **69,** 115–124.

Borg-Graham, L. J., Monier, C., and Frégnac, Y. (1998). Visual input evokes transient and strong shunting inhibition in visual cortical neurons. *Nature* **393,** 369–373.

Bullier, J., and Henry, G. H. (1979). Laminar distribution of first-order neurons and afferent terminals in cat striate cortex. *J. Neurophysiol.* **42,** 1271–1281.

Carandini, M., and Heeger, D. J. (1994). Summation and division by neurons in primate visual cortex. *Science* **264,** 1333–1336.

Carandini, M., Heeger, D. J., and Movshon, J. A. (1997). Linearity and normalization in simple cells of the macaque primary visual cortex. *J. Neurosci.* **17,** 8621–8644.

Carandini, M., Heeger, D. J., and Movshon, J. A. (1999). Linearity and gain control in V1 simple cells. In: *Cerebral cortex. Volume 13, Models of cortical circuits* (P. S. Ulinski, E. G. Jones, and A. Peters, Eds.), pp. 401–443. New York, Plenum Publishers.

Card, J. P. (1998). Exploring brain circuitry with neurotropic viruses: new horizons in neuroanatomy. *Anat. Rec.* **253,** 176–185.

Casanova, C., Nordmann, J. P., Ohzawa, I., and Freeman, R. D. (1992). Direction selectivity of cells in the cat's striate cortex: differences between bar and grating stimuli. *Vis. Neurosci.* **9,** 505–513.

Chance, F. S., Nelson, S. B., and Abbott, L. F. (1998). Synaptic depression and the temporal response characteristics of V1 cells. *J. Neurosci.* **18,** 4785–4799.

Chubb, S. J., and Sperling, G. (1988). Drift-balanced random stimuli: a general basis for studying non-Fourier motion perception. *J. Opt. Soc. Am. A* **5,** 1986–2007.

Chung, S., and Ferster, D. (1998). Strength and orientation tuning of the thalamic input to simple cells revealed by electrically evoked cortical suppression. *Neuron* **20,** 1177–1189.

Cleveland, S., Kuschmierz, A., and Ross, H. G. (1981). Static input-output relations in the spinal recurrent inhibitory pathway. *Biol. Cybern.* **40,** 223–231.

Creutzfeldt, O. D., Kuhnt, U., and Benevento, L. A. (1974). An intracellular analysis of visual cortical neurones to moving stimuli: responses in a co-operative neuronal network, *Exp. Brain Res.* **21,** 251–274.

Crook, J. M., Eysel, U. T., and Machemer, H. F. (1991). Influence of GABA-induced remote inactivation on the orientation tuning of cells in area 18 of feline visual cortex: a comparison with area 17. *Neuroscience* **40,** 1–12.

Crook, J. M., Kisvárday, Z. F., and Eysel, U. T. (1996). GABA-induced inactivation of functionally characterized sites in cat visual cortex (area 18): effects on direction selectivity. *J. Neurophysiol.* **75,** 2071–2088.

Crook, J. M., Kisvárday, Z. F., and Eysel, U. T. (1997). GABA-induced inactivation of functionally characterized sites in cat striate cortex: effects on orientation tuning and direction selectivity. *Vis. Neurosci.* **14,** 141–158.

Crook, J. M., Kisvárday, Z. F., and Eysel, U. T. (1998). Evidence for a contribution of lateral inhibition to orientation tuning and direction selectivity in cat visual cortex: reversible inactivation of functionally characterized sites combined with neuroanatomical tracing techniques. *Eur. J. Neurosci.* **10,** 2056–2075.

Cynader, M. S., and Chernenko, G. (1976). Abolition of directional selectivity in the visual cortex of the cat. *Science* **193,** 504–505.

Dean, A. F., and Tolhurst, D. J. (1986). Factors influencing the temporal phase of response to bar and grating stimuli for simple cells in the cat striate cortex. *Exp. Brain Res.* **62,** 143–151.

DeAngelis, G. C., Ohzawa, I., and Freeman, R. D. (1993a). Spatiotemporal organization of simple-cell receptive fields in the cat's striate cortex. I. General characteristics and postnatal development. *J. Neurophysiol.* **69,** 1091–1117.

DeAngelis, G. C., Ohzawa, I., and Freeman, R. D. (1993b). Spatiotemporal organization of simple-cell receptive fields in the cat's striate cortex. II. Linearity of temporal and spatial summation. *J. Neurophysiol.* **69,** 1118–1135.

DeAngelis, G. C., Ghose, G. M., Ohzawa, I., and Freeman, R. D. (1999). Functional micro-organization of primary visual cortex: receptive field analysis of nearby neurons. *J. Neurosci.* **19,** 4046–4064.

DeValois, R. L., and Cottaris, N. P. (1999). Inputs to directionally selective simple cells in macaque striate cortex. *Proc. Natl. Acad. Sci. U.S.A.* **95,** 14488–14493.

Douglas, R. J., and Martin, K. A. C. (1991). A functional microcircuit for cat visual cortex. *J. Physiol. (Lond.)* **440,** 735–769.

Douglas, R. J., Martin, K. A. C., and Whitteridge, D. (1991). An intracellular analysis of the visual responses of neurones in cat visual cortex. *J. Physiol. (Lond.)* **440,** 659–696.

Duysens, J., Orban, G. A., van der Glas, H. W., and de Zegher, F. E. (1982). Receptive field structure of area 19 as compared to area 17 of the cat. *Brain Res.* **231,** 279–291.

Emerson, R. C. (1997). Quadrature subunits in directionally selective simple cells: spatiotemporal interactions. *Vis. Neurosci.* **14,** 357–371.

Emerson, R. C., and Citron, M. C. (1992). Linear and nonlinear mechanisms of motion selectivity in simple cells of the cat's striate cortex. In: *Nonlinear vision: Determinants of neural receptive fields, function and networks* pp. 75–89. (R. B. Pinter and B. Nabet, Eds.), Boca Raton, CRC Press.

Emerson, R. C., and Gerstein, G. L. (1977a). Simple striate neurons in the cat. I. Comparison of responses to moving and stationary stimuli. *J. Neurophysiol.* **40,** 119–135.

Emerson, R. C., and Gerstein, G. L. (1977b). Simple striate neurons in the cat. II. Mechanisms underlying directional asymmetry and directional selectivity. *J. Neurophysiol.* **40,** 136–155.

Emerson, R. C., Citron, M. C., Vaughn, W. J., and Klein, S. A. (1987). Nonlinear directionally selective subunits in complex cells of cat striate cortex, *J. Neurophysiol.* **58,** 33–65.

Emerson, R. C., Bergen, J. R., and Adelson, E. H. (1992). Directionally selective complex cells and the computation of motion energy in cat visual cortex. *Vis. Res.* **32,** 203–218.

Eysel, U. T., Muche, T., and Worgotter, F. (1988). Lateral interactions at direction-selective striate neurones in the cat demonstrated by local cortical inactivation. *J. Physiol. (Lond.)* **399,** 657–675.

Ferster, D. (1986). Orientation selectivity of synaptic potentials in neurons of cat primary visual cortex. *J. Neurosci.* **6,** 1284–1301.

Ferster, D. (1992). The synaptic inputs to simple cells of the cat visual cortex. In: *Progress in Brain Research. Volume 90. GABA in the Retina and Central Nervous System,* (R. R. Mize, R. E. Marc, and A. M. Sillito, Eds.) pp. 423–441. New York, Elsevier.

Ferster, D., (1994). Linearity of synaptic interactions in the assembly of receptive fields in cat visual cortex, *Curr. Opin. Neurobiol.* **4,** 563–568.

Ferster, D., and Jagadeesh, B. (1992). EPSP-IPSP interactions in cat visual cortex studied with *in vivo* whole-cell patch recording. *J. Neurosci.* **12,** 1262–1274.

Ferster, D., and Lindstrom, S. (1983). An intracellular analysis of geniculo-cortical connectivity in area 17 of the cat. *J. Physiol. (Lond.)* **342,** 181–215.

Freund, T. F., Martin, K. A. C., Somogyi, P., and Whitteridge, D. (1985). Innervation of cat visual areas 17 and 18 by physiologically identified X- and Y-type thalamic afferents. II. Identification of postsynaptic targets by GABA immunocytochemistry and golgi impregnation. *J. Comp. Neurol.* **242,** 275–291.

Gabbiani, F., and Koch, C. (1998). Principles of spike train analysis. In: *Methods in Neuronal Modeling* pp. 313–360 (C. Koch and I. Segev, Eds.), Cambridge, MA, MIT Press.

Ganz, L., and Felder, R. (1984). Mechanism of directional selectivity in simple neurons of the cat's visual cortex analyzed with stationary flash sequences. *J. Neurophysiol.* **51,** 294–324.

Ghose, G. M., Ohzawa, I., and Freeman, R. D. (1994). Receptive-field maps of correlated discharge between pairs of neurons in the cat's visual cortex. *J. Neurophysiol.* **71,** 330–346.

Giaschi, D., Douglas, R., Marlin, S., and Cynader, M.S. (1993). The time course of direction-selective adaptation in simple and complex cells in cat striate cortex. *J. Neurophysiol.* **70,** 2024–2034.

Gil, Z., Connors, B. W., and Amitai, Y. (1999). Efficacy of thalamocortical and intracortical synaptic connections: quanta, innervation, and reliability. *Neuron* **23,** 385–397.

Gilbert, C. D. (1977). Laminar differences in receptive field properties of cells in cat primary visual cortex. *J. Physiol. (Lond.)* **268,** 391–421.

Gilbert, C. D. (1983). Microcircuitry of the visual cortex. *Annu. Rev. Neurosci.* **6,** 217–247.

Goodwin, A. W., and Henry, G. H. (1975). Direction selectivity of complex cells in a comparison with simple cells. *J. Neurophysiol.* **38,** 1524–1540.

Goodwin, A. W., Henry, G. H., and Bishop, P. O. (1975). Direction selectivity of simple striate cells: properties and mechanisms. *J. Neurophysiol.* **38,** 1500–1523.

Grzywacz, N. M., and Koch, C. (1987). Functional properties of models for direction selectivity in the retina. *Synapse* **1,** 417–434.

Hamada, T., Yamashima, M., and Kato, K. (1997). A ring model for spatiotemporal properties of simple cells in the visual cortex. *Biol. Cybern.* **77,** 225–233.

Hammond, P., and Kim, J. N. (1996). Role of suppression in shaping orientation and direction selectivity of complex neurons in cat striate cortex. *J. Neurophysiol.* **75,** 1163–1176.

Hammond, P., and MacKay, D. M. (1981). Modulatory influences of moving textured backgrounds on responsiveness of simple cells in feline striate cortex. *J. Physiol. (Lond.)* **319,** 431–442.

Hartveit, E., and Heggelund, P. (1992). The effect of contrast on the visual response of lagged and nonlagged cells in the cat lateral geniculate nucleus, *Vis. Neurosci.* **9,** 515–525.

Heeger, D. J. (1992). Normalization of cell responses in cat striate cortex. *Vis. Neurosci.* **9.** 181–197.

Heeger, D. J. (1993). Modeling simple-cell direction selectivity with normalized, half-squared, linear operators. *J. Neurophysiol.* **70,** 1885–1898.

Heggelund, P., and Hartveit, E. (1990). Neurotransmitter receptors mediating retinal input to cells in the cat lateral geniculate nucleus: I. Lagged cells. *J. Neurophysiol.* **63,** 1347–1360.

Hildreth, E. C., and Koch, C. (1987). The analysis of visual motion: from computational theory to neuronal mechanisms. In: *Annual Review of Neuroscience, Vol. 10,* (W. M. Cowan, Ed.), pp. 477–534. Palo Alto, Annual Reviews, Inc.

Hirsch, J. A., Alonso, J. M., Reid, R. C., and Martinez, L. M. (1998). Synaptic integration in striate cortical simple cells. *J. Neurosci.* **18,** 9517–9528.

Hoover, J. E., and Strick, P. L. (1999). The organization of cerebellar and basal ganglia outputs to primary motor cortex as revealed by retrograde transneuronal transport of herpes simplex virus type 1. *J. Neurosci.* **19,** 1446–1463.

Hubel, D. H., and Wiesel, T. N. (1959). Receptive fields of single neurones in the cat's striate cortex. *J. Physiol. (Lond.)* **148,** 574–591.

Hubel, D. H., and Wiesel, T. N. (1962). Receptive fields, binocular interaction and functional architecture in the cat's visual cortex. *J. Physiol. (Lond.)* **160,** 106–154.

Humphrey, A. L., and Murthy, A. (1999). Cell types and response timings in the medial interlaminar nucleus and C-layers of the cat lateral geniculate nucleus. *Vis. Neurosci.* **16,** 513–526.

Humphrey, A. L., and Saul, A. B. (1992). Action of brainstem reticular afferents on lagged and non-lagged cells in the cat lateral geniculate nucleus. *J. Neurophysiol.* **68,** 673–691.

Humphrey, A. L., and Saul, A. B. (1998). Strobe rearing reduces direction selectivity in area 17 by altering spatiotemporal receptive-field structure. *J. Neurophysiol.* **80,** 2991–3004.

Humphrey, A. L., and Weller, R. E. (1988a). Functionally distinct groups of X-cells in the lateral geniculate nucleus of the cat. *J. Comp. Neurol.* **268,** 429–447.

Humphrey, A. L., and Weller, R. E. (1988b). Structural correlates of functionally distinct X-cells in the lateral geniculate nucleus of the cat. *J. Comp. Neurol.* **268,** 448–468.

Humphrey, A. L., Sur, M., Uhlrich, D. J., and Sherman, S. M. (1985). Projection patterns of individual X- and Y-cell axons from the lateral geniculate nucleus to cortical area 17 in the cat. *J. Comp. Neurol.* **233,** 159–189.

Humphrey, A. L., Saul, A. B., and Feidler, J. C. (1998). Strobe rearing prevents the convergence of inputs with different response timings onto area 17 simple cells. *J. Neurophysiol.* **80,** 3005–3020.

Jacobson, L. D., Gaska, J. P., Chen, H., and Pollen, D. A. (1993). Structural testing of multi-input linear-nonlinear cascade models for cells in macaque striate cortex. *Vis. Res.* **33,** 609–626.

Jagadeesh, B., Wheat, H. S., Kontsevich, L. L., Tyler, C. W., and Ferster, D. (1997). Direction selectivity of synaptic potentials in simple cells of the cat visual cortex. *J. Neurophysiol.* **78,** 2772–2789.

Jones, J. P., and Palmer, L. A. (1987). The two-dimensional spatial structure of simple receptive fields in cat striate cortex. *J. Neurophysiol.* **58,** 1187–1211.

Kato, H., Bishop, P. O., and Orban, G. A. (1978). Hypercomplex and the simple/complex cell classification in cat striate cortex. *J. Neurophysiol.* **41,** 1071–1095.

Kisvárday, Z. F. (1992). GABAergic networks of basket cells in the visual cortex. In: *Progress in Brain Research, Volume 90, GABA in the Retina and Central Nervous System* (R. R. Mize, R. E. Marc, and A. M. Sillito, Eds.) pp. 385–405. New York, Elsevier.

Kisvárday, Z. F., and Eysel, U. T. (1993). Functional and structural topography of horizontal inhibitory connections in cat visual cortex. *Eur. J. Neurosci.* **5,** 1558–1572.

Kisvárday, Z. F., Kim, D.-S., Eysel, U. T., and Bonhoeffer, T. (1994). Relationship between lateral inhibitory connections and the topography of the orientation map in cat visual cortex. *Eur. J. Neurosci.* **6,** 1619–1632.

Kontsevich, L. L. (1995). The nature of the inputs to cortical motion detectors. *Vis. Res.* **19,** 2785–2793.

Kwon, Y. H., Esguerra, M., and Sur, M. (1991). NMDA and non-NMDA receptors mediate visual responses of neurons in the cat's lateral geniculate nucleus. *J. Neurophysiol.* **66,** 414–428.

Langmoen, I. A., and Andersen, P. (1983). Summation of excitatory postsynaptic potentials in hippocampal pyramidal cells. *J. Neurophysiol.* **50,** 1320–1329.

Levitt, J. B., Lund, J. S., and Yoshioka, T. (1996). Anatomical substrates for early stages in cortical processing of visual information in the macaque monkey. *Behav. Brain Res.* **76,** 5–19.

Livingstone, M. S. (1998). Mechanisms of direction selectivity in macaque V1. *Neuron* **20,** 509–526.

Maex, R., and Orban, G. A. (1996). Model circuit of spiking neurons generating directional selectivity in simple cells. *J. Neurophysiol.* **75,** 1515–1545.

Maffei, L., Fiorentini, A., and Bisti, S. (1973). Neural correlates of perceptual adaptation to gratings. *Science* **182,** 1036–1038.

Malpeli, J. G. (1983). Activity of cells in area 17 of the cat in absence of input from layer A of lateral geniculate nucleus. *J. Neurophysiol.* **49,** 595–610.

Malpeli, J. G., Lee, C., Schwark, H. D., and Weyand, T. G. (1986). Cat area 17: I. pattern of thalamic control of cortical layers. *J. Neurophysiol.* **56,** 1062–1073.

Marlin, S., Hasan, S.J., and Cynader, M.S. (1988). Direction-selective adaptation in simple and complex cells in cat striate cortex. *J. Neurophysiol.* **59,** 1314–1330.

Marr, D., and Ullman, S. (1981). Direction selectivity and its use in early visual processing. *Proc. R. Soc. Lond. B* **211,** 151–180.

Martin, K. A. C., and Whitteridge, D. (1984). From, function, and intracortical projections of spiny neurones in the striate cortex of the cat. *J. Physiol. (Lond.)* **353,** 463–504.

Martin, K. A. C., Somogyi, P., and Whitteridge, D. (1983). Physiological and morphological properties of identified basket cells in the cat's visual cortex. *Exp. Brain Res.* **50,** 193–200.

Mastronarde, D. N. (1987a). Two classes of single-input X-cells in cat lateral geniculate nucleus. I. Receptive-field properties and classification of cells. *J. Neurophysiol.* **57,** 357–380.

Mastronarde, D. N. (1987b). Two classes of single-input X-cells in cat lateral geniculate nucleus. II. Retinal inputs and the generation of receptive-field properties. *J. Neurophysiol.* **57,** 381–413.

Mastronarde, D. N., Humphrey, A. L., and Saul, A. B. (1991). Lagged Y cells in the cat lateral geniculate nucleus. *Vis. Neurosci.* **7,** 191–200.

McLean, J., and Palmer, L. A. (1989). Contribution of linear spatiotemporal receptive field structure to velocity selectivity of simple cells in area 17 of cat. *Vis. Res.* **29,** 675–679.

McLean, J., Raab, S., and Palmer, L. A. (1994). Contribution of linear mechanisms to the specification of local motion by simple cells in areas 17 and 18 of the cat. *Vis. Neurosci.* **11,** 271–294.

Mel, B. W. (1993). Synaptic integration in an excitable dendritic tree. *J. Neurophysiol.* **70,** 1086–1101.

Movshon, J. A., Thompson, I. D., and Tolhurst, D. J. (1978). Spatial summation in the receptive fields of simple cells in the cat's striate cortex. *J. Physiol. (Lond.)* **283,** 53–77.

Murthy, A., and Humphrey, A. L. (1999). Inhibitory contributions to spatiotemporal receptive field structure and direction selectivity in simple cells of cat area 17. *J. Neurophysiol.* **81,** 1212–1224.

Murthy, A., Humphrey, A. L., Saul, A. B., and Feidler, J. C. (1998). Laminar differences in the spatiotemporal structure of simple cell receptive fields in cat area 17. *Vis. Neurosci.* **15,** 239–256.

Nakayama, K. (1985). Biological image motion processing: a review. *Vis. Res.* **25,** 625–660.

Nelson, S. B., Toth, L., Sheth, B., and Sur, M. (1994). Orientation selectivity of cortical neurons during intracellular blockade of inhibition. *Science* **265,** 774–777.

Orban, G. A. (1984). *Neuronal operations in the visual cortex.* New York, Springer-Verlag.

Orban, G. A. (1994). Motion processing in monkey striate cortex. In: *Cerebral Cortex,* Volume 10, Primary Visual Cortex in Primates (A. Peters and K. S. Rockland, Eds.) pp. 413–442. New York, Plenum Press.

Orban, G. A., Kennedy, H., and Maes, H. (1981). Response to movement of neurons in areas 17 and 18 of the cat: direction selectivity. *J. Neurophysiol.* **45,** 1059–1073.

Pasternak, T., Schumer, R. A., Gizzi, M. S., and Movshon, J. A. (1985). Abolition of visual cortical direction selectivity affects visual behavior in cats. *Exp. Brain Res.* **61,** 214–217.

Pei, X., Vidyasagar, T. R., Volgushev, M., and Creutzfeldt, O. D. (1994). Receptive field analysis and orientation selectivity of postsynaptic potentials of simple cells in cat visual cortex. *J. Neurosci.* **14,** 7130–7140.

Peters, A., and Payne, B. R. (1993). Numerical relationships between geniculocortical afferents and pyramidal cell modules in cat primary visual cortex. *Cerebral Cortex* **3,** 69–78.

Pettigrew, J. D., Nikara, T., and Bishop, P. O. (1968). Responses to moving slits by single units in cat striate cortex. *Exp. Brain Res.* **6,** 373–390.

Pollen, D. A., and Ronner, S. F. (1981). Phase relationships between adjacent simple cells in the visual cortex. *Science* **212,** 1409–1411.

Reichardt, W. (1961). Autocorrelation, a principle for the evaluation of sensory information by the central nervous system. In: *Sensory communication* (W. A. Rosenblith, Ed.), pp. 303–317. New York, Wiley.

Reid, R. C., and Alonso, J. M. (1995). Specificity of monosynaptic connections from thalamus to visual cortex. *Nature* **378,** 281–284.

Reid, R. C., and Alonso, J. M. (1996). The processing and encoding of information in the visual cortex. *Curr. Opin. Neurobiol.* **6,** 475–480.

Reid, R. C., Soodak, R. E., and Shapley, R. M. (1987). Linear mechanisms of directional selectivity in simple cells of cat striate cortex. *Proc. Natl. Acad. Sci. U.S.A.* **84,** 8740–8744.

Reid, R. C., Soodak, R. E., and Shapley, R. M. (1991). Directional selectivity and spatiotemporal structure of receptive fields of simple cells in cat striate cortex. *J. Neurophysiol.* **66,** 505–529.

Rose, D. (1977). On the arithmetical operation performed by inhibitory synapses onto the neuronal soma. *Exp. Brain Res.* **28,** 221–223.

Rosenquist, A. C. (1985). Connections of visual cortical areas in the cat. In: *Cerebral Cortex, Volume 3, Visual Cortex* (A. Peters and E. G. Jones, Eds.) pp. 81–113. New York, Plenum Press.

Saul, A. B. (1995). Adaptation aftereffects in single neurons of cat visual cortex: response timing is retarted by adapting. *Vis. Neurosci.* **12,** 191–205.

Saul, A. B. (1999). Visual cortical cells: who inhibits whom. *Vis. Neurosci.* **16,** 667–676.

Saul, A. B., and Cynader, M. S. (1989a). Adaptation in single units in visual cortex: the tuning of after-effects in the spatial domain. *Vis. Neurosci.* **2,** 593–608.

Saul, A. B., and Cynader, M. S. (1989b). Adaptation in single units in visual cortex: the tuning of aftereffects in the temporal domain. *Vis. Neurosci.* **2,** 609–620.

Saul, A. B., and Humphrey, A. L. (1990). Spatial and temporal response properties of lagged and non-lagged cells in the cat lateral geniculate nucleus. *J. Neurophysiol.* **64,** 206–224.

Saul, A. B., and Humphrey, A. L. (1992a). Evidence of input from lagged cells in the lateral geniculate nucleus to simple cells in cortical area 17 of the cat. *J. Neurophysiol.* **68,** 1190–1207.

Saul, A. B., and Humphrey, A. L. (1992b). Temporal frequency tuning of direction selectivity in cat visual cortex. *Vis. Neurosci.* **8,** 365–372.

Schwark, H. D., Malpeli, J. G., Weyand, T. G., and Lee, C. (1986). Cat area 17. II. Response properties of infragranular layer neurons in the absence of supragranular layer activity, *J. Neurophysiol.* **56,** 1074–1087.

Shmuel, A., and Grinvald, A. (1996). Functional organization for direction of motion and its relationship to orientation maps in cat area 18. *J. Neurosci.* **16,** 6945–6964.

Sillito, A. M. (1975). The effectiveness of bicuculline as an antagonist of GABA and visually evoked inhibition in the cat's striate cortex. *J. Physiol. (Lond.)* **250,** 287–304.

Sillito, A. M. (1977). Inhibitory processes underlying the directional specificity of simple, complex, and hypercomplex cells in the cat's visual cortex. *J. Physiol. (Lond.)* **271,** 775–785.

Sillito, A. M., Kemp, J. A., Milson, J. A., and Berardi, N. (1980). A re-evaluation of the mechanisms underlying simple cell orientation selectivity. *Brain Res.* **194,** 517–520.

Somogyi, P., Martin, K. A. C., and Whitteridge, D. (1983). Synaptic connections of morphologically identified and physiologically characterized large basket cells in the striate cortex of cat. *Neuroscience* **10,** 261–294.

Spear, P. D. (1991). Functions of extrastriate visual cortex in non-primate species. In: *The neural basis of visual function* (A. Leventhal, Ed.), pp. 339–370. Basingstroke, England, McMillan Press.

Stratford, K. J., Tarczy-Hornoch, K., Martin, K. A. C., Bannister, N. J., and Jack, J. J. B. (1996). Excitatory synaptic inputs to spiny stellate cells in cat visual cortex. *Nature* **382,** 258–261.

Suarez, H., Koch, C., and Douglas, R. (1995). Modeling direction selectivity of simple cells in striate visual cortex within the framework of the canonical microcircuit. *J. Neurosci.* **15,** 6700–6719.

Tal, D., and Schwartz, E. L. (1997). Computing with the leaky integrate-and-fire neurons: logarithmic computation and multiplication. *Neural Comput.* **9,** 305–318.

Tanaka, K. (1983). Cross-correlation analysis of geniculostriate neuronal relationships in cats. *J. Neurophysiol.* **49,** 1303–1318.

Thorson, J. (1966). Small-signal analysis of a visual reflex in the locust. II. Frequency dependence. *Kybernetik* **3,** 53–66.

Tolhurst, D. J., and Dean, A. F. (1991). Evaluation of a linear model of directional selectivity in simple cells of the cat's striate cortex. *Vis. Neurosci.* **6,** 421–428.

Tolhurst, D. J., and Heeger, D. J. (1997a). Contrast normalization and a linear model for the directional selectivity of simple cells in cat striate cortex. *Vis. Neurosci.* **14,** 19–25.

Tolhurst, D. J., and Heeger, D. J. (1997b). Comparison of contrast-normalization and threshold models of the responses of simple cells in cat striate cortex. *Vis. Neurosci.* **14,** 293–309.

Tolhurst, D. J., Dean, A. F., and Thompson, I. D. (1981). Preferred direction of movement as an element in the organization of cat visual cortex. *Exp. Brain Res.* **44,** 340–342.

Tonegawa, S., Tsien, J. Z., McHugh, T. J., Huerta, P., Blum, K. I., and Wilson, M. A. (1996). Hippocampal CA1-region-restricted knockout of NMDAR1 gene disrupts synaptic plasticity, place fields, and spatial learning. *Cold Spring Harbor Symp. Quant. Biol.* **61,** 225–238.

Torre, V., and Poggio, T. (1978). A synaptic mechanism possibly underlying directional selectivity to motion. *Proc. R. Soc. Lond. B* **102,** 409–416.

Toyama, K., Kimura, M., and Tanaka, K. (1981). Cross-correlation analysis of interneuronal connectivity in cat visual cortex. *J. Neurophysiol.* **46,** 191–201.

Tsumoto, T., Eckart, W., and Creutzfeldt, O. D. (1979). Modification of orientation sensitivity of cat visual cortex neurons by removal of GABA-mediated inhibition. *Exp. Brain Res.* **34,** 351–363.

Ulinski, P. S. (1999). Neural mechanisms underlying the analysis of moving visual stimuli, In: *Cerebral cortex, Volume 13, Models of cortical circuits* (P. S. Ulinski, E. G. Jones, and A. Peters, Eds.) pp. 283–299. New York, Plenum Publishers.

Usrey, W. M., and Reid, R. C. (1999). Synchronous activity in the visual system. *Annu. Rev. Physiol.* **61,** 435–456.

Van Santen, J. P. H., and Sperling, G. (1985). Elaborated Reichardt detectors. *J. Opt. Soc. Am. A.* **2,** 300–321.

Vautin, R. G., and Berkley, M. A. (1977). Responses of single cells in cat visual cortex to prolonged stimulus movement: neural correlates of visual aftereffects. *J. Neurophysiol.* **40,** 1051–1065.

Vidyasagar, T. R., and Heide, W. (1986). The role of GABAergic inhibition in the response properties of neurones in cat visual area 18. *Neuroscience* **17,** 49–55.

Watson, A. B., and Ahumada, A. J., Jr. (1985). Model of human visual-motion sensing. *J. Opt. Soc. Am. A* **2,** 322–341.

Weliky, M., Bosking, W. H., and Fitzpatrick, D. (1996). A systematic map of direction preference in primary visual cortex. *Nature* **379,** 725–728.

Wolf, W., Hicks, T. P., and Albus, K. (1986). The contribution of GABA-mediated inhibitory mechanisms to visual response properties of neurons in the kitten's visual cortex. *J. Neurosci.* **6,** 2779–2795.

Worgotter, F., and Eysel, U. T. (1991). Topographical aspects of intracortical excitation and inhibition contributing to orientation specificity in area 17 of the cat visual cortex. *Eur. J. Neurosci.* **3,** 1232–1244.

Worgotter, F., and Holt, G. (1991). Spatiotemporal mechanisms in receptive fields of visual cortical simple cells: a model. *J. Neurophysiol.* **65,** 494–510.

Worgotter, F., and Koch, C. (1991). A detailed model of the primary visual pathway in the cat: comparison of afferent excitatory and intracortical inhibitory connection schemes for orientation selectivity. *J. Neurosci.* **11,** 1959–1979.

Worgotter, F., Niebur, E., and Koch, C. (1991). Isotropic connections generate functional asymmetrical behavior in visual cortical cells. *J. Neurophysiol.* **66,** 444–459.

Zhou, Y. X., and Baker, C. L., Jr. (1996). Spatial properties of envelope-responsive cells in area 17 and 18 neurons of the cat. *J. Neurophysiol.* **75,** 1038–1050.

10

Long-Range Intrinsic Connections in Cat Primary Visual Cortex

Kerstin E. Schmidt

Max-Planck-Institute for Brain Research, Frankfurt-am-Main, Germany

Siegrid Löwel

Leibniz-Institute for Neurobiology, Magdeburg, Germany

INTRODUCTION

One of the main tasks of our visual system is to extract features of a scene that belong to the same object and to segregate them from background. It is generally accepted that the activity of neurons in the primary visual cortex (area 17 or V1) is not only dependent on a stimulus appearing in their classical receptive field, but can be dramatically modulated by more global characteristics of a visual scene such as the contours and surfaces within which a stimulus is embedded. While the visual system is composed of a multitude of areas (like other sensory systems), all representing particular but different aspects of a visual stimulus, there is growing evidence that part of the process of spatial integration occurs in primary visual cortex. Long-range tangential connections within area 17 are hypothesized to be the anatomical substrate of these integrative capabilities. As the name indicates, long-range connections span large distances (up to 8 mm) within a cortical area. They are a characteristic feature not only of the primary visual cortex but also of other neocortical areas such as auditory, somatosensory, motor, and even prefrontal cortex. Neuronal computations within an area may be modulated by both feedforward (incoming sensory input) and "top-down" feedback projections, but they are also critically dependent on the layout of the intrinsic microcircuitry. This chapter focuses on the detailed layout of long-range horizontal connections in the primary visual cortex of cats, their topographic relation to functional cortical maps, learning- and activity-induced modifiability, and possible functions.

HISTORICAL OVERVIEW

Long-range horizontal connections in the visual cortex were probably observed indirectly more than 200 years ago by the Italian medical student Francesco Gennari (1782) and by Vicq d'Azyr (1786). Both described a prominent stripe in the human occipital cortex, the stria of Gennari, which is visible macroscopically because of its strong myelination (Fig. 10-1). The stria corresponds to a fiber plexus in upper layer IV of the six-layered neocortex and was later called the external band of Baillarger. As inferred from lesions in human brain tissue, the stria is mainly of intracortical origin, as it does not show signs of degeneration after destruction of the underlying white matter (Braitenberg, 1962; see also Szentágothai, 1973).

METHODOLOGICAL DEVELOPMENTS

Long-range intrinsic connections are currently recognized as an important and characteristic feature of neocortical circuitry, but they did not become the subject of scientific investigation until 1970 for both methodological and conceptual reasons. The Golgi stain (Golgi, 1873), which has been used in neuroanatomical

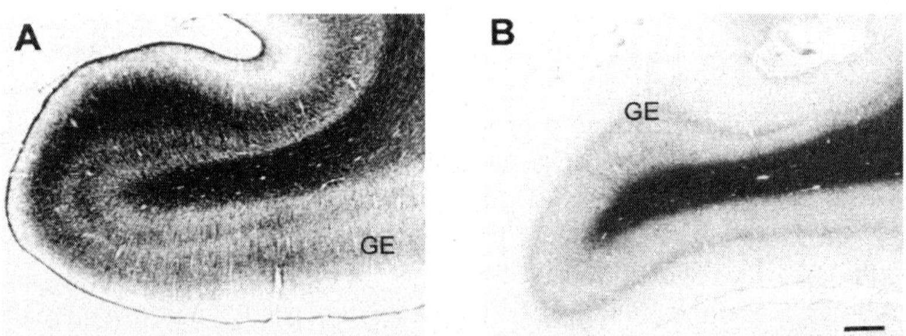

FIGURE 10-1. Myelin stained coronal sections of the cat's **(A)** and the human's **(B)** primary visual cortex (area 17, also termed striate cortex) demonstrating the stria of Gennari (GE). **(A)** Gulyas staining of the cat's primary visual cortex. The upper (also called external band of Baillarger) of the two dark stripes is situated in lower layer III and upper layer IV. Its lower part probably corresponds to the classic stria of Gennari. The lower stripe corresponds mainly to upper layer V. It is also called the internal band of Baillarger. After cortical lesions, horizontal degeneration was predominantly observed within these two stripes. **(B)** The prominent streak (GE) corresponds to upper layer IV and contains many fibers indicated by dark colour in the myelin stain. (Tissue prepared by W. Schlote). (Creutzfeld et al., 1975). Scale bars: 1 mm. (Also see Figs. 1-7 and 1-26, this volume.)

studies for more than a century, labels both dendritic and axonal fibers. This technique was used to detail and classify a variety of neuron types mainly on the basis of their dendritic morphology, and not their axon arbors, because axons are usually much thinner than dendrites and only stain in their proximal unmyelinated part immediately after leaving the soma (Ramón y Cajal, 1922; Lorente de Nó, 1922; O'Leary, 1941; Lund, 1973). Even so, Golgi studies revealed axon fragments that ran parallel to the cortical lamination, but they did not draw the anatomists' attention, because it seemed almost impossible to follow axons for long distances in the tangential plane, parallel to cortical layers. When the initial parts of axons were encountered, they were described as leaving the soma vertical to the cortical lamination and descending. From Golgi and myelin stains it was suggested that both the external and the less striking internal band of Baillarger (which is situated in layer Vb) contain oblique and horizontally running axon collaterals derived mainly from pyramidal cells of cortical layers III and Va (Clark and Sunderland, 1939; Braitenberg, 1962; for review see Braak, 1984). However, as soon as sensitive degeneration techniques became available, the extent of lateral axonal connections was immediately realized from the widespread extent of the degeneration induced by small electrical or mechanical lesions of cortex (Nauta and Gygax, 1953). In area 17 of the cat, the combination of Golgi stains, degeneration techniques, and electron microscopy finally revealed that the stria contained both terminals and preterminal passing fibers of specific afferents and terminals of intracortical origin, such as basket cell axons and collaterals of pyramidal neurons (Szentágothai, 1973). In macaque monkeys, and later also in cat visual cortex, horizontal axonal degeneration was observed up to a radius of 3.5 mm from the lesion site (Fisken et al., 1975; Creutzfeldt et al., 1977). However, degeneration studies could not differentiate between extrinsic and intrinsic cortical connections.

At the end of the 1970s, intracellular filling of single neurons allowed investigators to reconstruct all dendritic and axonal processes of a cell, revealing for the first time the detailed morphology of long axon collaterals (Gilbert and Wiesel, 1979; 1981; Lin et al., 1979). Finally, modern tracing techniques like intracellular and extracellular injections of horseradish peroxidase (HRP), injections of fluorescent latex microspheres, biocytin, and dextran amines visualized extensive horizontal connections travelling distances of up to 8 mm within specific layers, without entry into the white matter. These connections were termed long-range intrinsic, horizontal, tangential or intralaminar connections. They are especially prominent in supragranular layers II/III and in layer V.

CONCEPTUAL DEVELOPMENTS

The other reason long-range horizontal connections came late as a prime subject of neuroscientific investigations is the long-standing dominance of the columnar concept of the neocortical organization, which emphasizes vertical connections. The classical view was mainly derived from Golgi studies. For example, in the somatosensory cortex, Mountcastle (1957) observed that neurons were grouped into columns extending vertically through all layers within bundles of separated vertical connections. Just a few years later, Hubel and Wiesel (1962) demonstrated that neurons in the primary visual cortex were also grouped into vertical domains according to their receptive field properties. In vertical electrode penetrations, neuronal response properties such as receptive field position, ocular dominance, orientation, and direction selectivity did not change significantly, so that neurons responding to similar visual stimuli are arranged in columns extending from layer I to layer VI. In contrast, tangential penetrations traveling for long distances parallel to the cortical lamination revealed more or less continuous changes in response properties (see Chapter 1). In a plane parallel to the cortical surface, neuronal selectivities vary systematically, and it is now known that columns or domains of similar functional properties form highly organized and periodic patterns (see Chapters 2 and 3).

Additionally, a strong vertical connection loop indicated multiple streams for parallel signal processing within the columnar structures. Vertical connections extend predominantly from layer IV, which receives the main thalamocortical input to supragranular (I–III) and infragranular (V, VI) layers (Lund et al., 1979) (see Figs 1-9 and 1-33) "backward" to subcortical structures and "forward" to extrastriate cortical areas (for review see Salin and Bullier, 1995). Studies that have investigated these connections in detail indicate that most of them (except for connections between area 17 and a few subcortical structures) (Salin et al., 1989) exhibit a point-to-point (vertical) topology by interconnecting locations within and between the different cortical areas representing retinotopically matching parts of the visual field. This is compatible with the view of a strong vertical processing stream within the cortical hierarchy. However, when intrinsic horizontal connections were first described, it appeared that they covered more than a "hypercolumn" of cortical distance (Rockland and Lund, 1982; Gilbert and Wiesel, 1989) and it was not readily appreciated how they contributed to cortical operations. However, it was recognized that they could allow intrinsic connections to combine signals of neurons with nonoverlapping receptive fields within primary visual cortex. In addition, there is now evidence of considerable convergence and divergence within the corticocortical connections in the visual system (Salin and Bullier, 1995). However, the issue of whether the distances in visual field angle spanned by the long-range intrinsic connections match or exceed those of extrinsic feedback connections is still an open question.

LAYOUT OF LONG-RANGE HORIZONTAL CONNECTIONS

Even though long-range horizontal connections had been described some 25 years ago, interest in their analysis was boosted only after the discovery of their patchy layout. In 1982 Rockland and Lund demonstrated for the first time that

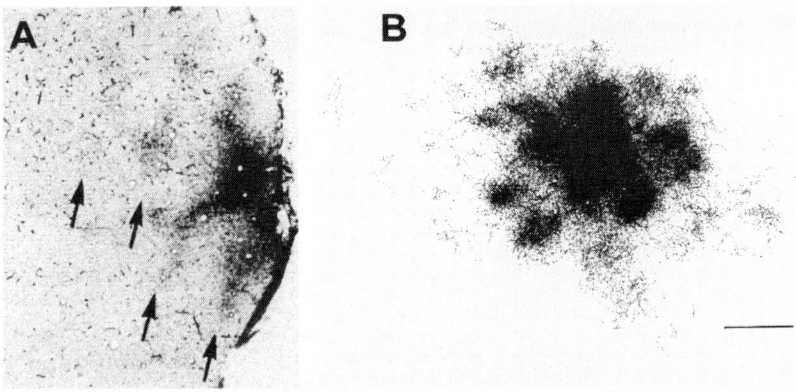

FIGURE 10-2. Long-range intrinsic connections in the primary visual cortex of the tree shrew **(A)** and the cat **(B)**. **(A)** The discovery of the patchiness of intrinsic connections. Horizontal section through tree shrew visual cortex illustrating patches (arrows) of anterogradely labeled terminals and retrogradely labeled cell bodies resulting from an extracellular injection of horseradish peroxidase (HRP). Magnification ×35. (Modified from Anatomical banding of intrinsic connections in striate cortex of tree shrews *(Tupaia glis)*, Rockland, Lund and Humphrey, *Journal of Comparative Neurology*, copyright © 1982 A. R. Liss inc., reproduced by permission of Wiley-Liss, Inc.) **(B)** Reconstruction of lateral connections in cat area 17. Distribution of synaptic boutons labeled anterogradely from a biocytin injection site (dark region in the center). Each dot represents one single bouton; excitatory and inhibitory boutons are plotted together. *Note that at least 20 distinct patches can be discriminated but interpatch regions are also heavily innervated.* Scale bar: 1000 μm. (Reproduced from Kisvárday, Toth, Rausch and Eysel, [1997], *Cerebral Cortex*, 7, by permission of Oxford University Press.)

extracellular injections of the tracer HRP into tree shrew visual cortex revealed a stripelike pattern of label in a plane tangential to the cortical surface (Rockland and Lund, 1982; Rockland et al., 1982) (Fig. 10-2). Dense patches of HRP were visible extending 2 to 3 mm around the injection sites, and these patches were most prominent in layers II and III A. The patches consisted of both anterogradely labeled terminals and retrogradely labeled neurons, mainly pyramidal cells. In coronal sections, the patches measured about 230 μm in diameter and were interleaved by label-free spaces at a center-to-center distance of 450 to 500 μm. Long horizontally and sometimes obliquely traveling axons were observed between the patches, probably giving rise to clustered axon terminals. In contemporaneous studies, intracellular injections of the same tracer permitted visualization and reconstruction of single axons emanating from pyramidal and spiny stellate cells. These studies were in the cat visual cortex (Gilbert and Wiesel, 1983; Martin and Whitteridge, 1984; Kisvárday and Eysel, 1992). The axons ran parallel to the cortical layers and were seen to give rise to terminal bundles at regularly spaced intervals (Fig. 10-3).

Early experiments using HRP injections in the tree shrew and the ferret indicated that the patches look more like periodically beaded stripes than round patches. This was true for both coronally cut sections and for horizontal sections

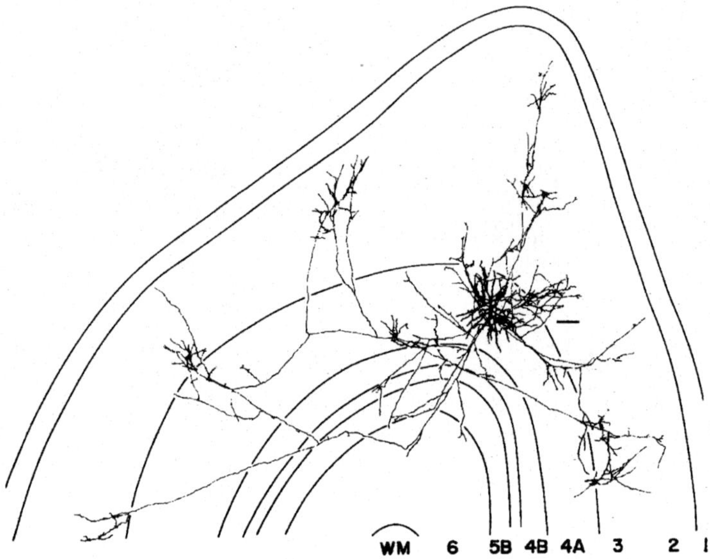

FIGURE 10-3. Camera lucida reconstruction of a HRP-filled spiny stellate cell of cortical layer IV displaying an extensive and patchy axonal distribution. The cell body of this neuron is located in layer 4A, whereas most of the collateral branches are restricted to cortical layers 2 and 3. Frontal section through the primary visual cortex of a cat. Scale bar: 100 μm. (Modified from Martin and Whitteridge [1984], *Journal of Physiology,* reproduced by permission of The Physiological Society.)

of some regions of primary visual cortex (Rockland et al., 1982; see also Ruthazer and Stryker, 1996). Reconstructions from serial sections revealed more or less regular rows of patches, sometimes with blind endings or twistings. Finally, both shapes seem to occur depending on the area, the species, and the technique used. In primate V1 and V2, large HRP injections revealed lattice-like or reticular patterns (Rockland and Lund, 1983, Rockland, 1985), whereas small injections labeled patches surrounded by unlabeled cortex (Livingstone and Hubel, 1984b). In the cat, the intrinsic connections are arranged in both irregular beaded bands and curved rows of isolated patches, both circular and oval-shaped (Luhmann et al., 1986, 1991; Gilbert and Wiesel, 1989; Callaway and Katz, 1990; Löwel and Singer, 1992; Schmidt et al., 1997a; Kisvárday et al., 1997; Yousef et al., 1999).

To date, patchy patterns of intrinsic connections have been observed in the primary visual cortices of several species including tree shrew (Rockland and Lund, 1982; Rockland et al., 1982; Fitzpatrick, 1996; Bosking et al., 1997), cat (Gilbert and Wiesel, 1983; 1989; Martin and Whitteridge, 1984; Luhmann et al., 1986, 1991; Kisvárday et al., 1986; 1994; 1997; Kisvárday and Eysel, 1992, 1993; Callaway and Katz, 1990; Albus et al., 1991; Katz and Callaway, 1992; Löwel and Singer, 1992; Lübke and Albus, 1992; Galuske and Singer, 1996; Schmidt et al., 1997a, 1997b; Buzás et al., 1998; Yousef et al., 1999), ferret (Rockland, 1985;

Durack and Katz, 1996; Ruthazer and Stryker, 1996) and squirrel monkey and macaque monkey (Rockland and Lund, 1983; Livingstone and Hubel, 1984a; Amir et al., 1993; Malach et al., 1993, Lund et al., 1993; Yoshioka et al., 1996). Patchiness has also been observed in extrastriate visual areas (Price, 1986; Matsubara et al., 1985, 1987; Yoshioka et al., 1992; Amir et al., 1993; Levitt et al., 1994; Malach et al., 1994; Kisvárday et al., 1997; Malach et al., 1997) and for corticocortical connections between different visual cortical areas (Gilbert and Kelly, 1975; Wong-Riley, 1979; Montero, 1980; Tigges et al., 1981; Rockland and Lund, 1982; Gilbert and Wiesel, 1983, 1989; Bullier et al., 1984; Ferrer et al., 1988, 1992; Price et al., 1994) and between the two hemispheres of the brain (Houzel et al., 1994; Boyd and Matsubara, 1994; Schmidt et al., 1997a; for earlier citations see review of Innocenti, 1986). Finally, clustered intrinsic and corticocortical connections are also a common feature of cat and monkey somatosensory and motor cortex (Jones et al., 1978; Matsubara and Phillips, 1988; Keller, 1993), auditory cortex (Imig and Brugge, 1978; Imig and Reale, 1981; DeFelipe et al., 1986), monkey prefrontal (Pucak et al., 1996) and frontal cortex (Goldmann and Nauta, 1977), and human visual (Burkhalter and Bernardo, 1989) and temporal cortex (Galuske et al., 2000), and thus a general feature of cortical organization.

TYPES OF NEURONS FORMING LONG-RANGE HORIZONTAL CONNECTIONS

Most long-range horizontal connections arise from pyramidal or spiny stellate cells, the two types of cortical excitatory neurons (Gilbert, 1983; Gilbert and Wiesel, 1983; Martin and Whitteridge, 1984). Pyramidal cells constitute the main cell type in the neocortex, and their somata are distributed in all cortical layers, except layer I. They are characterized by a triangular cell body, several basal dendrites, and a large apical dendrite, which is directed radially toward the pial surface. Pyramidal cell dendrites are covered with spines in high density, dendritic protrusions that receive at least one excitatory synapse (Peters and Kaiserman-Abramof, 1969). Spiny stellate cells lack a dominant apical dendrite but also have spines (see Chapter 1). They are present almost exclusively in layer IV of primary sensory areas (Lund, 1973; Valverde, 1986) and are the major recipients of thalamocortical afferents terminating in layer IV (Gilbert, 1983 (see Chapter 8)). It has been estimated that at least half of the pyramidal and spiny stellate neurons in the cortex contribute to long-range clustered connections. A few supragranular pyramidal neurons even seem to participate only in intrinsic circuits and do not extend their axon into the white matter (Gilbert and Wiesel, 1983).

About 20% of all cortical neurons are immunopositive for γ-aminobutyric acid (GABA), the major inhibitory neurotransmitter in the cerebral cortex (Gabbott and Somogyi, 1986). Compared with the excitatory network, the inhibitory connections have a much more restricted lateral extent. Nevertheless, the degeneration study of Fisken et al. (1975) showed that more than 10% of all degenerating symmetrical terminals were located as far as 2 to 3 mm from the center of a cortical lesion. Among the GABA-immunopositive neurons, only large basket cells, large multipolar cells, and dendrite-targeting cells provide long-range axon collaterals extending up to 2 mm (Somogyi et al., 1983; Kisvárday et al., 1994, 1997; see Chapter 1). Their axons are relatively straight and surround the somata of both pyramidal and nonpyramidal neurons (Kisvárday and Eysel, 1993; see also Somogyi et al., 1983). The other inhibitory cell types have predominantly local axon collaterals (LeVay, 1988; Matsubara, 1988). The percentage of GABA-positive inhibitory neurons in areas 17 and 18 that give rise to long lateral connections is in the range of 3–10% depending on which tracers and immunohistochemical protocols are combined. Since HRP might travel transsynaptically, studies using this tracer might have overestimated the number of inhibitory neurons participating in long-range connections (Matsubara and Boyd, 1992). On the other hand, studies using a fluorescent tracer and peroxidase-based immunohistochemical detection of GABA (LeVay, 1988; Albus and Wahle, 1994) might underestimate the number of double-labeled neurons, because the peroxidase reaction might quench the fluorescence. The actual percentage of long-range projecting neurons which are GABA immunoreactive is probably about 5%. A study of Albus et al. (1991) indicated that approximately 70% of all GABAergic neurons send out projections shorter than 1 mm; the remaining 30% occasionally give rise to connections with a lateral spread of 1–2.5 mm. GABAergic neurons with projections longer than 1 mm seem to be more numerous in infragranular layers (Matsubara and Boyd, 1992; see also Fisken et al., 1975).

Inhibitory neurons labeled by retrograde tracers seem to be more or less homogenously distributed in the primary visual cortex. Only about 60% of GABA immunopositive neurons reside within the dense clusters produced by retrograde fluorescent tracing (Albus et al., 1991; Albus and Wahle, 1994). Taking into account that they constitute only about 20% of all cortical neurons, inhibitory cells do not contribute critically to the clustering of intrinsic connections.

SYNAPTIC TARGETS OF LONG-RANGE INTRINSIC CONNECTIONS

Most of the long-range axon collaterals in primary visual cortex contact other pyramids. The majority of axons make asymmetrical type 1 synapses that are not GABA-immunoreactive and therefore thought to be excitatory (Kisvárday et al., 1986; area 18: LeVay, 1988; McGuire et al., 1991). The most frequent postsynaptic targets are dendritic spines (84–87%) probably of pyramidal cells, because the main projecting layers of intrinsic axons, namely cortical layers II, III and V, are basically free of spiny stellate neurons (Lund, 1973). The remaining synapses are with dendritic shafts of both pyramidal and GABA-immunoreactive nonpyramidal neurons (Kisvárday et al., 1986; McGuire et al., 1991). In summary, only about 5% of the postsynaptic structures of long-range intrinsic connections in cat primary visual cortex are GABAergic dendritic shafts and, rarely, somata (Kisvárday et al. 1986; area 18: LeVay, 1988). In the macaque monkey, Fisken et al. (1975) had reported that about 9% of all degenerating synapses in a column of 3 mm radius surrounding a focal cortical lesion were symmetrical and, therefore, presumably inhibitory. Later studies demonstrated that long-range connections in the monkey contacted the dendrites of spiny and nonspiny cells in the proportion to which these cell types occur in the cortex (spiny:nonspiny = 80%:20%) (Hendry et al., 1987; McGuire et al., 1991), so that there was no evidence for preferential connectivity between excitatory neurons. Nevertheless, given that about 80% of all cortical neurons are excitatory, the primary role of the long-range intrinsic connections is the activation of other excitatory cells.

DIVERGENCE AND CONVERGENCE OF LONG-RANGE HORIZONTAL CONNECTIONS AT THE ULTRASTRUCTURAL LEVEL

Long-range intrinsic connections are formed by myelinated collaterals that leave the main (vertically) descending axon trunk before it enters the white matter. Collaterals can arise from the same axon in more than one layer (Gilbert and Wiesel, 1983; Kisvárday et al., 1986). They do not always run strictly parallel to the lamination but may also ascend and/or descend (Gilbert and Wiesel, 1983). In any case, patches of terminal branches are in register in supragranular and infragranular layers (Gilbert and Wiesel, 1983; Kisvárday et al., 1986), with the most prominent projections extending within layers II/III and V. In the literature, the term *cluster* or *patch* is commonly used to describe a group of labeled terminals or neurons that are separated from each other by more than 500 μm. These patches are variable in size but usually cover an average area of about 300 to 400 μm in diameter when labeled with HRP (Kisvárday et al., 1986) and 500 to 600 μm when labeled with biocytin (Kisvárday et al., 1997). Only about 5% (Albus and Wahle, 1994) of all neurons in the cortical volume covered by a patch participate in projecting to another distinct columnar volume. Average interpatch distances are in the range of 800 to 1100 μm in cat area 17 (Gilbert and Wiesel, 1983, using HRP: 800 μm; Martin and Whitteridge, 1984, using HRP, 1000 μm; Kisvárday and Eysel, 1992, using HRP, 1100 μm; Galuske and Singer, 1996, using DiI: 900 μm; Kisvárday et al., 1997, using biocytin: 900μm) and slightly larger in cat area 18 (Kisvárday et al., 1997, using biocytin: 1200 μm). Up to 10% of the retrogradely labeled excitatory neurons in a distance of more than 1000 μm from the injection site center are located between clusters (Albus and Wahle, 1994).

Estimates based on reconstructed biocytin-labeled axons indicate that about 80 of 300 to 1200 synapses of individual pyramidal neurons are localized within a single patch. This suggests that each pyramidal cell contributes afferents to four to eight different patches. Because only up to 4 of about 4800 (about 0.1%) excitatory inputs to a pyramidal neuron are from the same axon, a single pyramidal cell axon might contact from 20 to as many as 80 different pyramidal cells in a distant patch. Another estimate was made for the overall influence of a patchy axon on the total cell population of a single patch (400 μm diameter). The results showed that, on average, only about 1–3% of the cells in a given patch are contacted by the same axon originating from a remote site (Kisvárday and Eysel, 1992). On the other hand, inhibitory neurons seem to have much more target cells: one basket cell makes about 2000–3000 synaptic contacts with 300 to 600 neurons (Kisvárday et al., 1993; Eysel et al., 1999a).

The density of anterogradely labeled boutons (within the patches) decreases with increasing distance from the pyramidal cell soma. Proximal patches lying within the dendritic field of the parent soma have more than twice as many boutons as do remote ones outside the dendritic field of the parent soma (Kisvárday and Eysel, 1992) (Fig. 10-2B). This suggests, on a semiquantitative level, that the impact of shorter-range connections on neocortical circuitry is much higher than that of long-range connections.

Another interesting feature of the network of long-range intrinsic connections is reciprocity. Injections of HRP had revealed a patchy distribution of both retrogradely labeled cell somata and anterogradely labeled axon terminals (Rockland and Lund, 1982). The assumption that a reciprocal network of clustered connections is responsible for these observations was confirmed by Kisvárday and Eysel (1992). Using biocytin as both an anterograde and a retrograde tracer, they visualized a patchy network of 10 interconnected pyramidal neurons in supragranular layers of cat area 17 (Fig. 10-4). Indeed, several remote pyramidal cells participated in the very same patchy network, with their terminal patches overlapping. The connections between the patches were predominantly reciprocal and made by contacts with distal segments of basal and apical dendrites. The largest distance observed between reciprocally interconnected patches in cat primary visual cortex was 2.7 mm (Kisvárday and Eysel, 1992).

TOPOGRAPHIC RELATIONS BETWEEN LONG-RANGE INTRINSIC CONNECTIONS AND FUNCTIONAL CORTICAL MAPS

MODULAR SELECTIVITY

When the clustering or patchiness of long-range connections was first observed, it was debated whether the pattern reflected that only a certain subpopulation of neurons gave rise to and received long-range connections ("fixed lattice hypothesis") or that different subsets of neurons (with different morphology and/or function) were specifically interconnected. The characteristic spacing of the observed clusters suggested a relationship to the functional architecture of the primary visual cortex, namely to orientation and/or ocular dominance columns, consisting of neurons with similar functional properties (Hubel and Wiesel, 1962)

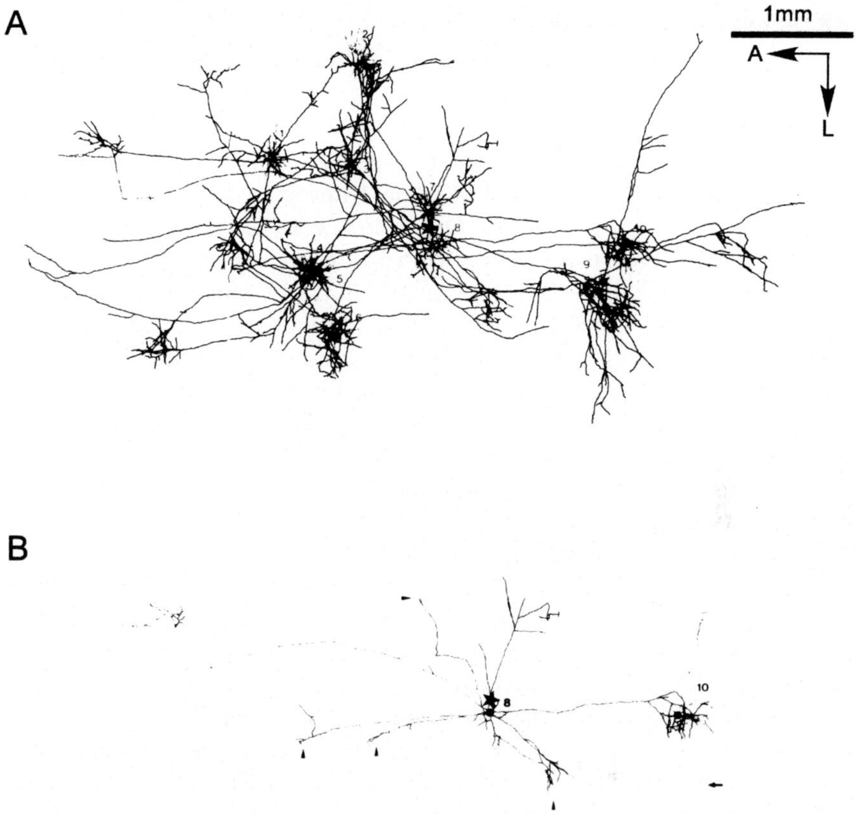

FIGURE 10-4. Reciprocity of the long-range lateral network. **(A)** Reconstruction of the complete axonal (black) and dendritic (gray) arborizations of 10 pyramidal neurons labeled from the same biocytin injection (black star) in cat area 17. Many of the patches receive overlapping axons from up to five pyramidal cells at different sites (e.g., the patch at the site of cell No. 3). **(B)** Some of the cells (e.g., cells No. 8 and 10) provide clustered axonal terminals to each other's dendritic field indicating the reciprocity of intrinsic connections. (Modified from *Neuroscience,* 46(2), Kisvárday and Eysel, Cellular organization of reciprocal patchy networks in layer III of cat visual cortex [area 17], p275–286, copyright 1992, with permission from Elsevier Science.)

(see also Chapter 2 and 3). Theoretical considerations from Mitchison and Crick (1982) predicted that neurons sharing similar functional properties such as orientation preference are selectively interconnected, which argued for a continuous system rather than a fixed lattice of patchy connections. The first trial to solve these issues by comparing the layout of both clustered long-range connections and orientation columns in the tree shrew visual cortex, however, showed only that the basic features of the two patterns are similar, but it did not reveal a systematic topographical relationship (Rockland et al., 1982).

Intrinsic Connections and Blobs

Historically, a systematic relationship between clustered long-range connections and a cortical functional system was first shown in the primary visual cortex of macaque and squirrel monkeys. Regions of high cytochrome oxidase activity (so-called cytochrome oxidase blobs) corresponded to neuronal domains with nonoriented, monocular, color- and low spatial fequency specific responses (Horton and Hubel, 1981; Livingstone and Hubel, 1984a; Tootell et al., 1988a, 1988b; Shoham et al., 1997) and received distinct thalamocortical input in both monkeys (Livingstone and Hubel, 1982; Fitzpatrick et al., 1983; Hendry and Yoshioka, 1994; for review see Casagrande, 1994) and cats (Boyd and Matsubara, 1996 (see Chapter 5)). Although cytochrome oxidase blob and interblob regions were initially described to be independent but reciprocally interconnected subcompartments (Livingstone and Hubel, 1984b), it is still a matter of debate whether blobs really represent a functionally separate compartment.

Biocytin tracing studies in both V1 and V2 of macaque and squirrel monkeys confirmed Livingstone and Hubel's original observations (Malach et al., 1993, 1994; Levitt et al., 1994; Yoshioka et al., 1996). Quantitative analysis showed that 68% of the intrinsic connections extended between cytochrome oxidase-rich blob regions, which occupied only about 20% of the cortical surface. When injections were made into nonblob regions (80% of cortical surface), patches of labeled neurons and terminals were located in nonblob compartments only with a probability of 73%, which seems to be random. A distinct border area between blob and nonblob territories (discriminable on the basis of its intrinsic connectivity) could not be specified (Yoshioka et al., 1996). In the cat, a colocalization of cytochrome oxidase blobs and patchy connections has so far been demonstrated for both callosal (Boyd and Matsubara, 1994) and feedforward connections from areas 17/18 to lateral susprasylvian cortex (Boyd and Matsubara, 1999) but *not* for long-range intrinsic connections (see Chapter 5).

Excitatory Long-Range Intrinsic Connections and Orientation Domains

First experimental evidence for the original hypothesis of Mitchison and Crick (1982) was obtained by cross-correlation studies (Michalski et al., 1983; Nelson and Frost, 1985; Ts'o et al., 1986). It was shown that the probablilty of synchronized spiking between neurons was higher when they had similar orientation preference, even if the neurons were separated by up to 2 mm (Ts'o et al., 1986; Ts'o and Gilbert, 1988). Because the ability to synchronize spikes on a short time scale is taken as evidence for a direct synaptic connection, it was concluded that neurons with similar orientation preference are selectively connected by long intracortical fibers (Ts'o et al., 1986).

The first direct alignments of anatomical connectivity patterns and functional domains in cat area 18 painted a different picture. Combining the patterns of HRP-labeled neurons with electrophysiologically recorded functional maps, Matsubara and colleagues observed connections predominantly between sites of different, frequently even orthogonal orientations (Matsubara et al., 1985, 1987). As discussed earlier, however, results from experiments using large extracellular injections of HRP are difficult to interpret because of distinct methodological difficulties (i.e., transneuronal transport of HRP). Two years later, Gilbert and Wiesel (1989) demonstrated that long-range intrinsic connections preferentially connected domains activated by the same orientation; they used a combination of 2-deoxyglucose (2-DG) labeling and fluorescent tracing. At that time, a new retrograde tracer, latex microspheres coupled with either rhodamine or fluorescein as the chromophore, the so-called beads, had become available (Katz et al., 1984; Katz and Iarovici, 1990). One of the big advantages of beads, compared with other retrograde tracers, is that they do not diffuse and thus allow small and localized injection sites. Gilbert and Wiesel (1989) injected beads into electrophysiologically characterized orientation preference domains. Subsequently, the pattern of iso-orientation domains was visualized by labeling with 2-DG while the animals were stimulated with moving gratings of the "injected" orientation. Post mortem, the retrogradely labeled neurons and the 2-DG domains were superimposed in the same sections of flat-mounted visual cortices. There was a strong tendency of the neurons to be located in the 2-DG-labeled domains and thus in regions preferring the same stimulus orientation as the injection site. However, no quantification of the data was delivered. Using the same experimental design in a different context (the analysis of connections in strabismic animals) our own analyses allowed a crude estimate of the actual percentage of retrogradely labeled neurons in the 2-DG-labeled orientation domains. In area 17 of a normally raised cat, about 56% (60% in strabismic cats) of all retrogradely labeled neurons were localized in domains of the same orientation preference (± 27°) as at the injection site, but only about 21% of the cells were localized in cross-oriented domains (± 27°) (Schmidt et al., 1997a). Thus the density of retrogradely labeled neurons in iso-orientation compartments as assessed with 2-DG (the compartment was defined to cover 30% of the cortical surface) was significantly higher than outside as can also be seen in the data of Gilbert and Wiesel (1989) (Fig. 10-5).

Kisvárday et al. (1997) gave a detailed quantification of the specificity of long-range intrinsic connections by combining anterograde biocytin tracing with electrophysiological mapping of cat areas 17 and 18. The distribution of labeled synaptic boutons was superimposed on orientation preference maps interpolated from the electrophysiological data. In area 17, 53% of the excitatory boutons were located in iso-orientation domains (± 0–30° difference from the orientation preference at the injection site), 30% in oblique (± 30–60°) and 17% in cross-oriented (± 60–90°) domains. The bouton distributions in area 18 were very similar (iso, 59%; oblique, 30%; cross, 11%). The selectivity for iso-orientation domains was higher in the close vicinity (distances less than 500 μm) of the injection site than in more remote patches. The advantage of that study over previous ones is that the finest details of excitatory connections, the boutons making the synaptic contacts, are matched with the functional maps of an area. While retrograde tracing studies can provide numbers of neurons participating in a particular pro-

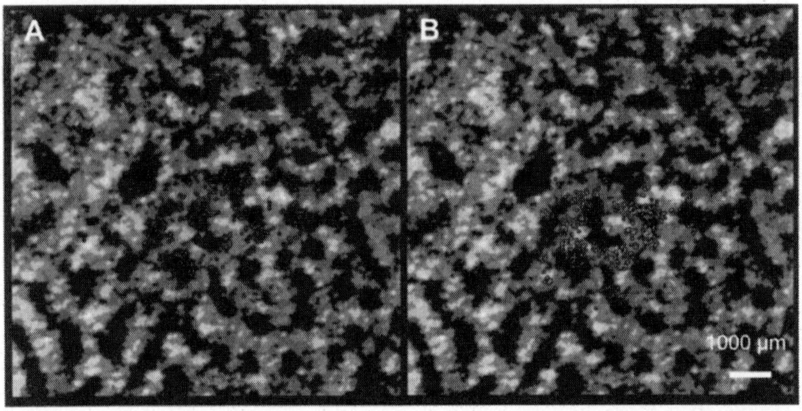

FIGURE 10-5. Superposition of retrogradely labeled neurons and 2-DG labeled horizontal orientation columns in cat primary visual cortex. **(A)** An injection of red microspheres was made into a column preferring horizontal contours. Of the labeled neurons, 54.6% (n = 414) are localized in iso-orientation domains. **(B)** Cell distribution after an injection of green microspheres in a column preferring vertical contours. Only 22% of the labeled neurons (n = 360) are localized in the 2-DG labeled horizontal orientation domains. (From Schmidt, Kim, Singer, Bonhoeffer and Löwel, 1997, *Journal of Neuroscience* 17, p5480–5492, copyright 1997 by the Society for Neuroscience.) See color insert for color reproduction of this figure.

jection, the quantification of axonal boutons additionally allows to estimate the functional strength of these connections, when it is assumed that individual synapses have the same functional impact.

The latter type of studies became possible with the development of optical imaging of intrinsic signals to visualize functional cortical maps (Bonhoeffer and Grinvald, 1996) (see Chapter 2). By making it possible to record all kinds of functional maps in a certain cortical area in vivo, intrinsic signal imaging overcomes the major limitation of the 2-DG technique, namely the visualization of one, maximally two (see Chapter 3), different activity patterns (e.g., orientation domains) in one brain region (e.g., visual cortex). Optically recorded functional maps can be used for both targeting of precise tracer injections into identified domains (Kisvárday et al., 1994; Schmidt et al., 1997a, 1997b; Yousef et al., 1999) and for superimposing of labeled neuron and bouton distributions with the functional architecture (e.g., Malach et al., 1993, 1994; Kisvárday et al., 1994; Yoshioka et al., 1996; Bosking et al., 1997). In monkey V1, superimposition of biocytin-labeled patches with orientation preference maps visualized by optical imaging, demonstrated that connections were as orientation specific as in the cat (Malach et al., 1993). Similarly, in the primary visual cortex of the tree shrew, 57% of biocytin-labeled boutons contacted sites in the optically imaged orientation preference map with the same orientation preference as the injection site (± 35°; Bosking et al., 1997) (Fig. 10-6).

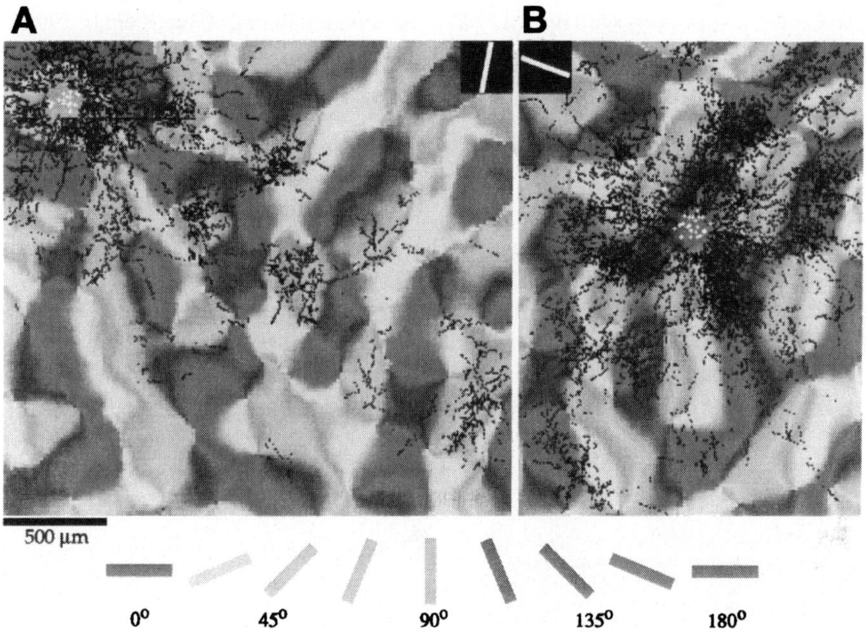

FIGURE 10-6. Topographic relationship between long-range intrinsic connections and iso-orientation domains in tree shrew primary visual cortex. The functional architecture of the cortex was visualized using optical imaging of intrinsic signals. In these maps, the preferred orientation for every region of the imaged cortex is color-coded according to the scheme below the figure (e.g., red codes for 0° and blue codes for 90° orientation preference). Biocytin was injected extracellulary into a domain preferring contours of 80° (light blue) **(A)** or 160° (red) **(B).** Both retrogradely labeled neurons (white symbols) and anterogradely labeled boutons (black dots) are superimposed with the optically recorded orientation preference map. Near the injection site, labeled boutons are found at sites with all orientation preferences, whereas at longer distances, boutons are located preferentially at sites preferring the same orientation as the injection site. (From Bosking, Zhang, Schofield and Fitzpatrick, 1997, *Journal of Neuroscience* 17, p2112–2127, copyright 1997 by the Society for Neuroscience.) See color insert for color reproduction of this figure.

More recently, Yousef et al. (1999) quantitatively analyzed the degree of orientation selectivity of long-range intrinsic connections with respect to the different cortical layers. Using a combination of optical imaging and injections of both latex microspheres and biocytin they analyzed connections in supragranular, granular, and infragranular layers of cat area 18. Layer IV lateral networks are in general much shorter (about 50%) than layer III networks and display a less clear patchy pattern. Moreover, long range (> 500 μm) connections in layer IV were distributed almost equally across orientations (iso, 35%; oblique, 34%; cross, 31%), suggesting that the long-range layer IV circuitry has a different functional role from that of the iso-orientation biased layer II/III circuitry.

To date, the most elaborate protocol for the detailed analysis of intrinsic circuitry is to reconstruct single intracellularly filled (with biocytin) neurons and to superimpose the reconstructions with optically imaged functional maps (Buzás et al., 1998). This technique makes it possible to analyze the relationship between the 3-dimensional anatomy of individual neurons, their receptive field properties (measured in vivo), and the functional map in which the neuron is embedded (Fig. 10-7).

So far, only injections into orientation domains (i.e. regions of low orientation gradient) have been considered. It is possible, however, that lateral connections in regions of high orientation gradient such as orientation (or pinwheel) centers are organized differently. Some evidence in this direction was recently obtained in cat area 18. Orientation center injections resulted in patchy labeling, although the patches were less remarkable than after injections into orientation domains (Kisvárday et al., 1996, 1998). Quantitative analyses of bouton distributions with optically recorded orientation preference maps revealed striking differences between the two experimental paradigms. Injections into orientation centers pro-

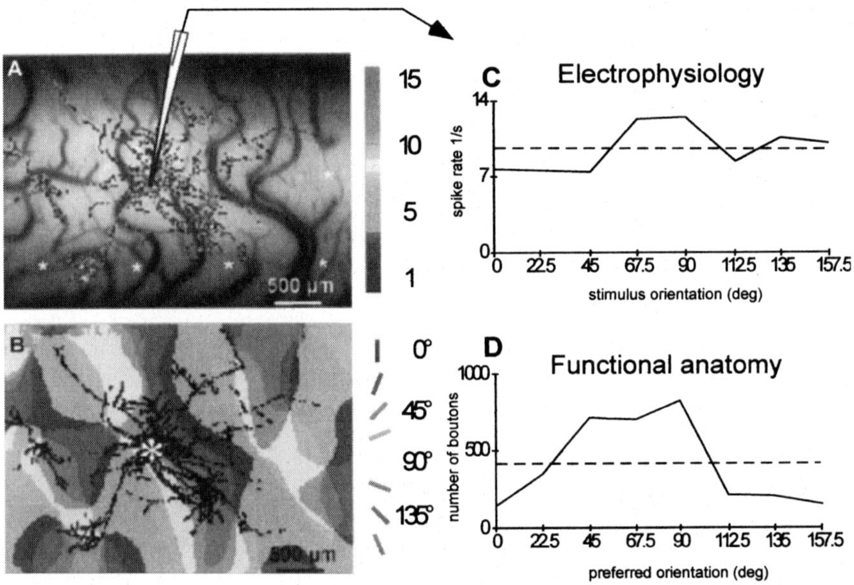

FIGURE 10-7. Relationship between a neuron's axonal projections, its receptive field properties, and the functional map in which the neuron is embedded. **(A)** Superposition of a density map of all axon terminals (colored dots) of a biocytin-filled pyramidal neuron with the vascular image containing five reference penetrations (white stars). Color scheme indicates the number of terminals per image-pixel. **(B)** Superposition of the bouton distribution (shown in black) with the optically visualized orientation map. The preferred orientation for every region of the imaged cortex is color-coded according to the scheme on the right side of the figure. Panels **(C)** and **(D)** provide a quantitative comparison between the electrophysiologically measured orientation tuning of the neuron **(C)** and the orientation distribution of all labeled boutons **(D).** (Modified from *Brain Research Protocols,* 3, Buzás, Eysel, and Kisvárday, Functional topography of single cortical cells: an intracellular approach combined with optical imaging, p199–208, copyright 1998, with permission from Elsevier Science.) See color insert for color reproduction of this figure.

duced boutonal patches whose "orientation" distribution varied from each other and from the injection site, whereas injections into orientation domains produced patches of axonal boutons with "orientation" preferences rather similar to the injection site. The combination of all patch distributions into a single one yielded a distribution similar to the one of the injections site. This kind of relationship was found applicable also for other cases regardless of the exact location of the injection site (Kisvárday et al., 1996). However, further experiments in area 17 are needed to establish the generality of this observation.

Methodologically, it is important to keep in mind that studies counting boutons and relating them to functional maps might underestimate the specificity of long-range connections, because synaptic contacts are also made on dendrites that can be located up to several hundred microns away from the parent soma (and thus in a different orientation domain). On the other hand, "simply" outlining patches after mass tracer injections might overestimate the specificity of the long-range circuitry because neurons connecting interpatch regions are usually omitted.

Inhibitory Connections and Orientation Domains

Most of the studies analyzing the topographic relationship between functional maps and intrinsic connections did not explicitly discriminate between excitatory and inhibitory connections. Because HRP, latex microspheres, and biocytin label both excitatory and inhibitory neurons and terminals, inhibitory connections must be identified by either GABA-immunohistochemistry or single cell reconstructions (Albus and Wahle, 1994; Kisvárday et al., 1994, 1997; Yousef et al., 1999). Although, their number is relatively small, Kisvárday et al. (1994, 1997) managed to reconstruct biocytin-labeled large basket cells, the main substrate of long-range inhibitory connections in cat visual cortex. Basket cell axons are characterized by segments with clusters of boutons aligned in rows separated by larger bouton-free segments (see Chapter 1). The inhibitory network is about one third to one half the extent of the excitatory network (Kisvárday et al., 1997; Crook et al., 1998) and involves more neurons. Inhibitory and excitatory terminals labeled from the same injection site do not overlap extensively. In cat areas 17 and 18, the density of inhibitory boutons is highest in the surround of excitatory terminal clusters outside the injection site core. In all, 48% of inhibitory boutons were located in iso-orientation domains (± 30°), 28% in oblique, and 24% in cross-oriented domains (Kisvárday et al., 1997). These data indicate that there seem to be slightly fewer inhibitory connections between iso-orientation domains than excitatory ones.

Relationship of Intrinsic Connections to Ocular Dominance

In the primary visual cortex of cats, long-range intrinsic connections extend between domains of left and right eye dominance with equal probability (Löwel and Singer, 1992; Schmidt et al., 1997a, see also Matsubara et al., 1987), so that there is no specific relationship between the two systems.

The situation is different in macaque monkey V1, in which intrinsic connections seem to have a preference for same-eye targets. First evidence was provided by cross-correlation studies showing interactions between cells with matched orientation and eye preference, at varying horizontal separations (Ts'o and Gilbert, 1988). More recently, two studies using anterograde biocytin labeling and optical imaging showed that the percentage of connections between same eye territories is on average 60–65% (Malach et al., 1993) thus, despite strong inter-individual variability, significantly greater than that of connections between opposite eye territories (Yoshioka et al., 1996). This difference in ocular dominance "selectivity" of the long-range intrinsic connections may be due to the different ocular dominance distribution in supragranular layers of the monkey compared with the cat primary visual cortex. In macaque monkeys, ocular dominance columns in the supragranular layers I–III are relatively obvious, and they are in precise register with the termination zones of thalamocortical afferents in layer IVc (Hubel and Wiesel, 1969; Hendrickson and Wilson, 1979; Horton and Hubel, 1981) in which monocular neurons predominate (Hubel and Wiesel, 1968). In contrast, the majority of supragranular neurons in normally raised cats are binocular, and they display only slight preferences for one eye or the other. However, when cats are reared with artificial strabismus, the ocular dominance distribution in area 17 changes dramatically and comes to resemble that in monkey V1. In strabismic cats, the segregation of thalamocortical afferents in layer IV is enhanced compared with normally raised animals (Shatz et al., 1977; Löwel, 1994) and two almost exclusively monocular subpopulations of neurons develop in supragranular and infragranular layers, in addition to the monocular population in layer 4 (Hubel and Wiesel, 1965). This experience-dependent change has a dramatic influence on the selectivity of the long-range intrinsic connections. Monocular injections of fluorescent beads combined with 2-DG autoradiography in area 17 of strabismic cats revealed that up to 85% of the retrogradely labeled neurons were located in the domains of the same eye preference than at the injection site (Löwel and Singer, 1992; for review see Löwel and Singer, 2000).

Other Columnar Systems

A relationship between long-range intrinsic connections and other functional systems such as direction selectivity or spatial frequency domains has so far not been reported in the cat. However, there is evidence from the ferret primary visual cortex that local, and to a lesser extent distant (up to ≈ 2 mm), excitatory synaptic inputs are iso-direction tuned, indicating the existence of a network of direction-specific, long-range horizontal connections (Roerig and Kao, 1999).

AXIAL SELECTIVITY

As originally postulated by Mitchison and Crick (1982), the pattern of long-range intrinsic connections is not only related to the system of orientation domains but also to the topography of visual space. The reconstruction of single axons in cat primary visual cortex revealed that the axonal fields are not circular but elliptically elongated, extending for a greater distance along one cortical axis compared with the orthogonal one (Gilbert and Wiesel, 1983; Kisvárday and Eysel, 1992). The observed anisotropy could not fully be explained by a different cortical magnification factor. Gilbert and Wiesel (1983) had already speculated that the axis of a cell's axonal field elongation is related to its receptive field orientation translated into cortical coordinates, but experimental evidence was missing. Studies in the visual cortex of both tree shrews (Fitzpatrick, 1996; Bosking et al., 1997) and cats (Schmidt et al., 1997b) later confirmed this suggestion, showing that in addition to the modular selectivity, long-range horizontal fibers exhibit axial specificity (i.e., there is a systematic relationship between a neuron's orientation preference and the distribution of its axon arbors across the cortical map of visual space). In tree shrews, biocytin-labeled horizontal axons extended for longer distances and gave rise to a larger number of terminal boutons along an axis of the visual field map that corresponded to the neuron's preferred orientation. In area 17 of cats use of latex microspheres to label interconnected neurons revealed a tendency for cell distributions to be elongated along the cortical axis corresponding to the orientation preference of the injection site (Schmidt et al., 1997b). Tracer injections into domains preferring horizontal contours resulted in distributions of retrogradely labeled cells elongated along an axis parallel to the cortical representation of the horizontal meridian; injections into domains preferring vertical contours resulted in cell distributions elongated parallel to the cortical representation of the vertical meridian at the 17/18 border (Fig. 10.8). Thus, long-range horizontal connections preferentially link neurons with co-oriented and co-axially aligned (colinear) receptive fields.

The axial selectivity of intrinsic connections is hypothesized to be the anatomical substrate for the physiological finding that responses to optimally oriented stimuli in the classical receptive field of a neuron are enhanced when colinearly aligned contours are presented outside the classical receptive field (Nelson and Frost, 1985; Kapadia et al., 1995).

Tracer injections into cat area 18 or monkey V2 revealed distributions of retrogradely labeled neurons elongated predominantly along a cortical axis parallel to the 17/18 (V1/V2) border (macaque and squirrel monkey: Rockland, 1985; cat: Matsubara et al., 1987). In that area, the cortical magnification factor is twice as large for the vertical compared with the horizontal orientation (e.g., Cynader et al., 1987) (see Chapter 1). Elongation of intrinsic networks in a direction parallel to the 17/18 border, therefore, might at least partly compensate for the large asymmetry in cortical magnification. Thus a possible axial selectivity of the horizontal network in area 18 is difficult to detect, given the pronounced asymmetry in the visual field representation. The same reasoning applies in the case of large extracellular tracer injections into monkey V1, which generally result in elliptical anterograde labeling. Microinjections of biocytin into layer III result in an asymmetrical field (average anisotropy = ratio long axis/short axis = 1.8) of labeled axon terminal clusters in layers I-III, with the longer axis of the label oriented orthogonal to the rows of blobs and ocular dominance domains, parallel to the V1/V2 border (Yoshioka et al., 1996; see also Malach et al., 1993). In contrast to

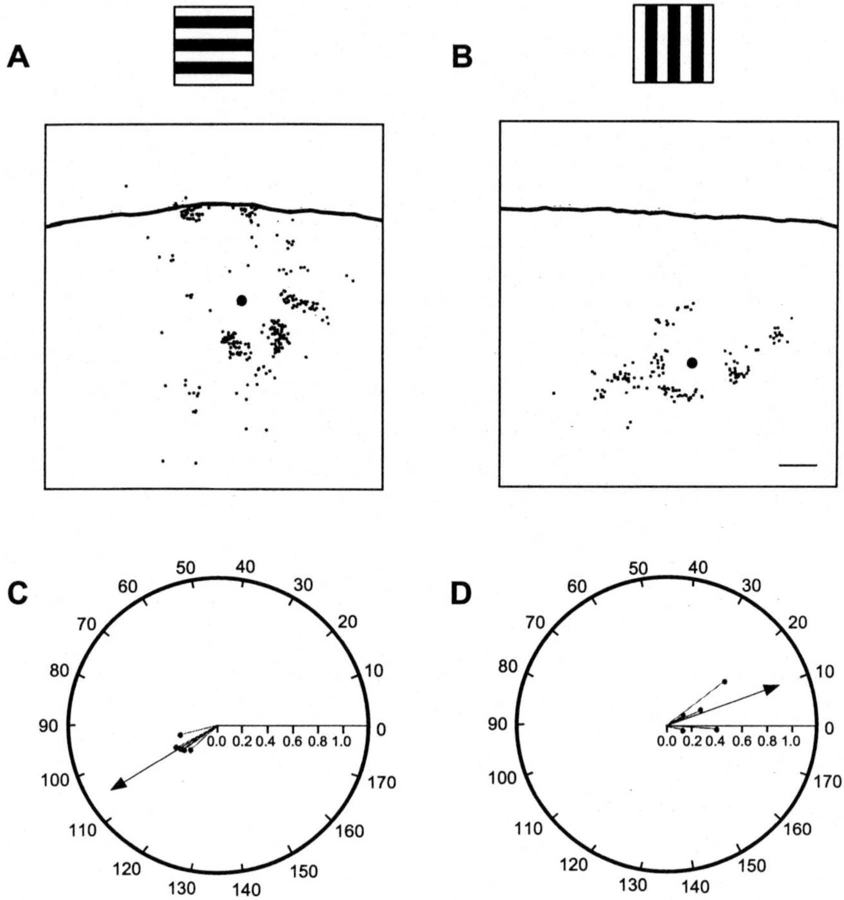

FIGURE 10-8. Axial selectivity of long-range intracortical connections in cat area 17. **(A, B)** Distribution of retrogradely labeled neurons resulting from extracellular injections of fluorescent latex microspheres in domains preferring horizontal **(A)** or vertical **(B)** contours. The continuous line in the upper parts of the pictures represent the 17/18 border (i.e., the representation of the vertical meridian). Labeled neurons are indicated by small dots and injection sites by large dots. *Note that cell distributions are elongated orthogonal to (A) or along (B) the vertical cortical axis.* **(C)** and **(D)** are vector plots of the distributions of labeled neurons in the different directions reduced to a range of 180°. Gray dots are the end points of section mean vectors reflecting the mean amount, distance, and direction of labeled neurons by the same injection site in five representative sections. The black arrows are the resulting vectors of all sections for an injection in a domain preferring horizontal **(C)** or vertical contours **(D).** *Note, that the vectors are separated by 95° and orthogonal to each other.* (Modified from Schmidt et al. [1997b]. The perceptual grouping criterion of colinearity is reflected by anisotropies of connections in the primary visual cortex, *European Journal of Neuroscience*, with permission from Blackwell Science Ltd.)

the cat, monkey V1 has an anisotropic cortical magnification factor (Van Essen et al., 1984; Dow et al., 1985), whereby the cortical representation of the horizontal meridian measures only half the representation of the vertical meridian (Malach et al., 1993; Grinvald et al., 1994). Nevertheless, there is some evidence for axial selectivity of long-range intrinsic connections also in the monkey (squirrel monkey: Sincich and Blasdel, 1995).

SUMMARY: FUNCTIONAL TOPOGRAPHY OF LONG-RANGE INTRINSIC CONNECTIONS

There is general agreement that excitatory long-range horizontal connections preferentially link neurons with similar orientation preference and colinearly aligned receptive fields. In contrast, inhibitory long-range connections are both shorter than excitatory ones and less orientation selective. Furthermore, the functional preference of long-range intrinsic connections may depend both on the layer (supragranular and infragranular layers I-III/V-VI versus layer IV) and on the exact position of the injections site within a functional map (linear region versus pinwheel-center). Published data further indicate that there is massive cross-talk between domains of different orientation preference both in excitatory and inhibitory networks.

Rockland and Lund (1982, 1983) originally suggested a lattice-like, but fixed, pattern of connections that link neurons only at certain cortical locations. This hypothesis has not been supported by experimental evidence, as almost every tracer injection into visual cortex resulted in a patchy pattern of labeling. Furthermore, double tracer injections into immediately adjacent cortical territories revealed both interdigitating and overlapping patch systems, depending on the relation between the functional properties of the cells at the two injection sites (Matsubara et al., 1987; Schmidt et al., 1997a). These observations indicate a rather continuous system of patchy connections between subsets of orientation-selective neurons.

POSSIBLE FUNCTIONS

Long-range horizontal connections in the primary visual cortex span a cortical region much larger (up to 8 mm) than that corresponding to the classical receptive field of a neuron (the part of visual space that must be stimulated to evoke a rate modulation of a neuron's spiking activity). This fact, along with what is known about their specific functional topography, suggests that horizontal connections may be important for the integration of information from widely distant points in the visual field and for context-dependent modifications of neuronal responsivity. Contrary to long-standing belief, recent physiological and anatomical evidence suggests that part of this process of spatial integration occurs at the level of the primary visual cortex. The relevance of this "early stage" modulation for further visual processing is exemplified by links to perceptual phenomena. It has been known since the work of the Gestalt psychologists that certain grouping criteria, such as neighborhood, color, direction of motion, velocity, orientation, and colinearity, make contours perceptually more salient (Wertheimer, 1938; Grossberg and Mignolla, 1985; Field et al., 1993). The topology, excitatory nature, and receptive field properties of the interconnected neurons make long-range intrinsic connections ideally suited to serve as a substrate for perceptual grouping by enhancing the saliency of adjacent, colinear, and similarly orientated contours that are processed by distributed neurons with nonoverlapping classical receptive fields. The following example demonstrates how the Gestalt criterion of colinearity supports perceptual grouping.

Figure 10.9 illustrates that our visual system tends to group contour segments into one coherent figure if they have the same orientation and are additionally colinearly aligned along the contour path rather than orthogonal to it (Field et al., 1993; Polat and Sagi, 1993, 1994; Kapadia et al., 1995; Schmidt et al., 1997b). To achieve this, neuronal responses to colinearly aligned contours must be distinguished from physically identical but noncolinearly arranged contours in a context-dependent way. Obviously, both the perceived location and the form of the individual contour elements are not influenced by the context. The only information that can be used for the context-dependent processing is the topological relation between the location and orientation of the contour elements. Selection of elements composing the object, therefore, must be done at a level of visual processing where neurons have orientation-selective receptive fields whose size is equal to or smaller than that of the constituting elements of the picture. Such receptive fields are present in early visual areas such as area 17 (V1) and area 18 (V2). Long-range intrinsic connections in those areas are well suited to perform

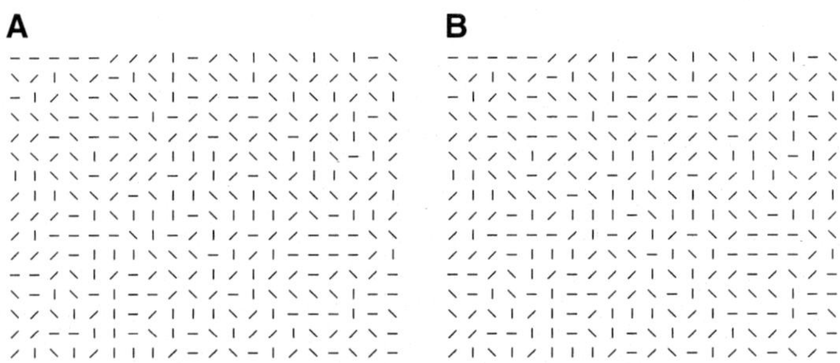

FIGURE 10-9. An example of perceptual grouping on the basis of colinearity. The colinearly arranged line segments in **(A)** that define the edges of a rhombus are grouped together and stand out from the randomly distributed line segments of the background. **(B)** The line segments are arranged in the same way as in **(A)** except those that defined the outline of the rhombus, which are now parallel to each other and hence orthogonal to the corresponding segments in **(A)** (Modified from Schmidt et al. [1997b]. The perceptual grouping criterion of colinearity is reflected by anisotropies of connections in the primary visual cortex, *European Journal of Neuroscience,* with permission from Blackwell Science Ltd.)

the context-dependent grouping operations given their functional architecture. Because they extend between neurons of similar orientation preference and colinearly aligned receptive fields, they could selectively enhance the saliency of neuronal responses to distributed elements of an object such as that shown in Fig. 10.9, according to criteria such as colinearity and orientation.

In a similar way, long-range intrinsic connections could be thought as being the substrate of other psychophysical phenomena in which perception is strongly influenced by the context in which objects are embedded (Gibson and Radner, 1937; Westheimer et al., 1976; Badcock and Westheimer, 1985; Westheimer, 1986; Polat and Sagi, 1993, 1994). They could enable the visual system to identify simple features from a noisy background (Wertheimer, 1938; Dresp, 1993; Grossberg and Mingolla, 1985; Field, et al., 1993; Kapadia et al., 1995) at an early level of cortical visual processing. In principle, long-range intrinsic connections can also contribute to the emergence of illusory contours and mediate perceptual filling-in of artificial or real scotomas (Kanizsa, 1979; Yarbus, 1957; Krauskopf, 1961; Crane and Piantanida, 1983; Ramachandran and Gregory, 1991). At present, however, the importance of long-range connections for the latter psychophysical phenomena is purely speculative. Many aspects of scene segmentation are probably also influenced by feedback connections from so-called higher cortical areas in which receptive fields are usually larger than in V1. It is likely that scenes are segmented as a result of a dynamic interplay of sensory input and intrinsic and feedback projections. In particular, when ambiguities in a visual scene cannot be solved by evaluating the local relations of an object's elements, feedback from areas with larger receptive fields is likely to play an important role.

Taken together, available evidence supports the hypothesis that long-range intrinsic connections in primary visual cortex are the anatomical substrate of context-dependent modifications and thus are important for visual scene analysis. It follows that the criteria for perceptual grouping might be partially determined by the functional architecture of these connections. Evidence in favor of this hypothesis comes from investigations in strabismic cats in which the changed perceptual capabilities are associated with both a change in the layout of the long-range intrinsic network (Löwel and Singer, 1992; Schmidt et al., 1997a) and modified neuronal interactions (König et al., 1993; for a review see Löwel and Singer, 2000).

Several electrophysiological findings in V1 have been attributed to intrinsic excitatory and inhibitory connections.

1. Besides the thalamocortical and vertically running intracortical circuits, they have been ascribed to contribute to orientation and direction selectivity (Eysel et al., 1987, 1990; Crook et al., 1997, 1998).
2. Mirroring spatial integration on the physiological level and hardly explainable by other cortical circuits are the generation of large composite receptive fields within V1 under certain circumstances (Singer and Tretter, 1976; Gilbert an Wiesel, 1985; Bolz and Gilbert, 1990; Schwarz and Bolz, 1991).

3. Also in this regard are the direct inhibitory and substreshold excitatory effects from beyond the classical receptive field (Blakemore and Tobin, 1971; Nelson and Frost, 1978; Morrone et al., 1982; Allman et al., 1985).
4. Similarly, orientation specific intrinsic connections are also a likely candidate for the synchronization of responses of spatially separated neurons as a function of stimulus coherence (Gray et al., 1989).
5. They may also be responsible for adaptive topographic reorganization of cortical maps after deafferentiation where the effects are too large to be mediated by rearrangements of the thalamocortical input (Kaas et al., 1990; Heinen and Skavenski, 1991; Gilbert and Wiesel, 1992; Darian-Smith and Gilbert, 1994; see also Florence et al., 1998).

PLASTICITY OF LONG-RANGE CONNECTIONS IN THE ADULT

The functional role of intrinsic connections in a particular situation may vary depending on the time course of the mediated function and the modifiability of the connections. Perceptual decisions probably operate on a millisecond time scale. The filling-in of an artificial scotoma requires longer time periods and might be followed by processes summarized by the term *learning*. Complete removal of the visual input leads to larger scale topographic changes in cortical maps with a delay of days to weeks.

During postnatal development intrinsic long-range connections are modifiable, whereby both spontaneous neuronal activity and sensory experience have been shown to play crucial roles in shaping the network (Callaway and Katz, 1991; Katz and Callaway, 1992; Ruthazer and Stryker, 1996; for review see Schmidt et al., 1999; Löwel and Singer, 2000). Unexpected just a few years ago, anatomical evidence now shows that these connections also remain plastic during adult life. Learning-induced or use-dependent changes may be mediated not only by long-range intrinsic connections but also by thalamocortical, local, feedforward, and feedback interareal connections. However, as outlined in the previous sections, their extent and characteristic functional topography make long-range intrinsic connections well suited to transmit feature-specific information over a large region of visual space in primary visual cortex. The next section reviews anatomical and physiological evidence indicating that long-range intrinsic connections in primary sensory cortices mediate both short-term dynamics and long-term modifications (of receptive fields) involving permanent reorganization of cortical representations.

SHORT-TERM DYNAMIC CHANGES OF RECEPTIVE FIELD SIZE AND LOCATION

Based on a number of studies showing that receptive fields can change within minutes to hours, the receptive field of a neuron is understood as a highly dynamic rather than a fixed neuronal property. Pettet and Gilbert (1992) stimulated primary visual cortical neurons with an artificial scotoma (a mask) covering their receptive field. The surround of the visual stimulus consisted of moving oriented bars. In cases in which the orientation of the bars matched the orientation preference of the neuron, the original receptive field size expanded up to fivefold within minutes and collapsed again when the center of the scotoma was stimulated with a small oriented stimulus. Because an important prerequisite of the receptive field expansion was the stimulation of the surround, it was concluded that dynamic changes of receptive field properties are context-dependent. The dynamic receptive field expansion of a binocular neuron can also be induced by stimulation of the nonconditioned eye (Volchan and Gilbert, 1995). Because afferents of the two eyes first converge onto single cells in the primary visual cortex, this was assumed to be the neural site of the dynamic changes. Also, because the effect was dependent on stimulus orientation, long-range intrinsic connections were discussed as a likely candidate to mediate the contextual influences.

To test whether dynamic short-term plasticity indeed involved lateral connections Das and Gilbert (1995a) quantified the strength of intracortical connections between two conditioned neurons. In their experiment, a mask covered the receptive fields of two neurons with similar response properties. Cross-correlograms were computed from spike trains evoked by a moving bar covering a common part of both receptive fields. Concomitant with the expansion of receptive field sizes, the effective connectivity between the coconditioned neurons increased and returned to baseline when the region of the artificial scotoma was visually stimulated. Other receptive field parameters such as orientation preference did not change. No significant cross-correlation developed between neurons that did not correlate before conditioning either because of a large spatial separation or a large difference in orientation preference. These data indicate that the observed modifications occurred within the existing framework of intrinsic long-range connections and are probably based on changes in synaptic efficacy initiated by phenomena such as long-term potentiation (LTP) or long-term depression (see page 415).

LONG-TERM CHANGES OF RECEPTIVE FIELDS AND CORTICAL MAPS

First evidence for adult cortical plasticity came from studies of the somatosensory system by Merzenich and colleagues. Cutting the median nerve in a monkey lead to expansion of the representation of surrounding skin surfaces into the regions formerly representing the denervated parts (Merzenich et al., 1983a, 1983b). Thereby, the cortical territory receiving input from the remaining intact skin parts adjacent to the deafferented field was expanded. The representation of some skin surfaces were even shifted to completely new locations. At the same time, the receptive field sizes decreased as if to compensate for the magnification factor of the newly gained representation. Similarly, in the motor cortex, amputation of one digit led to occupation of the initially silenced cortical region by inputs from adjacent digits (Merzenich et al., 1984).

Further experiments demonstrated that representational plasticity did not necessarily require amputation. Artificially fusing two independent fingers (syndactyly) so that they could only be used together caused the border of representations to smear. In addition, responses to movements of either digit could be elicited in both of the formerly strictly separated cortical representations (Clark et al., 1988). When adult owl monkeys were trained to use a particular digit repetitively for a behavioral task that was repeated several thousand times, the cortical representation of that digit expanded at the cost of the other, less often used digits (Jenkins et al., 1990). Thus, practice alone was sufficient to enlarge the region of cortex containing neurons that were activated during the repetitive behavioral task. This emphasizes the highly dynamic nature of cortical representations, even in the adult brain. Moreover, the observations indicate that perceptual processes themselves must be based on and supported by dynamic neuronal circuitries. In summary, basic features of somatosensory and motor cortical maps, such as receptive field sizes, representation of skin surfaces, and the boundaries between different submodalities, are dynamically maintained throughout life in a use-dependent way.

In the visual system of cats, adult plasticity was first observed in the lateral geniculate nucleus (LGN). Receptive fields of neurons located in the part of an LGN-layer corresponding to a monocular retinal lesion were displaced compared to the normal topography in the other (intact eye's) layer (Eysel et al., 1980, 1981). About a month or more after lesioning, the silenced geniculate cells shifted their receptive field center up to 5° of visual field angle into the immediate surround of the retinal lesion, depending on the eccentricity of the lesion. Excitatory spread within the retina from intact to lesioned areas was histologically excluded. The visual cortex as the origin of a new input via corticofugal connections could also be discarded on the basis of latency measurements and the nature of the responses. However, the optic tract fibers displayed considerable overlap within a LGN layer. Therefore, the lateral spread of activity accounting for the changes was assumed to result from intralaminar reorganization within the LGN. As a possible mechanism, an unmasking of previously ineffective excitatory inputs and/or intralaminar collateral sprouting was considered (Eysel et al., 1980).

It is now well established that representations in the primary visual cortex can also be reorganized reversibly as well as permanently after visual conditioning, restricted deafferentiation, or electrical stimulation. Visual cortical map reorganization was first demonstrated after retinal lesions in both monkey V1 and cat area 17. When V1/area 17 neurons were deprived of all their normal feedforward input by binocular lesions affecting corresponding retinal loci, they acquired new (or modified) receptive fields driven by inputs from intact retinal regions surrounding the lesions (cat: Kaas et al., 1990; Gilbert and Wiesel, 1992; Chino et al., 1992, 1995; Darian-Smith and Gilbert, 1995; monkey: Heinen and Skavenski, 1991; for review, see Kaas, 1991; Buonomano and Merzenich, 1998; Gilbert, 1998). Other reports demonstrated that reorganization also occurs after monocular lesions, indicating that both eyes are capable of inducing cortical plasticity independently

(Schmid et al., 1996; Calford et al., 1999). Rearranged fields typically appeared at the edge of the scotoma.

In the primary visual cortex of cats, substantial reorganization on a small spatial scale occurred within hours after the lesions (Chino et al., 1992; Gilbert and Wiesel, 1992), similar to observations in the somatosensory system (Calford and Tweedale, 1988; Kaas, 1991; Kelahan and Doetsch, 1984; Merzenich et al., 1983a, 1983b; Wall and Cusick, 1984), whereby receptive fields adjacent to the lesion boundary increased in size and shifted their location, but neuronal responses were sluggish and fatigued easily (Darian-Smith and Gilbert, 1995). After longer recovery periods of more than 2 to 3 months, between about 50% (Heinen and Skavenski, 1991) and 100% of all neurons located in an initially silenced cortical region (measuring up to 6–10 mm in diameter) acquired new receptive fields if the retinal lesion covered a visual field angle smaller than 5° (Chino et al., 1995; Darian-Smith and Gilbert, 1995). In the long run, orientation and direction selectivity recovered to normal levels, but maximal response amplitudes and contrast thresholds (= responsiveness) remained reduced (Chino et al., 1995). In addition, recovered receptive fields were sometimes larger and less defined (Heinen and Skavenski, 1991; Gilbert and Wiesel, 1992).

As originally shown in monkey and human motor cortex (Clark et al., 1988; Jenkins et al., 1990; Karni et al., 1995), representational plasticity in the visual system can also be produced by long-lasting activity changes without peripheral (retinal) lesions. Sugita (1996) fitted adult monkeys with prisms reversing the visual field. After a few months, neurons in the primary visual cortex developed novel receptive fields in the ipsilateral hemifield that normally only activates neurons in the contralateral cortex. These results indicated that (1) visual cortical neurons can acquire new inputs from distant areas (probably involving callosal connections) as well as from neighboring retinal areas and (2) visual input changes—not necessarily lesions—are sufficient to induce rearrangements in the visual field map.

WHERE DOES REORGANIZATION TAKE PLACE?

One of the most interesting questions resulting from these studies is: Where in the brain does reorganization take place? In the case of retinal lesions that cover all layers including the ganglion cell layer, a modification at the retinal level is unlikely to contribute to large-scale cortical rearrangements. There is limited evidence for a contribution of thalamic nuclei and spreading corticothalamic connections to cortical reorganization in the adult somatosensory and motor system (Rasmusson and Nance, 1986; Wells and Tripp, 1987; Garraghty and Kaas, 1991). After ligating the ulnar and median nerve in a squirrel monkey, Garraghty and Kaas (1991) observed a reorganization in the ventrobasal complex of the thalamus corresponding to the cortical changes. On the other hand, a more recent study demonstrated that a large part of thalamic plasticity in the somatosensory system is cortically mediated (Krupa et al., 1999; for a review see Kaas, 1999). In the "visual" thalamus, the lateral geniculate nucleus, the actual scotoma remained silent even months after the retinal lesion and thus at a time at which cortical reorganization was already completed (Gilbert and Wiesel, 1992; Darian-Smith and Gilbert, 1995). Anatomical analysis demonstrated that the thalamocortical afferents did not sprout into the cortical scotoma (Darian-Smith and Gilbert, 1995). Because the physical extension of geniculate arbors within the cortex is in the range of 1 to 3 mm (Humphrey et al., 1985), these arbors may account for short-term changes, which are in the range of 3 mm (Chino et al., 1992). Although the lateral geniculate nucleus itself exhibits a limited increase of its lateral excitatory spread in response to removal of visual input (up to 250 μm beyond normal; Eysel et al., 1980), this fact alone cannot explain cortical changes observed after long-term deafferentiation.

Neurons in the cortical scotoma can display new receptive fields separated by 6 to 10 mm in cortical distance from their original position (Kaas et al., 1990, Gilbert and Wiesel, 1992; Darian-Smith and Gilbert, 1995). Deafferentiating the upper limb in a monkey also led to cortical reorganization over more than 10 mm (Pons et al., 1991). The only known system that could mediate these large-scale modifications are the long-range intracortical connections (Gilbert and Wiesel, 1979, 1989, Luhmann et al., 1991; Katz and Callaway, 1992; Kisvárday et al., 1992, 1997; Galuske and Singer, 1996; for review see Löwel and Singer, 2000). Some of the characteristics displayed by reorganized receptive fields match features of the functional layout of these connections. Two remote parts of one receptive field that were separated by 10° of visual field angle by the scotoma displayed the same orientation and spatial frequency tuning (Chino et al., 1995). Similar observation were made by Das and Gilbert (1995b) using optical imaging of intrinsic signals to determine long-term adaptive changes after retinal lesions. The area of cortex that remained silent in the optical images immediately after the lesion recovered after 5 months. At that time, receptive fields of the neurons in the initially silenced area had shifted their receptive fields to outside the scotoma, but the orientation preference map within the cortical scotoma resembled the map recorded before lesioning the retina. Because neurons in this part of the cortical map do not receive feed-forward input (input from the retina), the recorded responses must be mediated by cortical inputs. The pattern of reorganization reflecting the extent and specificity of long-range horizontal connections suggests that they are the neuronal substrate of the plastic modifications. It cannot be excluded that feedback connections also take part in the reorganization of primary visual cortex; however, the observations that (1) most of the reorganized neuronal responses, although sluggish, were typical for V1 receptive fields and (2) feed-forward input to higher cortical areas is silenced at topographical locations corresponding to the scotoma site make this possibility rather unlikely.

SHORT-TERM CELLULAR MECHANISMS AND LONG-TERM REARRANGEMENTS OF LONG-RANGE INTRINSIC CONNECTIONS

Short-term plastic changes may be triggered by dynamic adjustments of excitation and inhibition and are most probably achieved by rapid modulations of the effectiveness of preexisting connections, whereby subthreshold excitatory inputs may become suprathreshold by either suppressing inhibitory or potentiating excitatory connections (Chino et al., 1992; Gilbert and Wiesel, 1992). The efficacy of existing synapses can be strengthened by mechanisms such as LTP in a use-dependent manner according to Hebbian rules (Hebb, 1949). On a longer time scale, new synapses can be formed, thereby establishing entirely new connections.

Immediately after a retinal lesion, excitability in a local region surrounding the cortical scotoma was increased (Eysel et al., 1999b), whereas an imbalance between excitation and inhibition was observed 2 weeks later (Rosier et al., 1995; Arckens et al., 1997). These rather acute changes disappeared after longer postlesion periods. Cortical lesions evoke qualitatively similar rearrangements in the cortical region outside the lesion as peripheral deafferentiation (Eysel and Schweigart, 1999). Specifically, NMDA receptor-mediated excitatory postsynaptic potentials, known to be involved in synaptic plasticity, were increased in the region surrounding the scotoma (Mittmann et al., 1994). LTP involving NMDA receptors has been demonstrated in visual cortical slices of both cats (Hirsch and Gilbert, 1993) and rats (Artola and Singer, 1987; Artola et al., 1990; Kirkwood and Bear, 1994).

A number of findings link the phenomenon of synaptic plasticity at the cellular level with that of topographical reorganization at the cortical map level. In cat visual cortex, a variety of conditioning paradigms (pairing protocols) demonstrated that use-dependent synaptic changes can be elicited in vivo. The orientation selectivity of visual cortical neurons was shifted by pairing the presentation of oriented bars with either iontophoretically applied currents (Frégnac et al., 1988, 1992), direct glutamate application (Greuel et al., 1988), or, more recently, with electrical stimulation of the reticular formation (Galuske et al., 1997). Moreover, coupling strength measured as the cross-correlation peak between neurons with nonoverlapping receptive fields was also enhanced by pairing visual stimulation with electrical stimulation of the reticular formation, which is known to increase cortical acetylcholine levels and facilitate use-dependent synaptic gain changes (Herculano-Houzel et al., 1999). Modifications of orientation preference maps in adult cats were induced by pairing the presentation of oriented whole-field gratings with reticular stimulation (Galuske et al., 1997). Within hours, iso-orientation domains preferring the conditioned orientation expanded at the cost of nonconditioned domains, whereas conditioning without pairing induced only large-scale habituation. Similar changes in orientation maps were observed after intracortical microstimulation (Godde and Dinse, 1998, 1999). Although, the majority of these changes were within the range of the local connectivity, the involved mechanisms might also apply for long-range synaptic gain changes.

One possibility to translate short-term synaptic gain changes into long-term alterations of cortical topography is the outgrowth of new connections, a characteristic phenomenon of early postnatal development (for review see Schmidt et al., 1999; Löwel and Singer, 2000). As early as 1960, experiments in adult rabbits demonstrated the ingrowth of horizontal fibers into destroyed cortical laminae that were devoid of surviving cells (Rose et al., 1960). More than 30 years later, morphological changes of long-range intrinsic connections were demonstrated in adult cats by using extracellular biocytin injections into cortex just outside the boundary of the original cortical scotoma induced by binocular retinal lesions. The comparison of the density of lateral projections into reorganized and nondeprived visual cortex revealed that axon collaterals from cortical neurons surrounding the visual cortical scotoma predominantly branched into the deprived as opposed to the normal cortical area (Darian-Smith and Gilbert, 1994, 1995). Axon fibers were 57–88% denser within reorganized than within normal cortex. The increase in connectivity was accompanied by a reduced bouton density per single axon in the reorganized region. Because the newly grown fibers did not extend beyond 4 mm, it was concluded that reorganization is mediated by modifications in the preexisting framework of long-range intrinsic connections rather than by an extension of fibers beyond their normal level (Darian-Smith and Gilbert, 1994). Similarly, functional reorganization in the rat motor cortex was mediated and constrained by the given layout of intrinsic connectivity (Huntley, 1997). However, a recent study in the somatosensory system indicated that sprouting may occur beyond the framework of pre-existing connections. By analyzing the distributions of thalamic and cortical connections in macaque monkeys with long-standing accidental trauma to a forelimb, it was shown that thalamocortical projections were relatively normal, whereas connections in the somatosensory cortex (area 3b and 1) were markedly more widespread than in normal animals (Florence et al., 1998).

Thus outgrowth of new connections, previously thought to be restricted to so-called "critical periods" in early life, also occurs in the adult brain and most probably mediates long-term long-range representational plasticity in the cortex.

CONCLUSIONS

1. The majority of long-range intrinsic connections of excitatory cortical neurons in the primary visual cortex of cats link neurons with similar response properties (e.g., orientation preference and possibly also direction preference) according to the rule: "like connects to like."
2. There is extensive cross-talk between different orientation domains.
3. There are clear differences in the layout of long-range excitatory and inhibitory connections.
4. The selectivity of the long-range connections may depend on the relative position of a neuron within a functional map (e.g., projections extending from pinwheel centers are less orientation selective than projections originating from linear zones of an orientation preference map).
5. Both the selectivity and maximal lateral extent of long-range connections depend on the layer (layer IV connections are probably less orientation-selective than supragranular and infragranular projections and much shorter, about 50%, than layer III connections).

6. It may well be that the functional specificity of long-range intrinsic connections depends on a variety of functional properties, of which only a minority have been analyzed to date.
7. Many different functions have been ascribed to long-range intrinsic connections. In particular, there is growing evidence that they are important for perceptual integration.

ACKNOWLEDGMENTS

We thank Peter Buzás, William Bosking, Kevan Martin, Rainer Goebel, Kathleen Rockland and Zoltan Kisvárday for providing us with original versions and/or permission for reproduction of their published figures. We also thank Ralf Galuske and James Waltz for helpful comments on the manuscript.

REFERENCES

Albus, K., Wahle, P., Lübke, J., and Matute, C. (1991). The contribution of GABA-ergic neurons to horizontal intrinsic connections in upper layers of the cat's striate cortex. *Exp. Brain Res.* **85,** 235–239.

Albus, K., and Wahle, P. (1994). The topography of tangential inhibitory connections in the postnatally developing and mature striate cortex of the cat. *Eur. J. Neurosci.* **6,** 779–792.

Allman, J., Miezin, F., and McGuiness, E. (1985). Stimulus specific responses from beyond the classical receptive field: neurophysiological mechanisms for local-global comparisons in visual neurons. *Annu. Rev. Neurosci.* **7,** 407–430.

Amir, Y., Harel, M., and Malach, R. (1993). Cortical hierarchy reflected in the organization of intrinsic connections in macaque monkey visual cortex. *J. Comp. Neurol.* **334,** 19–46.

Arckens, L., Qu, Y., Wouters, G., Pow, D., Eysel, U. T., Orban, G., and Vandesande, F. (1997). Changes in glutamate immunoreactivity during retinotopic reorganization of cat striate cortex. *Soc. Neurosci. Abstr.* **23,** 2362.

Artola, A., and Singer, W. (1987). Long-term potentiation and NMDA receptors in rat visual cortex. *Nature* **330,** 649–652.

Artola, A., Bröcher, S., and Singer, W. (1990). Different voltage-dependent thresholds for inducing long-term depression and long-term potentiation in slices of rat visual cortex. *Nature* **347,** 69–72.

Badcock, D. R., and Westheimer, G. (1985). Spatial location and hyperacuity: the center-surround localization has two subsrates. *Vis. Res.* **25,** 1259–1269.

Blakemore, C., and Tobin, E. A. (1971). Lateral inhibition between orientation detectors in the cat's visual cortex. *Exp. Brain Res.* **15,** 439–440.

Bolz, J., and Gilbert, C. D. (1990). The role of horizontal connections in generating long receptive fields in the cat visual cortex. *Eur. J. Neurosci.* **1,** 263–268.

Bonhoeffer, T., and Grinvald, A. (1996). Optical imaging based on intrinsic signals: the Methodology. In: Brain mapping: The methods (A. Toga, and J. C. Mazziotta, Eds.), pp. 55–97. San Diego, Academic Press.

Bosking, W. H., Zhang, Y., Schofield, B., and Fitzpatrick, D. (1997). Orientation selectivity and the arrangement of horizontal connections in tree shrew striate cortex. *J. Neurosci.* **17,** 2112–2127.

Boyd, J., and Matsubara, J. (1994). Tangential organization of callosal connectivity in the cat's visual cortex. *J. Comp. Neurol.* **347,** 197–210.

Boyd, J. D., and Matsubara, J. A. (1996). Laminar and columnar patterns of geniculocortical projections in the cat: relationship to cytochrome oxidase. *J. Comp. Neurol.* **365,** 659–682.

Boyd, J. D., and Matsubara, J. A. (1999). Projections from V1 to lateral suprasylvian cortex: an efferent pathway in the cat's visual cortex that originates preferentially from CO blob columns. *Vis. Neurosci.* **16,** 849–860.

Braak, H. (1984). Architectonics as seen by lipofuscin stains. In: Cerebral cortex (A. Peters, and E. G. Jones, Eds.), Vol. 1, pp. 59–104. New York, Plenum Press.

Braitenberg, V. (1962). A note on myeloarchitectonics. *J. Comp. Neurol.* **118,** 141–151.

Bullier, J., Kennedy, H., and Salinger, W. (1984). Branching and laminar origin of projections between visual cortical areas in the cat. *J. Comp. Neurol.* **228,** 329–341.

Buonomano, D. V., and Merzenich, M. M. (1998). Cortical plasticity: from synapses to maps. *Annu. Rev. Neurosci.* **21,** 149–186.

Burkhalter, A., and Bernardo, K. L. (1989). Organization of corticocortical connections in human visual cortex. *Proc. Natl. Acad. Sci. U.S.A.* **86,** 1071–1075.

Buzás, P., Eysel, U. T., and Kisvárday, Z. F. (1998). Functional topography of single cortical cells: an intracellular approach combined with optical imaging. *Brain Res. Brain Res. Protoc.* **3,** 199–208.

Calford, M. B., and Tweedale, R. (1988). Immediate and chronic changes in responses of somatosensory cortex in adult flying-fox after digit amputation. *Nature* **332,** 446–448.

Calford, M. B., Schmid, L. M., and Rosa, M. G. (1999). Monocular focal retinal lesions induce short-term topographic plasticity in adult cat visual cortex. *Proc. R. Soc. Lond. B. Biol. Sci.* **266,** 499–507.

Callaway, E. M., and Katz, L. C. (1990). Emergence and refinement of clustered horizontal connections in cat striate cortex. *J. Neurosci.* **10,** 1134–1153.

Callaway, E. M., and Katz, L. C. (1991). Effects of binocular deprivation on the development of clustered horizontal connections in cat striate cortex. *Proc. Natl. Acad. Sci. U.S.A.* **88,** 745–749.

Casagrande, V. A. (1994). A third parallel visual pathway to primate area V1. *Trends Neurosci.* **17,** 305–310.

Chino, Y. M., Kaas, J. H., Smith, E. L., Langston, A. L., and Cheng, H. (1992). Rapid reorganization of cortical maps in adult cats following restricted deafferentiation in retina. *Vis. Res.* **32,** 789–796.

Chino, Y. M., Smith, E. L., Kaas, J. H., Sasaki, Y., and Cheng, H. (1995). Receptive field properties of deafferentiated visual cortical neurons after topographic map reorganization in adult cats. *J. Neurosci.* **15,** 2417–2433.

Clark, S. A., Allard, T., Jenkins, W. M., and Merzenich, M. M. (1988). Receptive fields in the body-surface map in adult cortex defined by temporally correlated inputs. *Nature* **332,** 444–445.

Clark, W. E. L., and Sunderland, S. (1939). Structural changes in the isolated visual cortex. *J. Anat.* **73,** 563–574.

Crane, H. D., and Piantanida, T. P. (1983). On seeing reddish green and yellowish blue. *Science* **221,** 1078–1079.

Creutzfeldt, O. D., Garey, L. J., Kuroda, R., and Wolff, J.-R. (1977). The distribution of degenerating axons after small lesions in the intact and isolated visual cortex of the cat. *Exp. Brain Res.* **27,** 419–440.

Crook, J. M., Kisvárday, Z. F., and Eysel, U. T. (1997). GABA-induced inactivation of functionally characterized sites in cat striate cortex: effects on orientation tuning and direction selectivity. *Vis. Neurosci.* **14,** 141–158.

Crook, J. M., Kisvárday, Z. F., and Eysel, U. T. (1998). Evidence for a contribution of lateral inhibition to orientation tuning and direction selectivity in cat visual cortex: reversible inactivation of functionally characterized sites combined with neuroanatomical tracing techniques. *Eur. J. Neurosci.* **10,** 2056–2075.

Cynader, M. S., Swindale, N. V., and Matsubara, J. A. (1987). Functional topography in cat area 18. *J. Neurosci.* **7,** 1401–1413.

Darian-Smith, C., and Gilbert, C. D. (1994). Axonal sprouting accompanies functional reorganization in adult cat striate cortex. *Nature* **368,** 737–740.

Darian-Smith, C., and Gilbert, C. D. (1995). Topographic reorganization in the striate cortex of the adult cat and monkey is cortically mediated. *J. Neurosci.* **15,** 1631–1647.

Das, A., and Gilbert, C. D. (1995a). Receptive field expansion in adult visual cortex is linked to dynamic changes in strength of cortical connections. *J. Neurophysiol.* **74,** 779–792.

Das, A., and Gilbert, C. D. (1995b). Long-range horizontal connections and their role in cortical reorganization revealed by optical recording of cat primary visual cortex. *Nature* **375,** 780–784.

DeFelipe, J., Hendry, S. H. C., and Jones, E. G. (1986). A quantitative electron microscopic study of basket cells and large GABAergic neurons in the monkey sensory-motor cortex. *Neuroscience* **17,** 991–1009.

Dow, B. M., Vautin, R. G., and Bauer, R. (1985). The mapping of visual space onto foveal striate cortex in the macaque monkey. *J. Neurosci.* **5,** 890–902.
Dresp, B. (1993). Bright lines and edges facilitate the detection of small line targets. *Spatial Vision* **7,** 213–225.
Durack, J. C., and Katz, L. C. (1996). Development of horizontal projections in layer 2/3 of ferret visual cortex. *Cerebral Cortex* **6,** 178–183.
Eysel, U. T., Gonzalez-Aguilar, F., and Mayer, U. (1980). A functional sign of reorganization in the visual system of adult cats: lateral geniculate neurons with displaced receptive fields after lesions of the nasal retina. *Brain Res.* **181,** 285–300.
Eysel, U. T., Gonzalez-Aguilar, F., and Mayer, U. (1981). Time-dependent decrease in the extent of visual deafferentation in the lateral geniculate nucleus of adult cats with small retinal lesions. *Exp. Brain Res.* **41,** 256–263.
Eysel, U. T., Wörgötter, F., and Pape, H.-C. (1987). Local cortical lesions abolish lateral inhibition at direction selective cells in cat visual cortex. *Exp. Brain Res.* **68,** 606–612.
Eysel, U. T., Crook, J. M., and Machemer, H. F. (1990). GABA-induced remote inactivation reveals cross-orientation inhibition in the cat striate cortex. *Exp. Brain Res.* **80,** 626–630.
Eysel, U. T., and Schweigart, G. (1999). Reorganization of receptive fields at the border of chronic visual cortical lesions. *Cerebral Cortex* **9,** 101–109.
Eysel, U. T., Crook, J. M., and Kisvárday, Z. (1999a). Sehen, wie das Gehirn sieht. *Physiologie* **13,** 12–20.
Eysel, U. T., Schweigart, G., Mittmann, T., Eyding, D., Qu, Y., Vandesande, F., Orban, G., and Arckens, L. (1999b). Reorganization in the visual cortex after retinal and cortical damage. *Restor. Neurol. Neurosci.* **15,** 153–164.
Ferrer, J. M., Price, D. J., and Blakemore, C. (1988). The organization of corticocortical projections from area 17 to area 18 of the cat's visual cortex. *Proc. R. Soc. Lond. B. Biol. Sci.* **233,** 77–98.
Ferrer, J. M., Kato, N., and Price, D. J. (1992). Organization of association projections from area 17 to areas 18 and 19 and to suprasylvian areas in the cat's visual cortex. *J. Comp. Neurol.* **316,** 261–278.
Field, A. J., Hayes, A., and Hess, A. F. (1993). Contour integration by the human visual system: evidence for a local "association field." *Vis. Res.* **33,** 171–193.
Fisken, R. A., Garey, L. J., and Powell, T. P. S. (1975). The intrinsic, association and commissural connections of area 17 of the visual cortex. *Philos. Trans. R. Soc. London Ser. B* **272,** 487–536.
Fitzpatrick, D., Itoh, K., and Diamond, I. T. (1983). The laminar organization of the lateral geniculate body and the striate cortex in the squirrel monkey *(Saimiri sciureus). J. Neurosci.* **3,** 673–702.
Fitzpatrick D. (1996). The functional organization of local circuits in visual cortex: insights from the study of tree shrew striate cortex. *Cerebral Cortex* **6,** 329–341.
Florence, S. L., Taub, H. B., and Kaas, J. H. (1998). Large-scale sprouting of cortical connections after peripheral injury in adult macaque monkeys. *Science* **282,** 1117–1121.
Frégnac, Y., Shulz, D., Thorpe, S., and Bienenstock, E. (1988). A cellular analogue of visual cortical plasticity. *Nature* **333,** 367–370.
Frégnac, Y., Shulz, D., Thorpe, S., and Bienenstock, E. (1992). Cellular analogs of visual cortical epigenesis. I. Plasticity of orientation selectivity. *J. Neurosci.* **12,** 1280–1300.
Gabbott, P. L., and Somogyi, P. (1986). Quantitative distribution of GABA-immunoreactive neurons in the visual cortex (area 17) of the cat. *Exp. Brain Res.* **61,** 323–331.
Galuske, R. A., and Singer W. (1996). The origin and topography of long-range intrinsic projections in cat visual cortex: a developmental study. *Cerebral Cortex* **6,** 417–430.
Galuske, R. A. W., Singer, W., and Munk, M. H. J. (1997). Reticular activation facilitates use-dependent plasticity of orientation preference maps in the cat visual cortex. *Soc. Neurosci. Abstr.* **23,** 2059.
Galuske, R. A. W., Schlote, W., Bratzke, H., and Singer, W. (2000). Interhemispheric asymmetry of the modular structure in human temporal cortex. *Science,* **289,** 4946–4945.
Garraghty, P. E., and Kaas, J. H. (1991). Functional reorganization in adult monkey thalamus after peripheral nerve injury. *Neuroreport* **2,** 747–750.
Gennari, F. (1782). *De Peculiari Structura Cerebri Nonnulisque Ejus Morbis. Pacae aliae. Anatom. Observ. Accedunt.* Parma, Regio Typographeo.

Gibson, J. J., and Radner, M. (1937). Adaptation, after-effects and contrasts in the perception of tilted lines. *J. Exp. Psychol.* **20,** 453–467.

Gilbert, C. D., and Kelly, J. P. (1975). The projections of cells in different layers of the cat's visual cortex. *J. Comp. Neurol.* **163,** 81–105.

Gilbert, C. D., and Wiesel, T. N. (1979). Morphology and intracortical projections of functionally characterized neurons in the cat visual cortex. *Nature* **280,** 120–125.

Gilbert, C. D., and Wiesel, T. N. (1981). Laminar specialization and intracortical connections in cat primary visual cortex. In: The organization of the cerebral cortex (F. O. Schmitt, F. G. Worden, G. Edelman, and S. G. Dennis, Eds.), pp. 163–191. Cambridge, Mass. and London, MIT Press.

Gilbert, C. D. (1983). Microcircuitry of the visual cortex. *Annu. Rev. Neurosci.* **6,** 217–247.

Gilbert, C. D., and Wiesel, T. N. (1983). Clustered intrinsic connections in cat visual cortex. *J. Neurosci.* **3,** 1116–1133.

Gilbert, C. D., and Wiesel, T. N. (1985). Intrinsic connectivity and receptive field properties in visual cortex. *Vis. Res.* **25,** 365–374.

Gilbert, C. D., and Wiesel, T. N. (1989). Columnar specificity of intrinsic horizontal and corticocortical connections in cat visual cortex. *J. Neurosci.* **9,** 2432–2442.

Gilbert, C. D., and Wiesel, T. N. (1992). Receptive field dynamics in adult primary visual cortex. *Nature* **356,** 150–152.

Gilbert, C. D. (1998). Adult cortical dynamics. *Physiol. Rev.* **78,** 467–485.

Godde, B., and Dinse, H. R. (1998). Plastic reorganization of orientation maps in area 18 of adult cats induced by intracortical microstimulation. *Soc. Neurosci. Abstr.* **24,** 1875.

Godde, B., and Dinse, H. R. (1999). Intracortical microstimulation is efficient to induce systematic plastic reorganization of orientation maps in area 18 of adult cats. Proceedings of Human Brain Mapping HBM'99. *Neuroimage* **9**(6) 286.

Goldmann, P. S., and Nauta, W. J. H. (1977). Columnar distribution of cortico-cortical fibers in the frontal association, limbic and motor cortex of the developing rhesus monkey. *Brain Res.* **122,** 393–413.

Golgi, C. (1873). Sulla sostanza grigia del cervello. *Gaz. Med. Ital. Lombardia Rep. II* **7,** 1–69. (Quoted from *Opera Omnia,* 1903, Vol. 1, pp. 99–111, Milan, Hoepli.)

Gray, C. M., König, P., Engel, A. K., and Singer, W. (1989). Oscillatory responses in cat visual cortex exhibit inter-columnar synchronization which reflects global stimulus properties. *Nature* **338,** 334–337.

Greuel, J. M., Luhmann, H. J., and Singer, W. (1988). Pharmacological induction of use-dependent receptive field modifications in the visual cortex. *Science* **242,** 74–77.

Grinvald, A., Lieke, E. E., Frostig, R. D., and Hildesheim, R. (1994). Cortical point-spread function and long-range lateral interactions revealed by real-time optical imaging of macaque monkey primary visual cortex. *J. Neurosci.* **14,** 2545–2568.

Grossberg, S., and Mignolla, E. (1985). Neural dynamics of perceptual grouping: textures, boundaries and emergent segmentations. *Percept. Psychophys.* **38,** 141–171.

Hebb, D. O. (1949). *The organization of behaviour. A neuropsychological theory.* New York, Wiley.

Heinen, S. J., and Skavenski, A. A. (1991). Recovery of visual responses in foveal V1 neurons following bilateral foveal lesions in adult monkey. *Exp. Brain Res.* **83,** 670–674.

Hendrickson, A. E., and Wilson, J. R. (1979). A difference in [14C]deoxyglucose autoradiographic patterns in striate cortex between Macaca and Saimiri monkeys following monocular stimulation. *Brain Res.* **170,** 353–358.

Hendry, S., H., Schwark, H. D., Jones, E. G., and Yan, J. (1987). Numbers and proportions of GABA-immunoreactive neurons in different areas of monkey cerebral cortex. *J. Neurosci.* **7,** 1503–1519.

Hendry, S. H., and Yoshioka, T. (1994). A neurochemically distinct third channel in the macaque dorsal lateral geniculate nucleus. *Science* **264,** 575–577.

Herculano-Houzel, S., Munk, M. H., Neuenschwander, S., and Singer, W. (1999). Precisely synchronized oscillatory firing patterns require electroencephalographic activation. *J. Neurosci.* **19,** 3992–4010.

Hirsch, J. A., and Gilbert, C. D. (1993). Long-term changes in synaptic strength along specific intrinsic pathways in the cat visual cortex. *J. Physiol. (Lond.)* **461,** 247–262.

Horton, J. C., and Hubel, D. H. (1981). Regular patchy distribution of cytochrome oxidase staining in primary visual cortex of macaque monkey. *Nature* **292,** 762–764.

Houzel, J. C., Milleret, C., and Innocenti, G. (1994). Morphology of callosal axons interconnecting areas 17 and 18 of the cat. *Eur. J. Neurosci.* **6,** 898–917.

Hubel, D. H., and Wiesel, T. N. (1962). Receptive fields, binocular interaction and functional architecture in the cat's visual cortex. *J. Physiol. (Lond.)* **160,** 106–154.

Hubel, D. H. and Wiesel, T. N. (1965). Binocular interaction in striate cortex of kittens reared with artificial squint. *J. Neurophysiol.* **28,** 1041–1059.

Hubel, D. H., and Wiesel, T. N. (1968). Receptive fields and functional architecture of monkey striate cortex. *J. Physiol. (Lond.)* **195,** 215–243.

Hubel, D. H., and Wiesel, T. N. (1969). Anatomical demonstration of columns in the monkey striate cortex. *Nature* **221,** 747–750.

Humphrey, A. L., Sur, M., Uhlrich, D. J., and Sherman, S. M. (1985). Projection patterns of individual X and Y cell axons from the lateral geniculate nucleus to cortical area 17 in the cat. *J. Comp. Neurol.* **233,** 159–189.

Huntley, G. W. (1997). Correlation between patterns of horizontal connectivity and the extent of short-term representational plasticity in rat motor cortex. *Cerebral Cortex* **7,** 143–156.

Imig, T. J., and Brugge, J. F. (1978). Sources and termination of callosal axons related to binaural and frequency maps in primary auditory cortex of the cat. *J. Comp. Neurol.* **182,** 637–660.

Imig, T. J., and Reale, R. A. (1981). Ipsilateral corticocortical projections related to binaural columns in cat primary auditory cortex. *J. Comp. Neurol.* **203,** 1–14.

Innocenti, G. M. (1986). General organization of callosal connections in cerebral cortex. In: Cerebral cortex (A. Peters, and E. G. Jones, Eds.), Vol. 5, pp. 291–353. New York, Plenum Press.

Jenkins, W. M., Merzenich, M. M., Ochs, M. T., Allard, T., and Guic-Robles, E. (1990). Functional reorganization of primary somatosensory cortex in adult owl monkeys after behaviorally controlled tactile stimulation. *J. Neurophysiol.* **63,** 82–104.

Jones, E. G., Coulter, J. D., and Hendry, S. H. C. (1978). Intracortical connectivity of architectonic fields in the somatic sensory, motor and parietal cortex of monkeys. *J. Comp. Neurol.* **181,** 291–348.

Kaas, J. H., Krubitzer, L. A., Chino, Y. M., Langston, A. L., Polley, E. H., and Blair, N. (1990). Reorganization of retinotopic cortical maps in adult mammals after lesions of the retina. *Science* **248,** 229–321.

Kaas, J. H. (1991). Plasticity of sensory and motor maps in adult mammals. *Annu. Rev. Neurosci.* **14,** 137–167.

Kaas, J. H. (1999). Is most of neural plasticity in the thalamus cortical? *Proc. Natl. Acad. Sci. U.S.A.* **96,** 7622–7623.

Kanizsa, G. (1979). *Organization in vision: Essays on Gestalt perception.* New York, Praeger.

Kapadia, M. K., Ito, M., Gilbert, C. D., and Westheimer, G. (1995). Improvement in visual sensitivity by changes in local context: Parallel studies in human observers and in V1 of alert monkeys. *Neuron* **15,** 843–856.

Karni, A., Meyer, G., Jezzard, P., Adams, M. M., Turner, R., and Ungerleider, L. G. (1995). Functional MRI evidence for adult motor cortex plasticity during motor skill learning. *Nature* **377,** 155–158.

Katz, L. C., Burkhalter, A., and Dreyer, W. J. (1984). Fluorescent latex microspheres as a retrograde neuronal marker for in vivo and in vitro studies of visual cortex. *Nature* **310,** 498–500.

Katz, L. C., and Iarovici, D. M. (1990). Green fluorescent latex microspheres: a new retrograde tracer. *Neuroscience* **34,** 511–520.

Katz, L. C., and Callaway, E. M. (1992). Development of local circuits in mammalian visual cortex. *Annu. Rev. Neurosci.* **15,** 31–56.

Kelahan, A., and Doetsch, G. S. (1984). Time-dependent changes in the functional organization of somatosensory cerebral cortex following digit amputation in adult racoons. *Somatosens. Mot. Res.* **2,** 49–81.

Keller, A. (1993). Intrinsic connections between representation zones in the cat motor cortex. *Neuroreport* **4,** 515–518.

Kirkwood, A., and Bear, M. F. (1994). Homosynaptic long-term depression in the visual cortex. *J. Neurosci.* **14,** 3404–3412.
Kisvárday, Z. F., Martin, K. A., Freund, T. F., Magloczky, Z., Whitteridge, D., and Somogyi, P. (1986). Synaptic targets of HRP-filled layer III pyramidal cells in the cat striate cortex. *Exp. Brain Res.* **64,** 541–552.
Kisvárday, Z. F., and Eysel, U. T. (1992). Cellular organization of reciprocal patchy networks in layer III of cat visual cortex (area 17). *Neuroscience* **46,** 275–286.
Kisvárday, Z. F., and Eysel, U. T. (1993). Functional and structural topography of horizontal inhibitory connections in cat visual cortex. *Eur. J. Neurosci.* **5,** 1558–1572.
Kisvárday, Z. F., Beaulieu, C., and Eysel, U. T. (1993). Network of GABAergic large basket cells in cat visual cortex (area 18). Implications for lateral disinhibition. *J. Comp. Neurol.* **327,** 398–415.
Kisvárday, Z. F., Kim, D.-S., Eysel, U. T., and Bonhoeffer, T. (1994). Relationship between lateral inhibitory connections and the topography of the orientation map in cat visual cortex. *Eur. J. Neurosci.* **6,** 1619–1632.
Kisvárday, Z. F., Bonhoeffer, T., Kim, D.-S., and Eysel, U. T. (1996). Functional topography of horizontal neuronal networks in cat visual cortex (area 18). In *Brain theory: Biological basis and computational theory of vision.* Proceedings of the 5th International Meeting on Brain Theory, Trento, Italy (A. Aertsen A., and V. Braitenberg, Eds.), pp. 97–122. Amsterdam, The Netherlands, Elsevier.
Kisvárday, Z. F., Tóth, E., Rausch, M., and Eysel, U. T. (1997). Orientation-specific relationship between populations of excitatory and inhibitory lateral connections in the visual cortex of the cat. *Cerebral Cortex* **7,** 605–618.
Kisvárday, Z. F., Buzas, P., and Eysel, U., T. (1998). Comparison between orientation domains and centre projections in the cat visual cortex. *Eur. J. Neurosci. Suppl.* **10,** 234.
König, P., Engel, A. K., Löwel, S., and Singer, W. (1993). Squint affects synchronization of oscillatory responses in cat visual cortex. *Eur. J. Neurosci.* **5,** 501–508.
Krauskopf, J. (1961). Heterochromatic stabilized images: a classroom demonstration. *Am. J. Psychol.* **80,** 632–637.
Krupa, D. J., Ghazanfar, A. A., and Nicolelis, M. A. L. (1999). Immediate thalamic sensory plasticity depends on corticothalamic feedback. *Proc. Natl. Acad. Sci. U.S.A.* **96,** 8200–8205.
LeVay, S. (1988). The patchy intrinsic projections of visual cortex. *Prog. Brain Res.* **75,** 147–161.
Levitt, J. B., Yoshioka, T., and Lund, J. S. (1994). Intrinsic cortical connections in macaque visual area V2: evidence for interaction between different functional streams. *J. Comp. Neurol.* **342,** 551–570.
Lin, C. S., Friedlander, M. J., and Sherman, S. M. (1979). Morphology of physiologically identified neurons in the visual cortex of the cat. *Brain Res.* **172,** 344–348.
Livingstone, M. S., and Hubel, D. H. (1982). Thalamic inputs to cytochrome oxidase-rich regions in monkey visual cortex. *Proc. Natl. Acad. Sci. U.S.A.* **79,** 6098–6101.
Livingstone, M. S., and Hubel, D. H. (1984a). Specificity of intrinsic connections in primate primary visual cortex. *J. Neurosci.* **4,** 2830–2835.
Livingstone, M. S., and Hubel, D. H. (1984b). Anatomy and physiology of a color system in the primate visual cortex. *J. Neurosci.* **4,** 4309–356.
Lorente de Nó, R. (1922). La corteza cerebral del ratón (Primera contribución-La corteza acústica). *Trab. Lab. Invest. Biol. Madrid* **20,** 41–78.
Löwel, S., and Singer, W. (1992). Selection of intrinsic horizontal connections in the visual cortex by correlated neuronal activity. *Science* **255,** 209–212.
Löwel, S., and Singer, W. (1993). Monocularly induced 2-deoxyglucose patterns in the visual cortex and lateral geniculate nucleus of the cat. II. Awake animals and strabismic animals. *Europ. J. Neurosci.* **5,** 857–869.
Löwel, S. (1994). Ocular dominance column development: strabismus changes the spacing of adjacent columns in cat visual cortex. *J. Neurosci.* **14,** 7451–7468.
Löwel, S., and Singer, W. (2001). Plasticity of intracortical connections. In: *Perceptual learning* (M. Fahle, and T. Poggio, Eds.), MIT Press.
Lübke, J., and Albus, K. (1992). Rapid rearrangement of intrinsic tangential connections in the striate cortex of normal and dark-reared kittens: lack of exuberance beyond the second postnatal week. *J. Comp. Neurol.* **323,** 42–58.

Luhmann, H. J., Martínez-Millán, L., and Singer, W. (1986). Development of horizontal intrinsic connections in cat striate cortex. *Exp. Brain Res.* **63,** 443–448.

Luhmann, H. J., Singer, W., and Martínez-Millán, L. (1991). Horizontal interactions in cat striate cortex. I. Anatomical substrate and postnatal development. *Eur. J. Neurosci.* **2,** 344–357.

Lund, J. S. (1973). Organization of neurons in the visual cortex, area17, of the monkey *(Macaca mulatta). J. Comp. Neurol.* **147,** 455–495.

Lund, J. S., Henry, G. H., MacQueen, C. L., and Harvey, A. R. (1979). Anatomical organization of the primary visual cortex (area 17) of the cat. A comparison with area 17 of the macaque monkey. *J. Comp. Neurol.* **184,** 599–618.

Lund, J. S., Yoshioka, T., and Levitt, J. B. (1993). Comparison of intrinsic connectivity in different areas of macaque monkey cerebral cortex. *Cerebral Cortex* **3,** 148–162.

Malach, R., Amir, Y., Harel, M., and Grinvald, A. (1993). Relationship between intrinsic connections and functional architecture revealed by optical imaging and in vivo targeted biocytin injections in primate striate cortex. *Proc. Natl. Acad. Sci. U.S.A.* **90,** 10469–10473.

Malach, R., Tootell, R. B., and Malonek, D. (1994). Relationship between orientation domains, cytochrome oxidase stripes, and intrinsic horizontal connections in squirrel monkey area V2. *Cerebral Cortex* **4,** 151–165.

Malach, R., Schirman, T. D., Harel, M., Tootell, R. B., and Malonek, D. (1997). Organization of intrinsic connections in owl monkey area MT. *Cerebral Cortex.* **7,** 386–393.

Martin, K. A., and Whitteridge, D. (1984). Form, function and intracortical projections of spiny neurones in the striate visual cortex of the cat. *J. Physiol. (Lond.)* **353,** 463–504.

Matsubara, J., Cynader, M., Swindale, N. V., and Stryker, M. P. (1985). Intrinsic projections within visual cortex: evidence for orientation-specific local connections. *Proc. Natl. Acad. Sci. U.S.A.* **82,** 935–939.

Matsubara, J. A., Cynader, M. S., and Swindale, N. V. (1987). Anatomical properties and physiological correlates of the intrinsic connections in cat area 18. *J. Neurosci.* **7,** 1428–1446.

Matsubara, J. A. (1988). Local, horizontal connections within area 18 of the cat. *Prog. Brain Res.* **75,** 163–172.

Matsubara, J. A., and Phillips, D. P. (1988). Intracortical connections and their physiological correlates in the primary auditory cortex (AI) of the cat. *J. Comp. Neurol.* **268,** 38–48.

Matsubara, J. A., and Boyd, J. D. (1992). Presence of GABA-immunoreactive neurons within intracortical patches in area 18 of the cat. *Brain Res.* **583,** 161–170.

McGuire, B. A., Gilbert, C. D., Rivlin, P. K., and Wiesel, T. N. (1991). Targets of horizontal connections in macaque primary visual cortex. *J. Comp. Neurol.* **305,** 370–392.

Merzenich, M. M., Kaas, J. H., Wall, J., Nelson, R. J., Sur, M., and Felleman, D. (1983a). Topographic reorganization of somatosensory cortical areas 3b and 1 in adult monkeys following restricted deafferentation. *Neuroscience* **8,** 33–55.

Merzenich, M. M., Kaas, J. H., Wall, J. T., Sur, M., Nelson, R. J., and Felleman, D. J. (1983b). Progression of change following median nerve section in the cortical representation of the hand in areas 3b and 1 in adult owl and squirrel monkeys. *Neuroscience* **10,** 639–665.

Merzenich, M. M., Nelson, R. J., Stryker, M. P., Cynader, M. S., Schoppmann, A., and Zook, J. M. (1984). Somatosensory cortical map changes following digit amputation in adult monkeys. *J. Comp. Neurol.* **224,** 591–605.

Michalski, A., Gerstein, G. L., Czarkowska, J., and Tarnecki, R. (1983). Interactions between cat striate cortex neurons. *Exp. Brain Res.* **51,** 97–107.

Mitchison, G., and Crick, F. (1982). Long axons within the striate cortex: their distribution, orientation, and patterns of connection. *Proc. Natl. Acad. Sci. U.S.A.* **79,** 3661–3665.

Mittmann, T., Luhmann, H. J., Schmidt-Kastner, R., Eysel, U. T., Weigel, H., and Heinemann, U. (1994). Lesion-induced transient suppression of inhibitory function in rat neocortex in vitro. *Neuroscience* **60,** 891–906.

Montero, V. M. (1980). Patterns of connections from the striate cortex to cortical visual areas in superior temporal sulcus of macaque and middle temporal gyrus of owl monkey. *J. Comp. Neurol.* **189,** 45–59.

Morrone, M. C., Burr, D. C., and Maffei, L. (1982). Functional implications of cross-orientation inhibition of cortical visual cells. I. Neurophysiological evidence. *Proc. R. Soc. Lond. [Biol.]* **B216,** 335–354.

Mountcastle, V. B. (1957). Modality and topographic properties of single neurons in cat's somatic sensory cortex. *J. Neurophysiol.* **20,** 408–434.

Nauta, W. J. H., and Gygax, P. A. (1953). Silver impregnation of degenerating axons in the central nervous system: a modified technique. *Stain Technol.* **29,** 91–93.

Nelson, J. I., and Frost, B. J. (1978). Orientation-selective inhibition from beyond the classical receptive field. *Exp. Brain Res.* **139,** 359–365.

Nelson, J. I., and Frost, B. J. (1985). Intracortical facilitation among co-oriented, co-axially aligned simple cells in cat striate cortex. *Exp. Brain Res.* **61,** 54–61.

O'Leary, J. L. (1941). Structure of the area striata in the cat. *J. Comp. Neurol.* **75,** 131–164.

Peters, A., and Kaiserman-Abramof, I. R. (1969). The small pyramidal neuron of the rat cerebral cortex. The synapses upon dendritic spines. *Z. Zellforsch. Mikrosk. Anat.* **100,** 487–506.

Pettet, M. W., and Gilbert, C. D. (1992). Dynamic changes in receptive-field size in cat primary visual cortex. *Proc. Natl. Acad. Sci. U.S.A.* **89,** 8366–8370.

Polat, U., and Sagi, D. (1993). Lateral interactions between spatial channels: suppression and facilitation revealed by lateral masking experiments. *Vis. Res.* **33,** 993–999.

Polat, U., and Sagi, D. (1994). The architecture of perceptual spatial interactions. *Vis. Res.* **34,** 73–78.

Pons, T., Garraghty, P. E., Ommaya, A. K., Kaas, J. H., Taub, E., and Mishkin, M. (1991). Massive cortical reorganization after sensory deafferentiation in adult macaques. *Science* **252,** 1857–1860.

Price, D. J., Ferrer, J. M., Blakemore, C., and Kato, N. (1994). Functional organization of corticocortical projections from area 17 to area 18 in the cat's visual cortex. *J. Neurosci.* **14,** 2732–2746.

Pucak, M. L., Levitt, J. B., Lund, J. S., and Lewis, D. A. (1996). Patterns of intrinsic and associational circuitry in monkey prefrontal cortex. *J. Comp. Neurol.* **376,** 614–630.

Ramachandran, V. S., and Gregory, T. L. (1991). Perceptual filling-in of artificially induced scotomas in human vision. *Nature* **350,** 699–702.

Ramón y Cajal, S. (1922). Studien über die Sehrinde der Katze. *J. Psychol. Neurol.* **29,** 161–181.

Rasmusson, D. D., and Nance, D. M. (1986). Non-overlapping thalamocortical projections for separate forepaw digits before and after cortical reorganization in the racoon. *Brain Res. Bull.* **16,** 399–406.

Rockland, K. S., and Lund, J. S. (1982). Widespread periodic intrinsic connections in the tree shrew visual cortex. *Science* **215,** 1532–1534.

Rockland, K. S., Lund, J. S., and Humphrey, A. L. (1982). Anatomical binding of intrinsic connections in striate cortex of tree shrews *(Tupaia glis). J. Comp. Neurol.* **209,** 41–58.

Rockland, K. S., and Lund, J. S. (1983). Intrinsic laminar lattice connections in primate visual cortex. *J. Comp. Neurol.* **216,** 303–318.

Rockland, K. S. (1985). Anatomical organization of primary visual cortex (area 17) in the ferret. *J. Comp. Neurol.* **241,** 225–236.

Roerig, B., and Kao, J. P. Y. (1999). Organization of intracortical circuits in relation to direction preference maps in ferret visual cortex. *J. Neurosci.* **19,** 1–5.

Rose, J. E., Malis, L. I., Kruger, L., and Baker, C. P. (1960). Effects of heavy ionizing, monoenergetic particles on the cerebral cortex II. Histological appearance of laminar lesions and growth of nerve fibers after laminar destructions. *J. Comp. Neurol.* **115:** 243–295.

Rosier, A. M., Arckens, L., Demeulemeester, H., Orban, G. A., Eysel, U. T., Wu, Y. J., and Vandesande, F. (1995). Effect of sensory deafferentation on immunoreactivity of GABAergic cells and on GABA receptors in the adult cat visual cortex. *J. Comp. Neurol.* **359,** 476–489.

Ruthazer, E. S., and Stryker, M. P. (1996). The role of activity in the development of long-range connections in area 17 of the ferret. *J. Neurosci.* **16,** 7253–7269.

Salin, P. A., Bullier, J., and Kennedy, H. (1989). Convergence and divergence in the afferent projections to cat area 17. *J. Comp. Neurol.* **283,** 486–512.

Salin, P. A., and Bullier, J. (1995). Corticocortical connections in the visual system: structure and function. *Physiol. Rev.* **75,** 107–154.

Schmid, L. M., Rosa, M. G., Calford, M. B., and Ambler, J. S. (1996). Visuotopic reorganization in the primary visual cortex of adult cats following monocular and binocular retinal lesions. *Cerebral Cortex* **6,** 388–405.

Schmidt, K. E., Kim, D.-S., Singer, W., Bonhoeffer, T., and Löwel, S. (1997a). Functional specificities of long-range intrinsic and interhemispheric connections in the visual cortex of strabismic cats. *J. Neurosci.* **17,** 5480–5492.

Schmidt, K. E., Goebel, R., Löwel, S., and Singer, W. (1997b). The perceptual grouping criterion of colinearity is reflected by anisotropies of connections in the primary visual cortex. *Eur. J. Neurosci.* **9,** 1083–1089.

Schmidt, K. E., Galuske, R. A. W., and Singer, W. (1999). Matching the modules: cortical maps and long-range intrinsic connections in visual cortex during development. *J. Neurobiol.* **41,** 10–17.

Schwarz, C., and Bolz, J. (1991). Functional specificity of a long-range connection in cat visual cortex: a cross-correlation study. *J. Neurosci.* **11,** 2995–3007.

Shatz, C. J., Lindström, S., and Wiesel, T. N. (1977). The distribution of afferents representing the right and left eyes in the cat's visual cortex. *Brain Res.* **131,** 103–116.

Shoham, D., Hübener, M., Schulze, S., Grinvald, A., and Bonhoeffer, T. (1997). Spatio-temporal frequency domains and their relation to cytochrome oxidase staining in cat visual cortex. *Nature* **385,** 529–533.

Sincich, L., and Blasdel, G. G. (1995). Lateral connections and orientation preference in layers II/III of squirrel monkey striate cortex. *Soc. Neurosci. Abstr.* **21,** Part 1, 393.

Singer, W., and Tretter, F. (1976). Unusually large receptive fields in cats with restricted visual experience. *Exp. Brain Res.* **26,** 171–184.

Somogyi, P., Kisvárday, Z. F., Martin, K. A., and Whitteridge, D. (1983). Synaptic connections of morphologically identified and physiologically characterized large basket cells in the striate cortex of cat. *Neuroscience* **10,** 261–294.

Sugita, Y. (1996). Global plasticity in adult visual cortex following reversal of visual input. *Nature* **380,** 523–526.

Szentágothai, J. (1973). Synaptology of the visual cortex. In: Handbook of sensory physiology: Central visual information (R. Jung, Ed.), Vol. 7, pp. 269–324. Berlin, Springer.

Tigges, J., Tigges, M., Anschel, S., Croos, N. A., Letbetter, W. D., and McBride, R. L. (1981). Areal and laminar distribution of neurons interconnecting the central visual cortical areas 17, 18, 19 and MT in squirrel monkey *(Saimiri). J. Comp. Neurol.* **202,** 539–560.

Tootell, R. B., Silverman, M. S., Hamilton, S. L., De Valois, R. L., and Switkes, E. (1988a). Functional anatomy of macaque striate cortex. III. Color. *J. Neurosci.* **8,** 1569–1593.

Tootell, R. B., Silverman, M. S., Hamilton, S. L., Switkes, E., and De Valois, R. L. (1988b). Functional anatomy of macaque striate cortex. V. Spatial frequency. *J. Neurosci.* **8,** 1610–1624.

Ts'o, D., Gilbert, C. D. and Wiesel, T. N. (1986). Relationships between horizontal interactions and functional architecture in cat striate cortex as revealed by cross-correlation analysis. *J. Neurosci.* **6,** 1160–1170.

Ts'o, D. and Gilbert, C. D. (1988). The organization of chromatic and spatial interactions in the primate striate cortex. *J. Neurosci.* **8,** 1712–1727.

Valverde, F. (1986). Intrinsic neocortical organization: some comparative aspects. *Neuroscience* **18,** 1–23.

Van Essen, D. C., Newsome, W. T., and Maunsell, J. H. (1984). The visual field representation in striate cortex of the macaque monkey:asymmetries, anisotropies, and individual variability. *Vis. Res.* **24,** 429–448.

Vicq d'Azyr, F. (1786). *Traité d'anatomie et de physiologie.* Paris, Didot.

Volchan, E., and Gilbert, C. D. (1995). Interocular transfer of receptive field expansion in cat visual cortex. *Vis. Res.* **35,** 1–6.

Wall, J. T., and Cusick, C. G. (1984). Cutaneous responsiveness in primary somatosensory (S-I) hindpaw cortex before and after partial hindpaw deafferentiation in adult rats. *J. Neurosci.* **4,** 1499–1515.

Wells, J., and Tripp, L. N. (1987). Time course of reactive synaptogenesis in the subcortical somatosensory system. *J. Comp. Neurol.* **255,** 466–475.

Wertheimer, M. (1938). *Laws of organization in perceptual forms.* New York, Harcourt, Brace, Jovanovich.

Westheimer, G., Shimamura, K., and McKee, S. (1976). Interference with line orientation sensitivity. *J. Opt. Soc. Am.* **66,** 332–338.

Westheimer, G. (1986). Spatial interaction in the domain of diparity signals in human stereoscopic vision. *J. Physiol. (Lond.)* **370,** 619–629.

Wong-Riley, M. (1979). Columnar cortico-cortical interconnections within the visual system of the squirrel and macaque monkeys. *Brain Res.* **162,** 201–217.

Yarbus, A. L. (1957). The perception of an image fixed with respect to the retina. *Biophysics* **2,** 683–690.

Yoshioka, T., Levitt, J. B., and Lund, J. S. (1992). Intrinsic lattice connections of macaque monkey visual cortical area V4. *J. Neurosci.* **12,** 2785–2802.

Yoshioka, T., Blasdel, G. G., Levitt, J. B., and Lund, J. S. (1996). Relation between patterns of intrinsic lateral connectivity, ocular dominance, and cytochrome oxidase-reactive regions in macaque monkey striate cortex. *Cerebral Cortex* **6,** 297–310.

Yousef, T., Bonhoeffer, T., Kim, D.-S., Eysel, U. T., Tóth, E., and Kisvárday, Z. F. (1999). Orientation topography of layer 4 lateral networks revealed by optical imaging in cat visual cortex (area 18). *Eur. J. Neurosci.* **11,** 4291–4308.

11

PHARMACOLOGICAL STUDIES ON RECEPTIVE FIELD ARCHITECTURE

ULF T. EYSEL

Department of Neurophysiology, Ruhr-University, Bochum, Germany

RECEPTIVE FIELD ARCHITECTURE IN CAT STRIATE CORTEX

A visual receptive field (RF) is that area on the retina (or the corresponding part of the visual field) that, when stimulated, changes the activity (by excitation or inhibition) of a visual neuron. The first descriptions of cat striate cortex cell RFs were published 40 years ago by Hubel and Wiesel (1959, 1962). They used moving and flashing light stimuli to test the spatial properties of the RFs. They classified the RFs of visual cortical cells as simple and complex. The simple cell receptive fields are characterized as composed from elongated, spatially separated ON and OFF subfield regions that can be excited by light or dark stimuli, respectively (Fig.11-1 A,B). Complex cells, with comparatively larger RF areas, respond to ON and OFF stimuli as well; however, the respective subfields are superimposed, not separated in space. Both cell types were found to possess new properties that were not present in the afferent geniculocortical fibers. Figure 11-1 E,F shows the two most studied of these new features, orientation specificity and direction selectivity (Hubel and Wiesel, 1962). Hypercomplex cells with higher order specificity (e.g., angles, corners) were initially described in nonstriate areas of the cat (Hubel and Wiesel, 1965), but later it was found that they are also present in striate cortex and that the basic property of hypercomplexity is "end-inhibition" (Rose, 1977; Gilbert 1977; Kato et al., 1978), which can be present in simple as well as in complex cells. Primarily it was assumed that the RFs of cells

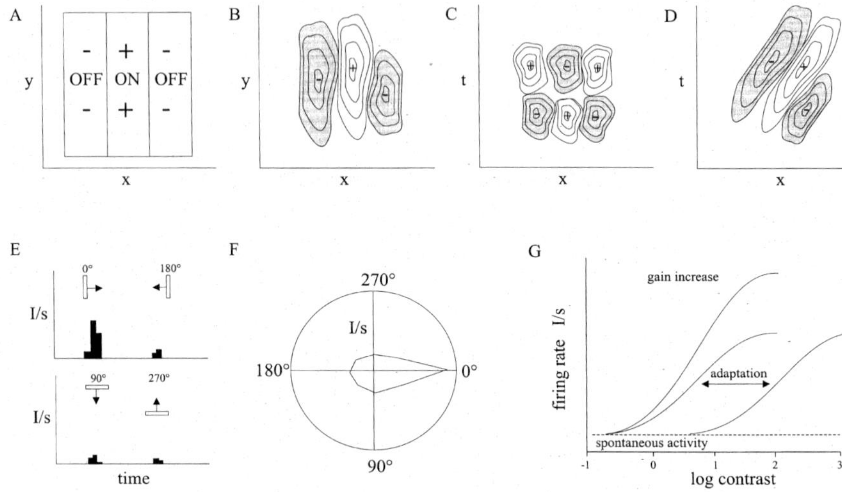

FIGURE 11-1. Receptive fields and diagrams of receptive field functions. **(A)** Textbook drawing of the RF of a simple cell with elongated, vertically oriented, parallel bands of inhibitory (OFF) and excitatory (ON) regions. **(B)** Realistic spatial (xy-) profile of a simple cell with properties corresponding to the textbook example in **(A)** obtained with small flashing bars of vertical orientation. The 443 profile of the same cell. The response is cut into "time-slices" and for each time slice the y-values are integrated at value of the x-axis. This way the temporal profile of the RF becomes visible. Early ON responses are followed by OFF responses and early OFF responses are followed by OFF responses. This simple cell xt-profile is space-time separable. **(D)** Same methodical procedure as in **(C),** but this time the ON- and OFF regions travel continuously in space and time and form a simple cell RF-profile that is inseparable in the spatiotemporal domain. **(E)** Conventional peristimulus time histograms (PSTH) obtained by averaging responses to light bars moving across the RF in different orientations. The vertical bar (upper histogram) elicits the strongest response when moving from left to right. The orthogonally oriented bar (lower histogram) elicits very weak responses in both directions of motion. This RF shows orientation specificity for vertical bars and direction selectivity for motion to the right. F. Polar diagram of the responses of the cell in **(E)** for eight different orientations. The length of the vectors indicates the average activity for the given orientation. The convention shown here (0° corresponds to a vertical bar moving to the right) is used in all polar diagrams in this chapter representing responses to moving bars. **(G)** Diagram showing fundamental terms of contrast adaptation. typical contrast-response curves are of sigmoid shape, beginning at the spontaneous activity level and running into saturation at the other end. Contrast adaptation is a shift of the operating point along the contrast axis. Contrast gain changes are characterized by changes in the slope of the contrast-response curve.

in the cat primary visual cortex are constructed by simple feed-forward excitation from geniculocortical inputs (Hubel and Wiesel, 1962); however, the RF architecture is further elaborated by intracortical excitatory and inhibitory networks, both within a column (Douglas et al., 1989; Douglas and Martin, 1991) and laterally across substantial distances (Maffei and Fiorentini, 1976; Allman et al., 1985; Ts'o et al., 1986; Gilbert and Wiesel, 1989). RFs show characteristic space-time profiles that can be separable (Fig. 11-1C) or inseparable (Fig. 11-1D) (for a review see DeAngelis et al., 1995). When talking about architecture in this article, we are well aware that the striate cortex RF is not static, but that its spatial organization is highly dynamic (Dinse et al., 1990; Eckhorn et al., 1993; Shevelev et al., 1992, 1993; DeAngelis et al., 1993a), and that structure and properties of the RFs are dependent on context of the stimuli as well as the intrinsic state of the system (Wörgötter and Eysel, 2000).

In striate cortex, most of the excitatory synapses are glutamatergic, and at the inhibitory synapses γ-aminobutyric acid (GABA) is used as a transmitter. Excitation is mediated through α-amino 3-hydroxy-5-methyl-4-isoxazole-proprionic acid (AMPA) and kainate (non-NMDA) as well as through *N*-methyl-D-aspartate (NMDA) receptor gated channels (Fox et al., 1989, 1990). Inhibition is mediated through both $GABA_A$ and $GABA_B$ receptor-gated channels (Sillito, 1975a; Baumfalk and Albus, 1987; Allison et al., 1996). In adddition to these main transmitter systems, modulation by ascending influences from the brainstem and basal forebrain couple striate cortex cell RF properties to states of the brain such as arousal or attention (Wörgötter et al., 1998; Wörgötter and Eysel, 2000). These modulatory systems use acetylcholine (ACh) and catecholamines as transmitters.

The specificity of excitatory and inhibitory transmission at striate cortex cells and its modulation contribute to the functional architecture of the RFs and can be investigated using pharmacological tools such as specific agonists or blockers. All these studies have been first performed in the cat visual cortex. Later some of these approaches have been repeated in the primate, whereas others have not yet been transferred to other visual model systems.

This chapter reviews a selection of pharmacological studies that have helped to shed light into basic mechanisms that contribute to the structure of RFs and response properties of cells in the striate cortex.

Two fundamentally different approaches were exploited to study visual cortical functions pharmacologically: (1) interference with synaptic transmission at a given cell and its close neighbors by micro-iontophoretical application of neurotransmitters or agonists or antagonists of specific receptors (see next section) and (2) interference with network functions by inactivation of specific parts of the intrinsic striate cortex circuitry (see page 447 and following pages).

SPECIFIC LOCAL SYNAPTIC INPUT SYSTEMS TO STRIATE CORTEX CELLS

The specific functional architecture of cat striate cortex cells is based on the different cell types, specific local and long-range connectivity, and different synaptic transmitter systems and specific neurotransmitter receptors (Fig. 11-2).

After visual or electrical stimulation of the geniculocortical pathway in the cat, glutamate (GLU) and aspartate (ASP) levels were increased in perfusates analyzed with high-performance liquid chromatography. Further investigation showed that glutamate was synaptically released by the geniculocortical system,

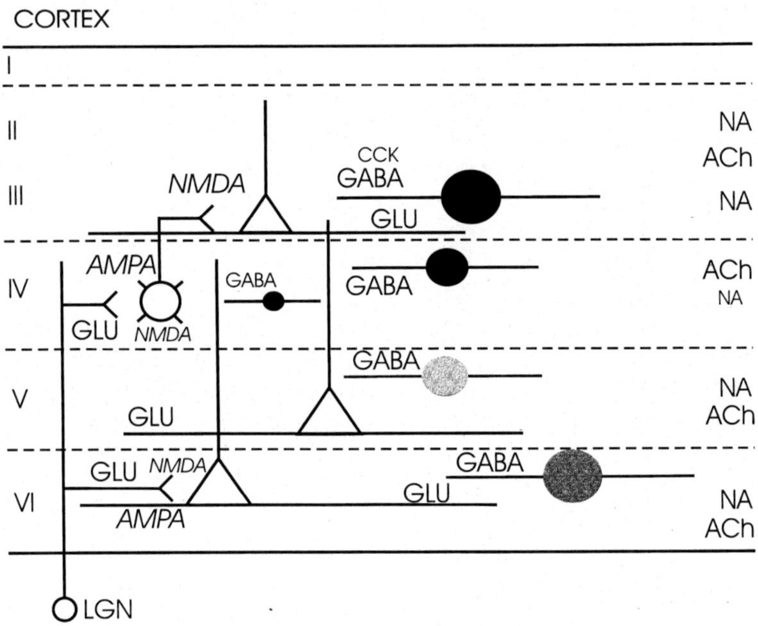

FIGURE 11-2. Summary diagram of excitatory (white symbols) and inhibitory (filled symbols) cell types in the striate cortex and of the neurotransmitters/neuromodulators and receptors (in italics) addressed in this chapter. Triangles symbolize pyramidal cell bodies with their apical dendrite, open circles stellate cells, filled circles GABAergic cells. The cortical layers I–IV from the pia mater to the white matter are indicated to the left. Abbreviations: ACh, acetylcholine; AMPA, α-amino3-hydroxy-5-methyl-4-isoxazole-proprionic acid; CCK, cholecystokinine; GABA, γ-aminobutyric acid; GLU, glutamate; NA, noradrenaline; NMDA, *N*-methyl-D-aspartate. The main inputs from the LGN (lateral geniculate nucleus) to layers IV and VI are schematically shown.

whereas aspartate was more likely released by intracortical connections, possibly via collaterals of corticogeniculate cells (Tamura et al., 1990). In an earlier study, Tsumoto et al. (1986) applied kynurenic acid, a broad-spectrum antagonist of excitatory amino acid receptors, to elucidate the possible role of GLU and ASP as transmitters in the cat visual cortex. In fact, kynurenic acid eliminated the visual responses of the vast majority of visual cortical neurons. The most reliable suppression was seen in simple cells and in cells situated in layer IV and the upper part of layer VI, whereas for cells in other layers, transmission was not blocked at all or only incompletely suppressed (see Chapter 7).

The excitatory input from the lateral geniculate nucleus to cortical cells in layer 4 and from cortical cells in layer VI to layer IV uses glutamate as neurotransmitter and it acts primarily at α AMPA receptors (Fox et al., 1989), and intracortical excitation from layer 4 to layer 2/3 cells also uses glutamate. However, the AMPA as well as the NMDA receptors are involved. An additional functional influence of NMDA receptors on geniculocortical transmission in cat cortex has been shown by several authors (Miller et al. 1989; Sillito et al., 1990; Kwon et al., 1991). When non-NMDA receptors are blocked with CNQX (6-cyano-7-nitroquinoxaline-2,3-dione) or NMDA-receptors are blocked with D-amino-5-phosphonovalerate (APV), the visual responses are reduced to 11.7% or 45% of the control, respectively. This indicates a leading role for non-NMDA receptors in geniculocortical transmission and an additional involvement of NMDA receptors (Sato et al., 1999).

About 20% of all neurons in the cat striate cortex are inhibitory and use GABA as their neurotransmitter (Gabbot and Somogyi, 1986). The distribution of these cells is not uniform across all cortical layers. The density of GABAergic cells was lowest in layer V and somewhat higher in layers II,III and the upper part of layer VI. The highest density of GABAergic cells is found in upper layer IV and lower layer III, which contain about 50% of all GABA-immunopositive cells in the striate cortex. Notably, this is the region where the majority of geniculocortical terminals are located (see Chapter 1). This finding suggests that fast inhibitory interactions occur early in visual processing in cat primary visual cortex. That conjecture is further supported by the demonstration of direct geniculocortical inputs to GABAergic cells (Freund et al., 1985).

Noradrenaline (NA) is the most abundant catecholamine within the visual cortex. It plays important roles in modulating the sensitivity of cortical neurons to visual stimuli. A bilateral projection from the locus coeruleus to cat striate cortex has been shown anatomically (Sato et al., 1989). When a polyclonal antibody against dopamine-β-hydroxylase was used to localize noradrenaline-containing afferents in visual cortex of adult cats, noradrenergic fibers were found to be present throughout all cortical layers and exhibited higher densities below and above the layer IV, and a band of lower staining in layer IV. While tangential fibers predominated in layers I, V and VI, relatively straight radial fibers traversed layers II and III (Liu and Cynader, 1994). Both α- and β-adrenergic receptors are present (Liu et al., 1993). Whereas the β-adrenergic inputs seem to provide facilitatory and suppressive effects, the α-adrenergic system acts facilitatory either by a direct facilitation or disinhibition (Sato et al., 1989).

Acetylcholine is the neurotransmitter of the cholinergic projections that reach the striate cortex from the magnocellular nuclei of the basal forebrain, in particular from the diagonal band of Broca and the substantia innominata (Albus, 1981; Bear et al., 1985; Stichel and Singer, 1985). These inputs provide terminals in all cortical layers and thus can influence the activity of all cortical cell types (deLima and Singer, 1986), which makes the cholinergic system another candidate to change the functional RF architecture, especially in relation to attention and the state of arousal.

Many other modulators act on striate cortical cells in the cat. One is cholecystokinin (CCK), which is colocalized with GABA (Somogyi et al., 1984) and most frequently found in cells of layers II/III that form symmetrical synapses at somata, dendritic shafts, or spines of pyramidal cells (Freund et al., 1986). These patterns show a developmental dependence (Shaw et al., 1986). Another modulator present in cat striate cortex is substance P (SP). Labeling for that tachykinin is most prominent in neuronal processes and some neurons of layer V, with a gradually decreasing density of labeled elements in layers I, II, III, VI, and with lowest density in layer IV (Gu et al., 1994).

The possible influences of all these neurotransmitters and neuromodulators in the cat striate cortex can be examined by local application of the respective receptor agonists or specific blockers. In the next section, their contributions to different aspects of RF architecture are discussed.

PHARMACOLOGICAL STUDIES OF THE GENERAL EXCITABILITY OF STRIATE CORTEX CELLS

Excitatory amino acids (EAA) are the main excitatory transmitters in the striate cortex. Application of their agonists increases the activity of the neurons, and application of the antagonists leads to a decrease in activity. This is evident from studies using blockers as well as agonists. Kynurenic acid, a blocker of glutamate and aspartate receptors, eliminates visual responses in 83% of striate cortex cells (Tsumoto et al., 1986). When the glutamatergic receptors of striate cortex cells are pharmacologically stimulated by glutamate microiontophoresis, the vast majority of cells (69%) show excitatory responses accompanied by enhanced visual responses (Kraszewski and Michalski, 1989). In a smaller population (18%), the same authors observed suppressive effects of glutamate iontophoresis. This suppressive effect could be explained by excitatory action of glutamate on inhibitory cells.

Microiontophoretic application of GABA has a strong inhibitory action at all visual cortical cells, excitatory as well as inhibitory cells (Wallingford et al., 1973; Sillito, 1975a). Because of this silencing effect it has been used for inactivation of parts of the striate cortex in many studies (see page 449). More information about the general effects of GABA was gained with application of the specific $GABA_A$ antagonist bicuculline. With systemic application of bicuculline, a differential effect of blockade of inhibition at simple and complex cells was reported (Pettigrew and Daniels, 1973). Whereas simple cells showed a depressed response, complex cells had facilitated responses and hypercomplex cells lost their end-inhibition. However, systemic application affects GABAergic inhibition not only in the striate cortex but also in all other parts of the visual system: the retina, the lateral geniculate nucleus, and higher visual areas. Local microiontophoretic application of bicuculline led to more consistent results. All striate cortex cells showed increased spontaneous activity as well as visual excitability when GABAergic inhibition was blocked (Sillito, 1975a). Similar effects were observed in 11 to 28-day old kittens, where 63% of the cells showed a strong increase of the firing rate, both during visually evoked responses and during periods without stimulation. In 25% of neurons the discharge frequency was more than doubled (Wolf et al., 1986).

Activation of the noradrenergic system by electrical stimulation of the locus coeruleus revealed differential general effects on the striate cortex cells. In 52%

of the cells the amplitude of the responses to visual stimuli decreased, whereas in 36% of neurons the responses became stronger. Suppressive effects prevailed in the upper layers II, III, and IV, whereas facilitation was found most often in layers V and VI. These effects were also clearly cell-type specific. In simple cells the suppressive effect dominated throughout all layers; the previously mentioned laminar pattern was seen in complex cells (Sato et al., 1989). Local microiontophoretic application of NA revealed about an equal proportion of neurons whose responses were depressed, facilitated, or not affected. Independent of the direction of the effect on the response amplitude, the signal-to-noise ratio was invariably improved during the action of NA owing to a reduction in spontaneous activity (Kasamatsu and Heggelund, 1982).

Microiontophoretical application of ACh to the striate cortex had a facilitatory effect on the responses of the majority of neurons (74%) and a suppressive effect in some cells (16%). These effects were mediated by muscarinic ACh receptors, as both the facilitatory and inhibitory effects of ACh were abolished by application of atropine, the specific antagonist of the muscarinic receptors. Facilitatory effects of ACh were most typical for cells in supragranular and infragranular layers (II–III and V–VI). In layer IV about half the cells showed facilitation. During facilitation, the spontaneous activity remained mostly unaffected, whereas the response amplitudes to visual stimulation increased, which resulted in an improvement of the signal-to-noise ratio in the majority of facilitated cells. (Sato et al., 1987).

Microiontophoresis of CCK at striate cortical cells revealed very inhomogeneous effects. Facilitation was observed in 31%, and suppression in 24% of the cells. Mixed, biphasic, or dose-dependent effects were seen in 20% of the neurons, and in the remaining cells no effects were detected. The excitatory effects were mainly seen in simple cells and were rare in complex cells (7%). More complex cells (31%) showed suppressive effects. Facilitation prevailed in layers II through IV, whereas suppression was more frequently found in layers V and VI.

SP and other tachykinins affected the responses of only 24% of the investigated striate cortex cells (Albus et al., 1992). Of these, in about two thirds of the cells, mostly from layer V, the responses were enhanced; in the others, mainly seen in layer IVc, depression was observed.

SPATIOTEMPORAL ASPECTS OF RF ARCHITECTURE

Neurons in the striate cortex have RFs of different subfield structure and size. Simple cells have smaller RFs with adjacent bands of ON and OFF subfields. Complex cells are characterized by larger RFs at a given eccentricity in the visual field and overlapping, spatially nonseparable regions with ON and OFF responses, and hypercomplex cells (simple or complex-like) are defined by the presence of additional end-inhibitory zones (Hubel and Wiesel, 1959, 1962, 1965; Jones and Palmer, 1987a, 1987b; Jones et al., 1987; Rose, 1977; Gilbert, 1977; Kato et al., 1978). These spatial characteristics of the striate cortex RFs appear in part generated, or at least significantly sculptured, by the balance between excitatory and inhibitory synaptic inputs, as can be shown with pharmacological studies of RF architecture.

When glutamate receptors were blocked with CNQX or APV, the visual responses were strongly reduced, and as a rule the excitatory receptive field size diminished (Sato et al., 1999). During the action of CNQX as well as APV, the responses of cortical cells to moving stimuli of constant velocity became weaker and were elicited from a smaller region in the visual field, whereby the decrease of the receptive field size and the response amplitude occurred in parallel. This change of spatial RF parameters is similar to a "tip of the iceberg" effect in which more or less of the response moves above or below some threshold level of activity or visibility. Application of the antagonist of GABAergic inhibition bicuculline led to the opposite effects. The most pronounced effect here was a significant increase of the dimensions of both ON and OFF subfields (Pernberg et al. 1998; Frégnac and Shulz, 1999). However, ON and OFF regions expanded predominantly within the outer borders of the RFs, so that ON and OFF subfields often became inseparable in space leading to ON-OFF responses in regions that had either ON or OFF responses before when tested with single flashing light bars (Sillito, 1975b; Eysel and Shevelev, 1994). As a result, the total size of the RF increased only slightly. These observations suggest that the OFF responses in ON subfields and the ON responses in OFF subfields are normally suppressed by GABAergic inhibition, and thus the characteristic clear structure of simple RFs is, at least partially, based on intracortical inhibition. Similar changes of the spatiotemporal organization of cortical RFs occurred during application of bicuculline in kittens ranging in age from 11 to 28 days (Wolf et al., 1986). The spatiotemporal structure of RFs in area 17, when estimated from responses to presentations of counterphase gratings, was also shown to change under blockade of $GABA_A$ inhibition (Murthy and Humphrey, 1999), indicating that not only spatial but also temporal properties of the RF are sculptured by inhibition. This, in turn, influences the direction selectivity of striate cortex RFs (see page 441).

Application of ACh affects RF dimensions in accordance with suppressive and facilitatory effects on a cell's activity. The minimal discharge field became smaller when ACh suppressed the responses of a cell, and increased in size when ACh enhanced single cell activity (Sato et al., 1987).

Spatiotemporal characteristics of RFs were also influenced by NA. Microiontophoretic application of NA led to enhancement of the selectivity for stimulus speed (Mc Lean and Waterhouse, 1994) and to changes in direction selectivity (see page 442).

CONTRAST GAIN CONTROL AND ADAPTATION

When considering effects of the contrast on response characteristics of neurons, one must discriminate between two properties, contrast gain and contrast adaptation. Contrast gain characterizes sensitivity to rapid changes of the stimulus contrast and is defined as the slope of the contrast-response function. Alterations of the contrast gain are reflected as changes of the steepness of the contrast-response function. Contrast adaptation is characterized by an alteration of the range of sensitivity to changes of stimulus contrast and is reflected by a parallel shift of the sigmoid contrast response function along the contrast axis (Fig. 11-1G).

To investigate the effects of EAA on contrast gain, contrast-response curves were measured under control conditions and under the influence of the glutamate receptor agonists NMDA and quisqualate as well as their antagonists APV and CNQX, respectively (Fox et al., 1990). Levels of NMDA that only marginally increased the spontaneous activity significantly increased the contrast-gain (Fig. 11-3A), whereas quisqualate mainly increased the spontaneous activity and induced a parallel upward shift of the contrast-response curves without significantly changing the slope (Fig. 11-3B). Quite in line with these results, the NMDA antagonist APV reduced the contrast gain in the supragranular layers of the cat striate cortex. In layer IV cells the contrast gain was not affected either by NMDA or by APV; here the only effect was an increase of decrease of spontaneous activity. CNQX also reduced the contrast gain in some neurons. However, it has to be kept in mind that blockade of AMPA receptors hinders the initial depolarization, which is necessary to release the Mg^{2+} block of the NMDA receptors and thus can indirectly reduce the NMDA response. These findings indicate that the voltage sensitivity of the NMDA receptors represents a basis for proportional response amplification from threshold to saturation.

However, not only contrast gain but also contrast adaptation are under control of excitatory amino acid receptors (McLean and Palmer, 1996). After an adaptation period, during which a drifting grating with higher contrast was presented, the responses to iontophoretic application of glutamate or NMDA were reduced, but the responses to application of AMPA remained unchanged. Blockade of the postsynaptic ionotropic glutamate receptors did not interfere with this effect, but blockade of the presynaptic metabotropic glutamate receptors with α-methyl-4-carboxyphenylglycine (MCPG) reduced the contrast adaptation effect. These results suggest a contribution of presynaptic glutamatergic mechanisms to contrast adaptation. An additional postsynaptic NMDA-dependent effect might also be involved, as immediately after a high-contrast adaptation, there seemed to be a decrease in the NMDA-mediated component of the visual response.

Although most RF properties are multifactorially controlled by glutamatergic and GABAergie mechanisms and contrast adaptation was thought to be related to inhibition (Ohzawa and Freeman, 1982; Ohzawa et al., 1985), there is little support of this notion in the literature. In one early study $GABA_A$ receptors were blocked with *N*-methyl bicuculline (NMB) applied locally by micro-iontophoresis (DeBruyn and Bonds, 1986). The effects were comparable to the effects of quisqualate (i.e., an upward shift of the contrast sensitivity function) (Fig. 11-3). However, in addition to an increase in spontaneous activity during blockade of $GABA_A$ receptors with NMB, the range of contrast adaptation, measured as the spacing of the contrast-response curves along the contrast axis with 10%, 20%,

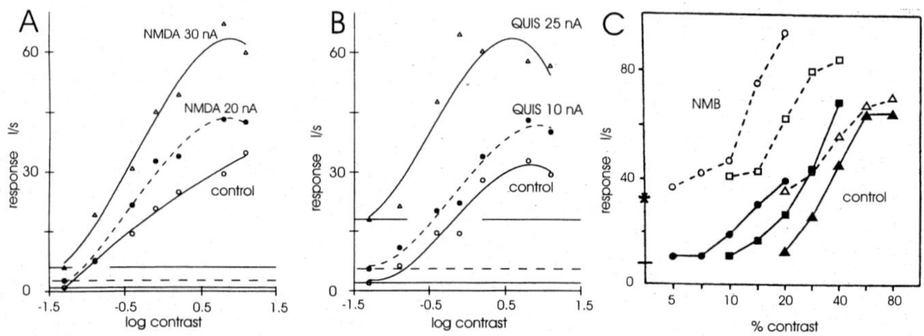

FIGURE 11-3. Influence of NMDA and quisqualate (QUIS) on contrast gain of a layer III cell **(A,B)** and the effect of *N*-methyl bicuculline (NMB) on contrast adaptation of a complex cell **(C)**. NMDA applied microiontophoretically with different ejection currents as indicated increases the spontaneous activity only marginally but leads t an increase in contrast gain **(A)**, and quisqualate increases the spontaneous activity and responses in a dose dependent manner without changing the contrast gain. (A and B reproduced with permission from Fox et al., 1990.) **(C)** Contrast-response curves at three different adaptation levels (10% circles, 20% squares, and 40% triangles) before (filled symbols) and during application of *N*-methyl bicuculline (NMB). The response level is increased during blockade of $GABA_A$ receptors, contrast adaptation is qualitatively unchanged. (C reproduced with permission from DeBruyn and Bonds, 1986.)

and 40% adapting contrasts, increased as well (Fig. 11-3C). At the same time, the contrast gain (slope of the contrast response function) as a rule did not increase. On the basis of these results, the authors concluded that GABAergic transmission is not strongly involved in contrast adaptation. A subsequent study on this topic (McLean and Palmer, 1996) also concluded that $GABA_A$ receptors have little influence on contrast adaptation. In that study the authors also found no appreciable effects of application of the β-adrenergic antagonist propranolol and ACh on contrast adaptation. Without doubt EAA receptors play the leading role (McLean and Palmer, 1996), but given the fact that the amplification of responses by the voltage-sensitive NMDA receptors can be easily controlled by inhibition (reviewed in Daw et al., 1993), one might still ask whether GABAergic inhibition can be indirectly involved in the mechanisms underlying contrast adaptation and contrast gain control.

ORIENTATION TUNING

Orientation tuning is one of the characteristic features of cells in the primary visual cortex (Fig. 11-1E, F). This property is not, or only faintly, present in the lateral geniculate nucleus, one level below the striate cortex. Hubel and Wiesel (1962) proposed that in cat striate cortex, orientation selectivity is established by the spatial organization of the excitatory input from the dorsal lateral geniculate nucleus (dLGN) to simple cells in layer IV. Simple cell orientation selectivity was thought to reflect excitatory convergence from geniculate afferents with receptive fields aligned in rows (see Chapter 8). Excitatory convergence of this type may play a role in the generation of orientation selectivity (Ferster, 1987; Chapman et al., 1991; Pei et al., 1994). At the time Hubel and Wiesel carried out their pioneering studies, it was not known that only a small fraction (5–20%) of the excitatory synapses in layer IV (the main geniculocortical recipient zone) derive from geniculate afferents (Garey and Powell, 1971; LeVay and Gilbert, 1976; LeVay, 1986; Peters and Payne, 1993; Ahmed et al., 1994). The major excitatory input is derived from a massive convergent input from other cortical cells. The cortical input has the potential to enhance orientation selectivity via excitatory interactions both within the same orientation "column" (Douglas and Martin, 1991) and between "columns" with similar orientation preference (Ts'o et al., 1986; Gilbert and Wiesel, 1989) (see also Chapters 10 and 12). As a consequence of intracortical processing, orientation tuning shows considerable dynamics (Dinse et al., 1990; Shevelev et al. 1993). Furthermore, intracellular studies (Creutzfeldt et al., 1974; Innocenti and Fiore, 1974; Volgushev et al., 1993; Pei et al., 1994) and results from experiments using localized blockade of $GABA_A$-mediated intracortical inhibition (Sillito, 1975b, 1977a, 1979; Tsumoto et al., 1979; Sillito et al., 1980) established the importance of inhibitory processes for generating and shaping orientation and direction selectivity in area 17, although the way in which intracortical inhibition contributes to orientation tuning is still a matter of intense debate (Ferster and Koch, 1987; Douglas and Martin, 1991; Sillito, 1992; Vidyasagar et al., 1996).

In this context, local pharmacological studies appear to be a useful tool to investigate the possible mechanisms underlying orientation selectivity. Some of the pharmacological studies have supported the excitatory feedforward hypothesis of Hubel and Wiesel (e.g., Chapman et al., 1991; Nelson et al., 1994), whereas, others aimed at elucidating the role of intracortical inhibition have essentially extended the original hypothesis and have assigned major contributions to intracortical inhibitory processes in the sharpening of orientation tuning.

The classical micro-iontophoretical studies with the $GABA_A$ receptor antagonist bicuculline (Sillito, 1975b) convincingly demonstrated the reduction, or even the loss, of orientation tuning of the responses to moving bars in many cells in cat striate cortex (Fig. 11-4A-D). Similar results were obtained by Tsumoto et al. (1979), who reported that orientation tuning was substantially reduced during micro-iontophoresis of bicuculline, however, it was never abolished, and only additional systemic application of the GABA synthesis inhibitor 3-mercaptopropionic acid led to a complete loss of orientation tuning in 4 of 13 cells. When studying the effects of bicuculline on the responses of complex cells in more detail, Sillito (1979) found two categories of cells. Complex cells of one class lost orientation selectivity completely and thus received, most probably, nonoriented excitatory inputs. In neurons of another group, orientation tuning was only weakened. Further analysis showed that the strongest inhibitory effect seemed to arise from orientations on either side of the optimal. This led to the conclusion that intracortical connections between cortical columns representing different orienta-

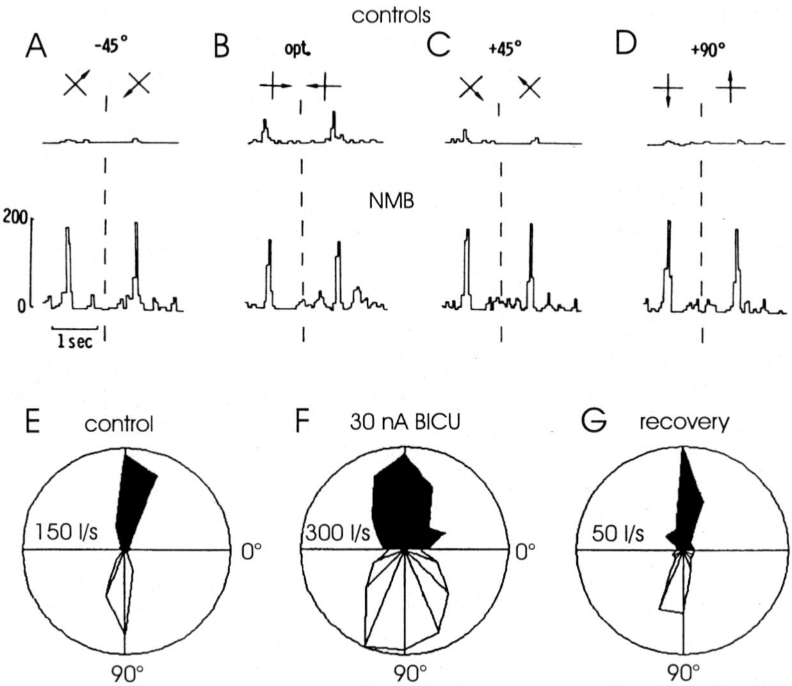

FIGURE 11-4. The effects of $GABA_A$ receptor blockade on orientation tuning of striate cortex cells. **(A–D)** Controls and PSTHs obtained from an area 17 simple cell during microiontophoresis of *N*-methyl bicuculline (NMB). Orientation of stimuli and direction of movement is shown above. In the control situation vertical is the optimal orientation, during NMB application all orientations elicit similar responses. (A–D reproduced with permission from Sillito, 1984.) **(E)** Polar plot from an area 17 complex cell stimulated with an oriented flashing bar. OFF responses upwards in black, ON responses downwards. **(F)** During bicuculline micro-iontophoresis with 30 nA, the orientation tuning of both the ON and OFF responses widens, and returns to normal during recovery **(G).** (Data from Eysel et al., 1998.)

tions would underlie the observed sharpening of orientation specificity (see Sillito, 1984; Sillito and Murphy, 1988 for earlier reviews). A quantitative reevaluation of bicuculline effects on orientation tuning with flashing light bars in cat striate neurons (Fig. 11-4E-G) revealed significant changes in all characteristics of the tuning: its width increased; the selectivity (ratio of maximal discharge frequencies at optimal and worst orientation) diminished by a factor of 1.5; and tuning quality, a factor combining tuning width and selectivity, dropped by a factor of 2.5 (Eysel et al., 1998). Bicuculline had quite comparable effects on orientation tuning in kittens aged 11 to 28 days, where orientation selectivity was eliminated in 18% of the neurons and was decreased in an additional 40% of the studied cells (Wolf et al., 1986). When drifting oriented gratings were used for stimulation instead of moving single bars, it became evident that different phases of responses are differentially affected by $GABA_A$ receptor blockade (Pfleger and

Bonds, 1995). In an early response phase of about 120 ms after onset of the movement, orientation tuning was significantly reduced, and the results were comparable to those obtained with moving bars, but the following steady state response retained significant orientation selectivity. Thus, while confirming that early orientation selective inhibition is mediated by $GABA_A$ receptors, these results led to the suggestion that if GABAergic inhibition plays a role in sharpening of orientation tuning of sustained responses, receptors other than $GABA_A$ receptors should be involved. Indeed, the orientation tuning of the sustained response was proved to depend strongly on $GABA_B$ receptor function. This was demonstrated with micro-iontophoretical application of the $GABA_B$ antagonist 2-hydroxy-saclofen (Allison et al., 1996). This specific $GABA_B$ antagonist significantly reduced orientation tuning in the late part of the response to a continuously drifting grating without affecting the early transient phase after drift onset. This indicates that both fast $GABA_A$ and delayed $GABA_B$ responses are involved in the inhibitory sharpening of orientation tuning. Another study used the so-called time-slice analysis combined with micro-iontophoretical application of bucuculline to investigate the temporal properties of $GABA_A$ receptor mediated inhibition with respect to the sharpening of orientation tuning in striate cortex cells (Eysel and Shevelev, 1994). In accordance with the preceding results, fast GABAergic inhibition acted between 30 and 100 to 150 ms after stimulus onset and sharpened and stabilized orientation tuning in a large group of cells. During $GABA_A$ receptor blockade, orientation tuning widened and preferred orientation sometimes changed in time (see also page 445). However, orientation tuning of the late responses was no different from that measured under control conditions. This similarity was established after approximately 120 ms, a time when $GABA_A$ receptor mediated inhibition is followed by the later and longer lasting action of the $GABA_B$ receptors (Pfleger and Bonds, 1995; Allison et al., 1996).

Although the data from a large number of studies with local extracellular applications of GABA receptor antagonists resulted in a convergent picture of intracortical inhibitory contributions to the improvement of orientation selectivity, one study with an alternative approach, namely intracellular blockade of $GABA_A$ inhibition at striate cortex cells by solutions containing the chloride-channel blockers picrotoxin or DIDS (4,4′-diisorhiocyanarostilbene-2,2′-disulfonic acid) and potassium channel blocker cesium, arrived at the opposite conclusion (Nelson et al., 1994). The use of this pipette solution in patch clamp recordings resulted in an effective blockade of the response to application of the $GABA_A$ agonist muscimol in vitro as well as to electrically evoked cortical inhibition in vivo. However, the orientation tuning of spike responses remained unchanged in all investigated cells. From these results the authors concluded that orientation selectivity is generated primarily by recurrent intracortical excitation. Subsequent computer simulations demonstrated that recurrent excitation is indeed capable of creating a high degree of selectivity of the final responses from poorly selective inputs (Somers et al., 1995) (Chapter 10). This view of the cortical circuitry is in accord with earlier models assigning a major role for the emerging properties of striate cortex RF architecture to exatatory intracortical feedback systems (Douglas and Martin, 1991). In light of these findings it can be suggested that the effects on RF architecture observed in the orientation domain with local application of $GABA_A$ antagonists are due to disinhibition not necessarily at the cell under study but in the general local cortical network.

An additional influence of the cholinergic input from the nucleus basalis of Meynert acts on the same circuitry and appears suited to enhance orientation selectivity and quality of the tuning as defined previously by facilitating the specific responses without changing the background activity (Sillito et al., 1985). This notion was questioned by other results where responses to optimal as well as nonoptimal orientations were enhanced and orientation tuning width was widened. A similar action seems to be taken by the tachykinins. SP was found to generally modulate excitability (see page 433), but not to change receptive field properties such as orientation selectivity (Albus et al., 1992).

DIRECTION SPECIFICITY

Direction specificity, like orientation tuning, is a property that emerges for the first time in cat primary visual cortex. And like orientation tuning, direction selectivity was initially suggested to be based on the spatial organization of the excitatory input from ON and OFF center geniculate cells to simple cells in layer IV (Hubel and Wiesel, 1962). Synergistic and/or antagonistic interactions between these subregions, when activated by stimuli moving in different directions, were thought to determine direction preference and selectivity. It was later claimed that simple-cell direction selectivity is largely independent of interactions between ON and OFF subregions (Goodwin et al., 1975; Heggelund, 1984; Peterhans et al., 1985). However, in a more recent study the detailed analysis revealed RFs that had a so-called inseparable spatiotemporal structure. Profiles of these RFs, plotted in the space-time domain, are tilted, thus the spatial outline of the RF is systematically shifted during the development of the response to visual stimuli (Fig. 11-1D). It has been suggested, that such a spatiotemporal organization can be related to directional specificity of the RFs (DeAngelis et al., 1993b; McLean et al., 1994).

Based on pharmacological experiments, specific effects on receptive field architecture and especially direction selectivity have been postulated for NMDA and AMPA receptors at cat visual cortex cells (Rivadulla et al. 1999). AMPA and NMDA receptors—in a delicate interplay with GABAergic inhibition—are involved in the generation of direction selectivty. Micro-iontophoretical application of APV leaves only the AMPA receptors effective and shows a linear reduction of responses without significantly changing the degree of direction selectivity. Micro-iontophoresis of CNQX (now only NMDA receptors remain active) shows an extreme suppression of responses in the nonpreferred direction that is released when GABAergic inhibition is antagonized. This effect uses the NMDA receptor property of the depolarization dependent Mg^{2+} block and is most closely related to the crucial role of GABAergic inhibition in the generation of direction selectivity.

When non-NMDA receptors were blocked with CNQX no change in direction selectivity of striate cortex cells was observed (Sato et al., 1999). This is not in accordance with the data described before and might be due to the fact that cells with complete suppression of the response in one direction (direction index = 1) were excluded from the analysis.

The distinctive role of GABAergic inhibition for the direction selectivity was shown several years ago (Sillito, 1975b, 1977a; Tsumoto et al., 1979). In many cases, even selectivity of the highest degree, when a cell responded to only one direction of motion of a bar, was completely lost when $GABA_A$ receptors were blocked with bicuculline (Fig. 11-5A). Complete loss of the direction selectivity during bicuculline application was observed in many simple cells in all cortical layers, and in some complex cells in the upper cortical layers (Fig. 11-5B). In other complex cells found in all cortical layers, the direction selectivity was insensitive to bicuculline microiontophoresis (Fig. 11-5C) (Sillito, 1977a). Direction selectivity of hypercomplex cells of the upper cortical layers was also usually resistant to bicuculline application. In contrast to orientation tuning (see page 439 for comparison), direction selectivity was also reduced by intracellular blockade of chloride conductances (Nelson et al., 1994). This result strongly supports the

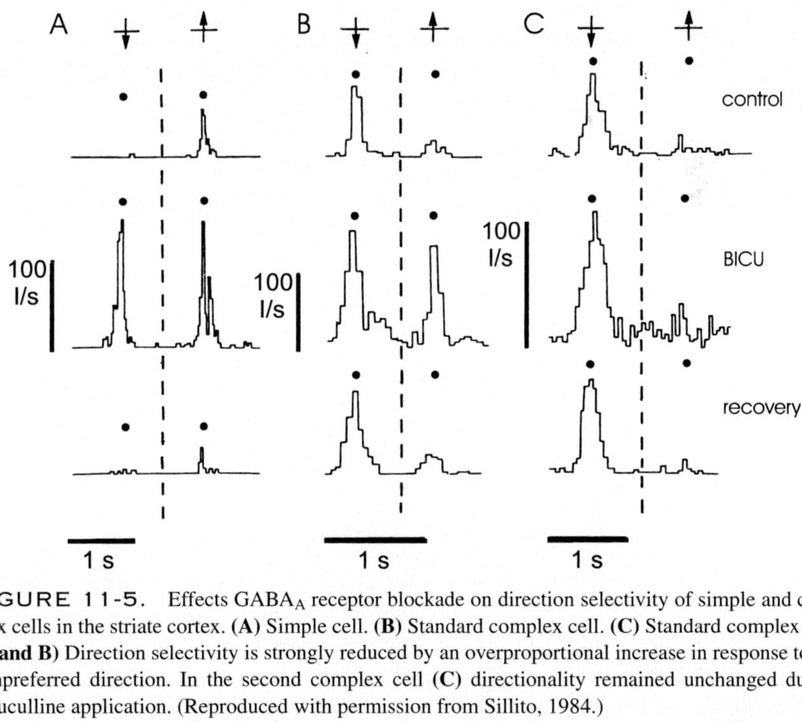

FIGURE 11-5. Effects $GABA_A$ receptor blockade on direction selectivity of simple and complex cells in the striate cortex. **(A)** Simple cell. **(B)** Standard complex cell. **(C)** Standard complex cell. **(A and B)** Direction selectivity is strongly reduced by an overproportional increase in response to the nonpreferred direction. In the second complex cell **(C)** directionality remained unchanged during bicuculline application. (Reproduced with permission from Sillito, 1984.)

involvement of inhibitory mechanisms in the creation of direction selectivity in cat striate cortex (see Chapter 9).

Direction selectivity in many simple cells of cat striate cortex is related to the spatiotemporal structure of their RFs. In these cells response timing changes gradually across the RF, and when a stimulus moves along this temporal gradient, it produces stronger responses than movement in the opposite direction (Movshon et al., 1978a; McLean and Palmer 1989; Reid et al., 1991). It has been demonstrated that this effect accounts for about half the directional tuning of simple cells in layer IV of striate cortex (Murthy et al., 1998). Micro-iontophoretical application of bicuculline methiodide (BMI) indeed led to a parallel reduction of the direction selectivity and a decrease of the spatiotemporal gradient of RFs in layer IV of cat striate cortex, pointing to the close relation between these two parameters (Murthy and Humphrey, 1999). However, the same analysis applied to layer VI cells revealed a reduction of direction selectivity that was not correlated to the changes of the spatiotemporal RF structure. These results illustrate the diversity of inhibitory mechanisms that can lead to direction selectivity (see pages 458–462). In 11 to 28-day old kittens bicuculline micro-iontophoresis led to a reduction (17%) or loss (58%) of direction specificity (Wolf et al., 1986).

Noradrenergic modulation influenced the selectivity for stimulus speed and direction of movement (McLean and Waterhouse, 1994). In about half the studied cells, velocity tuning became more narrow and the direction index increased. Thus, in addition to improving the signal-to-noise ratio (see page 433) NA seems to modulate the spatiotemporal tuning of RFs in cat area 17.

Activation of the cholinergic system had an enhancing effect on direction selectivity in striate cortex cells, irrespective of whether the firing rate of a cell was suppressed or facilitated by ACh (Sillito and Kemp, 1983; Murphy and Sillito, 1991). This strengthening of specificity is discussed in relation to different actions of ACh in the visual cortex (Murphy and Sillito, 1991). An M_1-receptor-mediated facilitation at pyramidal cells by reduction of potassium conductances associated with the M-current and the Ca^{2+}-dependent current or a M_2-receptor mediated excitation of inhibitory interneurons may lead to the strengthening of inhibitory input to pyramidal cells and thus to a reduction of their excitability due to a decrease of input resistance. These bidirectional influences of ACh on cortical cells may be one of the reasons for discrepancy with a report of reduced direction selectivity and orientation tuning (Sato et al., 1987).

LENGTH-TUNING AND RESPONSES TO COMPLEX STIMULI

In the original description length-tuning was a property of "hypercomplex cells" (Hubel and Wiesel, 1965). The basic mechanism behind this property is the end-stopping or end-inhibition seen in hypercomplex cells, which seems to rely on a powerful intracortical inhibition (Sillito, 1977b). Experiments that studied hypercomplex cells in the superficial layers of the striate cortex during application of the $GABA_A$ receptor antagonist bicuculline revealed that the strong inhibitory side-band effects were often weakened, but could not completely be eliminated (Sillito and Versiani, 1977). Given the fact that length tuning is not a unique property of complex cells, but is also found in simple cells of layer IV (Rose, 1977; Gilbert, 1977; Kato et al., 1978), this residual length selectivity could be explained by convergence of layer IV on layer III cells. Unfortunately, the effect of blocking GABAergic inhibition at end-stopped simple cells in layer IV has not been investigated in the cat striate cortex.

In two-thirds of cells studied in cat primary visual cortex, responses to complex oriented stimuli-like crosses differ significantly from responses to a single oriented bar (Shevelev et al., 1995), owing either to depression or facilitation by the crosslike figures. In a micro-iontophoretical study using bicuculline to block $GABA_A$ receptors a significant contribution of intracortical inhibition was demonstrated (Shevelev et al., 1998b). One-third of the studied cells that showed stronger responses to crosses than to single bars under control conditions lost their specific sensitivity to crosses during application of bicuculline (Fig. 11-6A).

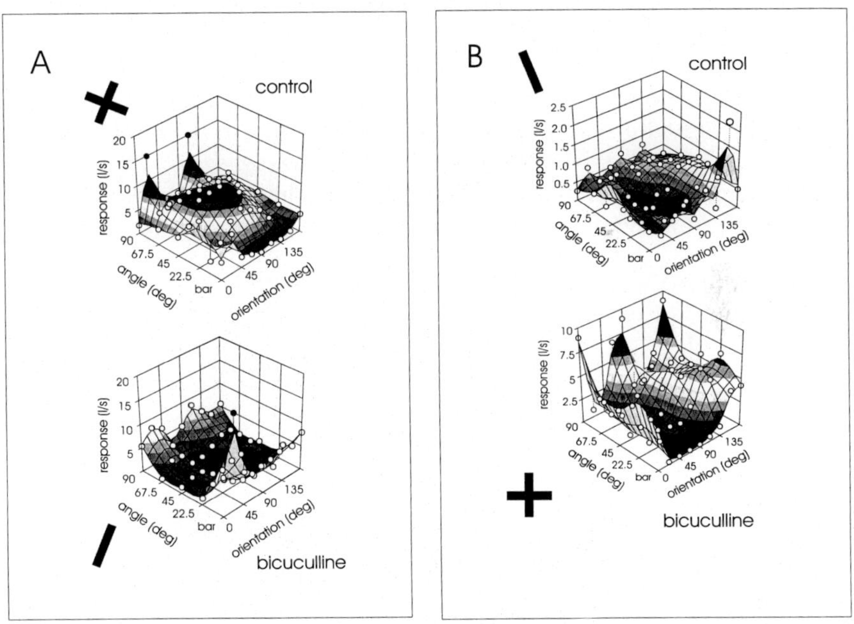

FIGURE 11-6. Effect of $GABA_A$ receptor blockade on the responses striate cortex cells to complex, crosslike figures. The three-dimensional graphs shows the orientation of one bar on the x-axis, the angle between bars on the y-axis, and the response on the z-axis. An angle of 0° corresponds to a single bar of the given orientation. **(A)** Simple cell shows higher sensitivity to a cross of two bars with 22.5° and 112.5° orientation than to the single bar at 22.5° under control conditions. With bicuculline microiontophoretically applied (lower diagram) a higher sensitivity to the singe bar at 22.5° is observed. **(B)** Complex cell with optimal response to a single bar of 157.5° under control conditions that develops a higher sensitivity to a cross with bars of 0° and 90° when bicuculline is applied. (Data from Shevelev et al., 1998.)

Another one third had no cross sensitivity before (i.e., the response to an optimally oriented bar was stronger than to any kind of crosslike figure) but became more sensitive to crosses during blockade of $GABA_A$ receptors (Fig. 11-6B). From these experiments with complex stimuli, one can conclude, that intracortical inhibition can modify RF architecture either in a way that achieves highly reliable tuning to the orientation of a single bar or a higher order selectivity for more complex stimulus patterns.

RF DYNAMICS

Receptive field architecture is not static but changes in time after stimulus onset. RFs are highly dynamic structures: ON and OFF subfields, excitation, and inhibition emerge in time, move, and disappear again (Dinse et al. 1990; Eckhorn et al. 1993, Shevelev et al. 1992; deAngelis et al., 1993a, 1993b; deAngelis et al., 1995).

GABAergic inhibition has been shown to play a major role for RF dynamics. The spatiotemporal RF architecture of simple cells in cat striate cortex (area 17) and area 18 is completely comparable (deAngelis et al., 1993a, 1993b; Pernberg et al., 1998). During blockade of $GABA_A$ receptors, the subfields of many cells widened in a way that simple cells became complex-like (Pernberg et al., 1998; Frégnac and Shulz, 1999) with spatially superimposed ON and OFF regions (Fig. 11-7). At the same time, the temporal dynamics of the fields changed. In about half the investigated cell RFs, the inverse late responses (Fig. 11-1C) that are related to stimulus offset were completely abolished (Fig. 11-7). This indicates that GABAergic inhibition under normal conditions introduces both spatial and temporal contrast in the activity of simple cells in the visual cortex.

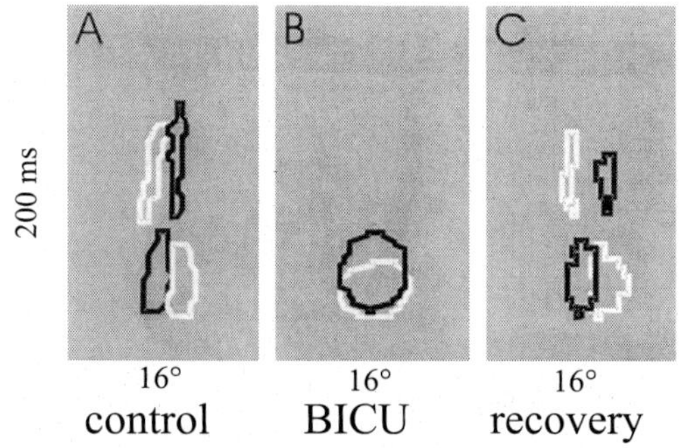

FIGURE 11-7. Influence of bicuculline on the xt-profile of a simple cell. **(A)** Control xt-profile before application of bicuculline (BICU) shows a space-time separable RF with an ON (white) and an OFF (black) subfield typical for a simple cell. **(B)** During microiontophoresis of BICU with 20 nA the early ON and OFF response fields completely overlapped in space as normally typical for complex cells and the late, inverse responses were abolished. **(C)** After recovery the original properties of the space-time separable field were regained. (Data from Pernberg et al., 1998.)

The fast topographical reorganization of excitatory and inhibitory subfields in the RF might be directly related to the dynamics of orientation tuning that was observed in striate cortex cells (Dinse et al. 1990; Shevelev et al., 1993). The new finding was that preferred orientation is not a static property in all cells, but in some neurons it was changed with time after stimulus onset as visible in polar plots obtained with the so-called time-slice method (Fig. 11-8A, B). In a study with the $GABA_A$ receptor antagonist bicuculline, it was shown that this dynamic property was differentially related to GABAergic inhibition in two groups of striate cortex cells. The first group of cells was characterized by sharp orientation tuning and stable preferred orientation during the whole response. During blockade of inhibition, these cells typically developed orientation tuning dynamics between 30 and 150 ms after stimulus onset when GABAergic inhibition was blocked (Fig. 11-8C). The other group showed significant dynamics of orientation tuning under control conditions that were reduced or lost during bicuculline micro-iontophoresis (Fig. 11-8D) (Eysel and Shevelev, 1994; Shevelev et al., 1998a). As in the findings with complex stimuli (crosslike figures), inhibition seems to provide twofold and bidirectional contributions to RF architecture. In most cells inhibition increases contrast and stability of RF architecture, which leads to sharpening of orientation tuning in the spatial domain and its stabilization in the temporal domain. In other cells, inhibition enhances dynamics of RF architecture and facilitates changes of the preferred orientation during the time course of the response. Different functional tasks can be performed by the two sets of neurons. The first group can reliably resolve the orientation of contours with high angular resolution, and the second group could use temporal orientation coding and act as orientation scanning devices (Shevelev et al., 1993).

RF PLASTICITY

Characteristic features of RF architecture such as ocular dominance, orientation selectivity, RF size, and ON and OFF response balance are subject to substantial plasticity during early postnatal life (critical period) and can also be modified in an activity-dependent way in the adult cat striate cortex (Frégnac et al., 1988, 1992; Shulz and Frégnac, 1992; Eysel et al., 1998; Frégnac and Shulz, 1999). These experiments have provided in vivo evidence that the modifiability of mature visual cortical connections follows the Hebbian rule, which requires both presynaptic and postsynaptic activity within a defined time window (Hebb, 1949). It also provides evidence that in vivo plasticity may exploit mechanisms involved in the long-term changes of synaptic transmission long-term potentiation [LTP] and long-term depression [LTD] observed in vitro in the rat and cat visual cortex (Artola and Singer, 1987; Kimura et al., 1989; Hirsch and Gilbert, 1993; Kirkwood and Bear, 1994).

Some of the basic mechanisms underlying this modifiability of RFs by specific use or training of cells in the striate cortex have also been assessed in vivo by pharmacological studies. Pairing of visual stimuli with microiontophoretical

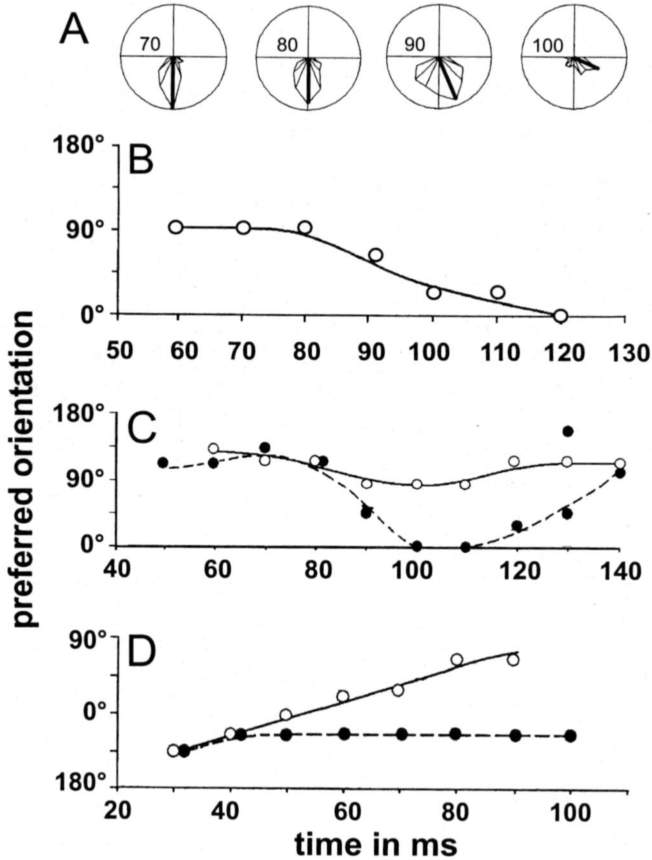

FIGURE 11-8. Effects of $GABA_A$ receptor blockade on orientation tuning dynamics of cells in the cat striate cortex. **(A)** Example of a simple cell in area 17 of the cat that shows orientation tuning dynamics under normal conditions. The polar diagrams show tuning of ON-responses to a flashing oriented bar in 10 ms time slices at different times after stimulus onset (starting point of each time slice indicated in ms). The preferred orientation is drawn as thick line in each polar plot. **(B)** Shift of the optimal orientation in degrees (ordinate) after stimulus onset (abscissa) for the simple cell shown in **(A).** Within 40 ms the preferred orientation of this cell travels from 90° to 0° between 80 and 120 ms after stimulus onset. **(C)** Preferred orientation versus time after stimulus onset for a simple cell with negligible orientation tuning dynamics under control condition (open symbols). When bicuculline was microiontophoretically applied (+50 nA), this cell developed significant tuning dynamics (filled symbols) between 70 ms and 140 ms after stimulus onset. The preferred orientation moved within 30 ms from 135° to 0° and after 10 ms of stability within another 30 ms back from 0° to 112.5°. C. Complex cell in area 17 that displayed orientation tuning dynamics ranging over the full range of 180° between 30 ms and 100 ms after stimulus onset under control conditions (open symbols). When bicuculline was applied (+15 nA), this dynamics was completely lost (filled symbols). (Data from Shevelev et al., 1998a.)

application of ACh, NA, glutamate, or one of its receptor agonists (NMDA) led to long-lasting (40 and more minutes) changes of RF architecture of cortical cells (Greuel et al., 1988). Specifically, ocular dominance was shifted in RFs of 4 to 6-week-old kittens but not in adult cats when stimulation of the initially less effective (nondominant) eye was combined with iontophoresis of NA, ACh, NMDA, or glutamate either in a "cocktail" or separately. As mentioned previously, all of these drugs activate visual cortex cells and can facilitate activation of NMDA- and voltage-dependent Ca^{2+}-conductances that are involved in long-term changes of synaptic efficacy (recently reviewed by Malenka and Nicoll, 1999). Indeed, the intracellular Ca^{2+}-level can be directly related to the induction of LTP or LTD (Hansel et al., 1996).

Using a similar associative conditioning paradigm exclusively in adult cats with pharmacological pairing with either glutamate or GABA, significant changes in orientation tuning were found (McLean and Palmer, 1998). As in many other studies, only about half of the cells changed their responses after the conditioning procedure. The outcome was not a shift in preferred orientation toward the positively (with glutamate) or negatively (with GABA) reinforced orientations, but an additional strengthening of the response to the positively conditioned orientation or a weakening of the response to the negatively conditioned orientation (Fig. 11-9A). This result shows that positive reinforcement widens orientation tuning, whereas sharpening of orientation tuning is only possible when inhibition is involved. Consequently a change of the preferred orientation is possible only when the initially preferred orientation is negatively reinforced by inhibition and the new preferred orientation is positively reinforced by coexcitation. In a similar proportion of cells, spatial phase tuning curves could be modified by conditioning (McLean and Palmer, 1998). In these experiments, after conditioning with spatial phases differing by 90° from the initial optimum, the preferred response was shifted to spatial phases 30–40° offset from the initially preferred phase (Fig. 11-9B). These results clearly demonstrate the ability of cells in the adult cat striate cortex to change their RF architecture in a use-dependent way. Receptive fields are definitively not fixed, static morphological and functional entities, their architecture can be changed throughout life and adapted to functional needs. Such changes of RFs on early stages of the visual system can be interpreted in relation to perceptual learning (for a review see Eysel, 2000).

STRIATE CORTICAL NETWORK EFFECTS ON RF PROPERTIES

Pharmacological methods can be applied to reversibly alter the functional connectivity in the cortical network and to get further insight into the architecture of receptive fields. Micro-iontoporesis can be applied to locally activate or block receptors at a cell under study. This kind of approach has been described above (pages 432–447). These pioneering experiments addressed in a very direct way

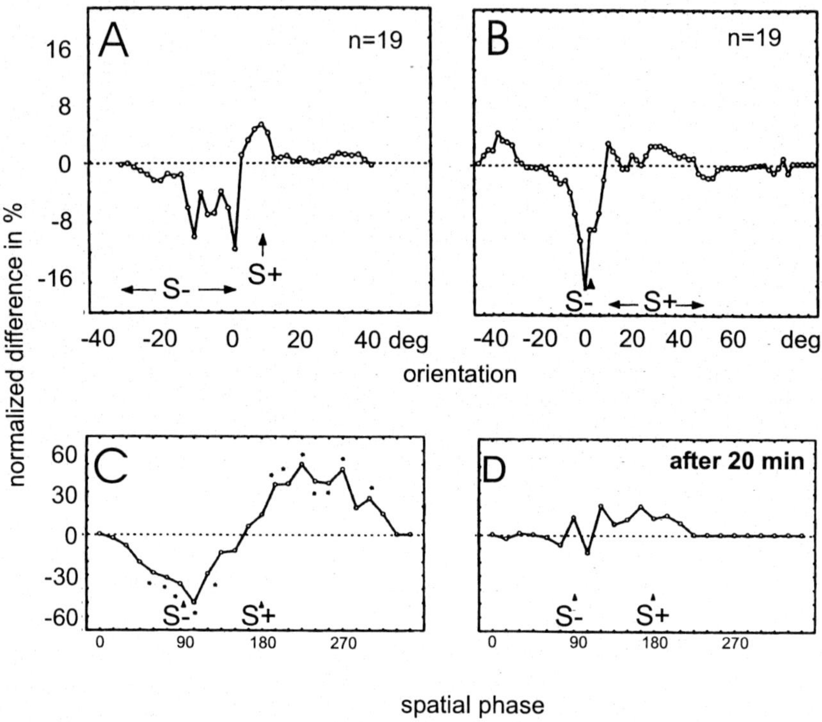

FIGURE 11-9. Plasticity of orientation tuning and spatial phase in the adult cat striate cortex after conditioning with stimuli of different orientation or phase paired with microiontophoresis of Glutamate (S+) or GABA (S–). **(A,B)** Conditioning with different orientations (average data from 19 cells). The difference of the normalized firing rate before and after conditioning (%) is plotted against orientation (degrees). **(A)** Curves aligned to the positively conditioned orientation (S+) with variable negatively conditioned orientations (S–, range indicated by arrows). **(B)** Curves aligned to the orientation of the negatively conditioned stimulus (S–) with accordingly variable S+ (range indicated by arrows). **(C)** This figure shows a single example of an area 17 simple cell positively and negatively conditioned with different spatial phase. Normalized difference of firing rate is plotted against spatial phase. Clear facilitatory effects of S+ and depressive effects by S– are present shortly after conditioning. The effects decrease substantially within 20 minutes after conditioning as shown in **D.** (Reproduced with permission from McLean and Palmer, 1999.)

the problem of the *functional anatomy* underlying the cortical receptive field. Later, a quite independent approach was developed to investigate influences from defined remote regions on RF architecture by reversible inactivation of the striate cortex (Bolz and Gilbert, 1986; Eysel et al., 1987). This was primarily done by injection of the inhibitory transmitter GABA, which completely suppresses the activity of all cell types in the striate cortex. This method was used to study the columnar connectivity (see page 449) and its influence on RF architecture by silencing infragranular layers and recording the responses of layer IV and supragranular cells. In addition to the columnar (vertical) network, the striate cortex is also characterized by intercolumnar long-range lateral connections (see Chapter 10), which contribute to RF architecture and can be studied by inactivation of laterally remote sites (see pages 452–462). It turned out that the two independent approaches—direct influence on the transmitter receptors at a cell under study and dissection of the intracortical network by silencing of the remote cells contributing to the RF architecture—were quite useful to provide new as well as complementary evidence (for earlier reviews on inactivation studies see Bolz et al., 1989; Eysel, 1992; Eysel and Wörgötter, 1992; Eysel et al., 1994).

PHARMACOLOGICAL DISSECTION OF VERTICAL (COLUMNAR) CORTICAL NETWORKS

The vertical columnar network has been studied in detail and the intracortical pathway of information processing can be described as follows: the geniculocortical input innervates layer IV (with a small projection to layer VI as well), layer IV projects to layers II/III, which in turn projects to layer V. Layer V projects to layer VI (and sparsely back to the supragranular layers), and layer VI cells project back to layer IV (Gilbert and Wiesel, 1979) (see also Chapters 1 and 7). This circuit has been further functionally elaborated and led to the proposal of a "canonical microcircuit" (Douglas et al., 1989; Douglas and Martin, 1991), which has since proven useful to explain many aspects of intracortical processing related to the emergence of RF architecture.

In some inactivation studies, single striate cortex cells were recorded while the activity was blocked in individual layers of the same cortical column. In this way a pharmacological dissection of the columnar cortical circuit was possible (for an earlier review see Bolz et al., 1989). Many of the characteristics of RF architecture that have been pharmacologically investigated with specific agonists and antagonists are specifically related to the intracortical connectivity between different cortical layers. The inactivation method is most specific if the blocking agent allows reversibility and only affects cells localized in the vicinity of an injection site but not fibers of passage. The inhibitory neurotranmsitter GABA and its longer acting analog muscimol are most suited for local, reversible inactivation; the action of the local anesthetic lidocaine is also reversible but silences both somata and fibers of passage.

End-inhibition

End-inhibition (Bishop and Henry, 1972; Rose, 1977) has been ascribed to interlaminar connections predominantly mediating inhibition from cells in layer VI to cells in layer IV (Gilbert, 1977). In the first study that used GABA for localized inactivation, the role of columnar connectivity in determining end-inhibition of layer IV cells has been assessed with localized pressure injection of GABA into layer VI of the same cortical columns below or micro-iontophoresis (Bolz and Gilbert, 1986). End-inhibited simple cells in layer IV (as well as supragranular complex cells) reversibly lost their end-inhibition when layer VI was inactivated by GABA injection (Fig. 11-10A). The authors suggested that the larger RFs of

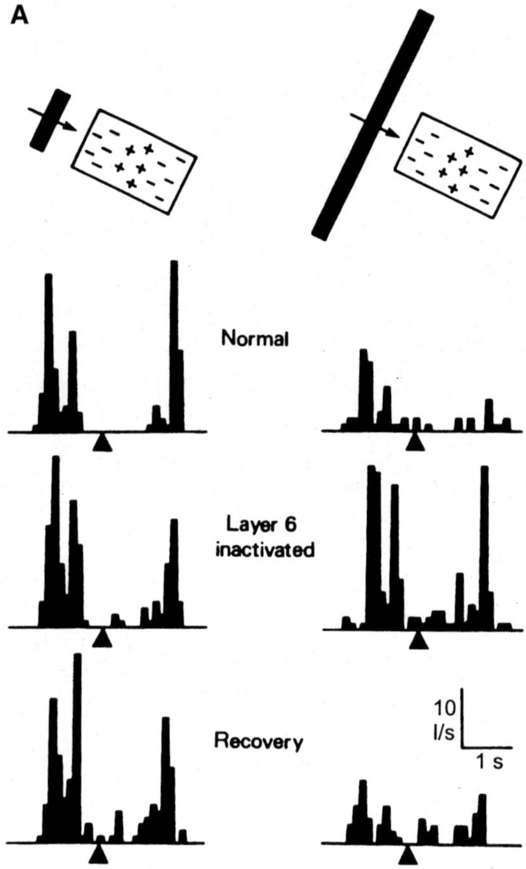

FIGURE 11-10. Dissection of columnar cortical circuits by local inactivation with GABA in the lower cortical layers and recordings from single cells above the inactivation site. **(A)** Inactivation in layer VI (pressure injection of 0.1 μl GABA) abolished end-inhibition in a layer IV simple cell. The cell showed the maximal response for both directions of stimulus movement across the RF (reversal of movement direction indicated by black triangles) when the stimulus bar (1° length) was adjusted to the dimensions of the RF (1°× 1.75°), and this response remained unchanged during layer VI inactivation (middle row). The right column shows significant end-inhibition (54%) in the normal control condition and direction preference to the right when the RF was stimulated with a long bar (8° length). The end-inhibition as well as the direction preference was abolished during inactivation of layer VI and returned after recovery (bottom row). (Reproduced with permission from Bolz and Gilbert, 1986.) **(B)** Orientation tuning curves obtained from a supragranular cell before, during and after inactivation of the infragranular layers with GABA micro-iontophoresis. The complex cell shows a symmetrical widening of tuning width. **(C)** Example of a supragranular layer simple cell with asymmetrical widening of orientation tuning width during inactivation of the infragranular layers. (Reproduced with permission from Allison and Bonds, 1994.)

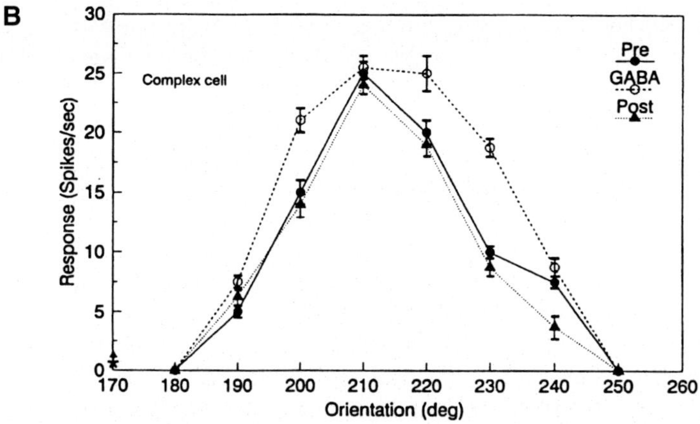

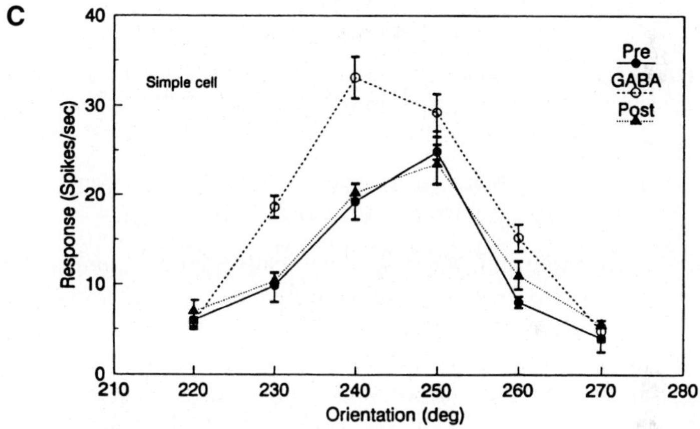

FIGURE 11-10 *(Continued)*

layer VI cells are well suited to introduce end-inhibition at the smaller RFs of layer VI simple cells and that the complex cells in superficial layers get the "hypercomplex property" from the layer IV simple cells (Bolz and Gilbert, 1986; Bolz et al., 1989). Although this study showed the direct influence of layer VI cells on layer IV end-inhibition, further investigation showed that the corticofugal feedback to the lateral geniculate nucleus is involved as well (Murphy and Sillito, 1987).

Spatial RF Organization

The two-dimensional RF architecture was studied in the upper layers of the cat striate cortex and layers V–VI were inactivated below by GABA micro-iontophoresis from two barrels of a micropipette (Sun and Bonds, 1994). The RF size increased, RF substructure was often disturbed, and the noise in the RF area increased reversibly during inactivation of the lower cortical layers with GABA. Such results clearly demonstrate vertically organized mechanisms contributing to the normal RF architecture. Not only spatial RF organization but also orientation tuning could be shown to depend on the intact connectivity between infragranular and supragranular layers in the striate cortex of the cat.

Orientation Tuning

When the lower cortical layers were inactivated by GABA micro-iontophoresis, the orientation tuning bandwidth of about half the investigated neurons increased by roughly 50% (Allison and Bonds, 1994). Changes of orientation tuning were either symmetrical to the originally preferred orientation (Fig. 11-10B) or asymmetrical at only one side of the tuning curve (Fig. 11-10C). The symmetry or asymmetry of the changes in the tuning curves depended on the vertical alignment of recording and inactivation sites with symmetrical widening detected when the two sites were perfectly aligned, whereas asymmetrical widening and eventual small changes in preferred orientation (by $6.67 \pm 5.77°$) were observed when cells were not in vertical registration.

PHARMACOLOGICAL DISSECTION OF HORIZONTAL (INTERCOLUMNAR) CORTICAL NETWORKS

Similar to inactivating and recording in vertically aligned columns, the GABA inactivation method also proved useful to study the contributions of lateral projecting axonal systems in cat striate cortex.

Long-ranging axonal systems that run parallel to the cortical surface form a morphological basis for lateral processing in the cat visual cortex, especially in the upper and lower layers were initially shown by degeneration methods (Creutzfeldt et al., 1977) (see also Chapter 10). After the demonstration of clusters of terminals emitted at regular intervals by intracellularly labeled pyramidal cell axons spanning several millimeters of horizontal distance (Gilbert and Wiesel, 1979, 1983; Martin and Whitteridge, 1984; Kisvárday et al., 1986) and the observation of periodic patchy labeling (Matsubara et al., 1985; Luhmann et al., 1986; LeVay, 1988; Gilbert and Wiesel, 1989), increasing interest was directed toward horizontal interactions within area 17 (see Chapter 10). This interest was focused on the long-range axonal systems of pyramidal cells and predominantly on excitatory functions (Nelson and Frost, 1985) as shown with cross-correlation analysis of recordings from pairs of cells with similar orientation specificity (T'so et al., 1986). More details about the lateral excitatory network were disclosed by 2-deoxyglucose labeling combined with local injections of rhodamine latex beads (Gilbert and Wiesel, 1989), which revealed connections between iso-orientation domains in area 17. On the other hand, inhibitory lateral interactions have been observed between cells with dissimilar orientation preferences in cross-correlation histograms obtained from pairs of cells in the cat visual cortex (Hata et al., 1988). To connect orientation columns of all different preferred orientations, connections have to travel laterally over at least 1 mm, which is the average distance between iso-orientation bands (Löwel et al., 1987) (see Chapter 3). Excitatory as well as inhibitory connections can bridge this distance, although their maximal possible horizontal spread is markedly different. Axons of pyramidal cells span up to about 3 mm in the neonatal (Luhmann et al., 1986) and 5 mm in the adult cat visual cortex (Kisvárday and Eysel, 1992), whereas the axons of GABAergic cells give rise to axons over distances of 0.5–2 mm (Martin et al., 1983; Somogyi et al., 1983; Albus et al., 1991; Albus and Wahle, 1994). These axons have been shown to link cells with similar as well dissimilar preferred orientations (Kisvárday and Eysel, 1993; Kisvárday et al., 1994, 1997).

Lateral excitatory and inhibitory interactions could play a significant role in generating or sharpening specific response properties of visual cortical neurons. Lateral iso-orientation inhibition can produce selectivity for direction of motion (Bishop et al., 1971; Creutzfeldt et al., 1974; Innocenti and Fiore, 1974). An orientation bias might be primarily introduced in different ways: by aligned geniculate inputs (Hubel and Wiesel, 1962) (see also Chapter 8), by an orientational bias already present in retina (Levick and Thibos, 1980; Leventhal and Schall, 1983; Thibos and Levick, 1985) and LGN (Daniels et al., 1977; Vidyasagar and Urbas, 1982; Soodak et al., 1987; Shou and Leventhal, 1989), or by inhibition between nonoriented cells with laterally displaced receptive fields (Heggelund, 1981). Inhibition between cells with different preferred orientations ("cross-orientation inhibition") can further refine the orientation tuning (Ferster and Koch, 1987).

Given the differential properties of long-ranging lateral connections and the fact that local inactivation does not selectively inactivate excitatory or inhibitory cells, the interpretation of studies that silenced lateral input systems is not always simple. Nevertheless, a certain specificity in local inactivation can be achieved by restricting GABA injections to special layers or by varying the distance across the columnar system of the visual cortex. In particular the knowledge of the functional properties (orientation preference, direction selectivity) at recording and inactivation sites proved especially helpful for understanding the results. As exemplified next, our studies have demonstrated that the orientation tuning and/or directionality of single cells in area 17 can be substantially modified by localized inactivation of laterally remote, visuotopically corresponding sites in the same area, and that the results are to a large extent predictable on the basis of the relative orientation and direction preference at the recording and inactivation sites.

RF Size of Layer VI Cells

The generation of long RFs is one possible role of long range horizontal excitatory connections in striate cortex. When layer V was inactivated by small pressure injections of GABA lateral to a cell under study in layer VI, the layer IV cell lost a part of its elongated RF (Fig. 11-11) corresponding to the topography of the layer V RFs (Bolz and Gilbert, 1989). The long-ranging axons of layer V pyramidal cells give rise to synapses on layer VI cells (Gilbert and Wiesel, 1979, 1983;

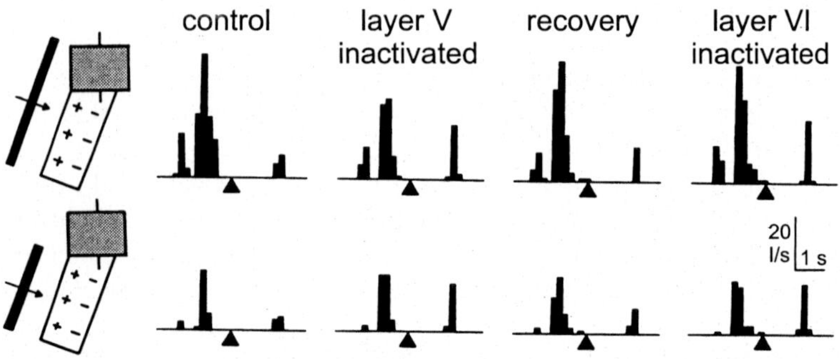

FIGURE 11-11. Effect of inactivation of layer V lateral from the recording site of a layer VI simple cell. The pictograms to the left show RF size and preferred orientation (small lines) at the layer V inactivation site. The significantly larger RF of the cell recorded in layer VI is shown with ON and OFF regions. The first column of histograms shows the responses to a long bar covering the full length of the layer VI cell (above) and the response to a shorter bar that was adjusted to that part of the layer VI cell RF that was not shared by the RF of the layer V inactivation site. The larger response in the upper histogram indicated length summation in this cell. When layer V was inactivated by pressure injection of 0.5 μl of GABA, the response to the long bar was decreased whereas the response to the shorter bar remained unaffected. The effect showed full recovery in the next column of histograms and could not be obtained when layer VI was inactivated with the same inactivation micropipette advanced into layer VI (rightmost column of histograms). (Reproduced with permission of Bolz and Gilbert, 1989.)

Martin and Whitteridge, 1984) and thus can contribute to the long RF of the layer VI cell. In about 20% of the cells studied with inactivation in layer V directionality was strongly reduced, as in the example shown in Fig. 11-11. This probably indicates that lateral inhibitory interactions are present in the infragranular cortical circuitry similar to those extensively discussed for the supragranular layers and layer IV (see pages 458–462).

Orientation Tuning

Orientation selectivity and direction selectivity (Hubel and Wiesel, 1962) were originally described as typical properties of visual cortical cells in response to moving bar-shaped stimuli. Ever since Sillito and others demonstrated the loss of both specificities in many cells during blockade of GABAergic transmission with the $GABA_A$ antagonist bicuculline (Sillito, 1975b, 1977a, 1979; Sillito et al., 1980; Tsumoto et al., 1979; Wolf et al., 1986), the possible contributions of GABA-mediated inhibitory interactions to these cortical response specificities have remained a matter of continuous debate (for a review see Vidyasagar et al., 1996). Orientation specificity was originally ascribed to spatially oriented subcortical inputs (Hubel and Wiesel, 1962), but the previously mentioned effects of bicuculline led to the assumption that intracortical inhibition must be involved (Sillito, 1979; Sillito et al., 1980). Regarding the question whether inhibition from iso- or cross-orientations prevails, some intracellular recordings disclosed exclusively iso-orientation excitation and inhibition (Ferster, 1986, 1987; Douglas et al., 1991; Sato et al., 1991), whereas others reported also cross-orientation inhibitory inputs (Creutzfeldt et al., 1974; Innocenti and Fiore, 1974; Sato et al., 1991; Volgushev et al., 1993; Pei et al., 1994). Several extracellular studies demonstrated inhibition from a broader range of orientations with respect to a target cell (Blakemore and Tobin, 1972; Morrone et al., 1982; Ramoa et al., 1986; Matsubara et al., 1987; Hata et al., 1988). Such interactions imply lateral connections between columns of different orientation preferences.

When we first tested orientation tuning with remote lateral inactivation in the supragranular layers of the cat striate cortex (Eysel et al., 1990), we used an array of four inactivation pipettes with an independently moving recording electrode in the center (Fig. 11-12A). The average distance between recording and inactivation sites was 500 μm, a distance corresponding to half the hypercolumnar size in area 17, where dissimilar orientation preferences can be expected. Although inactivation was performed in the supragranular layers, in all cortical depths orientation tuning was significantly widened in 33 of 54 and lost in 3 of 54 cells. The widening of orientation tuning was due to a significant increase of responses to previously nonpreferred orientations (Fig. 11-12B, C).

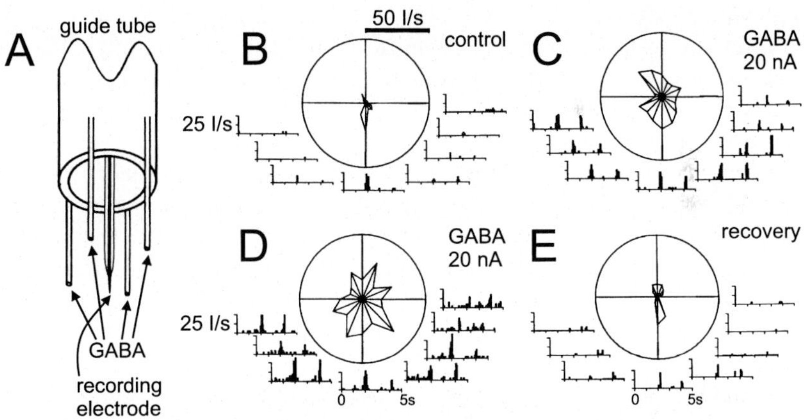

FIGURE 11-12. Effects of inactivation in the supragranular layers of striate cortex on orientation tuning of a simple cell recorded before during and after GABA micro-iontophoresis at four laterally remote sites. **(A)** Schematical drawing of the circular inactivation-micropipette array. The inactivation pipettes were glued to a guide tube of 1 mm diameter. The recording electrode was moved through the center of the guide tube. The lateral distance between each inactivation site and the recording site was 500 μm, accordingly. **(B)** The polar plot shows narrow orientation tuning before inactivation, the small PSTHs show individual responses to back and forth movement of the stimulus bar at the corresponding eight orientations between 0° and 157.5°. **(C,D)** Continuous GABA micro-iontophoresis (+20 nA) at all four inactivation sites led to widening of orientation tuning (5 to 9 minutes **(C)** and 15 to 19 minutes **(D)** after the beginning of microiontophoresis. **(E)** The effects returned to normal 10 minutes after discontinuation of the application. (Data from Eysel et al., 1990.)

The influence of distance of inactivation in terms of prevailing blockade of excitatory or inhibitory contributions to orientation specificity was assessed in a different study with varying distances (0.4 to 2.9 mm) between recording and inactivation sites (Wörgötter and Eysel, 1991). GABA inactivation was most effective in reducing orientation tuning owing to disinhibition of nonoptimal orientations when the distance of the inactivation site was about 0.5 mm. Inversely, a reduction in orientation tuning resulting from reduced excitation with the optimal orientation occurred most effectively from a distance of 1 mm.

These findings were then more specifically addressed when we started to discriminate iso- and cross-inactivation sites regardless of distance from the recording site (Crook et al., 1997, 1998). It turned out that the relation of orientation preference at the recording and inactivation sites is the first crucial parameter for the outcome of an inactivation experiment. This can be seen most impressively in an example in which the inactivation site displayed an approximately horizontal optimal orientation with strong preference for downward movement (Fig. 11-13A). Three different cells were recorded at two different recording sites, two with the same optimal orientation, one with the same (Fig. 11-13B) and one with the opposite preferred direction (Fig. 11-13C), and one with the orthogonal, "cross-oriented" optimal orientation (Fig. 11-13D). Inactivation at one and the same site had differential effects on the different cells. The cell with the same direction preference reversed its direction selectivity owing to a loss of excitation in its previously preferred direction (Fig. 11-13B), the cell with the opposite preferred direction lost directionality owing to disinhibition of the originally suppressed response in the nonpreferred direction (Fig. 11-13C), and the cell with orthogonal optimal orientation widened its orientation tuning in response to the moving bar (Fig. 11-13D).

As a rule inactivation at iso-orientation sites did not have an effect on orientation tuning except for 2 of 40 (5%) of the cells that showed an increase of tuning width accompanied by a decrease of the response in the optimal orientation (preferred direction). By contrast, cross-orientation inactivation induced a significant decrease in orientation specificity in 21 of 35 cells (60%) with an increase in mean tuning half-width of 126% from 23.5° (± 8.2°) to 53.0° (± 16.2°). Across all investigated cells, cross-orientation inactivation resulted in a 78% increase in mean tuning half-width, whereas iso-orientation inactivation had negligible influence on orientation tuning (increase in mean tuning half-width by about 7%). This difference was statistically highly significant.

Simple cells were more strongly affected by cross-orientation inactivation than C-cells. The analysis of significant inactivation effects on cells recorded from different layers (histologically identified) revealed an effectiveness of 29% in layer II, 71% in layer III, and 82% in layer IV. The inactivation sites were in layer III and in upper layer IV. The most prominent effect leading to a decrease in orientation specificity was an increase in responses to nonoptimal stimuli elicited by cross-orientation inactivation. In view of the orientation specificity of lateral excitatory and inhibitory connections, the most plausible explanation for these effects is that they reflected the loss of a cross-orientation inhibitory input (Sillito, 1979; Morrone et

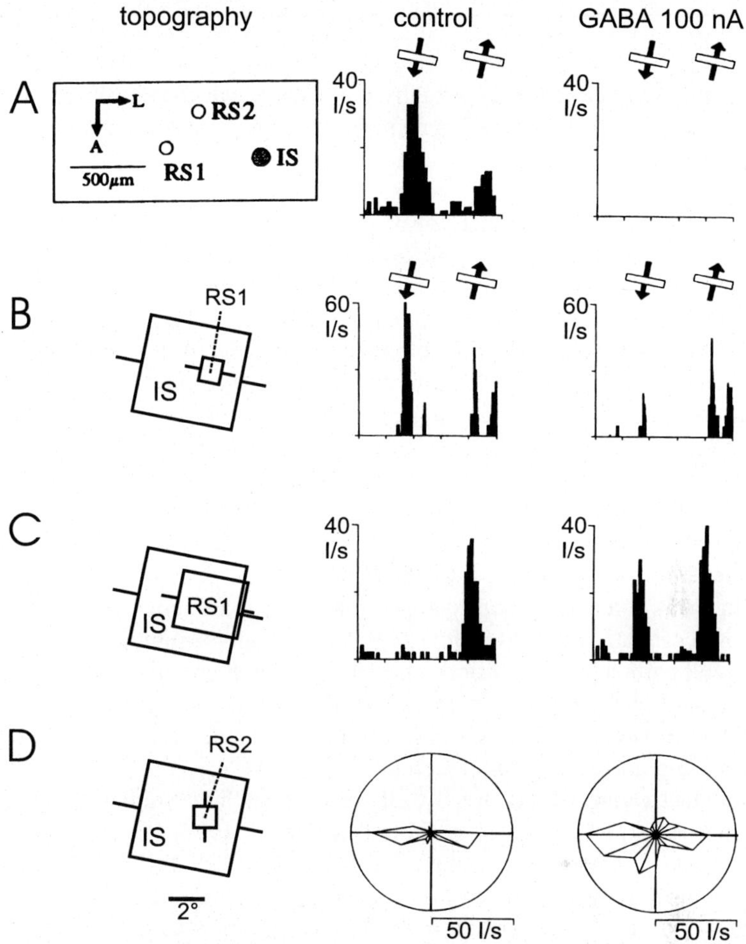

FIGURE 11-13. Specific inactivation effects on directionality at an iso-orientation recording site and orientation tuning at a cross-oriented recording site can be elicited from one and the same inactivation site in area 17. **(A)** The drawing in the box shows the topography of the inactivation site (IS) and recording sites (RS1, RS2) on the surface of area 17. Inactivation was performed with GABA micro-iontophoresis in the upper half of layer IV. The distance to the recording sites is 650 μm to RS1 and 520 μm to RS2. The middle column comprises the control recordings. In the multiunit recording at the inactivation site downward motion is preferred. During GABA iontophoresis with +100 nA, the activity at the inactivation site is completely silenced (right column). **(B)** The simple cell recorded in upper layer IV had a small RF completely overlapping with the RF of the inactivation site (IS) and the same orientation preference (left column). A slight direction preference for downward motion under control conditions (middle column) was completely lost due to strongly reduced excitation in this direction during GABA micro-iontophoresis at the inactivation site while the response to the opposite direction remained unchanged (right column). **(C)** A complex cell recorded in lower layer IV again shared the orientation preference with the inactivation site and had a larger RF that nearly completely overlapped with the RF of the inactivation site. This cell preferentially responded to upward motion (control). During inactivation at IS the suppression of the null-direction was released and directionality nearly lost (right column). **(D)** The completely overlapping RF of a simple cell recorded at the other recording site (RS2) was approximately orthogonally tuned in relation to the cells at the inactivation site. The sharp orientation tuning shown by the polar plot in the middle column was significantly widened during inactivation at IS. All of the effects were measured 4 minutes after onset and were completely reversible within 4 minutes after offset of GABA micro-iontophoresis (not shown). (Data from Crook et al., 1997.)

al., 1982; Ramoa et al., 1986; DeAngelis et al., 1992), which contributes primarily to orientation tuning by suppressing responses to nonoptimal orientations.

A straightforward explanation for the increase in response to nonoptimal stimuli observed after laterally remote inactivation is a loss of a monosynaptic inhibitory input to a recorded cell from cells at the inactivation sites. This is supported by combined anatomical/physiological experiments in areas 17 and 18 (Crook et al., 1998), which always revealed direct inhibitory projections from an inactivation site to a recording site in cases where cross-orientation inactivation caused broadening of orientation tuning. Increases in response to nonoptimal stimuli were elicited from inactivation sites in layer IV or layers III + IV in cells recorded in layers III–IV, or from an inactivation site in layer III in cells recorded in layers II–IV (Crook et al., 1997). In principle, all of these effects could have been mediated directly via the inactivation of large basket cells in layers III or IV, which have long-range tangential projections in the same or neighboring layers (Martin et al., 1983; Somogyi et al., 1983) and innervate both iso-orientation and cross-orientation sites (Kisvárday and Eysel, 1993). The interpretation of the effects in terms of a loss of monosynaptic inhibition is also consistent with results from cross-correlation studies in area 17 that demonstrated inhibitory connections between cell pairs with similar or dissimilar orientation preferences over horizontal distances of up to ~600 μm, both within and between layers II/III and IV (Hata et al., 1991). However, the effects may also have been mediated via indirect pathways involving one or more interneurons.

These inactivation results agree well with intracellular studies in which stationary flashed stimuli were used (Volgushev et al., 1993; Pei et al., 1994). In most cells that were driven monosynaptically from the dLGN, the excitatory input was initially weakly biased for orientation or nonoriented, and sharp orientation tuning developed as a result of inhibition at cross-orientations and facilitation around the optimum orientation. In cells that were driven indirectly from the dLGN, orientation tuning was determined largely by the excitatory input, but in some cases it was sharpened by cross-orientation inhibition. These results, together with those obtained with inactivation, support models of cortical orientation selectivity based on weak afferent geniculate biases and broadly tuned cross-orientation inhibition (Vidyasagar, 1987; Wörgötter and Koch, 1991).

Direction Selectivity

Lateral inhibition in visual cortex cells has been shown by intracellular recordings (Benevento et al., 1972; Innocenti and Fiore, 1974), and its participation in direction selectivity has been supported by a number of detailed studies (Goodwin et al., 1975; Emerson and Gerstein, 1977; Ganz and Felder, 1984). Several models have assumed that inhibition generates direction selectivity in the visual cortex (Sillito, 1977a; Barlow, 1981); however, excitatory convergence has also been considered to contribute to direction selectivity in complex cells (Movshon et al., 1978b, see also Chapter 9). Using intracortical local inactivation we have shown a loss of direction selectivity in cells of layers II–VI when laterally remote areas were silenced by GABA micro-iontophoresis (Eysel et al., 1987, 1988).

The influence of distance of inactivation in terms of prevailing blockade of inhibitory contributions to direction specificity was assessed by inactivating at lateral distances between 0.4 and 2.9 mm (Wörgötter and Eysel, 1991). GABA inactivation of cells about 1 mm lateral to the recording site most effectively induced a loss of directionality owing to disinhibition of the response to the nonpreferred direction of motion (i.e., lateral inhibition over a cortical distance of 1 mm was statistically most effective in generating direction specificity).

When we also discriminated the functional properties of recording and inactivation sites (Crook et al., 1997) of 40 inactivations from iso-orientation sites, 26 (65%) produced an effect on directionality, whereas only 6 (17%) of 35 cells subjected to cross-orientation inactivation showed an effect on directionality. This difference was statistically highly significant. The effects were due to an increase or decrease in response to the preferred direction, or to an increase in response to the nonpreferred direction (Fig. 11-13 B, C). Overall, iso-orientation inactivation had a significantly greater effect on the direction index (DI) than did cross-orientation inactivation.

When iso-orientation sites were effectively inactivated, by far the most frequent effect was a reduction of directionality in the recorded cell (89%). This was most frequently due to an increase of the response to the nonpreferred direction after inactivation of an iso-orientation site with a preferred response opposite to that of the recorded cell (58%), less often directionality decreased due to a decrease of the preferred response after inactivation of an iso-orientation site with the same preferred response (31%), and in a few cases (11%) directionality increased after inactivation of an iso-orientation site with the same preferred response (Fig. 11-14).

A decrease in direction selectivity resulting from an increase in response to the nonpreferred direction was reliably elicited by inactivation of iso-orientation sites with opposite direction preference. These effects must involve the loss of an inhibitory input that normally plays a major role in sharpening direction selectivity. Previous evidence for a contribution of intracortical inhibition to direction selectivity derived from intracellular studies revealing inhibitory postsynaptic potentials (IPSPs) in response to the nonpreferred direction of motion (Creutzfeldt et al., 1974; Innocenti and Fiore, 1974; Sato et al., 1991), and the pharmacological studies referred to previously in which blockade of the $GABA_A$-receptors during bicuculline microiontophoresis led to an increased response to the nonpreferred direction (Sillito, 1975b, 1977a; Tsumoto et al., 1979). Suppressive influences feeding forward through the receptive field in the nonpreferred direction were also observed in quantitative receptive-field studies (Emerson and Gerstein, 1977; Ganz and Felder, 1984; Emerson et al., 1987). The reversible inactivation experiments in the cortical network close to the recorded cells clearly show that directionality can be decreased or abolished by a reduction of intracortical inhibition (Eysel et al., 1987, 1988). The experiments also show that the intracortical inhibitory input that suppresses a cell's response to its nonpreferred direction is not dependent on the topographical relationship between the recording and inactivation sites and derives primarily from cells with approximately the

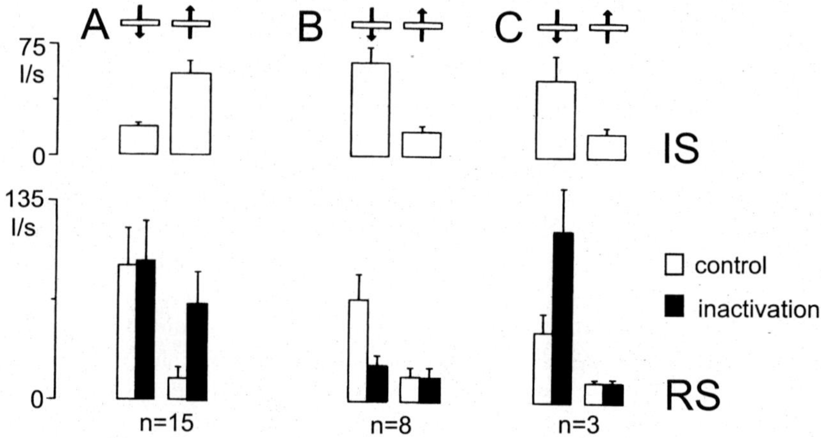

FIGURE 11-14. Predictability of the effects of iso-orientation inactivation on directionality. The upper row shows the average responses at the inactivation sites, the lower row the responses at the recording sites before (open bars) and during GABA inactivation (black bars) at the inactivation site. **(A)** Opposite preferred direction at inactivation and recording sites (n = 15). In this situation the predominant effect is a loss of directionality due to an increase of the response in the nonpreferred direction. **(B, C)** Same direction preferred at inactivation and recording sites (n = 11). The predominant effect of inactivation is shown in **(B),** where directionality decreases due to a reduction of the response in the preferred direction. **(C)** Only few cells showed an increase of directionality due to larger responses to the preferred direction when inactivation was performed at a site with the same preferred direction. (Reproduced with permission from Crook et al., 1997.)

same receptive field position and orientation preference, but opposite direction preference (Crook et al., 1997). Combining $GABA_A$ inactivation with the use of a masking paradigm (Crook et al., 1996) revealed that lateral inhibition contributing to direction selectivity operates over small distances within a cell's excitatory discharge region as had been suggested by earlier detailed RF studies (Emerson and Gerstein, 1977; Ganz and Felder, 1984; Emerson et al., 1987).

A decrease of direction selectivity owing to a decrease in response to the preferred direction was most frequently elicited by inactivation of an iso-orientation site with the same direction preference as the recorded cell. This decreases in response reflects a loss of excitation within the cortex. Anatomical studies have shown that even in layer IV (the main geniculocortical input zone), the number of intracortical excitatory contacts is much larger than that of the geniculate input (Garey and Powell, 1971; LeVay and Gilbert, 1976; LeVay, 1986; Peters and Payne, 1993; Ahmed et al., 1994). The functional dominance of intracortical excitatory inputs is also indicated by the reduced excitation after intracortical inactivation. The horizontally extending axonal arborizations of supragranular layer pyramidal cells and the laterally directed projections of layer IV spiny stellate cells that innervate layer IV and/or layers II–III (Gilbert and Wiesel, 1983; Martin and Whitteridge, 1984) are a suitable substrate for the effects of iso-orientation inactivation, especially because their patchy pattern of innervation (Gilbert and Wiesel, 1983; Martin and Whitteridge, 1984; Kisvárday and Eysel, 1992) displays the highest density at iso-oriented sites in striate cortex (Gilbert and Wiesel, 1989). The comparatively massive effect in response to inactivation of a single cortical site might be due to the fact that this local inactivation can indirectly cause a loss of mutual excitation among a cluster of highly interconnected excitatory cells that have iso-orientation/direction connections with each other, and a recorded cell. Dense local interconnections exist in the horizontal axonal systems of supragranular-layer pyramidal cells and layer IV spiny stellate cells (Kisvárday et al., 1986; Ahmed et al., 1994; Anderson et al., 1994). These interconnections could give rise to recurrent excitation (Douglas et al., 1995) among cells within iso-direction domains. The patchy horizontal fiber system thought to mediate excitatory coupling between remote iso-orientation domains (Ts'o et al., 1986; Gilbert and Wiesel, 1989; Engel et al., 1990) shows a high degree of reciprocity (Boyd and Matsubara, 1991; Kisvárday and Eysel, 1992) (see also Chapters 5 and 10). Silencing one cluster of cells in this system of reciprocal connections between cells with co-oriented, co-axially aligned receptive fields could cause a reduction in the recurrent excitation within this network of laterally remote iso-direction domains with the same direction preference and could thus cause the observed massive decrease of excitation at a recorded cell.

Seldom was an increase in direction selectivity elicited by an increase of the response to the preferred direction during inactivation of an iso-orientation site with the same direction preference. This effect proves that there is inhibition not only between cells with opposite direction preferences, but also between cells with the same preferred direction of motion. This inhibitory effect could be mediated monosynaptically via laterally directed iso-orientation inhibitory projections from remote cortical sites to the recorded cells as revealed by cross-correlation analysis (Hata et al., 1991) or via indirect pathways that involve at least one inhibitory interneuron on the way to a recorded cell. The functional role of the inhibition revealed by this effect might be the regulation of cortical amplification (Douglas and Martin, 1991) by iso-orientation inhibition.

In summary, intracortical contributions to directionality seem to consist of two main mechanisms, a major contribution of inhibition from sites with opposite preferred direction and a less frequent excitation from sites with the same preferred direction of motion.

The significance of intracortical inhibition for shaping responses in the striate cortex are continuously under debate. The magnitude of inhibition tuned nonoptimal orientations and nonpreferred directions as shown in intracellular studies (Creutzfeldt et al., 1974; Innocenti and Fiore, 1974; Sato et al., 1991; Volgushev et al., 1993; Pei et al., 1994) was considered insufficient by a number of authors to suppress strong excitation like that evoked by optimal stimuli (Douglas et al., 1988; Douglas and Martin, 1991; Berman et al., 1992). The alternative possibility of shunting inhibition was not considered to be relevant in the visual cortex (Douglas et al., 1988), although in a recent intracellular study, effective shunting inhibition was demonstrated to act during the early phase of visual cortical processing (Borg-Graham et al., 1998).

With a completely independent approach, the inactivation studies have shown contributions of intracortical inhibition to sharpening or even generation of orientation tuning and direction selectivity comparable to the results with $GABA_A$-receptor antagonists. This exemplifies that intracortical processing is involved in suppressing responses to nonoptimal stimuli in the cat striate cortex. It is independent from the question as to whether the pronounced increases in response to nonoptimal stimuli are due to the direct inactivation of an inhibitory input to a recorded cell (Creutzfeldt et al., 1974; Innocenti and Fiore, 1974; Sato et al., 1991; Volgushev et al., 1993; Pei et al., 1994), as would be supported by the GABA-antagonist studies, or to a more indirect inhibitory action on the relatively weak geniculocortical excitatory input to a recurrent cortical network (Douglas and Martin, 1991; Somers et al., 1995; Suarez et al., 1995) (see Chapter 12).

ACKNOWLEDGMENTS

The scientific contributions of my colleagues Prof. Shevelev, Drs. Crook, Kisvàrday, Pernberg, Volgushev and Wörgötter that substantially contributed to this chapter are gratefully acknowledged. Our own work was continuously supported by the Deutsche Forschungsgemeinschaft.

REFERENCES

Ahmed, B., Anderson, J. C., Douglas, R. J., Martin, K. A. C., and Nelson, J. C. (1994). Polyneuronal innervation of spiny stellate neurons in cat visual cortex. *J. Comp. Neurol.* **341,** 39–49.

Albus, K. (1981). Hypothalamic and basal forebrain afferents to the cat's visual cortex: a study with horseradish *Neurosi. Lett.* **24,** 117–121.

Albus, K., Chao, H. H. A., and Hicks, T. P. (1992). Tachykinins preferentially excite certain complex cells in the infragranular layers of feline striate cortex. *Brain Res.* **587,** 353–357.

Albus K., and Wahle, P. (1994). The topography of tangential inhibitory connections in the postnatally developing and mature striate cortex of the cat. *Eur. J. Neurosci.* **6,** 779–792.

Albus, K., Wahle, P., Lübke, J., and Matute, C. (1991). The contribution of GABA-ergic neurons to horizontal intrinsic connections in upper layers of the cat's striate cortex. *Exp. Brain Res.* **85,** 235–239.

Allison, J. D., and Bonds, A. B. (1994). Inactivation of the infragranular striate cortex broadens orientation tuning of supragranular visual neurons in the cat. *Exp. Brain Res.* **101,** 415–426.

Allison, J. D., Kabara, J. F., Snider, R. K., Casagrande, V. A., and Bonds, A. B. (1996). $GABA_B$-receptor-mediated inhibition reduces the orientation selectivity of the sustained response of striate cortical neurons in cats. *Vis. Neurosci.* **13,** 559–566.

Allman, J., Miezin, F., and McGuiness, E. (1985). Stimulus specific responses from beyond the classical receptive field: neurophysiological mechanisms for local-global comparisons in visual neurons. *Annu. Rev. Neurosci.* **8,** 407–430.

Anderson, J. C., Douglas, R. J., Martin, K. A. C., and Nelson, J. C. (1994). Synaptic output of physiologically identified spiny stellate neurons in cat visual cortex. *J. Comp. Neurol.* **341,** 16–24.

Artola, A., and Singer, W. (1987). Long-term potentiation and NMDA receptors in rat visual cortex. *Nature* **330,** 649–652.

Barlow, H. B. (1981). Critical limiting factors in the design of the eye and visual cortex. *Proc. R. Soc. Lond. B.* **212,** 1–34.

Baumfalk, U., and Albus, K. (1987). Baclofen inhibits the spontaneous and visually evoked responses of neurones in the striate cortex of the cat. *Neurosci. Lett.* **75,** 187–192.

Bear, M. F., Carnes, K. M., and Ebner, F. F. (1985). An investigation of cholinergic circuitry in cat striate cortex using acetylcholinesterase histochemistry. *J. Comp. Neurol.* **234,** 411–430.

Benevento, L. A., Creutzfeldt, O. D., and Kuhnt, U. (1972). Significance of intracortical inhibition in the visual cortex. *Nat. New Biol.* **238,** 124–126.

Berman, N. J., Douglas, R. J., and Martin, K. A. (1992). GABA-mediated inhibition in the neural networks of visual cortex. *Prog. Brain Res.* **90,** 443–476.

Bishop, P. O., Coombs, J. S., and Henry, G. H. (1971). Interaction effects of visual contours on the discharge frequency of simple striate neurones. *J. Physiol.* **219,** 659–687.

Bishop, P. O., and Henry, G. H. (1972). Striate neurons: receptive field concepts. *Invest. Ophthalmol.* **11,** 346–354.

Blakemore, C., and Tobin, E. A. (1972). Lateral inhibition between orientation detectors in the cat's visual cortex. *Exp. Brain Res.* **15,** 439–440.

Bolz, J., and Gilbert, C. D. (1986). Generation of end-inhibition in the visual cortex. *Nature* **320,** 362–365.

Bolz, J., and Gilbert, C. D. (1989). The role of horizontal connections in generating long receptive fields in the cat visual cortex. *Eur. J. Neurosci.* **1,** 263–268.

Bolz, J., Gilbert, C. D., and Wiesel, T. N. (1989). Pharmacological analysis of cortical circuitry. *TINS* **12,** 292–296.

Borg-Graham, L. J., Monier, C., and Fregnac, Y. (1998). Visual input evokes transient and strong shunting inhibition in visual cortical neurons. *Nature* **393,** 369–373.

Boyd, J., and Matsubara, J. (1991). Intrinsic connections in cat visual cortex: a combined anterograde and retrograde tracing study. *Brain Res.* **560,** 207–215.

Chapman, B., and Stryker, M. P. (1992). Origin of orientation tuning in the visual cortex. *Curr. Opin. Neurobiol.* **2,** 498–501.

Chapman, B., Zahs, K. R., and Stryker, M. P. (1991). Relation of cortical cell orientation selectivity to alignment of receptive fields of the geniculocortical afferents that arborize within a single orientation column in ferret visual cortex. *J. Neurosci.* **11,** 1347–1358.

Creutzfeldt, O. D., Garey, L. J., Kuroda, R., and Wolff, J. R. (1977). The distribution of degenerating axons after small lesions in the intact and isolated visual cortex of the cat. *Exp. Brain Res.* **27,** 419–440.

Creutzfeldt, O. D., Kuhnt, U., and Benevento, L. A. (1974). An intracellular analysis of visual cortical neurones to moving stimuli: responses in a co-operative neuronal network. *Exp. Brain Res.* **21,** 251–274.

Crook, J. M., and Eysel, U. T. (1992). GABA-induced inactivation of functionally characterized sites in cat visual cortex (area 18): effects on orientation tuning. *J. Neurosci.* **12,** 1816–1825.

Crook, J. M., Eysel, U. T., and Machemer, H. F. (1991). Influence of GABA-induced remote nactivation on the orientation tuning of cells in area 18 of feline visual cortex: a comparison with area 17. *Neuroscience* **40,** 1–12.

Crook, J. M., Kisvárday, Z. F., and Eysel, U. T. (1996). GABA-induced inactivation of functionally characterized sites in cat visual cortex (area 18): local determinants of direction selectivity. *J. Neurophysiol.* **75,** 2071–2088.

Crook, J. M., Kisvárday, Z. F., and Eysel, U. T. (1997). GABA-induced inactivation of functionally characterized sites in cat striate cortex: effects on orientation tuning and direction selectivity. *Vis. Neurosci.* **14,** 141–158.

Crook, J. M., Kisvárday, Z. F., and Eysel, U. T. (1998). Evidence for a contribution of lateral inhibition to orientation tuning and direction selectivity in cat visual cortex: reversible inactivation of functionally characterized sites combined with neuroanatomical tracing techniques. *Eur. J. Neurosci.* **10,** 2056–2075.

Daniels, J. D., Norman, J. L., and Pettigrew, J. D. (1977). Biases for oriented moving bars in lateral geniculate nucleus neurons of normal and stripe-reared cats. *Exp. Brain Res.* **29,** 155–172.

Daw, N. W., Stein, P. G. S., and Fox, K. (1993). The role of NMDA receptors in information processing. *Annu. Rev. Neurosci.* **16,** 207–222.

De Lima, A. D., and Singer, W. (1986). Cholinergic innervation of the cat striate cortex: a choline acetyltransferase immunocytochemical. *J. Comp. Neurol.* **250,** 324–338.

DeAngelis, G. C., Ohzawa, I., and Freeman, R. D. (1993a). Spatiotemporal organization of simple-cell receptive fields in the cat's striate cortex. I. General characteristics and postnatal development. *J. Neurophysiol.* **69,** 1091–1117.

DeAngelis, G. C., Ohzawa, I., and Freeman, R. D. (1993b). Spatiotemporal organization of simple-cell receptive fields in the cat's striate cortex. II. Linearity of temporal and spatial summation. *J. Neurophysiol.* **69,** 1118–1135.

DeAngelis, G. C., Ohzawa, I., and Freeman, R. D. (1995). Receptive-field dynamics in the central visual pathways. *TINS* **18,** 451–458.

DeAngelis, G., Robson, J. G., Ohzawa, I., and Freeman, R. D. (1992). Organization of suppression in receptive fields of neurons in cat visual cortex. *J. Neurophysiol.* **68,** 144–163.

DeBruyn, E. J., and Bonds, A. B. (1986). Contrast adaption incat visual cortex is not mediated by GABA. *Brain Res.* **383,** 339–342.

Dinse, H. R., Krüger, K., and Best, J. (1990). A temporal structure of cortical information processing. *Concepts Neurosci.* **1,** 199–238.

Douglas, R. J., Koch, C., Mahowald, M., Martin, K. A. C., and Suarez, H. H. (1995). Recurrent excitation in neocortical circuits. *Science* **269,** 981–985.

Douglas, R. J., and Martin, K. A. C. (1991). A functional microcircuit for cat visual cortex. *J. Physiol.* **440,** 735–769.

Douglas, R. J., Martin, K. A. C., and Whitteridge, D. (1988). Selective responses of visual cortical cells do not depend on shunting inhibition. *Nature* **332,** 642–644.

Douglas, R. J., Martin, K. A. C., and Whitteridge, D. (1989). A canonical microcircuit for neocortex. *Neural Comput* **1,** 480–488.

Eckhorn, R., Krause, F., and Nelson, J. I. (1993). The RF-cinematogram. *Biol. Cybern.* **69,** 37–55.

Emerson, R. C., Citron, M. C., Vaughn, W. J., and Klein, St. (1987). Nonlinear directionally selective subunits in complex cells of cat Striate cortex. *J. Neurophysiol.* **58,** 33–65.

Emerson, R. C., and Gerstein, G. L. (1977). Simple striate neurons in the cat. II. Mechanisms underlying directional asymmetry and directional selectivity. *J. Neurophysiol.* **40,** 136–155.

Engel, A. K., König, P., Gray, Ch., and Singer, W. (1990). Stimulus-dependent neuronal oscillations incat visual cortex: inter-columnar interaction as determined by cross-correlation analysis. *Eur. J. Neurosci.* **2,** 588–606.

Eysel, U. T. (1992). Lateral inhibitory interactions in areas 17 and 18 of visual cortex. In: Progress in brain research (R. R. Mize, R. E. Marc, and A. M. Sillito, Eds.), Vol. 90, pp. 407–422, Amsterdam, Elsevier.

Eysel, U. T. (2000). Plasticity of receptive fields on early stages of the adult visual system. In: Perceptual learning (M. Fahle, Ed.), in press. MIT Press, Cambridge.

Eysel, U. T., Crook, J. M., and Machemer, H. F. (1990). GABA-induced remote inactivation reveals cross-orientation inhibition in the cat striate cortex. *Exp. Brain Res.* **80,** 626–630.

Eysel, U. T., Eyding, D., and Schweigart, G. (1998). Repetitive visual stimulation elicits LTP-like receptive field plasticity in adult cat visual cortex. *Neuroreport* **9,** 949–954.

Eysel, U. T., Kisvárday, Z. F., Wörgötter, F., and Crook, J. M. (1994). Large basket cells and lateral inhibition in cat visual cortex. In Structural and functional organization of the neocortex (B. Albowitz, B., Albus, K., Kuhut, U., Northdurft, H.-Ch., and Wahle, P., Eds.), *Exp. Brain Res.* Series 24, pp. 212–220, Heidelberg, Springer.

Eysel, U. T., Muche, T. H., and Wörgötter, F. (1988). Lateral interactions at direction selective striate neurones in the cat demonstrated by local cortical inactivation. *J. Physiol.* **399,** 657–675.

Eysel, U. T., and Shevelev, I. A. (1994). Time-slice analysis of inhibition in cat striate cortical neurones. *Neuroreport* **5,** 2033–2036.

Eysel, U. T., Shevelev, I. A., Lazareva, N. A., and Sharaev, G. A. (1998). Orientation tuning and receptive field structure in cat striate cortex neurons during local blockade of intracortical inhibition. *Neuroscience* **84,** 25–36.

Eysel, U. T., and Wörgötter, F. (1992). Horizontal intracortical contributions to functional specificity in cat visual cortex. In Information processing in the cortex: Experiments and theory (A. Aertsen, and V. Braitenberg, Eds.), pp. 301–323, Berlin-Heidelberg-New York, Springer.

Eysel, U. T., Wörgötter, F., and Pape, H.-Chr. (1987). Local cortical lesions abolish lateral inhibition at direction selective cells in cat visual cortex. *Exp. Brain Res.* **68,** 606–612.

Ferster, D. (1986). Orientation selectivity of postsynaptic potentials in neurons of cat primary visual cortex. *J. Neurosci.* **6,** 1284–1301.

Ferster, D. (1987). Origin of orientation-selective EPSP in simple cells of cat visual cortex. *J. Neurosci.* **7,** 1780–1791.

Ferster, D., and Koch, Ch. (1987). Neuronal connections underlying orientation selectivity in cat visual cortex. *TNS* **10,** 487–492.

Fox, K., Sato, H., and Daw, N. (1989). The location and function of NMDA receptors in cat and kitten visual cortex. *J. Neurosci.* **9,** 2443–2454.

Fox, K., Sato, H., and Daw, N. (1990). The effect of varying stimulus intensity on NMDA-receptor activity in cat visual cortex. *J. Neurophysiol.* **64,** 1413–1428.

Frégnac, Y., and Shulz, D. E. (1999). Activity-dependent regulation of receptive field properties of cat area 17 by supervised Hebbian learning. *J. Neurobiol.* **41,** 69–82.

Frégnac, Y., Shulz, D., Thorpe, S., and Bienenstock, E. (1988). A cellular analogue of visual cortical plasticity. *Nature* **333,** 367–370.

Frégnac, Y., Shulz, D., Thorpe, S., and Bienenstock, E. (1992). Cellular analogs of visual cortical epigenesis. I. Plasticity of orientation selectivity. *J. Neurosci.* **12,** 1280–1300.

Freund, T. F., Maglóczky, Z., Soltész, I., and Somogyi, P. (1986). Synaptic connections axonal and dendritic patterns of neurons immunoreactive for cholecystokinin in the visual cortex of the cat. *Neuroscience* **19,** 1133–1159.

Freund, T. F., Martin, K. A. C., Somogyi, P., and Whitteridge, D. (1985). Innervation of cat visual areas 17 and 18 by physiologically identified X- and Y-type thalamic afferents. II.

Identification of postsynaptic targets by GABA immunocytochemistry and Golgi imp. *J. Comp. Neurol.* **242,** 275–291.

Gabbott, P. L. A., and Somogyi, P. (1986). Quantitative distribution of GABA-immunoreactive neurons in the visual cortex (area 17) of the cat. *Exp. Brain Res.* **61,** 323–331.

Ganz, L., and Felder, R. (1984). Mechanism of directional selectivity in simple neurons of the cat's visual cortex analyzed with stationary flash sequences. *J. Neurophysiol.* **51,** 294–324.

Garey, L. J., and Powell, T. P. (1971). An experimental study of the termination of the lateral geniculocortical pathway in the cat and monkey. *Proc. R. Soc. Lond. B Biol. Sci.* **179,** 41–63.

Gilbert, C. D. (1977). Laminar differences in receptive field properties of cells in cat primary visual cortex. *J. Physiol.* **268,** 391–421.

Gilbert, C. D., and Wiesel, T. N. (1979). Morphology and intracortical projections of functionally characterised neurones in the cat visual cortex. *Nature* **280,** 120–125.

Gilbert, C. D., and Wiesel, T. N. (1983). Clustered intrinsic connections in cat visual cortex. *J. Neurosci.* **3,** 1116–1136.

Gilbert, C. D., and Wiesel, T. N. (1989). Columnar specificity of instrinsic horizontal and corticocortical connections in cat visual cortex. *J. Neurosci.* **9,** 2432–2442.

Goodwin, A. W., Henry, G. H., and Bishop, P. O. (1975). Direction selectivity of simple striate cells: properties and mechanism. *J. Neurophysiol.* **38,** 1500–1523.

Greuel, J. M., Luhmann, H. J., and Singer, W. (1988). Pharmacological induction of use-dependent receptive field modifications in the visual cortex. *Science* **242,** 74–77.

Gu, Q., Liu, Y., and Cynader, M. S. (1994). A study of tachykinin-immunoreactivity in the cat visual cortex. *Brain Res.* **640,** 336–340.

Hansel, C., Artola, A., and Singer, W. (1996). Different threshold levels of postsynaptic [Ca2+]i have to be reached to induce LTP and LTD in neocortical pyramidal cells. *J. Physiol. (Paris)* **90,** 317–319.

Hata, Y., Tsumoto, T., Sato, H., Hagihara, K., and Tamura, H. (1988). Inhibition contributes to orientation selectivity in visual cortex of cat. *Nature* **335,** 815–817.

Hata, Y., Tsumoto, T., Sato, H., and Tamura, H. (1991). Horizontal interactions between visual cortical neurones studied by cross-correlation analysis in the cat. *J. Physiol.* **441,** 593–614.

Hebb, D. O. (1949). *The organization of behavior.* New York, Wiley.

Heggelund, P. (1981). Receptive field organization of simple cells in cat striate cortex. *Exp. Brain Res.* **42,** 89–98.

Heggelund, P. (1984). Direction asymmetry by moving stimuli and static receptive field plots for simple cells in cat striate cortex. *Vis. Res.* **24,** 13–16.

Hicks, T. P., Albus, K., Kaneko, T., and Baumfalk, U. (1993). Examination of the effects of cholecystokinin 26–33 and neuropeptide Y on responses of visual cortical neurons of the cat. *Neuroscience* **52,** 263–279.

Hirsch, J. A., and Gilbert, C. D. (1993). Long-term changes in synaptic strength along specific intrinsic pathways in the cat visual cortex. *J. Physiol.* **461,** 247–262.

Hubel, D. H., and Wiesel, T. N. (1959). Receptive fields of single neurones in the cat's striate cortex. *J. Physiol.* **148,** 574–591.

Hubel, D. H., and Wiesel, T. N. (1962). Receptive fields, binocular interaction and functional architecture in the cat's visual cortex. *J. Physiol.* **160,** 106–154.

Hubel, D. H., and Wiesel, T. N. (1965). Receptive fields and functional architecture in two nonstriate visual areas (18 and 19) of the cat. *J. Neurophysiol.* **41,** 229–289.

Innocenti, G. M., and Fiore, L. (1974). Post-synaptic inhibitory components of the responses to moving stimuli in area 17. *Brain Res.* **80,** 122–126.

Jones, J. P., and Palmer, L. A. (1987a). The two-dimensional spatial structure of simple receptive fields in cat striate cortex. *J. Neurophysiol.* **58,** 1187–1211.

Jones, J. P., and Palmer, L. A. (1987b). An evaluation of the two-dimensional gabor filter model of simple receptive fields in cat striate cortex. *J. Neurophysiol.* **58,** 1233–1258.

Jones, J. P., Stepnoski, A., and Palmer, L. A. (1987). The two-dimensional spectral structure of simple receptive fields in cat striate cortex. *J. Neurophysiol.* **58,** 1212–1232.

Kasamatsu, T., and Heggelund, P. (1982). Single cell responses in cat visual cortex to visual stimulation during iontophoresis of noradrenaline. *Exp. Brain Res.* **45,** 317–327.

Kato, H., Bishop, P. O., and Orban, G. A. (1978). Hypercomplex and simple/ complex cell classification in cat striate cortex. *J. Neurophysiol.* **41,** 1071–1095.

Kimura, F., Nishigori, A., Shirokawa, T., and Tsumoto, T. (1989). Long-term potentiation and *N*-methyl-D-aspartate receptors in the visual cortex of young rats. *J. Physiol.* **414,** 125–144.

Kirkwood, A., and Bear, M. F. (1994). Hebbian synapses in visual cortex. *J. Neurosci.* **14,** 1634–1645.

Kisvárday, Z. F., and Eysel, U. T. (1992). Cellular organization of reciprocal patchy networks in layer III of cat visual cortex (area 17). *Neuroscience* **46,** 275–286.

Kisvárday, Z. F., and Eysel, U. T. (1993). Functional and structural topography of horizontal inhibitory connections in cat visual cortex. *Eur. J. Neurosci.* **5,** 1558–1572.

Kisvárday, Z. F., Kim, D. S., Eysel, U. T., and Bonhoeffer, T. (1994). Relationship between lateral inhibitory connections and the topography of the orientation mpa in cat visual cortex. *Eur. J. Neurosci.* **6,** 1619–1632.

Kisvárday, Z. F., Martin, K. A. C., Freund, T. F., Maglóczky, Z. S., Whitteridge, D., and Somogyi, P. (1986). Synaptic targets of HRP-filled layer III pyramidal cell in the cat striate cortex. *Exp. Brain Res.* **64,** 541–552.

Kisvárday, Z. F., Tóth, E., Rausch, M., and Eysel, U. T. (1997). Orientation-specific relationship between populations of excitatory and inhibitory lateral connections in the visual cortex of the cat. *Cerebral Cortex* **7,** 605–618.

Kraszewski, K., and Michalski, A. (1989). Modulation of single cell responses and neuronal interactions produced by iontophoresis of glutamate in adult cat visual cortex. *Acta Neurobiol. Exp. (Warsz).* **49,** 171–192.

Kwon, Y. H., Esguerra, M., and Sur, M. (1991). NMDA and non-NMDA receptors mediate visual responses of neurons in the cat's lateral geniculate nucleus. *J. Neurophysiol.* **66,** 414–428.

LeVay, S. (1986). Synaptic organization of claustral and geniculate afferents to the visual cortex of the cat. *J. Neurosci.* **6,** 3564–3575.

LeVay, S. (1988). Patchy intrinsic projections in visual cortex, area 18, of the cat: morphological and immunocytochemical evidence for an excitatory function. *J. Comp. Neurol.* **269,** 265–274.

LeVay, S., and Gilbert, C. D. (1976). Laminar patterns of geniculocortical projection in the cat. *Brain Res.* **113,** 1–19.

Leventhal, A. G., and Schall, J. D. (1983). Structural basis of orientation sensitivity of cat retinal ganglion cells. *J. Comp. Neurol.* **220,** 465–475.

Levick, W. R., and Thibos, L. N. (1980). Orientation bias of cat retinal ganglion cells. *Nature* **286,** 389–390.

Liu, Y., and Cynader, M. (1994). Postnatal development and laminar distribution of noradrenergic fibers an cat visual cortex. *Dev. Brain Res.* **82,** 90–94.

Liu, Y., Jia, W., Strosberg, A. D., and Cynader, M. (1993). Development and regulation of beta adrenergic receptors in kitten visual cortex: an immunocytochemical and autoradiographic study. *Brain Res.* **31,** 274–286.

Löwel, S., Freeman, B., and Singer, W. (1987). Topographic organization of the orientation column system in large flat-mounts of the cat visual cortex. A 2-deoxyglucose study. *J. Comp. Neurol.* **255,** 401–415.

Luhmann, H. J., Millan Martinez, and W., Singer (1986). Development of horizontal intrinsic connections in cat striate cortex. *Exp. Brain Res.* **63,** 443–448.

Maffei, L., and Fiorentini, A. (1976). The unresponsive regions of visual cortical receptive fields. *Visi. Res.* **16,** 1131–1139.

Malenka, R. C., and Nicoll, R. A. (1999). Long-term potentiation: a decade of progress? *Science* **285,** 1870–1874.

Martin, K. A., Somogyi, P., and Whitteridge, D. (1983). Physiological and morphological properties of identified basket cells in the cat's visual cortex. *Exp. Brain Res.* **50,** 193–200.

Martin, K. A. C., and Whitteridge, D. (1984). Form, function and intracortical projections of spiny neurones in the striate visual cortex of the cat. *J. Physiol.* **353,** 463–504.

Matsubara, J. A., Cynader, M. S., and Swindale, N. V. (1987). Anatomical properties and physiological correlates of the intrinsic connections in cat area 18. *J. Neurosci.* **7,** 1428–1446.

Matsubara, J., Cynader, M., Swindale, N. V., and Stryker, M. P. (1985). Intrinsic projections within visual cortex: evidence for orientation-specific local connections. *Proc. Natl. Acad. Sci. U.S.A.* **82,** 935–939.

McLean, J., and Palmer, L. A. (1989). Contribution of linear spatiotemporal receptive field structure to velocity selectivity of simple cells in area 17 of cat. *Vis. Res.* **29,** 675–679.

McLean, J., and Palmer, L. A. (1996). Contrast adaptation and excitatory amino acid receptors in cat striate cortex. *Vis. Neurosci.* **13,** 1069–1087.

McLean, J., and Palmer, L. A. (1998). Plasticity of neuronal response properties in adult cat striate cortex. *Vis. Neurosci.* **15,** 177–196.

McLean, J., Raab, S., and Palmer, L. A. (1994). Contribution of linear mechanisms to the specification of local motion by simple cells in areas 17 and 18 of the cat. *Vis. Neurosci.* **11,** 271–294.

McLean, J., and Waterhouse, B. D. (1994). Noradrenergic modulation of cat area 17 neuronal responses to moving visual stimuli. *Brain Res.* **667,** 83–97.

Miller, K. D., Chapman, B., and Stryker, M. P. (1989). Visual responses in adult cat visual cortex depend on *N*-methyl-D-aspartate receptors. *Proc. Natl. Acad. Sci. U.S.A.* **86,** 5183–5187.

Morrone, M. C., Burr, D. C., and Maffei, L. (1982). Functional implications of cross-orientation inhibition of cortical visual cells. I. Neurophysiological evidence. *Proc. R. Soc. Lond.* **216,** 335–354.

Movshon, J. A., Thompson, I. D., and Tolhurst, D. J. (1978a). Spatial summation in the receptive field of simple cells in the cat's striate cortex. *J. Physiol.* **283,** 53–77.

Movshon, J. A., Thompson, I. D., and Tolhurst, D. J. (1978b). Receptive field organization of complex cells in the cat's striate cortex. *J. Physiol.* **283,** 79–99.

Murphy, P. C., and Sillito, A. M. (1987). Corticofugal feedback influences the generation of length tuning in the visual pathway. *Nature* **329,** 727–729.

Murphy, P. C., and Sillito, A. M. (1991). Cholinergic enhancement of direction selectivity in the visual cortex of the cat. *Neuroscience* **40,** 13–20.

Murthy, A., and Humphrey, A. L. (1999). Inhibitory contributions to spatiotemporal receptive-field structure and direction selectivity in simple cells of cat area 17. *J. Neurophysiol.* **81,** 1212–1224.

Murthy, A., Humphrey, A. L., Saul, A. B., and Feidler, J. C. (1998). Laminar differences in the spatiotemporal structure of simple cell receptive fields in cat area 17. *Vis. Neurosci.* **15,** 239–256.

Nelson, J. I., and Frost, B. J. (1985). Intracortical facilitation among co-oriented, co-axially aligned simple cells in cat striate cortex. *Exp. Brain Res.* **61,** 54–61.

Nelson, S., Toth, L., Sheth, B., and Sur, M. (1994). Orientation selectivity of cortical neurons during intracellular blockade of inhibition. *Science* **265,** 774–777.

Ohzawa, I., and Freeman, R. D. (1982). Contrast gain control in the cat visual cortex. *Nature* **298,** 266–268.

Ohzawa, I., Sclar, G., and Freeman, R. D. (1985). Contrast gain control in the cat's visual system. *J. Neurophysiol.* **54,** 651–667.

Pei, X., Vidyasagar, T. R., Volgushev, M., and Creutzfeldt, O. D. (1994). Receptive field analysis and orientation selectivity of postsynaptic potentials of simple cells in cat visual cortex. *J. Neurosci.* **14,** 7130–7140.

Pernberg, J., Jirmann, K.-U., and Eysel, U. T. (1998). Structure and dynamics of receptive fields in the visual cortex of the cat (area 18) and the influence of GABAergic inhibition. *Eur. J. Neurosci.* **10,** 3596–3606.

Peterhans, E., Bishop, P. O., and Camarda, R. M. (1985). Direction selectivity of simple cells in cat striate cortex to moving light bars. *Exp. Brain Res.* **57,** 512–522.

Peters, A., and Payne, B. R. (1993). Numerical relationships between geniculocortical afferents and pyramidal cell modules in cat primary visual cortex. *Cerebral Cortex* **3,** 69–78.

Pettigrew, J. D., and Daniels, J. D. (1973). Gamma-aminobutyric acid antagonism in visual cortex: different effects on simple, complex, and hypercomplex neurons. *Science* **182,** 81–83.

Pfleger, B., and Bonds, A. B. (1995). Dynamic differentiation of GABAA-sensitive influences on orientation selectivity of complex cells in the cat striate cortex. *Exp. Brain Res.* **104,** 81–88.

Ramoa, A. S., Shadlen, M., Skottun, B. C., and Freeman, R. D. (1986). A comparison of inhibition in orientation and spatial frequency selectivity of cat visual cortex. *Nature* **321,** 237–239.

Reid, R. C., Soodak, R. E., and Shapley, R. M. (1991). Directional selectivity and spatiotemporal structure of receptive fields of simple cells in cat striate cortex. *J. Neurophysiol.* **66,** 505–529.

Rivadulla, C., Sharma, J., and Sur, M. (1999). Role of NMDA and AMPA glutamate receptors in orientation and direction selectivity of V1 neurons. *Soc. Neurosci. Abstr.* **25,** 677.

Rose, D. (1977). Responses of single units in cat visual cortex to moving bars of light as a function of bar length. *J. Physiol.* **271,** 1–23.

Sato, H., Daw, N. W., and Fox, K. (1991). An intracellular recording study of stimulus-specific response properties in cat area 17. *Brain Res.* **544,** 156–161.

Sato, H., Fox, K., and Daw, N. W. (1989). Effect of electrical stimulation of locus coeruleus on the activity of neurons in the cat visual cortex. *J. Neurophysiol.* **62,** 946–958.

Sato, H., Hata, Y., Masui, H., and Tsumoto, T. (1987). A functional role of cholinergic innervation to neurons in the cat visual cortex. *J. Neurophysiol.* **58,** 765–780.

Sato, H., Hata, Y., and Tsumoto, T. (1999). Effects of blocking non-*N*-methyl-D-aspartate receptors on visual responses of neurons in the cat visual cortex. *Neuroscience* **94,** 697–703.

Shaw, C., Wilkinson, M., Cynader, M., Needler, C., Aoki, C., and Hall, S. E. (1986). The laminar distributions and postnatal development of neurotransmitter and neuromodulator receptors in cat visual cortex. *Brain Res. Bull.* **16,** 661–671.

Shevelev, I. A., Eysel, U. T., Lazareva, N. A., and Sharaev, G. A. (1998a). The contribution of intracortical inhibition to dynamics of orientation tuning in cat striate cortex neurons. *Neuroscience* **84,** 11–23.

Shevelev, I. A., Jirmann, K. U., Sharaev, G. A., and Eysel, U. T. (1998b). Contribution of GABAergic inhibition to sensitivity to cross-like figures in striate cortex. *Neuroreport* **9,** 3153–3157.

Shevelev, I. A., Novikova, R. V., Lazareva, N. A., Tikhomirov, A. S., and Sharaev, G. A. (1995). Sensitivity to cross-like figures in the cat striate neurons. *Neuroscience* **69,** 51–57.

Shevelev, I. A., Sharaev, G. A., Lazareva, N. A., Novikova, R. V., and Tikhomirov, A. S. (1993). Dynamics of orientation tuning in the cat striate cortex neurons. *Neuroscience* **56,** 865–876.

Shevelev, I. A., Volgushev, M. A., and Sharaev, G. A. (1992). Dynamics of responses of V1 neurons evoked by stimulation of different zones of receptive field. *Neuroscience* **51,** 445–450.

Shou, T., and Leventhal, A. G. (1989). Organized arrangement of orientation-sensitive relay cells in the cat's dorsal lateral geniculate nucleus. *J. Neurosci.* **9,** 4287–4302.

Shulz, D., and Frégnac, Y. (1992). Cellular analogs of visual cortical epigenesis. II. Plasticity of binocular integration. *J. Neurosci.* **12,** 1301–1318.

Sillito, A. M. (1975a). The effectiveness of bicuculline as an antagonist of gaba and visually evoked inhibition in the cat's striate cortex. *J. Physiol.* **250,** 287–304.

Sillito, A. M. (1975b). The contribution of inhibitory mechanisms to the receptive field properties of neurones in the striate cortex of the cat. *J. Physiol.* **250,** 305–329.

Sillito, A. M. (1977a). Inhibitory processes underlying the directional specificity of simple, complex and hypercomplex cells in the cat's visual cortex. *J. Physiol.* **271,** 699–720.

Sillito, A. M. (1977b). The spatial extent of excitatory and inhibitory zones in the receptive field of superficial layer hypercomplex cells. *J. Physiol.* **273,** 791–803.

Sillito, A. M. (1979). Inhibitory mechanisms influencing complex cell orientation selectivity and their modification at high resting discharge levels. *J. Physiol.* **289,** 33–53.

Sillito, A. M. (1984). Functional considerations of the operation of GABAergic inhibitory processes in the visual cortex. In: Cerebral cortex (E. G. Jones and A. Peters, Eds.), Vol. 2, pp. 91–117, New York, Plenum Press.

Sillito, A. M. (1992). GABA mediated inhibitory processes in the function of the geniculostriate system. *Prog. Brain Res.* **90,** 349–384.

Sillito, A. M., and Kemp, J. A. (1983). Cholinergic modulation of the functional organization of the visual cortex. *Brain Res.* **289,** 143–155.

Sillito, A. M., Kemp, J. A., Milson, J. A., and Berardi, N. (1980). A re-evaluation of the mechanisms underlying simple cell orientation selectivity. *Brain Res.* **194,** 517–520.

Sillito, A. M., and Murphy, P. C. (1988). GABAergic processes in the central visual system. In: Neurotransmitters and cortical function (M. Avoli, T. A. Reader, R. W. Dykes, and P. Gloor, Eds.), pp. 167–185, New York, Plenum Publishing Corporation.

Sillito, A. M., Murphy, P. C., Salt, T. E., and Moody, C. I. (1990). Dependence of retinogeniculate transmission in cat on NMDA receptors. *J. Neurophysiol.* **63,** 347–355.

Sillito, A. M., Salt, T. E., and Kemp, J. A. (1985). Modulatory and inhibitory processes in the visual cortex. *Vis. Res.* **25,** 375–381.

Sillito, A. M., and Versiani, V. (1977). The contribution of excitatory and inhibitory inputs to the length preference of hypercomplex cells in layers II and III of the cat's striate cortex. *J. Physiol.* **273,** 775–790.

Somers, D. C., Nelson, S. B., and Sur, M. (1995). An emergent model of orientation selectivity in cat visual cortical simple cells. *J. Neurosci.* **15,** 5448–5465.

Somogyi, P., Hodgson, A. J., Smith, A. D., Nunzi, G., Gorio, A., and Wu, J. Y. (1984). Different populations of gabaergic neurons in the visual cortex and hippocampus of cat contain somatostatin- or cholecystokinin-immunoreactive material. *J. Neurosci.* **4,** 2590–2603.

Somogyi, P., Kisvárday, Z. F., Martin, K. A. C., and Whitteridge, D. (1983). Synaptic connections of morphologically identified and physiologically characterized large basket cells in the striate cortex of cat. *Neuroscience* **10,** 261–294.

Soodak, R. E., Shapley, R. M., and Kaplan, E. (1987). Linear mechanism of orientation tuning in the retina and lateral geniculate nucleus of the cat. *J. Neurophysiol.* **58,** 267–275.

Stichel, C. C., and Singer, W. (1985). Organization and morphological characteristics of choline acetyltransferase-containing fibers in the visual thalamus and striate cortex of the cat. *Neurosci. Lett.* **53,** 155–160.

Suarez, H., Koch, C., and Douglas, R. (1995). Modeling direction selectivity of simple cells in striate visual cortex within the framework of the canonical microcircuit. *J. Neurosci.* **15,** 6700–6719.

Sun, M., Bonds, A. B. (1994). Two-dimensional receptive-field organization in striate cortical neurons of the cat. *Vis. Neurosci.* **11,** 703–720.

Tamura, H., Hicks, T. P., Hata, Y., Tsumoto, T., and Yamatodani, A. (1990). Release of glutamate and aspartate from the visual cortex of the cat following activation of afferent pathways. *Exp. Brain Res.* **80,** 447–455.

Thibos, L. N., and Levick, W. R. (1985). Orientation bias of brisk-transient y-cells of the cat retina for drifting and alternating gratings. *Exp. Brain Res.* **58,** 1–10.

Ts'o, D. Y., Gilbert, C. D., and Wiesel, T. N. (1986). Relationships between horizontal interactions and functional architecture in cat striate cortex as revealed by cross-correlation analysis. *J. Neurosci.* **6,** 1160–1170.

Tsumoto, T., Eckart, W., and Creutzfeldt, O. D. (1979). Modification of orientation sensitivity of cat visual cortex neurons by removal of GABA-mediated inhibition. *Exp. Brain Res.* **34,** 351–363.

Tsumoto, T., Masui, H., and Sato, H. (1986). Excitatory amino acid transmitters in neuronal circuits of the cat visual cortex. *J. Neurophysiol.* **55,** 469–483.

Vidyasagar, T. R. (1987). A model of striate response properties based on geniculate anisotropies. *Biol. Cybern.* **57,** 11–23.

Vidyasagar, T. R., Pei, X., and Volgushev, M. (1996). Multiple mechanisms underlying the orientation selectivity of visual cortical neurones. *TINS* **19,** 272–277.

Vidyasagar, T. R., and Urbas, J. V. (1982). Orientation sensitivity of cat LGN neurones with and without inputs from visual cortical areas 17 and 18. *Exp. Brain Res.* **46,** 157–169.

Volgushev, M., Pei, X., Vidyasagar, T. R., and Creutzfeldt, O. D. (1993). Excitation and inhibition in orientation selectivity of cat visual cortex neurons revealed by whole-cell recording in vivo. *Vis. Neurosci.* **10,** 1151–1155.

Wallingford, E., Ostdahl, R., Zarzecki, P., Kaufman, P., and Somjen, G. (1973). Optical and pharmacological stimulation of visual cortical. *Nat. New Biol.* **242,** 210–212.

Wolf, W., Hicks, T. P., and Albus, K. (1986). The Contribution of GABA-Mediated Inhibitory Mechanisms to Visual Response Properties of Neurons in the Kitten's Striate Cortex. *J. Neurosci.* **6,** 2779–2795.

Wörgötter, F., and Eysel, U. T. (1991). Topographical aspects of intracortical excitation and inhibition contributing to orientation specificity in area 17 of the cat visual cortex. *Eur. J. Neurosci.* **3,** 1232–1244.

Wörgötter, F., and Eysel, U. T. (2000). External and internal influences on the dynamic behavior of visual cortical receptive fields. *TINS,* **23,** 497–503.

Wörgötter, F., and Koch, C. (1991). A detailed model of the primary visual pathway in the cat: comparison of afferent excitatory and intracortical inhibitory connection schemes for orientation selectivity. *J. Neurosci.* **11,** 1959–1979.

Wörgötter, F., Suder, K., Zhao, Y., Kerscher, N., Eysel, U. T., and Funke, K. (1998). State-dependent receptive field restructuring in the visual cortex. *Nature* **396,** 165–168.

12

Orientation Selectivity and Its Modulation by Local and Long-Range Connections in Visual Cortex

David Somers

Department of Psychology, Boston University, Boston, Massachusetts

Valentin Dragoi and Mriganka Sur

Department of Brain and Cognitive Sciences, Massachusetts Institute of Technology, Cambridge, Massachusetts

OVERVIEW AND INTRODUCTION

Orientation selectivity of neurons in the primary visual cortex (V1) is one of the most thoroughly investigated receptive field properties in all of neocortex; however, its underlying neural mechanisms are still debated (see Das 1996; Sompolinsky and Shapley, 1997; Ferster and Miller, 2000). The earliest and simplest proposal for the generation of orientation selectivity was made by Hubel and Wiesel (1962). They proposed that a cortical simple cell receives input from a row of neurons in the lateral geniculate nucleus (LGN) whose receptive fields are aligned along the axis of orientation of the cortical receptive field (see Fig. 8-1 in Chapter by Reid et al.). This feedforward model, which postulates essentially linear summation of geniculate inputs, has had substantial experimental support (see for a recent review, Ferster and Miller, 2000).

At the same time, several kinds of evidence suggest that such a model does not capture the nonlinear complexity of cortical receptive fields and responses. In particular, V1 cells exhibit three significant kinds of nonlinear summation. First, the orientation selectivity of cortical cells is remarkably independent of stimulus contrast or the level of afferent drive. A purely feedforward model would predict that orientation tuning would broaden as contrast increases. Additions to the feedforward model have therefore been required. One proposal is that contrast gain control occurs by pooling the activity of cortical neurons and dividing or normalizing the geniculate input (Albrecht and Geisler, 1991; Heeger, 1992). Although recent evidence indicates that cortical neurons can exhibit the large changes in conductance that are required for such divisive normalization (Borg-Graham et al., 1998; Hirsch et al., 1998), the time course of changes is too brief for them to play a meaningful role in contrast gain control.

A second nonlinear feature of cortical cell responses is that they integrate inputs from a wide region of space that extends beyond the classical receptive field: stimuli in the receptive field surround can either enhance or suppress responses to stimuli in the receptive field center (Knierim and Van Essen, 1992; Grinvald et al., 1994; Sillito et al., 1995). We and others have shown that the effect of the surround on the center depends critically on the level of center drive (Toth et al., 1996; Sengpiel et al., 1997, 1998; Polat et al., 1998). Furthermore, stimulating the surround with stimuli that have a significantly different orientation than the cell's preferred orientation can enhance responses to optimally oriented center stimuli and generate supraoptimal responses (Sillito et al., 1995; Levitt and Lund, 1997). Third, orientation selectivity is not a static feature of V1 cells but can be altered dynamically by spatial and temporal context. For instance, adapting a cell with a stimulus of a given orientation causes a short-term change in the cell's preferred orientation (Dragoi et al., 2000). Responses on the flank of the tuning curve near the adapting orientation are suppressed, whereas responses on the far flank are actually enhanced, leading to a shift in orientation preference away from the adapting orientation. Such active nonlinear dynamics of orientation tuning are consistent with a progressive evolution of orientation tuning in V1 neurons observed on a rapid time scale using reverse correlation techniques (Ringach et al., 1997). More generally, altering the balance of excitation and inhibition in cortical columns alters the orientation tuning of adjacent columns (Toth et al., 1997a).

In this review, we argue that a model that takes into account not only thalamocortical excitation but also intracortical excitation between neurons in a cortical local circuit provides a fuller explanation of orientation selectivity. We have demonstrated that such a model provides a natural explanation for the contrast invariance of orientation tuning (Somers et al., 1995a). We summarize our model later and address some recent data that appear to contradict the model or support it. By adding long-range excitatory connections to neurons that are locally connected, we have shown that the paradoxical enhancement/suppression of center responses by surround stimulation can be explained (Somers et al., 1998). Such spatial context effects are actually part of a broader range of spatial receptive field dynamics governed by stimulus contrast, including length summation in receptive fields and supersaturation of responses. Next we describe the effects of long-range connections on local recurrent networks. Recurrent connections between excitatory neurons alone cannot explain supraoptimal responses accompanying surround stimulation at nonoptimal orientations. We have shown that recurrent connections between inhibitory neurons provide a sufficient explanation for this phenomenon (Dragoi and Sur, 2000), and we summarize the model. Finally, we have recently described the temporal dynamics and short-term plasticity of orientation tuning in V1 neurons following adaptation to oriented stimuli (Dragoi et al., 2000). These data provide strong support for the hypothesis that recurrent excitation and inhibition between local circuit neurons mediate orientation tuning. Together, these models and results provide a unifying explanation for orientation selectivity and for several of the key nonlinear phenomena that accompany orientation tuning—phenomena that are beyond the purview of feedforward models alone.

CONTRIBUTIONS OF LOCAL CORTICAL EXCITATION TO THE GENERATION OF ORIENTATION SELECTIVITY: THE EMERGENT MODEL

INTRODUCTION

Models of visual cortical orientation selectivity can be broadly divided into three categories: feedforward, inhibitory, and recurrent excitatory (Fig. 12-1). In feedforward models, all first-order cortical neurons receive converging input from a population of LGN neurons whose receptive fields are aligned in strongly oriented fashion in visual space. The bandwidth or sharpness of a cortical cell's orientation tuning is determined by the aspect ratio of its LGN projection. Many inhibitory models use a mild feedforward bias to establish the initial orientation preference of cortical neurons and inhibitory inputs—from cortical neurons preferring different orientations—to suppress nonpreferred responses. In recurrent models, cortical excitation among cells preferring similar orientations, combined with iso-orientation inhibition from a broader range of orientations, integrates and amplifies a weak thalamic orientation bias, which is distributed across the cortical columnar population. In this section we present our emergent model, which was one of the first recurrent models of orientation selectivity (Somers et al., 1995a; Ben-Yishai et al., 1995). Subsequently, a significant number of models have incorporated similar principles and in some cases expanded or refined these ideas (Troyer et al., 1998; McLaughlin et al., 2000; Pugh et al., 2000; Chance et al., 1999).

Feedforward models have received renewed experimental support since our original model was published (Ferster et al., 1996; Reid and Alonso, 1995; Chung

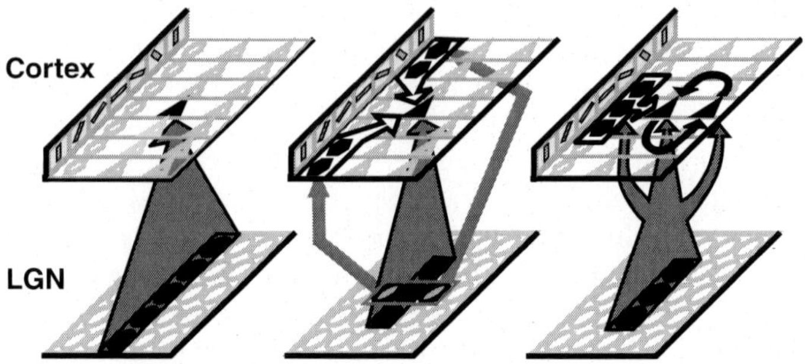

FIGURE 12-1. Models of visual cortical orientation selectivity. In feedforward models all "first-order" cortical neurons (triangle = excitatory, hexagon = inhibitory) receive converging input (gray arrow) from a population of LGN neurons that cover a strongly oriented region of visual space. A cortical cell's orientation tuning bandwidth is determined by the aspect ratio of its LGN projection. Many inhibitory models use a mild feedforward bias and use inhibitory inputs (white arrows), from cortical neurons preferring different orientations, to suppress nonpreferred responses. Here, we present a model in which recurrent cortical excitation (black arrows) among cells preferring similar orientations, combined with iso-orientation inhibition from a broader range of orientations, integrates and amplifies a weak thalamic orientation bias.

and Ferster, 1998; Anderson et al., 2000); however, problems linger with feedforward models (Ringach et al., 1997; Gardner et al., 1999). The key idea of feedforward models, that cortical neurons obtain orientation selectivity from elongated patterns of converging LGN inputs (Fig. 12-1 left) (Hubel and Wiesel, 1962; Ferster, 1987), is supported *qualitatively* by experiments. Thalamic ON cells (or OFF cells) are generally in close spatial register with the ON subfields (or OFF subfields) of cortical simple cells to which they provide input (Reid and Alonso, 1995). On average, cortical simple cell subfields are elongated along an axis parallel to the preferred response orientation (Jones and Palmer, 1987; Chapman et al., 1991). But for many simple cells, subfield length-to-width ratios (aspect ratios) are *quantitatively* insufficient to account for the sharp orientation selectivity exhibited (Watkins and Berkley, 1974; Jones and Palmer, 1987). Reports of low mean aspect ratios (Chapman et al., 1991; Pei et al., 1994; Reid and Alonso, 1995) are also inconsistent with feedforward models of orientation selectivity. Weakly biased feedforward inputs can be sharpened by using high firing thresholds (the "iceberg" effect, Creutzfeldt et al., 1974b), but this mechanism incorrectly predicts broadening of orientation tuning with increasing stimulus contrast (Sclar and Freeman, 1982; Wehmeier et al., 1989). Pure feedforward models also cannot account for the loss of orientation selectivity under iontophoresis of bicuculline, a $GABA_A$ antagonist that reduces inhibition over a localized population of cortical neurons (Sillito, 1975; Tsumoto et al., 1979; Sillito et al., 1980) (see Chapter 11).

Mechanisms that use shunting ("divisive") inhibition (e.g., Koch and Poggio, 1985; Carandini and Heeger, 1994) or hyperpolarizing ("subtractive") inhibition at nonpreferred orientations (Fig. 12-1 middle) (e.g., Wehmeier et al., 1989; Worgotter and Koch, 1991) can sharpen tuning in cells which have mildly oriented thalamocortical inputs; such models can also produce contrast-invariant orientation tuning and can account for bicuculline-induced tuning loss. However, inhibitory models are inconsistent with other experimental data. Although shunting inhibition has recently been rediscovered in cortex (Borg-Graham et al., 1998; Hirsch et al., 1998), it occurs only very transiently and appears insufficient to account for orientation selectivity (Douglas et al., 1988; Berman et al., 1991; Dehay et al., 1991; Ferster and Jagadeesh, 1992; Anderson et al., 2000). Inhibitory postsynaptic potentials (IPSPs) are evoked strongly by stimuli presented at the preferred orientation, whereas stimuli at cross-orientations evoke weak IPSPs (Ferster 1986; Douglas et al., 1991). Such inhibitory tuning conflicts with cross-orientation inhibitory models (Bishop et al., 1973; Morrone et al., 1982), and strong iso-orientation suppression poses problems even for models that use other forms of hyperpolarizing inhibition to sharpen thalamocortical input (e.g., Worgotter and Koch, 1991). Furthermore, results from our laboratory conflict with all orientation models that rely heavily on direct inhibitory input. Intracellular blockade of inhibition had *negligible* effect on sharpness of orientation tuning of blocked cells (Nelson et al., 1994). These results also appear to conflict with reports that orientation tuning can be abolished by bicuculline-induced extracellular inhibitory blockade (Sillito et al., 1980; Nelson, 1991) (see Chapter 11).

In this section, we demonstrate that this apparent paradox can be resolved by considering the effects the two inhibitory blockades have on the tuning of cortical excitatory inputs. Our computer simulations also demonstrate that local, recurrent cortical excitation can generate sharp, contrast-invariant orientation tuning in circuits that have strong iso-orientation inhibition and weakly oriented thalamocortical excitation (Fig. 12-1 right). This model primarily addresses the circuitry within a single cortical "hypercolumn"; effects of adding long-range cortical connections to this circuitry are addressed in the next section.

KEY ASSUMPTIONS OF THE MODEL

Although the model incorporated significant biological detail, only three assumptions were critical to this model. First, converging LGN inputs must provide *some* orientation bias at the columnar population level. Consistent with experiment (Creutzfeldt et al., 1974a; Watkins and Berkley, 1974; Jones and Palmer, 1987; Chapman et al., 1991; Pei et al., 1994), this bias may be weak and distributed across a population with many cells that receive unoriented input. The second assumption of the model—that local (< 1 mm horizontal distance) intracortical inhibitory connections must arise from cells with a broader distribution of orientation preferences than do intracortical excitatory connections—differs from prior inhibitory models in that it is consistent with experimental evidence for strong iso-orientation inhibition (Ferster, 1986; Douglas et al., 1991; Anderson et al., 2000). Narrowly tuned iso-orientation excitation and more broadly tuned iso-orientation inhibition can be realized by a simple difference-of-gaussians like structure in the orientation domain. This idea is supported by cross-correlation data (Hata et al., 1988; Michalski et al., 1983) and is consistent with a key hypothesis of many models of the development of orientation selectivity (e.g., Rojer and Schwartz, 1990; Miller, 1992, 1994; Swindale, 1992; Grossberg and Olson, 1994). The original model (Somers et al., 1995a) used a somewhat large difference in the input ranges of excitatory and inhibitory inputs. Below (Fig. 12-4C) we show that the inhibitory inputs can be much narrower than we originally claimed; inhibitory inputs need be only slightly broader than excitatory cortical inputs. This seems consistent with recent experimental reports (Anderson et al., 2000). The final assumption is that cortical inhibition must approximately balance cortical excitation. Too much inhibition produced low response rates and too little inhibition permitted nonselective amplification of all stimulus responses. However, many sets of parameters satisfied the "balance" requirement. This hypothesis is consistent with reports that excitatory postsynaptic potentials (EPSP) and inhibitory postsynaptic potentials (IPSP) strengths roughly covary across orientations (Ferster, 1986; Douglas et al., 1991; however, see Pei et al., 1994 for a differing view). Balanced cortical excitation and inhibition have also been invoked by computational models of cortical response variability (Softky and Koch, 1993; Shadlen and Newsome, 1994).

MODEL STRUCTURE AND IMPLEMENTATION

The model addresses the generation of orientation selectivity in cortical neurons that receive direct thalamic input. The circuit represents layer IV simple cells that lie under a 1700 μm by 200 μm patch of visual cortical area 17 of the cat. This small cortical area contains orientation columns representing a full set of preferred orientations (e.g., a hypercolumn), as well as neighboring columns. The model implemented 2205 cortical neurons, of which 20% were inhibitory and the rest excitatory (Gabbott and Somogyi, 1986). In total, the network contained more than 180,000 synapses.

Cortical neurons were organized into 21 orientation columns, spanning the length dimension of the cortical patch. An initial, but weak orientation structure was generated in cortex by providing orientation-biased converging LGN input to each cortical column. Simple cell subfield aspect ratios varied between 1:1 (unoriented) and 3:1 (moderately oriented). Cortical neurons received synaptic inputs from lateral geniculate (excitatory), cortical excitatory, and cortical inhibitory neurons. The model assumed LGN inputs contribute 20% of all synapses, and that cortical excitatory and inhibitory neurons, respectively, receive 20% and 10% of their synapses from cortical inhibitory neurons. Model cortical excitatory cells provided 60% and 70% synapses onto excitatory and inhibitory neurons, respectively.

To generate iso-orientation excitation and inhibition, cortical excitatory and inhibitory inputs were chosen from normal distributions, each with maximal connectivity probability *within* the cortical column of the postsynaptic cell. Inhibitory inputs were drawn from a broader range of orientations than the excitatory inputs. Inhibitory inputs had a range of ± 60° (~± 350 μm), whereas excitatory inputs had an effective range of ± 15° (~± 100 μm). We show later in this section (see Fig. 12-4c) that inhibitory inputs need be only modestly broader than cortical excitatory inputs to provide this degree of cortical sharpening effect. Effects of long-range connections are addressed in the next section.

Excitatory and inhibitory cortical neurons were modeled separately using experimentally reported input resistances, membrane time constants, and firing characteristics of regular-spiking (RS) and fast-spiking (FS) neurons (Connors et al., 1982; McCormick et al., 1985). Each cortical neuron was modeled as a single voltage compartment in which the membrane potential, V, was given by:

$$C_m \, dV_i/dt = -\sum_{j=1}^{k} g_{ji}\,(t - t_{ji})\,(V_i\,(t) - E_{Excit}) - \sum_{j=k+1}^{k+l} g_{ji}\,(t - t_{ji})\,(V_i\,(t) - E_{Inhib})$$

$$- g_{Leak}\,(V_i\,(t) - E_{Leak}) - g_{AHP}\,(t - t_{spike})\,(V_i\,(t) - E_{AHP})$$

where the synaptic conductances generated at each postsynaptic cell i by the spiking of each presynaptic cell j (excitatory if j <= k; inhibitory if k < j <= l) were given by:

$$g_{ji}\,(t) = \bar{g}_{ji} \sum_{p}^{S_j} (t - t_p)\,(e/\tau_{peak})\,\exp\,[-\,(t - t_p)/\tau_{peak}]$$

where t_{ji} and $\bar{g}_{ji}$ describe the delay and maximal conductance change produced for the synapse between cell j and cell i. The numbers of excitatory and inhibitory synapses received by cell i were k and l, respectively. AMPA receptor-mediated excitatory synaptic effects and $GABA_A$ receptor-mediated synaptic inhibitory effects were implemented as linear conductance changes. After-hyperpolarization effects were spike-triggered (with delay t_{spike} = 1 ms). In general, parameter choices were very conservative; see Somers et al. (1995a) for further discussion.

MODEL PERFORMANCE

Response Behavior

Orientation tuning properties of the model were thoroughly investigated. Both excitatory and inhibitory neurons were sharply selective. Mean HW tuning of excitatory cells was 17.1° ± 0.6° (SD), and mean HW tuning of inhibitory cells was 20.5° ± 0.7° (SD). The sharpness of tuning is consistent with typical physiological values for cortical simple cells (Watkins and Berkley, 1974; Orban, 1984). Figure 12-2A displays a simulated intracellular trace of a typical model cortical excitatory cell in response to flashed bar stimuli oriented at 0°, 22.5°, 45°, and 90°. As is true for all cells in the column, this cell is sharply selective for the 0° stimulus.

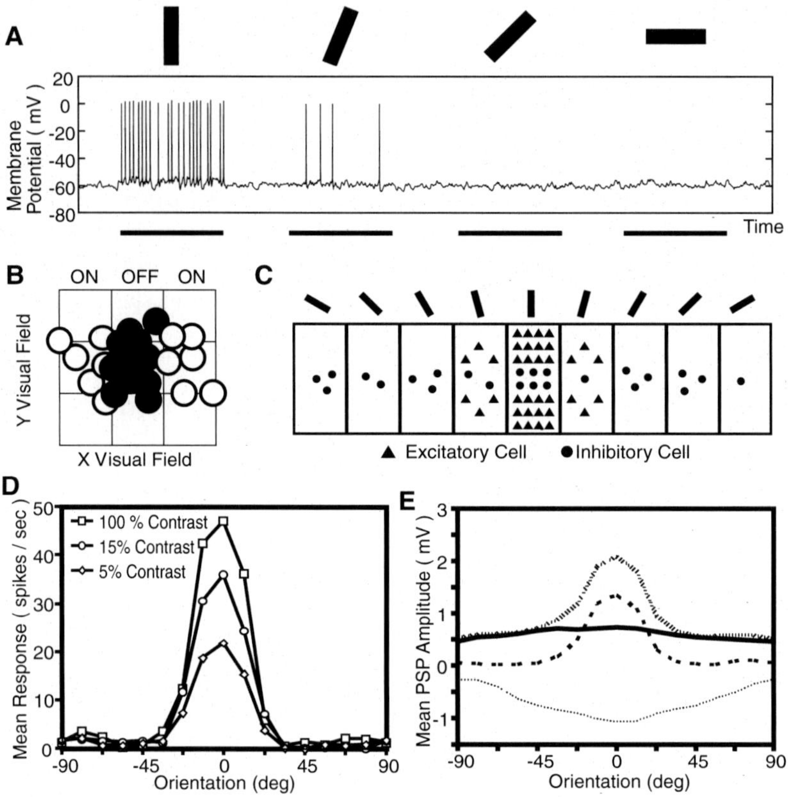

FIGURE 12-2. Response of a cell selective for 0° stimuli. **(A)** Simulated intracellular trace from one cell in the model network in response to flashed dark bars oriented at 0°, 22.5°, 45°, and 90°. Black bars on time axis indicate 500 msec stimulus presentation. **(B)** Thalamic and **(C)** cortical input fields of this cell. ON (white) and OFF (black) thalamic subfields exhibited only a mild orientation bias for 0° stimuli. Cortical excitatory (triangles) and inhibitory (circles) inputs arose most densely from cells within the 0° column. Inhibitory distribution was broader than the excitatory distribution. **(D)** As contrast increased, peak response increased but selectivity did not broaden. **(E)** Postsynaptic potentials evoked in the example cell by thalamic excitatory (solid), cortical excitatory (thick dashed), and cortical inhibitory (thin dashed) synaptic inputs.

Sharpness of orientation selectivity remained approximately constant across stimulus contrasts values. As contrast increased, peak responses increased but selectivity did not broaden. This contrast-invariance of orientation tuning in the model replicates experimental findings (Sclar and Freeman, 1982). Figure 12-2D displays orientation-response curves at three different contrasts for an example cell.

The mechanisms underlying orientation tuning of the model were investigated by measuring the postsynaptic potentials (PSPs) contributed by different synaptic input sources. Both excitatory (EPSP) and inhibitory (IPSP) PSPs were strongest at the preferred orientation (Fig. 12-2E). Notably, stimulus-evoked IPSPs (in excess of spontaneous levels) were, on average, 8.3 times as strong for the preferred orientation as for the orthogonal or cross-orientation (90°) stimulus. These PSP tuning properties of the model are consistent with intracellular reports of weak cross-orientation IPSPs and strong iso-orientation IPSPs (Ferster, 1986; Douglas et al., 1991).

In the model, the EPSP tuning resulted from a combination of broadly tuned thalamocortical input and sharply tuned corticocortical excitation. Broad tuning of thalamocortical input resulted from the low length to width ratios of the (regions of thalamic convergence onto) cortical subfields (Fig. 12-2C). Sharp tuning of cortical EPSPs reflected input from well-tuned cortical excitatory cells with similar orientation preferences. Cortical inhibitory inputs were also drawn most heavily from within the preferred orientation column (Fig. 12-2B), and these cells were also well tuned. Cortical inhibitory inputs were drawn from a broader range of orientations than cortical excitatory inputs; therefore cortical inhibition was more broadly tuned than cortical excitation. We show later (see Fig. 12-4C) that the critical orientation sharpening property can be obtained with much narrower cortical inhibition than was used here. Net EPSPs and IPSPs were evoked over a similar range of orientations. EPSP and IPSP tuning curves differed primarily in their slope at oblique orientations.

Claim of Emergence, Speed of Emergence

The PSP records revealed that cortical excitatory inputs were the source of the largest and best-tuned orientation signal (see Fig. 12-4). Therefore, cortical excitation was the leading cause of sharp orientation selectivity. This finding may appear paradoxical, as sharp tuning of cortical neurons was both the "effect" and the "cause" of the effect; however, in a feedback circuit such an explanation is not tautological. Rather, it implies that the effect is an *emergent* property of the recurrent circuitry.

Because sharp orientation selectivity resulted from cortical feedback, rapid sharpening of EPSP tuning can be observed in the model's intracellular records at the beginning of a response. Broadly tuned thalamocortical EPSPs are quickly joined by sharply tuned intracortical EPSPs. Pei et al. (1994) reported a similar sharpening in PSP tuning in vivo. Similar rapid sharpening effects have also been reported in anesthetized and awake monkeys (Ringach et al., 1997; Dragoi et al., 1999) and in optical recordings of cats (Shoham et al., 1999). However, the temporal evolution of tuning is less apparent in the model's extracellular traces. For nearly all model cells, sharp orientation selectivity emerged by the first or second spike of the response. This is consistent with experimental reports (Vogels and Orban, 1991). Because of the integrative properties of membranes, response latencies were generally shortest for preferred stimuli (c.f., Dean and Tolhurst, 1986). This "head-start" for preferred orientations likely contributed to the rapid emergence of sharp tuning in the model. Although not incorporated in this model, brief but strong shunting inhibition that has recently been observed might also play an important role in controlling transient responses (Borg-Graham et al., 1998; Hirsch et al., 1998).

Analysis of Orientation Tuning

To investigate further the mechanisms underlying orientation selectivity in the model, the maximal synaptic conductances (strengths) for all cortical inhibitory and/or cortical excitatory synapses were manipulated. The precise set of network connections, however, was otherwise unaltered. Inactivation of cortical excitatory and cortical inhibitory synapses revealed tuning effects of the converging thalamocortical inputs. LGN inputs alone generated broad tuning with a large variance in tuning across the population of cortical neurons (Fig. 12-3a). The most sharply tuned neuron in the feedforward circuit was more broadly tuned than the most broadly tuned cell in the full network. This demonstrates that the role of intracortical excitation was not simply to "convey" sharp tuning from a few cells, which receive highly oriented thalamocortical inputs, to other neurons.

Simulations were also performed on a variation of the model with normal intracortical inhibition and normal thalamocortical excitation, but inactivated intracortical excitation. This network ("inhib 1") exhibited tuning sharper than that of the feedforward network, but significantly broader than that of the full feedback network. Mean excitatory cell HW tuning was 38.4° ± 22.7° SD. This network also exhibited lower response rates than either the full network or the feedforward network. Mean peak response of excitatory neurons was 9.8 spikes/sec ± 2.3 sp/s SD. (compared with 47 sp/s for the full network). The tuning advantage of this inhibitory network over the feedforward network can be attributed to an "iceberg" effect; relatively unoriented hyperpolarizing inhibition is similar to raising the threshold. To investigate more thoroughly iceberg-type effects in this model, the maximal inhibitory synaptic conductances were doubled and cortical excitation remained. This ("inhib 2") network exhibited only a mild improvement in orientation tuning over the "inhib 1" network. The increase in inhibition also resulted in a further reduction of response rates. In contrast, the full recurrent network exhibited both physiologically sharp orientation tuning and robust responses. These results demonstrate that the orientation tuning properties of the model cannot be accounted for by either the thalamocortical or intracortical inhibitory connections it utilized. Thus these results demonstrate the utility of local, recurrent, cortical excitatory connections in the generation of sharp orientation selectivity by the model.

To explore further the robustness of the model, orientation behavior was explored for a broad range of cortical excitatory and inhibitory synaptic strengths. Mean response and orientation tuning of the summed responses of all excitatory neurons in the 0° column were measured for each parameter set (Fig. 12-3b). This analysis revealed a diagonal band of "balanced" excitatory and inhibitory strengths that yield both sharp tuning and robust responses. In the region below this diagonal, inhibition dominated, causing reduction of response. In the region above the diagonal band, excitation dominated, causing nonselective amplifica-

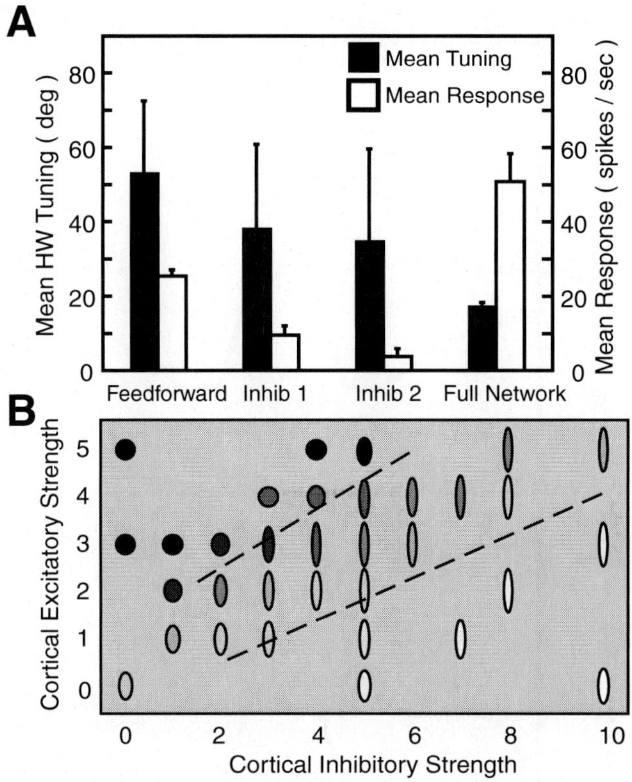

FIGURE 12-3. **(A)** Recurrent network tuning properties compared to those of the network components. **(B)** Sensitivity of the model to cortical excitatory and inhibitory strengths. Average peak response (indicated by symbol shading; darker = stronger response) and orientation tuning bandwidth (indicated by symbol shape; more elliptic = sharper tuning) of averaged responses of cortical excitatory neurons, for different cortical synaptic strengths (in nS). The model exhibited sharp orientation tuning and strong responses provided that cortical excitation and inhibition were approximately balanced. The region between the two lines defines the region of parameter space that satisfied this "balance" requirement. Where inhibition dominated, response rates fell; where excitation dominated, tuning broadened.

tion of *all* responses and thus a broadening of tuning. Explorations of networks with different spatial spreads of cortical connections also revealed similar (but shifted) diagonal bands of "balanced" excitation and inhibition for which sharp tuning was observed.

Additional analysis was performed to further illuminate the mechanism by which this model achieved sharp orientation tuning and robust responses (despite utilizing poorly oriented thalamocortical inputs and iso-orientation inhibition). Figure 12-4A displays the mean orientation response curve for cells in the thalamocortical network and an example of a "desired" orientation response curve

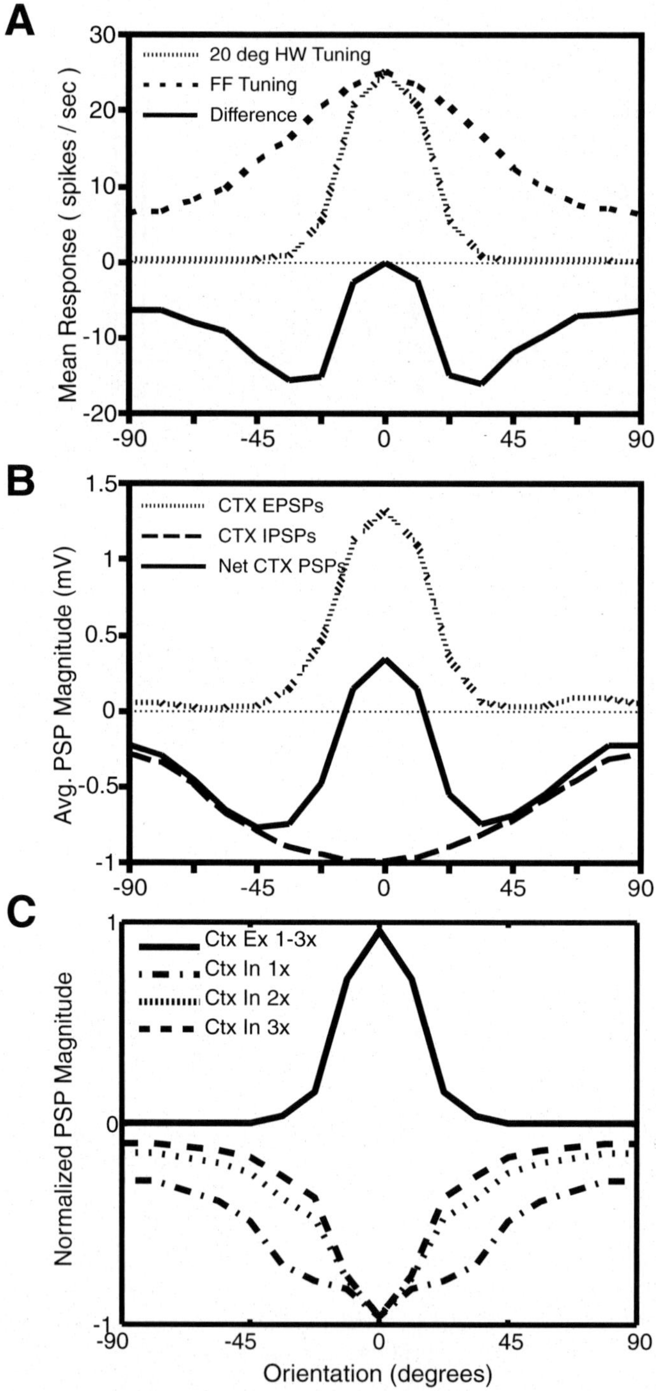
A
20 deg HW Tuning
FF Tuning
Difference
Mean Response (spikes / sec)
30
20
10
0
-10
-20
-90
-45
0
45
90
B
CTX EPSPs
CTX IPSPs
Net CTX PSPs
Avg. PSP Magnitude (mV)
1.5
1
0.5
0
-0.5
-1
-90
-45
0
45
90
C
Ctx Ex 1-3x
Ctx In 1x
Ctx In 2x
Ctx In 3x
Normalized PSP Magnitude
1
0
-1
-90
-45
0
45
90
Orientation (degrees)

with HW tuning of 20° (typical tuning for simple cells; Orban, 1984). The desired tuning curve was scaled so that the peak responses of the two curves were equal. To achieve 20° HW tuning, the difference between the desired orientation tuning response and the thalamocortical response must be provided by cortical inputs. Therefore, the difference curve (Fig. 12-4A) reveals the shape of orientation tuning of the net cortical contribution required to achieve sharp orientation selectivity. This curve indicates that net cortical inhibition should be strongest at approximately 20° to 40° from the preferred orientation. Modest net cortical inhibition is required at the cross-orientation.

The net cortical tuning curve indicates that little or no iso-orientation inhibition is required; an iso-orientation specific increase in inhibition would both reduce responsiveness and broaden tuning. Since experimental intracellular recordings indicate that cortical inhibition is actually strongest at the iso-orientation (Ferster, 1986; Douglas et al., 1991; Anderson et al., 2000), the net cortical orientation tuning curve cannot be accounted for solely by the cortical inhibitory inputs. However, the net cortical curve *can* be matched by combination of somewhat broad iso-orientation cortical inhibition and iso-orientation cortical excitation. Figure 12-4B displays the average cortical excitatory and inhibitory inputs (PSPs) to excitatory cortical cells in the full model. The net cortical orientation tuning curve produced by their sum exhibits the same center-surround structure displayed in the difference curve of Fig. 12-4A.

This analysis demonstrates that narrowly tuned cortical excitation can provide the "missing link" between the combination of broadly tuned thalamocortical excitatory inputs and iso-orientation inhibitory inputs on the one hand and sharply tuned orientation output responses on the other hand. The "center-surround" orientation tuning mechanism described here can be generated by any of a family of pairs of excitatory and inhibitory cortical inputs in which excitation and inhibition are approximately balanced and inhibition is more broadly tuned than excitation.

FIGURE 12-4. Cortical requirements for sharp orientation selectivity. **(A)** Comparison of orientation responses produced by thalamocortical inputs alone (FF tuning) with typical physiological tuning (20° HW). The difference between these curves must be contributed by intracortical inputs. Note that net inhibition is most strongly required approximately 20° to 40° from the preferred orientation. **(B)** Average cortical PSPs received by excitatory cells in the 0° column. Narrow iso-orientation excitation and broader iso-orientation inhibition combined to yield "center-surround" cortical orientation tuning with net excitation in response to preferred stimulus orientations and net inhibition for nonpreferred stimuli. Net cortical tuning satisfied inhibitory requirements of the difference curve (in part A) and amplified preferred responses. **(C)** Family of cortical excitatory and inhibitory PSP curves that produce the net cortical PSP curve required in panel A. As the strength of net cortical excitation increases relative to net thalamocortical excitation, the shape of the required IPSP curve becomes narrower. Normalized curves are shown for net cortical EPSPs 1, 2, and 3 times net thalamocortical EPSPs (based on preferred orientation amplitude). Solid line shows a normalized EPSP curve (always the same shape). Broken lines shows IPSP curve for net cortical EPSPs 1,2, and 3 times as strong as net thalamocortical.

Such a family of balanced excitatory and inhibitory cortical inputs was generated in the model as stimulus contrast was varied. This family also admits much more narrowly tuned cortical inhibitory inputs (Fig. 12-4C) than we proposed earlier. As the size of the cortical inputs increases relative to the size of the thalamocortical inputs, the shape of the "required" inhibitory curve narrows substantially. If the net cortical excitatory inputs at the preferred orientation are two or three times as strong as the net thalamocortical inputs at the preferred orientation, the cortical inhibitory input curve narrows to be very similar to the cortical excitatory curve. The exact ratio of thalamocortical and intracortical input strengths is still under experimental investigation. Individual thalamocortical synapses appear to be stronger than individual intracortical excitatory synapses (Stratford et al., 1996; Gil et al., 1999), but cortical excitatory synapses in layer IV outnumber thalamocortical synapses by as much as a factor of 15 (LeVay, 1986; Peters and Payne, 1993; Ahmed et al., 1994). Given these values, the curves shown in Fig. 12-4C seem quite plausible (Stratford et al, 1996). In the case of the narrowest curves, the excitatory and inhibitory curves are so similar as to be difficult to distinguish experimentally. This appears consistent with a recent estimate of cortical orientation currents (Anderson et al., 2000).

Inhibitory Blockade Studies

A direct paradox in the experimental literature on orientation selectivity lies in the dramatic effects observed under different forms of intracortical inhibitory blockade. Extracellular iontophoretic application of the $GABA_A$ antagonist bicuculline methiodide reduced inhibitory synaptic transmission across a local population of cells and disrupted orientation tuning. Sufficiently large bicuculline doses *abolished* orientation selectivity. Bicuculline application also substantially increased response rates. These results argue for a critical role for intracortical inhibition in orientation selectivity (e.g., Sillito et al., 1980). In contrast, our laboratory (Nelson et al., 1994) reported that intracellular blockade of inhibition had negligible effect on orientation tuning (Fig. 12-5A,B). In these experiments whole-cell pipettes were used to deliver CsF-DIDS (4,4′-diisothiocyanostilbene-2,2′-disulfonic acid) solution intracellularly to silence inhibitory voltage conductances (Cl^-, K^+). A mild, fixed hyperpolarizing current was injected to compensate for the increase in spontaneous firing rate. The critical difference between the two sets of inhibitory blockade experiments appears to be the number of cells that lost inhibitory inputs. Disruption of orientation selectivity requires long bicuculline ejection times (Sillito et al., 1980; Nelson, 1991). This suggests that the drug effects spread across a local population of neurons. In contrast, intracellular blockade affects only the recorded neuron. This issue was explored in the model. Bicuculline application was simulated by a 50% reduction in cortical inhibitory synaptic strength for all neurons in the 0° column; synaptic connections onto neurons in other columns were unaffected. CsF-DIDS application was simulated by a 100% reduction in inhibitory synaptic strength (Cl-channels) and an 80% reduction in afterpolarization conductance in a single neuron. The DIDS cell was either

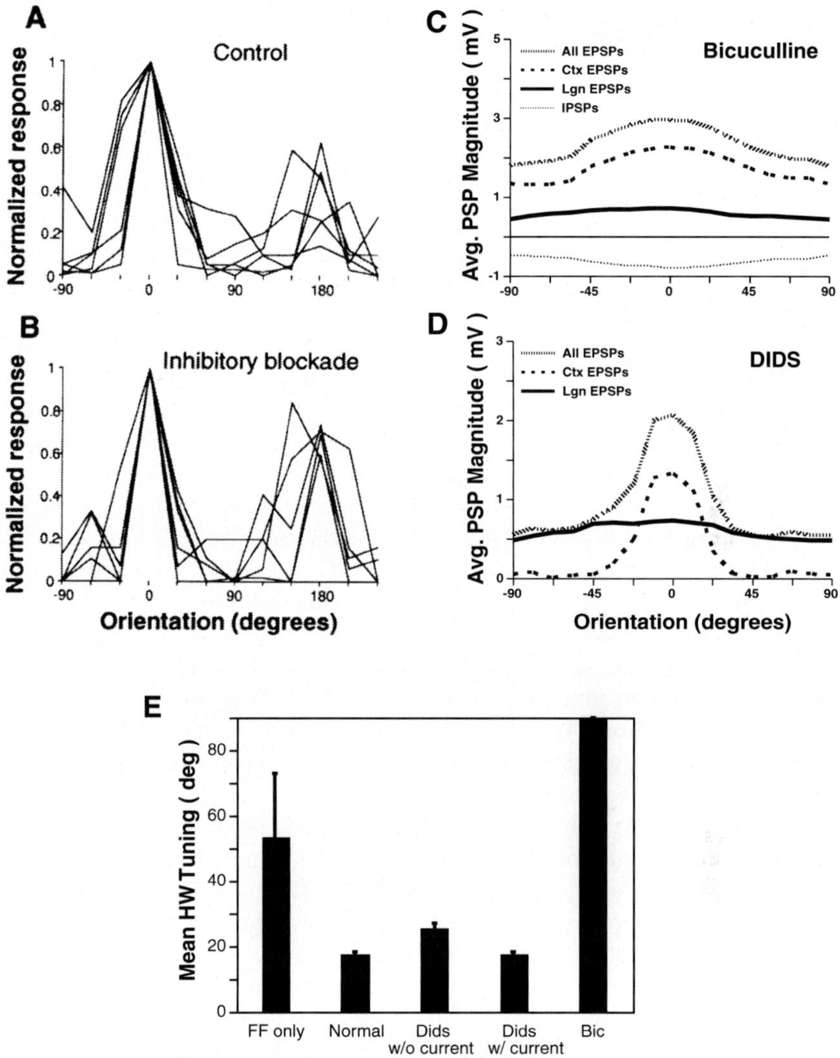

FIGURE 12-5. Inhibitory blockade experiments and model results. **(A)** and **(B):** Experimental intracellular blockade of inhibition in single neurons (shown in B) resulted in minimal change in orientation tuning versus normal (shown in A). **(C–E)** Model results. **(C)** Simulated bicuculline application reduced inhibition across the column and resulted in a loss of cortical EPSP tuning. Thus bicuculline disrupted both net EPSP tuning and response tuning, despite the fact that strong inhibitory PSPs were still present. In contrast, **(D)** silencing of inhibition in a single cell had negligible effect on the tuning of other cells. Therefore, cortical EPSPs and net EPSPs retained their orientation selectivity under CsF-DIDS. **(E)** Summary of model inhibitory blockade results.

injected with a fixed (for all orientations) –0.3 nA hyperpolarizing current or received no injected current.

Under bicuculline application, all cells in the 0° column became unoriented and peak excitatory responses increased by 87% (Fig. 12-5C). Both effects are consistent with experimental reports (Sillito et al., 1980; Nelson, 1991). With inhibitory efficacy reduced, the columnar population amplified responses to all stimuli and thus disrupted tuning. Bicuculline firing rates were higher than the normal firing rates, but they were far below cellular saturation levels (maximum firing rate > 300 sp/s). In agreement with experiment (Sillito et al., 1980; Nelson, 1991), response rates and tuning disruption effects decreased with bicuculline dosage.

DIDS simulations were performed individually for 10 neurons (all excitatory) both with and without injected current. With injected current the 10 cells exhibited mean HW tuning of 17° (Fig. 12-5D), which is identical to the mean tuning of these cells in the normal network. Under DIDS application and without current injection, these 10 cells exhibited mean HW tuning of 25°. Thus the simulations, like experiment, indicate little loss of orientation selectivity with blockade of direct inhibitory inputs. In the model, DIDS cells were sharply orientation selective because they received sharply tuned cortical excitatory input. The iceberg effect caused by the hyperpolarizing current contributed only mildly to the DIDS tuning.

In the model the fundamental difference between the two inhibitory blockade paradigms was the tuning of the intracortical excitatory inputs (Fig. 12-5C,D). Blockade of inhibition in a single cell had a negligible effect on the tuning of other cells in the network. Therefore, the cortical excitatory inputs to the blocked cell retained their sharp orientation tuning and the net input was well tuned (Fig. 12-5D). In contrast, the reduction in inhibition across the local population that occurs with bicuculline dramatically altered the tuning of intracortical excitatory inputs. With insufficient "balancing" inhibition, the recurrent orientation tuning mechanism failed to "emerge" and instead became disruptive. The columnar population amplified responses to all orientations and cortical excitatory cells lost their sharp tuning (Fig. 12-5C).

The inhibitory blockade simulation results are summarized in Fig. 12-5E. Note that bicuculline tuning was substantially worse than the thalamocortical tuning. In comparison, inhibitory models predict that bicuculline tuning should be no worse than the thalamocortical tuning. Thus consideration of intracortical excitation not only provides a mechanism for the persistence of tuning under direct inhibitory blockade of single cells but also provides a more complete explanation than do inhibitory models for the abolition of tuning under extracellular inhibitory blockade.

DISCUSSION

This work demonstrates that recurrent cortical excitatory and inhibitory inputs can rapidly sharpen orientation tuning even in cells that receive unoriented thalamocortical input. The model requires that the orientation column population receive some thalamocortical bias in orientation that iso-orientation inhibition be somewhat more broadly tuned than iso-orientation excitation, and that cortical excitation and inhibition balance each other. Of these requirements, only the second has proved controversial. How much broader must inhibition be than excitation? Although we initially believed the difference to be substantial, we show here that only a modest difference is required. This appears consistent with recent data (Anderson et al., 2000), although experimental methods with sufficient precision will be required to address this question conclusively.

More potent criticism of this model has come in the form of experimental evidence that has resurrected the linear feedforward model of Hubel and Wiesel. Efforts to silence cortical inputs have revealed little change in the sustained component of orientation tuning of many layer IV cells (Ferster et al., 1996; Chung and Ferster, 1998). These experiments demonstrate clearly that feedforward inputs play a significant role in orientation tuning, but also raise other issues. Intracellular records in the cortical silencing experiments (Ferster et al., 1996; Chung and Ferster, 1998) reveal tuning that appears to be substantially broader than that observed in normal extracellular recordings. Intracellular recordings also appear not to reveal the constant DC component across orientation that is predicted by the Hubel-Wiesel model (Carandini and Ferster, 2000). Although careful mapping of thalamic inputs to simple cell subfields has reported a close spatial correlation (Reid and Alonso, 1995) (see Chapter 8), it has not revealed the high aspect ratio of subfields that the linear model requires. A recent study (Gardner et al., 1999) reported that the linear component only accounts for about half of orientation tuning; the rest is nonlinear. In addition, several different groups have reported that orientation tuning rapidly sharpens at the beginning of responses (Ringach et al., 1997; Shoham et al., 1999; Dragoi et al., 1999), which is a key prediction of our model.

Shunting inhibition is another potential mechanism that has been revealed by recent experiments (Borg-Graham et al, 1998; Hirsch et al., 1998). Although the shunting appears far too brief to support the central hypothesis of normalization models (Heeger, 1992; Carandini and Heeger, 1994), the strength and timing of the shunting suggest that it may play a powerful role in shaping cortical responses. One potential utility is in controlling the strength of recurrent excitatory inputs. The implications of this use are just starting to be addressed (Chance and Abbott, 2000). At this time, it appears likely that orientation selectivity will not be accounted for by a single mechanism, but rather by a combination of feedforward, cortical inhibitory, cortical excitatory, and spike threshold mechanisms. A newer model (Troyer et al., 1998) invokes recurrent excitation between neurons to amplify phase-selective responses and strong inhibition from spatially opponent subfields to generate contrast invariance in orientation selectivity. However, key assumptions of the model, such as a population of neurons that respond in contrast-dependent manner to all orientations, remain untested. Other evidence suggests that feedback from cortex to thalamus (Murphy et al., 1999) and cellular adaptations may also contribute (Anderson et al., 2000b). All of this suggests that this particular cortical computation is far more complex than the field had anticipated. At the same time, orientation tuning is fundamentally a nonlinear property of V1 neurons, and it appears parsimonious to invoke nonlinear mechanisms to explain its major nonlinear features.

EFFECT OF LONG-RANGE CONNECTIONS ON ORIENTATION-SPECIFIC RESPONSES

INTRODUCTION

In this section, we present modeling (Somers et al., 1998) and experimental results (Toth et al., 1996) on dynamic local and long-range receptive field effects. We present a recurrent cortical model suggesting how the net effect on a cell, of stimulating one portion of the receptive field, can switch between excitatory and inhibitory in different stimulus contexts, without requiring dynamic cellular or synaptic properties. We also present experimental results supporting this notion. This model extends the model from the prior section by incorporating long-range (> 1.5 mm) intracortical excitatory connections (Rockland and Lund, 1982; Gilbert and Wiesel, 1983) (see Chapters 10 and 11).

The dynamic response modeling ideas arose out of consideration of cortical gain control issues (Somers et al., 1995b). To ensure the stability of a recurrent excitatory network, inhibition must be relatively strong at high drive levels (to prevent run-away positive feedback); yet, inhibition cannot be allowed to be too strong at low, suprathreshold drive levels (or else responses will be squashed). These same gain control principles can also produce dynamic receptive field effects with fixed circuitry. It turns out that this computational idea has interesting pieces of data to support it, including three previously unconnected phenomena: supersaturation, contrast-dependent length summation, and surround facilitation and suppression. In all three examples, receptive field regions shift from having excitatory influences to inhibitory influences as the contrast (and hence not excitatory drive) increases for the classical receptive field component of the stimulus.

Increasing the contrast of a stimulus in the receptive field can, at high contrast, actually cause responses of many V1 cells to decline or "supersaturate" (Li and Creutzfeldt, 1984) (Fig. 12-6A). This effect has been observed at contrast levels for which LGN responses have not saturated or supersaturated (Albrecht and Hamilton, 1982; Bonds 1991). A related phenomenon concerns the receptive field length of end-stopped cells. These cells typically respond maximally to high contrast stimulus bars of a certain length (and orientation), and responses decline for longer bars (e.g., Hubel and Wiesel, 1965), presumably owing to encroachment of the stimulus on inhibitory zones located at the ends of excitatory regions. End zones have maximal inhibitory effect when activated by stimuli of the preferred orientation (Orban et al., 1979; DeAngelis et al., 1994). However, these inhibitory zones can become excitatory as stimulus contrast is decreased, so that low contrast (preferred orientation) bars have longer optimal lengths than do high contrast

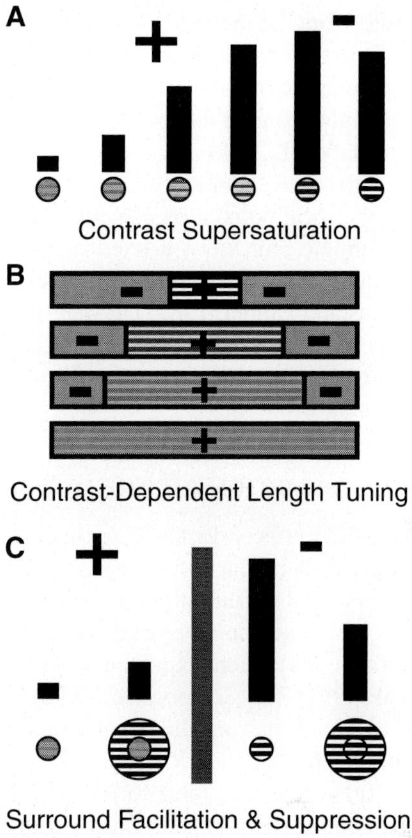

FIGURE 12-6. Schematic representation of different receptive field influences that shift from excitatory to inhibitory as central stimulus contrast increases. **(A)** A high-contrast surround stimulus facilitates the response of a low-contrast center stimulus, but the same surround-stimulus suppresses responses when the center stimulus is of high contrast. **(B)** Increasing the contrast of a center stimulus alone can decrease responses at very high contrast levels ("contrast supersaturation"). This suggests that the excitatory portion of the classical RF can become inhibitory. **(C)** Decreasing the contrast of a stimulus can cause the region of end-stopping to shift outward over the receptive field. At low contrasts, length summation can extend several degrees into the "inhibitory end-zones." Thus, the border between excitatory and inhibitory regions appears to shift with stimulus contrast. For all three receptive field effects, modulatory stimuli tend to have both the strongest facilitatory and strongest suppressive effects when oriented at the preferred orientation of the classical receptive field.

bars (Fig. 12-6B). This effect was first reported briefly (Jagadeesh and Ferster, 1990; Jagadeesh, 1993) but has since been replicated (Sceniak et al., 1999). A stimulus placed outside the classical receptive field can sometimes enhance responses (Maffei and Fiorentini, 1976; Knierim and Van Essen, 1992) and sometimes suppress responses (Blakemore and Tobin 1972; Nelson and Frost, 1978; Gulyas et al., 1987; Born and Tootell, 1991; Knierim and Van Essen, 1992;

Grinvald et al., 1994) (Fig. 12-6C). However, this seeming paradox had not been fully explored. In response to the model results, our laboratory performed a number of studies to investigate this dynamic behavior (Toth et al., 1996). Work by others has also supported and expanded this notion (Levitt and Lund, 1997; Sengpiel et al., 1997, 1998; Polat et al., 1998).

This work suggests that the relevant unit of receptive field integration is not the single neuron, but rather the local population of cells within a column of 300 to 500 μm diameter. In the single unit view, a fixed input to a fixed circuit will always exert a fixed influence on responses (Hartline, 1940; Kuffler, 1953; Movshon et al., 1978; Jones and Palmer, 1987; DeAngelis et al., 1995); however, in the population view, the net influence of a fixed input can change, provided that the integrating population consists of inhibitory and excitatory subpopulations with different (static) response properties.

KEY ASSUMPTIONS OF THE MODEL

The dynamic receptive field effects demonstrated here rely on an asymmetry between excitatory and inhibitory inputs in the local cortical circuitry. The critical property is that net inhibitory local circuit influences grow stronger at a greater rate than that of local excitatory influences as external drive levels increase. This idea is also supported by electrical stimulation experiments in slices of visual cortex (Hirsch 1995; Weliky et al., 1995). We (Somers et al., 1998) have proposed two specific nonlinear mechanisms for achieving this local circuit asymmetry; however, other plausible circuit nonlinearities could achieve the same functional results. The first (which we review here) is to utilize inhibitory neurons that, as a population, have higher response gain and higher contrast threshold than excitatory neurons. Based on intracellular data (McCorrmick et al., 1985), it is known that fast-spiking or inhibitory cells have much higher response gains than regular-spiking or excitatory neurons. However, the requirement for a higher contrast threshold has not been examined experimentally, and this assumption represents one experimental prediction of our model. Because this threshold difference need occur only at a population level, it is possible to achieve this result if multiple classes of inhibitory neurons, some with high thresholds and some with low thresholds (or a continuum) exist. Lund and colleagues (c.f., Lund et al., 1995) have identified several morphological classes of inhibitory neurons in V1 (see Chapter 1). Our analysis suggests that high threshold inhibitory neurons play a central role in gain control.

Synaptic physiology studies (Abbott et al., 1997; Tsodyks and Markram, 1997) indicate that rapid activity-dependent changes in synaptic efficacy at excitatory-excitatory cortical synapses may profoundly influence cortical processing. We suggest that if efficacy changes at excitatory-inhibitory synapses have the opposite sign, as is suggested by the data of Thomson and Deuchars (1994), then modulation of short-range synapses is sufficient to support contextual switching effects for both local and long-range inputs to the local cortical circuitry. Because this circuit with short-term synaptic plasticity cannot be called "fixed," we have omitted the result in this chapter and refer the interested reader to Somers et al. (1998). The important point is that the central requirements of the model can be achieved by diverse local circuit mechanisms.

MODEL STRUCTURE AND IMPLEMENTATION

The model (Somers et al., 1998) extends our prior local circuit model of orientation selectivity (Somers et al., 1995a) by incorporating long-range intracortical excitatory connections. Cortical circuitry under a 3.5 mm × 7 mm patch of primary visual cortex was represented by a model with 20,250 spiking cortical neurons and more than 1.3 million cortical synapses. Whenever possible, known anatomical values, ratios, and constraints were imposed on the model. In scaling up such a nonlinear system, parameter modifications were unavoidable; however, the essential behaviors of the earlier model (i.e., orientation sharpening, and amplification) persist in the present model. Neurons were organized into a 45 × 90 grid of "mini-columns" (Peters and Yilmaz, 1993) based on an orientation map obtained by optical recording of intrinsic signals of cat visual cortex (Toth et al., 1996) (Fig. 12-7).

Both excitatory and inhibitory neurons made short-range intracortical connections, whereas only excitatory neurons made long-range connections (see Chapter 10). Each type of connection targeted both excitatory and inhibitory postsynaptic neurons (Beaulieu and Somogyi, 1990; McGuire et al., 1991; Anderson et al., 1994). Short-range connection probabilities fell linearly with distance. Long-range excitatory neurons preferentially targeted neurons with orientation preferences similar to their own (Gilbert and Wiesel, 1989; McGuire et al., 1991; Malach et al., 1993). Connection probabilities declined linearly with the orientation difference between presynaptic and postsynaptic cells, from $\rho_{peak} = 0.005$ at $\varphi=0°$ to $\rho_{edge}=0.001$ at $\varphi=90°$.

We used two differences between the two cell classes as one way to create an asymmetry in the local circuitry that provides a local gain control mechanism. Fast-spiking (FS) or inhibitory neurons have higher input resistances and thus have higher current gains (greater slope of the frequency versus current plot) than those of regular striking (RS) or excitatory neurons (McCormick et al., 1985). Current thresholds for FS and RS neurons are similar, but inhibitory neurons receive substantially fewer synapses and presumably less synaptic current than excitatory neurons. Thus, we hypothesize that FS neurons have higher functional thresholds (e.g., contrast thresholds) than RS neurons. The combination of higher gain and threshold for a population of inhibitory neurons is one mechanism used to achieve a generalized gain control mechanism in the local circuitry. The essential nonlinearity here is that the ratio of local excitatory currents to local inhibitory currents evoked by a stimulus configuration should be high for low stimulus intensities and should decrease at higher stimulus intensities.

Stimuli consisted of circular ("center") and annular ("surround") gratings of differing contrasts, orientations, and radii. Central RF studies used only a central grat-

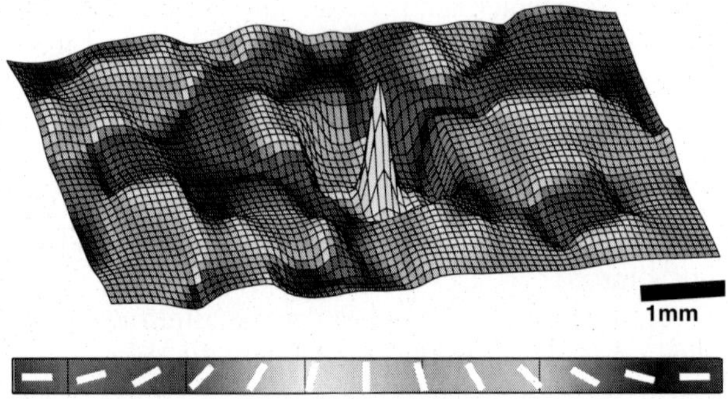

FIGURE 12-7. Connectivity of cortical circuitry in the model. The color map represents the orientation preference of each cortical mini column. The pattern of intracortical connections to cells in the central (Yellow) mini-column is represented by the surface amplitude, which codes the net (Σ Excit – Σ Inhib) strength of intracortical connections from each column to the cells of the central column. Three classes of intracortical connections are included in the model: short-range excitatory, short-range inhibitory, and long-range excitatory. Short-range connections are densest in the vicinity of the presynaptic cells and fall off with distance. Short-range excitatory connections are more numerous, but more spatially restricted than short-range inhibitory connections. Long-range excitatory connections can span the entire circuit and preferentially target cells with similar orientation biases. All connections target both excitatory and inhibitory neurons. See color insert for color reproduction of this figure.

ing with a diameter approximately equal to the RF (1°). End-stopping studies varied this diameter. Surround studies combined a 1° diameter center stimulus with a surround annulus (1° inner diameter; 4° outer diameter). Thalamic neuron responses to these stimuli increased linearly with log contrast. LGN responses were independent of stimulus orientation, phase, and spatial frequency. Converging thalamocortical inputs to a column were biased for a particular stimulus orientation and location (Hubel and Wiesel, 1962), and orientation selectivity was enhanced by intracortical connections (Somers et al., 1995a). The average firing rate of the converging thalamocortical input to single cortical neurons was a function of stimulus orientation (θ_{cent}, θ_{surr}), contrast ($\%Cont_{cent}$, $\%Cont_{surr}$), size, and position.

MODEL PERFORMANCE

Contrast Saturation and Supersaturation

The model captures physiological responses to oriented grating stimuli of differing contrasts presented within the classical receptive field. Model behavior is analyzed for a population of neurons lying in the central orientation domain of the model (yellow peak of Fig. 12-7). Figure 12-8A shows mean responses of excitatory cortical neurons to oriented grating stimuli of different contrasts presented within the classical receptive field. Responses were averaged for the same set of neurons in 20 networks, each constructed with the same connectivity probability rules.

To isolate cortical effects, thalamic responses were designed to increase linearly with log contrast. The dotted lines in Fig. 12-8A show contrast response curves for the cortical neurons at two different stimulus orientations when intracortical synapses are shut off and only thalamocortical inputs are active; note the linear responses above threshold. In the full model, with intracortical synapses active, cortical responses (data points and solid lines in Fig. 12-8A) exhibit a saturation nonlinearity at higher contrasts. This occurs at contrast levels for which thalamic responses have not yet saturated. Contrast saturation in the model is achieved as a network property rather than as a cellular property. Saturation response levels vary with stimulus orientation. Consistent with our prior model (Somers et al., 1995a), cortical inputs sharpen orientation tuning (compare differences between 0° and 20° responses). Note that intracortical connections amplify responses to low contrast stimuli at the preferred orientation, but attenuate responses to high contrast stimuli. Contrast saturation in our model reflects a gain change in local cortical circuitry that results from an asymmetry between the properties of excitatory and inhibitory neurons. Inputs from inhibitory neurons grow proportionally stronger at high levels of input drive; population responses for inhibitory neurons saturate only when LGN responses saturate.

For a subpopulation of modeled neurons (18 of 36 cells), a decline in response at high contrast was observed (Fig. 12-8B). Li and Creutzfeldt (1984) described this behavior in detail, calling the decline "supersaturation," because it occurs at contrast levels beyond which normal contrast saturation can be observed. Those authors suggested that supersaturation might result from reduced drive from thalamic sources at high contrasts; here, we show that a purely cortical mechanism is sufficient. The supersaturation effect, although infrequently discussed, can also be seen in other experimental reports (Albrecht and Hamilton, 1982; Bonds, 1991). Thus, for many cells, even the classical excitatory receptive field can have inhibitory effects when stimulation is strong.

Length-tuning

The effects of varying stimulus length and contrast were also systematically explored within the simulations. Responses to a high contrast visual stimulus were observed to decline beyond a characteristic preferred size (see top curve of Fig. 12-8C). Thus the model exhibited a form of length tuning or end-stopping (Hubel and Wiesel, 1965). Consistent with experiments (Orban et al., 1979; DeAngelis et al., 1994), high contrast iso-orientation stimuli that extend into end-zone (or side band) regions have potent inhibitory effects on responses to the central stimulus. In agreement with experimental findings (Jagadeesh and Ferster, 1990; Jagadeesh 1993; Sceniak et al., 1999), the length of the excitatory receptive field increased with decreasing contrast, and at low-stimulus contrasts responses to optimal orientation gratings continued to increase monotonically. Both the experiment and simulations indicate that the borders between excitatory and inhibitory regions shift depending on the level of stimulus contrast.

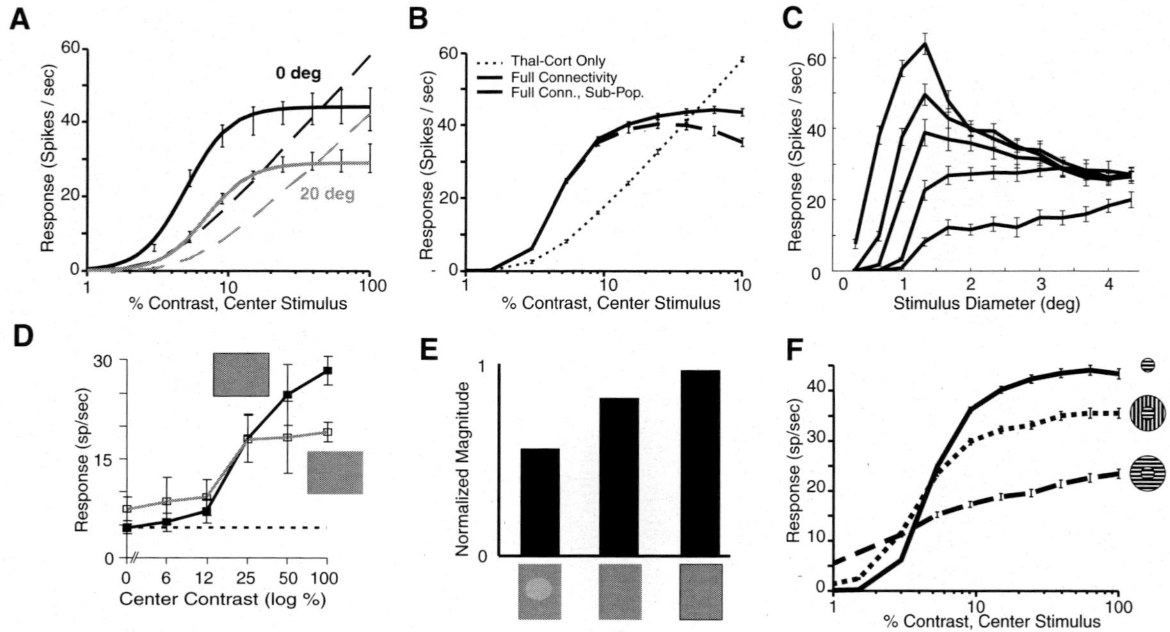
A
Response (Spikes / sec)
0 deg
20 deg
% Contrast, Center Stimulus
B
Thal-Cort Only
Full Connectivity
Full Conn., Sub-Pop.
Response (Spikes / sec)
% Contrast, Center Stimulus
C
Response (Spikes / sec)
Stimulus Diameter (deg)
D
Response (sp/sec)
Center Contrast (log %)
E
Normalized Magnitude
F
Response (sp/sec)
% Contrast, Center Stimulus

Facilitation Suppression

While preserving classical receptive field properties, the model also captures paradoxical extraclassical receptive field modulations (Knierim and Van Essen, 1992; Toth et al., 1996; Levitt and Lund, 1997; Sengpiel et al., 1997, 1998; Polat et al., 1998; Sillito et al., 1995). The contrast of a central grating was varied under three different surround stimulus conditions: no surround, a high contrast cross-orientation surround stimulus, and a high contrast iso-orientation surround stimulus. The modulatory influence of "surround" gratings on responses to optimal orientation "center" stimuli shifts from facilitatory to suppressive as center stimulus contrast increases (Fig. 12-8F). Compare this with experimental results from our laboratory (Toth et al., 1996) using single unit recording (Fig. 12-8D) and optical recording (Fig. 12-8E) techniques; a clear switch in receptive field influence is seen in both the data and in the model. Model simulations also obtain the result that both facilitation and suppression effects are strongest for iso-orientation surround stimuli (Knierim and Van Essen, 1992; Toth et al., 1996; Weliky et al., 1995; Sillito et al., 1995; Levitt and Lund, 1997). These effects emerge from the local intracortical interactions (as shown later) and do not require synaptic plasticity (cf. Hirsch and Gilbert, 1991; Gilbert et al., 1996) or complex cellular properties (cf., Bush and Sejnowski, 1994). This model also provides the first unified account of these classical and extraclassical RF modulations.

FIGURE 12-8. Model and experimental results on shifting excitatory and inhibitory regions of the receptive field. **(A)** Contrast response functions of model. Grating stimuli were shown at the preferred orientation (black curves) and 20° off of preferred (gray curves). Response of the full model circuit is shown with solid curves; response to thalamocortical inputs alone is shown with dashed lines. Cortical inputs strongly amplify responses to low suprathreshold contrasts, but attenuate responses to high contrasts. Cortical inputs also sharpen orientation selectivity. **(B)** For a subpopulation (18 of 36) of these neurons (dashed line), responses decline at high contrasts (supersaturation). **(C)** Response versus stimulus diameter at five different contrast levels (60%, 10%, 5%, 3%, and 2%) of the preferred orientation (10 trials, SD shown). For the highest contrast stimuli (top curve) responses are maximal for small stimuli (< 1.5°) and decrease for larger stimuli. As stimulus contrast decreases (lower curves), excitatory length summation occurs for progressively larger stimuli. Experimental results on surround facilitation and suppression from **(D)** single unit recordings and **(E)** optical recording of intrinsic signal. Central-grating stimuli were presented alone and again in conjunction with a high-contrast iso-orientation surround grating. Only the contrast of the central grating was changed. Resulting contrast response functions (CRFs) for the composite stimulus were higher at low center contrasts and lower at high center contrasts than center-only CRFs. Suppression and facilitation effects were largely unaffected by phase relation between center and surround stimuli. Optical recording data **(E)** show same effects, using only end points of CRFs. **(F)** Model results on facilitation and suppression produced by high-contrast surround stimuli. Responses of the same model neurons to varying contrast levels of a center stimulus under three fixed surround conditions: no surround (solid), high contrast cross-orientation surround (dotted), and high-contrast iso-orientation surround (dashed). Both surround stimuli increase responses to low-contrast centers but decrease responses to high-contrast center stimuli. Both facilitation and suppression effects are stronger for the iso-orientation surround.

Here, we have demonstrated one mechanism to produce the local circuit asymmetry: a population-level bias among inhibitory neurons to have higher gain and higher contrast thresholds than excitatory cells. Elsewhere we (Somers et al., 1998) have demonstrated a second mechanism: differential adaptation and enhancement at local intracortical excitatory synapses onto excitatory and inhibitory neurons, respectively (Thomson et al., 1993a, 1993b, 1995; Thomson and West, 1993; Thomson and Deuchars, 1994).

DISCUSSION

The functional asymmetry in the local cortical circuitry provides a generalized contrast gain control mechanism that can be accessed by all input sources. At low center-driving contrasts, local cortical excitation provides strong amplification of suprathreshold inputs. As center drive levels increase, local inhibition grows to reduce the cortical amplification factor. As a result, the model achieves saturating and supersaturating contrast response functions as a local circuit property. This result is consistent with experimental evidence that contrast saturation occurs when neither cellular firing rates nor synaptic inputs have reached plateau values. Thus, we suggest that the notion that excitation saturates is incorrect; rather, inhibition counterbalances excitation. The model makes the specific prediction that cortical amplification is strongest for low-contrast, suprathreshold stimuli (of the preferred orientation) and that amplification decreases and may even become attenuation as contrast increases.

This local circuit form of gain control, unlike cellular or input saturation ideas, also captures spatial contrast gain effects. Excitatory receptive field regions are spatially restricted when the total inputs are strong, but enlarge, as inputs become weaker. These expansion-contraction effects are adaptive in that they collect (excitatory) information more widely when the overall signal is weak but restrict their spatial spread when the signal from more central receptive field regions is stronger. It should be noted that the term *contrast gain control* has an ambiguous meaning in the literature, referring both to instantaneous contrast saturation as discussed here and slower acting contrast adaptation.

Sceniak et al. (1999) performed a linear, difference-of-gaussians analysis of their contrast-dependent dependent length summation data and found that their effects were best fit by allowing the excitatory space constant to shrink as stimulus contrast increased. This linear analysis suggested that surround strength effects were variable, increasing for some cells and decreasing for others. On this basis, these authors have criticized our long-range model because it suggests that excitatory and inhibitory changes are coupled; however, this analysis is problematic. Our model makes a nonlinear prediction about the excitatory and inhibitory strength parameters; this was not tested in their analysis. It would be interesting to see this prediction analyzed with their data. One might argue that a linear interpretation is simpler than a nonlinear one; however, in this case, the linear model *requires* that the functional connectivity of excitatory connections change with stimulus contrast, whereas the nonlinear model requires no such changes. One mechanism that they suggest for functional connectivity changes is synaptic depression at excitatory synapses, which is the second mechanism that we proposed and demonstrated (Somers et al., 1998).

Co-aligned, iso-orientation surround stimuli tend to facilitate responses for both high and low-contrast center stimuli (Kapadia et al., 1995; Polat et al., 1998). This effect may be related to perceptual contour integration (Grossberg and Mingolla, 1985; Field et al., 1993; Polat and Sagi, 1993) and may result from excitatory-excitatory connections between co-aligned, iso-orientation cell clusters (Bosking et al., 1997). Long-range connections in our model are not biased for spatial anisotropies. Our model could achieve alignment-specific facilitation by adding aligned long-range inputs that are more biased to synapse onto excitatory targets than are the present alignment nonspecific long-range inputs. Our model could similarly be extended to incorporate other specific patterns of long-range excitation, such as feedback from area V2 (Bullier et al., 1996).

SUPRAOPTIMAL RESPONSES AND DYNAMIC PROPERTIES OF RECURRENT INHIBITION

INTRODUCTION

We have described in the previous section how a functional asymmetry in the local cortical circuitry provides a sufficient explanation for the contrast dependency of contextual interactions. When the classical receptive field is optimally stimulated at high contrast, the presence of an iso-oriented surround stimulus increases the local inhibitory input to reduce the cortical gain, whereas when the center drive decreases, the local excitation provides strong amplification of suprathreshold inputs. However, it is unlikely that this mechanism would be able to explain cross-orientation surround facilitation (i.e., the fact that stimulating the surround with a grating whose orientation differs significantly from the cell's preferred orientation facilitates responses to optimal stimulation within the center) (Sillito et al., 1995; Levitt and Lund, 1997). In this case neurons respond beyond the level expected after stimulation with the optimal orientation (Fig. 12-9A). One prediction of our long-range model (Somers et al., 1998; also Stemmler et al., 1995) is that a cross-oriented surround would provide weak input to the local populations of excitatory and inhibitory neurons such that cortical responses are maintained around center-alone response levels.

To explore potential mechanisms that would explain the emergence of supraoptimal responses elicited by the cross-oriented surround, we reasoned that at high-contrast center and surround stimulation local inhibitory populations receive tonic inputs that show orientation specificity. Indeed, it has been reported (Weliky et al., 1995; Weliky and Katz, 1994) that activation of long-range horizontal connections evokes direct iso-orientation excitatory and multisynaptic

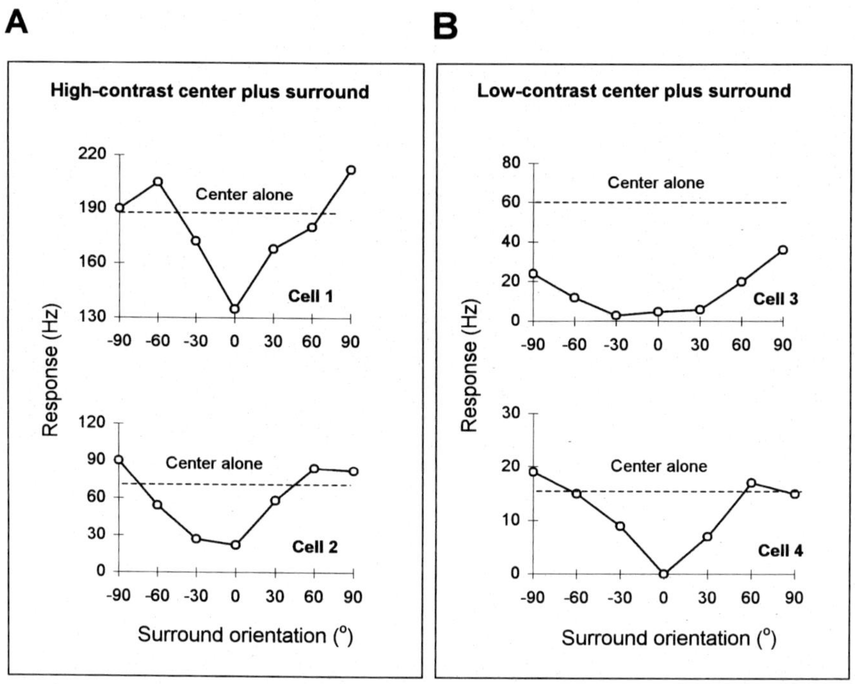

FIGURE 12-9. Contrast and orientation dependence of contextual effects (experimental data). **(A)** Experimental data: Responses to the high-contrast optimal center stimulus paired with high-contrast surround gratings of varying orientations. Cells 1 and 2 are adapted from Fig. 12-1 (cells A and D) of Levitt and Lund (1997). **(B)** Experimental data: Responses to the low-contrast optimal center stimulus paired with high-contrast surround gratings of varying orientations. Cells 3 and 4 are adapted from Fig. 12-1 (cells D and E) of Levitt and Lund (1997). Dashed lines in all panels indicate response to the optimal stimulus alone.

inhibitory responses from local pyramidal cells such that there is stronger activation of iso-orientation domains and gradually weaker activation of cross-orientation domains (see Chapters 10 and 11). It is thus possible that for certain center-surround orientation configurations, the stronger activation of selected local inhibitory neurons could trigger disinhibitory mechanisms that would facilitate responses in the presence of a cross-oriented surround. Disinhibitory interactions have a well-established role in other brain systems (e.g., Ito et al., 1968; Kelly and Renaud, 1974; Getting and Dekin, 1985; Hultborn et al., 1971), and yet the function of disinhibitory mechanisms in the visual cortex has not been explored. In this section we describe how recurrent inhibition in V1 can trigger disinhibitory interactions that help explain cross-orientation surround facilitation. Previous theoretical considerations, (Blomfield, 1974; Koch and Poggio, 1985) have suggested that inhibition could play a crucial vetoing role in the emergence and shaping of receptive field properties, such as direction (and possibly orientation) selectivity. However, these models have considered the properties of inhibitory projections to excitatory neurons without exploring the functional role of recurrent inhibition.

MODEL STRUCTURE AND IMPLEMENTATION

We have used a model of recurrent inhibition (Dragoi and Sur, 2000) to understand why surround stimuli presented at nonoptimal orientations drive neurons beyond optimal responses (obtained by high-contrast center stimulation at the preferred orientation), whereas when the CRF is stimulated at low contrast, the facilitatory effects disappear (Sillito et al., 1995; Levitt and Lund, 1997) (Fig. 12-9). The model incorporates short-and long-range interactions that describe the processing of information at two sequential stages: LGN and V1. The model configures 8,712 LGN neurons arranged on the array of 11×11 locations, with 72 cells per each location of the visual patch and 17,424 cortical neurons.

Cellular Models

LGN cells are modeled as single units whose mean rate of firing is given by:

$$dLGN_i/dt = -0.01\ LGN_i + R_i\ (1 - LGN_i)$$

where LGN_i represents the lateral geniculate cell. The retinal input, R_i, is set such that thalamic responses increase linearly with the log of stimulus contrast. R_i is maximal when the preferred orientation of the cortical cell that corresponds topographically to LGN_i matches the stimulus orientation, and it decays exponentially to zero with increasing orientation differences (Dragoi and Sur, 2000). The spread of geniculate inputs to the cortex ensures that each LGN cell synapses on a group of cortical cells with a broad range of orientations (with a spread of 60°). Cortical cells receive center stimulation as an oriented input stimulus applied to the receptive field center of LGN cells and surround stimulation as oriented stimuli applied to the surrounding hypercolumns.

Excitatory and inhibitory cortical neurons are modeled separately as single units whose mean rate of firing is given by:

Excitatory cells

$$dE_i/dt = -0.01\ E_i + (J^{fe}\ F_i + \Sigma_j\ J_{ij}^{ee}\ E_j + \Sigma_j\ J_{ij}^{me}\ E_j)\ (1 - E_i) - \Sigma_j\ J_{ij}^{ie}\ I_j\ E_i$$

Inhibitory cells

$$dI_i/dt = -0.01\ I_i + r\ (J^{fi}\ F_i + \Sigma_j\ J_{ij}^{ei}\ E_j + \Sigma_j\ J_{ij}^{mi}\ E_j)\ (1 - I_i) - \Sigma_j\ J_{ij}^{ii}\ I_j\ I_i$$

where E_i represents excitatory and I_i represents inhibitory cells. In agreement with experimental evidence (Connors et al., 1982; McCormick et al., 1985), model inhibitory cells have a higher firing rate than excitatory cells ($r = 3$). Each cortical cell receives stimulus-specific feedforward input, F_i, as the summed response of LGN cells centered at i with a spread of 60°. J^{fe} and J^{fi} are the strengths of feedforward connections and are equal for excitatory (fe) and inhibitory (fi) cells. J_{ij}^{ee} and J_{ij}^{ei} are the strengths of recurrent excitatory connections (ee) and excitatory projections to inhibitory cells (ei). J_{ij}^{ie} and J_{ij}^{ii} are the strengths of the inhibitory projections to excitatory cells (ie) and inhibitory cells (ii). J_{ij}^{me} and J_{ij}^{mi} are the strengths of long-range (modulatory) inputs to excitatory cells (me) and inhibitory cells (mi).

Short-range Intracortical Connections

Excitatory neurons are interconnected by recurrent excitatory synapses (Martin, 1988; Peters and Payne, 1993), whereas inhibitory neurons are interconnected by recurrent inhibitory synapses (Beaulieu and Somogyi, 1990; Kisvarday et al., 1993; Sik et al., 1995). In addition, local excitatory cells excite neighboring inhibitory cells, which in turn inhibit excitatory cells (Beaulieu and Somogyi, 1990; McGuire et al., 1991; Anderson et al., 1994a) (Fig. 12-10A). Short-range excitatory and inhibitory connection strengths decrease as cortical neurons become more widely separated in orientation (Fries et al., 1977; Nelson and Frost 1978; Miller, 1992). The strength of excitatory connections decays exponentially from 0.01 at distance zero to 75% of the peak value at 40° orientation difference between presynaptic and postsynaptic cells; the strength is 0 beyond 40°. Consistent with evidence from cross-correlation studies (Toyama et al. 1981; Michalski et al., 1983; Hata et al., 1988) and from combined imaging and intracellular recording (Tucker and Katz, 1998), intracortical inhibitory connections arise from cells with a broader distribution of orientation preferences than do intracortical excitatory connections. However, given the current disagreement on the exact orientation spread of inhibitory connections (Ferster, 1988; Hirsch and Gilbert, 1991; Tucker and Katz, 1998; Roerig and Katz, 1998), we have restricted inhibitory inputs to an orientation difference of 60° between presynaptic and postsynaptic cells. Inhibitory connection strengths are stronger than excitatory connections (Komatsu et al., 1988; Thomson and West, 1993; Thomson and Deuchars, 1994), and they decay exponentially with distance from a maximum value at distance 0 to 10% of the peak value at 60° orientation difference between presynaptic and postsynaptic cells. The strength of inhibitory connections is 0 beyond 60°.

Long-range Intracortical Connections

Long-range horizontal connections (Gilbert and Wiesel, 1979; Rockland and Lund, 1982a; Livingston and Hubel, 1984; Martin and Whitteridge, 1984) link cells across distinct regions of the visual field and spread across four orientation hypercolumns (or locations) in the model. Model long-range horizontal connections are excitatory and originate from pyramidal cells in the surround (cf. Gilbert and Wiesel, 1989) (Fig. 12-10A). These cells contact other pyramidal cells, as well as nearby inhibitory cells that are locally interconnected within a range of ± 60° (Kisvarday et al., 1986; McGuire et al., 1991). Model activation of horizontal connections evokes direct iso-orientation excitatory and multisynaptic inhibitory responses from local pyramidal cells in an orientation-dependent fashion: stronger activation of iso-orientation domains and gradually weaker activation of cross-orientation domains. We set the strengths of model long-range horizontal connections as being maximal when they connect cortical cells with the same orientation preference; the strengths gradually decrease with increasing relative orientation between cells (Weliky et al., 1995; Weliky and Katz, 1994). Long-range connection strengths decay exponentially from $J_i^{mi} = 0.03$ and $J_i^{me} = 0.01$ at distance 0 to 25% of the peak value at 60° orientation difference between presynaptic and postsynaptic cells.

MODEL PERFORMANCE

Analysis of Recurrent Inhibition

Our goal in the simulations is to understand how neuronal responses are modulated by changes in the balance between local excitation and inhibition as a result of surround stimulation. Therefore, the receptive field center was always stimulated at the optimal orientation, whereas surround orientation was varied systematically from 0° to 180° to fully investigate the orientation dependence of contextual effects. Figure 12-10B–E shows the changes in excitatory and inhibitory cortical responses in a representative population of cells with orientation preferences in the range [–30°, 75°]. The high-contrast surround stimulus is oriented either at 0° (maximum suppression) or 60° (maximum facilitation), whereas the orientation of the center stimulus is fixed at 0°. In all the simulations the center stimulus is either presented at 100% (high) or 15% (low) contrast level.

High-contrast Center Stimulation

When the surround is oriented at 60°, strongly it activates the inhibitory cells in the vicinity of iso-orientation domains (e.g., 60°) and more weakly the inhibitory cells in the vicinity of non-iso-orientation domains (e.g., 0°) (Fig. 12-10A). The surround modulation is orientation-dependent: pyramidal cells outside the CRF project more strongly to iso-orientation domains and only weakly to non-iso-orientation domains (e.g., the projection to the horizontal pyramidal cell in Fig. 12-10A) and to the neighboring inhibitory cells. Figure 12-10B–E represents the changes in inhibitory and excitatory responses relative to the center alone condition. Because of their higher firing rates, inhibitory cells with orientation preference close to that of the surround (i.e., the cells with orientation preferences near 60°) are able to fire continuously in response to the tonic center and surround stimuli and thus exert tonic inhibition on their postsynaptic targets (i.e., inhibitory and excitatory cells with orientation preferences near 0°). The effect of this interaction can be seen in Fig. 12-10B, which shows that the activity of inhibitory cells oriented away from the surround orientation (e.g., the 0° cells) diminishes below the center alone condition. The net effect of this removal of tonic inhibition, or disinhibition, from pyramidal cells in the vicinity of the 0°

domain is an increase in the strength of excitation relative to the center-alone condition (Fig. 12-10C). This explains the supraoptimal responses obtained when the surround is cross-oriented with respect to the center stimulus (Fig. 12-11A). However, when the surround is oriented at 0°, there is an increase in the inhibitory responses above the center-alone condition (Fig. 12-10B) that decreases the

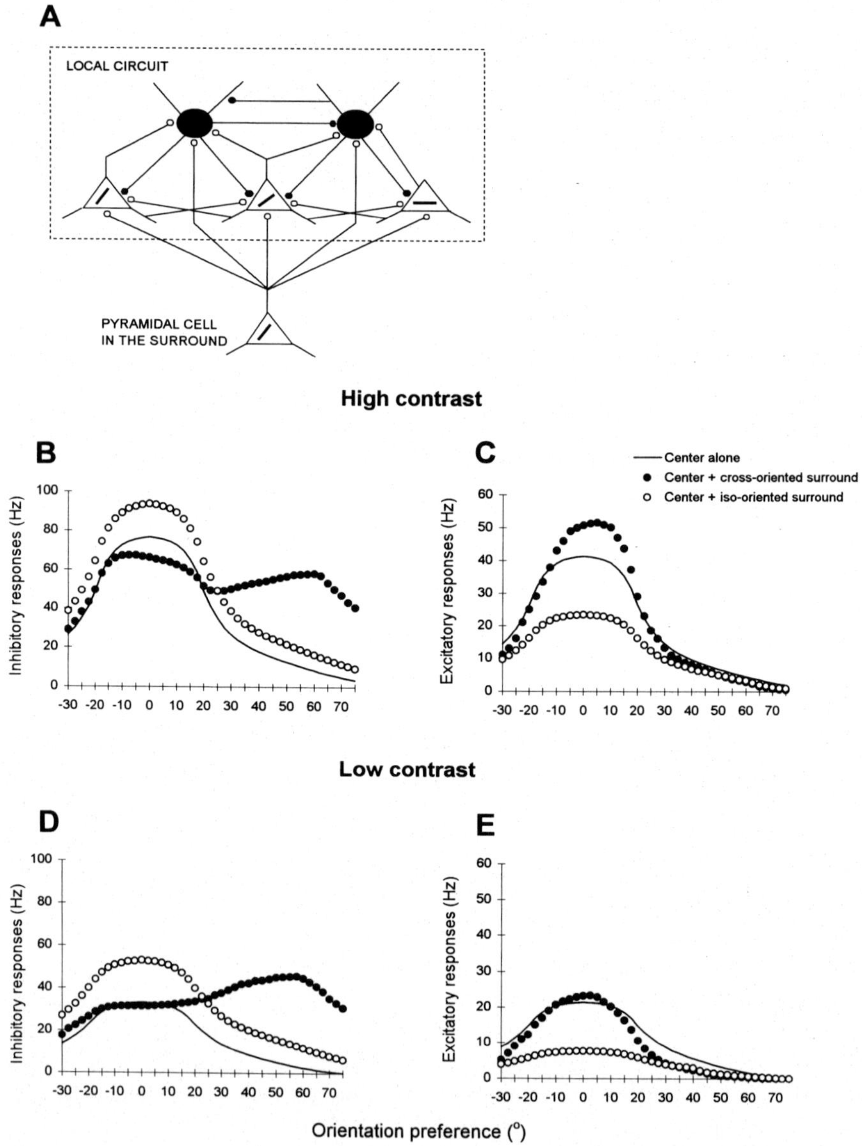

strength of excitation relative to the center-alone condition (Fig. 12-10C). This explains the iso-orientation surround suppression (Fig. 12-11A).

Low-contrast Center Stimulation

The mechanism just described is manifested when the receptive field is stimulated at high contrast. However, it becomes ineffective when the center is presented with a low-contrast stimulus. If the surround is cross-oriented at 60° and the center orientation is 0° and 15% contrast level, inhibitory neurons iso-oriented relative to the surround (i.e., preferred orientation 60°) decrease their response to suppress only weakly their postsynaptic targets, including other inhibitory cells. Indeed, Fig. 12-10D shows that, relative to the center-alone condition, inhibitory responses are suppressed to a lesser extent than in the high-contrast case. Therefore, the iso-oriented inhibitory neurons are no longer capable of sustaining the release from inhibition of the non-iso-orientation excitatory cells, and thus the total excitatory input to cortical cells does not differ substantially from the center-alone condition (Fig. 12-10E). The net effect of this interaction is the disappearance of surround cross-orientation facilitation (Fig. 12-11B), and a broader tuning of the suppressive effects.

In summary, the recurrent inhibition model demonstrates that orientation-dependent long- and short-range connections can have biphasic modulatory effects, depending on the relative orientation and contrast between center and surround. Thus, Fig. 12-11A shows that responses to the center stimulus are suppressed by an iso-oriented surround. However, responses to the same center stimulus become supraoptimal in the presence of an orthogonal or oblique surround. These results provide a good fit to the experimental data obtained in similar conditions (Levitt

FIGURE 12-10. Analysis of contrast-dependent center-surround interactions. **(A)** Schematic diagram of the interactions between representative cells in the superficial layers of V1 embedded in their local circuit (dashed rectangle). Pyramidal cells in the surround project to both excitatory (triangles) and inhibitory (filled circles) neurons. Empty circles, excitatory connections; filled circles, inhibitory connections. **(B)** The response of inhibitory cortical cells with orientation preferences between –30° and 75°. When the surround is cross-oriented (filled circles), cells oriented away from the surround orientation are released from local inhibition relative to the center-alone condition (solid line). When the surround is iso-oriented (empty circles) there is an overall increase in the local inhibition level. **(C)** The response of excitatory cortical cells with orientation preferences between –30° and 75°. When the surround is cross-oriented, the net effect of the removal of tonic inhibition from pyramidal cells in the vicinity of the 0° domain is an increase in the strength of excitation relative to the center-alone condition. When the surround is iso-oriented, there is a decrease in the response of excitatory cells relative to the center-alone condition. Center stimulus: 100% contrast level, orientation is fixed at 0° (panels C and D). **(D)** The response of inhibitory cortical cells with orientation preferences between –30 and 75°. When the surround is cross-oriented, inhibitory responses are suppressed to a lesser extent than in the high-contrast case. **(E)** The response of excitatory cortical cells with orientation preferences between –30 and 75°. When the surround is cross-oriented the total excitatory input to cortical cells does not differ substantially from the center-alone condition. Center stimulus: 15% contrast level, orientation is fixed at 0°.

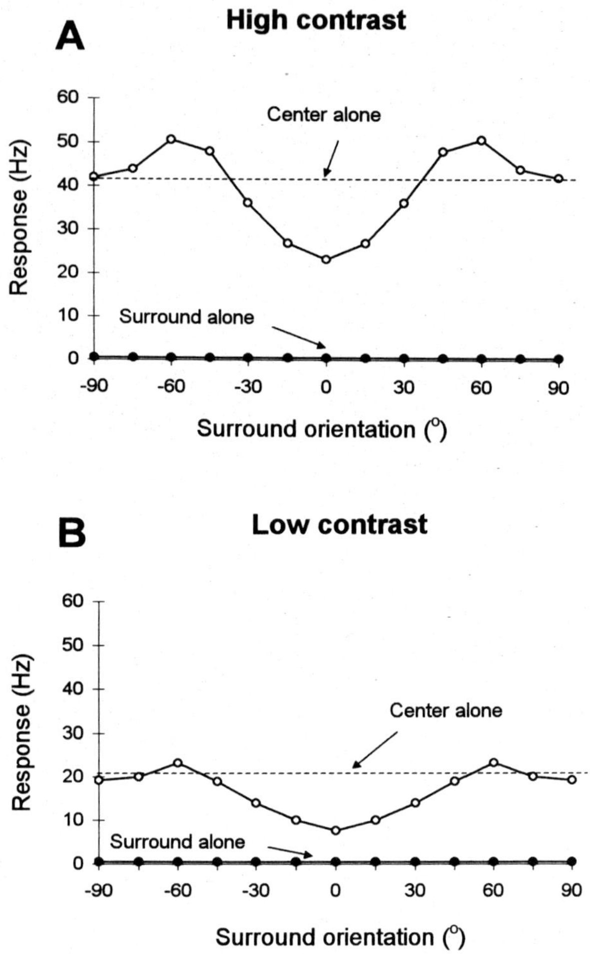

FIGURE 12-11. Contrast and orientation dependence of contextual effects (modeling results). **(A)** Model responses to the high-contrast optimal center stimulus (contrast 100%) paired with a high-contrast surround of varying orientation (empty circles) and to the surround stimulus alone (filled circles). Center contrast and surround orientation values are identical to those used by Levitt and Lund (1997). **(B)** Model responses to the low-contrast optimal center stimulus (contrast 15%) paired with a high-contrast surround of varying orientation (empty circles) and to the surround stimulus alone (filled circles). Center contrast and surround orientation values are identical to those used by Levitt and Lund (1997). Dashed lines in all panels indicate response to the optimal stimulus alone (high-center contrast value is 100% in panel A; low-center contrast value is 15% in panel B).

and Lund, 1997) (Fig. 12-9A). When the center stimulus is presented at low contrast, the facilitatory effects induced by cross-oriented surround stimuli disappear or become very small (Fig. 12-11B), and this determines a broader tuning of the suppressive effects (Levitt and Lund, 1997) (Fig. 12-9B). We have thus demonstrated that under some center-surround configurations, the responses of inhibitory interneurons can be completely reversed: iso-oriented stimuli in the surround increase the firing rate of local inhibitory cells that further suppress their postsynaptic pyramidal cells, and cross-oriented stimuli in the surround decrease the firing rate of local inhibitory cells that further disinhibit their postsynaptic pyramidal cells. In addition, the magnitude of the disinhibitory effect decreases with reductions in the center contrast level. Thus, the recurrent inhibition model advances a clear-cut prediction: measuring the total intracellular excitatory and inhibitory synaptic responses during in vivo presentations of different center-surround configurations would yield inhibitory responses below the center-alone condition and excitatory responses above the center-alone condition when the surround is cross-oriented, and inhibitory responses above the center-alone condition and excitatory responses below the center-alone condition when the surround is iso-oriented.

SHORT-TERM PLASTICITY OF ORIENTATION TUNING INDUCED BY PATTERN ADAPTATION

INTRODUCTION

The last two sections showed that selective stimulation of the receptive field center and surround explains nonlinear contextual effects in the contrast and orientation domains. These effects arise as a consequence of spatial interactions and persist as long as the center and surround stimuli are presented together. However, if changes in the gain of local circuitry are indeed responsible for changes in the response properties of cortical neurons and if orientation selectivity is an emergent property of recurrent networks, we would predict that altering the efficacy of intracortical orientation-specific inputs would induce *changes* in the tuning properties of neurons. One way to alter the efficacy of cortical inputs to a V1 neuron is to enhance or suppress the excitability of individual or small groups of neurons. We have shown that focal iontophoresis of bicuculline or GABA to a cortical column alters the orientation tuning properties of adjacent cortical columns (Toth et al., 1997). Another way to alter intracortical inputs is to adapt a cell to a stimulus of fixed orientation. It is known that adapting neurons to a potent stimulus can reduce responses to subsequent similar stimuli. In a recent study (Dragoi et al., 2000), we examined the changes in orientation tuning using pattern adaptation (Movshon and Lennie, 1979; Saul and Cynader, 1989; Carandini et al., 1998) as the induction procedure by analyzing how the entire profile of the orientation tuning curve changes after short- and long-term adaptation to a particular stimulus orientation.

RESULTS

Single Cell Responses

In contrast to the common view of adaptation as simply a passive process of response suppression, we found that exposure to particular orientations reveals an active process of plasticity by which responses on the flank of the tuning curve near the adapting orientation are depressed, whereas responses on the opposite flank of the tuning curve are enhanced. For instance, Fig. 12-12A shows how the preferred orientation of a representative cell changes after 2 minutes of exposure to one orientation located on one flank of the cell's tuning curve, followed by a period of recovery, subsequent adaptation to a different orientation located on the opposite flank with respect to the preferred orientation, and a final period of recovery. When the difference between the cell's preferred orientation and that of the adapting stimulus ($\Delta\theta$) is –22.5°, there is a shift in preferred orientation to the right, away from the adapting stimulus. In contrast, when the adapting stimulus is presented on the right flank of the tuning curve ($\Delta\theta = 45°$), the preferred orientation shifts to the left and then returns to the original value after 10 minutes of recovery. However, adaptation to stimuli orthogonal to the cell's preferred orientation ($\Delta\theta$ between approximately 60° and 90°) does not induce any change in preferred orientation. Figure 12-12B illustrates the behavior of one representative cell that exhibits a stimulus-dependent shift after 2 minutes of adaptation to a 22.5° stimulus, but the orientation preference remains unchanged when $\Delta\theta$ is 90°. In both postadaptation conditions, there is a decrease in response at the preadaptation preferred orientation and a broadening of tuning.

Interestingly, the shape of the orientation tuning curve undergoes profound changes when neurons are serially exposed to different adaptation periods. For instance, Fig. 12-12C,D shows one cell that exhibits significant shifts in orientation after adaptation for 10 seconds, 2 minutes, and 10 minutes to a stimulus oriented 45° away from the cell's peak orientation. Both the response reduction on the near flank (toward the adapting orientation) and facilitation on the far flank of the tuning curve (away from the adapting orientation) build up gradually in time: increasing the adaptation time from 10 seconds to 10 minutes shows a progressive depression of responses on the near flank and a progressive facilitation of responses on the far flank. For the largest adaptation period (10 minutes), we found that many cells increase their response at the new preferred orientation by a factor of 2 or more (Fig. 12-12D). The orientation at which the adapting stimulus is presented elicits weak or no responses under the various time periods (e.g., Fig. 12-12C,D). Yet adaptation induces a significant shift in optimal orientation, along with the reorganization of responses around the new preferred orientation. Thus, orientation plasticity involves an active process of network synaptic changes that lead to a new preferred orientation rather than simply a passive reduction of orientation-selective responses around the adapting orientation.

Optical Imaging Studies of Adaptation

We also investigated short-term plasticity in orientation tuning at the population level by carrying out optical imaging of intrinsic signals (Grinvald et al., 1986) from an expanse of V1 (Dragoi et al., 2000). Figure 12-13A shows three composite orientation preference maps from one animal, combining response images at eight different stimulus orientations, obtained during control, adaptation, and

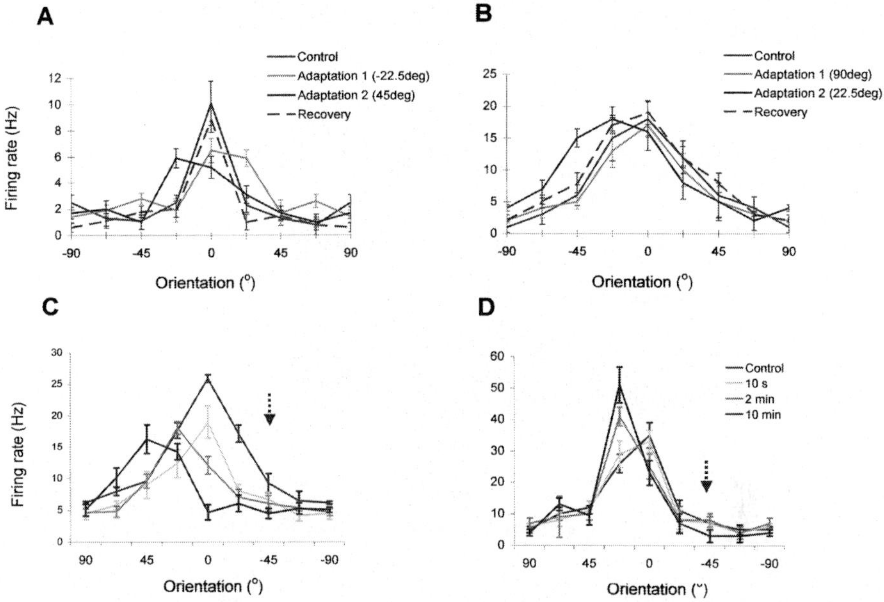

FIGURE 12-12. Plasticity of orientation tuning in V1 cells. **(A-B)** Orientation tuning curves of two representative cells that were successively adapted to two different orientations. Each graph represents orientation tuning during four conditions: control (red), adaptation to the first orientation (light blue), adaptation to the second orientation (dark blue), and recovery (red, dashed line). In our tuning curve display convention, the control optimal orientation is represented as 0°, and all subsequent tuning curves (during adaptation and recovery) are represented relative to the control condition. **(C-D)** Tuning curves of cells that show adaptation-induced response suppression on the near flank and response facilitation on the far flank. Each cell was serially exposed to different adaptation periods: 10 seconds, 2 minutes, and 10 minutes. Tuning curves were calculated in each of the four conditions: control (red), 10-second adaptation (light blue), 2-minute adaptation (blue), and 10-minute adaptation (dark blue). The adapting orientation is marked by the blue arrow. See color insert for color reproduction of this figure.

recovery conditions. The adapting orientation was fixed throughout the experiment at 135° (dark green bar in Fig. 12-13A). We determined the change in cortical responses induced by the adapting stimulus by computing the difference in orientation preference between control, adaptation, and recovery conditions for each pixel. If the adapting orientation shifts the preferred orientation of cells, pixels flanking the adapting orientation would change their vector angle away from the adapting orientation (dark green). For example, most pixels preferring 112.5° (light green) would shift toward 90° (yellow), and the 90° pixels would shift toward 67.5° (orange). At the opposite side of the adapting orientation, most pixels preferring 157.5° (light blue) would shift toward 180° (dark blue), and the 180° (or 0°) pixels would shift toward 22.5° (purple). Pixels whose orientation preference exactly matches that of the adapting stimulus (i.e., 135° [dark green] would show minimal changes in their angle; others with actual preferred orientation close to

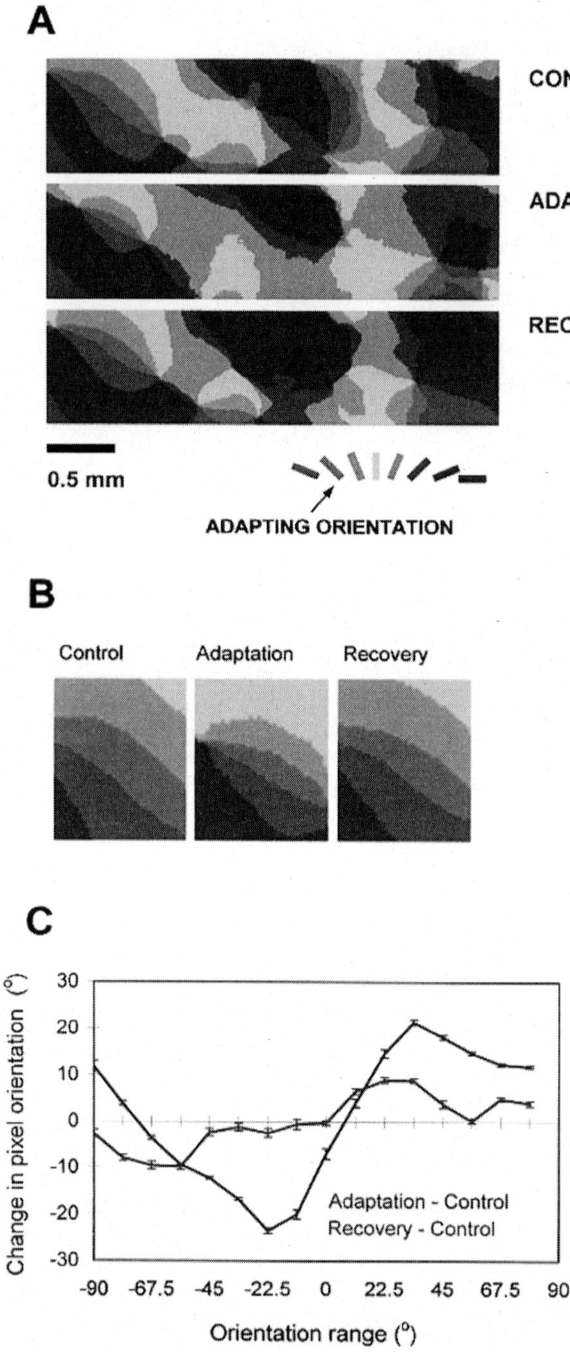
A
CONTROL
ADAPTATION
RECOVERY
0.5 mm
ADAPTING ORIENTATION
B
Control
Adaptation
Recovery
C
Change in pixel orientation (°)
30
20
10
0
-10
-20
-30
-90
-67.5
-45
-22.5
0
22.5
45
67.5
90
Orientation range (°)
Adaptation - Control
Recovery - Control

FIGURE 12-13. Changes in orientation tuning in an expanse of V1 demonstrated by optical imaging. **(A)** Composite maps of orientation angle obtained during control, adaptation, and recovery conditions. Data analysis was performed using original, unfiltered, single orientation maps in all three conditions. To obtain these composite maps we summed vectorially the response at each pixel to the eight single stimulus orientations (including both directions of motion) and displayed the resultant angle of preferred orientation in pseudocolor according to the key at bottom. Each map was smoothed using a low-pass filter, 5×5 kernel size. Adapting orientation is coded dark green. **(B)** Magnified portion from **(A),** showing the postadaptation repulsive shift in orientation and the recovery from adaptation. Depending on the difference between pixel orientation and that of the adapting stimulus ($\Delta\theta$) orientation domains exhibit repulsive shift toward neighboring orientations: for $\Delta\theta = -22.5°$ (light green) most pixels shift toward the "yellow" domain; for $\Delta\theta = 0°$ (dark green) many pixels are unchanged and others shift either toward the "light green" or "light blue" domains; for $\Delta\theta = 22.5°$ (light blue) most pixels shift toward the "dark blue" domain. **(C)** Change in pixel orientation during adaptation and recovery. Orientation changes were calculated for all pixels in the entire map shown in **(A)** by subtracting the pixel vector angle in the control map from that of the corresponding pixel in the adaptation or the recovery map. We divided all pixels into 16 bins, which are represented, relative to the adapting orientation (135°), as the following intervals: [–90°, –78.75°], [–78.75°, –67.5°] ... [78.75°, 90°]. Each of these bins is represented on the abscissa by the lower bound of each orientation interval. The numbers of the y-axis represent the average change in pixel angle. See color insert for color reproduction of this figure.

135° [135° ± 11.25°] would also shift repulsively with respect to the adapting orientation). Finally, if the shift is reversible, all pixels should revert to their initial orientation after recovery from adaptation. Figure 12-13B shows a magnified portion from Fig. 12-13A that captures the repulsive shift in orientation during adaptation and then shows that most pixels recover to their original orientation.

Figure 12-13C demonstrates quantitatively that a repulsive shift in orientation follows adaptation. The change in orientation for each pixel during adaptation and recovery was quantified by calculating the difference between the vector angle of each pixel before and after adaptation, and before adaptation and after recovery. There is a positive change in orientation when the difference between each pixel's orientation and the adapting orientation is positive, and a negative change in orientation when the difference between each pixel's orientation and the adapting orientation is negative. Also paralleling the single-unit data, Fig. 12-13C shows that as the orientation difference increases from 0 to about ± 22.5°, the repulsive shift increases, and this effect diminishes as the orientation difference approaches ± 90°. During recovery there is a reversal toward original (control) pixel values.

These results demonstrate that the orientation-selective responses of adult V1 cells are reorganized actively and nonlinearly by the temporal context of stimulation. Previous examination of the effect of temporal context pertains to studies that have demonstrated that adapting neurons to a potent stimulus can reduce responses to subsequent similar stimuli. This property has been characterized with respect to many stimulus dimensions, such as orientation (Blakemore and Campbell, 1969; Hammond et al., 1989; Nelson, 1991; Carandini et al., 1998), contrast (Movshon and Lennie, 1979; Ohzawa et al., 1982; Carandini and Ferster, 1997; Carandini et al., 1997), spatial frequency (Movshon and Lennie, 1979; Saul and Cynader, 1989), direction of motion (Maffei et al., 1973; Hammond et al., 1985; 1986), and velocity (Hammond et al., 1985). Saul and Cynader (1989) have demonstrated that pattern adaptation causes changes in the spatial frequency tuning of a small group of V1 neurons, an effect related to the changes in orientation preference reported here. However, it is generally assumed that in the orientation domain adaptation reduces responses at all orientations, the maximum reduction being obtained when the cell's preferred orientation and that of the adapting stimulus are the same (this property can be generalized to most stimulus attributes). A reduction of responses may result from mechanisms at the level of individual neurons, such as tonic hyperpolarization of the membrane potential of V1 cells (Carandini and Ferster, 1997), resulting possibly from synaptic depression (Abbott et al., 1997; Chance et al., 1998) or to slow hyperpolarizing Ca- and Na-activated potassium channels (Sanchez-Vives et al., 2000). The changes in orientation selectivity reported here (i.e., shifts in orientation preference by depression of responses on the near flank and facilitation of responses on the far flank) imply a network mechanism that reorganizes responses across a broad range of orientations, possibly through changes in the gain of local cortical circuits that mediate recurrent excitation and inhibition (Douglas et al., 1995; Somers et al., 1995a, 1998) and include disinhibitory mechanisms (Dragoi and Sur, 2000). For example, if the local cortical circuit includes broadly tuned orientation inhibition, hyperpolarization of neurons representing the adapting orientation could cause disinhibition of responses on the far flank of the tuning curve in the recorded neuron, an effect that could be further amplified via local excitatory interactions.

CONCLUDING REMARKS

The models and data presented here constitute a powerful argument for the proposal that orientation selectivity is generated and modulated by a combination of thalamocortical inputs to VI neurons and intracortical processing. Cortical neurons are embedded in diverse circuitry: each neuron receives not only excitatory thalamocortical inputs but also excitatory intracortical input from local neurons, inhibitory input from local interneurons, and excitatory long-range input from distant neurons. It would be surprising indeed if these diverse components of circuitry did not find expression in the responses of V1 cells. In fact these complex inputs are integrated in nonlinear fashion, and this integration is expressed in the various nonlinear response properties of V1 cells. A final feature of V1 neurons that is difficult to explain by feedforward connections alone is their responses to subjective or illusory contours and the systematic mapping of subjective contour orientation within V1 (Sheth et al., 1996). A model involving iso-orientation intracortical excitation to amplify LGN responses at line ends is able to reproduce the subjective contour responses and representations within V1

(Kalarickal and Sur, 1999). These response properties include orientation selectivity, modulation of orientation selective responses by spatial context or surround stimuli, and modulation as well by temporal context or pattern adaptation. A linear feedforward model of orientation selectivity was adequate for explaining the simplest, first-order, responses of V1 neurons, at a time when responses to unitary stimuli presented within the receptive field center were regarded as the only significant property of V1 cells. As our appreciation of the complexity of V1 responses has increased, our models for explaining such complexity need to keep pace with the known richness of circuitry. A description of V1 circuits that incorporates this knowledge has to include recurrent excitation and inhibition, and indeed such models provide a natural explanation for many of the complex nonlinear responses of V1 cells. They also provide hope that as we understand and incorporate even more aspects of cortical connections, we shall indeed be able to unravel the mechanisms behind even complex emergent properties of cortical neurons.

ACKNOWLEDGMENTS

We thank Jennifer Vazquez for help with preparing the manuscript. Supported by fellowships from Merck Inc. and the McDonnell-Pew Foundation (V.D.), and grants from the NIH (M.S., D.S. to Edward H. Adelson).

REFERENCES

Abbott, L. F., Varela, J. A., Sen, K., and Nelson, S. B. (1997). Synaptic depression and cortical gain control. *Science* **275,** 220–224.

Ahmed, B. A., Anderson, J. C., Douglas, R. J., Martin, K. A. C., and Nelson, J. C. (1994). Polyneuronal innervation of spiny stellate neurons in cat visual cortex. *J. Comp. Neurol.* **341,** 39–49.

Albrecht, D. G., and Geisler, W. (1991). Motion selectivity and the contrast-response function of simple cells in the striate cortex. *Vis. Neurosci.* **7,** 531–546.

Albrecht, D. G., and Hamilton, D. B. (1982). Striate cortex of monkey and cat: contrast response function. *J. Neurosci.* **48,** 217–237.

Alonso, J. M., Usrey, W. M., and Reid, R. C. (1996). Precisely correlated firing in cells of the lateral geniculate nucleus. *Nature* **383,** 815–819.

Anderson, J. C., Douglas, R. J., Martin, K. A. C., and Nelson, J. C. (1994a). Synaptic output of physiologically identified spiny stellate neurons in cat visual cortex. *J. Comp. Neurol.* **341,** 16–24.

Anderson, J. C., Douglas, R. J., Martin, K. A. C., and Nelson, J. C. (1994b). Map of the synapses formed with the dendrites of spiny stellate neurons of cat visual cortex. *J. Comp. Neurol.* **341,** 25–38.

Anderson, J. S., Carandini, M., and Ferster, D. (2000a). Orientation tuning of input conductance, excitation, and inhibition in cat primary visual cortex. *J. Neurophysiol.* **84,** 909–926.

Anderson, J., Lampl, I., Reichova, I., Carandini, M., and Ferster, D. (2000b). Stimulus dependence of two-state fluctuations of membrane potential in cat visual cortex. *Nat. Neurosci.* **3,** 617–621.

Beaulieu, C., and Somogyi, P. (1990). Targets and quantitative distribution of GABAergic synapses in the visual cortex of the cat. *Eur. J. Neurosci.* **2,** 296–303.

Ben-Yishai, R., Bar-Or, R. L., and Sompolinsky, H. (1995). Theory of orientation tuning in visual cortex. *Proc. Natl. Acad. Sci. U.S.A.* **92,** 3844–3848.

Berman, N. J., Douglas, R. J., Martin, K. A. C., and Whitteridge, D. (1991). Mechanisms of inhibition in cat visual cortex. *J. Physiol.* **440,** 697–722.

Bishop, P. O., Coombs, J. S., and Henry, G. H. (1973). Receptive fields of simple cells in the cat striate cortex. *J. Physiol.* **231,** 31–60.

Blakemore, C., and Tobin, E. A. (1972). Lateral inhibition between orientation detectors in the cat's visual cortex. *Exp. Brain Res.* **15,** 439–440.

Blakemore, C. and Campbell, F. W. J. (1969). Adaptation to spatial stimuli. *J. Physiol. (Lond.)* **200,** 11P.

Blomfield, S. (1974). Arithmetical operations performed by nerve cells. *Brain Res.* **9,** 115–124.

Bonds, A. B. (1989). Role of inhibition in the specification of orientation selectivity of cells in the cat striate cortex. *Vis. Neurosci.* **2,** 41–55.

Bonds, A. B. (1991). Temporal dynamics of contrast gain in single cells of the cat striate cortex. *Vis. Neurosci.* **6,** 239–255.

Bonds, A. B. (1993). The encoding of cortical contrast gain control. In: Contrast sensitivity (Shapley, R. M., and Lam, D-K., Eds.), pp. 215–230, Cambridge, MIT Press.

Borg-Graham, L. J., Monier, C., and Fregnac, Y. (1998). Visual input evokes transient and strong shunting inhibition in visual cortical neurons. *Nature* **393,** 369–373.

Born, R. T., and Tootell, R. B. H. (1991). Single-unit and 2-deoxyglucose studies of side inhibition in macaque striate cortex. *Proc. Natl. Acad. Sci. U.S.A.* **88,** 7071–7075.

Bosking, W. H., Zhang, Y., Schofield, B., and Fitzpatrick, D. (1997). Orientation selectivity and the arrangement of horizontal connections in tree shrew striate cortex. *J. Neurosci.* **17,** 2112–2127.

Bullier, J., Hupe, J. M., James, A., and Girard, P. (1996). Functional interactions between areas V1 and V2 in the monkey. *J. Physiol. (Paris)* **90,** 217–220.

Bush, P. C., and Sejnowski, T. J. (1994). Effects of inhibition and dendritic saturation in simulated neocortical pyramidal cells. *J. Neurophysiol.* **71,** 2183–2193.

Carandini, M., and Heeger, D. J. (1994). Summation and division by neurons in primate visual cortex. *Science* **264,** 1333–1336.

Carandini, M., and Ringach, D. (1997). Predictions of a recurrent model of orientation selectivity. *Vis. Res.* **37,** 3061–3071.

Carandini, M., and Ferster, D. (2000). Orientation tuning of membrane potential and firing rate in cat primary visual cortex. *J. Neurosci.* **20,** 470–484.

Carandini, M., Barlow, H. B., O'Keefe, O. P., Poirson, A. B., and Movshon, J. A. (1997). Adaptation to contingencies in macaque primary visual cortex. *Phil. Trans. R. Soc. Lond. B.* **352,** 1149–1154.

Carandini, M., and Ferster, D. (1997). A tonic hyperpolarization underlying contrast adaptation in cat visual cortex. *Science* **276,** 949.

Carandini, M., Movshon, J. A., and Ferster, D. (1998). Pattern adaptation and cross-orientation interactions in the primary visual cortex. *Neuropharmacology* **37,** 501.

Chance, F. S., and Abbott, L. F. (2000). Divisive inhibition in recurrent networks. *Network* **11,** 119–129.

Chance, F. S., Nelson, S. B., and Abbott, L.F. (1999). Complex cells as cortically amplified simple cells. *Nat. Neurosci.* **2,** 277–282.

Chance, F. S., Nelson, S. B., and Abbott L. F. (1998). Synaptic depression and the temporal response characteristics of V1 cells. *J. Neurosci.* **18,** 4785.

Chapman, B., Zahs, K. R., and Stryker, M. P. (1991). Relation of cortical cell orientation selectivity to alignment of receptive fields of the geniculocortical afferents that arborize within a single orientation column in ferret visual cortex. *J. Neurosci.* **11,** 1347–1358.

Chino, Y. M., and Kaplan, E. (1988). Abnormal orientation bias of LGN neurons in strabismic cats. *Invest. Ophthalmol. Vis. Sci.* **29,** 644–648.

Chung, S., and Ferster, D. (1998). Strength and orientation tuning of thalamic input to simple cells revealed by electrically evoked cortical suppression. *Neuron* **20,** 1177–1189.

Connors, B. W., Gutnick, M. J., and Prince, D. A. (1982). Electrophysiological properties of neocortical neurons in vitro. *J. Neurophysiol.* **48,** 1302–1320.

Creutzfeldt, O. D., Kuhnt, U., and Benevento, L. A. (1974a). An intracellular analysis of visual cortical neurons to moving stimuli: responses in a cooperative neuronal network. *Exp. Brain Res.* **21,** 251–274.

Creutzfeldt, O. D., Innocenti, G., and Brooks, D. (1974b). Vertical organization in the visual cortex (area 17). *Exp. Brain Res.* **21,** 315–336.
Crook, J. M., and Eysel, U. T. (1992). GABA-induced inactivation of functionally characterized sites in cat visual cortex (area 18): effects on orientation tuning. *J. Neurosci.* **12,** 1816–1825.
Das, A. (1996). Orientation in visual cortex: a simple mechanism emerges. *Neuron* **16,** 477.
Dean, A. F., and Tolhurst, D. J. (1986). Factors influencing the temporal phase of response to bar and grating stimuli for simple cells in the cat striate cortex. *Exp. Brain Res.* **62,** 143–151.
DeAngelis, G. C., Robson, J. G., Ohzawa, I., and Freeman, R. D. (1992). Organization of suppression in receptive fields of neurons in cat visual cortex. *J. Neurophysiol.* **68,** 144–163.
DeAngelis, G. C., Freeman, R. D., and Ohzawa, I. (1994). Length and width tuning of neurons in the cat's primary visual cortex. *J. Neurophysiol.* **71,** 347–374.
DeAngelis, G. C., Ohzawa, I., and Freeman, R. D. (1995). Receptive field dynamics in the central visual pathways. *Trends Neurosci.* **18,** 451–458.
Dehay, C., Douglas, R. J., Martin, K. A. C, and Nelson, C. (1991). Excitation by geniculocortical synapses is not "vetoed" at the level of dendritic spines in cat visual cortex. *J. Physiol.* **440,** 723–734.
Douglas, R. J., Martin, K. A. C., and Whitteridge, D. (1991). An intracellular analysis of the visual responses of neurones in cat visual cortex. *J. Physiol.* **44,** 659–696.
Douglas, R. J., Koch, C., Mahowald, M., Martin, K. A. C., and Suarez, H. H. (1995). Recurrent excitation in neocortical circuits. *Science* **269,** 981–985.
Douglas, R. J., and Martin, K. A. C. (1991a). Opening the grey box. *Trends Neurosci.* **14,** 286–293.
Douglas, R. J., and Martin, K. A. C. (1991b). A functional microcircuit for cat visual cortex. *J. Physiol.* **440,** 735–769.
Douglas, R. J., Martin, K. A. C., and Whitteridge, D. (1988). Selective responses of visual cortical neurones do not depend on shunting inhibition. *Nature* **332,** 642–644.
Douglas, R. J., Martin, K. A. C., and Whitteridge, D. (1989). A canonical microcircuit for neocortex. *Neural Comp.* **1,** 480–488.
Dragoi, V., Sharma, J., Miller, E. K. M., and Sur M. (1999). Dynamics of orientation adaptation in awake monkey primary visual cortex revealed by reverse correlation. *Soc. Neurosci. Abstr.* **25,** 1548.
Dragoi, V., and Sur, M. (2000). Dynamic properties of recurrent inhibition in primary visual cortex: contrast and orientation dependence of contextual effects. *J. Neurophysiol.* **83,** 1019–1030.
Dragoi, V., Sharma, J., and Sur, M. (2000). Adaptation-induced plasticity in adult visual cortex. *Neuron* (in press).
Dragoi, V., Sharma, J., and Sur, M. (2000). Adaptation-induced plasticity of orientation tuning in adult visual cortex. *Neuron.* **28,** 287–298.
Elias, S., and Grossberg, S. (1975). Pattern formation, contrast control, and oscillations in the short-term memory of shunting on-center off-surround networks. *Biol. Cybern.* **20,** 69–98.
Enroth-Cugell, C., Robson, J. G., Schweizer-Tong, D. E., and Watson, A. B. (1983). Spatio-temporal interactions in cat retinal ganglion cells showing linear spatial summation. *J. Physiol.* **341,** 279–307.
Eysel, U. T., Crook, J. M., and Machemer, H. F. (1990). GABA-induced remote inactivation reveals cross-orientation inhibition in the cat striate cortex. *Exp. Brain Res.* **80,** 626–630.
Ferster, D. (1986). Orientation selectivity of synaptic potentials in neurons of cat primary visual cortex. *J. Neurosci.* **6,** 1284–1301.
Ferster, D. (1987). Origin of orientation-selective EPSPs in simple cells of the cat visual cortex. *J. Neurosci.* **7,** 1780–1791.
Ferster, D. (1988). Spatially opponent excitation and inhibition in simple cells of the cat visual cortex. *J. Neurosci.* **8,** 1172–1180.
Ferster, D. (1994). Linearity of synaptic interactions in the assembly of receptive fields in cat visual cortex. *Curr. Opin. Neurobiol.* **4,** 563–568.
Ferster, D., and Koch, C. (1987). Neuronal connections underlying orientation selectivity in cat visual cortex. *Trends Neurosci.* **10,** 487–492.
Ferster, D., and Jagadeesh, B. (1992). EPSP-IPSP interactions in cat visual cortex studied with in vivo whole-cell patch recording. *J. Neurosci.* **12,** 1262–1274.

Ferster, D., Chung, S., and Wheat, H. (1996). Orientation selectivity of thalamic input to simple cells of cat visual cortex. *Nature* **380,** 249–252.

Ferster, D., and Miller, K. D. (2000). Neural mechanisms of orientation selectivity in the visual cortex. *Annu. Rev. Neurosci.* **23,** 441–471.

Field, D. J., Hayes, A., and Hess, R. F. (1993). Contour integration by the human visual system: evidence for a local association field. *Vis. Res.* **33,** 173–193.

Fregnac, Y., and Debanne, D. (1993). Potentiation and depression in visual cortical neurons: a functional approach to synaptic plasticity. In: 'Brain mechanisms of perception and memory: from neuron to behavior' (T. Ono, L. Squire, M. E. Raichle, D. I. Perrett, and M. Fukuda, Eds.), pp. 533–561. Oxford, UK, Oxford University Press.

Freund, T. F., Martin, K. A. C., Somogyi, P., and Whitteridge, D. (1985). Innervation of cat visual areas 17 and 18 by physiologically identified X- and Y-type thalamic afferents. II. Identification of postsynaptic targets by GABA immunocytochemistry and golgi impregnation. *J. Comp. Neurol.* **242,** 275–291.

Fries, W., Albus, K., and Creutzfeldt, O. D. (1977). Effects of interacting visual patterns on single cell responses in cat's striate cortex. *Vis. Res.* **17,** 1001–1008.

Gabbott, P. L. A., and Somogyi, P. (1986). Quantitative distribution of GABA-immunoreceptive neurons in the visual cortex (area 17) of the cat. *Exp. Brain Res.* **61,** 323–331.

Galarreta, M., and Hestrin, S. (1998). Frequency-dependent synaptic depression and the balance of excitation and inhibition in the neocortex. *Nat. Neurosci.* **1,** 587–594.

Gardner, J. L., Anzai, A., Ohzawa, I., and Freeman, R. D. (1999). Linear and nonlinear contributions to orientation tuning of simple cells in the cat's striate cortex. *Vis. Neurosci.* **16,** 1115–1121.

Ghose, G. M., Freeman, R. D., and Ohzawa, I. (1994). Local intracortical connections in the cat's visual cortex: postnatal development and plasticity. *J. Neurophysiol.* **72,** 1290–1303.

Gil, Z., Connors, B. W., and Amitai, Y. (1999). Efficacy of thalamocortical and intracortical synaptic connections: quanta, innervation, and reliability. *Neuron* **23,** 385–397.

Gilbert, C. D. (1992). Horizontal integration and cortical dynamics. *Neuron* **9,** 1–13.

Gilbert, C. D., Das, A., Ito, M., Kapadia, M., and Westheimer, G. (1996). Spatial integration and cortical dynamics. *Proc. Natl. Acad. Sci. U.S.A.* **93,** 615–622.

Gilbert, C. D., and Wiesel, T. N. (1983). Clustered intrinsic connections in cat visual cortex. *J. Neurosci.* **3,** 1116–1133.

Gilbert, C. D., and Wiesel, T. N. (1989). Columnar specificity of intrinsic horizontal and corticocortical connections in cat visual cortex. *J. Neurosci.* **9,** 2432–2442.

Gilbert, C. D., and Wiesel, T. N. (1979). Morphology and intracortical projections of functionally identified neurons in cat visual cortex. *Nature* **280,** 120–125.

Gilbert, C. D., and Wiesel, T. N. (1990). The influence of contextual stimuli on the orientation selectivity of cells in primary visual cortex of the cat. *Vis. Res.* **30,** 1689–1701.

Grinvald, A., Lieke, E. E., Frostig, R. D., Gilbert, C. D., and Wiesel, T. N. (1986). Functional architecture of cortex revealed by optical imaging of intrinsic signals. *Nature.* **324,** 361–364.

Grinvald, A., Lieke, E. E., Frostig, R. D., and Hildesheim, R. (1994). Cortical point-spread function and long-range lateral interactions revealed by real-time optical imaging of macaque monkey primary visual cortex. *J. Neurosci.* **14,** 2545–2568.

Grossberg, S. (1973). Contour enhancement, short-term memory, and constancies in reverberating neural networks. *Studies Appl. Math.* **52,** 217–257.

Grossberg, S. (1983). The quantized geometry of visual space: the coherent computation of depth, form, and lightness. *Behav. Brain Sci.* **6,** 625–692.

Grossberg, S., and Mingolla, E. (1985). Neural dynamics of perceptual grouping: textures, boundaries, and emergent segmentations. *Perception Psychophysics* **38,** 141–171.

Grossberg, S., and Olson, S. J. (1994). Rules for the cortical map of ocular dominance and orientation columns. *Neural Networks* **7,** 883–894.

Gulyas, B., Orban, G. A., Duysens, J., and Maes, H. (1987). The suppressive influence of moving textured backgrounds on responses of cat striate neurons to moving bars. *J. Neurophysiol.* **57,** 1767–1791.

Hammond, P., Mouat G. S., and Smith, A. T. (1985). Motion after-effects in cat striate cortex elicited by moving gratings. *Exp. Brain Res.* **60,** 411–416.

Hammond, P., Mouat G. S., and Smith, A. T. (1986). Motion after-effects in cat striate cortex elicited by moving texture. *Vis. Res.* **26,** 1055–1060.

Hammond, P., Pomfrett, C. J., and Ahmed, B. (1989). Neural motion after-effects in the cat's striated cortex: Orientation selectivity. *Vision Res. 29,* 1671–1683.

Hartline, H. K. (1940). The receptive fields of optic nerve fibers. *Am. J. Physiol.* **130,** 700–711.

Hata, Y., Tsumoto, T., Sato, H., Hagihara, K., and Tamura, H. (1988). Inhibition contributes to orientation selectivity in visual cortex of cat. *Nature* **335,** 815–817.

Heeger, D. J. (1992). Normalization of cell responses in cat striate cortex. *Vis. Neurosci.* **9,** 181–197.

Heeger, D. J. (1993). Modeling simple-cell direction selectivity with normalized, half-squared, linear operators. *J. Neurophysiol.* **70,** 1885–1898.

Hess, R., and Murata, K. (1974). Effects of glutamate and GABA on specific response properties of neurons in the visual cortex. *Exp. Brain Res.* **21,** 285–297.

Hestrin, S. (1992). Activation and desensitization of glutamate-activated channels mediating fast excitatory synaptic currents in the visual cortex. *Neuron* **9,** 991–999.

Hirsch, J. A. (1995). Synaptic integration in layer IV of the ferret striate cortex. *J. Physiol.* **483,** 183–199.

Hirsch, J. A., Alonso, J. M., Reid, R. C., and Martinez, L. (1998). Synaptic integration in striate cortical simple cells. *J. Neurosci.* **18,** 9517–9528.

Hirsch, J. A., Gilbert, C. D. (1991). Synaptic physiology of horizontal connections in the cat's visual cortex. *J. Neurosci.* **11,** 1800–1809.

Horton, J. C., and Sherk, H. (1984). Receptive field properties in the cat's lateral geniculate nucleus in the absence of on-center retinal input. *J. Neurosci.* **4,** 374–380.

Hubel, D. H., and Wiesel, T. N. (1962). Receptive fields, binocular interaction and functional architecture in the cat's visual cortex. *J. Physiol.* **165,** 559–568.

Hubel, D. H., and Wiesel, T. N. (1965). Receptive fields and functional architecture in two non-striate visual areas. *J. Neurophysiol.* **41,** 229–289.

Hubel, D. H., and Wiesel, T. N. (1977). Functional architecture of macaque monkey visual cortex. *Proc. R. Soc. (London) B* **198,** 1–59.

Humphrey, A. L., Sur, M., Uhlrich, D. J., and Sherman, S. M. (1985). Projection patterns of individual x-and y-cell axons from the lateral geniculate nucleus to cortical area 17 in the cat. *J. Comp. Neurol.* **233,** 159–189.

Jagadeesh, B., and Ferster, D. (1990). Receptive field lengths in cat striate cortex can increase with decreasing stimulus contrast. *Soc. Neurosci. Abstr.* **16,** 130.11.

Jagadeesh, B. (1993). The construction of receptive field properties of cells in the cat visual cortex from the synaptic inputs to the cortex. Ph.D. Dissertation. Northwestern University.

Jagadeesh, B., Wheat, H. S., and Ferster, D. (1993). Linearity of summation of synaptic potentials underlying direction selectivity of simple cells of the cat visual cortex. *Science* **262,** 1901–1904.

Jones, J. P., and Palmer, L. A. (1987). The two-dimensional spatial structure of simple receptive fields in cat striate cortex. *J. Neurophysiol.* **58,** 1187–1211.

Kalarickal, G. J., and Sur, M. (1999). Modeling orientation tuning of cortical neurons to subjective contours. *Soc. Neurosci. Abstr.* **25,** 677.

Kamphuis, W., and Lopez da Silva, F. H. (1990). The kindling model of epilepsy: the role of GABAergic inhibition. *Neurosci. Res. Commun.* **6,** 1–10.

Kapadia, M. K., Ito, M., Gilbert, C. D., and Westheimer, G. (1995). Improvement in visual sensitivity by changes in local context: parallel studies in human observers and in V1 of alert monkeys. *Neuron* **15,** 843–856.

Kaplan, E., Purpura, K., and Shapley, R. M. (1987). Contrast affects the transmission of visual information through the mammalian lateral geniculate nucleus. *J. Physiol.* **391,** 267–288.

Kisvarday, Z. F., Martin, K. A. C., Fruend, T. F., Magloczky, Z., Whitteridge, D., and Somogyi, P. (1986). Synaptic targets of HRP-filled layer III pyramidal cells in the cat striate cortex. *Exp. Brain Res.* **64,** 541–552.

Kisvarday, Z. F., Beaulieu, C., and Eysel, U. T. (1993). Network of GABAergic large basket cells in cat visual cortex (area 18): implication for lateral disinhibition. *J. Comp. Neurol.* **327,** 398–415.

Knierim, J. J., and Van Essen, D. C. (1992). Neuronal responses to static texture patterns in area V1 of the alert macaque monkey. *J. Neurophysiol.* **67,** 961–980.

Koch, C., Douglas, R. J., and Wehmeier, U. (1990). Visibility of synaptically induced conductance changes: Theory and simulations of anatomically characterized cortical pyramidal cells. *J. Neurosci.* **10,** 1728–1744.

Koch, C., and Poggio, T. (1985). The synaptic veto mechanism: does it underlie direction and orientation selectivity in the visual cortex? In "Models of the visual cortex" (D. R. Rose, and V. G. Dobson, Ed.), pp. 408–419, New York, J. Wiley.

Kohonen, T. (1984). *Self-organization and associative memory.* New York, Springer-Verlag.

Komatsu, Y., Nakajima, S., Toyama, K., and Fetz, E. (1988). Intracortical connectivity revealed by spike-triggered averaging in slice preparations of cat visual cortex. *Brain Res.* **442,** 359–362.

Kuffler, S. W. (1953). Discharge patterns and functional organization of the mammalian retina. *J. Neurophysiol.* **16,** 37–68.

LeVay, S. (1986). Synaptic organization of claustral and geniculate afferents to the visual cortex of the cat. *J. Neurosci.* **6,** 3564–3575.

Levitt, J. B., and Lund, J. S. (1997). Contrast dependence of contextual effects in primate visual cortex. *Nature* **387,** 73–76.

Li, C. Y., and Creutzfeldt, O. D. (1984). The representation of contrast and other stimulus parameters by single neurons in area 17 of the cat. *Pflugers Arch.* **401,** 304–314.

Linsenmeier, R. A., Frishman, L. J., Jakiela, H. G., and Enroth-Cugell, C. (1982). Receptive field properties of X and Y cells in the cat retina derived from contrast sensitivity measurements. *Vis. Res.* **22,** 1173–1183.

Livingston, M. S., and Hubel, D. H. (1984). Specificity of intrinsic connections in primate primary visual cortex. *J. Neurosci.* **4,** 2830–2835.

Lund, J. S., Yoshioka, T., and Levitt, J. B. (1993). Comparison of intrinsic connectivity in different areas of macaque monkey cerebral cortex. *Cerebral Cortex* **3,** 148–162.

Lund, J. S., Wu, Q., Hadingham, P. T., and Levitt, J. B. (1995). Cells and circuits contributing to functional properties in area V1 of macaque monkey cerebral cortex: bases for neuroanatomically realistic models. *J. Anat.* **187,** 563–581.

Maffei, L., and Fiorentini, A. (1976). The unresponsive regions of visual cortical receptive fields. *Vis. Res.* **16,** 1131–1139.

Maffei, L., Fiorentini, A., and Bisti, S. (1973). Neural correlate of perceptual adaptation to gratings. *Science.* **182,** 1036–1038.

Malach, R., Amir, Y., Harel, M., and Grinvald, A. (1993). Relationship between intrinsic connections and functional architecture revealed by optical imaging and in vivo targeted biocytin injections in primate striate cortex. *Proc. Natl. Acad. Sci. U.S.A.* **90,** 10469–10473.

Malpeli, J. G. (1983). Activity of cells in area 17 of the cat in absence of input from layer A of lateral geniculate nucleus. *J. Neurophysiol.* **49,** 595–610.

Malpeli, J. G., Lee, C., Schwark, H. D., and Weyand, T. G. (1986). Cat area 17. I. Pattern of thalamic control of cortical layers. *J. Neurophysiol.* **56,** 1062–1073.

Martin, K. A. C. (1988). From single cells to simple circuits in the cerebral cortex. *Q. J. Exp. Physiol.* **73,** 637–702.

Martin, K. A. C., and Whitteridge, D. (1984). Form, function, and intracortical projections of spiny neurons in the striate visual cortex of the cat. *J. Physiol.* **353,** 463–504.

Mason, A., Nicoll, A., and Stratford, K. (1991). Synaptic transmission between individual pyramidal neurons of the rat visual cortex in vitro. *J. Neurosci.* **11,** 72–84.

Matsubara, J. A., Cynader, M. S., and Swindale, N. V. (1987). Anatomical projections and physiological correlates of the intrinsic connections in cat area 18. *J. Neurosci.* **7,** 1428–1446.

McCormick, D. A., Connors, B. W., Lighthall, J. W., and Prince, D. A. (1985). Comparative electrophysiology of physiology of pyramidal and sparsely spiny stellate neurons of the neocortex. *J. Neurophysiol.* **54,** 782–806.

McGuire, B. A., Gilbert, C. D., Rivlin, P. K., and Wiesel, T. N. (1991). Targets of horizontal connections in macaque primary visual cortex. *J. Comp. Neurol.* **305,** 370–392.

McLaughlin, D., Shapley, R., Shelley, M., and Wielaard, D. J. (2000). A neuronal network model of macaque primary visual cortex (V1): Orientation selectivity and dynamics in the input layer 4Calpha. *Proc. Natl. Acad. Sci. U.S.A.* **97,** 8087–8092.

Michalski, A., Gerstein, G. I., Czarkowska, J., and Tarnecki, R. (1983). Interactions between cat striate cortex neurons. *Exp. Brain. Res.* **51,** 97–107.

Miller, K. D. (1992). Development of orientation columns via competition between on- and off-center inputs. *Neuroreport* **3,** 73–76.

Miller, K. D. (1994). A model for the development of simple cell receptive fields and the ordered arrangement of orientation columns through activity-dependent competition between on- and off-center inputs. *J. Neurosci.* **14,** 409–441.

Morrone, M. C., Burr, D. C., and Maffei, L. (1982). Functional implications of cross—orientation inhibition of visual cortical cells. I. Neurophysiological evidence. *Proc. R. Soc. Lond. B* **216,** 335–354.

Mountcastle, V. B. (1978). An organizing principle for the cerebral function: the unit module and the distributed system. In: 'The Mindful Brain' (G. M. Edelman, and V. B. Mountcastle, Eds.), pp. 7–50. Cambridge, MA, MIT Press.

Movshon, J. A., Thompson, I. D., and Tolhurst, D. J. (1978). Spatial summation in the receptive fields of simple cells in the cat's striate cortex. *J. Physiol.* **283,** 53–77.

Movshon, A., and Lennie, P. (1979). Pattern-selective adaptation in visual cortical neurones. *Nature* **278,** 850.

Murphy, P. C., Duckett, S. G., and Sillito A. M. (1999). Feedback connections to the lateral geniculate nucleus and cortical response properties. *Science* **286,** 1552–1554.

Nelson, J. I., and Frost, B. J. (1978). Orientation-selective inhibition from beyond the classic visual receptive field. *Brain Res.* **139,** 359–365.

Nelson, J. I., and Frost, B. J. (1985). Intracortical facilitation among co-oriented, co-axially aligned simple cells in cat striate cortex. *Exp. Brain Res.* **61,** 54–61.

Nelson, S. B. (1991). Temporal interactions in the cat visual system. III. Pharmacological studies of cortical suppression suggest a presynaptic mechanism. *J. Neurosci.* **11,** 369–380.

Nelson, S. B., Toth, L. J., Sheth, B., and Sur, M. (1994). Orientation selectivity of cortical neurons persists during intracellular blockade of inhibition. *Science* **265,** 774–777.

Ohzawa, I., Sclar, G., and Freeman, R. D. (1982). Contrast gain control in the cat visual cortex. *Nature.* **298,** 266–268.

Orban, G. A., Kato, H., and Bishop, P. O. (1979). Dimensions and properties of end-zone inhibitory areas in receptive fields of hypercomplex cells in cat striate cortex. *J. Neurophysiol.* **42,** 833–849.

Orban, G. A. (1984). *Neuronal operations in the visual cortex.* Berlin, Springer.

Pei, X., Vidyasagar, T. R., Volgushev, M., and Creutzfeldt, O. D. (1994). Receptive field analysis and orientation selectivity of postsynaptic potentials of simple cells in cat visual cortex. *J. Neurosci.* **14,** 7130–7140.

Peichl, L., and Wassle, H. (1979). Size, scatter and coverage of ganglion cell receptive field centers in the cat retina. *J. Physiol.* **291,** 117–141.

Peters, A., and Payne, B. R. (1993). Numerical relationships between geniculocortical afferents and pyramidal cell modules in cat primary visual cortex. *Cerebral Cortex* **3,** 69–78.

Peters, A., and Yilmaz, E. (1993). Neuronal organization in area 17 of cat visual cortex. *Cerebral Cortex* **3,** 49–68.

Peters, A., and Sethares, C. (1991). Organization of pyramidal neurons in area 17 of monkey visual cortex. *J. Comp. Neurol.* **306,** 1–23.

Peters, A., and Sethares, C. (1996). Myelinated axons and the pyramidal cell modules in monkey primary visual cortex. *J. Comp. Neurol.* **365,** 232–255.

Poggio, T., and Reichardt, W. E. (1976). Visual control of orienting behavior in the fly. II. Towards the underlying neural interactions. *Q. Rev. Biophys.* **9,** 377–438.

Polat, U., and Sagi, D. (1993). Lateral interactions between spatial channels: suppression and facilitation revealed by lateral masking experiments. *Vis. Res.* **33,** 993–999.

Polat, U., Mizobe, K., Pettet, M. W., Kasamatsu, T., and Morcia, A. M. (1998). Collinear stimuli regulate visual responses depending on cell's contrast threshold. *Nature* **391,** 580–584.

Press, W. H., Flannery, B. P., Teukolsky, S. A., and Vetterling, W. T. (1992). *Numerical recipes in C: the art of scientific computing,* 2nd ed., pp. 566–597. Cambridge, Cambridge University Press.
Pugh, M. C., Ringach, D. L., Shapley, R., and Shelley, M. J. (2000). Computational modeling of orientation tuning dynamics in monkey primary visual cortex *J. Comput. Neurosci.* **8,** 143–159.
Rakic, P. (1988). Specification of cerebral cortical areas. *Science* **241,** 170–176.
Ramoa, A. S., Shadlen, M., Skottun, B. C., and Freeman, R. D. (1986). A comparison of inhibition in orientation and spatial frequency selectivity of cat visual cortex. *Nature* **321,** 237–239.
Reid, R. C., and Alonso, J. M. (1995). Specificity of monosynaptic connections from thalamus to visual cortex. *Nature* **387,** 281–284.
Reid, R. C., Soodak, R. E., and Shapley, R. M. (1987). Linear mechanisms of direction selectivity in simple cells of cat striate cortex. *Proc. Natl. Acad. Sci. U.S.A.* **84,** 8740–8744.
Richter, J., and Ullman, S. (1982). A model for the temporal organization of X- and Y-type receptive fields in the primate retina. *Biol. Cybern.* **43,** 127–145.
Ringach, D. L., Hawken, M. J., and Shapely, R. (1997). Dynamics of orientation tuning in macaque primary visual cortex. *Nature* **387** 281–284
Rockland, K. S., and Lund, J. S. (1982a). Intrinsic laminate lattice connection in primate visual cortex. *Science* **215,** 1532–1534.
Rockland, K. S., and Lund, J. S. (1982b). Widespread periodic intrinsic connections in the tree shrew visual cortex. *Brain Res.* **169,** 19–40.
Roerig, B., and Katz, L. C. (1998). Relationships of synaptic input patterns to orientation preference maps in ferret visual cortex. *Soc. Neurosci. Abstr.* **24,** 766.
Rodieck, R. W., and Stone, J. (1965). Analysis of receptive fields of cat retinal ganglion cells. *J. Neurophysiol.* **28,** 833–849.
Roig, B. R., Kabara, J. F., Snider, R. K., and Bonds, A. B. (1996). Non-uniform influence from stimuli outside the classical receptive field on gain control of cat visual cortical neurons. *Invest. Opthalmol. Vis. Sci. Suppl.* **37,** 2198.
Rojer, A., and Schwartz, E. L. (1990). Cat and monkey cortical columnar patterns modeled by bandpass-filtered 2d white noise. *Biol. Cybern.* **62,** 381–391.
Salinas, E., and Abbott, L. F. (1996). A model of multiplicative neural responses in parietal cortex. *Proc. Natl. Acad. Sci. U.S.A.* **93,** 11956–11961.
Sanchez-Vives, M. V., Nowak, L. G., McCormick, D. A. (2000). Membrane mechanisms underlying contrast adaptation in cat area 17 in vivo. *J. Neurosci.* **20,** 4267–4285.
Saul, A. B., and Cynader, M. S. (1989). Adaptation in single units in visual cortex: the tuning of aftereffects in the spatial domain. *Vis. Neurosci.* **2,** 593.
Sceniak, M. P., Ringach, D. L., Hawken, M. J., and Shapley, R. (1999). Contrast's effect on spatial summation by macaque V1 neurons. *Nat. Neurosci.* **2,** 733–739.
Schiller, P. (1982). Central connections of the retinal ON and OFF pathways. *Nature* **297,** 580–583.
Sclar, G., and Freeman, R. D. (1982). Orientation selectivity in the cat's striate cortex is invariant with stimulus contrast. *Exp. Brain Res.* **46,** 457–461.
Sengpiel, F., Sen, A., and Blakemore, C. (1997). Characteristics of surround inhibition in cat area 17. *Exp. Brain Res.* **116,** 216–228.
Sengpiel, F., Baddeley, R. J., Freeman, T. C. B., Harrad, R., and Blakemore, C. (1998). Different mechanisms underlie three inhibitory phenomena in cat area 17. *Vis. Res.* **8,** 2067–2080.
Shadlen, M., and Newsome, W. (1994). Noise, neural codes and cortical organization. *Curr. Opin. Neurobiol.* **4,** 569–579.
Sherk, H., and Horton, J. C. (1984). Receptive field properties in the cat's area 17 in the absence of on-center geniculate input. *J. Neurosci.* **4,** 381–393.
Sheth, B. R., Sharma J., Rao, S. C., and Sur, M. (1996). Orientation maps of subjective contours in visual cortex. *Science* **274,** 2110–2115.
Shoham, D., Glaser, D. E., Arieli, A., Kenet, T., Wijnbergen, C., Toledo, Y., Hidersheim, R., and Grinvald, A. (1999). Imaging cortical dynamics at high spatial and temporal resolution with novel blue voltage-sensitive dyes. *Neuron* **24,** 791–802.
Shou, T., Softky, W. R., and Koch, C. (1993). The highly irregular firing of cortical cells is inconsistent with temporal integration of random EPSPs. *J. Neurosci.* **13,** 334–450.

Sik, A., Penttonen, M., Ylinen, A., and Buzsaki, G. (1995). Hippocampal CA1 interneurons: an in vivo intracellular labeling study. *J. Neurosci.* **15,** 6651–6665.

Sillito, A. M. (1975). The contribution of inhibitory mechanisms to the receptive field properties of neurones in the striate cortex of the cat. *J. Physiol.* **250,** 305–329.

Sillito, A. M., Grieve, K. L., Jones, H. E., Cudeiro, J., and Davis J. (1995). Visual cortical mechanisms deetcting focal orientation discontinuities. *Nature* **378,** 492–496.

Sillito, A. M., Kemp, J. A., Milson, J. A., and Berardi, N. (1980). A re-evaluation of the mechanisms underlying simple cell orientation selectivity. *Brain Res.* **194,** 517–520.

Soodak, R. E., Shapley, R. M., and Kaplan, E. (1987). Linear mechanism of orientation tuning in the retina and lateral geniculate nucleus of the cat. *J. Neurophysiol.* **58,** 267–275.

Somers, D. C., Nelson, S. B., and Sur, M. (1995a). An emergent model of orientation selectivity in cat visual cortical simple cells. *J. Neurosci.* **15,** 5448–5465.

Somers, D. C., Toth, L. J., Todorov, E., Rao, S. C., Kim, D-S., Nelson, S. B., Siapas, A. G., and Sur, M. (1995b). Variable gain control in local cortical circuitry supports context-dependent modulation by long-range connections. In: Lateral interactions in the cortex: structure and function (J. Sirosh, R. Miikkulainen, and Y. Choe, Eds.), Austin, Univ of Texas Press. WWW electronic book, *http://www.cs.utexas.edu/users/nn/web-pubs/htmlbook96/somers/.*

Somers, D. C., Todorov, E. V., Siapas, A. G., Toth, L. J., Kim, D. S., and Sur, M. (1998). A local circuit approach to understanding integration of long-range inputs in primary visual cortex. *Cerebral Cortex* **8,** 204–217.

Sompolinsky, H., and Shapley, R. (1997). New perspectives on the mechanisms for orientation tuning. *Curr. Opin. Neurobiol.* **7,** 514–522.

Stemmler, M., Usher, M., and Niebur, E. (1995). Lateral interactions in primary visual cortex: a model bridging physiology and psychophysics. *Science* **269,** 1877–1880.

Stratford, K. J., Tarcky-Hornoch, K., Martin, K. A. C., Bannister, N. J., and Jack, J. J. (1996). Excitatory synaptic inputs to spiny stellate cells in cat visual cortex. *Nature* **382,** 258–261.

Suarez, H., Koch, C., and Douglas, R. (1995). Modeling direction selectivity of simple cells in striate visual cortex within the framework of the canonical microcircuit. *J. Neurosci.* **15,** 6700–6719.

Swindale, N. V. (1992). A model for the coordinated development of columnar systems in primate striate cortex. *Biol. Cybern.* **66,** 217–230.

Thomson, A. M., and Deuchars, J. (1994). Temporal and spatial properties of local circuits in neocortex. *Trends Neurosci.* **17,** 119–126.

Thomson, A. M., and Deuchars, J. (1997). Synaptic interactions in neocortical local circuits: dual intracellular recordings in vitro. *Cerebral Cortex* **7,** 511–522.

Thomson, A. M., Deuchars, J., and West, D. C. (1993a). Large, deep layer pyramid-pyramid single axon EPSPs in slices of rat motor cortex display paired pulse and frequency-dependent depression, mediated presynaptically and self-facilitation, mediated postsynaptically. *J. Neurophysiol.* **70,** 2354–2369.

Thomson, A. M., Deuchars, J., and West, D. C. (1993b). Single axon excitatory postsynaptic potentials in neocortical interneurons exhibit pronounced paired pulse facilitation. *Neuroscience* **54,** 347–360.

Thomson, A. M., West, D. C., and Deuchars, J. (1995). Properties of single axon excitatory postsynaptic potentials elicited in spiny interneurons by action potentials in pyramidal neurons in slices of rat neocortex. *Neuroscience* **69,** 727–738.

Thomson, A. M., and West, D. C. (1993). Fluctuations in pyramid-pyramid excitatory postsynaptic potentials modified by presynaptic firing pattern and postsynaptic membrane potential using paired intracellular recordings in rat neocortex. *Neuroscience* **54,** 329–346.

Todorov, E. V., Siapas, A. G., and Somers, D. C. (1997a). A model of recurrent interactions in primary visual cortex. In: Advances in neural information processing systems, Vol. 9 (M. C. Mozer, M. I. Jordan, and T. Petsche, Eds.), pp. 118–124. Cambridge, MIT Press.

Todorov, E. V., Siapas, A. G., Somers, D. C., and Nelson, S. B. (1997b). Modeling visual cortical contrast adaptation effects. In: Computational neuroscience: trends in research 1997 (J. M. Bower, Ed.), pp. 525–531. New York, Plenum Press.

Tolhurst, D. J., and Dean, A. F. (1987). Spatial summation by simple cells in the striate cortex of the cat. *Exp. Brain Res.* **66,** 607–620.

Tolhurst, D. J., and Dean, A. F. (1990). The effects of contrast on the linearity of spatial summation of simple cells in the cat's striate cortex. *Exp. Brain Res.* **79,** 582–588.

Toth, L. J., Rao, S. C., Kim, D-S., Somers, D., and Sur, M. (1996). Subthreshold facilitation and suppression in primary visual cortex revealed by intrinsic signal imaging. *Proc. Natl. Acad. Sci. U.S.A.* **93,** 9869–9874.

Toth, L. J., Kim, D-S., Rao, S. C., and Sur, M. (1997a). Integration of local inputs in visual cortex. *Cerebral Cortex* **7,** 703–710.

Toth, K., Freund, T. F., and Miles, R. (1997b). Disinhibition of rat hippocampal pyramidal cells by GABAergic afferents from the septum. *J. Physiol.* **500,** 463–474.

Toyama, K., Kimura, M., and Tanaka, K. (1981). Organization of cat visual cortex as investigated by cross-correlation technique. *J. Neurophysiol.* **46,** 202–214.

Traub, R. D., and Miles, R. (1991). *Neuronal networks of the hippocampus.* Cambridge, Cambridge University Press.

T'so, D. Y., Gilbert, C. D., and Wiesel, T. N. (1986). Relationship between horizontal interactions and functional architecture in cat striate cortex as revealed by cross-correlation analysis. *J. Neurosci.* **6,** 1160–1170.

Troyer, T. W., Krukowski, A., Priebe, N. J., and Miller, K. D. (1998). Contrast-invariant orientation tuning in cat visual cortex: feedforward-tuning and correlation-based intracortical connectivity. *J. Neurosci.* **18,** 5908–5927.

Tsodyks, M. V., and Markram, H. (1997). The neural code between pyramidal neurons depends on neurotransmitter release probability. *Proc. Natl. Acad. Sci. U.S.A.* **94,** 719–723.

Tsumoto, T., Eckart, W., and Creutzfeldt, O. D. (1979). Modification of orientation sensitivity of cat visual cortex neurons by removal of GABA—mediated inhibition. *Exp. Brain Res.* **34,** 351–363.

Tucker, T. R., and Katz, L. C. (1998). Organization of excitatory and inhibitory connections in layer 4 of ferret visual cortex. *Soc. Neurosci. Abstr.* **24,** 1756.

Vidyasagar, T. R., and Urbas, J. V. (1982). Orientation sensitivity of cat LGN neurones with and without inputs from visual cortical areas 17, 18. *Exp. Brain Res.* **46,** 157–169.

Vogels, R., and Orban, G. A. (1991). Quantitative study of striate single unit responses in monkeys performing an orientation discrimination task. *Exp. Brain Res.* **84,** 1–11.

Volgushev, M., Vidyasagar, T. R., and Pei, X. (1995). Dynamics of the orientation tuning of postsynaptic potentials in the cat visual cortex. *Vis. Neurosci.,* **12,** 621–628..

Volgushev, M., Vidyasagar, T. R., and Pei, X. (1997). A linear model fails to predict orientation selectivity of cells in the cat visual cortex. *J. Physiol. (Lond)* **496,** 597–606.

Watkins, D. W., and Berkley, M. A. (1974). The orientation selectivity of single neurons in cat striate cortex. *Exp. Brain Res.* **19,** 433–446.

Wehmeier, U., Dong, D., Koch, C., and Van Essen, D. (1989). Modeling the visual system. In: Methods in neuronal modeling (C. Koch, and I. Segev, Eds.), pp. 335–359. Cambridge, MIT Press.

Weliky, M., Kandler, K., Fitzpatrick, D., and Katz, L. C. (1995). Patterns of excitation and inhibition evoked by horizontal connections in visual cortex share a common relationship to orientation columns. *Neuron* **15,** 541–552.

Weliky, M., and Katz, L. C. (1994). Functional mapping of horizontal connections in developing ferret visual cortex: experiments and modeling. *J. Neurosci.* **14,** 7291–7305.

Wilson, H. R., and Cowan, J. D. (1972). Excitatory and inhibitory interactions in localized populations of model neurons. *Biophys. J.* **12,** 1–24.

Worgotter, F., and Koch, C. (1991). A detailed model of the primary visual pathway in the cat: comparison of afferent excitatory and intracortical inhibitory connection schemes for orientation selectivity. *J. Neurosci.* **11,** 1959–1979.

13

Response Synchronization, Gamma Oscillations, and Perceptual Binding in Cat Primary Visual Cortex

WOLF SINGER

Max-Planck-Institute for Brain Research, Frankfurt-am-Main, Germany

INTRODUCTION

Since the seminal discovery by Hubel and Wiesel (1962) of feature selective neurons in cat primary visual cortex, it has been commonly held that the main function of this primary cortex is the extraction and encoding of elementary features of visual objects. It is assumed that features such as the orientation of contrast gradients or the motion of contour borders are detected by evaluating the spatial and temporal relations among the firing patterns in the array of retinotopically ordered thalamic input fibers and that these relations are specified by the selective convergence of these inputs onto individual cortical target cells (Chapman et al., 1991; Jagadeesh et al., 1993; Reid and Alonso, 1995) (see Chapter 8). More recent evidence, however, suggests that the functions of striate cortex are not confined to feature extraction. Thus, the responses of feature selective neurons have been shown to be exquisitely sensitive to the context in which these features are embedded. Responses to stimuli presented within the classical receptive field undergo drastic modifications when additional contours are present in regions surrounding the cell's receptive field. These effects can be remarkably strong and can even alter the cell's feature preference (Sillito et al., 1995; Das and Gilbert, 1999; Kapadia et al., 1995). Also, there is now evidence for attention-dependent modulation of responses in area 17 (Lamme and Spekreijse, 1998; Roelfsema et al., 1998), and experiments with reversible cooling of higher cortical areas indicate that responses of striate cortex neurons are modulated by top-down projections (Galuske et al., 2000; Hupé et al., 1998, 2001; Payne et al., 2000). Finally, neuronal responses in area 17 exhibit marked state-dependent modulation as is the case for neurons in all cortical areas (Contreras and Steriade, 1997a, 1997b; Steriade et al., 1996; Steriade, 1999; Livingstone and Hubel 1981). Thus, the response properties of striate cortex neurons are determined not only by the convergence patterns of thalamocortical feedforward connections but also by corticocortical projections originating both within striate cortex and other cortical areas, as well as by the more globally organized modulatory systems.

Together with the fact that striate cortex shares most of its organizational features with other areas of isocortex (Douglas et al., 1989), these dynamic characteristics are compatible with the view that the functions of striate cortex are not confined to feature extraction but cover the full range of functions commonly attributed to the neocortex. Apart from the selective recombination of input signals by neurons with feature-specific response properties, these functions could comprise synthetic operations such as the evaluation of context, the dynamic binding and segregation of responses during scene segmentation, and the establishment of new relationships between hitherto unbound features by learning. If so, one should expect that the neuronal codes utilized in area 17 ought to be the same as those postulated for the other cortical areas. Hence, the same catalog of unresolved questions listed here that are currently discussed with respect to cortical functions in general should also apply to striate cortex.

1. Are the results of the computations performed in a particular cortical area represented by sharply tuned, highly specific responses of "smart" neurons or by assemblies of broadly tuned neurons or are both strategies applied in parallel?
2. If assembly coding is exploited, is a particular feature always represented by the same population of neurons, or can a particular neuron participate at different times in the representation of quite different features?
3. Are representations of new contents, features, or their constellations generated by shaping the response preferences of individual "smart" neurons or by changing the composition of assemblies of cooperatively coupled broadly tuned neurons?
4. Are temporal covariations among the discharge patterns of distributed neurons exploited for the definition of relations among responses? If so, what is the temporal resolution with which such relations can be expressed and decoded?

5. Is there a need for the dynamic selection and grouping of responses for further joint processing and if so, is selection achieved solely by increasing the discharge rate of the selected responses and temporal summation in target cells, or can responses also be selected by synchronization and subsequent spatial summation?

These questions about the nature of neuronal codes and the structure of representations are often raised in the context of theories about higher cognitive functions such as the representation and recognition of perceptual objects. Thus, discussions focus on functions of higher cortical areas. However, if the principles of organization are as similar among cortical areas, as the many shared features suggest, these questions should also be pertinent for the understanding of striate cortex functions. Therefore, this chapter examines the extent to which current concepts about the nature of cortical codes and representations can be generalized to striate cortex functions.

TWO COMPLEMENTARY STRATEGIES FOR THE REPRESENTATION OF RELATIONS: SMART NEURONS AND ASSEMBLIES

Perceptual objects consist of unique, often highly complex, constellations of features, and they require for their adequate neuronal representation that their respective features and spatial and/or temporal relations be encoded in neuronal responses. Although the variety of basic feature dimensions that nervous systems exploit to classify perceptual objects is limited, the diversity of possible constellations is virtually unlimited. Thus, cognitive systems must explore a huge combinatorial space when searching for the consistent relations among features that define a perceptual object.

Similarly large combinatorial problems must also be solved for the programming of purposeful movements. Such a combinatorial problem emerges because there are large numbers of muscle fibers passing across many joint articulations, and specific subgroups of muscle fibers contract in specific spatial and temporal combinations to effect subcomponent actions of a complex movement repertoire. Because subcomponents may be used in different combinations, the simple stereotyped contraction of a given muscle fiber may participate in a virtually limitless number of complex movement repertoires.

The extraction of features in sensory processing is equivalent to the analysis and encoding of spatial and temporal relations. In the primary visual cortex of mammals, for example, temporal coincidence among the responses of colinearly aligned retinal ganglion cells is evaluated by the convergence of signals from these ganglion cells converging onto individual cortical neurons (Hubel and Wiesel, 1962; Chapman et al., 1991; Jagadeesh et al., 1993; Reid and Alonso, 1995) (see Chapter 8). In this way the feature orientation is extracted and represented. Iteration of this strategy in prestriate cortical areas leads to increasingly "smart" neurons that encode more and more complex spatial and temporal relationships (Tanaka et al., 1991) including those characteristic for stereotypes of real-world objects such as faces (Gross, 1992). Thus, one way of analyzing and representing relations is to selectively recombine signals by having subsets of input fibers converge onto target cells at subsequent processing stages (Fig. 13-1). This strategy of representing features and their constellations by the tuned

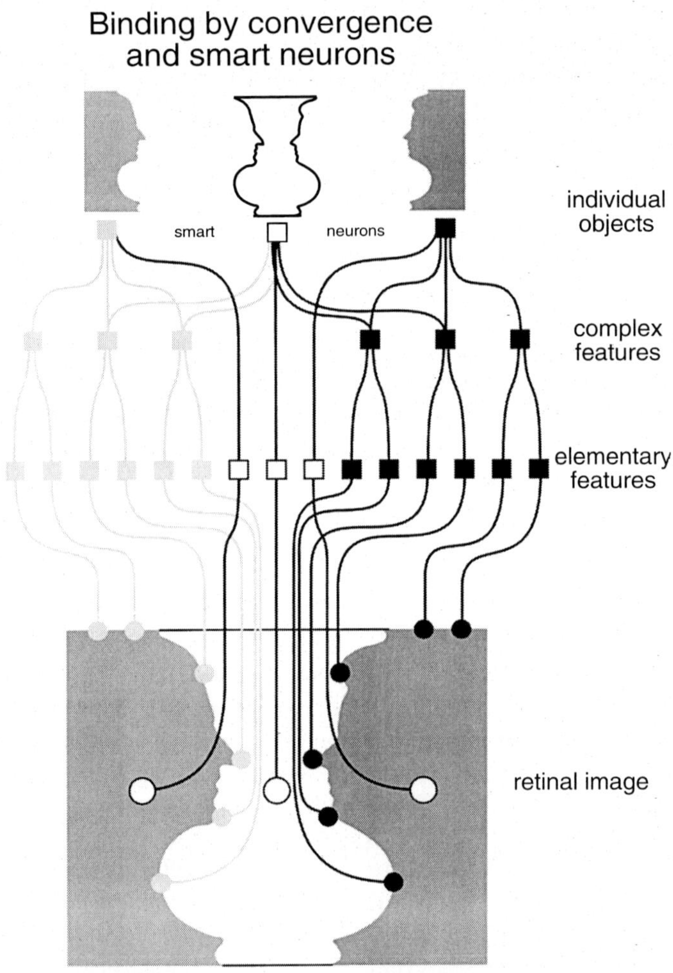

FIGURE 13-1. Schematic wiring diagram of a hierarchically organized feedforward network that generates smart neurons that respond selectively to different perceptual objects. Note that the smart neurons representing the faces and the vase, respectively, receive input from partially the same feature-specific neurons.

responses of individual cells (labeled line coding) is rapid and reliable because it can be realized in simple feedforward architectures. However, if used as the only representational strategy, it requires astronomical numbers of neurons to cope with the virtually infinite diversity of possible feature conjunctions (Sejnowski, 1986; Engel et al., 1992). Moreover, it is not easy to comprehend how such a strategy deals with the representation of novel objects at first encounter and how the system might cope with the representation of composite objects, categories, and semantic relations.

These constraints of labeled line coding have so far been discussed mainly in the context of higher cognitive functions such as the representation of complex perceptual objects. However, problems of very similar nature, especially the problem of the combinatorial explosion of representational units, arise also at lower levels of sensory processing. In the visual system this is the case for the encoding of elementary features such as the precise position, shape, and orientation of a contour segment—functions that; in all likelihood, are accomplished in primary visual cortex. Given the high resolution with which these features can be distinguished, one again requires a very large number of neurons if all possible conjunctions of these properties are encoded in a 1:1 relation by sharply tuned neurons.

To minimize neural circuitry, a complementary strategy is needed that permits sharing of neurons by representations of different contents and contexts. One such strategy is population coding, also known as coarse coding. In this concept information about a specific stimulus feature is distributed across large numbers of neurons and encoded in the graded responses of cells. The great advantage of this coding strategy is that a given cell can be recruited into different assemblies and can participate in the encoding of many different contents (Fig. 13-2). Accordingly neurons in such populations need to be broadly tuned, and they react with graded responses to variations of features along several dimensions, for example, orientation, location, contrast and length of a contour border. Hence, a particular stimulus always drives a large number of cells with overlapping preferences, and the precise nature and configuration of features needs to be assessed by interpolation from the population response. This coding strategy dramatically reduces the number of neurons required for the encoding of different features and appears to be applied at all levels of cortical processing, including primary visual cortex; however, it does, have a price.

A NEED FOR DYNAMIC RESPONSE SELECTION AND BINDING

As long as the populations activated by simultaneously presented stimuli do not overlap, population coding poses no special problems and can be realized in simple feedforward architectures. Decoding problems arise, however, when different, simultaneously present objects recruit partially overlapping populations. In this case the subsets of responses related to each object need to be identified, selected,

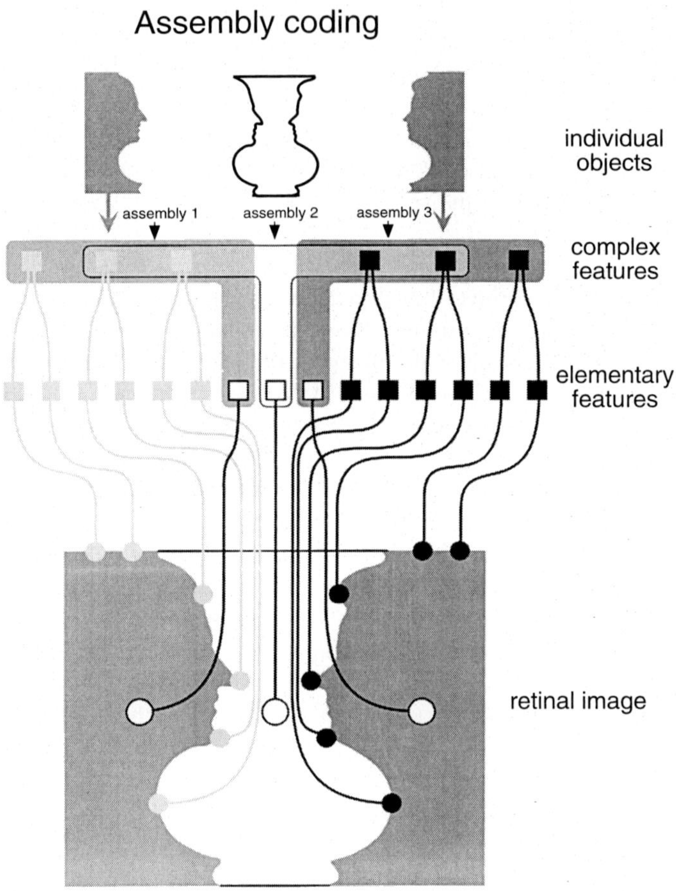

FIGURE 13-2. Schematic wiring diagram of neuronal architectures serving the representation of perceptual objects by assemblies. Note that the assembly representing the vase shares neurons with the assemblies representing the faces. To ensure stability of the respective assemblies, additional reciprocal connections among neurons constituting an assembly are required (shaded regions) that bind responses of neurons belonging to the same assembly.

and bound together for further joint processing without ambiguity and overlap with another neuron population. If this step does not occur, ambiguities arise as to whether a given neuron participates in the representation of one object or another. Such superposition problems are expected to occur whenever stimuli overlap either in euclidian or feature space, or both, and need to be represented simultaneously (Gray, 1999; von der Malsburg, 1999). Because of the broad tuning of cells participating in coarse coding, place codes do not provide suitable resolution of the problem. If place codes were called on, each feature or object requires representation by its own, unshared population of cells that annihilates all advantages of coarse coding. Moreover, it would require even more cells than coding with smart neurons. As coarse coding is applied at all levels of processing, this superposition problem is of a very general nature. Strategies to solve it therefore can be expected to be similar across the various stages of the cortical processing hierarchy.

Here is a firm example of the general problem adapted to the conditions in primary visual cortex. Owing to broad tuning (coarse coding) of neurons for position, orientation, length, width, contrast, and a few other stimulus features, a single elongated contour evokes graded responses in a very large number of neurons. Because the response amplitude of any one of these neurons is influenced by variations of the stimulus along any of these feature dimensions, individual responses convey little specific information. The precise configuration of the stimulus can be deduced only if a large number of the graded responses evoked by the stimulus are evaluated jointly. If only a single contour is present on a homogeneous background this poses no difficulty. Superposition problems arise, however, if several nearby or spatially overlapping contours are embedded in a textured background. In that case joint evaluation of responses needs to be restricted to responses evoked by the same contours.

A classical assumption is that there are neurons at subsequent processing levels that receive convergent input in various constellations from subsets of broadly tuned, low level neurons and thereby acquire selectivity for a particular constellation of features by binding signals by convergence. Population codes could then be fully disambiguated by such conjunction-specific binding units, but this solution is very expensive in terms of the number of required binding units if all possible conjunctions are represented in this explicit way. One would require as many binding units as there are distinguishable population states. This is clearly not an attractive strategy because it sacrifices the main advantage of coarse coding: the parsimonious use of neurons (Fig. 13-3).

To overcome these limitations of potential encoding strategies, it has been proposed that the cerebral cortex uses two coding strategies in parallel: first, partial disambiguation of coarse population codes by the implementation of conjunction units and, second, disambiguation of population codes by dynamic, context-dependent binding. In both cases the essence of the process is the representation of spatial and temporal relations. In the first case, this is achieved through binding by convergence; in the second case it is achieved through dynamic binding. The latter coding strategy cannot rely on simple feedforward architectures, but requires a complex network of highly specific reentry connections that link reciprocally cells distributed both within and across cortical areas. The role of these connections is to ensure dynamic and context-dependent association of neurons into functionally coherent assemblies. To distinguish between simple population codes that require only feedforward architectures and dynamic grouping that requires cooperative interactions through reentrant networks, the neuronal populations selected by dynamic grouping will henceforth be addressed as assemblies (for a detailed discussion of this distinction see Singer et al., 1997; Phillips and Singer, 1997; Singer, 1999).

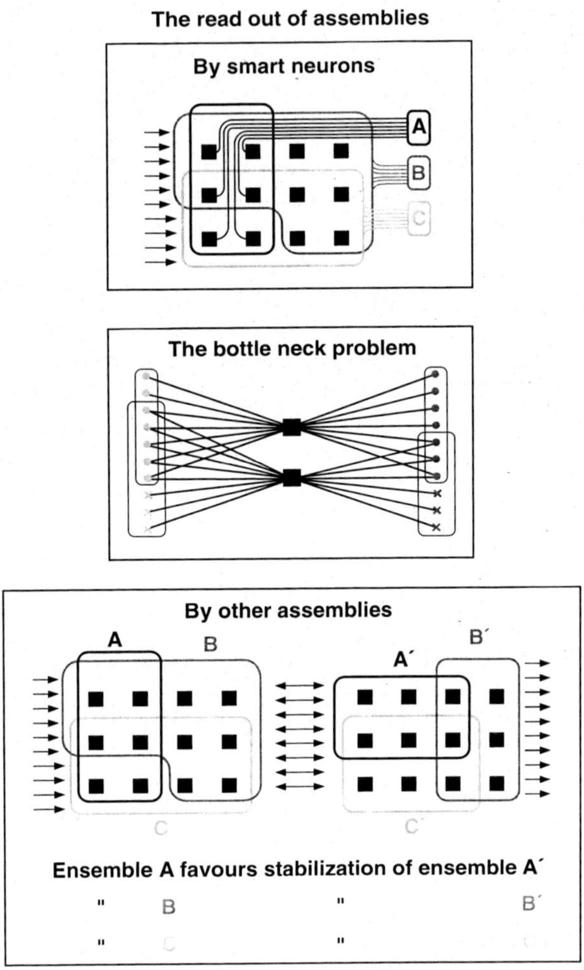

FIGURE 13-3. Schematic representation of solutions for the read-out of assemblies. The distributed responses of assemblies A to C could be read out (bound together) by smart neurons that receive convergent input from the members of each of the three assemblies (upper box). This solution is not attractive because it leads to a combinatorial explosion of the number of required read-out units. Moreover, it generates a bottle-neck problem (middle box) because the activity of individual read-out neurons needs to be redistributed onto large assemblies of neurons to orchestrate executive functions. The alternative option is the read-out of assemblies by assemblies (lower box). This requires parallel reentry connections between the respective cortical areas in which "sending" and "receiving" assemblies are configurated.

In the event that the cerebral cortex uses both smart neurons and assemblies to evaluate and represent spatial and temporal relationships, the following predictions should hold. Stereotyped, frequently occurring and behaviorally particularly relevant conjunctions should be represented by specific binding units, because this strategy is faster and less susceptible to binding errors. However, because these conjunction specific neurons cannot exhaust the full combinatorial space and cannot represent nonanticipated conjunctions, they should in turn remain broadly tuned and be recruitable into dynamically configurated assemblies that represent conjunctions for which there are no binding units. This dual coding strategy should be iterated throughout the processing hierarchy, whereby some feature conjunctions get represented explicitly by individual neurons and others by assemblies. Which conjunctions are represented explicitly by the binding units at the various processing stages can be deduced from the rate coded response properties of the respective neurons. Evidence indicates that these conjunctions become increasingly complex, abstract, and multimodal as one proceeds in the processing hierarchy until they shift progressively from stimulus related to movement-related features at the interface between sensory and motor systems.

At each processing level, novel conjunctions, for which specific binding units have not (yet) been implemented, can then be represented by dynamically configurated assemblies. Thus, despite the implementation of ever more sophisticated binding units, flexible and context-dependent grouping (binding) operations remain necessary at each processing level to represent relations that are not encoded by conjunction-specific neurons and to resolve superposition problems, if they occur.

In addition to its flexibility and virtually inexhaustible coding capacity, assembly coding has the further advantage that it circumvents bottle neck problems. Such problems arise if a complex content gets finally represented by only a few very smart neurons, as these have to orchestrate again the graded activity of myriads of distributed neurons to generate coordinated motor responses. Assembly coding circumvents this problem because computational results can be represented at all stages of processing by population codes that have the same format, and distributed representations (assemblies) can be mapped directly onto other distributed representations (Fig. 13-3).

In conclusion, if relations are encoded both by conjunction-specific neurons and dynamically associated assemblies of such neurons, rapid and flexible grouping operations must be accomplished at all levels of cortical processing, including primary visual cortex. To accomplish this binding, distributed but related neuronal responses must be labeled in a way that ensures joint processing in the absence of confounds with unrelated activity. Note, however, that the two strategies, although available at all stages of processing, need not always be used together. One can imagine conditions where the elementary features of a scene are all explicitly represented by conjunction-specific units and where the arrangement of contours poses no superposition problem, but where the objects defined by these features are novel and not covered by conjunction-specific cells of higher order. In this instance dynamic grouping may be necessary only at higher levels of processing. Conversely, a single object could be embedded in a complex background, allowing for many unfamiliar conjunctions between elements of the figure and elements of the background. In that instance extensive grouping operations are required at peripheral levels of processing to associate the correct set of elementary conjunction units, but no dynamic grouping would be necessary at higher levels provided there is no superposition problems.

DYNAMIC GROUPING MECHANISMS

Numerous theoretical studies have addressed the question of how neuronal responses can be grouped into assemblies on the basis of cooperative interactions within associative neuronal networks (for review see Hebb, 1949; Braitenberg, 1978; Edelman, 1987; Palm, 1990; Gerstein and Gochin, 1992), and they have provided solutions compatible with the functional architecture of the cerebral cortex (for review see Singer, 1995). The most critical question is how responses of cells that have been grouped into an assembly can be tagged as related. An unambiguous signature of relatedness is absolutely crucial for population codes because the meaning of responses changes with the context in which they are interpreted. Hence, false conjunctions are deleterious. Unfortunately, the risk of false conjunctions is constitutive in assembly coding. Assembly codes only economize on neuron numbers if the same cells can be bound into different assemblies at different times. Because complex scenes usually consist of spatially overlapping and contiguous objects, situations often arise in which the same group of neurons ought to be recruited simultaneously into different assemblies. As a matter of principle, assemblies that share common neurons but describe different objects cannot have activities that overlap in time (Fig. 13-4). They must be generated successively to avoid the merging of activities. Processing speed is thus critically limited by the rate at which different assemblies can be formed and dissolved. At peripheral levels of processing, where conjunctions need to be defined for many different often spatially contiguous features, the alternation rate between assemblies coding for different conjunctions of features must be considerably faster than the rate at which different objects can be perceived and represented. This must occur because the results of the various grouping operations must be interpreted jointly by higher processing stages for the evaluation of relations of higher order. Hence, the multiplexed results of low level grouping must alternate quickly enough to permit their association at higher processing stages even though they are transmitted sequentially.

One proposal is that dynamic binding of distributed responses for further joint processing is achieved by mechanisms of selective attention (Ghose and Maunsell, 1999; Reynolds and Desimone, 1999; Shadlen and Movshon, 1999; Wolfe and Cave, 1999). In vision it is assumed, that object-centered attention selectively enhances, via top-down projections, the discharge rate of neurons responding to features of the same contour or object and that this joint increase of saliency leads to joint processing of the selected responses at subsequent stages (e.g. see Lamme and Spekreijse, 1998; Roelfsema et al., 1998; Treisman, 1996) (Fig. 13-4). This mechanism has several difficulties if not supplemented by additional grouping operations: First, it presupposes that higher centers "know" which of the periph-

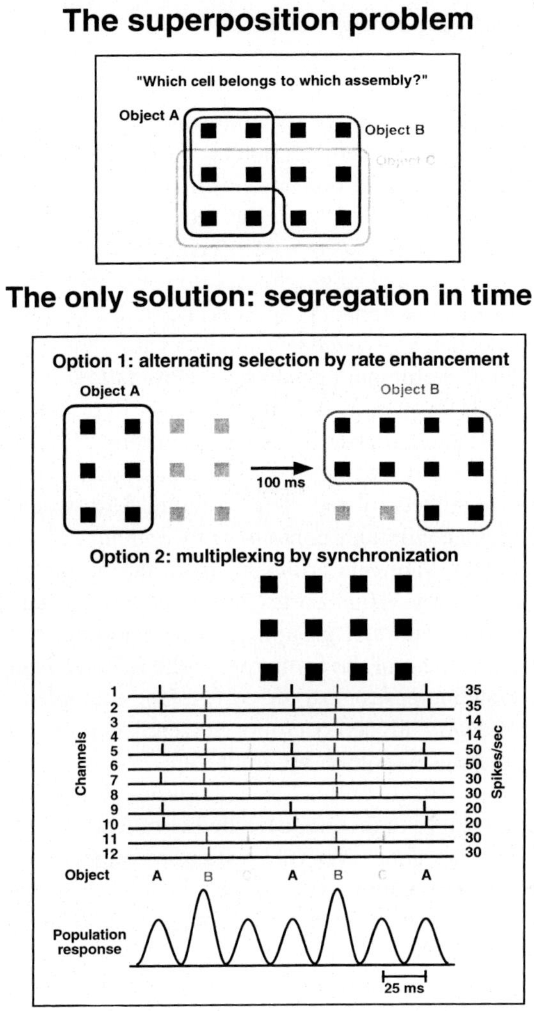

FIGURE 13-4. Options for the solution of the superposition problem. Superposition problems arise if perceptual objects are present whose corresponding assemblies share partly the same neurons (upper box). In this case, different assemblies need to be segregated in time to avoid false conjunctions. One option is to raise successively the saliency of responses belonging to the respective assemblies by enhancing the discharge rate of the corresponding responses (lower box, option 1). An alternative solution is to enhance the saliency of responses belonging to a particular assembly by making the discharges of the respective neurons coincident in time (option 2). This permits rapid multiplexing of the different assemblies because coincidence can be evaluated within short time intervals, as it does not require temporal summation. Here it is assumed that the different assemblies alternate at intervals of approximately 25 ms. Note that this temporal structure can be, but does not have to be, obvious in the discharge sequences of individual neurons (channels 1–12) but that the spike density function of the population response shows an oscillatory modulation in the 40 Hz range. Note also that the constellation of neurons contributing spikes to the oscillatory population response changes from cycle to cycle.

eral responses code for the contours of a particular object. However, this information is available only once an object is identified, which is often possible only after the population codes at lower levels have been disambiguated. Thus, it must first be clarified whether the many simultaneous responses in a low level area code for coherent or disjunct contours before one can decide whether one has to do with one or two objects. Second, it is difficult to comprehend how the top-down projections could enhance, with the required topological specificity, the responses of the cells coding for one of several overlapping contours, because these projections diverge over large cortical domains (Salin and Bullier, 1995), although the timing of their impact is appropriate (Hupé et. 2001).

These problems can be alleviated if responses within the respective processing stages undergo a first, preliminary grouping according to criteria that are adapted to the grouping problems encountered at each stage. In primary visual cortex this would be rather fundamental criteria, because neurons code for simple stimulus features. All that needs to be accomplished here is grouping of responses that are, in all likelihood, generated by the same contour. This would be the case for responses evoked by a continuous contour, or by contour segments that are colinear, or by contours that share similarities in any of the simple feature domains for which neurons in primary visual cortex are responsive. Analogous to the proposed mechanism of attentional grouping, these grouping operations could be based on joint enhancement of the firing rate of the selected neurons. And there is evidence for such rate increases in striate cortex. One example is the enhancement of responses to colinearly aligned contours (see e.g., Wörgötter et al., 1991).

Grouping of responses solely by joint rate enhancement, however, may encounter problems. First, it can lead to ambiguities. It may not always be easy for other processing stages to distinguish whether rate increases are due to grouping or to variations in stimulus properties such as position, orientation, or contrast changes by uneven illumination. Second, when objects overlap in euclidian or feature space, only responses to a single object can be grouped at any one moment. Otherwise it would be unclear which of the selected responses belong to which population code. Because evaluation of nonsynchronized rate changes requires integration of a minimal number of excitatory postsynaptic potentials (EPSPs) arriving successively from the selected cells, the pace at which different populations can be defined by rate enhancement is slow.

Both problems could be alleviated by introducing internal synchronization as an additional grouping mechanism. First, synchronization can bias the saliency of responses independently of rate fluctuations. Second, because it relies on coincidence detection and spatial summation rather than temporal summation, synchronization can define relations with sufficiently high temporal precision to permit rapid multiplexing (Fig. 13-4).

Grouping of responses by synchronization raises their impact and is therfore likely to enhance discharge rates of target cells in areas receiving synchronized input. Therefore, fast synchronization codes and more sustained rate codes likely coexist and complement one another. Sustained, rate-coded input from a given processing stage could then be rapidly disambiguated and bound together as grouped activity through synchronization, which would, in turn, lead to a specific pattern of sustained, rate-modulated responses at the next stage of signal processing. There, these activity patterns can again be subject to disambiguation by selective synchronization, and so on. Once grouping operations have converged the respective solutions may need to be stabilized. If they need to be kept separate, they could each engage in sustained synchronized oscillations with high intragroup and little intergroup synchrony, but if no superposition problems remain, rate enhancement alone should be sufficient. As discussed later there is evidence for both strategies.

PREDICTIONS

For internally generated synchronization to serve as a signature of relatedness, it should meet several criteria. First, to be compatible with known processing speed, synchronization must be achieved rapidly, maximally within a few tens of milliseconds (Rolls and Tovee, 1994; Thorpe et al., 1996). Second, changes in synchrony and changes in discharge rate should be adjustable independently, (i.e., there should be cases where synchrony among two neurons increases [decreases] without a concomitant increase [decrease] in discharge rate). Third, internal synchronization should be sufficiently precise to ensure that synchronous activity is more effective than nonsynchronized activity in driving target cells. Fourth, there should be systematic relations between the occurrence, dynamics, and topological distribution of synchronization patterns on the one hand and specific perceptual or motor processes on the other hand. These relations should be sufficiently consistent to permit predictions of behavior from measurements of synchronization patterns. Finally, the connections responsible for the internal generation of synchrony should be susceptible to use-dependent modifications of synaptic gain so that synchronization probability can be increased for groups of cells that have often been synchronously active in the past. This is required to install grouping criteria by learning and to form stable associations among frequently occurring feature constellations. The coincidence detecting mechanism that mediates these use-dependent gain changes should operate with the same temporal precision as the synchronizing mechanism. The next sections review experimental data supporting these predictions.

RESPONSE SYNCHRONIZATION IN STRIATE CORTEX

Evidence for the existence of internally generated response synchronization that meets the criteria defined previously was first obtained in cat primary visual cortex (Gray and Singer, 1987a, 1987b); since then, similar observations have been made in numerous brain structures in a variety of different species (for review see Singer, 1999).

Neurons in the visual cortex tend to synchronize their discharges with a precision in the millisecond range when activated with a single contour, but fail to do so when activated by different contours moving in different directions (Gray et al., 1989; Engel et al., 1991c). In addition, these stimulus-induced, context-dependent synchronization phenomena were associated with a conspicuous oscillatory modulation of cell firing in a frequency range between 30 and 50 Hz, the so-called γ frequency range. Two aspects make this synchronization interesting in the context of response selection and binding. First, it results from internal coordination of spike timing and is not simply caused by stimulus-locked changes in discharge rate. Second, synchronization probability changes in a systematic way when the perceptual coherence of stimulus constellations is modified. Thus, this type of synchrony is not a trivial reflection of anatomical connectivity such as shared input through bifurcating axons, but does result from context-dependent, dynamic interactions within the cortical network.

RESPONSE SYNCHRONIZATION, MECHANISMS AND PROPERTIES

Evidence indicates that the precise synchronization of cortical responses associated with oscillations in the γ frequency range results from intracortical interactions. This distinguishes them from the less precise synchronization phenomena that occur in association with oscillatory patterning of responses in the α (~10 Hz) or δ (< 4 Hz) frequency range and are due to intrathalamic or thalamocortical interactions (Contreras and Steriade, 1997a, 1997b; Steriade et al., 1996; Steriade, 1999).

The cortical origin of synchronization in the γ frequency range is suggested by several observations: First, isolated slices of the visual cortex can produce γ oscillations when appropriately stimulated pharmacologically (Buhl et al., 1998; Draguhn et al., 1998; Fisahn et al., 1998; Whittington et al., 1995; Tennigkeit, personal communication). Second, cortical networks contain at least two cell types with putative pace maker functions in the appropriate frequency range: non-pyramidal cells that exhibit an oscillatory fluctuation of their membrane potential in the γ frequency range (Llinas et al., 1991) and pyramidal cells that engage in rhythmic firing in the 40 Hz range (chattering cells) (Gray and McCormick, 1996) when sufficiently depolarized. Third, synchronization is mediated by corticocortical connections. This has been shown both for the intrinsic tangential connections that reciprocally link cells distributed across different columns (Löwel and Singer, 1992; König et al., 1993) (see Chapter 10) and for the long-range connections that mediate interactions between homologous areas in the two hemispheres via the corpus callosum (Engel et al., 1991a). Fourth, the network of reciprocally coupled inhibitory interneurons can maintain oscillatory activity in the γ frequency range even after blockade of ionotropic glutamate receptors if the interneurons are activated via metabotropic glutamate receptors (Beierlein et al., 2000; Tennigkeit unpublished observations). Whether this oscillatory activity depends solely on conventional synaptic interactions, or on the gap junctions (Gibson et al., 1999; Fukuda and Kosaka, 2000) that have recently been discovered among inhibitory interneurons in the neocortex as well is currently under investigation (Beierlein et al., 2000; Tamás et al., 2000). Fifth, local intracortical application of cholinergic agonists facilitates oscillatory activity in the γ frequency range and synchronization while blockade of muscarinic receptors has the reverse effect (Rodriguez et al., 2001). Sixth, analysis of the spatiotemporal patterning of oscillatory activity with multielectrodes suggests that the synchronization results from an intracortical self-organizing process that leads to an entrainment of distributed oscillators and is not caused by oscillatory subcortical input (Prechtl et al., 2000).

How the various cell populations and pace maker mechanisms interact to produce the oscillatory patterning of responses and their synchronization remains to be clarified. It is unclear whether oscillations are always synchronous across laminae or whether supragranular and infragranular oscillations can dissociate (see Chapter 7). It is clear, however, that there is a close relation between the oscillatory patterning of responses in the γ frequency range and the occurrence of precise synchronization of discharge patterns, especially if synchronization occurs over longer distances between cells located in different functional columns or cortical areas (König et al., 1995b). Synchronization of discharges with close to zero time lag and a precision in the range of < 10 ms can occur in the absence of any obvious oscillatory patterning, especially among closely spaced neurons with overlapping receptive fields (König et al., 1995a). However, strong correlations and correlations over larger distances are usually associated with an oscillatory patterning of the individual responses in the γ frequency range. This suggests that oscillatory patterning facilitates entrainment of neuronal groups into coherently discharging assemblies, a notion supported by early simulation studies (König and Schillen, 1991; for review Singer, 1993). One reason is that an oscillatory modulation of the membrane potential permits temporal decoupling between excitatory synaptic input and the occurrence of the postsynaptic spike. When the membrane potential of a cell undergoes an oscillatory modulation, EPSPs with a *N*-methyl-D-aspartate (NMDA)-receptor-mediated component evoke spikes, not necessarily at the time of their occurrence, but only when the cell reaches the peak of the next depolarizing cycle. This occurs because NMDA receptors still occupied by glutamate are reactivated by the cyclic depolarizations, which remove the voltage-dependent magnesium block. Through this mechanism, spikes can be delayed by more than a half-cycle of the oscillation, which permits synchronization of a cell's discharge to the oscillatory activity provided by inputs from pacemaker circuits (Volgushev et al., 1998). Another reason is that inhibitory postsynaptic potentials (IPSPs) can effectively delay the occurrence of action potentials when occurring during the trough and rising phase of oscillatory membrane potential changes (Stiefel et al., 2001). This process is rapid and can synchronize responses within less than an oscillation cycle by appropriate shifting of spike latencies. Evidence from in vivo experiments points in the same direction. Neurons in the visual system can engage in synchronous activity at the same time that they increase their discharge rate in response to the light stimulus (Gray et al., 1992; Neuenschwander and Singer, 1996; Castelo-Branco et al., 1998).

Another important property of response synchronization is its independence of changes in discharge frequency. Although the occurrence of epochs of precisely synchronized (oscillatory) discharge sequences is greatly enhanced by light stimulation, there is no consistent relationship between the amplitude of the responses and the magnitude of oscillatory patterning and/or synchronization because changes in spike timing that leave average discharge rate unaffected can lead to drastic changes in synchrony (Fig. 13-4) (König et al., 1996). Note, however, that one could determine discharge rates on a spike-by-spike basis, defining instant frequencies as the inverse of the respective interspike intervals. In that case, some of the synchronized events might be interpreted as the result of coherent, ultrarapid rate fluctuations, but even then relations between rate terms and synchrony can break down because synchrony can be achieved by advancing a spike in one cell (rate increase) and delaying a spike in the other (rate decrease).

In this context it is important to note that analysis of single cell responses often fails to reveal that a cell participates in a synchronously oscillating assembly because individual cells usually do not fire with every cycle. Interspike interval distributions may appear poissonian even though the cell's discharges are precisely time locked to an oscillatory process and synchronized with other cells in the population. Such cycle skipping is a well-known phenomenon of oscillatory processes in the hippocampus (Buzsaki and Chrobak, 1995; Buzsaki, 1996) and may be of functional significance rather than a reflection of noise fluctuations. If cycles are skipped in a systematic way, the constellations of cells discharging in synchrony can change from cycle to cycle. In case of 40 Hz oscillations, different contents could then be encoded in time slices following one another at intervals of 25 ms (Jensen and Lisman, 1998; Luck and Vogel, 1997) (Fig. 13-4). Indications that such a coding strategy may actually be used have been obtained in the olfactory system of insects (Wehr and Laurent, 1996).

There are several ways to determine whether a cell participates in a synchronously active assembly. One possibility is to record the local field potential (LFP), which reflects the average synaptic currents produced in nearby cells, and to correlate the cell's spike discharges with these LFP fluctuations. If this leads to a modulated correlation function, it indicates that the cell's discharges exhibit a consistent temporal relation to the discharges of cells contributing to the LFP. If, in addition, the resulting function exhibits an oscillatory patterning, it proves that the cell's discharges are synchronized to the activity of an assembly of synchronously oscillating cells. This is what one usually finds in striate cortex. When responses synchronize in the γ frequency range, cells typically fire with high temporal precision around the negative peaks of the LFP fluctuations (Gray and Singer, 1989). Another possibility to disclose oscillatory firing patterns and synchronization is to record multiunit activity. This strategy greatly increases the chances of detecting oscillatory patterns resulting from denser sampling. If some of the simultaneously recorded cells participate in an oscillatory ensemble, autocorrelation functions are likely to exhibit an oscillatory modulation. Synchronous discharges, in the absence of oscillations, are also detectable in autocorrelations as a central peak whose width corresponds to the precision of synchronization. Alternatively, or in addition, the spikes from multiunit recordings can be sorted and subjected to cross-correlation analysis. Finally, the optimal approach is to use multiple electrodes and to record both multiunit and field potential responses from many sites simultaneously and to subject these signals to auto- and cross-correlation analysis. This strategy is the only option for the detection of nonlocal synchronization phenomena. All these analyses should be applied to nonaveraged time series obtained from the same trials, because averaging, or trial scrambling, renders detection of oscillatory patterning and/or synchronization impossible owing to the absence of time locks to the stimulus for internally generated episodes of synchronized firing. Moreover, oscillation frequencies may vary from trial to trial (Gray et al., 1992). Hence, both oscillatory patterning and synchronization of responses, if resulting from internal interactions, can be detected only in single trial analysis. As detailed later, in other instances responses become synchronized by phase locking to stimuli. Although this type of synchronization is of great relevance for functions such as perceptual grouping, it is discussed only after having explored the putative functions of internally generated synchrony.

RELATION BETWEEN RESPONSE SYNCHRONIZATION AND PERCEPTUAL PHENOMENA

PERCEPTUAL GROUPING

Based on the hypothesis that internal synchronization of discharges could serve to group responses for joint processing, a series of experiments have been performed in search for a correlation between dynamic changes of response synchronization and particular stimulus configurations. One prediction to be tested was that in early visual areas synchronization probability should reflect some of the basic Gestalt-criteria according to which the visual system grouped related features during scene segmentation. A consistent finding was that neurons distributed across different columns within the same or different visual areas and even across hemispheres synchronized their responses with close to zero phase lag when activated with a single contour, but fired independently when stimulated simultaneously with two different contours (Gray et al., 1989; Engel et al., 1991a, 1991b, 1991c; Kreiter and Singer, 1996; Freiwald et al., 1995). This experiment suggested that synchronization was the result of a context-dependent selection and grouping process. Analysis of the dependence of synchrony on receptive field and stimulus configurations revealed that the probability and strength of response synchronization reflected elementary Gestalt-criteria for perceptual grouping, such as continuity, proximity, similarity in the orientation domain, colinearity, and common fate (for review see Singer, 1993; Singer et al., 1997; Gray, 1999). These initial experiments were performed in anesthetized cats, but more recent multielectrode recordings from awake cats and monkeys indicated that the synchronization phenomena are not artifacts of anesthesia but are even more pronounced when the animals are awake and attentive (Fries et al., 1997a; Gray and Viana Di Prisco, 1997; Frien et al., 1994, Kreiter and Singer, 1992, 1996; Friedman-Hill et al., 2000; Maldonado et al., 2000). It is noteworthy that none of these systematic changes in synchronization probability have been associated with systematic changes of the neuron's discharge rate.

A particularly close correlation between neuronal synchrony and perceptual grouping has recently been observed in experiments with plaid stimuli. These stimuli are well suited for the study of dynamic binding mechanisms, because minor changes of the stimulus cause a binary switch in perceptual grouping. Two superimposed gratings moving in different directions (plaid stimuli) may be perceived either as two surfaces, one being transparent and sliding on top of the other (component motion), or as a single surface, consisting of crossed bars, that moves in a direction intermediate to the component vectors (pattern motion) (Adelson and Movshon, 1982; Stoner et al., 1990). Which percept dominates depends on the luminance of grating intersections, this variable defines the degree of transparency or on the duty cycle of the gratings because it determines changes in foreground-background assignments (Albright and Stoner, 1995). Component (pattern) motion is perceived when luminance conditions are compatible (incompatible) with transparency (Fig. 13-5A). Here is a case where local changes in stimulus properties cause global changes in perceptual grouping. For component motion, responses evoked by the two gratings must be segregated and only responses evoked by the contours of the same grating must be grouped to represent the two

→

FIGURE 13-5. **(A)** Two superimposed gratings that differ in orientation and drift in different directions are perceived either as two independently moving gratings (component motion) or as a single pattern drifting in the intermediate direction (pattern motion), depending on whether the luminance conditions at the intersections are compatible with transparency. **(B)** Predictions on the synchronization behavior of neurons as a function of their receptive field configuration (left) and stimulation conditions (right). **(C)** Changes in synchronization behavior of two neurons recorded simultaneously from areas 18 and PMLS that were activated with a plaid stimulus under component (upper graph) and pattern motion conditions (lower graph). The two neurons preferred gratings with orthogonal orientation (see receptive field configuration, top, and tuning curves obtained with component and pattern, respectively) and synchronized their responses only when activated with the pattern stimulus (compare cross-correlograms on the right). (Adapted from Castelo-Branco et al., 2000.)

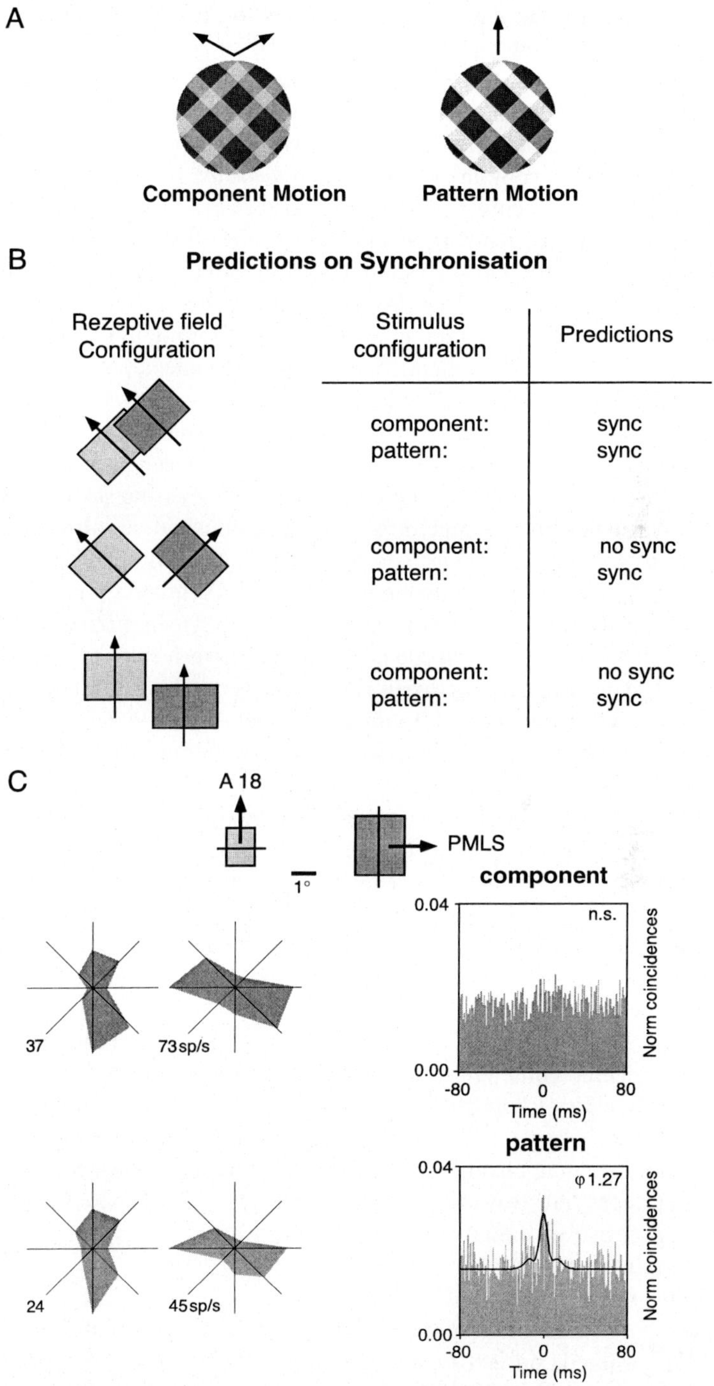
A
Component Motion
Pattern Motion
B
Predictions on Synchronisation
Rezeptive field Configuration
Stimulus configuration
Predictions
component: sync
pattern: sync
component: no sync
pattern: sync
component: no sync
pattern: sync
C
A 18
PMLS
1°
component
pattern
37
73sp/s
24
45sp/s
n.s.
φ1.27
0.04
0.00
-80
0
80
Time (ms)
Norm coincidences

surfaces, whereas for pattern motion, responses to all contours must be bound together to represent a single surface. If this grouping of responses is initiated by selective synchronization, three predictions must hold (Fig. 13-5B). First, neurons that prefer the direction of motion of one of the two gratings and have colinearly aligned receptive fields should always synchronize their responses, because they always respond to contours that belong to the same surface. Second, neurons that are tuned to the respective motion directions of the two gratings should synchronize their responses for pattern motion, because they then respond to contours of the same surface, but they should not synchronize for component motion because their responses are evoked by contours belonging to different surfaces. Third, neurons preferring the direction of pattern motion should synchronize only in the pattern and not in the component motion condition.

An important aspect of these predictions is that the expected changes in synchrony differ for different cell pairs, depending on the configuration of their receptive fields. Thus, when searching for relations between synchrony and cognitive functions, it is not only crucial to identify the processing stage at which one assumes a particular binding function to be accomplished but also to select the appropriate cell pairs. Averaging data across cell pairs with different receptive field configurations can mask dynamic changes in synchrony and is likely to reveal only the static anisotropies in the network of synchronizing connections. Such a difficulty may have contributed to the negative results of a recent study that failed to show a relation between perceptual grouping and internal synchronization in monkey striate cortex (Lamme and Spekreijse, 1998).

For the plaid stimuli, predictions were tested with multielectrode recordings from area 18 of primary visual cortex and area PMLS of lightly anesthetized cats. This was done after it had been confirmed with eye movement recordings in awake cats that the animals distinguished between component and pattern motion. Cross-correlation analysis of responses from cell pairs distributed solely within area 18 or area PMLS, or across the two areas, confirmed all three predictions. Cells synchronized their activity if they responded to contours that are perceived as belonging to the same surface (Castelo-Branco et al., 2000) (Fig. 13-5C). Analysis of the neuron's discharge rate confirmed that most of the cells in these visual areas respond preferentially to the component gratings of the plaids (component specific cells, Gizzi et al. 1990) and not to the pattern as a whole. However, in contrast to synchrony, variations in response amplitude failed to reflect the transition from component to pattern motion induced by transparency manipulation. Dynamic changes in synchronization could, thus, serve to encode in a context-dependent way the relations among the simultaneous responses to spatially superimposed contours and thereby bias their association with distinct surfaces. Future investigations are needed to clarify whether the populations of differentially synchronized neurons already serve as the final representation of the perceived surfaces, or whether for pattern motion additional assemblies are formed. They should consist of conjunction units tuned to the specific constellations of superimposed gratings, and their responses should be bound together to signal that they code for the same surface.

In all experiments referred to in this chapter, trials with different stimulus configurations were interleaved randomly to counterbalance possible effects of slow drifts in central state that may have caused covariations of discharge rate at a slow time scale. Moreover, phasic response components were excluded from correlation analysis to avoid artifactual correlations caused by nonstationarities in discharge rate (Brody, 1999a, 1999b). Accordingly, the shift predictors (correlograms between responses to the same stimulus configuration selected from different trials) were always flat, indicating that the changes in the correlations were caused by context-dependent changes of internal interactions and not by changes in stimulus locked or state-dependent covariations of discharge rate.

PERCEPTUAL SELECTION

The experiments reviewed in the previous section examined relations between synchronization and feature binding in the context of perceptual grouping, this section describes studies that deal with more global but still preattentive processes of stimulus selection. A close relation between response synchronization and stimulus selection has been found in cats who suffered from strabismic amblyopia, a developmental impairment of vision associated with suppression of the amblyopic eye, reduced visual acuity, and crowding. Crowding refers to the inability of amblyopic subjects to identify target stimuli if these are surrounded by nearby contours, and it is thought to result from false binding of responses evoked by the target and the embedding background. Quite unexpectedly, the light responses of individual neurons in the primary visual cortex were normal, and the neurons continued to respond vigorously to stimuli that the animals were unable to perceive through the amblyopic eye. The only significant correlate of amblyopia detected in this study was a reduction of response synchronization among neurons driven by the amblyopic eye. This reduction in synchrony became particularly pronounced when responses were evoked with gratings that had been identified in previous behavioral experiments as too fine to be resolvable by the animal through the amblyopic eye (Roelfsema et al., 1994). Impaired synchrony is likely to account for at least some of the perceptual deficits. By reducing the saliency of responses, it could explain why signals from the amblyopic eye cannot compete successfully with the well-synchronized responses from the normal eye and are excluded from supporting perception when both eyes are open. Recent recordings from area 21, a higher visual area, have indeed shown that the poorly synchronized responses evoked from the amblyopic eye drive neurons in area 21 less efficiently than the well-synchronized responses from the normal eye (Schröder et al., 1998). Poor synchronization is also likely to be responsible for the reduction in visual acuity and for crowding because it is expected to impair disambiguation of responses evoked by closely spaced stimuli.

A correlation between response synchronization and stimulus selection has also been documented in experiments on binocular rivalry that were again performed in strabismic animals (Fries et al., 1997a). Perception in strabismic subjects always alternates between the two eyes. This can be exploited to investigate how neuronal responses to constant stimuli change if they pass from being selected, and perceived to being suppressed and excluded from perception, and vice versa (Fig. 13-6). The outcome of these experiments was surprising because the responses of neurons in areas 17 and 18 were not attenuated when they were excluded from supporting perception. A close and highly significant correlation existed, however, between changes in the strength of response synchronization and the outcome of rivalry. Cells mediating responses of the eye that won in interocular competition increased the synchronicity of their responses on presentation of the rival stimulus to the other, losing eye, whereas the reverse was true for cells driven by the eye that became suppressed. In addition there was a marked increase in the oscillatory modulation of the synchronized discharges that supported perception. Thus, in primary visual cortex there are instances where internal selection of responses for further processing is associated with enhanced synchronization rather than with increased firing. This agrees with rivalry experiments in awake, behaving monkeys, which showed no systematic relation between the strength of visual responses and perception in early visual areas, but a clear correlation between perceptual suppression and loss of neuronal responses in higher visual areas (Logothetis and Schall, 1989; Leopold and Logothetis, 1996; Sheinberg and Logothetis, 1997). This is what one expects if the saliency of the responses from the two eyes is adjusted at early processing stages by modulating synchronization rather than discharge rates. These results do not, of course, exclude additional response selection by rate modulation. Thus, the outcome of rivalry can also be biased by changing the contrast of the patterns presented to the two eyes, and this manifests itself in rate changes.

DEPENDENCY ON CENTRAL STATES AND ATTENTION

A characteristic feature of response synchronization in the γ frequency range is its marked state dependency. It is particularly prominent when the cortex is in an active state, (i.e., when the electroencephalogram [EEG] is desynchronized and exhibits high power in the β and γ frequency range (Fig. 13-7) (Munk et al., 1996, Herculano-Houzel et al., 1999). Synchronization over long distances that requires oscillatory patterning of responses breaks down completely when the EEG becomes "synchronized" and exhibits high power in the low frequency range (< 10 Hz). This close correlation between the occurrence of response synchronization on the one hand and EEG states characteristic for the aroused and performing brain on the other hand is strong support for a functional role of precise

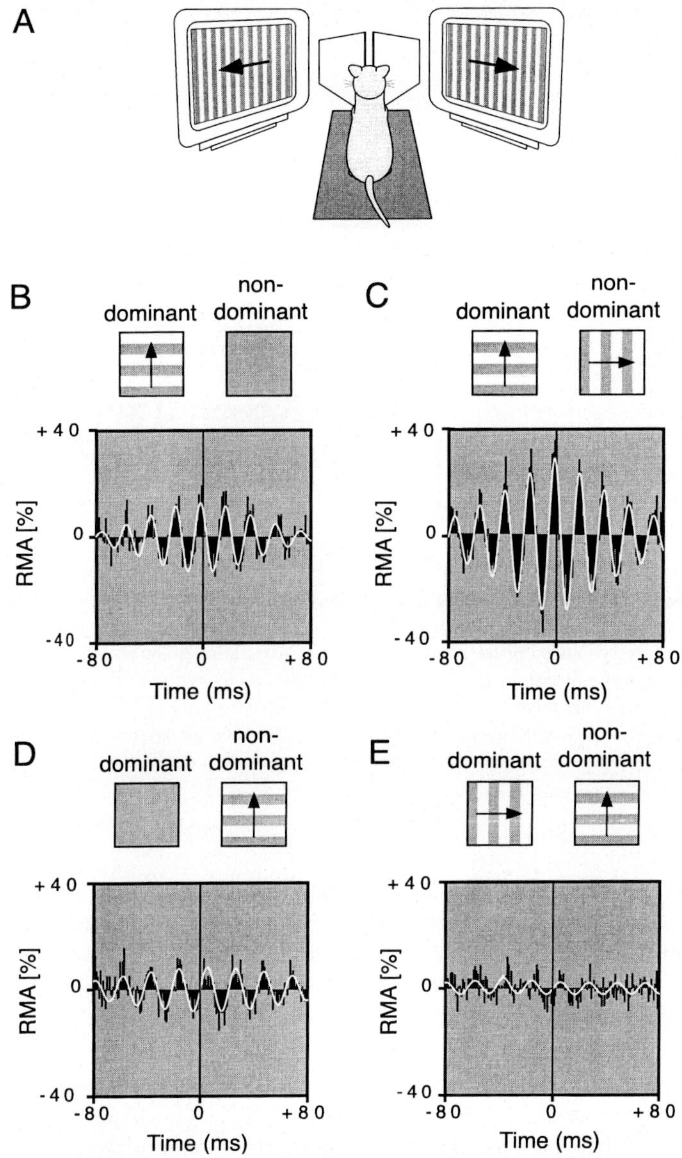

FIGURE 13-6. Neuronal synchronization under conditions of binocular rivalry. **(A)** Using two mirrors, different patterns were presented to the two eyes of strabismic cats. Panels **(B–E)** show normalized cross-correlograms for two pairs of recording sites activated by the eye that won **(B,C)** and lost **(D,E)** in interocular competition, respectively. Insets above the correlograms indicate stimulation conditions. Under monocular stimulation **(B),** cells driven by the winning eye show a significant correlation that is enhanced after introduction of the rivalrous stimulus to the other eye **(C).** The reverse is the case for cells driven by the losing eye (compare conditions **D** and **E**). The white continuous line superimposed on the correlograms represents a damped cosine function fitted to the data. RMA, relative modulation amplitude of the center peak in the correlogram, computed as the ratio of peak amplitude over offset of correlogram modulation. This measure reflects the strength of synchrony. (Modified from Fries et al., 1997a.)

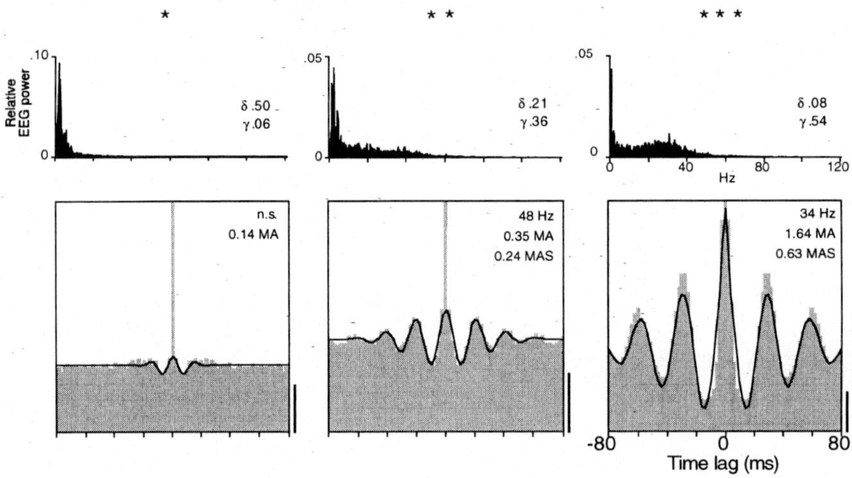

FIGURE 13-7. State dependence of response synchronization. *(Upper row)* Power distribution in the EEG recorded during three episodes from an anesthetized cat. *(Lower row)* Averaged cross-correlograms of multiunit responses evoked by a drifting grating and recorded from two different sites in A17 during corresponding episodes. Note that the synchronization of the responses and the oscillatory modulation in the γ-frequency range increase with increasing γ-activity in the EEG. Inserts in the power spectra give the relative power in the δ- and γ-frequency range. Inserts in the cross-correlograms give the oscillation frequency (Hz) and the relative modulation amplitude of the center peak (MA) and of the first side peak (MAS). (From Herculano-Houzel et al., 1999.)

synchrony in cortical processing. Precise response synchronization is indeed a much more sensitive indicator of changes in central state that are associated with a loss of perceptual functions than the response amplitudes of individual neurons. Discharge rates also fluctuate in a state-dependent way, but these fluctuations are neither related to changes in synchrony nor do they exhibit a close relation to changes in EEG power. It is not uncommon to observe an increase in discharge rate while oscillatory patterning and synchronization in the γ frequency range disappear and vice versa.

The magnitude and precision of synchronization in the γ frequency range reflect even fluctuations of attention in the fully aroused brain. Attention and task-dependent modulation of synchrony between areas 17 and 18 of the two hemispheres, but also between visual cortex and more frontally located cortical areas, has been observed in cats trained to perform a visually triggered motor response. The visual, association, somatosensory, and motor areas involved in the execution of the task synchronized their activity in the γ frequency range as soon as the animals focused their attention on the relevant stimulus. Consequently, the strength of synchronization among areas reflected precisely the coupling of these areas by corticocortical connections. Immediately after the appearance of the visual stimulus, synchronization increased further, and these coordinated activation patterns were maintained until the task was completed. However, once the reward was available and the animals engaged in consumatory behavior, these coherent patterns collapsed and gave way to low-frequency oscillatory activity that did not exhibit any consistent relations with regard to phase and areal topology (Roelfsema et al., 1997). These results suggest that an attention-related process had imposed a coherent temporal patterning on the activity of cortical areas required for the execution of the task. Here, attentional mechanisms, probably involving top-down projections and/or nonspecific thalamic projections (Ribary et al., 1991; Steriade, 1999), seem to prepare cortical areas, so that neurons can rapidly synchronize their responses both within and across areas once the stimulus appears, thereby accelerating selection and grouping of responses.

These anticipatory synchronization patterns would be particularly effective if they exhibited columnar specificity, because they could then act as a read out mechanism of the grouping criteria expressed in the architecture of tangential intracortical connections. Recent evidence supports such a scenario. Measurements of fluctuations in response latency revealed that self-generated γ activity exhibits a specific patterning that reflects the grouping criteria of vicinity and colinearity. Response latencies of neurons in striate cortex fluctuate in the range of about 20 ms for identical, repeatedly presented stimuli. The new finding is that these fluctuations of response latency are often correlated across cortical columns (Fries et al., 1997b, 2001). In this case, synchronization of the early response phase is better than expected from mere stimulus locking. Correlating response latencies of cells recorded from one site with the LFP fluctuations recorded from a site in the opposite hemisphere that exhibited latency covariations revealed that response latencies could be predicted by the polarity of the ongoing LFP oscillations, but only when these oscillations were in the γ frequency range. This suggests that the rapid synchronization of the very first spikes of a response is due to ongoing, oscillatory fluctuations of the cells' membrane potential and that these fluctuations can be coherent over large distances, in this case across hemispheres. These oscillations delay or advance responses to light stimuli depending on the timing of the stimulus relative to the phase of the ongoing oscillation, leading to synchronization of onset latencies in cells located in coherently oscillating columns. In this case, then, synchronization is achieved by latency adjustment, probably via the delay mechanism identified in the in vitro experiments (Volgushev et al., 1998). In oscillating cells the first discharges evoked by a barrage of EPSPs are likely to occur shortly after the peak of the respective next depolarizing cycle, and in cells that are already engaged in a synchronously oscillating network these first responses will thus be synchronous.

Analysis of the receptive field properties of cell pairs exhibiting correlated latency fluctuations revealed that the ongoing fluctuations of cortical activity are not random, and not simply noise, but exhibit columnar specificity: Fluctuations are coherent among cells with overlapping receptive fields, and, for cells with nonoverlapping fields, they are coherent only if these have similar orientation preference. Thus, the latency adjustments exhibit feature selectivity and could be exploited for rapid perceptual grouping. The grouping criteria that reside in the functional architecture of the intracortical association connections thus seem to be translated into specific spatiotemporal patterns of membrane potential fluctuations. Columns encoding groupable features oscillate in phase, and this activity enhances the temporal coherence of responses to nearby or colinear contours. Anatomical evidence indicates that the corticocortical association connections preferentially link columns coding for related features that tend to be grouped perceptually (Gilbert and Wiesel, 1989; Ts'o and Gilbert, 1988; Malach et al., 1993; Schmidt et al., 1997) (see Chapter 10). Further, physiological data suggest that the corticocortical connections contribute to the synchronization of spatially segregated groups of neurons (Engel et al., 1991a; Löwel and Singer, 1992; König et al., 1993). It appears likely then that the functional architecture of these association connections is continuously translated into coördinated fluctuations of neuronal excitability, which reflect dynamically the system's "knowledge" of grouping rules. Inevitably the pattern of these fluctuations is also modified by top-down influences from higher cortical areas and by immediately preceding changes of sensory input. Both effects would be equivalent to the functions commonly attributed to attentional mechanisms, the selection and binding of expected events, either as a consequence of bottom-up priming or of intentional top-down selection. The observed fluctuations could thus be equivalent with the system's updated expectancy that is determined by the fixed, locally installed grouping rules, by top-down influences and preceding sensory input that influences subsequent selection, grouping and binding of actual input patterns. Seen in this context, ongoing activity assumes the function of a predictor against which incoming activity is matched. One of the effects of matching these predictions with incoming signals is a rapid temporal regrouping of output activity (Fig. 13-8).

PLASTICITY OF SYNCHRONIZING CONNECTIONS

If synchronization of responses serves as a mechanism to group responses, synchronizing connections must be susceptible to use dependent modifications. These modifications are required to implement new grouping criteria by learning and to stabilize assemblies representing previously experienced conjunctions. Synchronization probability must increase durably for groups of cells that have been previously forced to engage repeatedly in highly synchronous firing. Further, the mechanism mediating these use-dependent changes in synchronization probability must be capable of distinguishing between synchronous and nonsynchronous firing with a temporal resolution that is in the same range as the temporal precision of observed synchrony. Both postulates have recently received experimental support. Herculano et al. (1997) have shown that synchronization probability increases between groups of neurons if they engage repeatedly in synchronous oscillatory firing in the γ frequency range. This enhanced tendency of neurons to synchronize their responses to coherent stimuli can be reduced again

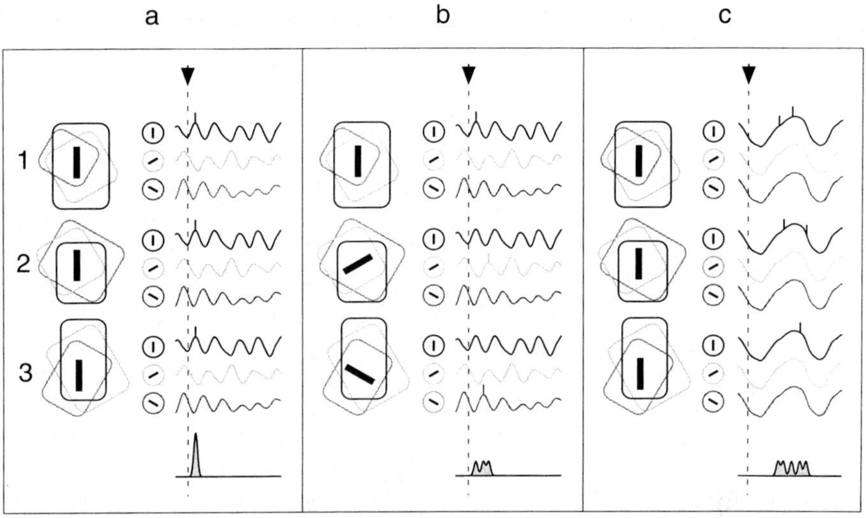

FIGURE 13-8. A model of rapid feature selective neuronal synchronization through correlated latency shifting. **(A)** Three groups of neurons are shown schematically as circles (middle column). The three groups have RFs at different locations in the visual field (left column) and each group contains neurons with three different orientation selectivities (small black bars inside circles). The graphs in the right column depict LFPs dominated by gamma frequency oscillations that reflect the neurons' membrane potential fluctuations. The model assumes that fluctuations are coherent for neurons of the same orientation preference, but incoherent for neurons preferring dissimilar orientation. When a stimulus array of three vertical bars (black bars over the RFs) is presented, only the neurons selective for vertical orientations respond. Arrow and broken vertical line over LFPs indicate the time when stimulus driven synaptic input reaches the recorded neurons. As their response onsets are shifted coherently by the coherent membrane potential fluctuations, their first spikes are correlated and add up to a strong population response, shown as spike density function in the lowest trace. Note that for the sake of clarity, we have focused on the effect of similarity of orientation selectivity and ignored RF overlap as a factor influencing the strength of latency correlations. **(B)** The same conventions as in **(A),** but stimulation with bars of different orientations. Neurons of different orientation selectivities are activated by the three bars. As in this case the neuronal membrane potentials fluctuate incoherently, the response onsets are not aligned and the population response is spread out in time. **(C)** The same conventions as in **(A),** but the LFP is dominated by low-frequency oscillations. These low-frequency oscillations are globally synchronous (coherence is not dependent on orientation selectivity), but because of their low frequency, they have no latency shifting effect. On stimulation with three aligned bars, the neurons fire more spikes than with γ-dominated LFP **(A),** but the latency is long, the spikes are not synchronized, and the reponses to nonaligned bars would exhibit the same temporal dispersion as the responses to aligned bars. (Adapted from Fries et al., 2001.)

by having the same cells engage repeatedly in oscillatory firing patterns that are decorrelated. These increases and decreases of synchronization probability appeared to depend on the precise phase relations of the oscillatory discharges during conditioning because on a coarser time scale (> 50 ms) the neurons' activity overlapped completely in both conditioning paradigms.

To obtain lasting modifications in synchronization probability in anesthetized cats, it was necessary to pair the light stimuli that induced synchronous (asynchronous) oscillatory firing of cortical cell pairs with electrical stimulation of the mesencephalic reticular formation (MRF). Activation of the MRF has several, probably causally related, effects. It desynchronizes the EEG, just as does arousal. This, in turn, enhances oscillatory modulation of visual responses in the γ frequency range and increases the magnitude and precision of synchronization (see previously). These effects are in all likelihood mediated by muscarinic receptors in the cortex that become activated by enhanced release of acetylcholine from projections originating in the basal forebrain. This is suggested by two observations: First, MRF stimulation enhances acetylcholine release in striate cortex and in nonrodents the only source of cholinergic afferents to striate cortex is the basal forebrain (Mesulam et al., 1983). Second, iontophoretic application of scopolamin in the striate cortex abolishes both γ oscillations and response synchronization (Rodriguez et al., 2001). The facilitation of lasting modifications of synchronization probability by MRF stimulation could thus have two quite different reasons. It could be due to enhanced synchrony and the resulting increase in cooperation among synchronously active synaptic inputs; but it could also be due to the plasticity-enhancing effect of acetylcholine. Both scenarios are supported by experimental evidence. Experiments on use-dependent synaptic plasticity in slices of the hippocampus and the visual cortex have revealed that induction of modifications such as long-term potentiation (LTP) and long-term depression (LTD) is facilitated both by increasing cooperativity among synaptic inputs (Gustafsson and Wigstroem, 1986), as well as by application of carbachol, the latter effect being mediated by muscarinic receptors (Bröcher et al., 1992). Both increased cooperativity and muscarinic activation enhance depolarization, the latter via blockade of K^+-currents. This, in turn, augments entry of calcium via NMDA-receptors and voltage-gated calcium channels, thus increasing the probability for the occurrence of LTP and LTD (Hansel et al., 1997). However, it is also possible that acetylcholine facilitates plasticity, not by its effect on membrane depolarization, but by its action on other intracellular signaling cascades. At present it is not possible to distinguish between these possibilities, but it will be important to find out whether any constellation that causes highly synchronized activity and strong depolarization induces lasting synaptic gain changes or whether the cholinergic afferents exert a more specific, activity independent gating function of plasticity.

The indications from the in vivo experiments of Herculano et al. (1997) that precise synchronization of neuronal discharges might play a critical role in use-dependent modifications of synaptic interactions are well supported by slice experiments. Varying the temporal relations between presynaptic and postsynaptic responses in simultaneously recorded coupled cortical cells revealed that LTP results when the EPSP precedes the postsynaptic spike within intervals of 10 ms or less, whereas the polarity of the modification reverses to LTD as soon as the EPSP follows the spike (Markram et al., 1997; Zhang et al., 1998). Thus, shifts of a few milliseconds in the timing relations between presynaptic and postsynaptic discharges suffice to invert the polarity of use-dependent synaptic modifications. The mechanism permitting such precise evaluation of the temporal contiguity of presynaptic and postsynaptic responses is probably the active dendritic response associated with the back-propagating spike (Magee and Johnston, 1997).

Under the sustained activation conditions applied in the in vivo experiments of Herculano et al. (1997), the high temporal selectivity of synaptic modifications most likely resulted from the oscillatory modulation of the neuronal responses. That an oscillatory patterning of activity can effectively narrow the temporal window for coincidence detection has been shown in vitro. Experiments in slices of the hippocampus indicate that the polarity of synaptic gain modifications depends critically on the phase relations between presynaptic and postsynaptic discharges if these overlap on a coarse time scale but exhibit an oscillatory modulation (Huerta and Lisman, 1996). The same holds for the visual cortex. Pyramidal cells in rat visual cortex slices were made to discharge tonically at 20 Hz by injecting sinusoidally modulated current through a patch pipette. Simultaneously, EPSPs were evoked, also at 20 Hz, by electrical stimulation of excitatory afferents. Changing the phase relations between presynaptic and postsynaptic activity revealed that the stimulated input tended to undergo LTP when the EPSPs were coincident with the spikes. On the other hand, afferents consistently underwent LTD when the EPSPs fell in the troughs of the membrane potential oscillations. Thus, although presynaptic and postsynaptic activation overlapped completely on a coarse time scale, phase shifts of less than 20 ms between individual EPSPs and spikes reversed the polarity of the synaptic modifications (Wespatat et al., 1999).

THE IMPACT OF SYNCHRONIZED RESPONSES

If response synchronization served response selection and grouping for further joint processing, the respective target networks must be able to differentiate between synchronized and nonsynchronized input; that is, synchronized input must have a stronger impact on postsynaptic responses than nonsynchronized input. Evidence indicates that this is the case both for the visual cortex itself and its target structures.

Simultaneous recordings from coupled neuron triplets along thalamocortical (Alonso et al., 1996; Usrey and Reid, 1999 (see Chapter 8) and intracortical pathways (Alonso and Martinez, 1998) in the visual system have revealed that EPSPs synchronized within intervals below 2 ms are much more effective than EPSPs dispersed over longer intervals. A similar conclusion is suggested by simulation studies (Niebur and Koch, 1994) and in vitro experiments (Stevens and Zador, 1998).

Evidence for the enhanced saliency of synchronized activity has also been obtained at a more global level with simultaneous recordings from several sites (area 18 and the posterior mediolateral suprasylvian sulcus, PMLS) of the cat visual cortex and retinotopically corresponding sites in the superior colliculus

(Brecht et al., 1998). The impact that a particular group of cortical cells has on target cells in the colliculus increased dramatically whenever the cortical cells synchronized their discharges with other cortical cell groups projecting to the same site in the tectum (Fig. 13-9). Finally, enhanced saliency of synchronized responses can also be inferred from the tight correlation between the strength of neuronal response synchronization and perception that was observed in experiments on binocular rivalry in cats (Fries et al., 1997a) and human subjects (Tononi and Edelman, 1998; Tononi et al., 1998) (see also later). Last but not least, simulation studies also indicate that neurons with conventional integrate and fire properties can be quite sensitive to the temporal dispersion of synaptic input. Large scale simulation of biologically inspired thalamocortical networks revealed that neurons exhibited a strong tendency to engage in oscillatory firing patterns and to synchronize their responses. When synchrony was artificially disrupted by introducing a jitter in spike timing, transmission across polysynaptic pathways was drastically reduced (Lumer et al., 1997a, 1997b).

Several mechanisms, some of which have been identified only recently, make synchronously arriving EPSPs more efficient than temporally dispersed EPSPs. First, because of their exponential decay, simultaneous EPSPs summate more

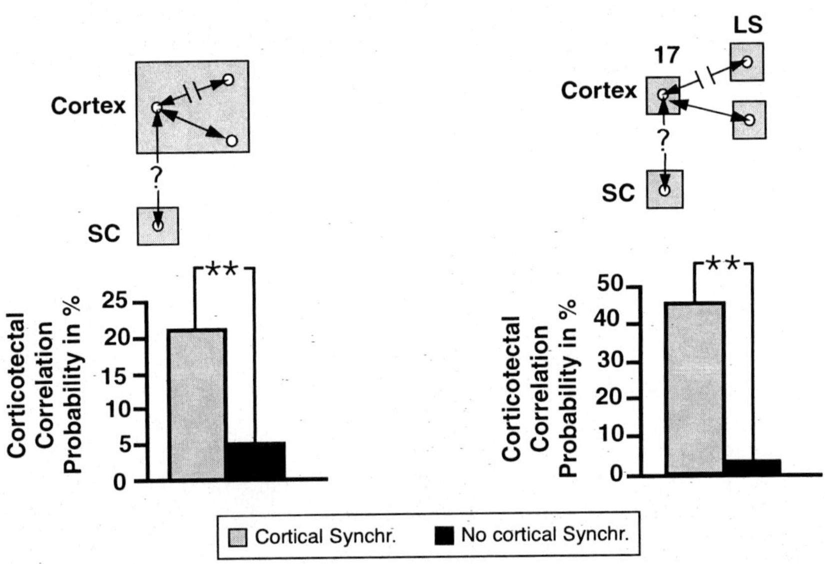

FIGURE 13-9. Dependence of corticotectal synchronization on intracortical synchronization within (left plots) and across cortical areas (right plots). **(Left)** Percentage of significant corticotectal correlations if cortical cells recorded from two sites in the same cortical area synchronize (shaded column) or do not synchronize (black column) their responses. **(Right)** Percentage of corticotectal interactions between A17 and the tectum if area 17 cells synchronize (shaded column) or do not synchronize (black column) their responses with those of cells in the lateral suprasylvian sulcus. (From Brecht et al., 1998.)

effectively than temporally dispersed EPSPs, and there is some evidence for supralinear summation resulting from voltage-gated dendritic conductances. Second, firing threshold is sensitive to the rising slope of the depolarization and lowers for fast rising depolarizations (Gray, personal communication). Third, the effect of EPSPs is dramatically enhanced when these coincide with a back-propagating dendritic spike and hence with the input that generated this spike (Larkum et al., 1999). All three mechanisms are sensitive to dispersions of EPSPs in the range of a few milliseconds.

GAMMA OSCILLATIONS AND VISUAL PERCEPTION IN HUMAN SUBJECTS

Although noninvasive measurements of cortical activity in human subjects with EEG or magnetoencephalogram (MEG) recordings do not allow one to distinguish between activity in striate and prestriate cortex, for two reasons these data deserve inclusion in this chapter. First, the fact that stimulation related γ activity can at all be recorded from the scalp indicates that very large populations of neurons engage in synchronous firing during visual perception. Second, these studies in human subjects reveal relations between the occurrence of response synchronization in the γ frequency range and conscious perception that can be analyzed only with great difficulties in trained animals.

Results from a series of pioneering EEG studies suggest that the engagement in cognitive tasks that require figure-ground distinctions and feature binding (Tallon-Baudry et al., 1996, 1997) perceptual switching during perception of bistable figures (Keil et al., 1999), or the retention of visual patterns in short-term memory (Tallon-Baudry et al., 1998, 1999) are associated with transient increases of γ oscillations in cortical areas involved in the respective tasks. These findings have recently been extended by the demonstration that during a face recognition task, populations of neurons not only synchronize locally on the basis of γ oscillations, but also phase lock across large distances with zero phase lag. A particularly interesting finding was that these large-scale synchronization patterns dissolve and settle into new configurations at exactly the time when subjects had recognized the pattern and prepared the execution of the motor response (Rodriguez et al., 1999).

A close correlation has also been found between the establishment of coherent γ oscillations across cortical areas and the acquisition of a visuotactile association in classical conditioning. Once subjects have learned about the association between a visual and a tactile stimulus, the visual stimulus evokes γ activity over the visual cortex that is coherent with γ activity over the somatosensory cortex contralateral to the stimulated finger, and this coherence breaks down again on extinction (Miltner et al., 1999). Finally, evidence has been obtained in an MEG study on binocular rivalry that activity evoked from the eye that dominates perception during rivalry is better synchronized than activity evoked by the suppressed eye (Tononi and Edelman, 1998; Tononi et al., 1998). This agrees with the rivalry experiments in cats (Fries et al., 1997a).

Consistent with a functional role of synchronized γ oscillations is the observation that these are closely related to arousal and attention. In human subjects γ oscillations are enhanced in response to attended, as compared to nonattended, stimuli (Tiitinen et al., 1993). They disappear in deep anesthesia (Madler and Pöppel, 1987), but are present during paradoxical sleep, during which dreaming (i.e., reactivation of memories) is thought to occur (Steriade, 1999; Llinas and Ribary, 1993).

STIMULUS LOCKED SYNCHRONIZATION AND PERCEPTUAL GROUPING

Simultaneously appearing visual stimuli evoke synchronous responses of neurons in striate cortex and periodically flickering stimuli can artificially induce synchronizations that are as high as internally generated synchrony (Rager and Singer, 1998). Thus, if synchronous neuronal responses are interpreted by the visual system as related and bound together, one expects perceptual grouping of simultaneously appearing pattern elements. Recent psychophysical experiments support this prediction.

These studies indicate that spatially distributed contour elements are bound perceptually and interpreted as elements of a coherent figure if they appear or change synchronously, whereas elements that follow different time courses are perceived as unrelated (Leonards et al., 1996; Usher and Donnelly, 1998; Alais et al., 1998; Lee and Blake, 1999). The temporal resolution of this grouping mechanism is surprisingly high. Temporal offsets between the respective appearances of figure and ground elements of less than 10 ms still support perceptual grouping (Leonards et al., 1996).

Interestingly the short temporal offsets between stimuli that support segmentation are no longer perceivable, suggesting a dissociation between the perceptibility of small differences in the time course of stimuli on the one hand and the effect of such differences on perceptual grouping on the other. This dissociation may be related to the fact that, in vision, the temporal cues supporting grouping as a function of stimulus synchrony are mediated mainly by the magnocellular pathway, whereas the other, nontemporal grouping cues are mediated by both the magnocellular and the parvocellular pathways (Leonards and Singer, 1998). These two pathways interact at multiple levels but subserve somewhat different functions. The magnocellular pathway is exquisitely sensitive to temporal features and can signal stimulus transients with high temporal resolution, whereas the parvocellular pathway operates with low temporal but high spatial and spectral resolution (for review of the extensive literature see Leonards and Singer, 1998). This functional dichotomy is relevant in the present context because it is a likely basis for the ability of the visual system to use both temporal and spatial cues in parallel for perceptual grouping. If, within the same matrix of line elements, one figure is defined by the synchronous onset of elements (temporal cue) and another spatially overlapping figure by differences in the orientation of the respective line elements (spatial cue), either the temporally or the spatially defined figure is perceived, depending on the relative saliency of the two cues (temporal offset versus orientation difference) (Leonards and Singer, 1998). This indicates that spatial and temporal grouping cues are processed in parallel and if conflicting, the less salient cue is disregarded.

How exactly the two grouping mechanisms interact is unknown. One possibility is that activity conveyed by the magnocellular system can interfere with internally generated synchrony and depending on the congruence of temporal and textural features, it either overrides or enhances internally generated temporal patterns.

CONCLUSIONS

The experimental evidence reviewed in this chapter supports the idea that the functions of primary visual cortex are not confined to feature extraction, but are an integral part of the constructivistic processes that underly visual perception. In addition to the elaboration of response selectivity for elementary features and some of their conjunctions, striate cortex seems to play as active a role in state-, context-, and attention-dependent selection and grouping of distributed responses as any other cortical area. Striate cortex also appears to share with other cortical areas the mechanisms of synaptic plasticity required for the experience-dependent formation of durable associations between distributed responses. The analysis of precise temporal relations between simultaneously recorded responses of several neurons also revealed the existence of highly dynamic, context-sensitive synchronization phenomena. Their numerous and close relations to perceptual processes suggest that they are not epiphenomenal but play an essential role in cortical processing and coding. These dynamic phenomena and their systematic relation to cognitive processes cannot be adequately assessed by evaluating the rate changes of single cells, because the firing rates of individual cells exhibit no systematic relation with response synchronization. Even pairwise analysis of temporal relations among distributed responses, as performed in most of the studies reviewed here, is probably revealing only the surface structures of immensely complex and ever changing dynamic states. These states, however, will eventually need to be analyzed more fully to obtain a deeper understanding of the cortical functions that go beyond sequential extraction and recombination of features in feedforward architectures. Analysis of these dynamic states requires massive parallel recording of spontaneous and evoked activity with high temporal resolution and the development of appropriate mathematical tools for the assessment of higher order correlations between distributed responses. Because the structure and function of the primary visual cortex is one of the best understood of all of the cerebral cortical areas, it offers itself once again as a highly suitable model for the analysis of general cortical functions.

REFERENCES

Adelson, E. H., and Movshon, J. A. (1982). Phenomenal coherence of moving visual patterns. *Nature* **300,** 523–525.

Alais, D., Blake, R., and Lee, S.-H. (1998). Visual features that vary together over time group together over space. *Nat. Neurosci.* **1,** 160–164.

Albright, T. D., and Stoner, G. R. (1995). Visual motion perception. *Proc. Natl. Acad. Sci. U.S.A.* **92,** 2433–2440.

Alonso, J.-M., and Martinez, L. M. (1998). Functional connectivity between simple cells and complex cells in cat striate cortex. *Nat. Neurosci.* **1,** 395–403.

Alonso, J.-M., Usrey, W. M., and Reid, R. C. (1996). Precisely correlated firing in cells of the lateral geniculate nucleus. *Nature* **383,** 815–819.

Beierlein, M., Gibson, J. R., and Connors, B. W. (2000). A network of electrically coupled interneurons drives synchronized inhibition in neocortex. *Nat. Neurosci.* **3,** 906–910.

Braitenberg, V. (1978). Cell assemblies in the cerebral cortex. In: *Architectonics of the cerebral Cortex.* Lecture Notes in Biomathematics 21, Theoretical Approaches in Complex Systems. (R. Heim, and G. Palm, Eds.), pp. 171–188. Berlin, Springer-Verlag.

Brecht, M., Singer, W., and Engel A. K. (1998). Correlation analysis of corticotectal interactions in the cat visual system. *J. Neurophysiol.* **79,** 2394–2407.

Brody, C. D. (1999a). Disambiguating different covariation types. *Neural Comp.* **11,** 1527–1535.

Brody, C. D. (1999b). Correlation without synchrony. *Neural Comp.* **11,** 1537–1551.

Bröcher, S., Artola, A., and Singer, W. (1992). Agonists of cholinergic and noradrenergic receptors facilitate synergistically the inductionof long-term potentiation in slices of rat visual cortex. *Brain Res.* **573,** 27–36.

Buhl, E. H., Tamás, G., and Fisahn, A. (1998). Cholinergic activation and tonic excitation induce persistent gamma oscillations in mouse somatosensory cortex in vitro. *J. Physiol.* **513,** 117–126.

Buzsaki, G. (1996). The hippocampo-neocortical dialogue. *Cerebral Cortex* **6,** 81–92.

Buzsaki, G., and Chrobak, J. J. (1995). Temporal structure in spatially organized neuronal ensembles: a role for interneuronal networks. *Curr. Opin. Neurobiol.* **5,** 504–510.

Castelo-Branco, M., Neuenschwander, S., and Singer, W. (1998). Synchronization of visual responses between the cortex, lateral geniculate nucleus, and retina in the anesthetized cat. *J. Neurosci.* **18,** 6395–6410.

Castelo-Branco, M., Goebel, R., Neuenschwander, S., and Singer, W. (2000). Neural synchrony correlates with surface segregation rules. *Nature* **405,** 685–689.

Chapman, B., Zahs, K. R., and Stryker, M. P. (1991). Relation of cortical cell orientation selectivity to alignment of receptive fields of the geniculocortical afferents that arborize within a single orientation column in ferret visual cortex. *J. Neurosci.* **11,** 1347–1358.

Contreras, D., and Steriade, M. (1997a). Synchronization of low-frequency rhythms in corticothalamic networks. *Neuroscience* **76,** 11–24.

Contreras, D., and Steriade, M. (1997b). State-dependent fluctuations of low-frequency rhythms in corticothalamic networks. *Neuroscience* **76,** 25–38.

Das, A., and Gilbert, C. D. (1999). Topography of contextual modulations mediated by short-range interactions in primary visual cortex. *Nature* **399,** 655–661.

Douglas, R. J., Martin, K. A. C., and Whitteridge, D. (1989). A canonical microcircuit for neocortex. *Neural Comput.* **1,** 480–488.

Draguhn, A., Traub, R. D., Schmitz, D., and Jefferys, J. G. R. (1998). Electrical coupling underlies high-frequency oscillations in the hippocampus in vito. *Nature* **394,** 189–192.

Edelman, G. M. (1987). *Neural darwinism: The theory of neuronal group selection.* New York, Basic Books.

Engel, A. K., König, P., Kreiter, A. K., and Singer, W. (1991a). Interhemispheric synchronization of oscillatory neuronal responses in cat visual cortex. *Science* **252,** 1177–1179.

Engel, A. K., Kreiter, A. K., König, P., and Singer, W. (1991b). Synchronization of oscillatory neuronal responses between striate and extrastriate visual cortical areas of the cat. *Proc. Natl. Acad. Sci. U.S.A.* **88,** 6048–6052.

Engel, A. K., König, P., and Singer, W. (1991c). Direct physiological evidence for scene segmentation by temporal coding. *Proc. Natl. Acad. Sci. U.S.A.* **88,** 9136–9140.

Engel, A. K., König, P., Kreiter, A. K., Schillen, T. B., and Singer, W. (1992). Temporal coding in the visual cortex: new vistas on integration in the nervous system. *Trends Neurosci.* **15,** 218–226.

Fisahn, A., Pike, F. G., Buhl, E. H., and Paulsen, O. (1998). Cholinergic induction of network oscillations at 40Hz in the hippocampus in vitro. *Nature* **394,** 186–189.

Freiwald, W. A., Kreiter, A. K., and Singer, W. (1995). Stimulus dependent intercolumnar synchronization of single unit responses in cat area 17. *Neuroreport* **6,** 2348–2352.

Friedman-Hill, S., Maldonado, P. E., and Gray, C. M. (2000). Temporal dynamics of neuronal activity in the striate cortex of alert macaque: I. Incidence and stimulus-dependence of oscillations. *Cerebral Cortex,* **10,** 1105–1116.

Frien, A., Eckhorn, R., Bauer, R., Woelbern, T., and Kehr, H. (1994). Stimulus-specific fast oscillations at zero phase between visual areas V1 and V2 of awake monkey. *Neuroreport* **5,** 2273–2277.

Fries, P., Roelfsema, P. R., Engel, A. K., König, P., and Singer, W. (1997a). Synchronization of oscillatory responses in visual cortex correlates with perception in interocular rivalry. *Proc. Natl. Acad. Sci. U.S.A.* **94,** 12699–12704.

Fries, P., Roelfsema, P. R., Singer, W., and Engel, A. K. (1997b). Correlated variations of response latencies due to synchronous subthreshold membrane potential fluctuations in cat striate cortex. *Soc. Neurosci. Abstr.* **23,** 499.

Fries, P., Neuenschwander, S., Engel, A. K., Goebel, R., and Singer, W. (2001). Rapid feature selective neuronal synchronization through correlated latency shifting. *Nature Neuroscience.* **4(2),** 194–200.

Fukuda, T., and Kosaka, T. (2000). Gap junctions linking the dendritic network of GABAergic interneurons in the hippocampus. *J. Neurosci.* **20,** 1519–1528.

Galuske, R. A. W., Schmidt, K. E., Goebel, R., Lomber, S. G., and Payne, B. R. (2000). Feedback control of orientation and direction maps in primary visual cortex. *Soc. Neurosci. Abstr.* **26,** 309.2.

Gerstein, G. L., and Gochin, P. M. (1992). Neuronal population coding and the elephant. In: *Information processing in the cortex, experiments and theory.* (A. Aertsen, and V. Braitenberg, Eds.), pp. 139–173. Berlin, Springer-Verlag.

Ghose, G. M., and Maunsell, J. (1999). Specialized representations in visual cortex: a role for binding? *Neuron* **24,** 79–85.

Gibson, J. R., Beierlein, M., and Connors, B. W. (1999). Two networks of electrically coupled inhibitory neurons in neocortex. *Nature* **402,** 75–79.

Gilbert, C. D., and Wiesel, T. N. (1989). Columnar specificity of intrinsic horizontal and cortico-cortical connections in cat visual cortex. *J. Neurosci.* **9,** 2432–2442.

Gizzi, M. S., Katz, E., Schumer, R. A., and Movshon, J. A. (1990). Selectivity for orientation and direction of motion of single neurons in cat striate and extrastriate visual cortex. *J. Neurophysiol.* **63,** 1529–1543.

Gray, C. M. (1999). The temporal correlation hypothesis of visual feature integration: still alive and well. *Neuron* **24,** 31–47.

Gray, C. M., and Singer, W. (1987a). Stimulus-dependent neuronal oscillations in the cat visual cortex area 17. *IBRO Abstr. Neurosci. Lett. Suppl.* **22,** 1301P.

Gray, C. M., and Singer, W. (1987b). Stimulus-dependent neuronal oscillations in the cat visual cortex: a cortical functional unit. *Soc. Neurosci. Abstr.* **13,** 404.3.

Gray, C. M., and Singer, W. (1989). Stimulus-specific neuronal oscillations in orientation columns of cat visual cortex. *Proc. Natl. Acad. Sci. U.S.A.* **86,** 1698–1702.

Gray, C. M., König, P., Engel, A. K., and Singer, W. (1989). Oscillatory responses in cat visual cortex exhibit inter-columnar synchronization which reflects global stimulus properties. *Nature* **338,** 334–337.

Gray, C. M., Engel, A. K., König, P., and Singer, W. (1992). Synchronization of oscillatory neuronal responses in cat striate cortex-temporal properties. *Vis. Neurosci.* **8,** 337–347.

Gray, C. M., and McCormick, D. A. (1996). Chattering cells: superficial pyramidal neurons in the visual cortex. *Science* **274,** 109–113.

Gray, C. M., and Viana Di Prisco, G. (1997). Stimulus-dependent neuronal oscillations and local synchronization in striate cortex of the alert cat. *J. Neurosci.* **17,** 3239–3253.

Gross, C. G. (1992). Representation of visual stimuli in inferior temporal cortex. *Philos. Trans. R. Soc. Lond. (Biol.)* **335,** 3–10.

Gustafsson, B., and Wigstroem, H. (1986). Hippocampal long-lasting potentiation produced by pairing single volleys and brief conditioning tetani evoked in separate afferents. *J. Neurosci.* **6,** 1575–1582.

Hansel, C., Artola, A., and Singer, W. (1997). Relation between dendritic Ca2+ levels and the polarity of synaptic long-term modifications in rat visual cortex neurons. *Eur. J. Neurosci.* **9,** 2309–2322.

Hebb, D. O. (1949). *The organization of behavior: A neuropsychological theory.* New York, Wiley.

Herculano, S., Munk, M. H. J., and Singer, W. (1997). Use-dependent reversible modification in the composition of synchronously firing ensembles in cat visual cortex. *Soc. Neurosci. Abstr.* **23,** 13.5.

Herculano-Houzel, S., M. H. J. Munk, S. Neuenschwander, and Singer, W. (1999). Precisely synchronized oscillatory firing patterns require electroencephalographic activation. *J. Neurosci.* **19,** 3992–4010.

Hubel, D. H., and Wiesel, T. N. (1962). Receptive fields, binocular interaction and functional architecture in the cat's visual cortex. *J. Physiol. (Lond.)* **160,** 106–154.

Huerta, P. T., and Lisman, J. E. (1996). Low-frequency stimulation at the throughs of Θ-oscillation induces long-term depression of previously potentiated CA1 synapses. *J. Neurophysiol.* **75,** 877–884.

Hupé, J. M., James, A. C., Payne, B. R., Lomber, S. G., Girard, P., and Bullier, J. (1998). Cortical feedback improves discrimination between figure and background by V1, V2 and V3 neurons. *Nature* **394,** 784–787.

Hupé, J.-M., James, A. C., Girard, P., Lomber, S. G., Payne, B. R., and Bullier, J. (2001). Feedback connections act on the early part of the responses in monkey visual cortex. *J. Neurophysiol.* in press.

Jagadeesh, B., H. S. Wheat, and Ferster, D. (1993). Linearity of summation of synaptic potentials underlying direction selectivity in simple cells of the cat visual cortex. *Science* **262,** 1901–1904.

Jensen, O., and Lisman, J. E. (1998). An oscillatory short-term memory buffer model can account for data on the Sternberg task. *J. Neurosci.* **18,** 10688–10699.

Kapadia, M. K., Ito, M., Gilbert, C. D., and Westheimer, G. (1995). Improvement in visual sensitivity by change in local context: parallel studies in human observers and in V1 of alert monkeys. *Neuron* **15,** 843–856.

Keil, K., Müller, M. M., Ray, W. J., Gruber, T., and Elbert, T. (1999). Human gamma band activity and perception of a Gestalt. *J. Neurosci.* **19,** 7152–7161.

König, P., and Schillen, T. B. (1991). Stimulus-dependent assembly formation of oscillatory responses: I. Synchronization. *Neural Comput.* **3,** 155–166.

König, P., Engel, A. K., Löwel, S., and Singer, W. (1993). Squint affects synchronization of oscillatory responses in cat visual cortex. *Eur. J. Neurosci.* **5,** 501–508.

König, P., Engel, A. K., Roelfsema, P. R., and Singer, W. (1995a). How precise is neuronal synchronization? *Neural Comput.* **7,** 469–485.

König, P., Engel, A. K., and Singer, W. (1995b). Relation between oscillatory activity and long-rang synchronization in cat visual cortex. *Proc. Natl. Acad. Sci. U.S.A.* **92,** 290–294.

König, P., Engel, A. K., and Singer, W. (1996). Integrator or coincidence detector? The role of the cortical neuron revisted. *Trends Neurosci.* **19,** 130–137.

Kreiter, A. K., and Singer, W. (1992). Oscillatory neuronal responses in the visual cortex of the awake macaque monkey. *Eur. J. Neurosci.* **4,** 369–375.

Kreiter, A. K., and Singer, W. (1996). Stimulus-dependent synchronization of neuronal responses in the visual cortex of the awake macaque moneky. *J. Neurosci.* **16,** 2381–2396.

Lamme, V. A. F, and Spekreijse, H. (1998). Neuronal synchrony does not represent texture segregation. *Nature* **396,** 362–366.

Larkum, M. E., Zhu, J. J., and Sakmann, B. (1999). A new cellular mechanism for coupling inputs arriving at different cortical layers. *Nature* **398,** 338–341.

Lee, S.-H., and Blake, R. (1999). Visual form created solely from temporal structure. *Science* **284,** 1165–1168.

Leonards, U., Singer, W., and Fahle, M. (1996). The influence of temporal phase differences on texture segmentation. *Vis. Res.* **36,** 2689–2697.

Leonards, U., and Singer, W. (1998). Two segmentation mechanisms with differential sensitivity for colour and luminance contrast. *Vis. Res.* **38,** 101–109.

Leopold, D. A., and Logothetis, N. K. (1996). Activity changes in early visual cortex reflect monkeys' percepts during binocular rivalry. *Nature* **379,** 549–553.

Llinas, R. R., Grace, A. A., and Yarom, Y., 1991, In vitro neurons in mammalian cortical layer 4 intrinsic oscillatory activity in the 10- to 50-Hz frequency range. *Proc. Natl. Acad. Sci. U.S.A.* **88,** 897–901.

Llinas, R., and Ribary, U. (1993). Coherent 40 Hz oscillation characterizes dream state in humans. *Proc. Natl. Acad. Sci. U.S.A* **90,** 2078–2081.

Livingstone, M. S., and Hubel, D. H. (1981). Effects of sleep and arousal on the processing of visual information in the cat. *Nature* **291,** 554–561.

Logothetis, N. K., and Schall, J. D. (1989). Neuronal correlates of subjective visual perception. *Science* **245,** 761–763.

Löwel, S., and Singer, W. (1992). Selection of intrinsic horizontal connections in the visual cortex by correlated neuronal activity. *Science* **255,** 209–212.

Luck, S. J., and Vogel, E. K. (1997). The capacity of visual working memory for features and conjunctions. *Nature* **390,** 279–281.

Lumer, E. D., Edelman, G. M., and Tononi, G. (1997a). Neural dynamic in a model of the thalamocortical system. I. Layers, loops and the emergence of fast synchronous rhythms. *Cerebral Cortex* **7,** 207–227.

Lumer, E. D., Edelman, G. M., and Tononi, G. (1997b). Neural dynamics in a model of the thalamocortical system. II. The role of neural synchrony tested through perturbations of spike timing. *Cerebral Cortex* **7,** 228–236.

Madler, C., and Pöppel, E. (1987). Auditory evoked potentials indicate the loss of neuronal oscillations during general anaesthesia. *Naturwissenschaften* **74,** 42–43.

Magee, J. C., and Johnston, D. A. (1997). A synaptically controlled, associative signal for Hebbian plasticity in hippocampal neurons. *Science* **275,** 209–213.

Malach, R., Amir, Y., Harel, M., and Grinvald, A. (1993). Relationship between intrinsic connections and functional architecture revealed by optical imaging and in vivo targeted biocytin injections in primate striate cortex. *Proc. Natl. Acad. Sci. U.S.A.* **90,** 10469–10473.

Maldonado, P. E., Friedman-Hill, S. R., and Gray, C. M. (2000). Temporal dynamics of neuronal activity in the striate cortex of alert macaque: II. Short and long-range temporally-correlated activity. *Cerebral Cortex,* **10,** 1117–1131.

Markram, H., Lübke, J., Frotscher, M., and Sakmann, B. (1997). Regulation of synaptic efficacy by coincidence of postsynaptic APs and EPSPs. *Science* **275,** 213–215.

Mesulam, M. M., Mufson, E. J., LeVey, A. I., and Wainer, B. H. (1983). Cholinergic innervation of cortex by the basal forebrain: cytochemistry and cortical connections of the septal area, diagonal band nuclei, nucleus basalis (substantia innominata) and hypothalamus in the rhesus monkey. *J. Comp. Neurol.* **214,** 170–197.

Miltner, W. H. R., Braun, C, Arnold, M., Witte, H., and Taub, E. (1999). Coherence of gamma-band EEG activity as a basis for associative learning. *Nature* **397,** 434–436.

Munk, M. H. J., Roelfsema, P. R., König, P., Engel A. K., and Singer, W. (1996). Role of reticular activation in the modulation of intracortical synchronization. *Science* **272,** 271–274.

Neuenschwander, S., and Singer, W. (1996). Long-range synchronization of oscillatory light responses in the cat retina and lateral geniculate nucleus. *Nature* **379,** 728–733.

Niebur, E., and Koch, C. (1994). A model for the neuronal implementation of selective visual attention based on temporal correlation among neurons. *J. Comput. Neurosci.* **1,** 141–158.

Palm, G. (1990). Cell assemblies as a guideline for brain research. *Concepts Neurosci.* **1,** 133–147.

Payne, B. R., Lomber, S. G., Schmidt, K. E., and Galuske, R. A. W. (2000). Positional and laminar origins of feedback influences over primary visual cortex. *Soc. Neurosci. Abstr.* **26,** 53.6.

Phillips, W. A., and Singer, W. (1997). In search of common foundations for cortical computation. *Behav. Brain Sci.* **20,** 657–722.

Prechtl, J. C., Bullock, T. H., and Kleinfeld, D. (2000). Direct evidence for local oscillatory current sources and intracortical phase gradients in turtle visual cortex. *Proc. Natl. Acad. Sci. U.S.A.* **97,** 877–882.

Rager, G., and Singer, W. (1998). The response of cat visual cortex to flicker stimuli of variable frequency. *Eur. J. Neurosci.* **10,** 1856–1877.

Reid, R. C., and Alonso, J. M., (1995). Specificity of monosynaptic connections from thalamus to visual cortex. *Nature* **378,** 281–284.

Reynolds, J. H., and Desimone, R. (1999). The role of neural mechanisms of attention in solving the binding problem. *Neuron* **24,** 19–29.

Ribary, U., Joannides, A. A., Singh, K. D., Hasson, R., Bolton, J. P. R., Lado, F., Mogilner, A., and Llinas, R. (1991). Magnetic field tomography of coherent thalamocortical 40 Hz oscillations in humans. *Proc. Natl. Acad. Sci. USA* **88,** 11037–11041.

Rodriguez, E., George, N., Lachaux, J.-P., Martinerie, J., Renault, B., and Varela, F. J. (1999). Perception's shadow: long-distance gamma band synchronization of human brain activity. *Nature* **397,** 430–433.

Rodriguez, R., Kallenbach, U., Singer, W., and Munk, M. H. J. (2001). Enhancement of gamma-oscillatory responses after cholinergic stimulation in cat visual cortex. Proceedings of the 4th Meeting of the German Neuroscience Society. 28th Göttingen Neurobiology Conference, Vol. II:574.

Roelfsema, P. R., König, P., Engel, A. K., Sireteanu, R., and Singer, W. (1994). Reduced synchronization in the visual cortex of cats with strabismic amblyopia. *Eur. J. Neurosci.* **6,** 1645–1655.

Roelfsema, P. R., Engel, A. K., König, P., and Singer, W. (1997). Visuomotor integration is associated with zero time-lag synchronization among cortical areas. *Nature* **385,** 157–161.

Roelfsema, P. R., Lamme, V. A. F., and Spekreijse, H. (1998). Object-based attention in the primary visual cortex of the macaque monkey. *Nature* **395,** 376–381.

Rolls, D. T., and Tovee, M. J. (1994). Processing speed in the cerebral cortex and the neurophysiology of visual masking. *Proc. R. Soc. Lond. B* **257,** 9–15.

Salin, P. A., and Bullier, J. (1995). Corticocortical connections in the visual system: structure and function. *Physiol. Rev.* **75,** 107–154.

Schmidt, K. E., Goebel, R., Löwel, S., and Singer, W. (1997). The perceptual grouping criterion of colinearity is reflected by anisotropies of connections in the primary visual cortex. *Eur. J. Neurosci.* **9,** 1083–1089.

Schröder, J.-H., Fries, P., Roelfsema, P. R., Singer, W., and Engel, A. K. (1998). Correlates of strabismic amblyopia in cat extrastriate visual areas. *Eur. J. Neurosci. Abstr.* **10,** 237.

Sejnowski, T. R. (1986). Open questions about computation in cerebral cortex. In: *Parallel distributed processing,* Vol. 2. (J. L. McClelland, and D. E. Rumelhart, Eds.), pp. 372–389. Cambridge, MIT Press.

Shadlen, M. N., and Movshon, J. A. (1999). Synchrony unbound: a critical evaluation of the temporal binding hypothesis. *Neuron* **24,** 67–77.

Sheinberg, D. L., and Logothetis, N. K. (1997). The role of temporal cortical areas in perceptual organization. *Proc. Natl. Acad. Sci. U.S.A.* **94,** 3408–3413.

Sillito, A. M., Grieve, K. L., Jones, H. L., Cuderio, J., and Davis, J. (1995). Visual cortical mechanisms detecting focal orientation discontinuities. *Nature* **378,** 492–496.

Singer, W. (1993). Synchronization of cortical activity and its putative role in information processing and learning. *Annu. Rev. Physiol.* **55,** 349–374.

Singer, W. (1995). Development and plasticity of cortical processing architectures. *Science* **270,** 758–764.

Singer, W. (1999). Neuronal synchrony: a versatile code for the definition of relations? *Neuron* **24,** 49–65.

Singer, W., Engel, A. K., Kreiter, A. K., Munk, M. H. J., Neuenschwander, S., and Roelfsema, P. R. (1997). Neuronal assemblies: necessity, signature and detectability. *Trends Cogn. Sci.* **1,** 252–261.

Steriade, M. (1999). Coherent oscillations and short-term plasticity in corticothalamic networks. *Trends Neurosci.* **22,** 337–345.

Steriade, M., Contreras, D., Amzica, F., and Timofeev, I. (1996). Synchronization of fast (30–40Hz) spontaneous oscillations in intrathalamic and thalamocortical networks. *J. Neurosci.* **16,** 2788–2808.

Stevens, C. F., and Zador, A. M. (1998). Input synchrony and the irregular firing of cortical neurons. *Nat. Neurosci.* **1,** 210–217.

Stiefel, K. M., Wespatat, V., Singer, W., and Tennigkeit, F. (2001). A computational model of the interaction of membrane potential oscillations with inhibitory synaptic input in cortical cells. *Neurocomputing,* in press.

Stoner, G. R., Albright, T. D., and Ramachandran, V. S. (1990). Transparency and coherence in human motion perception. *Nature* **344,** 153–155.

Tallon-Baudry, C., Bertrand, O., Delpuech, C., and Pernier, J. (1996). Stimulus specificity of phase-locked and non-phase-locked 40 Hz visual responses in human. *J. Neurosci.* **16,** 4240–4249.

Tallon-Baudry, C., Bertrand, O., Delpuech, C., and Pernier, J. (1997). Oscillatory γ-band (30–70-Hz) activity induced by a visual search task in humans. *J. Neurosci.* **17,** 722–734.

Tallon-Baudry, C., Bertrand, O., Peronnet, F., and Pernier, J. (1998). Induced γ-band activity during the delay of a visual short-term memory task in humans. *J. Neurosci.* **18,** 4244–4254.

Tallon-Baudry, C., Kreiter, A. K., and Bertrand, O. (1999). Sustained and transient oscillatory responses in the gamma and beta bands in a visual short-term memory task in humans. *Vis. Neurosci.* **16,** 449–459.

Tamás, G., Buhl, E. H., Lörincz, and Somogyi, P. (2000). Proximally targeted GABAergic syapses and gap junctions synchronize cortical interneurons. *Nat. Neurosci.* **3,** 366–371.

Tanaka, K., Saito, H., Fukada, Y., and Moriya, M. (1991). Coding visual images of objects in the inferotemporal cortex of the macaque monkey. *J. Neurophysiol.* **66,** 170–189.

Thorpe, S., Fize, D., and Marlot, C. (1996). Speed of processing in the human visual system. *Nature* **38,** 520–522.

Tiitinen, H., Sinkkonen, J., Reinikainen, K., Alho, K., Lavikainen, J., and Naatanen, R. (1993). Selective attention enhances the auditory 40-Hz transient response in humans. *Nature* **364,** 59–60.

Tononi, G., and Edelman, G. M. (1998). Consciousness and complexity. *Science* **282,** 1846–1851.

Tononi, G., Srinivasan, R., Russell, D. P., and Edelman, G. M. (1998). Investigating neural correlates of conscious perception by frequency-tagged neuromagnetic responses. *Proc. Natl. Acad. Sci. U.S.A.* **95,** 3198–3203.

Treisman, A. (1996). The binding problem. *Curr. Opin. Neurobiol.* **6,** 171–178.

Ts'o, D. Y., and Gilbert, C. D. (1988). The organization of chromatic and spatial interactions in the primate striate cortex. *J. Neurosci.* **8,** 1712–1727.

Usher, M., and Donnelly, N. (1998). Visual synchrony affects binding and segmentation in perception. *Nature* **394,** 179–182.

Usrey, W. M., and Reid, R. C. (1999). Synchronous activity in the visual system. *Annu. Rev. Physiol.* **61,** 435–456.

Volgushev, M., Chistiakova, M., and Singer, W. (1998). Modification of discharge patterns of neocortical neurons by induced oscillations of the membrane potential. *Neuroscience* **83,** 15–25.

von der Malsburg, C. (1999). The what and why of binding: the modeler's perspective. *Neuron* **24,** 95–104.

Wehr, M., and Laurent, G. (1996). Odour encoding by temporal sequences of firing in oscillating neural assemblies. *Nature* **384,** 162–166.

Wespatat, V., Tennigkeit, F., and Singer, W. (1999). Oscillations and long-term synaptic plasticity in rat visual cortex. In: *From Molecular Neurobiology to Clinical Neuroscience.* Proc. 1st Göttingen Conf. of the German Neurosci. Soc.; Vol. 1. 27th Göttingen Neurobiol. Conf., p. 102. (N. Elsner, and U. Eysel Eds.), Stuttgart, Thieme-Verlag.

Whittington, M. A., Traub, R. D., and Jefferys, J. G. R. (1995). Synchronized oscillations in interneuron networks driven by metabotropic glutamate receptor activation. *Nature* **373,** 612–615.

Wörgötter, F., Niebur, E., and Koch, C. (1991). Isotropic connections generate functional asymmetrical behavior in visual cortical cells. *J. Neurophysiol.* **66,** 444–459.

Wolfe, J. M., and Cave, K. R. (1999). The psychophysical evidence for a binding problem in human vision. *Neuron* **24,** 11–17.

Zhang, L. I., Tao, H. W., Holt, C. E., Harris, W. A., and Poo, M. (1998). A critical window for cooperation and competition among developing retinotectal synapses. *Nature* **395,** 37–44.

14

THE SPECIAL RELATIONSHIP BETWEEN β RETINAL GANGLION CELLS AND CAT PRIMARY VISUAL CORTEX

BERTRAM R. PAYNE AND
R. JARRETT RUSHMORE

Department of Anatomy and Neurobiology, Boston University School of Medicine, Boston, Massachusetts

INTRODUCTION

Primary visual cortex is a structure central to processing of visual signals. These signals are derived from a host of retinal ganglion cell types that have come to be recognized in the last 40 years. Two ganglion cell types that were discovered first and that appear to be the most prominent are the X and Y types. The X and Y cell types were designated on the basis of their physiological characteristics, and correspond to the morphologically identified β and α retinal ganglion cells, respectively. Both types supercede the ON and OFF functional dichotomy that was described by Kuffler (1953). These two ganglion cell types have different biologies that are expressed in their structure and connections. Consequently, the two ganglion cell types contribute to visual processing in fundamentally dissimilar ways. Similar conclusions can be reached for the remaining ganglion cell types, although it should be noted that their characteristics are much less well understood; these other cell types are frequently grouped together under the single rubric of γ (W) cells. One critical difference makes β cells stand out from the remaining cell types. Immature β cells have a dependence for survival on the integrity of primary visual cortex and its intermediate relay structure, the lateral geniculate nucleus (LGN). This dependence and its basis form the topic of this chapter.

According to Cowan (1972), the identification of neuronal dependencies in brain systems can be traced back to the work of Gudden and others in the latter part of the 19th century (Gudden, 1869, 1880, 1884; Ganser, 1882; Von Monakow, 1882, 1889), and the early part of the 20th century (Münzer and Wiener, 1902; Nissl, 1913; Winkler, 1918). Much of this early work was neglected or disputed at the time because of the threat it posed to the emerging neuron doctrine (Polyak, 1957; Shepherd, 1998), and it was not until later, with the acceptance of the doctrine, that the idea started to be acknowledged that some neurons had dependencies on second-order structures. For example, Walsh (1947) and Haddock and Berlin (1951) recognized a late optic nerve atrophy in patients who had sustained damage of visual cortex. This atrophy was interpreted as secondary degeneration of retinal ganglion cells that ensues from the primary retrograde degeneration of neurons in the LGN. Such studies reveal the trans-synaptic, or transneuronal, dependency of certain ganglion cells on primary visual cortex. This conclusion has been verified by a small number of studies on cats and monkeys in the succeeding half century (Tong et al., 1982; Payne et al., 1984b; Cowey and Stoerig, 1989; Weller and Kaas, 1989; Cowey et al., 1999; Johnson and Cowey, 2000).

After the concept of secondary, or transneuronal, retrograde degeneration of retinal ganglion cells consequent to lesions of primary visual cortex was established, it became possible to reinterpret older studies and recognize that Ganser (1882) and von Monakow (1889) were the first to describe the phenomenon, and to do so in both cats and dogs. Both investigators described a thinning of the optic nerve fiber layer of the retina in cats (Ganser) and dogs (von Monakow) after extensive lesions of cerebral cortex in young kittens and puppies. However, the topic remained largely dormant for more than 75 years, except for minor mention by noted investigators such as Klüver (1937) who detected degeneration of retina after lesions of the optic radiations or visual cortex. The first systematic study of the phenomenon was made by Van Buren (Van Buren, 1963a, 1963b) who described the appearance of the ganglion cell and inner nuclear layers of retina in monkey and human material following lesions of the optic nerve and visual cortex. In one monkey, surgical section of optic fibers at the chiasm triggered the death of large numbers of ganglion cells in nasal retina when measured 20 months later (Van Buren, 1963a, 1963b; Gills and Wadsworth, 1967). There was no evidence of change in the inner plexiform layer, although degeneration in the inner nuclear layer was evident. The outer plexiform and outer nuclear layers appeared to be unaffected by the lesion. In a similar way, examination of retinae 48 months after occipital lobectomy revealed a distinct loss of ganglion cells from the homonymous halves of the two retinae with a concomitant 50% reduction in cross-sectional area of the optic nerves. Neither degeneration nor shrinkage of the other retinal layers was detected. Congruent observations have since been made by Cowey (1974) and Weller and Kaas (1989) for monkeys, with additional observations that young retinal ganglion cells are more vulnerable than mature ganglion cells (Weller and Kaas, 1989), and that lesions that include more occipital cortex or delve deeper into the optic radiation induce greater ganglion cell degeneration than smaller or more superficial lesions (Cowey et al., 1999). As noted by Cowan (1972), "...the visual system represents an ideal site in which to investigate the cellular changes which occur during retrograde degeneration..." However, only limited work has been carried out on the effects of age and duration of the post-lesion period.

Some of the most revealing work has been carried out on cats. After ablation of primary visual cortical areas 17 and 18 there is a selective degeneration of β (X) retinal ganglion cells (Payne et al., 1984b). This dependency of β (X) cells is a characteristic of the immature brain, because they die in large numbers. Mature β (X) cells are immune to the cortical lesion, even with long post-lesion periods. Consequently, they survive the lesion of visual cortex and the considerable retrograde degeneration in the LGN. This Chapter summarizes the characteristics of cat ganglion cells that are vulnerable and those that appear to be largely insensitive to the cortical lesion. This is a worthwhile activity because elimination of β (X) cells has a potentially profound impact on visual processing in addition to the absence of primary visual cortex. The three most significant factors linked to the survival, and death, of cat retinal ganglion cells following lesions of primary visual cortex are (1) maturational status, (2) patterns of connectivity, and (3) rate of degeneration of LGN neurons. The attainment of these conclusions had a significant prerequisite, the elucidation of the anatomy and function of the multiple streams emanating from retina and destined for numerous brain targets, including the dorsal LGN and primary visual cortex. We review the relevant features of retinal ganglion structure, function, contributions to vision, and connectivity before describing the impact of primary visual cortex lesions on ganglion cells and visually guided behavior.

IIDENTIFICATION OF SUBSYSTEMS AND CONNECTIONS

OPTIC NERVE FIBERS

The existence of more than one class of retinal ganglion cell was first hinted in analyses of conduction velocity studies on cat optic nerve. Bishop and O'Leary (1938) noted that stimulation of the optic nerve with brief electric pulses evoked complex compound action potentials in the optic tract. These potentials consisted of two sharp peaks that were differentially sensitive to the intensity of the stimulating current and had different latencies. Simple arithmetic, based on the distance between the stimulating and recording electrodes and the latency of the responses, indicated conduction velocities in the region of 60 m/sec and 25 m/sec. These observations were subsequently confirmed and extended by Bishop et al. (1953) to suggest an additional broad-spectrum conduction group with transmission velocities well below 25 m/sec and exhibiting a wide peak. Based on theories of the relationship of fiber diameter and conduction velocity (Rushton, 1951), it was speculated that there were three major populations of axons in cat optic nerve: a large-diameter group that transmitted action potentials quickly and intermediate and smaller diameter groups with correspondingly slower velocities.

X and Y Systems

Single cell recording and the desire to measure contrast sensitivity ushered in the X and Y dichotomy of visual system components. Enroth-Cugell and Robson (1966) used a psychophysical approach that was normally applied to investigations of human vision and applied it to analyses of individual cat retinal ganglion cells. They used an oscilloscope-based visual display of light and dark bars. In one dimension the light intensity is constant, whereas in the orthogonal direction light intensities varies sinusoidally. The amplitude of the sinusoid could be varied to increase or decrease the contrast between regions. Moreover, the width of the bars could be varied to increase or decrease the total number of bars overlying the receptive field, and hence change the spatial frequency of the stimulus. The grating pattern could also be moved at a predetermined speed, thus allowing control over the temporal frequency of the stimulus.

Using these stimuli, Enroth-Cugell and Robson (1966) recorded from a minority of cells that were highly predictable in their responses to brightening and darkening of the receptive field. For these so-called X cells, influences within the receptive field determined the activity of the cell in a linear fashion. To determine linearity, the response of the cell to a particular grating stimulus is measured; this grating is introduced to the receptive field of the cell such that half the receptive field is brightened and the other half is dimmed relative to a uniform background. The cell does not respond to the presence of such a stimulus, even if the grating was introduced, modulated, or removed, because each aspect of the stimulus is integrated by the cell in a linear fashion. The stimulus, therefore, is said to be in a null position.

For the majority of cells recorded, termed Y cells, a null position of the grating stimulus could not be found. These cells sum influences within their receptive field in a nonlinear fashion. The cells always respond to onset and cessation of the stimulus, and they respond to presentation and removal of gratings at twice the rate of stimulation. These seminal results established the existence of parallel functional subsystems within the visual system, and these subsystems superceded the ON and OFF dichotomy already described by Kuffler (1953).

Subsequent studies extended the X/Y dichotomy to include time course and response to stimuli placed within the center of the receptive field (Cleland et al., 1971). One population of cells exhibited a substantial but transient increase in firing rate in response to the change in illumination and standing contrast. By comparison, the second population of cells gave an initial transient response that was followed by a sustained, elevated level of discharge (Fig. 14.1). The cells exhibiting transient discharges were equated with Y-responses, and the sustained discharges were equated with X-responses. However, these distinctions are not always readily evident because the level of adaptation contributes to the shaping of the overall response (Jakiela and Enroth-Cugell, 1976). Even so, Fukada (1971) provided evidence that the transient Y-cells transmitted action potentials distinctly more quickly than the sustained X-cells, and Cleland et al. (1971) provided evidence that action potential transmission of sustained cells took about twice as long as action potential transmission by transient cells. These studies were carried out in an era of intense investigative activity that was initially dominated by the Australian groups led by Stone in Sydney and Cleland and Levick in Canberra, although there were numerous significant contributions by groups led by Enroth-Cugell and Shapley in the United States.

Two major advances are highly relevant to this chapter. The first is the recognition that at all retinal locations X-cells, compared with Y-cells, respond to

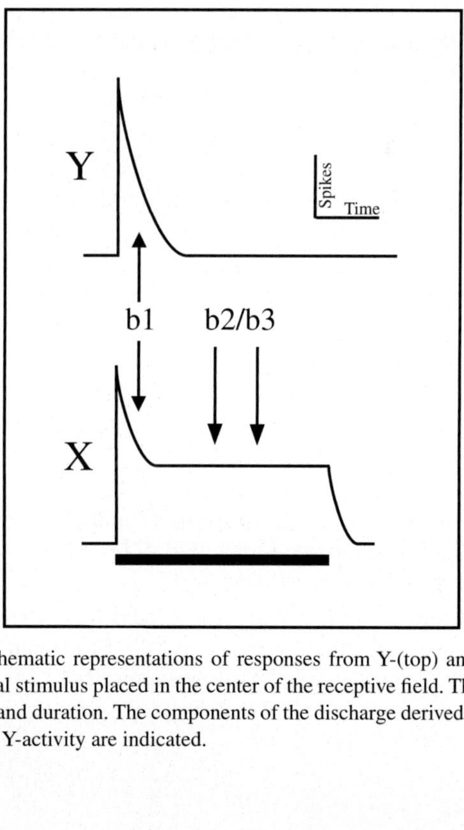

FIGURE 14-1. Schematic representations of responses from Y-(top) and X-(bottom) retinal ganglion cells to an optimal stimulus placed in the center of the receptive field. The horizontal bar represents stimulus presence and duration. The components of the discharge derived from b1, b2, b3 cone bipolar cells to the X- and Y-activity are indicated.

higher spatial frequencies at high contrasts and high levels of illumination (Cleland et al., 1979; Stone, 1983). The second is that X-cells are highly sensitive to low contrasts at mid-range spatial frequencies. Both of these sensitivities are at the expense of temporal sensitivity, which is low. X-cells are able to report the position of small stimuli with great accuracy and minute movements of small stimuli with even greater accuracy (Shapley and Victor, 1986). Moreover, they have a relatively high density in retina (Wässle and Boycott, 1991; Stein et al., 1996), and they are implicated in local aspects of vision and visual acuity, features in accord with their small receptive field size and high density (Peichl and Wässle, 1979).

Y-cells have larger receptive fields, and they respond uniquely to low spatial frequencies, exhibit higher contrast sensitivity at those lower frequencies (Troy, 1987), and possess high temporal sensitivity to rapid transients in illumination and to high stimulus velocities. Y-cells are also sensitive to high spatial frequencies by virtue of nonlinear subunits within the receptive field that result in activation without spatial resolution (Hochstein and Shapley, 1976; Shapley and Victor, 1979). However, their large receptive fields and sparse distribution (see page 570) in the retina report the position of such stimuli only coarsely. They are implicated in coarse, more global aspects of spatial vision, in the detection of rapid visual transients in illumination and movement, and they are excellent at stimulus detection.

Both X- and Y-cells give brisk responses and, under high-contrast conditions, Y-cells generate the dominant signals until the limit of α-cell acuity is reached. Thereafter, X-cell activity dominates (Shapley and Victor, 1978; 1979; 1980; 1981; Shapley and Perry, 1986; Troy, 1987; Freeman, 1991; Wässle and Boycott, 1991). For low-contrast stimuli, both X- and Y-cells detect and signal contrast to the brain (Troy and Enroth-Cugell, 1993), and both the X and Y cells are highly sensitive in their favored spatial and temporal frequency domains (Troy and Enroth-Cugell, 1993). However, the greater numerical superiority of the X-system (see later) affords it a greater overall sensitivity even at low temporal and spatial frequencies because of the convergence of signals on target neurons (Troy, 1983). Based on these multiple criteria, it became clear that the visual system contained two functional subsystems that each transmits signals about different aspects of the visual scene.

W System

During the initial years of the work on X- and Y-systems, other cells were identified in electrophysiological studies that did not fit neatly into the X- and Y-dichotomy. They had diverse properties. Relative to the X- and Y-responses, they responded rather sluggishly to visual stimuli, and they transmitted action potentials at a slower pace (Cleland and Levick, 1947b; Stone and Fukuda, 1974). They had either ON, OFF, or mixed ON-OFF receptive field centers, and gave either sustained or transient responses (Table 14-1).

TABLE 14-1 Major Characteristics of Retinal Ganglion Cells

	Alpha (α)	Beta (β)	Gamma (γ): Delta (δ)	Gamma (γ): Epsilon (ε)	Gamma (γ): Eta (η)	Gamma (γ): Zeta (ζ)	Gamma (γ): Theta (θ)
Soma							
Form:	Angular	Round	Oval-Angular	Oval-Angular	Round	Round	Ovoid-Round
Size:							
Center:	300–400 μm^2	100–250 μm^2	100–225 μm^2	100–225 μm^2	75–220 μm^2	100–200 μm^2	75–200 μm^2
Periphery:	>600 μm^2	300–500 μm^2	150–300 μm^2	300–450 μm^2	75–225 μm^2	100–225 μm^2	100–250 μm^2
Dendritic Field Area	Large	Small	Medium	Large	Small/Medium	Small/Medium	Medium
Center	0.01 mm^2	.001–.003 mm^2	0.02–.07 mm^2	0.03–0.12 mm^2	0.01–0.07 mm^2	0.01–0.03 mm^2	0.003–0.017 mm^2
Periphery	>0.4 mm^2	0.004–.01 mm^2	0.12–0.3 mm^2	0.12–0.6 mm^2	0.07–0.2 mm^2	0.01–0.2 mm^2	0.015–0.035 mm^2
Sholl ('56) Analysis on Dendrites*							
Total Intersections†	High	Low	Medium-High	Low	Medium	Low	High
Form	Broad, complex	Compact, complex	Broad, simple	Broad, sparse	Medium, complex	Low, asymmetric	Compact, Highly overlapping
IPL Stratum §	Inner or outer (Paramorphic)	Inner or outer (Paramorphic)	Inner	Outer	Inner	Border of inner and outer	Bistratified–inner and outer
Functional Type and Vigor	Y, on, off; Transient; Brisk	X, on, off; Sustained; Brisk	W, Off; Sustained; Sluggish	W, On Sustained Sluggish	W, Off Transient Sluggish	W, On-off; Transient Sluggish	W, On-off; Transient Sluggish
Frequency							
All Targets	5%	57%			38%		
LGN Target	6%	71%			23%‡		

Many features of ganglion cells in peripheral retina are larger than those in central retina. Even so, the ordinal size of the characteristics for the different ganglion cells remain essentially constant at any given location. The exceptions are small γ cells (η, ζ, θ); their cell bodies do not increase appreciably in size with increasing eccentricity. Central retina (< 2 mm from visual axis). P = Peripheral retina (> 2 mm from visual axis). Data from: Berson et al. (1998, 1999), Boycott and Wässle (1974, 1999), Dacey (1989); Dann et al. (1988); Fukuda et al. (1984), Isayama et al. (2000); Kolb et al. (1981); Leventhal et al. (1980, 1985); Pu et al. (1994); Ramoa et al. (1988); Saito (1983), Stanford (1987a,b), Stein et al. (1996); Stone (1983), Stone and Clarke (1980), Stone et al. (1980), Wässle and Boycott (1991) amongst others. IPL sublaminae inner = *a*, outer = *b*. Frequency represents estimate across retina, and exact fractions may vary by location.

* In some instances actual measurements are not known and data need to be collected for intact cats.

† = sum of intersections of all rings.

‡ Approximately 50% of γ ganglion cells do not project to LGN. Exclusion of these cells modifies the fraction of cells labeled by retrograde tracers injected into LGN.

§ Paramorphic = Identical *en face* dendritic arborizations that differ in level of stratification in IPL.

COVARIATION OF MORPHOLOGY, PHYSIOLOGY, AND INPUT

X (β) and Y (α) Systems

The physiological distinctiveness of X and Y cell responses was correlated with distinct morphologies of retinal ganglion cells. Boycott and Wässle (1974), using the Golgi impregnation technique on retinal flat mounts, recognized that at any given retinal location there are two obvious types of ganglion cells. The α-ganglion cells are prominent because they possess dendrites, somata, and axons that have greater diameters than the equivalent structures of the numerically more numerous β cells (Fig. 14-2). These variables are summarized for central and peripheral retina by ganglion cell type in Table 14-1. Moreover, α dendrites are radiate and arborize over a much larger area of retina than the bushy β dendrites. These differences in the dendrites of α and β cells are readily appreciable and can be demonstrated in a quantitative way using a Sholl analysis. Sholl (1956) developed a scheme to quantify size and complexity of dendritic fields of neurons in cerebral cortex by counting the number of times dendrites crossed concentric rings, centered on the cell body, and placed at increasingly greater distances from it. The increments for sequential rings were usually fixed. Examples of Sholl analyses are presented in Fig. 14-3 for one α cell and one β cell from the same

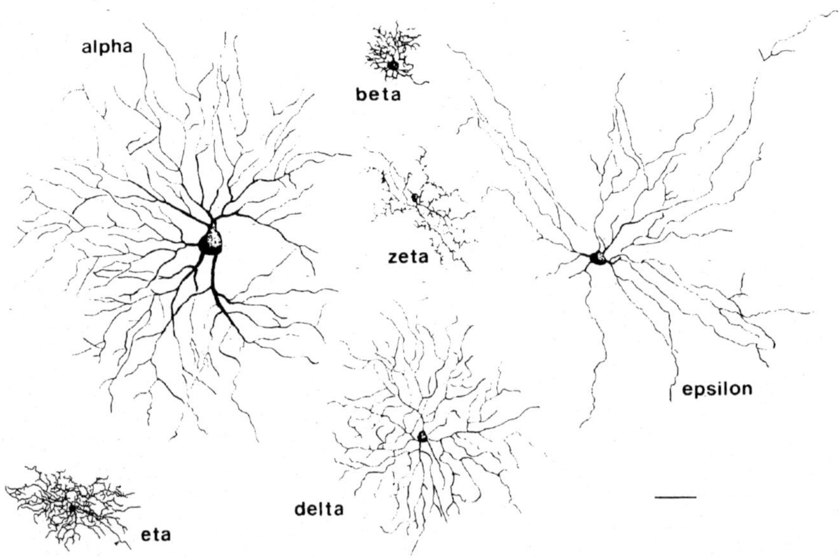

FIGURE 14-2. Camera lucida drawings of the major ganglion cell types. The δ (delta), ε (epsilon), η (eta), and ζ (zeta) cells compose the γ category. The scale bar is 100 μm. (Figure reproduced from Berson et al. [Berson et al., 1998; Berson et al., 1999], Journal of Comparative Neurology, Copyright © 1998, 1991. Reprinted by permission of Wily-Liss Inc., a subsidiary of John Wiley & Sons, Inc.)

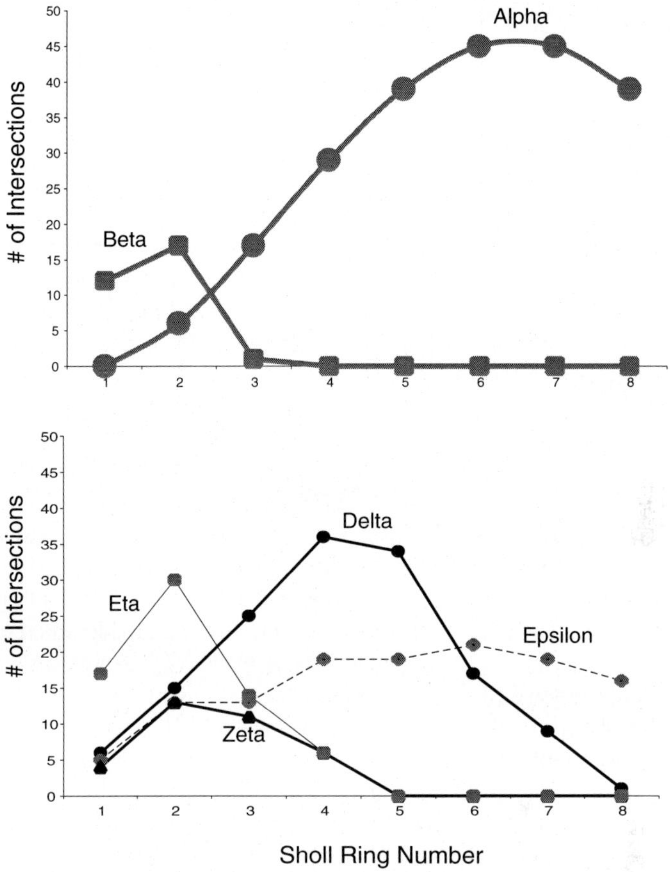

FIGURE 14-3. Sholl analysis reveals class specific differences in retinal ganglion cell dendritic fields. **(Upper)** α- and β-ganglion cells are readily distinguishable from each other. **(Lower)** The Sholl analysis of four γ-type retinal ganglion cells also reveals unique profiles for each class. α and β cells are also readily distinguishable from all γ cell types.

retinal location, and they reveal the greater size and greater overall branching of α cells compared to β cells.

Based on comparisons of the diameters of the dendritic trees and receptive field center diameters, Boycott and Wässle (1974) speculated that the α cells are the structural counterpart of functionally identified Y, or transient, units, and β cells are the structural counterpart of functionally identified X, or sustained, units (Boycott and Wässle, 1974; Levick, 1975; Peichl and Wässle, 1979). This speculation was confirmed by studies that combined electrophysiological analyses of neuron responses to characterized stimuli with subsequent injection of markers into the cell that permitted direct visualization of morphology (Saito, 1983; Stanford and Sherman, 1984; Stanford, 1987b).

An important insight of Boycott and Wässle (1974) was the recognition that structure of a given ganglion cell type is not isotropic across the retinal surface. For all α and β cells, dendritic, soma, and axon sizes are smaller in central retina than in the periphery and, indeed, many features of peripheral β cells are larger than the equivalent features of central α cells. Even so, at any given retinal location, all features of α cells are larger than the equivalent features of β cells, and the astuteness of Boycott and Wässle's observations lay in the recognition of gradients in morphological structure. It was soon recognized that there were equivalent gradients in physiological responsiveness (Cleland and Levick, 1974a; Levick, 1975; Peichl and Wässle, 1979; Stone, 1983).

W (γ) System

Although a neat correspondence between the X/Y and β/α cells was established, the same could not be written for the multicomponent W (γ) class of cells. Detailed analyses eluded good correspondences between the physiological responses and the anatomy until the recent work of Berson and colleagues. They extended the classification of retinal ganglion cells and fractionated the γ class into its component types. The initial studies of Boycott and Wässle (1974) already identified one class of cell, which they termed δ (delta), that did not fit the α/β dichotomy, and their observations were followed by Leventhal (1979), who identified ε (epsilon) cells. Berson's group (Berson et al., 1998, 1999; Isayama et al., 2000) added the η (eta), ζ (zeta) and θ (theta) cells that had been recalcitrant to identification for so many years. Examples of δ, ε, η, and ζ cells are shown in Fig. 14-2; and their morphological features are summarized in Table 14-1. They are all clearly different from α and β cells (Fig. 14-3). Berson correlated and more recently bridged the morphological data across to the physiological properties and receptive field characteristics for the cells that thus far compromise the γ-group (Berson et al., 1998, 1999; O'Brien et al., 1999; Isayama et al., 2000). As shown in Table 14-1, each of the γ subgroups has distinctive receptive field properties, but all can be collected into a category that is characterized by relatively sluggish responses.

NUMBERS

After some initial discrepancies based on sample counts of retinal ganglion cells and optic nerve fibers that ranged from about 120,000 (Stone, 1978; Stone and Campion, 1978) to about 200,000 or more (Hughes, 1975; Hughes and Wässle, 1976), it is now generally recognized that the actual total of ganglion cells in cat retina is about 160,000 per eye. These estimates were reached from counts made in retinal flat mounts (Hughes, 1981; Illing and Wässle, 1981) and by counts of axons in the optic nerve (Williams et al., 1986). Subsequent estimates that rely on the morphological features described previously indicate that about 7,000 to 8,000 are α-cells, or about 5% of the total population (Wässle et al., 1975, 1981a, 1981b, 1981c). In contrast, there are about 90,000 β-cells, and they account for 55% of the total ganglion cell population (Wässle et al., 1975, 1981a, 1981b, 1981c; Illing and Wässle, 1981; Williams et al., 1986; Stein et al., 1996). The remaining 65,000, or 40%, of ganglion cells fall into the multiple γ-subtypes (δ, ε, η, and ζ; [Boycott and Wässle, 1974; Berson et al., 1998, 1999]) that are each only poorly represented in numerical terms.

Even though it was thought that these fractions for α, β, and γ ganglion cells are relatively constant across retina (Illing and Wässle, 1981), new evidence from Berson's group suggests that there is some variability from one retinal location to another. Berson's group note that β cells compose up to 67% of all ganglion cells in central and temporal retina and about half the population in peripheral nasal retina (Stein and Berson, 1995; Stein et al., 1996).

SAMPLING OF TYPES

The anatomical evidence shows that contrary to the earliest electrophysiological studies on the X/Y dichotomy, Y cells are not the dominant ganglion cell type. The anatomical studies cited previously reported that at all retinal locations β-ganglion cells outnumber α-ganglion cells about 10:1. The difference between the anatomical and electrophysiological data reflects the sampling biases associated with use of electrodes to assay neural activity and infer frequency of occurrence. The discrepancy arises because both larger cells and larger axons generate action currents that spread through a greater volume of tissue than do smaller cells and axons. Thus, the probability of positioning an electrode tip within the detectable action-current field of a large cell or axon is much greater than the probability of positioning an electrode tip within the detectable action-current field of a smaller cell or axon (Humphrey, 1979). Thus, the detectability of larger cells or axons is increased disproportionally over detectability of smaller cells and axons—hence the discrepancy between the relative proportions of X/Y and β/α cells. Identical factors also account for the very low encounter rate of cells with sluggish response properties that compose the W-class. The impact of this factor is reduced by the use of finer, higher impedance electrodes, but it is not abolished (Levick and Cleland, 1974; Stone, 1974), and Levick (1975) warned against the uncritical acceptance of recording percentages of different classes of neurons as direct indicators of numerical preponderance.

CONNECTIONS

Afferents

Regardless of retinal location, the differences in the structure and function of α and β cells are paralleled by differences in the composition of the presynaptic inputs (Cohen et al., 1994). β-cell input is derived primarily from glutamatergic cone bipolar cells, whereas α-cell input is derived primarily from amacrine cells (Kolb, 1979; McGuire et al., 1986; Freed and Sterling, 1988; Kolb and Nelson,

1993), although some distinct departures from this general view have been noted (Weber et al., 1991). Moreover, b1 cone bipolar cells transmit the transient component of the responses to both α and β cells, whereas the sustained component of the β-cell response is a product of inputs from b2 and b3 cone bipolar cells (Fig. 14-1) (Freed and Sterling, 1988; Freed, 2000). As far as we know, no analyses have been made of the afferent connectivity of any of the cell types that compose the γ group.

Targets in Brain

Retinal ganglion cells have differential projections to target structures in brain, with the overwhelmingly dominant projections being to the dorsal LGN (dLGN), pretectum, and superior colliculus. Projections to the nuclei comprising the accessory optic system are extremely small by comparison (Farmer and Rodieck, 1982) and are not considered further, as is the case for cells that project to the ventral LGN (vLGN).

Illing and Wässle (1981) using retrograde labeling methods determined that all α- and β-retinal ganglion cells project to layers A and A1 of the dLGN, and all α and 43% of γ cells project to the C layers and the medial interlaminar nucleus (MIN) component of dLGN. In addition, all α and all γ cells project to the superior colliculus. About 10% of β cells project to the pretectum. Although subsequent studies using similar approaches have described slightly different patterns of innervation and fractional contributions of γ innervation of LGN and superior colliculus (Leventhal et al., 1985; Stein and Berson, 1995; Stein et al., 1996), the variability is largely accounted for by variations in the proportions of the three major classes of retinal ganglion cells in different regions of the retina, and by the inclusion of cell types that project specifically to the dLGN wing and to vLGN in the numerical estimates. Even so, all investigators concur that α cells have highly collateralized projections to all major visual structures; the vast majority of β cells project virtually exclusively to layers A and A1 of LGN with only a minor number possessing axon collaterals that reach the pretectum and possibly the superior colliculus (Sawai et al., 1985; Sur et al., 1987; Hada and Hayashi, 1990); and all members of the composite γ group project to the midbrain with about half of them also projecting axons to dLGN. Intra-axonal injection of opaque markers following response characterization (Sur and Sherman, 1982; Bowling and Michael, 1984; Sur et al., 1987; Tamamaki et al., 1995), as well as electrophysiological analyses, concur with the values presented for α- and β-cells (Cleland and Levick, 1974a). Moreover, the intra-axonal studies show that the termination fields of the X axons in LGN are much smaller (150 μm in mediolateral extent) than the termination fields of Y-axons (300 μm). Accordingly, X-terminations bear many fewer (500 to 600) boutons than Y-terminations (1000), even within a single A or A1 lamina. The cell bodies and axons of γ cells are too fine to permit comparable analyses to be carried out on them. These patterns of innervation are summarized in Fig. 14-4.

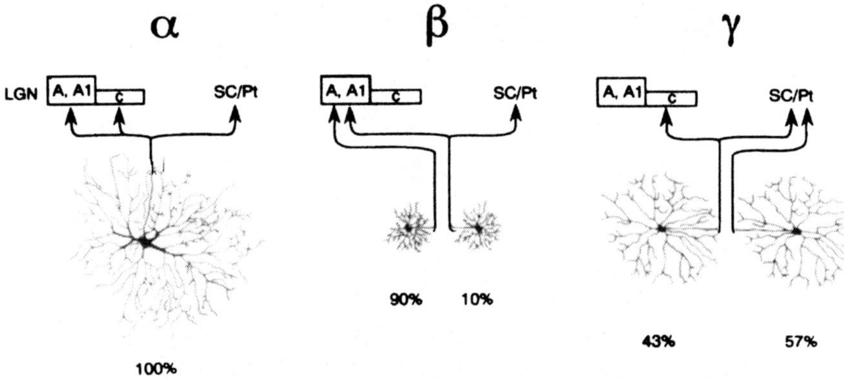

FIGURE 14-4. Patterns of projections of α, β, and γ cells to layers A and A1 and to the C-complex of LGN and to Superior Colliculus (SC) / Pretectum (Pt). (Figure taken from Payne and Cornwell [1984] and reproduced with permission from Elsevier.)

Relay to Primary Visual Cortex

Anatomy

The projections from LGN to cortex are also highly specific and were eloquently summarized by Holländer and Vanegas (1977) following their analyses using retrograde tracers. Their view, with updates, is schematized in Fig. 1-8 and are only summarized here (Fig. 14-5). Within the laminated portion of LGN, most of the neurons that project to area 17 are located in the A-laminae with additional projections from neurons in the C-laminae, and none in MIN. Most of the neurons that project to area 18 are located in layers A1 and C and in MIN. Most cells that project to area 19 are located in the C-laminae and in MIN. That is, as one goes from area 17 to 18 and then on to area 19, the source of the projection shifts within LGN from the A-laminae through the C-laminae into MIN. The source of projections to area 18 covers all components of LGN, and the projecting cells are, on the whole, larger than the neurons that project to either area 17 or to area 19. In addition, there is some evidence that the largest LGN neurons also project to cortex forming the medial bank and fundus of the middle suprasylvian sulcus (Kalil et al., 1991; Lomber et al., 1995; MacNeil et al., 1997). As expected, there is considerable topographic order in the form of the projections in the areas 17 and 18 (see Fig. 1-2). The broad patterns described by Holländer and Vanegas were largely confirmed in subsequent studies (Maciewicz, 1974, 1975; Leventhal, 1979; Geisert, 1980; Niimi et al., 1981), with the added information, gleaned from highly sensitive retrograde tracing studies, that the small cells in the C layers project to a wide expanse of cortex beyond areas 17, 18, and 19 (Raczkowski and Rosenquist, 1980, 1983).

Retrograde labeling methods that use two labels have shown that some projections to cortex are via axons that branch to innervate more than one cortical area

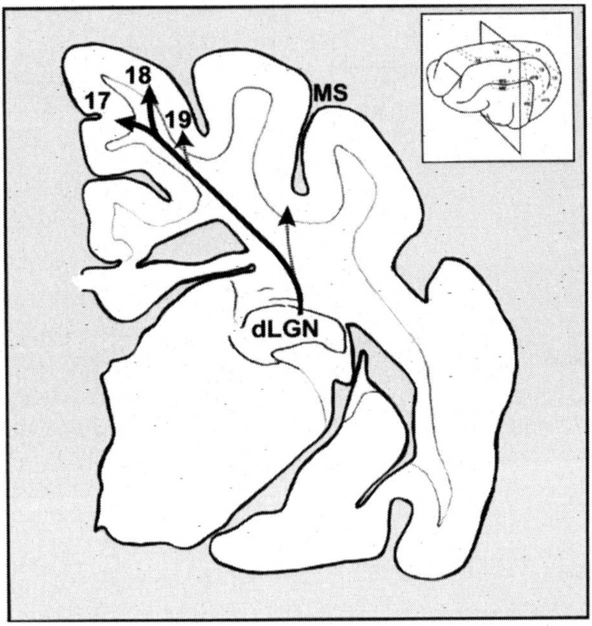

FIGURE 14-5. **(Inset)** A dorsolateral view of the cat cerebral cortex with visual areas delineated. The rectangle represents a coronal plane of section shown in the larger panel. The larger panel shows the location of the dorsal division of the lateral geniculate nucleus (dLGN) and its major projections to areas 17 and 18 and minor projections to area 19 and middle suprasylvian (MS) cortex (see also Fig. 1-4).

(Geisert, 1980; Bullier et al., 1984; Kaufman and Rosenquist, 1985; Birnbacher and Albus, 1987). These studies have been replicated in experiments that fill axons with opaque markers following physiological characterization (Freund et al., 1985; Humphrey et al., 1985), and they show that, just like their forebears in the optic nerve, LGN Y-axons have substantial collateralized projections and innervate both areas 17 and 18. In contrast, LGN X-axons have very focused projections, like their functional counterparts in the optic nerve, and they send projections exclusively to area 17. Within area 17 the axons usually terminate in a single continuous clump with an overall dimension in the range of 0.5 to 0.8 mm^2. In contrast, Y-axons may arborize over an area of 2.0 to 2.8 mm^2. Because the intra-axonal injections were made within the white matter below areas 17 and 18, it is not known if the Y-axons also send collateral projections to other cortical areas, as some retrograde tracing studies cited previously suggest. As in the optic nerve, axons transmitting W signals have diameters that are too small to impale and inject with markers. However, double label retrograde tracing methods indicate that some, and possibly all, project axons to more than one target in cortex (Kaufman and Rosenquist, 1985).

From these anatomical studies, we learn that axons in the visual radiation, like their functional counterparts in the optic nerve, exhibit varying degrees of axon collateralization to innervate multiple cortical areas. Many W- and Y-axons innervate more than one area, whereas X-axons innervate area 17 only. When combined with the retinal innervation of LGN, it becomes clear that β-retinal ganglion cells are the origin of a highly restricted pathway to area 17, whereas α- and γ-ganglion cells are the origin of highly divergent pathways to cerebral cortex and elsewhere.

Anatomical studies reveal five morphological types of neurons in LGN (Guillery, 1966; Updyke, 1979). As shown in Fig. 14-6, Class I cells are characterized by a large polygonal to round cell body with thick uncomplicated dendrites that extend from six or more radially disposed primary dendrites that may cross LGN lamina boundaries. A small number of dendritic appendages may be present. Types II and III neurons are characterized by small-to-medium-sized oval cell bodies. They have a bipolar configuration with tufts of tapering dendrities emerging from each pole and an axis oriented perpendicular to the laminar borders of LGN. The dendritic fields are bilobed. Type II neurons are distinguished from type III neurons by the form and position of dendritic appendages: type II neurons have grapelike appendages positioned at or near dendrite branch points, whereas type III neurons have complex appendages of variable form scattered along the lengths of the dendrites. Dendrites are usually confined to a single LGN lamina. Type IV cells are characterized by a rather heterogeneous group of small-to-medium cell

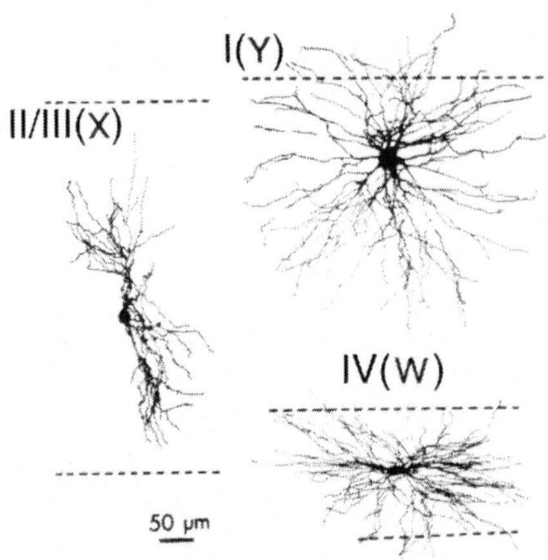

FIGURE 14-6. dLGN cells. Camera lucida drawings of the three major dLGN cell types from Stanford et al. (1983). The type I cells transmit Y-signals, the types II/III cells carry X-signals, and the type IV cells relay W-signals. Type II and III cells are distinguished by the presence of grapelike appendages at dendritic branch points (type II) and the presence of complex dendritic appendages along the length of the dendrites (type III). (Figure reproduced with permission of the American Physiological Society, and modified according to the anatomical conventions of Guillery [1966].)

bodies with long axes oriented parallel to LGN lamina borders. Dendrites may be thin and sinuous or varicose. Type V neurons (not shown) possess medium-sized cell bodies with large, yet sparse and smooth, dendritic trees (Updyke, 1979).

Each cell type has a characteristic distribution across the layers of LGN, and each is related to transmission of specific types of visual signals. Types I, II, and III neurons are located in the magnocellular layers A, A1, and C of LGN. Type I neurons, and a minor number of type II neurons that have dendrites that extend beyond a single LGN layer, transmit Y signals (Friedlander et al., 1981). The remaining type II and the type III neurons transmit X signals. Type IV neurons are the dominant neurons type in the parvocellular LGN layers, and they transmit W signals (Stanford et al., 1981, 1983). Type V neurons are scattered throughout LGN, and are believed to be interneurons because of an absence of an axon that reaches distant targets. The discrepancies identified for type II neurons suggest that either the morphological features appropriate for linking to the physiological X- and Y-characteristics have not been identified, or that the functional tests that have been applied to separate LGN cells do not provide a perfect match to the commonly used morphological features.

Signal Amplification

It is worth noting that the anatomy of the retino-geniculo-cortical pathway reveals a greater numerical amplification of the Y-system than the X-system (Peters and Payne, 1993). Based on the 2:1 incidence of X- and Y-neurons in LGN (Friedlander and Stanford, 1984), the 360,000 relay cells in LGN (Peters and Payne, 1993), and the foregoing numbers of retinal ganglion cells types, we calculate an approximately threefold amplification of the X-system and an approximately 15-fold amplification of the Y-system. Of course, we need to keep in mind the bases for sampling biases in numerical estimates based on electrophysiological measures noted earlier. Even so, Friedlander and Stanford (1984) assure us that the ratios they obtained in their electrophysiological studies match the ratios of X- and Y-LGN neurons obtained in their anatomicofunctional study of LGN neurons in which neurons were impaled and injected with opaque markers following electrophysiological characterization as X- or Y-type responses (Friedlander et al., 1981). Regardless of the actual numbers, the amplification is likely to be real because of the collateralization Y-axons undergo as they approach LGN and the greater number of terminal boutons each Y-axon forms in LGN (Bowling and Michael, 1984; Sur et al., 1987; Tamamaki et al., 1995). As noted in Chapter 1, evidence suggests that the amplification of the Y-system continues up to primary visual cortex.

Signal Transfer

Even though there is a numerical amplification in the number of signals transmitted to primary visual cortex, the amplification is not without modification of transmitted signals. The characterization of the transgeniculate transfer of signals to visual cortex as a "relay" is a misnomer. Ample numbers of studies have shown that action potential transfer from optic axon to LGN neuron is imperfect because LGN neurons generate fewer action potentials to a given stimulus than do ganglion cells afferent to them (Cleland et al., 1971; Cleland and Lee, 1985). Consequently, signal transfer should be viewed as probabilistic and not as a strict relay (Usrey et al., 1999). Moreover, convergence of signals from multiple ganglion cells, and local and more distant recurrent circuits, have considerable potential to modify LGN-neuron signal strength and properties. For example, LGN neurons are characterized by stronger antagonistic surrounds, and many possess an additional synergistic region beyond the traditional antagonistic immediate surround (Hammond, 1973). At present, it is not known if the lower activity is an inherent property of the neurons or reflects the presence of a stronger surround inhibition (Hammond, 1973; Bullier and Norton, 1979a, 1979b; So and Shapley, 1979; Cleland and Lee, 1985; Wilson et al., 1988). Inhibition of X-LGN neurons is greater than for Y LGN neurons (Mitzdorf and Singer, 1977; Bullier and Norton, 1979b). Even so, in broad terms, neurons in the magnocellular LGN layers have receptive field properties largely akin to those elucidated for retina (Hubel and Wiesel, 1961; Usrey et al., 1999).

It is necessary to inspect the view that signal transfer by type through LGN is unblemished by mixing with other types. There is robust evidence that many LGN cells respond in ways that characterize them unequivocally as being of the X- or Y-type (Mastronarde, 1987a, 1987b; Humphrey and Weller, 1988a; Saul and Humphrey, 1990; Mastronarde et al., 1991; Mastronarde, 1992) (see Chapter 9). However, there are numerous reports in the literature of LGN cells that respond in ways that do not equivocally classify them as X- or Y-cells (Cleland et al., 1971; Bullier and Norton, 1977, 1979a, 1979b; Kratz et al., 1978; Derrington and Fuchs, 1979; So and Shapley, 1979), but little notice was paid to them because, at best, their existence obscured the emerging primary dichotomy of visual function, or, at worst, they reflected poor technique in classification.

However, in 1987 Mastronarde (1987a, 1987b) provided clear evidence for a subdivision of LGN X-cells into two subtypes termed lagged and nonlagged X-cells. These cells are differentiated by the rapidity of their response to visual stimuli. The nonlagged group of cells respond promptly to the stimulus with a latency to spot stimulation of about 50 to 60 ms, whereas the lagged group exhibits an initial decrease in on-going activity before activation, which only becomes manifest 100 ms, or more, after stimulus onset (see Chapter 9). The latter group also exhibited a transient increase in activity at stimulus offset. This differentiation was soon replicated by Humphrey and Weller (1988a), who also provided evidence for morphological differences between the two types of cells (Humphrey and Weller, 1988b), as did Saul and Humphrey (1990), and Bowling (1989). Subsequently, Mastronarde, Humphrey and Saul (Mastronarde et al., 1991) reported on lagged and nonlagged LGN Y-cells.

Now that the X- and Y-streams have been fractionated, evidence is emerging that the distinction between the X- and Y-systems in LGN may not be as pure as was once thought. For example, of the conventional, non-lagged X- and Y-cells there also appear to be two sub-types in each class: those that receive input from a single retinal ganglion cell and those that receive converging inputs from two or more ganglion cells, which may be of the same or of another functional class (Mastronarde, 1987a, 1987b, 1992). Moreover, Wolfe and Palmer (1998) provide evidence based on space-time plots for a continuous, broad, yet modal, distributions of temporal responses in the cat LGN that blur the distinctions between the functional streams that emerge from retina. Even so, at a coarser level of analysis it seems that the dominant LGN input to area 17 is by the X-system, and the dominant LGN input to area 18 is by the Y-system (Chapter 1).

Retino-Geniculo-Cortical Coupling

The substantial physiological coupling between X- and Y-optic axons and their counterparts is demonstrable with anatomical techniques. For example, tritiated amino acids, sugars, or wheat germ agglutinin injected into the eyes are transported trans-synaptically through LGN to areas 17 and 18 and demarcate the X- and Y-terminations in layer 4 (Shatz et al., 1977; Anderson et al., 1988). The absence of label from other layers and from other areas show that the W-retino-geniculate system is much smaller and/or coupled much more weakly (see Chapter 1).

These results are supported by the retrograde trans-synaptic transport of label from areas 17 and 18 back to retina. LeVay and Voigt (1990) injected wheat germ agglutinin conjugated to horseradish peroxidase into areas 17 and 18. Injections into area 17 labeled neurons at visuotopically corresponding positions in LGN and, after retrograde trans-synaptic transport, in retina. Labeled ganglion cells contained dense reaction product, and all cells were classifiable on the basis of soma size and initial portions of their dendritic trees into α and β types. No cells of the γ group were labeled. Similar injections into area 18 also labeled LGN neurons and both α- and β-retinal ganglion cells at visuotopically matching regions, although labeling of α cells was much denser than the labeling of the β cells. The labeling of β cells is surprising because it is thought that β cells do not transmit signals that reach area 18 directly from LGN (see previously and Chapter 1). At this point in our understanding of the retino-geniculo-cortical pathway, it is not known if this surprising result reflects indiscriminate spread of tracer from LGN neurons to terminals of optic axons arising from both α and β cells and unrelated to cortical target locus. In that event, the result is uninteresting, as it reveals a limitation of this retrograde trans-synaptic tracing technique. Alternatively, the label in β-ganglion cells may be a true representation of coupling between optic axons and LGN neurons, a view that is leant some credence by the data on convergence of X- and Y-signals in LGN cited previously. Regardless of interpretation, there is no doubt that retinal α- and β-ganglion cells are strongly coupled with one or other, or both, areas 17 and 18. This knowledge of retino-geniculo-cortical circuitry allows us to interpret the repercussions of visual cortex lesions on LGN and retina. These are the topics of subsequent sections, but first we summarize the status of visual connections in the newborn cat.

VISUAL SYSTEM CONNECTIONS AND FUNCTION IN THE NEWBORN

This section summarizes the status of pathways and connections present within the visual system at about the time of birth to provide an interpretive baseline for the subsequent descriptions of the repercussions of visual cortex lesions incurred shortly after birth.

Pathway tracing studies have revealed that the basic patterns of connections present in the adult are already present at birth. For example, the basic pattern of projections of retina to LGN is indistinguishable from the pattern exhibited by the mature cat (Shatz, 1983; Sretavan and Shatz, 1986). The contralateral and ipsilateral eyes project in alternating sequence to the various layers of LGN and its MIN component in a dorsoventral sequence that accords with the sequence of the mature cat. Moreover, both areas 17 and 18 receive projections from LGN, and the topographic relationships of the two structures mimics the topography present in the adult cat (see Chapter 1) (Cornwell et al., 1984; Shatz and Luskin, 1986; Payne et al., 1988). However, it is important to note that LGN continues its developmental rotation through birth (Kalil, 1978), and the characteristic coronal and sagittal images of LGN of the mature cat differ in the newborn. Many neurons in the C-complex of LGN project to regions of cortex beyond areas 17 and 18, just as they do in the mature cat (Cornwell et al., 1984). Worthy of note are a small population of neurons in layers A and A1 that have projections to middle suprasylvian cortex in the newborn cat (Cornwell et al., 1984; Bruce and Stein, 1988; Tong et al., 1991) and thus exhibit a pattern of projection that differs from that exhibited by the mature brain. Accordingly, LGN axons in the neonate terminate in continuous bands throughout both areas 17 and 18, as in the adult (Shatz and Luskin, 1986). However, the axons terminate in all cortical layers, with the densest terminations in layers 1 and 4 (Kato et al., 1984; Shatz and Luskin, 1986).

Even though the visual pathways are in place in newborn kittens, the system is far from being structurally and functionally mature. In retina, α and β cells are identifiable by the relative size of the morphological features, although both bear transient dendritic appendages that are a hallmark of immature neurons (Dann et al., 1988; Ramoa et al., 1988; Ault and Leventhal, 1994). Moreover, cells are smaller than in the adult, and synapse number is much below mature levels (Cragg, 1975; Maslim and Stone, 1986; Crooks and Morrison, 1989). Corresponding measures in peripheral retina show that it is less mature than central retina. In LGN, optic axons are not fully elaborated, synapses with target neurons are small and low in number, and the target neurons themselves are small and have poorly developed dendrites (Garey et al., 1973; Kalil, 1978; Winfield et al., 1980; Mason, 1982, 1983; Sur et al., 1984; Friedlander et al., 1985). In visual cortex, all cortical neurons are immature, in particular the most recent arrivals in the cortical plate subjacent to layer 1. These cells are distinctly bipolar and have not yet started to develop basal dendrites that radiate in the laminar domain. Many neurons destined for layer 2/3 continue to arrive from the periventricular generation zone, and the upper cellular layers are incompletely formed (Luskin and Shatz, 1985); the low presence of these cells accounts for the broad translaminar spread of LGN terminations. Axon arbors are extremely immature (Kato et al., 1984) and synaptic density is 7.5% of the mature density (Cragg, 1975). When growth of cortex and individual neurons are taken into account, this percentage translates to 0.8% to 3% of the adult number of synapses per neuron (Cragg, 1975; Winfield, 1981). Few axons in the optic tract and optic radiation are wrapped in myelin (Moore et al., 1976)

Functional connections in the visual pathway to cerebral cortex have been demonstrated at birth using electrical stimulation and recording techniques. Conduction velocities of fibers are uniformly slow and in the range of 0.5 to 1.0 m/sec (Beckmann and Albus, 1982) and reflect the small size and almost complete absence of myelination. Moreover, electrical stimulation of the optic radiation or LGN evokes surface-positive, deep-negative potentials that reflect current sinks in the upper layers and current sources in the deeper layers, and the presence of functional synapses between fibers in the visual radiation and cortical neurons (Kato et al., 1983; Komatsu et al., 1985). However, neither the laminar distribution nor magnitude of the synaptic currents resembles their counterparts in the mature cat (Mitzdorf and Singer, 1978). In parallel, spatial organization of receptive fields is poorly developed and cortical responses are sluggish and fatigue readily (Hubel and Wiesel, 1963; Albus and Wolf, 1984). These are all features that reinforce the view that the visual pathway remains functionally immature at birth.

VISUAL CORTEX LESIONS

LESIONS IN ADULTHOOD

Degeneration of LGN

Removal of a large portion of area 17 induces medium and small cells in LGN layers A and A1 to degenerate, whereas lesions restricted to area 18 do not result in any localized, severe degeneration in LGN (Garey and Powell, 1967). However, combined destruction of both areas 17 and 18 induces large, medium, and small cells in layers A, A1, and C to degenerate. Further expansion of the lesion to include area 19 appears to induce more complete degeneration of LGN that includes more neurons in the C layers and the medial interlaminar subdivision of LGN (Doty, 1973). In short, the increase in lesion size from areas 17 through 18, to include area 19, results in progressively greater degeneration in LGN, which spreads throughout compartments and layers in accord with the patterns of connectivity revealed by Holländer and Vanegas (1977) and indicated earlier. Even so, it is important to note that although the rate of degeneration may be relatively rapid and many cells show chromatolysis and atrophy within days or weeks of the cortical ablation, many cells may remain relatively healthy for several months (Chow and Dewson, 1966; Madarász et al., 1983).

Thus, while the initial degenerative effects of the lesion may progress rapidly, complete degeneration is reached only relatively slowly. An example of severe degeneration of LGN following a lesion of areas 17, 18, and 19 sustained 4 years earlier is shown in Fig. 14-7.

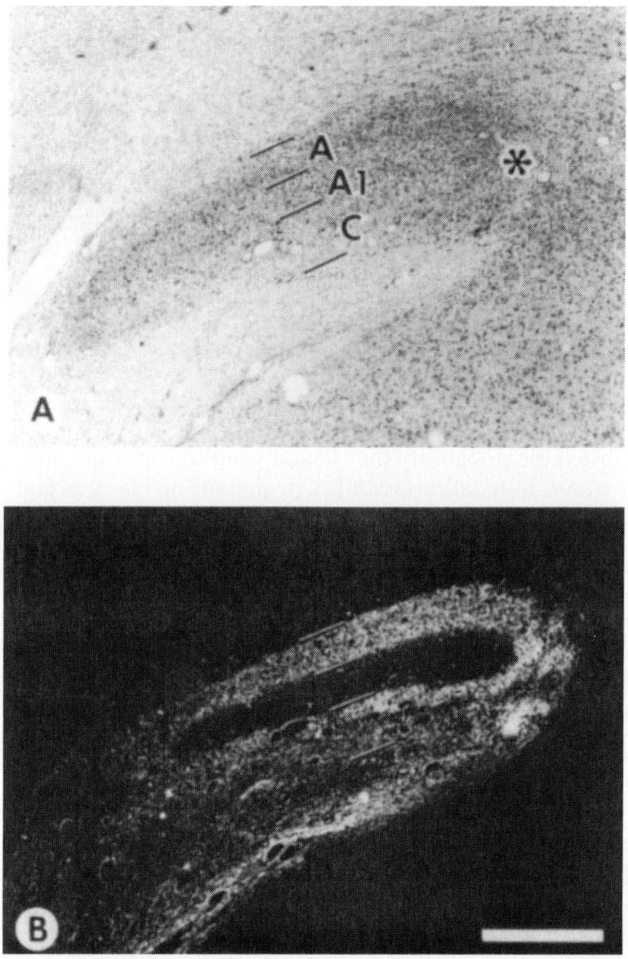

FIGURE 14-7. **(A)** Brightfield photomicrograph of LGN induced to degenerate by a lesion of areas 17, 18, and 19 incurred 4 years earlier in an adult cat. Note absence of clear definition of layers, and absence of neurons that are obvious in more medial portions of thalamus (right). **(B)** Darkfield photomicrograph of retinal projections to the same degenerated LGN. Note presence of label throughout layers A and C of the nucleus. Label is absent from layer A1 because it is innervated by the ipsilateral eye, which was not labeled. (Figure reproduced from Lomber et al. [1993], Journal of Comparative Neurology, Copyright © 1993. Reprinted by permission of Wily-Liss Inc., a subsidiary of John Wiley & Sons, Inc.)

Retinal Projections to LGN

Even 4 years after a lesion of areas 17, 18, and 19 and severe degeneration of LGN, dense retinal projections to LGN are maintained, and they have the same configuration as retinal projections to an intact LGN (Fig. 14-7) (Lomber et al., 1993). Overall, the density of innervation may even increase relative to levels identified in intact cats, but that increase can be explained by the partial collapse of overall dLGN structure without significant reduction in retinal input. This possibility was tested and confirmed in cats that underwent partial ablations of areas 17 and 18 by showing that projections to degenerated portions of LGN were denser than projections to intact portions of the same nucleus (Lomber et al., 1993). However, even though there is substantial evidence for maintenance of significant retinal projections to LGN following the cortical lesion, the bulk anterograde labeling method that was used cannot eliminate the possibility that individual retinal axon collaterals, or even whole axons, might retract from LGN, and that could signal the death of retinal ganglion cells. Nevertheless, it is worth noting that even removal of all contiguous visual areas and complete degeneration of LGN following long postlesion periods does not significantly moderate the density of innervation of LGN by retinal fibers.

Retinal Ganglion Cells

We have tested the possibility that retinal ganglion cells die after ablations of areas 17 and 18. We tested this possibility by measuring the sizes and determining the densities of ganglion cells in retinal flat mounts prepared 4 years after cortical lesion. We also injected tracers into the degenerated LGN to label ganglion cells by retrograde transport. This labeling strategy should enhance the probability that we should detect a loss of ganglion cells, because counts and measures were made on the subgroup of ganglion cells that project to LGN and did not include ganglion cells that project to other brain targets. Figure 14-8 (top) compares the labeling in a strip of midperipheral retina from an intact cat and from a cat with a longstanding lesion of areas 17, 18, and 19 (P180). There is no obvious decrease in the numbers of ganglion cells labeled. Moreover, even when the β cells are examined independently of the α and γ groups, there is no evidence for a lesion-induced decline (Fig. 14-8, bottom). Furthermore, we conclude that there is no axon withdrawal from LGN. A different picture emerges when equivalent procedures are carried out on retinae obtained from cats that sustained equivalent lesions shortly after birth. This is the topic of the next section.

LESION AT BIRTH

Degeneration of LGN

Lesions of visual cortex incurred by the newborn have severe degenerative repercussions on LGN, and the nucleus is reduced to a ghost of its normal size. However, as in the adult, lesions of visual cortex that are limited to area 17 induce

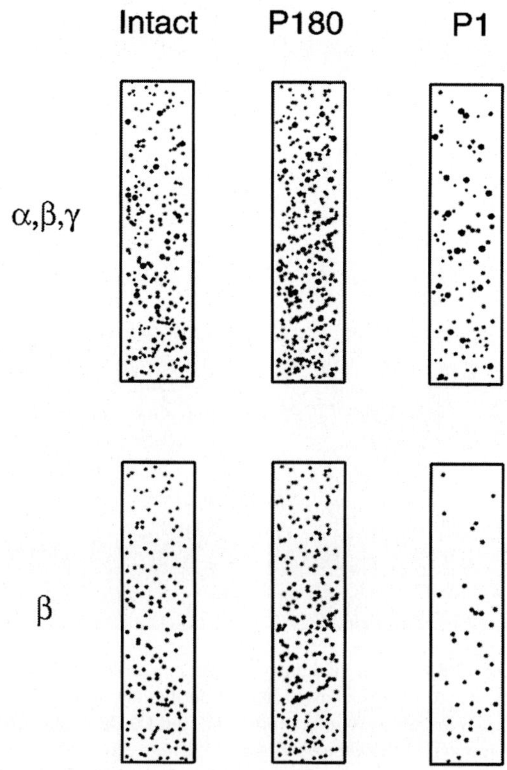

FIGURE 14-8. Distribution of all retinal ganglion cells **(top)** and β retinal ganglion cells alone **(bottom)** in 0.4 × 2 mm retinal strips in an intact cat, a cat with an equivalent lesion incurred in adulthood (P180), and a cat with a lesion of primary visual cortical areas incurred on the day of birth (P1). The bottom of each panel is closer to central retina; hence, there is a higher density of ganglion cells toward the bottom of each panel. Strips are from midperipheral retina, 2 mm from the area centralis. Postlesion periods were in excess of 4 years. See color insert for color reproduction of this figure.

less degeneration than lesions that include portions of area 18, and lesions that include area 19 in addition to areas 17 and 18 are particularly destructive on LGN (Fig. 14-9A). However, some neurons in the C layers and a population of particularly large neurons scattered throughout the nucleus survive (Doty, 1961; Spear et al., 1980; Lomber et al., 1993). These large neurons have been identified as type I and the smaller neurons as type IV (MacNeil et al., 1997). Some are hypertrophied. Types II and III neurons were not detected. Many of the surviving neurons identified maintain or establish projections to cortex on the medial bank of the middle suprasylvian sulcus (Kalil et al., 1991; Lomber et al., 1995; MacNeil et al., 1997).

Except for the scattered surviving type I neurons, the overall patterns of degeneration can be comprehended in terms of the patterns of projection of LGN

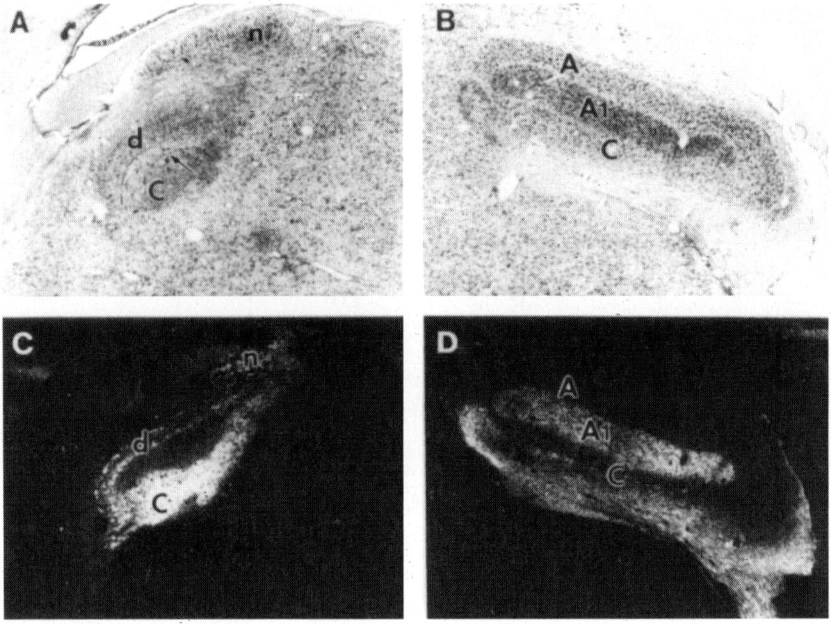

FIGURE 14-9. **(A and B)** Light field photomicrographs of dLGN in an adult cat after unilateral ablation of primary visual cortical areas 17 and 18 on the day of birth. **(A)** dLGN ipsilateral to cortical ablation. **(B)** dLGN ipsilateral to intact cortex. **(C and D)** Darkfield photomicrographs of the same sections shown in A and B to show the distribution of tritiated amino acids that had been injected into the eye. **(C)** Contralateral projections. **(D)** Ipsilateral projections. Layers A and A1 and the C-complex are indicated. In the intact dLGN note dominant retinal projections to magnocellular layer A1 and very weak retinal projections to parvocellular C layers. In the degenerated dLGN note the reverse: greatly reduced projections to the severely degenerated ghost magnocellular layers and a massive increase in projections to the residual neurons in the C complex. d = severely degenerated magnocellular layers A and A1. n = normal tail of dLGN that is connected to an intact, caudal portion of areas 17 and 18 in this instance. (Figure modified from Payne et al. [1984b], and reproduced with the permission of the Royal Society.)

neurons on to cortex at the time the lesion is incurred; in short, those neurons that project to the lesion are axotomized and they die. Neurons with projections to targets not included in the lesion survive. Included in this group is a significant population of neurons that establish projections to middle suprasylvian cortex. As discussed by Lomber et al. (1995), it is not known if these cells have projections to MS cortex at the time the lesion of areas 17 and 18 is incurred, as suggested by Bruce and Stein (1988) and Tong et al. (1991) and are naturally protected, or whether the parent axon is triggered by the lesion to establish a new projection to middle suprasylvian cortex, a region that lies above the optic radiation. Whatever the mechanism, it is clear that projections from LGN to cerebral cortex are somewhat greater in cats that sustained lesions of areas 17 and 18 shortly after birth compared with cats that sustained equivalent lesions in adulthood, even if the overall size of the degenerated nucleus is smaller. The same cells do not survive when the lesion extends laterally to include middle suprasylvian cortex in addition to areas 17, 18, and 19 (Lomber et al., 1993), and LGN becomes virtually unrecognizable. Finally, of note, is the observation that visual cortex lesions in the neonate induce a very rapid degenerative response by LGN and a clearance of debris within only a matter of days. This rate of degeneration stands in stark contrast to the prolonged degenerative response and long time course of elimination followed by mature LGN neurons in response to visual cortex lesions.

Retinal Projections to LGN

Examination of the retinal innervation of LGN shows a marked reduction in optic innervation of the severely degenerated layers A and A1 (Fig. 14-9C and D) (Payne et al., 1984b) to 17% or less of normal levels (Lomber et al., 1993). However, this density is clearly an underestimate of the impact of the lesion on LGN, because the nucleus undergoes a massive collapse to about one-tenth its normal size, or less, and we estimate that the real innervation of layers A and A1 is reduced to about 2% of normal levels, or less. However, there is a marked increase in the density of innervation of surviving neurons in the C-complex. It is not known whether this increase is merely a product of the collapse of the nucleus, sprouting of axons to synapse with surviving neurons, or a redeployment of some or all axons that normally innervate layers A and A1.

Retinal Ganglion Cells

We have tested the possibility that retinal ganglion cells withdraw axons from LGN, or even die after ablations of areas 17, 18, and 19 in the neonate. First, examination of flat-mounted retinae stained to show cell bodies reveal that 68% of medium-sized cells disappear from the retina (Pearson et al., 1981), and examination of flat-mounted retinae show a reduction of approximately 60% in the number of ganglion cells that could be labeled by injection of retrograde tracers into LGN (Fig. 14-8, top, P1). Counterstaining of the retinae to reveal all ganglion cell bodies showed that the decreased labeling reflected a real disappearance of neurons and not simply a failure to label a significant fraction of ganglion cells as a result of axon retraction from LGN (Payne et al., 1984b). Subsequent sorting of ganglion cells by type based on soma size and labeling of the proximal portions of the dendritic trees revealed that the degeneration was confined to the β type, of which about 90% die (Fig. 14-8 [P1, bottom] and 14-10) (Payne et al., 1984b). It is not known what it is about the target neurons in LGN that promotes survival of the remaining 10% of β-retinal ganglion cells or why possessing connections with afferent sources is insufficient to promote survival.

Temporal Retina

Degeneration in temporal retina may not be as severe as in nasal retina. The first studies on visual cortex lesion-induced retinal degeneration were carried out

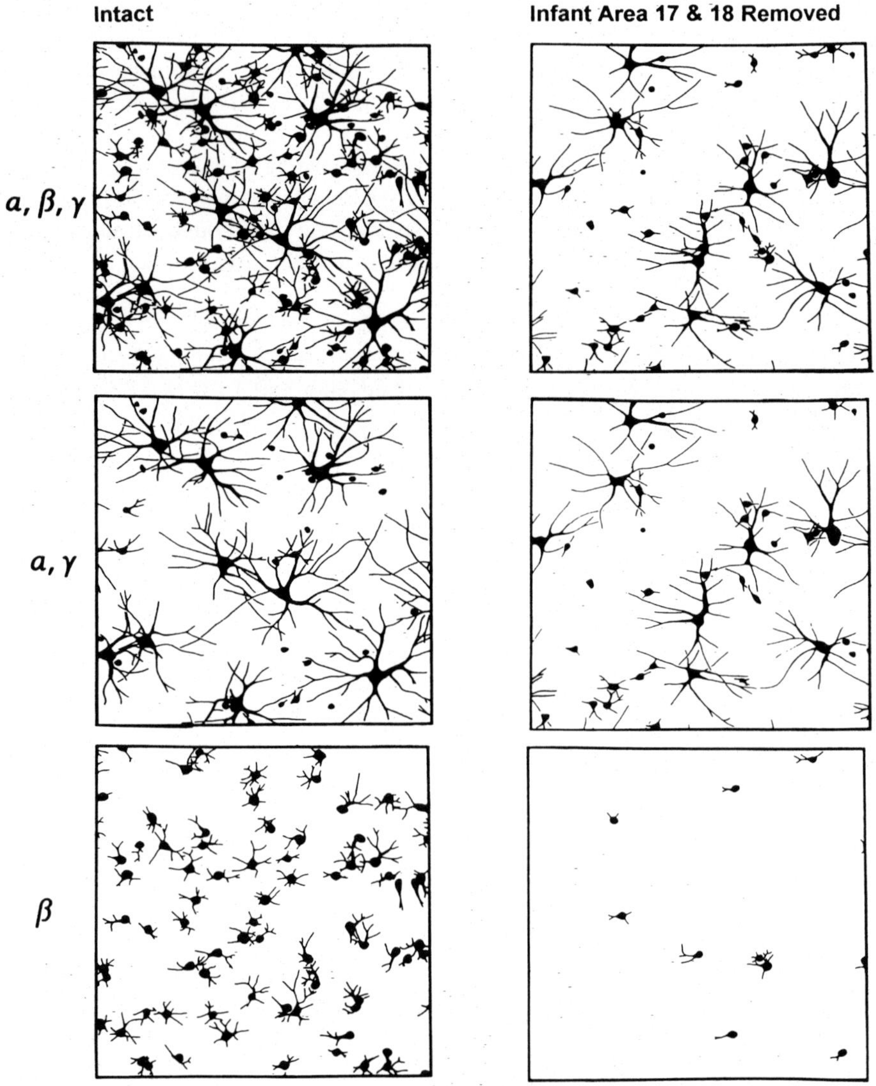

FIGURE 14-10. Drawing of retinal ganglion cells labeled by injection of HRP into an intact dLGN *(left)* and a dLGN of a cat which sustained a complete lesion of areas 17 and 18 shortly after birth *(right)*. *(Top row)* α, β and γ cells. *(Middle row)* α and γ cells. *(Bottom row)* β cells. *Note:* (1) In intact cats β cells are the most abundant cell type being ~3 times as common as γ cells and ~10 times as common as α cells. (2) After the early lesion ~90% of β cells have been eliminated, yet all α and γ cells remain. (3) Some α cells are smaller than normal, and some γ cells are larger than normal. (Figure modified from Payne et al. [1984b], and reproduced with the permission of the Royal Society.)

on nasal retina, and the magnitude and specificity of the degenerations have both been confirmed in subsequent studies using retrograde trans-synaptic transport of tracer injected into middle suprasylvian cortex. Wheat germ agglutinin conjugated to horseradish peroxidase injected into middle suprasylvian cortex labels large numbers of surviving neurons in the C-complex of LGN with smaller numbers labeled in the severely degenerated layers A & A1, and in retina. The labeled retinal ganglion cells are present in large swathes of nasal (contralateral) and temporal (ipsilateral) hemiretinae (Payne and Lomber, 1998). Embedded within these swathes are distinct foci of densely labeled cells (Fig. 14-11) at homonymous positions that match the visual coordinates at the tracer deposit site in cortex (Palmer et al., 1978; Grant and Shipp, 1991).

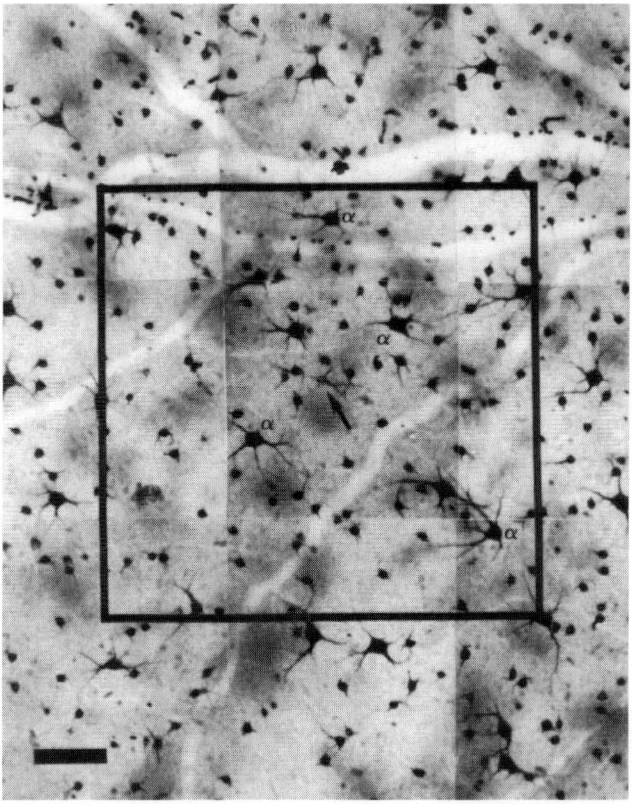

FIGURE 14-11. Photographic montage of a field of ganglion cells labeled in right temporal retina after injection of WGA-HRP into MS cortex in a cat that underwent bilateral ablation of areas 17, 18, and 19 on P1. Arrow indicates the position of a shrunken α cell. Selected α-retinal ganglion cells are indicated. Scale bar = 200 μm. (Figure reproduced from Payne and Lomber, [1998] with permission of Springer-Verlag, Copyright © 1998.)

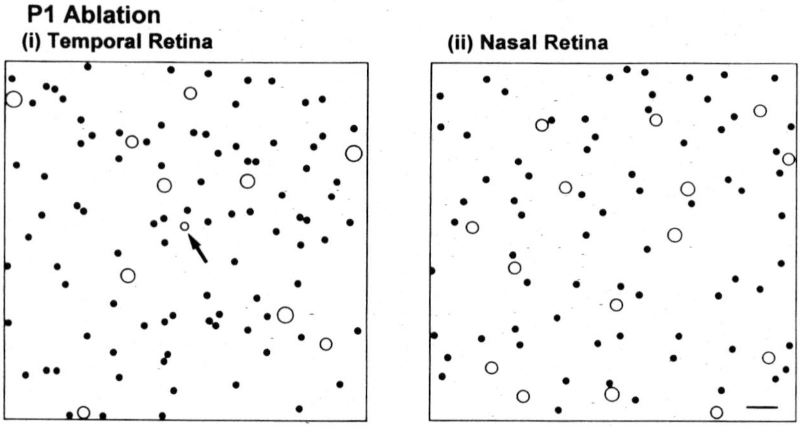

FIGURE 14-12. Plots of labeled and identified α (open circles) and non-α-retinal ganglion cells (filled circles) in temporal and nasal hemiretinae. Region shown is indicated by the square in Fig. 14.11. Scale bar = 100 μm. (Figure reproduced from Payne and Lomber [1998] with permission of Springer-Verlag, Copyright © 1998.)

In the dense foci α-ganglion cells are readily identified by the size of the soma and the prominent labeling of dendrites, and it is obvious that they are more abundant in nasal retina than in temporal retina (Fig. 14-12). Moreover, in nasal retina, the non-α cells outnumber α cells about 4:1, whereas the ratio in temporal retina is 8:1 (Payne and Lomber, 1998). These differences between the two hemiretinae are made up partly by a lower number of α cells labeled and partly by a higher number of non-α cells labeled in temporal retina than in nasal retina (Fig. 14-12). Some of these labeled ganglion cells may be of the β type. They are known to compose a greater fraction of the ganglion cell population in temporal retina than in nasal retina (Stein et al., 1996), and it may be that 20% or more of them might survive. Such a conclusion on the differential sensitivity of β-ganglion cells in nasal compared to temporal hemiretinae concurs with the results of Rowe (1988).

Overall, based on data for β-ganglion cell degeneration in nasal and temporal hemiretinae, we estimate that ~78,000 of the total 90,000 β population die as a result of the neonatal lesion of areas 17 and 18. In addition, Rowe (1988) claimed that a fraction of the γ population also degenerates, but that finding has not been substantiated following lesions of areas 17 and 18. Of the surviving cells, many α cells in nasal hemiretina are atrophied compared to normal α-cells, whereas some γ cells are hypertrophied (Payne et al., 1984b; Rowe, 1988).

ELECTROPHYSIOLOGY

Electrophysiological assessment of neuron types in retinae of cats that sustained lesions of areas 17 and 18 also provide evidence for reduced numbers of X (β) cells in the presence of normal numbers of Y (α) and W (γ) cells (Tong et al., 1982). In intact cats, electrical stimulation of the optic chiasm and antidromic activation of ganglion cells within a small region of retina produces a histogram that is trimodal, regardless of retinal location. The cells with the shortest latency are usually considered to belong to the Y (α) class, those with intermediate latencies belong to the X (β) class, and those with the longest latencies belong to the W (γ) class (Fukada, 1971; Cleland and Levick, 1974a, 1974b; Stone and Fukuda, 1974). After neonatal lesion of areas 17, 18, and 19, the peak with medium latency is greatly reduced in amplitude, suggesting a disappearance of substantial numbers of X (β) cells. A minor reduction in the number of X (β) ganglion cells was also detected following a cortical lesion incurred in adulthood (Tong et al., 1982).

Tong et al. (1982) reached a similar conclusion when their analyses were based on receptive field sizes. In intact cats, receptive fields in one region of the retina usually have a bimodal distribution, with the receptive fields of the Y (α) and W (γ) cells spanning the same medium-to-large diameter spectrum, and X (β) cells usually having the smallest diameters. Cortical lesions incurred in infancy greatly reduce the number of X (β) cells encountered, and lesions incurred in adulthood have a slight impact. Other useful measures to distinguish between retinal ganglion cell types, such as cutoff velocity and linearity of spatial summation led to equivalent conclusions. Based on encounter rates in intact cats and cats with lesions of primary visual cortex, Tong et al. (1982) concluded that 78% of X (β) cells were lost from the retinae of cats with early lesions, and that 22% were lost from retinae of cats with lesions sustained in adulthood.

The incidence of LGN cells with X-properties is also reduced from 62% to 15% by lesions of areas 17 and 18. The decrease is regardless of age of the cat at the time the lesion was sustained, which was without impact on cells with W- and Y-properties (Tumosa et al., 1989). The reduction in early-lesion cats is understandable, because there are so few β (X) cells in retina, and it is a result that concurs with the absence of types II and III neurons in LGN that transmit X signals to cortex (Friedlander et al., 1981; MacNeil et al., 1997). The low numbers of X-cells in cats with lesions of areas 17 and 18 sustained in adulthood is more difficult to explain. However, it may result from optic terminals disconnecting from axotomized LGN neurons, as occurs elsewhere in the nervous system for presynaptic terminals following axotomy of target neurons (Matthews and Nelson, 1975; Brenner and Johnson, 1976; Payne et al., 1984a; Eugene and Taxi, 1991). Alternatively, the physiological characteristics of LGN cells receiving both X- and Y-signals (the X/Y cells of Mastronarde) may shift in preference to Y-dominated after an adult lesion of primary visual cortex and thus manifest in a decrease in X-signals encountered. A similar phenomenon does not characterize the response of W and Y pathways to removal of areas 17 and 18.

The values obtained by Tong et al. (1982) and Tumosa et al. (1989) on the impact of visual cortex lesions on detectability of cells with X-characteristics in LGN and retina are slightly at odds with the values obtained by MacNeil et al. (1997) and Payne et al. (1984b) in anatomical studies. The bases for the differences are unclear, but they do remind us of earlier discrepancies between anatomical studies and electrophysiological studies, and they remind us of Levick's (1975) warning against uncritical acceptance of recording percentages of different classes of neurons as direct indicators of numerical preponderance. It seems that more accurate measures are obtained from the anatomy.

FACTORS LINKED TO SURVIVAL AND DEATH OF GANGLION CELLS

It is unlikely that a single factor governs whether a given ganglion cell lives or dies after damage of primary visual cortex, but rather it is a group of factors that govern a composite response. The relative importance of possible factors is unknown at present. However, we have suggested that maturational status, patterns of connectivity at the time the lesion of primary visual cortex is sustained, and rate of degeneration of LGN neurons subsequent to the lesion are all important factors linked to the survival and death of ganglion cells. In the following sections we review these linkages.

MATURATIONAL STATUS

The evidence shows that immature β-retinal ganglion cells are highly susceptible to lesions of primary visual cortex and the ensuing degeneration of LGN neurons, causing them to die. Clearly, immature β cells have a dependency on LGN and primary visual cortex. Mature β-ganglion cells are also susceptible to the cortical lesion, but in a moderated way because they merely become disconnected from axotomized LGN neurons. In the absence of significant contacts with the target neurons, they manage to survive, presumably because they have passed through some critical maturational step that attenuates their dependence on other neurons. However, exactly why immature β-ganglion cells are so susceptible is as much a mystery today as it was when the phenomenon was first described by Ganser (1882) and von Monakow (1889).

CONNECTIONS

With Brain

As summarized previously, the brain of the neonatal cat has many features in common with the mature brain. It has the same overall form and the same patterns of connections between eye and brain and between LGN and visual cortex. However, it differs because it is smaller than the mature brain, and both the quantity and details of connections are far below their adult levels. Moreover, many neurons are structurally and physiologically immature, and axon arbors are small and synaptic connectivity is very low. Even so, the patterns of degeneration induced by lesions of primary visual cortex of the neonate broadly parallel the known patterns of projections. However, there are some minor exceptions. For example, the small number of LGN neurons that manage to establish projections to middle suprasylvian cortex do not die (see page 585), whereas the massive numbers of neurons in layers A and A1 that normally project to areas 17 and 18 degenerate.

The majority of β cells that send massive projections to layers A and A1 of dLGN succumb when their target neurons degenerate. The degeneration of vast numbers of β-retinal ganglion cells is matched by a massive withdrawal of retinal fibers from the severely degenerated layers (Fig. 14-9). On the other hand, α- and γ-retinal ganglion cells survive in large numbers; and that survival can be linked to the presence of collateral or sole projections to neurons in the parvocellular layers of dLGN, superior colliculus, and pretectum (Fig. 14-4), where large numbers of healthy target neurons reside. Survival of all, or most, α and γ cells results in increased density of projections to the LGN C layers, as noted earlier. Furthermore, these observations on death and survival of ganglion cells are concordant with the absence of types II and III neurons (X cells) and presence of types I (Y cells) and IV neurons (W cells) in LGN.

It is not clear why small numbers of β-retinal ganglion cells survive damage of primary visual cortex. It may be that they are more mature than their counterparts and are resistant to the lesion-induced degeneration of LGN. They may be the β cells that have collateral projections to the midbrain (Fig. 14-4) (Cleland and Levick, 1974a; Sawai et al., 1985; Sur et al., 1987; Hada and Hayashi, 1990). Alternatively, they, like a small number of LGN neurons, may be capable of developing, or sustaining projections to targets outside the zone of degeneration. They may even be medium-sized cells of another type that have been misclassified, although this possibility is unlikely because the total number of degenerated and surviving β cells equal the number of β cells present in an intact retina.

It is also not known why some α cells shrink and some γ cells hypertrophy. The α-cell responses may reflect withdrawal of axons from the degenerated layers A and A1, and a reduction in total axon arbor; the γ-cell response may reflect an expansion of terminal arbors (Payne et al., 1984b), although larger lesions are likely to have greater repercussions on these two cell types than lesions confined to areas 17 and 18 (Ault et al., 1993). It is also not known why the responses of some cells in temporal retina differ from the responses of their counterparts in nasal retina. Reasonable speculation must await additional details on the differences in the normal anatomy and development of pathways emanating from nasal and temporal hemiretinae. Whatever differences emerge, anterograde transsynaptic tracing shows that both retinal regions contribute to expanded projections to C-layers of LGN and on to a 60 mm^2 region of middle suprasylvian cortex (Tong et al., 1984; Payne and Lomber, 1998). Moreover, the coupling between the surviving ganglion cells and their targets in LGN is increased and approaches the retinogeniculate coupling identified in the intact cat (Shatz et al., 1977; Shatz and Stryker, 1978; Mower et al., 1985; Payne and Lomber, 1998).

One factor remains unclear. If β-retinal ganglion cells are the origin of a highly restricted pathway to area 17, why is it necessary to remove area 18 to effect degeneration of β cells? A simple answer might suggest that the presence of any neurons in layers A and A1 that survive cortical ablation promotes the formation of connections and the survival of ganglion cells. However, a more realistic answer may lie in the propensity of many axons originating from β and α cells to converge onto individual neurons in LGN, as emerging evidence from the intact cat is starting to suggest (see page 575). Some of these cells project to area 17, some to area 18, and some to both areas, thus making it essential that both cortical targets be removed. Future studies, particularly on the connections of LGN type 1 neurons will likely bear importantly on this point.

With Outer Retina

As in brain, immature β-retinal ganglion cells have different connectivity from mature α and γ cells. Unfortunately, no detailed information is available on the precise connectivity in the neonate or on the connectivity pattern after central visual cortex lesion in the neonate. What is clear, however, is that there is considerable reorganization in the inner plexiform layer (IPL) of the retina in response to a central lesion. Paradoxically, in response to the loss of 80% or more of the population of β cells, representing overall a loss of half of all retinal ganglion cells, the IPL increases in thickness by 15–40% (Rowe, 1988).

This unexpected finding may be accounted for by several factors. First, the increased thickness of the IPL may be due to elaboration of processes from cells in the ganglion cell layer, the inner nuclear layer, or from both. The elaborated processes may include those from the remaining β cells (Weber et al., 1998); however, given the magnitude of β-cell loss, it is unlikely that this factor contributes to the thickening of the IPL. The β-cell processes may ordinarily check the growth and extent of expansion of non-β-ganglion cells; removal of these influences may cause the hypertrophy of some ganglion cells and the hypotrophy of others. This possibility is supported by observed changes in the morphology of non-β-ganglion cells (Payne et al., 1984b; Rowe, 1990). Alternatively, increases in the thickness of the IPL could be due to hypertrophy or elongation of the amacrine and cone bipolar cell processes that serve as ganglion cell inputs. In addition, since the neonatal lesion interrupts a pruning process that reduces exuberant ganglion cell dendritic appendages to the adult form, the increased thickness could be produced by enduring dendritic appendages and even elaboration with a subsequent increase in contacts with presynaptic elements. A final possibility is that the Müller support cells of the retina contribute to the increased thickness by invading the IPL after large-scale death of β cells. Of course, the actual outcome may be an amalgam of the listed possibilities.

RATE OF LGN DEGENERATION

The speed of the degenerative response to lesions of primary visual cortex lesions differs between mature and immature LGN neurons. In the mature brain, lesions of primary visual cortex induce a detectable atrophy in LGN within a few days to weeks (Garey and Powell, 1968). However, many neurons remain for long periods and well beyond 1 year (Chow and Dewson, 1966). In contrast, equivalent lesions in the neonatal cat induces a massive and overwhelming degenerative response in LGN, which is at its peak 2 days later and is essentially complete about 1 week later, including the removal of neuron debris (Labar et al., 1981; Kalil, 1984). The swift response is accompanied by a prompt removal of retinal input (Lomber et al., 1993). Thus, there is an almost catastrophic elimination of LGN neurons in the neonate but a much more protracted elimination in the mature cat. Could it be that the fast speed of repercussions on the LGN in the neonate contributes greatly to the degeneration of β-retinal ganglion cells, and that the more slowly emerging impact of the visual cortex lesion on LGN in the adult cat contributes greatly to the survival of β ganglion cells?

This possibility has been tested. Kainic acid was injected directly into the adult LGN to kill neurons over a time scale that mimics the response of LGN to visual cortex lesion in the neonate (Pearson et al., 1991). This toxin is selective for the LGN neurons and has no direct impact on optic axon arbors (Pearson et al., 1991). However, between 2 and 4 months later, retinal projections were absent from the degenerated region in LGN and retained in their normal density to intact regions (Pearson et al., 1992). This result showed a reduction in innervation density in response to the death of target neurons. It also showed that the removal of synaptic targets emerged over a period greater than 2 months but less than 4 months, although retraction continued through 6 months and possibly beyond. Subsequent analyses of retinal ganglion cells show that they first atrophy and then degenerate over a period that parallels the modification in retinal projections (Pearson and Stoffler, 1992; Pearson and Thompson, 1993). Moreover, cells in peripheral retina (Pearson and Stoffler, 1992) are more vulnerable than their counterparts in central retina (Pearson and Thompson, 1993) over the same period. However, it is not clear whether the reduction in retinal projections triggers the death of parent cells, or whether the degeneration of the parent cell body in retina results in the elimination of the optic axon and innervation of LGN. Regardless of sequence, what is not in doubt is that the most vulnerable cell group is β-retinal ganglion cells (Pearson and Stoffler, 1992), a result that shows that even mature β cells have a dependence on LGN target neurons for survival. However, their response to LGN-target removal remains quite slow. Finally, and parenthetically, expansion of the kainic acid lesion to include the C-complex in addition to layers A and A1 reveals an additional susceptibility of α- and γ-retinal ganglion cells to rapid degeneration of target neurons (Pearson and Stoffler, 1992). Thus, fast degeneration of neurons in both the A-layers and C-layers triggers withdrawal of retinal projections and the death of retinal ganglion cells.

Likewise, lesions of the optic nerve trigger preferential degeneration of mature β-retinal ganglion cells. As the lesion is more proximal than visual cortex lesion-induced degeneration of LGN, the degenerative process starts within a few days of the nerve sustaining the lesion. Cell nuclei exhibit pyknosis (Silveira et al., 1994), and there is frank degeneration present 1 to 3 weeks later, with the first responding and most vulnerable cells located in peripheral temporal retina (Holländer et al., 1984). Subsequently, ganglion cells in central and nasal regions start to degenerate. The loss is predominantly of medium-sized cells, many of which are β cells (Holländer et al., 1984; Silveira et al., 1994). A proportion of the α and γ cells also degenerate, and the process is marked by enhanced neurofibrillar staining and dissolution of Nissl substance but without apparent pyknosis (Silveira et al., 1994). Cell bodies degenerate before any changes are seen in the axon, and there is no evidence for a dying back process. It has been suggested by Silveira et al., (1994) that the β cells undergo an excitotoxic death as a consequence of the large glutamatergic input from bipolar cells. The severity of the degeneration can be reduced by intravitreal administration of the NMDA-glutamate receptor antagonist, MK-801 (Russelakis-Carneiro et al., 1996).

Clearly, mature β-retinal ganglion cells are highly sensitive to disconnection from target neurons in LGN, but what permits them to survive visual cortex lesion-induced degeneration of LGN? The greater maturity and elaboration of β-cell axon arbors in LGN and the contacts with multiple target neurons (see page 575) likely play significant roles. Moreover, besides the slower degenerative response of the LGN neurons and β-retinal ganglion cells per se, it may be that the longer time scale permits small incremental adjustments to be made by retinal axon terminals in LGN as target neurons initially atrophy and then degenerate. These small adjustments may establish or strengthen sustaining connections with interneurons or with alternate LGN neurons that may receive inputs from multiple ganglion cells and degenerate at slower rates (Kalil and Behan, 1987). These connections may be sufficient to promote continued β-cell survival, and it would be fair to suggest that these connections somehow interfere with the excitotoxicosis suggested previously.

Thus, it seems that three significant factors contribute to the dependence of β-retinal ganglion cells on primary visual cortex: (1) the absence of significant collateral projections to regions other than layers A and A1 of LGN, (2) the rapid degenerative response of LGN neurons, and (3) the immature status of the β cells at the time the lesion is incurred. If any one of these three factors does not hold, then ganglion cells are protected from the deleterious effects of visual cortex lesions, and they do not have a dependency of primary visual cortex, or the dependency is highly attenuated and does not result in neuron death.

PRIMATES

Monkeys and humans also contain significant populations of β-ganglion cells in retina, and they, too, have a dependency on primary visual (striate) cortex for survival. This dependency is even greater than that exhibited by the cat because visual cortex lesions sustained even in adulthood trigger β-cell degeneration. As in the cat, α- and γ-ganglion cells appear immune to visual cortex lesions. In the following subsections we summarize the repercussions of lesions of primary visual cortex that reveal the dependency of β-retinal ganglion cells.

LESIONS AND IMPACT ON LGN

Ablation of striate cortex at any age results in severe retrograde degeneration of LGN (Polyak, 1957; Van Buren, 1963a, 1963b), which reaches completion within about 12 weeks (Mihailovic et al., 1971), and all layers become ghosts of their former selves (Cowey et al., 1999). However, embedded and scattered throughout the glial cell fields of the parvocellular and magnocellular zones are a small number of solitary neurons that survive (Van Buren, 1963a, 1963b). Either these cells are GABAergic interneurons, or they extend axons into the prestriate field (Yukie and Iwai, 1981; Cowey and Stoerig, 1989).

IMPACT ON RETINAL PROJECTIONS

The collapse of LGN has severe repercussions on retinal projections. Throughout the degenerated nucleus, retinal innervation is reduced, with the reductions being slight in the ghosts of the magnocellular layers and virtually complete in the ghosts of the parvocellular layers (Dineen et al., 1982; Weller and Kaas, 1989; Kisvárday et al., 1991). Whatever retinal innervation remains does not appear to form synapses with the surviving projections neurons but rather with the surviving GABAergic interneurons (Kisvárday et al., 1991).

IMPACT ON RETINA

The severe degeneration of LGN triggers degeneration and disappearance of retinal ganglion cells. The greatest impact is most obvious in central retina. In this region of a normal retina, ganglion cells are normally stacked six or more deep. With increasing eccentricity, ganglion cell densities exhibit a decreasing density. Following the lesion of striate cortex and degeneration of LGN, the density of ganglion cells across the entirety of the retina becomes much more uniform (Cowey, 1974; Weller and Kaas, 1989; Niida et al., 1990) owing to the disappearance of about 65–80% of the ganglion cells in the central region (Cowey and Stoerig, 1989; Johnson and Cowey, 2000). The proportional loss from peripheral retina is much less (Cowey and Stoerig, 1989; Niida et al., 1990). Retrograde pathway tracing studies that also reveal details of ganglion cell dendritic morphology show that the disappearance is confined to the β group that comprise about 80% of the total retinal ganglion cell population. Alpha and γ cells are present in their normal numbers, and they appear to be unaffected by the cortical ablation (Cowey and Stoerig, 1989). The pattern and degree of cell survival and death parallel the pattern and degree of innervation of major visual structures in brain. In the intact monkey, all β cells project to the parvocellular layers of LGN; and it is presumed that all γ cells project to the midbrain (Perry and Cowey, 1984; Perry et al., 1984; Williams et al., 1995) and other targets including the LGN interlaminar zones, whereas α cells project to both the magnocellular division of LGN and to the superior colliculus (Perry and Cowey, 1984; Perry et al., 1984).

The collateralization of α cell projections accounts for the maintenance of retinal projections to the magnocellular ghosts.

AGE AT LESION

The repercussions of visual cortex lesions on retinal innervation of LGN and on survival and death of retinal ganglion cells described previously pertains to adult monkeys that endured postlesion periods of up to 8 years. These patterns appear to be attained more quickly when lesions are made earlier in life or are less complete when shorter postlesion periods are applied to mature monkeys. For example, retinal projections to the parvocellular layers following striate removal in young (~2 3-year-old) monkeys are greatly reduced (Dineen et al., 1982), and this is matched by degeneration of a lesser degree in retina (Dineen and Hendrickson, 1981). Cowey et al. (1999) estimate that maximal retinal degeneration is achieved very slowly over 4 to 5 years. On the other hand, striate lesions sustained during the first 9 weeks of life result in a more severe reduction in retinal innervation of LGN and a more complete degeneration of retinal ganglion cells, and it is achieved more rapidly than in the mature monkey (Dineen and Hendrickson, 1981; Weller and Kaas, 1989). In addition, lesions sustained early in life result in sparser innervation of the magnocellular layers (Weller and Kaas, 1989), suggesting susceptibility by a subpopulation of α cells. It is unclear if these are typical α cells as the anatomy suggests, or whether they are a different type of cell. For example, Kaplan and Shapley (1982) provided physiological evidence that neurons with X-physiological properties project to the magnocellular layers of LGN. In functional terms, they have properties, and hence morphologies, that are distinct from the properties of the cells that innervate the parvocellular layers. Identification of two morphological types of magnocellularly projecting ganglion cells would reconcile the observations made in monkey with those made in cat on the repercussions of visual cortex lesions. What is clear, though, is that the time course of repercussions is graded in an inverse, age-dependent way, occurring both more rapidly and more completely after lesions sustained earlier rather than later in life. Moreover, monkey β cells have a dependency on primary visual cortex that extends throughout life.

COMPARISON WITH THE CAT

We recognize that it is risky to draw too many parallels between the details of cat, monkey, and human visual systems and the dependencies of retinal ganglion cells on primary visual cortex. However, we think it appropriate to discuss them in broad terms because of the numerous and substantial similarities in the organization, function, and developmental program of the visual system of cats and monkeys and, by extension, humans (Rakic, 1977; Palmer et al., 1978; Payne, 1993; Williams et al., 1993; Payne et al., 1996). Nonetheless, it is important to both affirm that details, magnitude, and complexity of monkey and human visual systems exceed those of the cat and appreciate that the developmental program is substantially longer. Even so, the evidence allows us to conclude that the dependencies shown by β cells in adult and adolescent monkeys broadly parallel the dependencies exhibited by immature β-retinal ganglion cells in cats.

SUMMARY

1. The dependency of retinal cell populations on the structural integrity of the primary visual cortex was discovered in the latter part of the 19th century; however, the debate concerning the neuron doctrine posed a significant barrier to the acceptance of the transneuronal degeneration of retinal ganglion cells consequent to primary visual cortex.
2. After a considerable period of quiescence, renewed interest in the topic resulted in a confirmation of retinal cell dependencies on primary visual cortex and extension of the phenomenon to primate visual systems. The most detailed work has been done, and continues to be done, in the cat.
3. β (X)-retinal ganglion cells in the cat and monkey, and by extension humans, have a unique and signature array of morphological, physiological, and connectional characteristics that set them apart from all other ganglion cell types. They are also extremely fragile and depend for survival on primary visual cortex and its intermediate relay structure, the LGN, both some distance away.
4. Maturational status, patterns of connectivity, and rate of degeneration of dLGN neurons all contribute to the selective degeneration of β (X)-retinal ganglion cells.
5. The absence of approximately 90% of the β (X)-retinal ganglion cells is certain to put limits on visual function in the face of primary visual cortex ablation. Future avenues of research will focus on these limits. Research will also likely begin to shed light on the reasons that certain β (X)-retinal ganglion cells do not degenerate.

ACKNOWLEDGMENTS

The authors thank NINDS for supporting the research work and the preparation of this chapter. We also thank Dr. David Berson and the following publishers for permission to reproduce the following figures: Elsevier Science (Fig. 14-4); John Wiley & Sons, Inc. (Fig. 14-2 and 14-7); Springer-Verlag (Fig. 14-11 and 14-12); the American Psychological Society (Fig. 14-6); and the Royal Society (Fig. 14-9 and 14-10).

REFERENCES

Albus, K., and Wolf, W. (1984). Early post-natal development of neuronal function in the kitten's visual cortex: a laminar analysis. *J. Physiol. (London)* **348,** 153–185.

Anderson, P. A., Olavarria, J., and Van Sluyters, R. C. (1988). The overall pattern of ocular dominance bands in cat visual cortex. *J. Neurosci.* **8,** 2183–2200.

Ault, S. J., and Leventhal, A. G. (1994). Postnatal development of different classes of cat retinal ganglion cells. *J. Comp. Neurol.* **339,** 106–116.

Ault, S. J., Thompson, K. G., Zhou, Y., and Leventhal, A. G. (1993). Selective depletion of beta cells affects the development of alpha cells in cat retina. *Vis. Neurosci.* **10,** 237–245.

Beckmann, R., and Albus, K. (1982). The geniculocortical system in the early postnatal kitten: an electrophysiological investigation. *Exp. Brain Res.* **47,** 49–56.

Berson, D. M., Isayama, T., and Pu, M. (1999). The Eta ganglion cell type of cat retina. *J. Comp. Neurol.* **408,** 204–219.

Berson, D. M., Pu, M., and Famiglietti, E. V. (1998). The zeta cell: a new ganglion cell type in cat retina. *J. Comp. Neurol.* **399,** 269–288.

Birnbacher, D., and Albus, K. (1987). Divergence of single axons in afferent projections to the cat's visual cortical areas 17, 18, and 19: a parametric study. *J. Comp. Neurol.* **261,** 543–561.

Bishop, G. H., and O'Leary, J. L. (1938). Potential records from the optic cortex of the cat. *J. Neurophysiol.* **1,** 391–404.

Bishop, P. O., Jeremy, D., and Lance, J. W. (1953). The optic nerve. Properties of a central tract. *J. Physiol.* **121,** 415–432.

Bowling, D. B. (1989). Timing differences between the light responses of X cells recorded simultaneously in cat lateral geniculate nucleus. *Vis. Neurosci.* **2,** 383–389.

Bowling, D. B., and Michael, C. R. (1984). Terminal patterns of single, physiologically characterized optic tract fibers in the cat's lateral geniculate nucleus. *J. Neurosci.* **4,** 198–216.

Boycott, B., and Wässle, H. (1999). Parallel processing in the mammalian retina: the Proctor Lecture. *Invest. Ophthalmol. Vis. Sci.* **40,** 1313–1327.

Boycott, B. B., and Wässle, H. (1974). The morphological types of ganglion cells of the domestic cat's retina. *J. Physiol. (Lond)* **240,** 397–419.

Brenner, H. R., and Johnson, E. W. (1976). Physiological and morphological effects of post-ganglionic axotomy on presynaptic nerve terminals. *J. Physiol. (London)* **260,** 143–158.

Bruce, L. L., and Stein, B. E. (1988). Transient projections from the lateral geniculate to the posteromedial lateral suprasylvian visual cortex in kittens. *J. Comp. Neurol.* **278,** 287–302.

Bullier, J., Kennedy, H., and Salinger, W. (1984). Bifurcation of subcortical afferents to visual areas 17, 18, and 19 in the cat cortex. *J. Comp. Neurol.* **228,** 309–328.

Bullier, J., and Norton, T. T. (1979a). Comparison of receptive-field properties of X and Y ganglion cells with X and Y lateral geniculate cells in the cat. *J. Neurophysiol.* **42,** 274–291.

Bullier, J., and Norton, T. T. (1979b). X and Y relay cells in cat lateral geniculate nucleus: quantitative analysis of receptive-field properties and classification. *J. Neurophysiol.* **42,** 244–273.

Bullier, J. H., and Norton, T. T. (1977). Receptive-field properties of X-, Y- and intermediate cells in the cat lateral geniculate nucleus. *Brain Res.* **121,** 151–156.

Chow, K. L., and Dewson, J. H. (1966). Numerical estimates of neurons and glia in the lateral geniculate body during retrograde degeneration. *J. Comp. Neurol.* **128,** 63–74.

Cleland, B. G., Dubin, M. W., and Levick, W. R. (1971). Sustained and transient neurones in the cat's retina and lateral geniculate nucleus. *J. Physiol. (London)* **217,** 473–496.

Cleland, B. G., Harding, T. H., and Tulunay-Keesey, U. (1979). Visual resolution and receptive field size: examination of two kinds of cat retinal ganglion cell. *Science* **205,** 1015–1017.

Cleland, B. G., and Lee, B. B. (1985). A comparison of visual responses of cat lateral geniculate nucleus neurones with those of ganglion cells afferent to them. *J. Physiol. (London)* **369,** 249–268.

Cleland, B. G., and Levick, W. R. (1974a). Brisk and sluggish concentrically organized ganglion cells in the cat's retina. *J. Physiol. (London)* **240,** 421–456.

Cleland, B. G., and Levick, W. R. (1974b). Properties of rarely encountered types of ganglion cells in the cat's retina and an overall classification. *J. Physiol. (London)* **240,** 457–492.

Cohen, E. D., Zhou, Z. J., and Fain, G. L. (1994). Ligand-gated currents of alpha and beta ganglion cells in the cat retinal slice. *J. Neurophysiol.* **72,** 1260–1269.

Cornwell, P., Ravizza, R., and Payne, B. (1984). Extrinsic visual and auditory cortical connections in the 4-day-old kitten. *J. Comp. Neurol.* **229,** 97–120.

Cowan, W. M. (1972). Anterograde and retrograde transneuronal degeneration in the central and peripheral nervous system. In: *Contemporary research methods in neuroanatomy.* (W. J. H. Nauta, and S. O. E. Ebbessokn, Eds.), pp. 217–251, New York, Springer-Verlag.

Cowey, A. (1974). Atrophy of retinal ganglion cells after removal of striate cortex in a rhesus monkey. *Perception* **3,** 257–260.

Cowey, A., and Stoerig, P. (1989). Projection patterns of surviving neurons in the dorsal lateral geniculate nucleus following discrete lesions of striate cortex: implications for residual vision. *Exp. Brain Res.* **75,** 631–638.

Cowey, A., Stoerig, P., and Williams, C. (1999). Variance in transneuronal retrograde ganglion cell degeneration in monkeys after removal of striate cortex: effects of size of the cortical lesion. *Vis. Res.* **39,** 3642–3652.

Cragg, B. G. (1975). The development of synapses in the visual system of the cat. *J. Comp. Neurol.* **160,** 147–166.

Crooks, J., and Morrison, J. D. (1989). Synapses of the inner plexiform layer of the area centralis of kitten retina during postnatal development: a quantitative study. *J. Anat.* **163,** 33–47.

Dacey, D. M. (1989). Monoamine-accumulating ganglion cell type of the cat's retina. *J. Comp. Neurol.* **288,** 59–80.

Dann, J. F., Buhl, E. H., and Peichl, L. (1988). Postnatal dendritic maturation of alpha and beta ganglion cells in cat retina. *J. Neurosci.* **8,** 1485–1499.

Derrington, A. M., and Fuchs, A. F. (1979). Spatial and temporal properties of X and Y cells in the cat lateral geniculate nucleus. *J. Physiol. (London)* **293,** 347–364.

Dineen, J., Hendrickson, A., and Keating, E. G. (1982). Alterations of retinal inputs following striate cortex removal in adult monkey. *Exp. Brain Res.* **47,** 446–456.

Dineen, J. T., and Hendrickson, A. E. (1981). Age correlated differences in the amount of retinal degeneration after striate cortex lesions in monkeys. *Invest. Ophthalmol. Vis. Sci.* **21,** 749–752.

Doty, R. W. (1961). Functional significance of the topographic aspects of the retinocortical projection. In: *The visual system: neurophysiology and psychophysics.* (R. Jung and H. Kornhuber, Eds.), pp. 228–247, Berlin, Springer Verlag.

Doty, R. W. (1973). Ablation of visual areas in the central nervous system. In: *Central processing of visual information, part B.* (R. Jung, Ed.), VII, 3B, pp. 483–541, Berlin, Springer-Verlag.

Enroth-Cugell, C., and Robson, J. G. (1966). The contrast sensitivity of retinal ganglion cells of the cat. *J. Physiol. (London)* **187,** 517–552.

Eugene, D., and Taxi, J. (1991). Effects of axotomy on synaptic transmission and structure in frog sympathetic ganglia. *J. Neurocytol.* **20,** 404–419.

Farmer, S. G., and Rodieck, R. W. (1982). Ganglion cells of the cat accessory optic system: morphology and retinal topography. *J. Comp. Neurol.* **205,** 190–198.

Freed, M. A. (2000). Parallel cone bipolar pathways to a ganglion cell use different rates and amplitudes of quantal excitation. *J. Neurosci.* **20,** 3956–3963.

Freed, M. A., and Sterling, P. (1988). The ON-alpha ganglion cell of the cat retina and its presynaptic cell types. *J. Neurosci.* **8,** 2303–2320.

Freeman, A. W. (1991). Spatial characteristics of the contrast gain control in the cat's retina. *Vis. Res.* **31,** 775–785.

Freund, T. F., Martin, K. A., and Whitteridge, D. (1985). Innervation of cat visual areas 17 and 18 by physiologically identified X- and Y-type thalamic afferents. I. Arborization patterns and quantitative distribution of postsynaptic elements. *J. Comp. Neurol.* **242,** 263–274.

Friedlander, M. J., Lin, C. S., Stanford, L. R., and Sherman, S. M. (1981). Morphology of functionally identified neurons in lateral geniculate nucleus of the cat. *J. Neurophysiol.* **46,** 80–129.

Friedlander, M. J., Martin, K. A., and Vahle-Hinz, C. (1985). The structure of the terminal arborizations of physiologically identified retinal ganglion cell Y axons in the kitten. *J. Physiol. (London)* **359,** 293–313.

Friedlander, M. J., and Stanford, L. R. (1984). Effects of monocular deprivation on the distribution of cell types in the LGNd: a sampling study with fine-tipped micropipettes. *Exp. Brain Res.* **53,** 451–461.

Fukada, Y. (1971). Receptive field organization of cat optic nerve fibers with special reference to conduction velocity. *Vis. Res.* **11,** 209–226.

Fukuda, Y., Hsiao, C. F., Watanabe, M., and Ito, H. (1984). Morphological correlates of physiologically identified Y-, X-, and W-cells in cat retina. *J. Neurophysiol.* **52,** 999–1013.

Ganser, S. (1882). Über die periphere und zentrale Anordnung der Sehnervenfasern and über das corpus bigeminum anterius. *Arch. Psychiat. Nervenkr* **13,** 341–381.

Garey, L. J., Fisken, R. A., and Powell, T. P. (1973). Observations on the growth of cells in the lateral geniculate nucleus of the cat. *Brain Res.* **52,** 359–362.

Garey, L. J., and Powell, T. P. (1968). The projection of the retina in the cat. *J. Anat.* **102,** 189–222.

Garey, L. J., and Powell, T. P. S. (1967). The projection of the lateral geniculate nucleus upon the cortex in the cat. *J. Anat. London* **169,** 107–126.

Geisert, E. E., Jr. (1980). Cortical projections of the lateral geniculate nucleus in the cat. *J. Comp. Neurol.* **190,** 793–812.

Gills, J., and Wadsworth, J. (1967). Retrograde transsyanptic degeneration of the inner nuclear layer of the retina. *Invest. Ophthalmol.* **6,** 437–448.

Grant, S., and Shipp, S. (1991). Visuotopic organization of the lateral suprasylvian area and of an adjacent area of the ectosylvian gyrus of cat cortex: a physiological and connectional study. *Vis. Neurosci.* **6,** 315–338.

Gudden, B. (1869). Experimentaluntersuchungen über das periphäre unde Zentrale nervensystem. *Arch. Psychiat. Nervenkr.* **2,** 693–723.

Gudden, B. (1880). Beitrag zur Kenntnis des corpus mammilare und der sogennanten Schenkel des Fornix. *Arch. Psychiat. Nervenkr.* **11,** 428–452.

Gudden, B. (1884). Über das Corpus mammilare und der sogennanten Schenkel des Fornix. *Verslg. Dtsch. Naturforsch.* **57,** 126–157.

Guillery, R. W. (1966). A study of Golgi preparations from the dorsal lateral geniculate nucleus of the adult cat. *J. Comp. Neurol.* **128,** 21–50.

Hada, J., and Hayashi, Y. (1990). Retinal X-afferents bifurcate to lateral geniculate X-cells and to the pretectum or superior colliculus in cats. *Brain Res.* **515,** 149–154.

Haddock, J. N., and Berlin, L. (1951). Transsynaptic degeneration in the visual system. *Arch. Neurol. Psychiat. (Chic.)* **64,** 66–73.

Hammond, P. (1973). Contrasts in spatial organization of receptive fields at geniculate and retinal levels: centre, surround and outer surround. *J. Physiol. (London)* **228,** 115–137.

Hochstein, S., and Shapley, R. M. (1976). Linear and nonlinear spatial subunits in Y cat retinal ganglion cells. *J. Physiol. (London)* **262,** 265–284.

Holländer, H., Bisti, S., Maffei, L., and Hebel, R. (1984). Electroretinographic responses and retrograde changes of retinal morphology after intracranial optic nerve section. A quantitative analysis in the cat. *Exp. Brain Res.* **55,** 483–493.

Holländer, H., and Vanegas, H. (1977). The projection from the lateral geniculate nucleus onto the visual cortex in the cat. A quantitative study with horseradish-peroxidase. *J. Comp. Neurol.* **173,** 519–536.

Hubel, D. H., and Wiesel, T. N. (1961). Integrative action in the cat's lateral geniculate body. *J. Physiol. (London)* **155,** 385–398.

Hubel, D. H., and Wiesel, T. N. (1963). Receptive fields of cells in striate cortex of very young, visually inexperienced kittens. *J. Neurophysiol.* **26,** 994–1002.

Hughes, A. (1975). A quantitative analysis of the cat retinal ganglion cell topography. *J. Comp. Neurol.* **163,** 107–128.

Hughes, A. (1981). Population magnitudes and distribution of the major modal classes of cat retinal ganglion cell as estimated from HRP filling and a systematic survey of the soma diameter spectra for classical neurones. *J. Comp. Neurol.* **197,** 303–339.

Hughes, A., and Wässle, H. (1976). The cat optic nerve: fibre total count and diameter spectrum. *J. Comp. Neurol.* **169,** 171–184.

Humphrey, A. L., Sur, M., Uhlrich, D. J., and Sherman, S. M. (1985). Termination patterns of individual X- and Y-cell axons in the visual cortex of the cat: projections to area 18, to the 17/18 border region, and to both areas 17 and 18. *J. Comp. Neurol.* **233,** 190–212.

Humphrey, A. L., and Weller, R. E. (1988a). Functionally distinct groups of X-cells in the lateral geniculate nucleus of the cat. *J. Comp. Neurol.* **268,** 429–447.

Humphrey, A. L., and Weller, R. E. (1988b). Structural correlates of functionally distinct X-cells in the lateral geniculate nucleus of the cat. *J. Comp. Neurol.* **268,** 448–468.

Humphrey, D. R. (1979). Extracellular, single-unit recording methods. In: *Electrophysiological techniques.* pp. 199–259, Bethesda, MD, Society for Neuroscience.

Illing, R. B., and Wässle, H. (1981). The retinal projection to the thalamus in the cat: a quantitative investigation and a comparison with the retinotectal pathway. *J. Comp. Neurol.* **202,** 265–285.

Isayama, T., Berson, D. M., and Pu, M. (2000). Theta ganglion cell type of cat retina. *J. Comp. Neurol.* **417,** 32–48.

Jakiela, H. G., and Enroth-Cugell, C. (1976). Adaptation and dynamics in X-cells and Y-cells of the cat retina. *Exp. Brain Res.* **24,** 335–342.

Johnson, H., and Cowey, A. (2000). Transneuronal retrograde degeneration of retinal ganglion cells following restricted lesions of striate cortex in the monkey. *Exp. Brain Res.* **132,** 269–275.

Kalil, R. (1978). Development of the dorsal lateral geniculate nucleus in the cat. *J. Comp. Neurol.* **182,** 265–291.

Kalil, R. (1984). Removal of visual cortex in the cat: effects on the morphological developments of the retino-geniculo-cortical pathway. In: *Development of visual pathways in mammals.* (J. Stone, B. Dreher, and D. Rapaport, Eds.), **9** pp. 257–274, New York, Alan R. Liss.

Kalil, R. E., and Behan, M. (1987). Synaptic reorganization in the dorsal lateral geniculate nucleus following damage to visual cortex in newborn or adult cats. *J. Comp. Neurol.* **257,** 216–236.

Kalil, R. E., Tong, L. L., and Spear, P. D. (1991). Thalamic projections to the lateral suprasylvian visual area in cats with neonatal or adult visual cortex damage. *J. Comp. Neurol.* **314,** 512–525.

Kaplan, E., and Shapley, R. M. (1982). X and Y cells in the lateral geniculate nucleus of macaque monkeys. *J. Physiol. (London)* **330,** 125–143.

Kato, N., Kawaguchi, S., and Miyata, H. (1984). Geniculocortical projection to layer I of area 17 in kittens: orthograde and retrograde HRP studies. *J. Comp. Neurol.* **225,** 441–447.

Kato, N., Kawaguchi, S., Yamamoto, T., Samejima, A., and Miyata, H. (1983). Postnatal development of the geniculocortical projection in the cat: electrophysiological and morphological studies. *Exp. Brain Res.* **51,** 65–72.

Kaufman, E. F., and Rosenquist, A. C. (1985). Efferent projections of the thalamic intralaminar nuclei in the cat. *Brain Res.* **335,** 257–279.

Kisvárday, Z. F., Cowey, A., Stoerig, P., and Somogyi, P. (1991). Direct and indirect retinal input into degenerated dorsal lateral geniculate nucleus after striate cortical removal in monkey: implications for residual vision. *Exp. Brain Res.* **86,** 271–292.

Klüver, H. (1937). Certain effects of lesions of the occipital lobes in macaques. *J. Psychol.* **4,** 383–401.

Kolb, H. (1979). The inner plexiform layer in the retina of the cat: electron microscopic observations. *J. Neurocytol.* **8,** 295–329.

Kolb, H., and Nelson, R. (1993). OFF-alpha and OFF-beta ganglion cells in cat retina: II. Neural circuitry as revealed by electron microscopy of HRP stains. *J. Comp. Neurol.* **329,** 85–110.

Kolb, H., Nelson, R., and Mariani, A. (1981). Amacrine cells, bipolar cells and ganglion cells of the cat retina: a Golgi study. *Vis. Res.* **21,** 1081–1114.

Komatsu, Y., Fujii, K., Nakajima, S., Umetani, K., and Toyama, K. (1985). Electrophysiological and morphological correlates in the development of visual cortical circuitry in infant kittens. *Brain Res.* **354,** 305–309.

Kratz, K. E., Webb, S. V., and Sherman, S. M. (1978). Electrophysiological classification of X- and Y-cells in the cats lateral geniculate nucleus. *Vis. Res.* **18,** 1261–1264.

Kuffler, S. W. (1953). Discharge patterns and functional organization of mammalian retina. *J. Neurophysiol.* **16,** 37–68.

Labar, D. R., Berman, N. E., and Murphy, E. H. (1981). Short- and long-term effects of neonatal and adult visual cortex lesions on the retinal projection to the pulvinar in cats. *J. Comp. Neurol.* **197,** 639–659.

LeVay, S., and Voigt, T. (1990). Retrograde transneuronal transport of wheat-germ agglutinin to the retina from visual cortex in the cat. *Exp. Brain Res.* **82,** 77–81.

Leventhal, A. G. (1979). Evidence that the different classes of relay cells of the cat's lateral geniculate nucleus terminate in different layers of the striate cortex. *Exp. Brain Res.* **37,** 349–372.

Leventhal, A. G., Keens, J., and Tork, I. (1980). The afferent ganglion cells and cortical projections of the retinal recipient zone (RRZ) of the cat's pulvinar complex. *J. Comp. Neurol.* **194,** 535–554.

Leventhal, A. G., Rodieck, R. W., and Dreher, B. (1985). Central projections of cat retinal ganglion cells. *J. Comp. Neurol.* **237,** 216–226.

Levick, W. R. (1975). Form and function of cat retinal ganglion cells. *Nature* **254,** 659–662.

Levick, W. R., and Cleland, B. G. (1974). Selectivity of microelectrodes in recordings from cat retinal ganglion cells. *J. Neurophysiol.* **37,** 1387–1393.

Lomber, S. G., MacNeil, M. A., and Payne, B. R. (1995). Amplification of thalamic projections to middle suprasylvian cortex following ablation of immature primary visual cortex in the cat. *Cerebral Cortex.* **5,** 166–191.

Lomber, S. G., Payne, B. R., Cornwell, P., and Pearson, H. E. (1993). Capacity of the retinogeniculate pathway to reorganize following ablation of visual cortical areas in developing and mature cats. *J. Comp. Neurol.* **338,** 432–457.

Luskin, M. B., and Shatz, C. J. (1985). Studies of the earliest generated cells of the cat's visual cortex: cogeneration of subplate and marginal zones. *J. Neurosci.* **5,** 1062–1075.

Maciewicz, R. J. (1974). Afferents to the lateral suprasylvian gyrus of the cat traced with horseradish peroxidase. *Brain Res.* **78,** 139–143.

Maciewicz, R. J. (1975). Thalamic afferents to areas 17, 18 and 19 of cat cortex traced with horseradish peroxidase. *Brain Res.* **84,** 308–312.

MacNeil, M. A., Einstein, G., and Payne, B. R. (1997). Transgeniculate signal transmission to middle suprasylvian cortex in intact cats and following early removal of areas 17 and 18: a morphological study. *Exp. Brain Res.* **114,** 11–23.

Madarász, M., Somogyi, J., Silakov, V. L., and Hamori, J. (1983). Residual neurons in the lateral geniculate nucleus of adult cats following chronic disconnection from the cortex. *Exp. Brain Res.* **52,** 363–374.

Maslim, J., and Stone, J. (1986). Synaptogenesis in the retina of the cat. *Brain Res.* **373,** 35–48.

Mason, C. A. (1982). Development of terminal arbors of retino-geniculate axons in the kitten- -I. Light microscopical observations. *Neuroscience* **7,** 541–559.

Mason, C. A. (1983). Postnatal maturation of neurons in the cat's lateral geniculate nucleus. *J. Comp. Neurol.* **217,** 458–469.

Mastronarde, D. N. (1987a). Two classes of single-input X-cells in cat lateral geniculate nucleus. I. Receptive-field properties and classification of cells. *J. Neurophysiol.* **57,** 357–380.

Mastronarde, D. N. (1987b). Two classes of single-input X-cells in cat lateral geniculate nucleus. II. Retinal inputs and the generation of receptive-field properties. *J. Neurophysiol.* **57,** 381–413.

Mastronarde, D. N. (1992). Nonlagged relay cells and interneurons in the cat lateral geniculate nucleus: receptive-field properties and retinal inputs. *Vis. Neurosci.* **8,** 407–441.

Mastronarde, D. N., Humphrey, A. L., and Saul, A. B. (1991). Lagged Y cells in the cat lateral geniculate nucleus. *Vis. Neurosci.* **7,** 191–200.

Matthews, M. R., and Nelson, V. H. (1975). Detachment of structurally intact nerve endings from chromatolytic neurones of rat superior cervical ganglion during the depression of synaptic transmission induced by post-ganglionic axotomy. *J. Physiol. (London)* **245,** 91–135.

McGuire, B. A., Stevens, J. K., and Sterling, P. (1986). Microcircuitry of beta ganglion cells in cat retina. *J. Neurosci.* **6,** 907–918.

Mihailovic, L. T., Cupic, D., and Dekleva, N. (1971). Changes in the numbers of neurons and glial cells in the lateral geniculate nucleus of the monkey during retrograde cell degeneration. *J. Comp. Neurol.* **142,** 223–229.

Mitzdorf, U., and Singer, W. (1977). Laminar segregation of afferents to lateral geniculate nucleus of the cat: an analysis of current source density. *J. Neurophysiol.* **40,** 1227–1244.

Mitzdorf, U., and Singer, W. (1978). Prominent excitatory pathways in the cat visual cortex (A 17 and A 18): a current source density analysis of electrically evoked potentials. *Exp. Brain Res.* **33,** 371–394.

Moore, C. L., Kalil, R., and Richards, W. (1976). Development of myelination in optic tract of the cat. *J. Comp. Neurol.* **165,** 125–136.

Mower, G. D., Caplan, C. J., Christen, W. G., and Duffy, F. H. (1985). Dark rearing prolongs physiological but not anatomical plasticity of the cat visual cortex. *J. Comp. Neurol.* **235,** 448–466.

Münzer, E., and Wiener, H. (1902). Das Zwischen- und Mittelhirn des Kaninchensund die beziehungen des Tielerzum übrigen Centralnervensystem, mit besonderer Berücksichtigung der Pyramidenbahn und Schleife. *Mschr. Pschiat. Neurol.* **12,** 241–279.

Niida, T., Mukuno, K., Isikawa, S., and Iwai, E. (1990). Transneuronal retrograde degeneration in adult monkey retina following ablation of striate cortex. In: *Vision, memory and the temporal lobe.* (E. Iwai, and M. Mishkin, Eds.), pp 369–375, New York, Elsevier.

Niimi, K., Matsuoka, H., Yamazaki, Y., and Matsumoto, H. (1981). Thalamic afferents to the visual cortex in the cat studied by retrograde axonal transport of horseradish peroxidase. *Brain Behav. Evol.* **18,** 114–139.

Nissl, F. (1913). Die Groβhirnanteile des Kaninchens. *Arch Psychiat. Nervenkr.* **52,** 867–953.

O'Brien B. J., Isayama T., and Berson, D. M. (1999). Light responses of morphologically identified cat ganglion cells. *Invest. Ophthalmol. Vis. Sci.* S815.

Palmer, L. A., Rosenquist, A. C., and Tusa, R. J. (1978). The retinotopic organization of lateral suprasylvian visual areas in the cat. *J. Comp. Neurol.* **177,** 237–256.

Payne, B. R. (1993). Evidence for visual cortical area homologs in cat and macaque monkey. *Cerebral Cortex* **3,** 1–25.

Payne, B. R., and Lomber, S. G. (1998). Neuroplasticity in the cat's visual system. Origin, termination, expansion, and increased coupling of the retino-geniculo-middle suprasylvian visual pathway following early ablations of areas 17 and 18. *Exp. Brain Res.* **121,** 334–349.

Payne, B. R., Lomber, S. G., Macneil, M. A., and Cornwell, P. (1996). Evidence for greater sight in blindsight following damage of primary visual cortex early in life. *Neuropsychologia* **34,** 741–774.

Payne, B. R., Pearson, H. E., and Berman, N. (1984a). Deafferentation and axotomy of neurons in cat striate cortex: time course of changes in binocularity following corpus callosum transection. *Brain Res.* **307,** 201–215.

Payne, B. R., Pearson, H. E., and Cornwell, P. (1984b). Transneuronal degeneration of beta retinal ganglion cells in the cat. *Proc. R. Soc. Lond. B Biol. Sci.* **222,** 15–32.

Payne, B. R., Pearson, H. E., and Cornwell, P. (1988). Neocortical connections in fetal cats. *Neurosci. Res.* **5,** 513–543.

Pearson, H. E., Labar, D. R., Payne, B. R., Cornwell, P., and Aggarwal, N. (1981). Transneuronal retrograde degeneration in the cat retina following neonatal ablation of visual cortex. *Brain Res.* **212,** 470–475.

Pearson, H. E., Sonstein, W. J., and Stoffler, D. J. (1991). Selectivity of kainic acid as a neurotoxin within the dorsal lateral geniculate nucleus of the cat: a model for transneuronal retrograde degeneration. *J. Neurocytol.* **20,** 376–386.

Pearson, H. E., and Stoffler, D. J. (1992). Retinal ganglion cell degeneration following loss of postsynaptic target neurons in the dorsal lateral geniculate nucleus of the adult cat. *Exp. Neurol.* **116,** 163–171.

Pearson, H. E., Stoffler, D. J., and Sonstein, W. J. (1992). Response of retinal terminals to loss of postsynaptic target neurons in the dorsal lateral geniculate nucleus of the adult cat. *J. Comp. Neurol.* **315,** 333–343.

Pearson, H. E., and Thompson, T. P. (1993). Atrophy and degeneration of ganglion cells in central retina following loss of postsynaptic target neurons in the dorsal lateral geniculate nucleus of the adult cat. *Exp. Neurol.* **119,** 113–119.

Peichl, L., and Wässle, H. (1979). Size, scatter and coverage of ganglion cell receptive field centres in the cat retina. *J. Physiol. (London)* **291,** 117–141.

Perry, V. H., and Cowey, A. (1984). Retinal ganglion cells that project to the superior colliculus and pretectum in the macaque monkey. *Neuroscience* **12,** 1125–1137.

Perry, V. H., Oehler, R., and Cowey, A. (1984). Retinal ganglion cells that project to the dorsal lateral geniculate nucleus in the macaque monkey. *Neuroscience* **12,** 1101–1123.

Peters, A., and Payne, B. R. (1993). Numerical relationships between geniculocortical afferents and pyramidal cell modules in cat primary visual cortex. *Cerebral Cortex* **3,** 69–78.

Polyak, S. L. (1957). *The vertebrate visual system.* Chicago, University of Chicago Press,

Pu, M., Berson, D. M., and Pan, T. (1994). Structure and function of retinal ganglion cells innervating the cat's geniculate wing: an in vitro study. *J. Neurosci.* **14,** 4338–4358.

Raczkowski, D., and Rosenquist, A. C. (1980). Connections of the parvocellular C laminae of the dorsal lateral geniculate nucleus with the visual cortex in the cat. *Brain Res.* **199,** 447–451.

Raczkowski, D., and Rosenquist, A. C. (1983). Connections of the multiple visual cortical areas with the lateral posterior-pulvinar complex and adjacent thalamic nuclei in the cat. *J. Neurosci.* **3,** 1912–1942.

Rakic, P. (1977). Prenatal development of the visual system in rhesus monkey. *Philos. Trans. R. Soc. Lond. B Biol. Sci.* **278,** 245–260.

Ramoa, A. S., Campbell, G., and Shatz, C. J. (1988). Dendritic growth and remodeling of cat retinal ganglion cells during fetal and postnatal development. *J. Neurosci.* **8,** 4239–4261.

Rowe, M. H. (1988). Changes in inner plexiform layer thickness following neonatal visual cortical ablation in cats. *Brain Res.* **439,** 345–349.

Rowe, M. H. (1990). Evidence for degeneration of retinal W cells following early visual cortical removal in cats. *Brain. Behav. Evol.* **35,** 253–267.

Rushton, W. A. H. (1951). A theory of the effects of fiber size in medullated nerve. *J. Physiol. (London)* **115,** 101–122.

Russelakis-Carneiro, M., Silveira, L. C., and Perry, V. H. (1996). Factors affecting the survival of cat retinal ganglion cells after optic nerve injury. *J. Neurocytol.* **25,** 393–402.

Saito, H. A. (1983). Morphology of physiologically identified X-, Y-, and W-type retinal ganglion cells of the cat. *J. Comp. Neurol.* **221,** 279–288.

Saul, A. B., and Humphrey, A. L. (1990). Spatial and temporal response properties of lagged and nonlagged cells in cat lateral geniculate nucleus. *J. Neurophysiol.* **64,** 206–224.

Sawai, H., Fukuda, Y., and Wakakuwa, K. (1985). Axonal projections of X-cells to the superior colliculus and to the nucleus of the optic tract in cats. *Brain Res.* **341,** 1–6.

Shapley, R., and Perry, V. H. (1986). Cat and monkey retinal ganglion cells and their visual functional roles. *TINS* **9,** 229–235.

Shapley, R., and Victor, J. (1986). Hyperacuity in cat retinal ganglion cells. *Science* **231,** 999–1002.

Shapley, R. M., and Victor, J. D. (1978). The effect of contrast on the transfer properties of cat retinal ganglion cells. *J. Physiol. (London)* **285,** 275–298.

Shapley, R. M., and Victor, J. D. (1979). Nonlinear spatial summation and the contrast gain control of cat retinal ganglion cells. *J. Physiol. (London)* **290,** 141–161.

Shapley, R. M., and Victor, J. D. (1980). The effect of contrast on the non-linear response of the Y cell. *J. Physiol. (London)* **302,** 535–547.

Shapley, R. M., and Victor, J. D. (1981). How the contrast gain control modifies the frequency responses of cat retinal ganglion cells. *J. Physiol. (London)* **318,** 161–179.

Shatz, C. J. (1983). The prenatal development of the cat's retinogeniculate pathway. *J. Neurosci.* **3,** 482–499.

Shatz, C. J., Lindstrom, S., and Wiesel, T. N. (1977). The distribution of afferents representing the right and left eyes in the cat's visual cortex. *Brain Res.* **131,** 103–116.

Shatz, C. J., and Luskin, M. B. (1986). The relationship between the geniculocortical afferents and their cortical target cells during development of the cat's primary visual cortex. *J. Neurosci.* **6,** 3655–3668.

Shatz, C. J., and Stryker, M. P. (1978). Ocular dominance in layer IV of the cat's visual cortex and the effects of monocular deprivation. *J. Physiol. (London)* **281,** 267–283.

Shepherd, G. M. (1998). *Foundations of the neuron doctrine (History of neuroscience, number 6).* Oxford, Oxford University Press,

Sholl, D. A. (1956). *The organization of the cerebral cortex.* London, Methuen.

Silveira, L. C., Russelakis-Carneiro, M., and Perry, V. H. (1994). The ganglion cell response to optic nerve injury in the cat: differential responses revealed by neurofibrillar staining. *J. Neurocytol.* **23,** 75–86.

So, Y. T., and Shapley, R. (1979). Spatial properties of X and Y cells in the lateral geniculate nucleus of the cat and conduction veolcities of their inputs. *Exp. Brain Res.* **36,** 533–550.

Spear, P. D., Kalil, R. E., and Tong, L. (1980). Functional compensation in lateral suprasylvian visual area following neonatal visual cortex removal in cats. *J. Neurophysiol.* **43,** 851–869.

Sretavan, D. W., and Shatz, C. J. (1986). Prenatal development of cat retinogeniculate axon arbors in the absence of binocular interactions. *J. Neurosci.* **6,** 990–1003.

Stanford, L. R. (1987a). W-cells in the cat retina: correlated morphological and physiological evidence for two distinct classes. *J. Neurophysiol.* **57,** 218–244.

Stanford, L. R. (1987b). X-cells in the cat retina: relationships between the morphology and physiology of a class of cat retinal ganglion cells. *J. Neurophysiol.* **58,** 940–964.

Stanford, L. R., Friedlander, M. J., and Sherman, S. M. (1981). Morphology of physiologically identified W-cells in the C laminae of the cat's lateral geniculate nucleus. *J. Neurosci.* **1,** 578–584.

Stanford, L. R., Friedlander, M. J., and Sherman, S. M. (1983). Morphological and physiological properties of geniculate W-cells of the cat: a comparison with X- and Y-cells. *J. Neurophysiol.* **50,** 582–608.

Stanford, L. R., and Sherman, S. M. (1984). Structure/function relationships of retinal ganglion cells in the cat. *Brain Res.* **297,** 381–386.

Stein, J. J., and Berson, D. M. (1995). On the distribution of gamma cells in the cat retina. *Vis. Neurosci.* **12,** 687–700.

Stein, J. J., Johnson, S. A., and Berson, D. M. (1996). Distribution and coverage of beta cells in the cat retina. *J. Comp. Neurol.* **372,** 597–617.

Stone, J. (1974). Sampling properties of microelectrodes assessed in the cat's retina. *J. Neurophysiol.* **36,** 1071–1079.

Stone, J. (1978). The number and distribution of ganglion cells in the cat's retina. *J. Comp. Neurol.* **180,** 753–771.

Stone, J. (1983). *Parallel processing in the visual system.* New York, Plenum Press.

Stone, J., and Campion, J. E. (1978). Estimate of the number of myelinated axons in the cat's optic nerve. *J. Comp. Neurol.* **180,** 799–806.

Stone, J., and Clarke, R. (1980). Correlation between soma size and dendritic morphology in cat retinal ganglion cells: evidence of further variation in the gamma-cell class. *J. Comp. Neurol.* **192,** 211–217.

Stone, J., and Fukuda, Y. (1974). Properties of cat retinal ganglion cells: a comparison of W-cells with X- and Y-cells. *J. Neurophysiol.* **37,** 722–748.

Stone, J., Leventhal, A., Watson, C. R., Keens, J., and Clarke, R. (1980). Gradients between nasal and temporal areas of the cat retina in the properties of retinal ganglion cells. *J. Comp. Neurol.* **192,** 219–233.

Sur, M., Esguerra, M., Garraghty, P. E., Kritzer, M. F., and Sherman, S. M. (1987). Morphology of physiologically identified retinogeniculate X- and Y-axons in the cat. *J. Neurophysiol.* **58,** 1–32.

Sur, M., and Sherman, S. M. (1982). Retinogeniculate terminations in cats: morphological differences between X and Y cell axons. *Science* **218,** 389.

Sur, M., Weller, R. E., and Sherman, S. M. (1984). Development of X- and Y-cell retinogeniculate terminations in kittens. *Nature* **310,** 246–249.

Tamamaki, N., Uhlrich, D. J., and Sherman, S. M. (1995). Morphology of physiologically identified retinal X and Y axons in the cat's thalamus and midbrain as revealed by intraaxonal injection of biocytin. *J. Comp. Neurol.* **354,** 583–607.

Tong, L., Kalil, R. E., and Spear, P. D. (1984). Critical periods for functional and anatomical compensation in lateral suprasylvian visual area following removal of visual cortex in cats. *J. Neurophysiol.* **52,** 941–960.

Tong, L., Spear, P. D., Kalil, R. E., and Callahan, E. C. (1982). Loss of retinal X-cells in cats with neonatal or adult visual cortex damage. *Science* **217,** 72–75.

Tong, L. L., Kalil, R. E., and Spear, P. D. (1991). Development of the projections from the dorsal lateral geniculate nucleus to the lateral suprasylvian visual area of cortex in the cat. *J. Comp. Neurol.* **314,** 526–533.

Troy, J. B. (1983). Spatial contrast sensitivities of X and Y type neurones in the cat's dorsal lateral geniculate nucleus. *J. Physiol. (London)* **344,** 399–417.

Troy, J. B. (1987). Do Y geniculate neurons have greater contrast sensitivity than X geniculate neurons at all visual field locations? *Vis. Res.* **27,** 1733–1735.

Troy, J. B., and Enroth-Cugell, C. (1993). X and Y ganglion cells inform the cat's brain about contrast in the retinal image. *Exp. Brain. Res.* **93,** 383–390.

Tumosa, N., McCall, M. A., Guido, W., and Spear, P. D. (1989). Responses of lateral geniculate neurons that survive long-term visual cortex damage in kittens and adult cats. *J. Neurosci.* **9,** 280–298.

Updyke, B. V. (1979). A golgi study of the class V cell in the visual thalamus of the cat. *J. Comp. Neurol.* **186,** 603–620.

Usrey, W. M., Reppas, J. B., and Reid, R. C. (1999). Specificity and strength of retinogeniculate connections. *J. Neurophysiol.* **82,** 3527–3540.

Van Buren, J. M. (1963a). *The retinal ganglion cell layer.* Springfield, Illinois, Charles C. Thomas.

Van Buren, J. M. (1963b). Trans-synaptic retrograde degeneration in the visual system of primates. *J. Neurol., Neurosurg. Psych.* **26,** 402–409.

Von Monakow, C. (1882). Über einige durch Exstirpation circumscripter Hirnrindenregionen bedingte Entwickelungenshemmungen des Kaninchengehirns. *Arch. Psych. Nervenkr.* **12,** 141–156.

Von Monakow, C. (1889). Experimentelle und pathologisch-anatomische Untersuchungen über die optischen Centren und Bahnen. *Arch. Psychiatr. Nervenkr.* **20,** 714–787.

Walsh, F. B. (1947). *Clinical neuro-ophalmology.* Baltimore, Williams and Wilkins,

Wässle, H., and Boycott, B. B. (1991). Functional architecture of the mammalian retina. *Physiol. Rev.* **71,** 447–480.

Wässle, H., Boycott, B. B., and Illing, R. B. (1981a). Morphology and mosaic of on- and off-beta cells in the cat retina and some functional considerations. *Proc. R. Soc. Lond. B Biol. Sci.* **212,** 177–195.

Wässle, H., Levick, W. R., and Cleland, B. G. (1975). The distribution of the alpha type of ganglion cells in the cat's retina. *J. Comp. Neurol.* **159,** 419–438.

Wässle, H., Peichl, L., and Boycott, B. B. (1981b). Dendritic territories of cat retinal ganglion cells. *Nature* **292,** 344–345.

Wässle, H., Peichl, L., and Boycott, B. B. (1981c). Morphology and topography of on- and off-alpha cells in the cat retina. *Proc. R. Soc. Lond. B Biol. Sci.* **212,** 157–175.

Weber, A. J., Kalil, R. E., and Stanford, L. R. (1998). Dendritic field development of retinal ganglion cells in the cat following neonatal damage to visual cortex: evidence for cell class specific interactions. *J. Comp. Neurol.* **390,** 470–480.

Weber, A. J., McCall, M. A., and Stanford, L. R. (1991). Synaptic inputs to physiologically identified retinal X-cells in the cat. *J. Comp. Neurol.* **314,** 350–366.

Weller, R. E., and Kaas, J. H. (1989). Parameters affecting the loss of ganglion cells of the retina following ablations of striate cortex in primates. *Vis. Neurosci.* **3,** 327–349.

Williams, C., Azzopardi, P., and Cowey, A. (1995). Nasal and temporal retinal ganglion cells projecting to the midbrain: implications for "blindsight". *Neuroscience* **65,** 577–586.

Williams, R. W., Bastiani, M. J., Lia, B., and Chalupa, L. M. (1986). Growth cones, dying axons, and developmental fluctuations in the fiber population of the cat's optic nerve. *J. Comp. Neurol.* **246,** 32–69.

Williams, R. W., Cavada, C., and Reinoso-Suarez, F. (1993). Rapid evolution of the visual system: a cellular assay of the retina and dorsal lateral geniculate nucleus of the Spanish wildcat and the domestic cat. *J. Neurosci.* **13,** 208–228.

Wilson, J. R., Bullier, J., and Norton, T. T. (1988). Signal-to-noise comparisons for X and Y cells in the retina and lateral geniculate nucleus of the cat. *Exp. Brain. Res.* **70,** 399–405.

Winfield, D. A. (1981). The postnatal development of synapses in the visual cortex of the cat and the effects of eyelid closure. *Brain. Res.* **206,** 166–171.

Winfield, D. A., Hiorns, R. W., and Powell, T. P. (1980). A quantitative electron-microscopical study of the postnatal development of the lateral geniculate nucleus in normal kittens and in kittens with eyelid suture. *Proc. R. Soc. Lond. B. Biol. Sci.* **210,** 211–234.

Winkler, C. A. (1918). The olfactory tract of the rabbit. In: *Opera Omnia.* 5, pp. 397–413., Haarlem, Bohn Press.

Wolfe, J., and Palmer, L. A. (1998). Temporal diversity in the lateral geniculate nucleus of cat. *Vis. Neurosci.* **15,** 653–675.

Yukie, M., and Iwai, E. (1981). Direct projection from the dorsal lateral geniculate nucleus to the prestriate cortex in macaque monkeys. *J. Comp. Neurol.* **201,** 81–97.

15

PRIMARY VISUAL CORTEX WITHIN THE CORTICO-CORTICO-THALAMIC NETWORK

JACK W. SCANNELL AND MALCOLM P. YOUNG

Neural Systems Group, Newcastle University, Newcastle-upon-Tyne, United Kingdom

INTRODUCTION

What is the function of primary visual cortex, and how does its function relate to the structure of the network within which it is embedded? This chapter poses a question that it does not pretend to be able to answer satisfactorily. Rather, we aim to develop a framework that may help provide a better explanation than we have at present. En route, we examine function, structure, and structure-function relationships. We consider function and structure-function relationships in rather general terms, but discuss the structure of the network in which visual cortex is embedded more comprehensively. Three themes link the ideas presented in the chapter. First, the visual system is a specific biological adaptation, so is best studied as such. Second, the normal role of the visual cortex may be better described as "inference about" rather than the more conventional "analysis of" the visual world. Third, explanations of the function of individual cortical areas must consider wider aspects of the cortico-cortico-thalamic network in which the areas are embedded. The first two themes are important because they bear on the kind of tasks and stimuli that we should use to explore the function of the visual cortex. The third theme is important because it bears on the way we think about structure and hence structure-function relationships.

The idea that structure and function are intimately related has driven much neuroanatomy. It has motivated the development of a range of methods for exploring the connectional structure of the brain (Marchi and Algeri, 1895; Nauta and Gygax, 1954; Kristensson et al., 1971; Cowan et al., 1972; Gerfen and Sawchenko, 1984) and the application of these methods to the thalamus and cortex in several species (Le Gros Clark, 1932, 1942; Rose and Woolsey, 1948; Polyak, 1927, 1933). Over the last 25 years, retrograde and anterograde tracers have resulted in an explosion in our knowledge of brain connectivity (Zeki and Shipp, 1988, 1995; Felleman and Van Essen, 1991; Young, 1993; Scannell et al. 1995; Pandya and Yeterian, 1985). In parallel with anatomical advances has come great progress in our knowledge of the functional properties of the cortical areas and thalamic nuclei that the connections link. So, for example, there is now a great deal of information about the functional properties of neurons in areas 17 and 18; however, there is rather less information about the roles that areas 17 and 18 normally play in the brain's global information processing economy.

In some areas of neuroscience, the relationship between structural and functional explanations has been intimate and fertile. We can explain, for example, the electrical properties of single nerve cells in terms of their structural organisation (e.g., Holmes and Rall, 1992; Agmonsnir and Segev, 1993; Bush and Sejnowski, 1994). Similarly, in some simple nervous systems, important behaviors can be related to the structural organization of networks of neurons in a formal and rigorous way (Kristan and Shaw, 1997; Lewis and Kristan, 1998). Compared with these examples, attempts to explain the properties of cortical areas, and their contributions to the behavior of the animal, in terms of cortical connectivity, look relatively unsuccessful. Much of the difficulty, no doubt, stems from the great complexity of the system in question. However, a critic might argue that connectional neuroanatomy has run in parallel with, rather than in synergy with, systems level neurophysiology. The two approaches often study the same brain areas, frequently feature in the introduction or discussion section of papers, and even involve the same researchers; but their formal interdependence is largely unexplored. Functional properties are only occasionally predicted from connectional neuroanatomy (but see Zeki, 1969, 1971, 1974; Scannell et al., 1996), and anatomy is almost never predicted from physiology. Thus at the level of whole systems in the brain, we still do not know how to relate the extrinsic connections that link different brain structures to the functional properties of the network or of its constituent structures (Vanduffel et al., 1995, 1997a, 1997b; Lomber et al., 1996a, 1996b; Michalski et al., 1993, 1994).

With these problems in mind, we remember Teuber (1955), who wrote that "no degree of refinement of ... technique can substitute for clarity of concepts referring to structure and function. ... Unless we work on our concepts, the accumulation of facts will hinder rather than help". Therefore, before we engage in our own post-hoc discussion of structure and function in the visual cortex, we must sharpen our concepts about function. We first consider what we mean by function in neuroanatomical structure-function relationships and the tactics that visual cortex is presumed to use to perform its functions. Then, we provide a coarse-grained but comprehensive examination of the organization of the network of which primary visual cortex is an integral component. Finally, we review work that has explored the relationship between the structure and function of cat cortex at the systems level and explore possible strategies to complement current methods for investigating structure-function relationships.

FUNCTION

CLARIFYING FUNCTION OF PRIMARY VISUAL CORTEX

Clarity over the meaning of function is necessary before we can relate the structure of the brain to brain function in a rigorous way. Young et al. (2000) have identified at least five senses of function that are in frequent, and often confused, use in the neuroscience literature: (1) the evolutionary sense of function as survival function, f_e; (2) function as a discrete local property, f_l; (3) function in the context of the network, f_c; (4) the global or behavioral function of the whole animal, f_g; and (5) function in the formal mathematical sense of a mapping between input and output, f_m. These senses are not independent, but show considerable overlap. For example, in the natural environment, evolutionary function, f_e, and behavior, f_g, have a great deal in common. Function, in all five senses, can also apply to different levels of organization. For example, any behavioral output, f_g, can be explained in terms of smaller individual behaviors. Thus the meaning of function, applied to structure-function relationships in neuroscience, varies over at least two dimensions, a sense dimension and a scale dimension. If we want to relate structure to function, we have to be clear about our structure and about the sense and the scale of the particular function that we have in mind. We consider the different senses of function first.

Function (Evolutionary), f_e

Visual systems have evolved for the sole purpose of conferring an evolutionary fitness benefit on their owners. Evolutionary fitness, in terms of increased survival and reproduction, is their only ultimate goal. Therefore, function in the sense of evolutionary fitness benefit, f_e, is the dominant sense of function in biology. Other senses of function should be ultimately explicable in terms of their contribution to evolutionary fitness. Function f_e can be studied using analysis of visual behavior in naturalistic conditions. By exploiting within and/or between species variability in visual behavior, environment, and neuronal architecture, it is possible to obtain information about the costs and benefits of different visual systems. This approach has provided compelling accounts of differences in neural processing in the early stages of the visual systems of some flies (O'Carrol et al., 1996, 1997). Evolutionary function could, in principle, provide a strong set of constraints to theories concerning the visual cortical network. However, given the huge amount of research on the cat visual system, we are ignorant about the uses to which most cats (or most macaques) put their visual systems most of the time. We are also ignorant of the evolutionary history of the cat visual system.

Given the absence of hard evidence about the evolutionary function of the visual systems of cats (and macaques), there is a surprising degree of consensus about the purposes of visual systems. Many researchers presume that the primary purpose of visual systems has something to do with reconstructing the image (Aloimonos and Rosenfeld, 1991; Tsotsos, 1987) or building a three-dimensional model of the world containing information about objects and their locations (Lennie, 1998; Marr, 1982; Treisman, 1986; Barrow and Tenebaum, 1986; Cavanagh, 1989). These ideas, some of which owe their origins to machine vision and artificial intelligence, are largely untested and may even be substantially wrong when applied to biological visual systems (Churchland et al., 1993).

Given limited neural resources (Baddeley, 1996; Scannell, 1997; Cherniak, 1990, 1992; Laughlin et al., 1998), it is unlikely that any biological visual system will devote costly neural machinery to processing information that does not have direct survival value. For example, the "bug detector" found in retina of the frog (Barlow, 1953) is not suited to the general task of "building a three-dimensional model of the world containing information about objects and their locations." However, the "bug-detector" is good at telling the frog's brain about the small subset of objects that the frog might like to eat (Lettvin et al., 1959). Similarly, some spiders have retinae that are effective "mate detectors" but that would be rather bad at the more general task of "reconstructing the image" (Land and Barth, 1992). Within mammals and other groups, peripheral sensory neurons are exquisitely tuned to the ecological niche occupied by the animal that bears them (Osorio et al., 1997; Osorio and Vorobyev, 1996). They are not general purpose mechanisms for sampling and reconstructing the environment, but are designed to specifically sample those aspects of the environment that are biologically relevant. There is ample evidence that peripheral sensory specializations are reflected in the organization of the brain (Azzopardi and Cowey, 1993, 1996; Catania and Kaas, 1997a; 1997b). Therefore, even if one is interested in central mechanisms of vision, it is necessary to take the evolutionary function of visual systems seriously. With this aim in mind, a vague, but perhaps biologically plausible, statement about the evolutionary function of the visual system might be that the visual system is engaged in providing cost-effective information about those aspects of the visual world that have an impact on survival and reproduction. The extent to which this involves reconstructing the image, or building three-dimensional models of the world, are still open empirical questions as far as the cat is concerned.

Even the rather vague statement that the visual system is engaged in providing cost-effective information about those aspects of the visual world that have impact on survival and reproduction bears on the study of the visual system. For example, visual systems are unlikely to be passive processors whose neurons come simply to reflect statistical redundancies in the visual input. This is true for two reasons. First, visual input is only one of many sources of information used in visual behavior. Second, animals sample their environments in a biased way. Nor is the ultimate goal of any structure in the visual system likely to be the production of an efficient code, as neuronal coding strategies, although of fundamental importance, are means and not biological ends (Baddeley and Hancock, 1991; Laughlin et al., 1998, Van Hateren, 1992; Field, 1994).

To summarize, visual systems are not general-purpose computers designed to build general models of the world. They are specific biological adaptations designed to serve particular species-specific behaviors (Barton and Dean, 1993; Barton, 1998; Tooby and Cosmides, 1995). Consequently, functional studies relying on stimuli or behaviors for which visual systems are not adapted may be difficult to interpret.

Function (Local), f_l

This is the function, f_l, of a component of a system when considered as an isolated element outside the system in which it is normally embedded. If we were to remove a capacitor from the circuit board of a radio, for example, we might say that the capacitor's function, f_l, is to store charge. If we could dissect out and isolate a neuron from area 17, its function, f_l, might be to integrate e.g. (Abeles, 1990), or differentiate (Chance et al., 1998), its synaptic inputs to produce a spike train. Local function, f_l, is easy to apply to the example of an electronic circuit board, as solder can be applied at the interface between components and the rest of the system. However, local function may be a concept that is difficult to apply to cortical areas because extrinsic projections reach right into the area's local circuits and form an intrinsic part of them. Therefore, local circuits lack an interface at which imaginary neural solder could be applied. Above the level of the single neuron, therefore, f_l will be an approximation with limited validity.

Function (in Context), f_c

Function in context, f_c, is the function of a component of a system when it is in place within the system (Lewis, 1970). For example, if we were to take the capacitor whose local function, f_l, was to store charge and solder it back into the radio, its function in context, f_c, could be to act as a high-pass filter to help tune the radio to different stations. The function in context, f_c, of brain components depends on the wider structure of the network of which they are a part. Hence the function, f_c, of area 17 is determined by the nature of its inputs, its internal computations, the nature of its outputs, and the organization of the network that contributes the inputs and receives the outputs. Much work on function at the level of the cortical area implicitly uses a sense of function that is close to f_c (Pasternak et al., 1989, 1995; Pasternak and Maunsell, 1992; Lomber et al., 1996a, 1996b; Petersen et al., 1988; Poeppel, 1996). In this chapter, function is usually function in context, f_c.

Systems neuroscience has four major approaches to determining function in context, f_c, of brain areas and their contribution to the brain's overall processing strategy. The first approach infers f_c of the area from the receptive field properties of a sample of neurons (Lennie, 1998; Barlow et al., 1967; Nikahara et al.,

1968). The second approach infers an area's f_c by inactivating it, either by cooling (Lomber et al., 1996a; 1996b), with pharmacological agents (Martin and Ghez, 1999; Worgotter et al., 1998), or with lesions (Pasternak et al., 1989; 1995; Pasternak and Maunsell, 1992). Changes in global function, f_g, or in the f_c of other areas are presumed to reflect the f_c of the inactivated structure. The third approach infers function by stimulating neurons within the area (Tehovnik, 1996; Salzman et al., 1992), and then measuring the behavioral effects of the activation. The fourth approach examines how gross measures of neuronal activation, such as cerebral blood flow (Petersen et al., 1988; Poeppel et al., 1996; see also Chapter 4) or uptake of metabolic markers (Vanduffel et al., 1995, 1997b; see also Chapter 3) covary with sensory stimuli or behavioral tasks (Scannell and Young, 1999). Regions where gross blood flow or metabolic rate covary with the stimulus or behavior are presumed to have an f_c that involves processing the stimulus or generating the behavior.

Although these approaches have contributed nearly all our useful information about function, all can be problematic (Lewis, 1970; Cumming and Parker, 1997; Poeppel, 1996; Scannell and Young, 1999; Young et al., 2000). We do not consider single-unit recording, functional imaging, or stimulation here. However, lesions or inactivations are commonly used to impute function in context, f_c, to brain structures. Young et al. (2000) have found that conventional inference from inactivations or lesions, including double dissociation, can misattribute function to structure. Lesion effects could be used to recover detailed and reliable information about the way in which structures contributed to particular functions in a simple network, *only* when the connectivity of the network was known. Thus the function in context, f_c, of structures such as primary visual cortex must be framed in terms of the wider connectional network (see later).

Function (Global), f_g

This is function, f_g, in the sense of the behavior of the whole animal. Global function, f_g, can include biological behaviors closely related to evolutionary function f_e, such as catching prey or finding mates, and to psychophysical responses such as orienting to food items or detecting gratings. Because global function is relatively easy to measure, changes in global function that result from lesions or inactivation are often used to make inferences function in context, f_c, of brain structures (Pasternak et al., 1989, 1995; Pasternak and Maunsell, 1992; Lomber et al., 1996a, 1996b; Young et al., 2000).

Function (Mathematical), f_m

This is function, f_m, in the formal mathematical or computational sense (equation 1).

$$y = f_m(\mathrm{x}) \tag{1}$$

Function f_m is the mapping of an input vector onto an output vector and can be applied to any level of the nervous system. For example, a psychophysical task might measure behavioral output y, which is a function, f_m, of sensory input x. Alternatively, an intracellular electrode might measure membrane potential y, which is a function, f_m, of synaptic input x. One could even consider the dependencies of the outputs of all the neurons in area 17, represented by vector y, on all inputs to area 17, represented by vector x. Thus, function, f_m, could apply to the whole animal and any processor, set of processors, or sub-processor within the brain.

Scales of Function

Function, in all five senses outlined previously, can also apply to different levels of organization. This is illustrated for some of the senses of function in Table 15-1. For example, the ultimate evolutionary function of the visual system will be to "leave more healthy offspring," which can be explained in terms of subsidiary evolutionary functions, "detect prey," which, perhaps in the case of frogs (Barlow, 1953), can be explained in terms of "identify small moving dark patches against a light background." Similarly, any behavioral response of the animal, f_g, can be explained in terms of simpler behavioral responses. Whenever we talk about structure-function relationships, we should bear in mind which type of function we are relating to the structure. Slipping between different scales and senses of function is likely to cause confusion.

IS THE FUNCTION, F_C, OF VISUAL CORTEX TO ANALYZE OR INFER?

We now consider a high-level functional question whose relevance will become clearer when we discuss the connectional structure of network of which

TABLE 15-1. Senses and scales of function.[a]

Sense	High-level	Intermediate-level	Low-level
f_e	Leave more offspring	Catch prey	Identify small moving dark patches against light background.
f_c	Infer scene parameters using image data and experience. Pass parameters to other structures in network	Infer local image parameters using local data and data from nearby region. Pass parameters to other structures in network.	Infer local contrast using image data and contrast data from nearby regions. Pass parameters to other structures in network.
f_g	Catch prey	Move forward, open mouth…	Step with right leg, step with left leg…
f_m	$y = f(x)$	$y = f(x) = g(x)/h(x)$	$f(x) = g(x)/h(x) = 2x^2/3\sqrt{x}$

[a]Reading vertically shows different, but partially overlapping, senses of function (e.g. 'catch prey' can occur under f_e and f_g). Reading horizontally gives examples of different scales of function.

visual cortex is a component. That is, is the high-level function, f_c, of the visual cortex best regarded as analysis of the scene or as inference about the scene?

Much of the language of visual neuroscience is derived from an explicit model of "vision-as-analysis" of the scene or image (Lennie, 1998). The vision-as-analysis model supposes that most of the useful information about the visual world is present in the immediate retinal image. Feedforward connections convey the useful information from the retina to primary visual cortex and to then on to successively higher levels for further analysis. Feedback projections have important but modest roles, concerned with gating, modulation, providing contextual information, or improving signal-to-noise ratio (Gulyas et al., 1987, 1990, Murphy and Sillito, 1987; Hupé et al., 1998). Through successive processing stages, information in the retinal image is gradually decoded to provide representations of objects or spatial relationships.

Despite the popularity and success of the vision-as-analysis model, it is possible that much of what appears to be analysis is a consequence of the way that visual systems are typically studied in the laboratory (see later). We think that it may be better to think in terms of "vision-as-inference." Vision-as-inference supposes that perhaps only a little information is normally derived from the immediate and local retinal signal. However, a great deal of information comes from the expectations of the visual system's neurons, gained via evolutionary or direct experience of the world (e.g., sky is above; eyes are normally parts of faces; colinear contours are generally connected; objects tend to move together). We believe that vision-as-inference may provide a better explanation of the behavioral (Knill and Richards, 1996; Knill et al., 1996; Yuille and Bulthoff, 1996; Williams et al., 1998a, 1998b) physiological (Peterhans and van der Heydt, 1989; Coppola, 1998a, 1998b) and, as we argue later, anatomical properties of the visual system. Vision-as-analysis can be regarded as a special condition of vision-as-inference (Fig. 15-11).

When contrasting the analysis and inference models, we should remember that primary visual cortex performs its evolutionary, local, in-context, and mathematical functions within a network that is designed by evolution and tuned by experience to work in complicated natural scenes. Such scenes typically contain many objects, shadows, clutter, varying illumination, and even evolved camouflage. These scenes also usually contain redundancy, that is, several cues, simultaneously present, that the visual system could combine to reach the right answer. Suppose that the visual cortex has to decide about some property of such a complex scene by monitoring the activity of its neurons. Because both neurons and the world are noisy and variable, the activity of the neurons must be interpreted probabilistically (Rieke et al., 1997). This is *best* done by using bayesian inference to combine prior knowledge about the world, the activity of the neurons, and the dependencies of neurons on the world to compute the posterior probability of the scene given all the information (equation 2). This optimal statistical strategy is not dependent on any particular neural coding tactic (e.g., sparse vs. compact code), although it is possible that different coding tactics may have more or less efficient neural implementation.

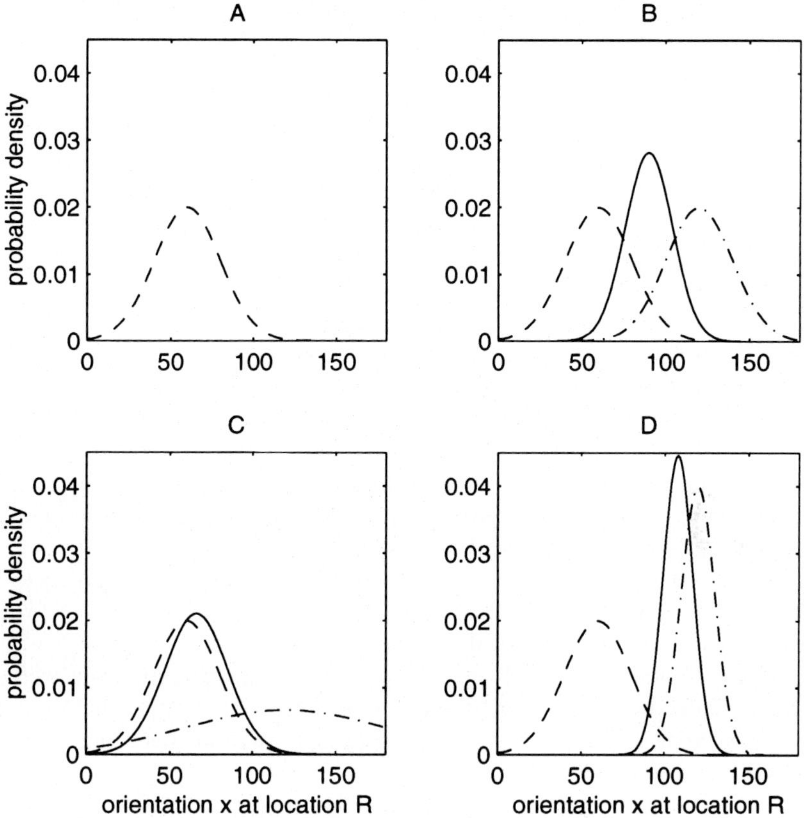

FIGURE 15-1. An illustrative example of analysis versus inference. Suppose neurons in the primary visual cortex estimate the orientation, x, of a contour lying within a small image region, R. **(A)** The cortical neurons' estimate of orientation using only a noisy signal from a sample of lateral geniculate cells looking directly at image region R. The y axis represents the likelihood of orientation, x. The most likely orientation is 60°, but the estimate is uncertain. **(B)** Bayesian inference. Here, the cortical neurons use a prior expectation (dots and dashes) about image region R, perhaps from neighboring parts of the scene, *and* the likelihood function from their geniculate input (dashes, as in **A**). The posterior probability (solid line) of orientation x is proportional to the product of the prior and the likelihood function and is the *best* estimate of orientation (equation 2). The posterior probability is less uncertain (i.e., narrower) than either the likelihood function or the prior expectation. **(C)** The results of bayesian inference when there is a weak prior expectation (dotted and dashed line) of the orientation at region R. A weak prior might be expected under typical neurophysiological conditions where there is no contextual information from outside the classical receptive field. Here, the posterior probability of local orientation (solid line) resembles the likelihood function (dashed line), so the decision about orientation depends primarily on analysis of the local image region R. **(D)** The results of bayesian inference with a very strong prior expectation (perhaps in a natural image with substantial contextual information). Here, the posterior probability (solid line) resembles the prior expectation (dotted and dashed line) and depends more weakly on the likelihood function (dashed line) from the local image region.

$$p(scene \mid x_n = K \times p(x_n \mid scene)p(scene) \qquad (2)$$

Here, x_n, is a vector describing the neurons' activity; $p(scene \mid x_n)$ is the posterior probability of the scene property, given that the neurons activity is x_n; $p(scene)$ is the prior probability of the scene property; $p(x_n \mid scene)$ is the likelihood function of the cells having activity levels, x_n, given the scene property; and K is a factor that ensures that the area under the posterior probability density function, $p(scene \mid x_n)$, is unity. The meaning of these possibly unfamiliar terms is explained in Fig. 15-1.

Concrete neurophysiological examples of inference as opposed to analysis comes from work that shows neurons early in the visual system that are sensitive to colinearity outside their receptive field. Such colinearity is associated with illusory contours (Peterhans and von der Heydt, 1989; Grosof et al., 1993) and "local association fields" (Field et al., 1993; Polat et al., 1998). Here we can think of the indirect input from outside the classical receptive field as establishing the prior expectation and the direct visual input from the cell's classical receptive field as generating the likelihood function. Given an illusory contour, the likelihood function is flat, as the input from the classical receptive field provides no useful information about the orientation or motion of the contour. However, the cell can still produce a tuned response as a consequence of the strong prior expectation (Fig. 15-1D) generated by experience of colinearity in the normal visual environment and the presence of oriented luminance contours outside the classical receptive field.

Further examples of the influence of prior expectations come from work in attention. Here, attention can be thought of as a prior expectation that modifies the way afferent signals (corresponding roughly to the likelihood function) are processed. For example, Treue and Maunsell (1996) presented two small moving objects simultaneously within the receptive fields of neurons in cortical areas MT and MST of the macaque. The two objects moved in opposite directions, one in the neuron's preferred direction and the other in the neuron's null direction. If the monkey paid attention to the object drifting in the neuron's preferred direction, then the neuron tended to respond at a high rate. However, if the monkey paid attention to the stimulus drifting in the neuron's null direction, then neuronal activity was strongly suppressed, typically by around 50%. This effect depended only on the attentional state of the animal, as the stimulus was exactly the same in both conditions. Similar phenomena occur in macaque area V4, where many neurons' orientation tuning depends on attentional state rather than retinal signal (Maunsell and Ferrera, 1995). These examples show that changes in the prior expectation, from either simultaneously present contextual information in the image, or from changes in the animals attentional state, have a large effect on the inferences that the visual system derives from the retinal signal. Indeed, when challenged with appropriate stimuli, the visual system produces responses dominated by the prior expectation and largely independent of analysis of the local image region.

In scenes with unusual or unnatural statistical properties (i.e., most stimuli used in vision research), there are two likely consequences. First, the visual system will have a weak prior expectation about the stimulus (most vision research stimuli are *designed* to do away with prior expectations altogether). Second, the visual system will have a wrong or ambiguous prior expectation about the stimulus. In the first case, the likelihood function, or analysis term, will be emphasized (e.g. Fig. 15-1A, C). The results of such studies will be consistent with the vision-as-analysis model. In the second case, we should expect optical illusions (Williams et al., 1998a; 199b) or bi-stable percepts such as the Necker cube. In familiar or natural scenes, for which the visual system has both strong prior expectations and substantial contextual information, the prior probability, or expectation term, will play a greater role (Fig. 15-1B, D).

Despite the bias of current vision research methods toward emphasizing analysis, there is evidence that prior expectations about the nature of the visual environment occur throughout the visual system. This is well known in invertebrates (Srinivasan et al., 1982; O'Carroll et al., 1996; 1997; Van Hateren, 1997). However, even in mammals, neural coding in the retina, lateral geniculate nucleus (LGN), and primary visual cortex all show evidence for prior knowledge of the visual environment (Atick, 1992; Atick and Redlich, 1992; Atick et al., 1992; Dan et al., 1996; Olshausen and Field, 1996; van Hateren and van der Schaaf, 1998; Roa and Ballard, 1999). Such prior expectations are typically thought of in terms of predictive coding. That is, the visual system exploiting a knowledge of the environment to maximize information transmission given a limited channel capacity (Shrininvasan et al., 1982), metabolic capacity (Baddeley, 1997; Baddeley et al., 1997), or the need for a sparse representation (Olshausen and Field, 1996). However, knowledge of the normal visual environment is also a necessary precondition for bayesian inference. As well as achieving coding efficiency, visual systems should exploit prior knowledge of the environment to make better decisions about biologically salient features of the world (Knill and Richards, 1996).

STRUCTURE

In this section, we move from function f_c of the visual system back to structure. First, we consider whether the machinery of visual cortex appears designed to directly analyze afferent input from the retina or whether it appears designed to pool ascending retinal signals with prior expectations from others within and outside the cortical area. Second, we explore the wider organization of the network that links the cortical areas and thalamic nuclei, concentrating on visual cortex.

EXTRINSIC AND INTRINSIC CONNECTIONS

Single Projections are Surprisingly Weak

Given the quantity and quality of research on the connections between the LGN and primary visual cortex, one might be forgiven for thinking that a large proportion of the synapses in primary visual cortex come from neurons in the LGN. In fact, this is not the case. Peters and Payne (1993) performed a quantitative analysis on the projection from the X and Y cells of the LGN to primary visual cortex. They estimated that geniculocortical axons account for only 5% of the total excitatory synapses on the average layer IV pyramidal neuron. This low value is even more surprising when we remember that layer IV receives more geniculate input than the other cortical layers (Boyd and Matsubara, 1996; see also Chapter 1). Consequently, well over 95% of area 17's excitatory synapses belong to neurons that originate not in the LGN, but from other cells in area 17, other cortical areas, and other thalamic nuclei (see Chapter 5).

This arrangement is not unique to the cat visual system. Similarly low numbers of synapses come from the projection from the LGN to the primary visual cortex of the macaque (fewer than 5% of synapses in layer IV) (Peters et al., 1994). Individual cortico-cortical projections provide similarly few synapses to their targets. The projection from macaque V1 to MT provides fewer than 5% of the excitatory synapses even in the densest target zones (Anderson et al., 1998), and the projection from V2 to V1 provides less than 6% of the excitatory synapses to upper layer pyramidal cells in V1 (Budd, 1998). Thus, the strongest thalamo-cortical and cortico-cortical projections contribute few excitatory synapses to their target areas. Even with of the differential amplification that is known to occur in some projections (Stratford et al., 1996), the anatomical data suggest that single thalamo-cortical or cortico-cortical projections will not dominate neural activity in their targets.

Sum of Inputs is Much Greater than Single Inputs

The strongest extrinsic thalamo-cortical or cortico-cortical projections rarely contribute more than 5% of the excitatory synapses to their densest target zones (Peters and Payne, 1993; Anderson et al., 1998; Budd, 1998). However, each cat cortical area receives, on average, approximately 15 cortico-cortical and 12 thalamo-cortical connections (Fig. 15-2) (Scannell et al., 1999). Scannell et al. (2000) have pooled quantitative retrograde tracing data from their own laboratories and

FIGURE 15-2. Quantitative distribution of connection densities. The graphs were computed from quantitative data from a large number of retrograde tracer injections in cortical areas PMLS, CGa. CGp, 6m, and PFC (Scannell et al., 2000; Vanduffel et al., 1997a; Olson and Musil, 1992; Musil and Olson, 1988a; 1988b; 1991). For each single tracer injection, afferent areas were ranked according to the proportion of labeled cells that they contained (x-axis). The mean and standard deviation of the proportion of labeled neurons was then computed for each rank. Thus, connection rank 1 represents the mean proportion of labeled neurons in the area or nucleus containing the most labeled cells, rank 2 represents the mean proportion of labeled neurons in the area or nucleus containing the second largest number of labeled cells, etc. The y-axis represents the proportion of total labeled extrinsic neurons in cortex or thalamus. **(A)** Frequency distribution of cortico-cortical connection densities. The graph shows that the area contributing the strongest cortico-cortical projection to the injected area typically contains approximately 30% of all labeled cortical neurons. Other areas

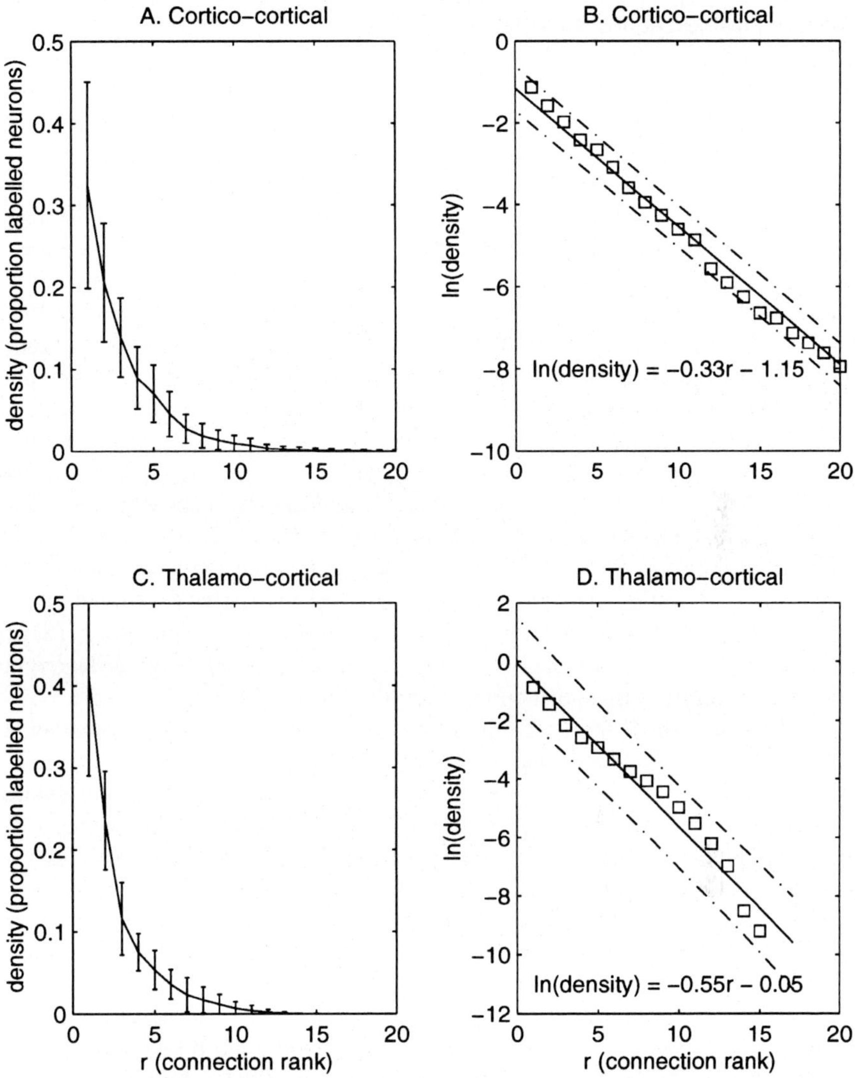

contribute approximately 70% of the labeled cells. **(B)** Same data as in **(A),** but log of connection density is plotted against rank (squares). The solid line shows a linear regression on the log of the data, and dashed lines show upper and lower 95% prediction limits. The straight line on log axes provides an excellent fit ($R^2 = 0.98$, $p < 0.001$), showing that connection density declines exponentially with rank. The regression equation is given in the bottom left corner. **(C)** Frequency distribution of thalamo-cortical connection densities. The graph shows that the nucleus contributing the strongest thalamic input to an area typically contains approximately 40% of all labeled thalamic neurons. All other nuclei contain approximately 60% of the labeled thalamic cells. **(D)** Same data as in **(C),** but log of connection density is plotted against rank (squares). The solid line shows a linear regression on the data, and dashed lines show upper and lower 95% prediction limits. The straight line on log axes provides an excellent fit ($R^2 = 0.94$, $p < 0.001$), showing that connection density declines exponentially with rank. The regression equation is given in the bottom left corner.

from the anatomical literature (Olson and Musil, 1992; Musil and Olson, 1988a, 1988b, 1991) to examine within and between individual variability in connection densities. The data make it possible to estimate the quantitative frequency distribution of cortico-cortical and thalamo-cortical connection densities to the typical cortical area (Fig. 15-2).

We can use Fig. 15-2 to estimate the proportion of extrinsic thalamo-cortical and cortico-cortical synapses in a typical cortical area that come from the strongest single projections if we make the following two assumptions. First, we assume that the strongest thalamo-cortical and cortico-cortical projections contribute similar numbers of synapses to their targets (Peters and Payne, 1993; Peters et al., 1994; Budd, 1998, Anderson et al., 1998). Second, we assume that the number of retrogradely labeled neurons in each extrinsic connection is proportional to the number of synapses that the extrinsic area contributes to the injected region. Given these assumptions, the strongest single cortico-cortical projection will contribute only 15% of extrinsic input and the strongest single thalamo-cortical projection will contribute only 20% of extrinsic input. Thus, it appears that a substantial majority of extrinsic thalamo-cortical and cortico-cortical synapses do not come from the strongest projections. These data suggest that it is important to look to the overall pattern of inputs, rather than any single input, to account for the functional properties of any given area. We note, however, that primary cortices (areas 17, 18, primary somatosensory and primary auditory context) may have slightly fewer, but stronger, connections that most cortical areas (see later).

In addition to the synapses from distant cortical areas and thalamic nuclei, an unknown proportion of synapses come from distant sites within the area. Long-range intrinsic projections are known to have highly specific topography linking, for example, regions of primary visual cortex that have similar orientation selectivity (Gilbert and Wiesel, 1983, 1989 (see Chapter 11)). Although we do not give these connections detailed consideration in this chapter, we think that they may also be important for storing prior expectations about the local structure of the visual world. This is illustrated by physiological examples of responses to colinear contours outside the classical receptive field (Peterhans and von der Heydt, 1989; Grosof et al., 1993; Polat et al., 1998), other contextual effects from outside the classical receptive field (Gilbert and Wiesel, 1990), and modulation of local activity by contrast gain control mechanisms (Heeger, 1992a, 1992b).

A consequence of the anatomy is that local computations are likely to be strongly dependent on the *overall pattern* of inputs and on the activity in a number of distant sites within the brain. Furthermore, function in context, f_c, and local function, f_l, cannot be clearly distinguished at the level of local cortical circuitry because long-range connections reach into the local circuitry and contribute a substantial proportion of its synapses. These features of cortical circuitry may explain why crude analyses of the gross area-to-area patterns of connections reveal relationships that look so much like those apparent to neurophysiologists (Young, 1992; Scannell et al., 1996). The anatomical features also explain why we should not be surprised when the properties of cortical areas are not predictable on the basis of single projections (Burns and Young, 2000). For example, macaque V4 and V5/MT exchange direct projections (Zeki and Shipp, 1995; Steele et al., 1991) but have very different "choruses" of inputs (Young, 1992; Young et al., 1995a, 1995b).

Anatomy shows that rather little of the machinery of the visual cortex is devoted to the propagation of feedforward signals. Cortical machinery appears designed to pool information from a wide range of sources, many of which have no access to the immediate retinal signal. These features argue against the vision-as-analysis model and for the vision-as-inference model. An anatomical arrangement that keeps afferent input from the retina in its proper informational place may help explain why the world looks so unlike the retinal signal (Young, 1994).

What about those occasions on which direct projections appear to provide a good account of some aspects of the function, f_c, of cortical areas or parts of the network? Even in these cases, we cannot be confident that we have found a structure-function, f_c, relationship that holds for normal visual processing. This is true for two reasons. First, areas with strong direct projections also tend to have very similar overall patterns of connections. Second, unfamiliar or unnatural stimuli may fail to engage some of the machinery normally involved in visual processing (Rao and Ballard, 1999). The first concern is illustrated in Fig. 15-3, which shows the correlation between direct and indirect connectivity.

The quantitative relationship shown in Fig. 15-3 can be illustrated with the example of structural projections and functional, f_c, interactions between area 17 and area 21a (Michalski et al., 1993, 1994). Activity in 17 and 21a could be related via the direct reciprocal projections that link the areas. However, activity could also be relayed via areas 18, 19, PLLS, PMLS, AMLS, 21b, LGLC1-3, LGLGW, LGLI, LPl, and PUL (Table 15-1). The 11 or more indirect pathways could well be more powerful that the single direct pathway (Fig. 15-2). Furthermore, any covariation in neuronal activity in 21a and area 17 could be mediated by common input from at least 11 other structures.

The second reason why we should be cautious when we think we have found structure-function relationship is that unfamiliar or unnatural stimuli may not engage some of the machinery normally involved in visual processing. In the previous section, we suggested that the stimuli typically used in vision research favor the vision-as-analysis model, as they may fail to trigger the prior expectations of the visual system. This idea has implications for studying the function of particular classes of connection. For example, Rao and Ballard (1999) proposed that feedback connections carry the higher area's statistical predictions of the activity of its inputs, while feedforward connections signal the difference between the prediction and the actual lower-level area's activity. Higher areas will not be able to make sensible statistical predictions about stimuli with which they have little experience. This may explain why back projections, which include some of the quantitatively strongest projections between brain areas (e.g., Budd, 1998), appear to have rather modest effects when studied with conventional neurophysiological stimuli (Gulyas et al., 1990; Murphy and Sillito, 1987; Hupé et al., 1998).

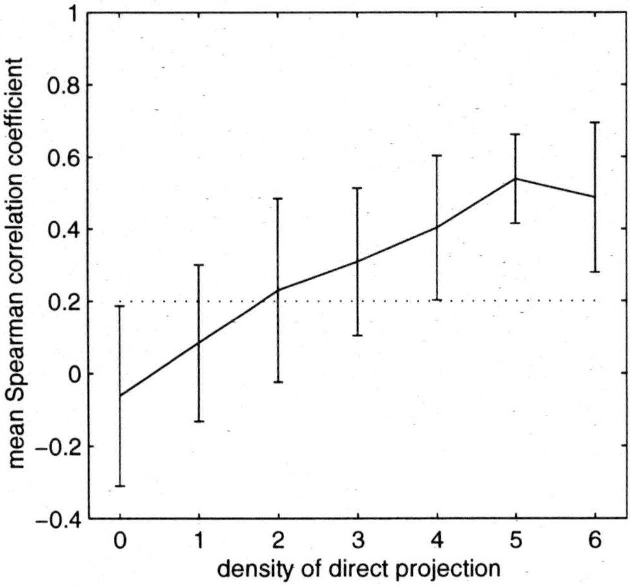

FIGURE 15-3. Cortical areas with strong direct projections also have similar patterns of connections with the rest of the network. The x-axis shows the sum the weights of direct connections between cortical areas (afferent plus efferent connection weights, from Table 15-2). The y-axis shows the mean Spearman rank correlation coefficient in connection pattern between areas computed from Table 15-1. Error bars are ± 1 standard deviation. The dotted line shows the critical value of rank correlation coefficient where $p > 0.05$. Areas or nuclei with strong direct projections tend to have significantly similar patterns of connections with the rest of the network, and areas or nuclei that lack direct projections tend to have uncorrelated patterns of connections with the rest of the network. Therefore, the strengths of direct and indirect influences between cortical areas covary strongly.

To summarize this section, single extrinsic connections contribute only a small fraction (< 5%) of synapses to their target area. However, there are many extrinsic cortico-cortical and thalamo-cortical connections, so the sum of extrinsic synapses is roughly six times the contribution of the strongest single projection. The overall pattern of projections is likely to be much more important than any individual projection. There is unlikely to be a clear division of labor between, on the one hand, local circuit microanatomy and physiology, and, on the other, gross extrinsic connectivity and global patterns of activation. Local computations will pool information from a wide variety of sources. Both local and global scales of processing should reflect similar dynamics. This is the design one would expect in a visual system in which the retinal signal as only one of many sources of information about the likely nature of the scene. Conventional neurophysiological stimuli may underestimate the importance of feedback projections by failing to invoke prior expectations about the visual environment.

BASIC AREA TO AREA PATTERN FOR VISUAL SYSTEM

Given a network in which the strongest single projections are several times weaker than the sum of all the extrinsic projections, it is necessary to consider the function, f_c, of brain structures in the context of the wider network in which they are embedded. A small first step along this path is to collect data on the organization of as much of the network as practical (Young, 1993; Scannell et al., 1993, 1995, 1999; Burns and Young, 2000; Stephan et al., 2000).

Table 15-2 shows such data and are drawn from a full database that provides more detailed information (http://www.psychology.ncl.ac.uk/jack/cor_thal.html). The connection strengths are on an ordinal scale that assumes *only* that the relationship between the ranks is monotonic; a 3 is stronger than a 2, which is stronger than a 1, which is stronger than a 0 (for clarity, 0 is represented in Table 15-2 as.). In quantitative terms, connections weighted 3 may be two or three orders of

TABLE 15-2A. Connections between cat cortical areas and thalamic nuclei.

A.	17	18	19	PLLS	PMLS	AMLS	ALLS	VLS	DLS	21a	21b	20a	20b	7	AES	PS	AI	AII	AAF	P	VPc	EPp	Tem	3a	3b	1	2	SII	SIV	4g	4	6l	6m	5Am	5Al	5Bm	5Bl	SSAi	SSAo	PFCMil	PFCMd	PFCL	Ia	Ig	Cga	CGp	RS	35	36	pSb	Sb	Enr	Hipp
17	.	3	3	1	3	3	.	.	.	3	2	3	.	.	.	.	.	.	.	.	.	.	.	.	.	.	.	.	.	.	.	.	.	.	.	.	.	.	.	.	.	.	.	.	.	.	.	.	.	.	.	.	.
18	3	.	3	2	2	3	.	1	.	3	2	2	.	1	.	.	.	.	.	.	.	.	.	.	.	.	.	.	.	.	.	.	.	.	.	.	.	.	.	.	.	.	.	.	.	.	.	.	.	.	.	.	.
19	3	3	.	1	3	2	1	1	.	3	1	3	1	2	.	1	.	.	.	.	.	.	.	.	.	.	.	.	.	.	.	.	.	.	1	1	1	.	.	.	.	.	.	.	.	.	.	.	.	.	.	.	.
PLLS	1	1	2	.	2	.	2	.	3	1	.	1	.	1	3	3	.	.	.	.	.	.	.	.	.	.	.	.	.	.	.	.	.	.	.	.	.	.	.	.	.	.	1	1	.	.	.	1	1	.	.	.	.
PMLS	3	3	3	1	.	3	.	3	.	1	1	3	.	.	1	1	.	.	.	.	.	.	.	.	.	.	.	.	.	.	.	.	.	.	.	.	.	.	.	.	.	.	.	.	.	.	.	1	1	.	.	.	.
AMLS	2	3	2	1	2	.	1	.	.	2	.	.	1	.	1	1	.	.	.	.	.	.	.	.	.	.	.	.	.	.	.	.	.	.	.	.	.	.	.	.	.	.	.	.	.	.	.	1	1	.	.	.	.
ALLS	.	1	1	2	.	1	.	.	.	.	.	.	.	.	2	.	.	.	.	.	.	.	.	.	.	.	.	.	.	.	.	.	.	.	.	.	.	.	.	.	.	.	1	1	.	.	.	1	1	.	.	.	.
VLS	2	1	2	1	3	.	1	.	2	.	1	2	.	.	.	.	.	.	.	.	.	.	.	.	.	.	.	.	.	.	.	.	.	.	.	.	.	.	.	.	.	.	.	.	.	.	.	.	.	.	.	.	.
DLS	.	.	2	3	.	.	2	2	.	.	1	.	.	.	2	.	.	.	.	.	.	.	.	.	.	.	.	.	.	.	.	.	.	.	.	.	.	.	.	.	.	.	.	.	.	.	.	.	.	.	.	.	.
21a	2	2	3	.	1	2	.	.	.	.	2	2	.	2	1	1	.	.	.	.	.	1	.	.	.	.	.	.	.	.	.	.	.	.	.	.	.	.	.	.	.	.	.	.	.	.	.	.	.	.	.	.	.
21b	2	2	1	.	1	.	.	.	1	2	.	2	.	1	.	1	.	.	.	.	.	1	.	.	.	.	.	.	.	.	.	.	.	.	.	.	.	.	.	.	.	.	.	.	.	.	.	.	.	.	.	.	.
20a	3	2	3	2	3	.	.	2	.	2	2	.	3	1	2	2	.	.	.	1	.	2	.	.	.	.	.	.	.	.	.	1	.	.	1	.	.	.	.	.	.	1	1	1	.	2	2	1	1	1	1	2	.
20b	.	.	1	.	.	1	.	.	.	.	.	3	.	.	2	2	.	.	.	1	.	2	.	.	.	.	.	.	.	.	.	1	.	.	1	.	.	.	.	.	.	1	.	.	.	2	2	1	1	1	1	2	.
7	.	1	2	1	.	2	1	.	.	2	2	2	2	.	.	1	.	.	.	.	.	1	.	.	.	.	.	.	.	.	.	2	3	1	2	2	2	1	2	.	1	1	.	1	.	2	.	1	.	.	.	.	.
AES	.	.	.	3	1	1	2	1	2	.	.	.	1	.	.	2	.	1	1	.	.	1	.	1	1	.	.	2	2	1	1	2	1	1	1	1	.	1	1	.	.	1	1	2	.	1	.	.	2	.	.	.	.
PS	.	.	.	3	.	.	.	.	.	.	.	.	2	.	2	.	.	.	.	.	.	2	.	.	.	.	.	.	.	.	.	.	.	.	.	.	.	.	.	.	1	1	1	3	.	.	.	.	2	.	.	.	.
AI	.	.	.	.	.	.	.	.	.	.	.	.	.	.	.	.	.	3	3	3	2	1	.	.	.	.	.	.	.	.	.	.	.	.	.	.	.	.	.	.	.	.	1	1	.	.	.	1	1	.	.	.	.
AII	.	.	.	.	.	.	.	.	.	.	.	.	.	.	.	.	2	.	2	3	2	1	2	.	.	.	.	.	.	.	.	.	.	.	.	.	.	.	.	.	.	.	1	1	.	.	.	2	2	.	.	.	.
AAF	.	.	.	.	.	.	.	.	.	.	.	.	.	.	.	.	3	2	.	1	1	.	.	.	.	.	.	.	.	.	.	.	.	.	.	.	.	.	.	.	.	.	.	.	.	.	.	.	.	.	.	.	.
P	.	.	.	.	.	.	.	.	.	.	.	.	.	.	.	.	2	2	2	.	2	2	.	.	.	.	.	.	.	.	.	.	.	.	.	.	.	.	.	.	.	.	.	.	.	2	2	2	.	.	.	.	.
VPc	.	.	.	.	.	.	.	.	.	.	.	.	.	.	.	.	2	.	1	2	.	2	.	.	.	.	.	.	.	.	.	.	.	.	.	.	.	.	.	.	.	.	.	.	.	.	.	.	.	.	.	.	.
EPp	.	.	1	1	.	.	1	1	2	1	1	2	2	2	1	.	1	2	1	2	2	.	2	.	.	.	.	.	.	.	.	1	1	.	.	.	.	.	.	.	1	1	2	2	.	2	.	1	2	.	.	.	.
Tem	.	.	.	.	.	.	.	.	.	.	.	.	.	.	.	.	1	1	.	1	2	.	.	.	.	.	.	.	.	.	.	.	.	.	.	.	.	.	.	.	.	.	.	.	.	.	.	1	3	.	.	.	.
3a	.	.	.	.	.	.	.	.	.	.	.	.	.	.	1	.	.	.	.	.	.	.	.	.	3	3	3	3	.	2	2	1	.	.	.	.	.	.	.	.	.	.	.	.	.	.	.	1	1	.	.	.	.
3b	.	.	.	.	.	.	.	.	.	.	.	.	.	.	1	.	.	.	.	.	.	.	.	3	.	3	3	3	1	2	2	1	.	.	.	.	.	1	1	.	.	.	.	.	.	.	.	1	1	.	.	.	.
1	.	.	.	.	.	.	.	.	.	.	.	.	.	.	.	.	.	.	.	.	.	.	.	3	3	.	3	3	.	2	2	.	.	.	.	.	.	.	.	.	.	.	.	.	.	.	.	1	1	.	.	.	.
2	.	.	.	.	.	.	.	.	.	.	.	.	.	.	.	.	.	.	.	.	.	.	.	3	3	3	.	3	.	2	2	.	.	.	.	.	.	.	.	.	.	.	.	.	.	.	.	1	1	.	.	.	.
SII	.	.	.	.	.	.	.	.	.	.	.	.	.	.	.	.	.	.	.	.	.	.	.	3	3	3	3	.	.	2	2	1	.	1	1	2	.	2	.	.	.	.	1	1	.	.	.	1	1	.	.	.	.
SIV	.	.	.	.	.	.	.	.	.	.	.	.	.	.	2	.	.	.	.	.	.	.	.	.	.	.	.	1	.	.	.	2	2	2	2	2	.	2	1	.	.	.	1	2	.	.	.	2	1	.	.	.	.
4g	.	.	.	.	.	1	.	.	.	.	.	.	.	1	2	.	.	.	.	.	.	.	.	3	2	2	2	3	1	.	2	.	1	2	2	2	2	2	2	.	.	.	.	1	.	.	.	.	.	.	.	.	.
4	.	.	.	.	.	1	.	.	.	.	.	.	.	.	1	.	.	.	.	.	.	.	.	2	2	1	1	3	1	2	.	.	.	2	1	2	1	1	1	.	.	.	.	1	.	.	.	.	.	.	.	.	.
6l	.	.	1	1	1	.	.	.	.	.	.	.	.	2	2	.	.	.	.	.	.	.	.	2	2	2	2	2	2	2	2	.	3	.	2	1	2	1	1	.	.	.	.	1	.	1	.	.	.	.	.	.	.
6m	.	.	.	1	.	1	1	.	.	.	.	1	1	3	2	.	.	.	.	.	.	2	.	.	.	1	1	1	2	2	.	3	.	1	1	1	2	.	2	2	.	.	2	2	.	3	.	2	1	.	.	1	.
5Am	.	.	.	.	.	1	.	.	.	.	.	.	.	1	2	.	.	.	.	.	.	.	.	3	2	3	2	2	2	3	2	.	1	.	2	3	.	2	3	.	.	.	1	1	1	1	.	.	.	.	.	.	.
5Al	.	.	1	.	.	2	.	.	.	1	1	1	1	2	2	.	.	.	.	.	.	.	.	3	3	1	2	1	2	3	1	2	1	1	.	2	3	2	3	.	.	1	2	2	1	1	.	.	.	.	.	.	.
5Bm	.	.	1	.	.	1	.	.	.	.	.	.	.	2	2	.	.	.	.	.	.	.	.	.	1	2	.	2	2	3	2	1	1	3	1	.	2	2	1	.	.	.	1	1	2	1	.	1	1	.	.	.	.
5Bl	.	1	2	1	2	2	2	.	.	2	2	.	.	3	.	.	.	.	.	.	.	.	.	.	1	.	2	.	.	2	1	2	2	.	2	2	.	1	2	.	.	1	1	1	1	1	.	.	.	.	.	.	.
SSAi	.	.	.	.	.	1	.	.	.	.	.	.	.	2	1	.	.	.	.	.	.	.	.	2	1	2	1	3	2	2	1	1	.	2	.	2	1	.	2	.	.	.	1	1	1	1	.	.	.	.	.	.	.
SSAo	.	.	.	.	.	.	.	.	.	.	.	.	.	2	2	.	.	.	.	.	.	.	.	1	1	1	2	.	1	2	1	1	2	2	2	2	2	1	.	.	.	.	1	1	2	1	.	.	.	.	.	.	.
PFCMil	.	.	.	.	.	.	.	.	.	.	.	.	.	.	.	.	.	.	.	.	.	.	.	.	.	.	.	.	.	.	.	1	1	.	.	.	.	.	.	.	3	2	3	.	1	1	.	2	2	1	3	3	.
PFCMd	.	.	.	.	.	.	.	.	.	.	.	.	.	.	.	.	1	.	.	.	.	1	1	.	.	.	.	.	.	.	.	2	2	.	.	.	.	.	.	3	.	3	3	1	3	3	2	2	2	1	2	2	.
PFCL	.	.	.	1	.	.	1	.	.	.	.	1	1	1	2	1	.	1	.	.	.	1	1	.	.	.	.	.	1	.	.	2	2	.	1	.	1	.	.	.	2	.	2	3	2	2	1	1	1	.	1	.	.
Ia	.	.	.	1	1	.	.	.	2	.	.	1	1	1	1	.	.	2	.	.	.	2	2	.	.	.	1	1	2	.	.	.	2	1	1	1	1	1	1	2	2	2	.	.	2	2	.	3	1	.	2	.	.
Ig	.	.	.	2	1	.	2	.	2	.	.	1	1	1	2	2	.	2	.	.	.	2	2	.	.	.	1	1	2	.	.	.	2	.	.	.	.	.	.	.	2	2	.	.	2	2	.	1	3	.	1	.	.
Cga	.	.	.	.	.	1	.	.	.	.	.	.	1	1	.	1	.	.	.	.	.	1	.	.	.	.	.	.	1	1	1	2	2	1	1	1	1	.	.	3	2	3	2	2	.	3	2	2	2	1	1	1	.
CGp	.	.	.	.	.	1	1	.	.	1	.	1	2	3	1	1	.	.	.	2	1	2	.	.	.	.	.	.	1	1	1	2	3	1	1	1	1	1	1	2	2	3	1	1	2	1	2	2	1	2	1	1	.
RS	.	.	.	.	.	.	.	.	.	.	.	2	2	2	.	.	.	.	.	2	.	.	.	.	.	.	.	.	.	.	.	.	.	.	.	.	.	.	.	.	1	1	.	.	2	2	.	1	1	.	2	1	.
35	.	.	.	1	1	.	.	.	.	2	2	2	2	1	1	1	1	2	.	2	.	2	1	.	.	.	.	1	1	.	1	.	1	.	.	.	.	.	.	2	.	.	3	2	.	1	2	.	3	2	.	3	1
36	.	.	.	.	.	.	.	.	.	1	1	1	1	.	1	2	.	1	2	.	1	1	3	.	.	.	.	1	2	.	1	.	1	.	.	.	.	.	.	1	.	.	1	3	.	1	1	3	.	2	.	2	.
pSb	.	.	.	.	.	.	.	.	.	.	.	.	.	.	.	.	.	.	.	.	.	.	.	.	.	.	.	.	.	.	.	.	.	.	.	.	.	.	.	.	.	.	.	.	.	1	.	2	2	.	2	2	2
Sb	.	.	.	.	.	.	.	.	.	.	.	.	.	.	.	.	.	.	.	.	.	.	.	.	.	.	.	.	.	.	.	.	.	.	.	.	.	.	.	2	.	.	.	.	.	.	.	2	.	.	.	2	2
Enr	.	.	.	.	.	.	.	.	.	.	.	.	.	.	.	.	.	.	.	.	.	.	.	.	.	.	.	.	.	.	1	.	1	.	.	.	.	.	.	2	1	.	1	1	.	1	2	3	2	3	2	.	3
Hipp	.	.	.	.	.	.	.	.	.	.	.	.	.	.	.	.	.	.	.	.	.	.	.	.	.	.	.	.	.	.	.	.	.	.	.	.	.	.	.	.	.	.	.	.	.	.	.	2	.	.	.	3	.

A. Connections between cortical areas. Connections weighted '3' are strong, connections weighted '2' are intermediate, connections weighted '1' are weak or sparse, and connections marked '.' have been explicitly reported as absent or are unreported. Reading horizontally in this part of the table, from a cortical structure, shows afferent connections and reading vertically shows efferent connections.

TABLE 15-2B. Connections between cat cortical areas and thalamic nuclei.

B.	17	18	19	PLLS	PMLS	AMLS	ALLS	VLS	DLS	21a	21b	20a	20b	7	AES	PS	AI	AII	AAF	P	VPc	EPp	Tem	3a	3b	1	2	SII	SIV	4g	4	6l	6m	5Am	5Al	5Bm	5Bl	SSSAi	SSAo	PFCMil	PFCMd	PFCL	Ia	Ig	Cga	CGp	RS	35	36	pSb	Sb	Enr	Hipp
MGV	.	.	.	.	.	.	.	.	.	.	.	.	.	.	.	.	3	.	2	3	2	1	.	.	.	.	.	.	.	.	.	.	.	.	.	.	.	.	.	.	.	.	.	.	.	.	.	.	.	.	.	.	.
MGDdd	.	.	.	.	.	.	.	.	.	.	.	.	.	.	.	.	2	3	2	2	2	2	2	.	.	.	.	.	.	.	.	.	.	.	.	.	.	.	.	.	.	.	2	2	.	.	.	.	.	.	.	.	.
MGDd	.	.	.	.	.	.	.	.	.	.	.	.	.	.	.	.	.	3	.	2	2	2	2	.	.	.	.	.	.	.	.	.	.	.	.	.	.	.	.	.	.	.	1	2	.	.	.	1	1	.	.	.	.
MGDds	.	.	.	.	.	.	.	.	.	.	.	.	.	.	.	.	.	3	.	.	.	.	1	.	.	.	.	.	.	.	.	.	.	.	.	.	.	.	.	.	.	.	.	.	.	.	.	.	.	.	.	.	.
MGvl	.	.	.	.	.	.	.	.	.	.	.	.	.	.	1	.	1	3	1	2	2	2	3	.	.	.	.	.	.	.	.	.	.	.	.	.	.	.	.	.	.	.	2	1	.	.	.	2	2	.	.	.	.
MGM	.	.	.	1	.	1	1	.	.	.	.	.	.	.	2	.	3	2	2	2	2	3	2	.	.	.	1	2	2	.	.	.	.	.	1	.	.	.	.	.	.	.	1	1	.	.	.	2	2	.	.	.	.
LGLA	3	3	.	.	1	.	.	.	.	.	.	.	.	.	.	.	.	.	.	.	.	.	.	.	.	.	.	.	.	.	.	.	.	.	.	.	.	.	.	.	.	.	.	.	.	.	.	.	.	.	.	.	.
LGLC	3	3	2	.	.	.	.	.	.	.	.	.	.	.	.	.	.	.	.	.	.	.	.	.	.	.	.	.	.	.	.	.	.	.	.	.	.	.	.	.	.	.	.	.	.	.	.	.	.	.	.	.	.
LGLC1-3	2	2	2	2	2	1	.	2	2	2	.	2	1	2	.	2	.	.	.	.	.	.	.	.	.	.	.	.	.	.	.	.	.	.	.	.	.	.	.	.	.	.	.	.	.	.	.	.	.	.	.	.	.
LGGW	1	2	2	1	2	2	.	.	.	2	.	.	.	.	.	.	.	.	.	.	.	.	.	.	.	.	.	.	.	.	.	.	.	.	.	.	.	.	.	.	.	.	.	.	.	.	.	.	.	.	.	.	.
LGI	2	2	2	1	2	2	.	.	.	2	1	.	1	.	.	.	.	.	.	.	.	.	.	.	.	.	.	.	.	.	.	.	.	.	.	.	.	.	.	.	.	.	.	.	.	.	.	.	.	.	.	.	.
POm	.	.	.	2	.	.	.	.	.	.	.	.	.	.	2	.	1	1	2	.	.	1	.	1	1	1	1	1	3	.	.	.	.	3	3	2	2	1	1	.	.	.	1	1	.	.	.	2	2	.	.	.	.
POl	.	.	.	.	.	.	1	.	.	.	.	.	.	.	2	.	2	1	2	2	2	2	.	.	.	.	1	2	2	.	2	.	.	2	.	.	.	2	2	.	.	.	.	2	1	1	.	.	.	.	.	.	.
SG	.	.	.	1	1	1	2	.	1	.	.	.	.	.	3	.	1	2	2	1	1	2	2	.	.	.	.	.	2	.	.	1	.	.	1	.	.	1	1	.	.	.	2	2	.	2	.	2	2	.	.	.	.
POi	.	.	.	2	.	1	2	.	.	.	.	.	.	.	3	.	.	.	.	.	.	.	.	.	.	.	.	.	3	.	.	.	.	.	.	.	.	.	.	.	.	.	.	2	.	.	.	3	3	.	.	.	.
LD	.	.	.	.	.	2	.	.	.	.	2	.	.	2	.	.	.	.	.	.	.	.	.	.	.	.	.	.	.	.	1	.	1	1	1	.	2	.	.	.	.	1	.	.	1	2	3	1	1	3	.	1	.
LPl	2	2	2	1	3	3	.	2	1	2	2	2	2	2	1	.	.	.	.	.	.	1	.	.	.	.	.	.	.	.	.	.	.	.	.	2	2	.	.	.	.	.	.	.	.	2	.	.	.	.	.	.	.
LPm	.	.	2	3	2	2	2	2	2	2	2	2	2	1	2	2	.	.	.	.	.	1	.	.	.	.	.	1	1	.	.	.	.	.	.	.	.	.	.	.	.	.	.	2	.	.	.	1	1	.	.	.	.
LM-Sg	.	.	.	2	1	1	2	.	.	.	.	.	.	.	3	.	.	2	.	.	.	2	1	.	.	.	.	1	1	.	.	2	2	.	.	1	1	.	.	2	2	1	2	3	.	2	.	2	2	.	.	.	.
PUL	1	1	2	1	2	2	.	.	.	2	2	2	2	2	.	.	.	.	.	.	.	.	.	.	.	.	.	.	.	.	.	.	.	.	.	1	2	2	2	.	.	.	.	1	1	1	1	1	1	1	.	.	.
Llc	.	.	.	.	.	.	.	.	.	.	.	.	.	3	.	.	.	2	.	.	.	.	.	.	.	.	.	1	1	.	.	.	.	1	1	2	2	.	.	1	1	1	.	1	.	.	.	.	.	.	.	.	.
Llo	.	.	.	.	.	.	.	.	.	.	.	.	.	.	.	.	.	2	.	.	.	.	.	2	2	.	2	1	1	1	2	.	.	1	2	3	2	.	.	.	.	.	.	1	.	.	.	.	.	.	.	.	.
VPL	.	.	.	.	.	.	.	.	.	.	.	.	.	.	.	.	.	.	.	.	.	.	.	3	3	3	3	3	1	1	1	.	.	2	.	.	.	2	2	.	.	.	.	.	.	.	.	.	.	.	.	.	.
VPM	.	.	.	.	.	.	.	.	.	.	.	.	.	.	1	.	.	.	.	.	.	.	.	3	3	3	3	3	1	1	1	.	.	.	1	.	.	2	2	.	.	.	.	1	.	.	.	.	.	.	.	.	.
VA-VL	.	.	.	2	1	1	.	.	.	1	1	1	1	2	1	.	.	.	.	.	.	.	.	2	2	2	2	1	2	3	3	2	3	3	3	2	2	.	.	2	2	2	1	2	1	1	1	.	.	1	.	.	.
VMB	.	.	.	.	.	.	.	.	.	.	.	.	.	.	.	.	.	.	.	.	.	.	.	.	.	.	.	.	.	.	.	2	.	.	.	.	.	.	.	2	1	2	2	1	.	.	1	2	.	.	.	1	.
VMP	.	.	.	.	.	.	.	.	.	.	.	.	.	1	2	.	.	2	2	.	.	.	1	.	.	2	.	1	2	3	3	2	3	1	1	.	.	.	.	3	3	3	3	3	1	1	1	1	1	.	.	1	.
SUM	.	.	.	.	.	.	.	.	.	.	.	.	.	.	.	.	.	.	.	.	.	.	.	.	.	.	.	.	.	.	.	1	1	.	.	1	.	.	.	1	1	2	.	1	1	1	1	.	.	1	.	.	.
MD	.	.	.	.	.	.	.	.	.	.	.	.	.	.	1	.	.	1	.	.	.	1	.	.	1	.	.	1	1	.	.	2	2	.	.	.	.	.	.	3	3	3	3	2	1	1	.	1	1	.	.	1	.
CMN	.	1	.	1	1	1	1	.	.	.	.	1	1	1	1	.	.	1	1	1	1	.	.	1	1	1	1	1	1	2	1	2	2	1	1	.	1	.	.	1	1	2	2	1	1	1	1	1	1	1	.	.	.
PAC	.	1	1	1	1	1	1	1	1	1	1	1	1	1	1	.	.	.	.	.	.	1	.	1	1	1	1	1	1	1	1	1	2	1	1	1	1	.	.	2	2	2	.	1	1	2	1	1	1	1	.	1	.
CLN	1	1	1	1	1	2	1	1	1	1	1	1	2	2	1	.	.	.	.	.	.	1	.	1	1	1	1	1	1	1	1	1	2	1	1	1	1	1	1	.	.	.	.	1	1	1	1	1	1	2	.	.	.
CM	.	1	1	1	1	1	1	1	1	.	.	1	1	1	1	.	.	.	.	.	.	.	.	.	.	.	.	1	.	1	1	1	1	1	.	1	1	.	.	.	.	.	1	1	.	.	.	.	.	2	.	1	.
PF	.	.	.	.	.	1	1	.	.	.	.	.	.	1	1	.	.	1	1	.	.	.	.	1	1	1	1	1	1	1	1	1	1	.	1	.	1	.	.	1	1	1	2	2	1	1	1	1	1	.	.	.	.
RH	.	.	.	.	.	.	.	.	.	.	.	.	.	.	1	.	.	1	.	.	.	.	.	.	.	.	.	.	.	1	1	1	1	.	.	.	.	.	.	1	1	1	1	1	1	.	1	2	2	2	.	2	.
PAT	.	.	.	.	.	.	.	.	.	.	.	.	.	.	.	.	.	.	.	.	.	.	.	.	.	.	.	.	.	.	.	.	.	.	.	.	.	.	.	2	2	2	2	1	1	.	.	1	1	.	.	.	.
PARA	.	.	.	.	.	.	.	.	.	.	.	.	.	.	.	.	.	.	.	.	.	.	.	.	.	.	.	.	.	.	.	.	.	.	.	.	.	.	.	2	1	.	.	.	1	.	.	2	2	1	2	2	.
PARP	.	.	.	.	.	.	.	.	.	.	.	.	.	.	.	.	.	.	.	.	.	.	.	.	.	.	.	.	.	.	.	.	.	.	.	.	.	.	.	1	1	.	.	.	1	.	.	2	2	1	2	2	.
MV-RE	.	.	.	.	.	.	.	.	.	.	.	.	.	.	.	.	.	.	.	.	.	.	.	.	.	.	.	.	.	.	.	.	1	.	.	.	.	.	.	2	1	.	2	1	2	1	2	2	2	2	3	3	3
AD	.	.	.	.	.	.	.	.	.	.	.	.	.	.	.	.	.	.	.	.	.	.	.	.	.	.	.	.	.	.	.	.	1	.	.	.	.	.	.	1	.	.	.	.	1	2	3	.	.	2	1	2	.
AM	.	.	.	.	.	.	.	.	.	.	.	.	.	.	.	.	.	.	.	.	.	.	.	.	.	.	.	.	.	.	.	.	1	.	.	.	.	.	.	2	2	.	.	.	3	3	3	.	.	1	2	2	.
AV	.	.	.	.	.	.	.	.	.	.	.	.	.	.	.	.	.	.	.	.	.	.	.	.	.	.	.	.	.	.	.	.	1	.	.	.	.	.	.	1	.	.	.	.	2	2	3	.	.	2	.	3	.

B. Connections between thalamus and cortex. We assume these are invariably reciprocal (Burton and Kopf, 1984b; Jones, 1985; Symonds and Rosenquist, 1984; Raczkowski and Rosenquist, 1983).

Anatomical abbreviations

1	area 1
17	area 17
18	area 18
19	area 19
2	area 2
20a	area 20a
20b	area 20b
21a	area 21a
21b	area 21b
35	area 35
36	area 36
3a	area 3a
3b	area 3b
4	areas 4f, 4sf, and 4δ
4g	area 4γ
5Al	lateral area 5A
5Am	medial area 5a
5Bl	lateral area 5B
5Bm	medial area 5B
6l	lateral area 6
6m	medial area 6
7	area 7
AAF	anterior auditory field
AD	anterodorsal nucleus
AES	anterior ectosylvian sulcus
AI	primary auditory field
AII	secondary auditory field
ALLS	anterolateral lateral suprasylvian area
AM	anteromedial nucleus
AMLS	anteromedial lateral suprasylvian area
AV	anteroventral nucleus
CGa	anterior cingulate cortex
CGp	posterior cingulate cortex
CLN	central lateral nucleus
CM	centre median nucleus
CMN	central medial nucleus
DLS	dorsolateral suprasylvian area
Enr	entorhinal cortex
EPp	posteror part of posterior ectosylvian gyrus
Hipp	hippocampus proper
Ia	agranular insula
Ig	granular insula
LD	lateral dorsal nucleus
LGLA	A laminae of the lateral geniculate
LGLC	magnocellular C laminae of the lateral geniculate
LGLC1-3	parvocellular C laminae of the lateral geniculate
LGGW	geniculate wing
LGI	medial interlaminar nucleus

LIc	caudal region of the lateral intermediate nucleus
LIo	oral region of the lateral intermediate nucleus
LM-Sg	lateral medial – suprageniculate complex
LPl	lateral nucleus of the lateral posterior complex
LPm	medial nucleus of the lateral posterior complex
MD	mediodorsal nucleus
MGDd	dorsal nucleus of the medial geniculate
MGDdd	deep dorsal nucleus of the medial geniculate
MGDds	superficial dorsal nucleus of the medial geniculate
MGM	magnocellular division of the medial geniculate
MGV	ventral division of the medial geniculate
MGvl	ventrolateral nucleus of the medial geniculate
MV-RE	medioventral-reuniens nucleus
P	posteror auditory field
PAC	paracentral nucleus
PARA	anterior paraventricular nucleus
PARP	posterior paraventricular nucleus
PAT	parataenial nucleus
PF	parafasicular nucleus
PFCl	lateral prefrontal cortex
PFCMd	dorsal medial prefrontal cortex
PFCMil	infralimbic medial prefrontal cortex
PLLS	posterolateral lateral suprasylvian area
PMLS	posteromedial lateral suprasylvian area
POi	intermediate region of the posterior complex
POl	lateral region of the posterior complex
POm	medial region of the posterior complex
PS	posterior suprasylvian area
pSb	presubiculum, parasubiculum and post subicular cortex
PUL	pulvinar
RH	rhomboid nucleus
RS	retrosplenial area
Sb	subiculum
SG	suprageniculate nucleus
SII	second somatosensory area
SIV	fourth somatosensory area
SSAi	inner (deep) suprasylvian sulcal region or area 5
SSAo	outer suprasylvian sulcal region of area 5
SUM	submedian nucleus
Tem	temporal auditory field
VA-VL	ventroanterior-ventrolateral complex
VLS	ventrolateral suprasylvian area
VMB	basal ventromedial nucleus
VMP	principal ventromedial nucleus
VPc	Ventroposterior auditory field
VPL	lateral part of the ventroposterior nucleus
VPM	medial part of the ventroposterior nucleus

magnitude stronger than connections weighted 1 (Musil and Olson, 1988a, 1988b; Scannell et al., 2000). Connections weighted 0 (. in Table 15-2) are simply weaker than those labeled 1 and, we have argued (Young et al., 1995a; Scannell et al., 1999), correspond to cases where the label density does not exceed the density that anatomists describe as background. We note that Fig. 15-2 shows that weak connections and background labeling almost certainly constitute a continuum. However, in line with anatomical conventions, in the following discussions we treat connections weighted as 0 as if they were absent.

Table 15-2A shows 826 cortico-cortical connections. Each cortical area connects with, on average, 15 other areas (standard deviation = 8). This level of connectivity corresponds to 28% of possible cortico-cortical connections. Table 15-2B shows 651 connections between thalamus and cortex, which we assume are wholly reciprocal (Burton and Kopf, 1984; Jones, 1985; Symonds and Rosenquist, 1984; Raczkowski and Rosenquist, 1983). On average, each thalamic nucleus projects to, and receives projections from, 15 cortical areas (standard deviation = 10). This constitutes 28% of all possible thalamo-cortical connections, were each nucleus to project to each cortical area. There are no

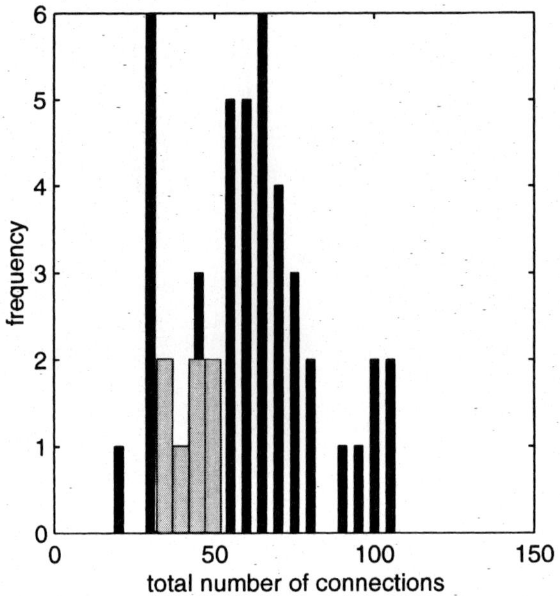

FIGURE 15-4. Primary sensory cortical areas connect with fewer other cortical areas and thalamic nuclei than nonprimary areas. The x-axis represents the total number of afferent and efferent projections of a given area. The y-axis shows the number of cortical areas making a given number of connection. Black bars represent nonprimary areas, and gray bars represent the primary sensory areas (primary visual cortex [17 and 18], primary auditory cortex [AI], primary somatosensory cortex [1, 2, 3a, 3b]. Primary sensory areas make significantly fewer connections than nonprimary areas ($p < 0.01$, Mann-Whitney U test).

direct extrinsic connections between the thalamic nuclei. The primary sensory cortices, including primary visual cortex, connect with fewer other cortical and thalamic structures than would be expected of typical cortical areas (Fig. 15-4) ($p < 0.01$, by Mann-Whitney U test.).

ANALYSIS OF CONNECTIVITY

As the strongest single extrinsic inputs are several times weaker than the sum of extrinsic inputs, function in context, f_c, and global function, f_g, are likely to depend on the overall pattern of connections. Because of the complexity of the data, attempts to relate structure and function at the level of the cortical network should aim for quantitative representation and analysis of connection data. Perhaps for different reasons, many anatomists have also recognized the need to represent large amounts of complex connection data in a compact and comprehensible way (McCulloch, 1944; Felleman and Van Essen, 1991; Nicolelis et al., 1990). A number of approaches have been developed for graphical representations and systematic analysis of connection data (Nicolelis et al., 1990; Young, 1992; Musil and Olson, 1991; Hilgetag et al., 1996; Graepel and Obermayer, 1999). These methods provide four major advantages over conventional intuitive methods for analysing the connectional organization of the cortical network. First, they simultaneously consider all the connections and nonconnections, a feat impossible for naked intuition. Second, they make explicit their input data and how the results are computed from the data. Therefore, given the same data and the same analysis methods, different researchers should produce the same results. Third, they allow models of anatomical organization (Young, 1992; Scannell, 1997), or models of structure-function relationships (Kötter and Sommer, 2000; Young, 1992), to be tested formally and quantitatively. Fourth, they can provide relatively comprehensive and relatively comprehensible models of the large-scale organization of neural systems (Scannell et al., 1996).

At present, however, these methods also have substantial limitations. First, like any analysis, they are only as good as the data they use. If the anatomical data are incomplete or inaccurate, the results of the analyses are likely to be misleading. Second, the present analysis methods consider only limited kinds or information (although, at present only rather limited information is available for the vast majority of connections as are known to exist). For example, nonmetric multidimensional scaling (NMDS) uses only the pattern of area-to-area connections and ignores all information on the microanatomy of the connections. Therefore, these methods represent a tradeoff. They are relatively comprehensive and rigorous, but at the cost of ignoring much of the richness and detail of the primary anatomical studies on which they are entirely dependent for data. In no way can any of the analysis methods be regarded as an alternative to good primary neuroanatomy. However, they may help us to sketch a rough map of a large and complex network.

The next section summarizes the results of three of the formal analysis methods: NMDS, optimal set analysis, and hierarchical analysis (Felleman and Van Essen, 1991; Rockland and Pandya 1979; Hilgetag and Young 1996).

Nonmetric Multidimensional Scaling

NMDS (Kruskal, 1964a, 1964b; Takane et al., 1977; Young et al., 1978) is the most widely used of the analysis methods, having been applied to databases for the macaque cortex, the cat thalamo-cortical and cortico-cortical network, and much of the rat brain (Young, 1993; Scannell and Young, 1993; Scannell et al., 1995; Young et al., 1995a; Stephan et al., 2000; Scannell et al., 1999; Burns and Young, 2000).

NMDS finds a configuration of points (in this case cortical areas and thalamic nuclei) that matches a set of experimentally measured proximities (in this case anatomical connections). The process is exactly analogous to deriving a map from a mileage chart in a road atlas. NMDS is useful for computing a rough map that locates brain areas, not in terms of their physical location, but in terms of their connections with the rest of the brain. If, as we suppose, function in context is closely tied to the overall pattern of connections, then the location of cortical areas or thalamic nuclei within these maps should provide a good guide to their function in context, f_C. The maps derived using NMDS can be compared quantitatively using a regression-like procedure, Procrustes' rotation (Schonemann and Carrol, 1970; Gower, 1971). This makes it relatively straightforward to test models of organization quantitatively (Young, 1992; Scannell et al., 1995).

However, NMDS analysis is not free from problems. First, there are many possible alternative algorithms and cost functions that tend to yield similar, but different, results. It is not clear which, if any, is the best to apply to connection data (Simmen et al., 1994; Young et al., 1994, 1995a; Goodhill et al., 1995). Therefore, it is necessary to apply a variety of methods to check that they yield consistent results. Only then can we be confident that we are looking at genuine features of the connection data and not simply the peculiarities of a particular NMDS method. Fortunately, for data shown in this chapter, a wide variety of different NMDS methods to produce highly congruent results (Scannell et al., 1999).

Second, NMDS, like the more familiar principal components analysis, is a form of data compression. A large and potentially high dimensional data set (in this case a connection matrix) is represented as a spatial configuration in a manageably small number of dimensions. For practical reasons we are limited to showing configurations in no more than three dimensions. Although this is probably reasonable for many of the analyses so far, it is likely that certain kinds of anatomical connection data cannot be represented adequately in this framework.

NMDS Representations of the Thalamo-cortical Network

The next section outlines the results of NMDS analyses of the thalamo-cortical network, focusing on the visual system. Although it is possible to use NMDS-derived configurations to test organizational models, we think that some of the most important results of the analyses are the graphical representations of connectivity themselves. In the next section, we review these representations and comment on the characteristics of the connectional architecture that appear robust and independent of the details of the NMDS method.

Figure 15-5 shows the two *least* similar NMDS configurations computed for the entire network using a variety of NMDS methods. The configurations are qualitatively and quantitatively similar, accounting for 78% of each other's variability, by Procrustes' rotation. Areas in Fig. 15-5 are color-coded according to the results of an alternative and independent method for analyzing connection data—optimal set analysis (see later). Scannell et al., (1999) found that the thalamo-cortico-cortical network is composed of three broadly distinct sensory/sensory-motor systems and a fourth frontolimbic system. This is in line with analyses of the isolated cortico-cortical system (Young, 1993, Scannell et al., 1995). In Fig. 15-5A, NMDS arranges points representing cortical areas and thalamic nuclei in a region of space that is roughly the shape of a triangular pyramid. Three of the corners correspond to the most topologically peripheral parts of the sensory and sensory/motor systems, whereas the fourth corner corresponds to the most

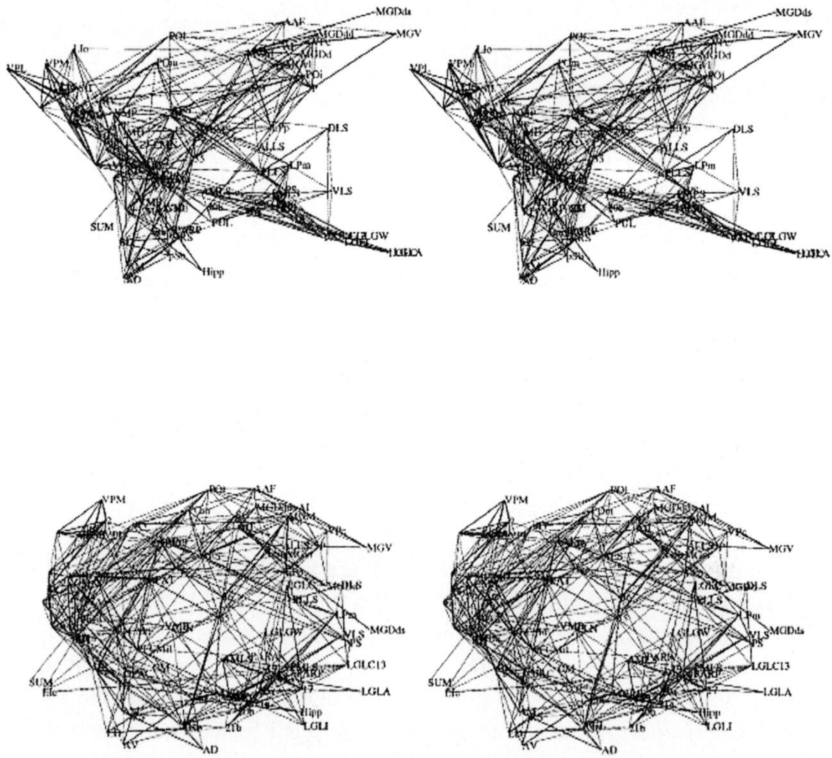

FIGURE 15-5. Connectional relationships in the thalamo-cortical network. The stereogram shows the two least similar three-dimensional configurations yielded by the variety of NMDS analyses of the connection data in Table 15-2 (Scannell et al., 1999). Different NMDS approaches arrange the points in different volumes of space, but the local organization within systems and the relationships between the systems are broadly similar. The *apparent* badness of fit (Young et al., 1995a, 1995b) and distance correlations for the upper and lower plots were 0.21 and 0.94 (upper plot) and 0.26 and 0.61 (lower plot). Areas are color coded according to the major sets independently identified by optimal set analysis (Fig. 15-7). The set of frontolimbic, multimodal, and higher order somatomotor structures (LM-Sg to PFCMil in Fig. 15-7) is colored brown; the set of auditory structures (MGM to MGDd in Fig. 15-7) is colored green; the set of lower order somatomotor structures (4g to 3b in Fig. 15-7) is colored red; The "hippocampal" set (Sb to Hipp in Fig. 15-7) is colored purple; the "anterior nuclei" set (LD to RS in Fig. 15-7) is colored blue/green; and the visual set (LP1 to PMLS in Fig. 15-7) is colored blue. The structures outside these sets are black. The figure was computed using all the connection data; but, for clarity, only strong reciprocal connections, with a sum of weights in both directions greater than 4, are drawn. The sets identified by optimal set analysis map neatly onto compact regions of the space identified by NMDS. This shows that NMDS and optimal set analysis agree with regard to the major features of the connectional organization of the thalamo-cortical network. Figures 15-5 and 15-6 may be free-fused to give three-dimensional images. Divergent viewing is easier for those used to "magic eye" pictures. To fuse divergently, hold the figure at arms length in front of the face, just below the line of sight. Look steadily at a distant object, just visible over the top of the page, then switch your view to the page without converging the eyes. Practice by drawing two dots with a felt-tipped pen about 4 cm apart at the very top of a piece of paper and repeat the procedure. At first, you should see four dots at the top of the paper just below the distant object on which you are fixating. If you switch your view from the distant object to the dots, the central two dots should move toward each other and then merge when you have achieved divergent fixation. For practice and hints, see Johnstone (1995). See color insert for color reproduction of this figure.

topologically central parts of the frontolimbic complex—the hippocampus and associated structures. The central region of the pyramid, where the four systems meet, contains those regions of the network where information from the different sensory modalities converges (e.g., 7, AES, LM-Sg, Ia, Ig, CGp, 36). The NMDS method in Fig. 15-5B arranges the points within a different volume of space, which in this case resembles a hollow shell. However, it also distinguishes the visual, auditory, somatomotor and frontolimbic structures and preserves similar local relations with systems.

Figure 15-5 illustrates the connectional architecture of the cat thalamo-cortical visual system (blue) within the overall connectional scheme. To emphasize the visual system, the visual areas in Fig. 15-5B are redrawn in Fig. 15-6. Signals from the retina enter the visual cortical network directly via the LGN, the geniculate wing, and LPI. Signals from the retina also reach the cortical network indirectly via the tectum and pretectum to PUL and LPm (Dreher, 1986; Berson and Graybiel, 1978, 1983; Updyke, 1981, 1983). Fig 15-5 and 15-6 show that areas 17, 18, and other lower order visual areas are in close association with the retinorecipient thalamic nuclei (components of the LGN and LPI). At higher levels, the visual system divergences somewhat into structures that appear biased toward the retinal component of the afferent signal and structures that appear more closely associated with the pretectal and tectal component of the afferent signal. In the former group are areas 20a, 20b, and PS, which lie between the lower order visual areas and frontal and limbic cortex. In the latter group are ALLS, AMLS, and PLLS, which lead toward visual areas and thalamic nuclei associated with somatomotor structures (AES, area 7). LPm and LM-Sg are associated with these visual areas.

Optimal Set Analysis

The limitations of NMDS and criticism of NMDS by Goodhill et al. (1995) and Simmen et al. (1994) encouraged the development of an entirely independent method for analyzing connection data. We call this method optimal set analysis (OSA). Optimal set analysis allows us to implement an attractive definition of a neural system: something containing components more connected with each other than with components of other systems. We designed OSA to arrange areas and nuclei into sets with as few as possible connections between areas in different sets and as few as possible nonconnections between areas within the same set. This lets the connection data itself decide objectively the number and composition of the sets of areas and nuclei. For example, the visual system might fall out of the analysis as a collection of cortical areas and thalamic nuclei that are highly interconnected with each other and relatively weakly connected with the other cortical systems.

OSA uses evolutionary optimization, a method based on simulated annealing (Aarts et al., 1988) to arrange neural structures into sets that best match our idea of neural systems (Hilgetag et al. 1996; 1997b; 1998; 2000; Scannell et al., 1999). We produced an optimal arrangement on the basis of two terms. The first was a repulsion term that counted nonexisting connections *within* the potential sets.

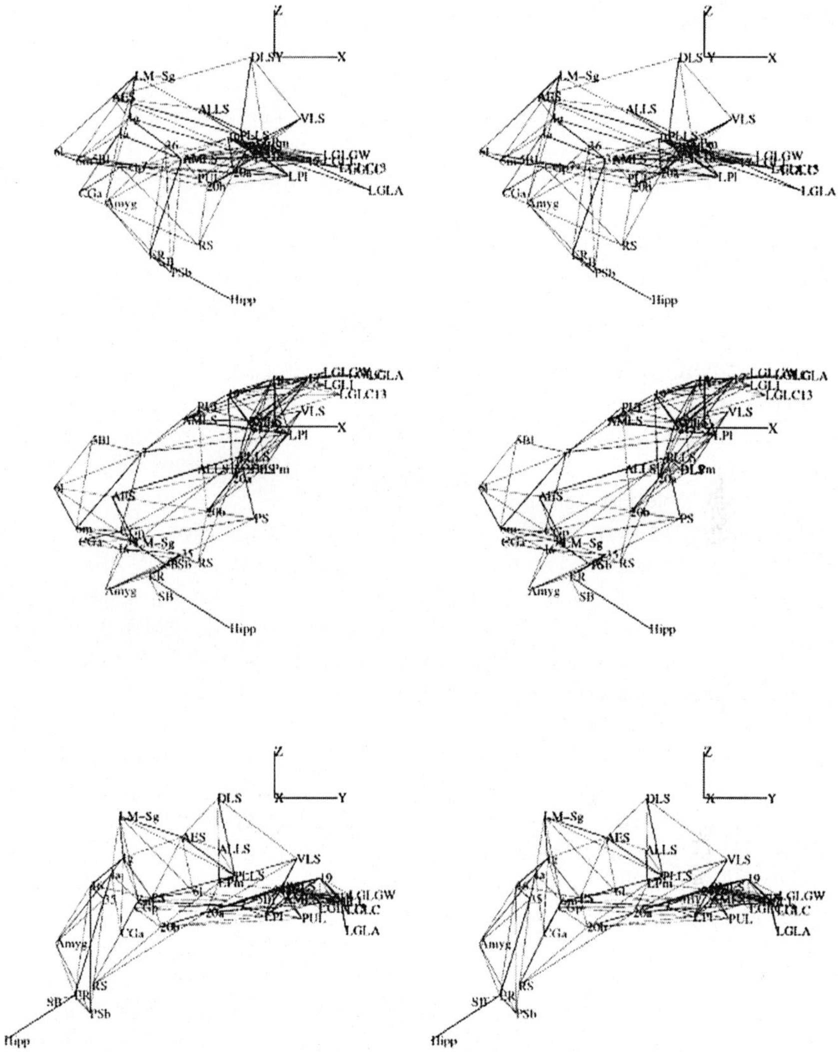

FIGURE 15-6. Connectional relationships in the visual system. Stereogram derived from the upper configuration in Fig. 15-5A computed using all the thalamo-cortico-cortical connection data. For clarity, the figure shows only the visual areas, and some frontolimbic structures with which the higher stations of the visual system are associated. The same configuration is shown from three different viewpoints. The *apparent* badness of fit (Young et al., 1995a) for scaling the wdsm1 matrix was 0.21, with a distance correlation of 0.94. Only strong reciprocal connections, with a sum of weights in both directions greater than 4, are drawn. See color insert for color reproduction of this figure.

Optimizing the repulsion component alone breaks larger clusters into smaller clusters. The second cost consisted of an attraction term counting existing connections *between* sets. Optimizing the attraction term alone will, in the limit case, result in one large cluster containing all the areas and nuclei. OSA lets us explore connectivity at a range of scales. For example, by having a low cost for nonconnections within sets and a high cost for connections between sets, we end up with a small number of sets, each of which tends to contain a large number of areas or nuclei. This means that the method can be used to look at large-scale structure within the network. Alternatively, by having a high cost for nonconnections within sets and a low cost for connections between sets, we can end up with a large number of small sets, each of which contains only the most connectionally associated structures. This means that the method can be used to look at particularly strong local connectional associations within the network. We can assess the statistical rarity of the goodness of fit of the sets yielded by optimal set analysis by comparing the cost of sets computed from connection data with the cost of sets computed from randomly shuffled connection matrices.

When compared with NMDS, OSA has several advantages. First, it avoids dimensional reduction. Second, it allows us to specify particular cost functions, in this case, functions that embody an attractive definition of a neural system. Third, by varying the cost function we can look for both local and global features of neural organization. The disadvantages of OSA are that it tends to produce a very large number of optimal solutions that then have to be summarized (Hilgetag et al., 1996). Sometimes the summaries are simple (Fig. 15-7), but they may be nearly as complex as the connection matrix itself (Fig. 15-8). In addition, the cost functions that we have implemented to date assume a metric relationship between connection densities, which is a major disadvantage when compared with NMDS.

Figure 15-7 shows a summary of the 31 solutions we found with costs within 1% of the single lowest cost solution at balanced attraction and repulsion. The balanced attraction and repulsion case is the most intuitively obvious. In this condition, the cost of a connection between sets is equal to the weight of the connection (1, 2, or 3), and the cost of a nonconnection within sets is equal to 1. The difference between the 31 solutions is small. All solutions agree on several clear and consistent sets (solid dark blocks in Fig. 15-7). From top to bottom of Fig. 15-3, these correspond to a set containing higher somatomotor and some multimodal and frontolimbic structures (LM-Sg to PFCMil), an auditory set (MGM to MGDd), a set of lower somatomotor structures (4g to 3b), two small sets of frontolimbic structures (one consisting of Hipp, Sb, and MV-Re, and the other containing pSb, LD, AD, AV, and RH), and a large set of visual structures (LPI to PS). Approximately 17 areas or nuclei have variable positions and do not consistently lie in any of these sets. Some of these structures have few connections and are pushed out of sets by the repulsive effect of many nonconnections (e.g., LGLA, LGLC). Other areas and nuclei may have substantial connections with structures in more than one of the major sets (e.g., 7, CMN, 36, Enr, POm, SG) and thus fit into different sets with little effect on the overall cost of the configuration. These aspects of the analyses would benefit from more experiments on the less studied

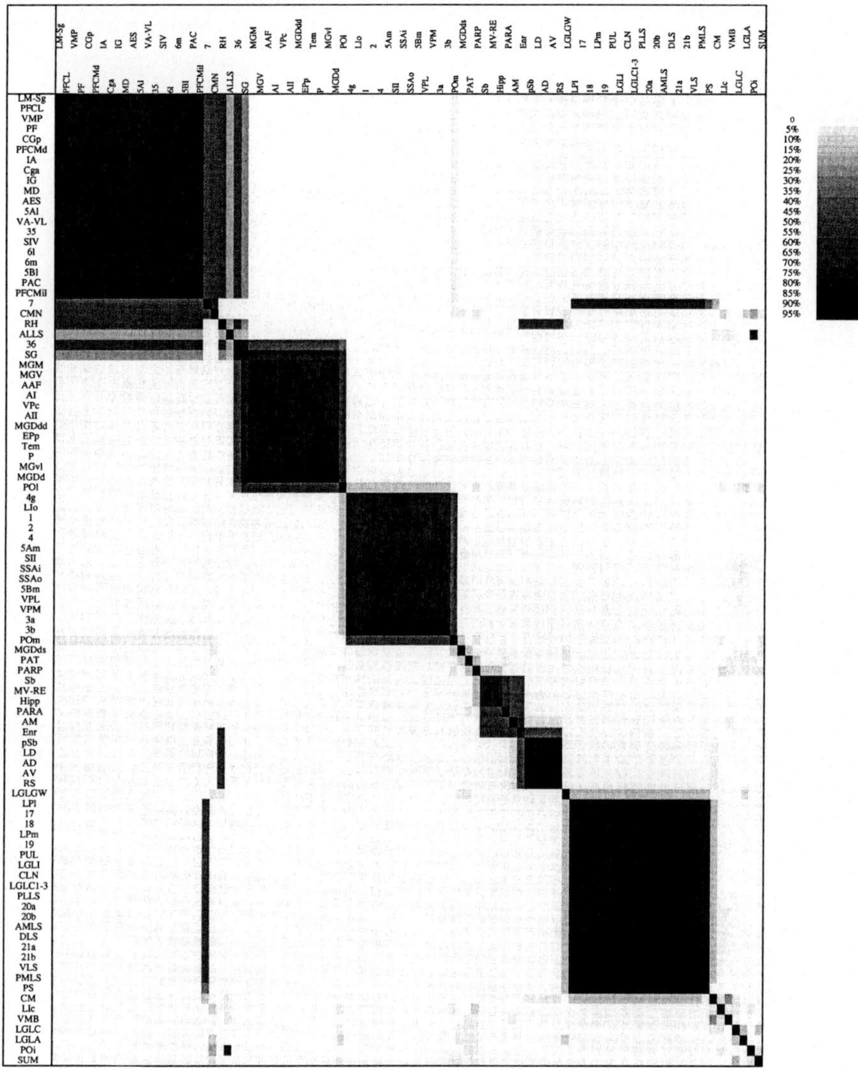

FIGURE 15-7. Optimal set analysis at balanced attraction and repulsion. Summary of the 31 solutions whose costs lie within 1% of the single lowest cost solution. Attraction (the factor multiplied by connection weight to compute the cost of connections between sets) was 1 and repulsion (the cost of nonconnections within sets) was 1. Cortical areas and thalamic nuclei are listed along the top and left of the table. The frequency with which areas or nuclei colocalized within the same set is given by the color (scale at top right of table). Darker colors show frequent colocalization, and light colors show rare colocalization. The analysis consistently finds a number of sets. From top to bottom of Fig. 15-3, these are a set of frontolimbic, multimodal, and higher order somatomotor structures (LM-Sg to PFCMil); a set of auditory structures (MGM to MGDd); a set of lower order somatomotor structures (4g to 3b); two small frontolimbic sets (Sb to Hipp, and pSb to RS); and a large visual set (LP1 to PMLS). Some structures do not have a strong affiliation with any of the large sets, either because of few known connections (e.g., LGLGW) or because of strong connections with more than one set (e.g., 7 and ALLS). The number of sets changes substantially when the attraction and repulsion parameters are at different values. We can be confident that the sets shown in the figure represent a good fit of the experimental data because the costs of the sets computed from connection data are approximately 20 z-scores lower than the mean cost of sets calculated with randomly shuffled connection matrices ($p < 10^{-12}$).

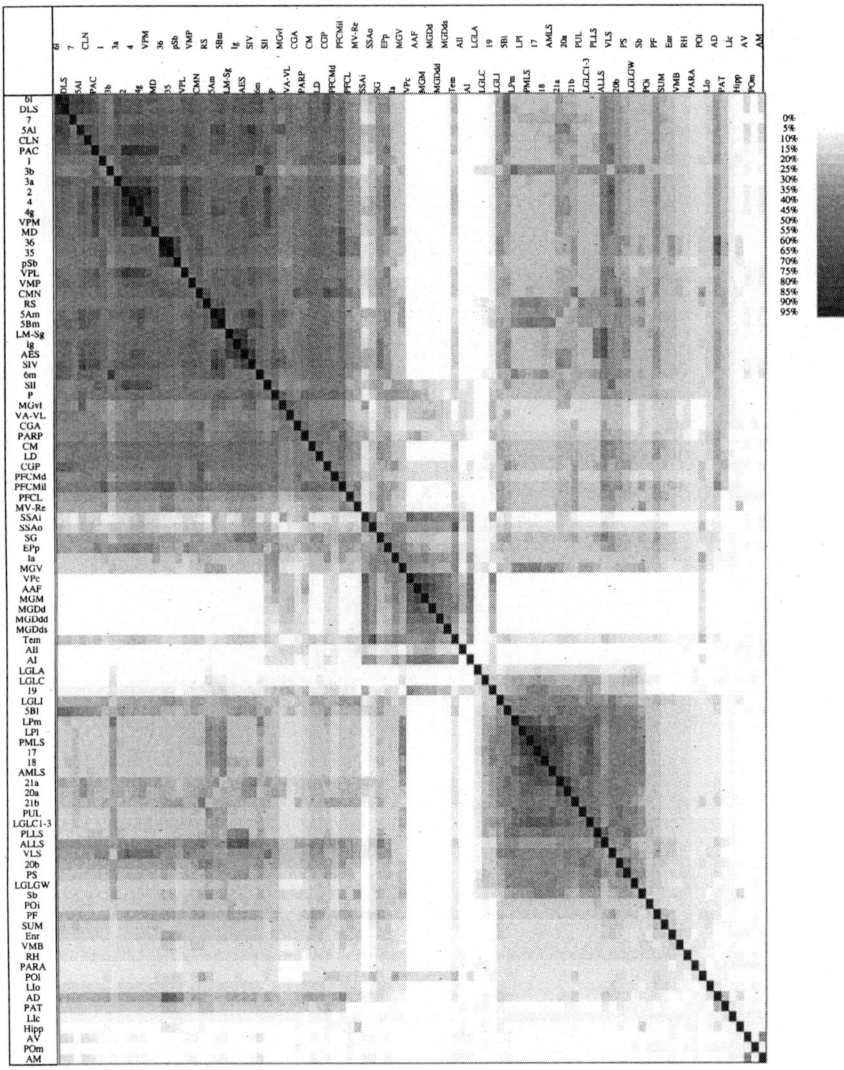

FIGURE 15-8. Summary of optimal set analysis with the attraction parameter ranging from 1 to 10 (repulsion set to 1) and repulsion parameter ranging from 2 to 30 (attraction set to 1). Cortical areas and thalamic nuclei are listed along the top and left of the table. The frequency with which areas or nuclei colocalized within the same set is given by the color (scale at top right of table). Darker colors show frequent colocalization within the same set, and light colors show rare colocalization.

areas and from quantitative connection data for all areas. Quantitative data, which would provide a more accurate cost function, are provided by some researchers (Sherk, 1986; Musil and Olson, 1991; Peters and Payne, 1993; Peters et al., 1994; MacNeil et al., 1997; Vanduffel et al., 1997a), but is unavailable for the vast majority of connections reported in the neuroanatomical literature.

OSA is in excellent agreement with NMDS over the main features of the connectional architecture of the cat thalamo-cortico-cortical network. Figure 15-5 plots the major sets illustrated in Fig. 15-7 on a stereogram showing three-dimensional configurations yielded by NMDS. Figure 15-5 shows that the sets (Fig. 15-7) map neatly and compactly onto the three-dimensional NMDS configurations with little stretching, fracturing, or overlap between the sets. It is clear that most of the unclassified structures lie either at the edge of the NMDS configurations (those with few connections) or else lie near the borders between the sets (those with substantial connections with more than one set). Thus there is excellent agreement between two entirely independent data analysis methods.

Over the full range of attraction and repulsion parameters, high attraction gave fewer, larger, clusters, identifying higher order associations between the major thalamo-cortical systems. High repulsion gave more, smaller clusters, identifying the most closely associated areas or nuclei. Figure 15-8 shows the pooled results from the battery of attraction and repulsion levels. Figures 15-7 and 15-8 are in good agreement, but the graded nature of Fig. 15-5 reveals both local and large-scale structure of the thalamo-cortical network and allows one to look across levels of organization. For example, the darkest squares show areas or nuclei that frequently colocalize across all levels of repulsion and therefore are strongly associated. The lightest squares show areas or nuclei that rarely colocalize, even at the highest levels of attraction and thus connectionally distant.

The results of OSA over range of attraction and repulsion parameters are in excellent agreement with the results of NMDS. Figure 15-8 shows that the higher order visual areas, anterior ectosylvian sulcus and area 7, are in close association with the somatomotor system. The figure also shows that areas ALLS, VLS, and, to a lesser extent, PLLS have the tendency to colocalize with somatomotor and frontolimbic structures that is not shared by most visual areas. Consistent with the findings of NMDS, area 5B1 is closely associated with visual structures. The A and magnocellular C layers of the LGN colocalize frequently only with other unimodal visual structures, identifying these nuclei as the most peripheral parts of the visual system. In global terms, areas 17 and 18 have similar patterns of connections and are closely associated with a large number of other visual structures, particularly LP1, PMLS, AMLS, 21a, 21b, 20a, PUL, and components of the LGN.

On the basis of the overall similarities or differences in connection patterns, we would expect activity in areas 17 and 18 to be most highly correlated with, and most predictable from, the activity in those area with which it most frequently colocalizes. This does not include the LGLA and LGLC, which, while strongly associated with 17 and 18, are not as connectionally similar to primary cortex as a number of other structures.

Hierarchical Analysis

Rockland and Pandya (1979) suggested rules to classify cortico-cortical connections as feedforward, feedback, or lateral. In recent interpretations (Felleman and Van Essen, 1991; Hilgetag et al., 1996; Crick and Koch, 1998), feedforward connections originate in the superficial or both deep and superficial layers and terminate predominantly in layer IV. Feedback connections originate in the deep or both deep and superficial layers and terminate in the deep and superficial layers, avoiding layer IV. Lateral connections arise in both the deep and superficial layers and terminate throughout the depth of the cortex.

By using data on the morphology of the origins and terminations of connections, it is possible to use the classification of connections as feedforward, feedback, or lateral to arrange cortical areas in a largely consistent hierarchy. When there are only a few areas and few connections, this is reasonably easy to do by hand. However, with a large number of areas and a large number connections, finding an arrangement that is most consistent with the connection data presents a formidable computational problem (Hilgetag et al., 1996). For example, 22 areas of the cat visual and frontolimbic system (Fig. 15-9) yield in the order 10^{24} potential solutions. The difficulty of the hierarchical ordering problem, and heroic attempts solve it by hand (e.g. Felleman and Van Essen, 1991), motivated Hilgetag et al (1996) to automate the process. In this case, the cost function reflected the number of connections whose directions are not consistent with the hierarchical relations of the areas that they link (e.g., connections with a feedforward morphology from an area that is higher in the hierarchy to an area that is lower in the hierarchy).

The study by Hilgetag et al. (1996) provided two major results. First, given current data, there is no single best hierarchy. Because connectivity is rather sparse, and most hierarchical relationships are only known at a low level of measurement (i.e., laminar origin and termination patterns tell us that A is above B, but not how far A is above B), hierarchies are necessarily indeterminate. Thus, one is left with a large number of equally good hierarchies, each violating the same number of, but different, hierarchical constraints. Second, although there is no best hierarchy, it is possible the fit the anatomical constraints rather well, although not perfectly, by arranging the areas into a serial hierarchy. In other words, there is a high degree of consistency in the relationships between areas. If area A sends a feedback projection to area B, which sends a feedback projection to area C, we can be confident that the projection from A to C is also of the feedback type.

Results of hierarchical analysis in the cat visual system are shown in Fig. 15-9. Hierarchical analysis was performed using data from 43 reliably classified feed-

FIGURE 15-9. Hierarchical analysis of the cat visual system. **(A)** Distribution of cat visual areas in 19 optimal hierarchies. The y-axis shows the level that areas occupied in hierarchy. The gray scale shows the frequency with which the area is occupied at a particular level. The black box indicates, for each area, the most frequent location (Hilgetag and Grant, personal communication). **(B)** The peak hierarchy for visual areas in the cat. The level of each area responds to each area's most frequent location. The lines show connections between areas. Connections whose hierarchical direction is violated by the arrangement of areas are shown by dashed lines.

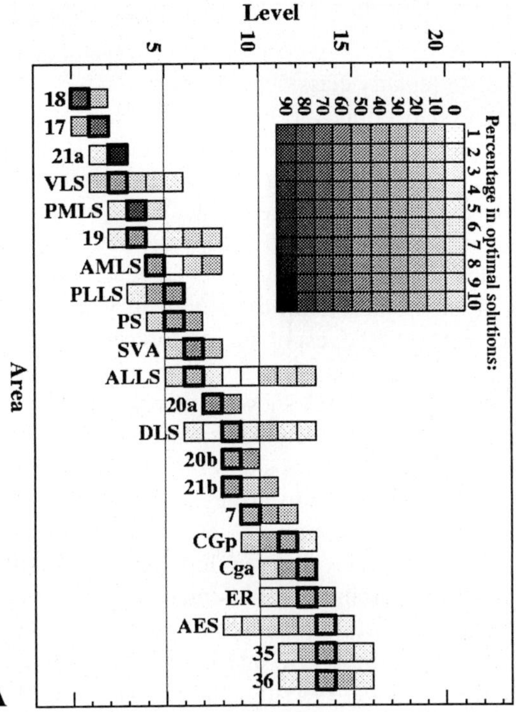
Level
5
10
15
20
Percentage in optimal solutions:
1 2 3 4 5 6 7 8 9 10
0
10
20
30
40
50
60
70
80
90
18
17
21a
VLS
PMLS
19
AMLS
PLLS
PS
SVA
ALLS
20a
DLS
20b
21b
7
CGp
Cga
ER
AES
35
36
Area

A

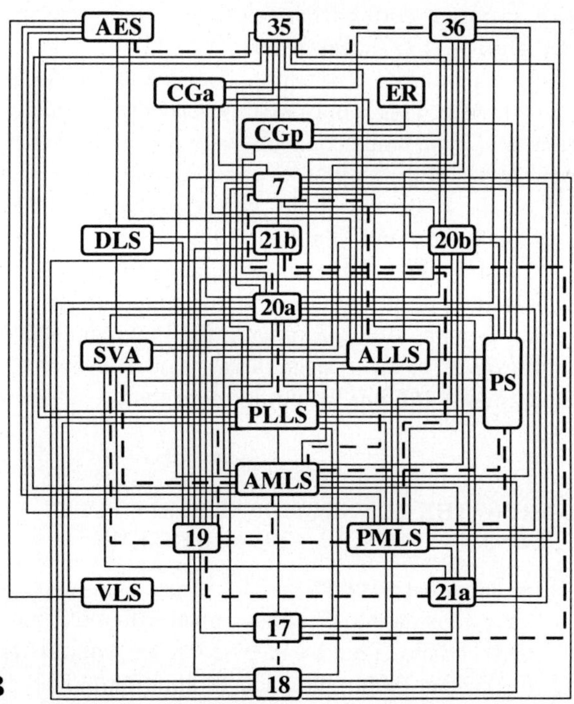
AES
35
36
CGa
ER
CGp
7
DLS
21b
20b
20a
SVA
ALLS
PS
PLLS
AMLS
19
PMLS
VLS
21a
17
18

B

forward connections, 43 reliably classified feedback connections, 22 less reliably classified feedforward connections, 38 less reliably classified feedback connections, 4 lateral connections, and 1 mixed feedforward or lateral connection. The arrangement of cortical areas according to these constraints produced 15 constraint violations. By the standards of hierarchical analysis, there were few optimal solutions—only 19. Areas 17 and 18 occupy the bottom levels. Interestingly, area 18 is usually below area 17.

One application of hierarchical analysis is to identify areas whose hierarchical constraints do not fit well with the rest of the network (Hilgetag et al., 1996). This can provide an insight into regions of cortex where cortical parcellation may be problematic. For example, Fig. 15-9A shows that area ALLS has a bimodal distribution of hierarchical levels. This would be expected if ALLS is really composed of two distinct areas, each of which has a different hierarchical level.

Using morphological constraints to arrange cortical areas into a multilevel hierarchy generates surprising controversy. For example, that onset latency is similar in cortical areas that occupy different levels in the hierarchy suggests that visual systems cannot be hierarchical (e.g., Schmolesky et al., 1998). In part, this controversy may be due to confusion between *hierarchy* in the strictly morphological sense of feedforward, feedback, and lateral projections and *hierarchy* in the more general sense of a linear system with each station being above or below its neighbors. We see no conflict between anatomical hierarchies computed from feedforward and feedback projections and broadly simultaneous onset latencies across many areas in visual cortex (Schmolesky et al., 1998; Schroeder et al., 1998; Nowak and Bullier, 1997). Onset latency is likely to be determined by, among other factors, the number of synapses between the retina and the cells in the cortical area in question. This is only weakly correlated with hierarchical level for at least two reasons. First, many connections between areas span a large number of hierarchical levels (Fig. 15-9B). Second, areas receive visual input that does not depend on the route via striate cortex (Table 15-2). (Dreher, 1986; Updyke, 1981, 1983; Berson and Graybiel, 1978, 1983; Graybiel and Berson, 1980). In the cat, for example, LPm receives input from the superficial layers of the superior colliculus and projects to most visual areas except 17 and 18 (Table 15-2); (Graybiel and Berson, 1980). Similarly, the Pulvinar receives input from the pretectum and projects to most visual areas, so that even the hierarchically highest areas such as AES (Figure 9) are no more than 3 steps from the retina (Table 15-2).

COMPARING THE CONNECTIONAL STRUCTURE OF CAT AND MACAQUE VISUAL SYSTEMS

NMDS and OSA show that a connectional divergence in higher visual areas is reflected in a functional division into areas concerned with motion and visuospatial functions (e.g., 7, AES) and those that are involved in static object vision (e.g., 20a, 20b) (Lomber et al., 1996a; 1996b; Scannell et al., 1996). This arrangement is analogous to the functional (Ungerleider and Mishkin, 1982) and connectional (Young, 1992) divergence into somewhat distinct object versus motion and visuospatial streams found in the macaque visual system. Cats and macaques diverged from a small insectivorous ancestor over 60 million years ago (Savage, 1986), so the similarities in visual system organization may well reflect convergent evolution rather than homology. This, in turn, could reflect a wiring efficiency advantage in separating form and motion vision (Jacobs and Jordan, 1992).

STRUCTURE-FUNCTION RELATIONSHIPS

The previous section focused on the analysis of cortical connectivity, arguing that data analytic methods are useful tools to help understand the pattern of connections between cortical areas and thalamic nuclei. The approach has been to describe the primary data as accurately and comprehensively as possible and then to analyze the data to find the patterns that may be there. This is distinctly different from synthetic modeling, which builds models to test the logical consequences of a set of ideas. However, to explore structure-function relationships, it is necessary to go beyond data analysis and to make explicit synthetic models of the way function is presumed to depend on structure. This section focuses on small number of recent studies that have used a synthetic approach to account for functional properties of brain networks in terms of the anatomical connections that link their stations.

STRUCTURE-ACTIVITY RELATIONSHIPS

Strychnine neuronography was an early physiological connection tracing method. (McCulloch, 1944; MacLean and Pribram, 1953). Saturated strychnine solution was applied to a 1 to 4 mm^2 patch of cortex, inducing local epileptic activity that was propagated via extrinsic connections to other cortical sites where it could be measured. The strength of the propagated activity was taken to indicate the strength of the neural connection between the sites (Kötter and Sommer, 2000).

Kötter and Sommer (2000) and Sommer and Kötter (1997) were interested to see if the pattern of connections between cortical areas in the cat (Scannell et al. 1995) could account for the pattern of activity spread observed in studies using strychnine neuronography (MacLean and Pribram, 1953). Kötter and Sommer devised a simple network model of the whole cortex, with each node representing a single cortical area, and with the connections between nodes representing cortico-cortical connections (Scannell et al. (1995). The model consisted of a simple excitatory network in which A_j (r + 1) is the activity of area j at time step $r + 1$. The area is active at time $r + 1$ if the sum of the activity of its inputs, A_i(r), at time r, multiplied by the connection weights, w_{ij}, exceeds a threshold Θ (equation 3). One node was "strychninized" by setting its activity to 1, and the threshold was adjusted to control the number of active areas.

$$A_j\,(r+1) = \begin{cases} 1 \text{ if } (\Sigma\, A_i(\text{r}) \times w_{i,j}) > \Theta \\ 0 \text{ if } (\Sigma\, A_i(\text{r}) \times w_{i,j}) < \Theta \end{cases} \tag{3}$$

We emphasize two results from Kötter and Sommer (2000). First, the network based on real anatomical connectivity reproduced the pattern of activity propagation from the strychnine experiments rather well. Network models with alternative patterns of connections, for example nearest neighbor connectivity, performed much worse. This confirms that the pattern of association fibers and low-level information on their relative strengths are both useful quantitative predictors of the global pattern of activity spread in cerebral cortex. Second, the model provides a good account of the strychnine data when activity spreads between areas via both direct and indirect pathways. Therefore, it is important to consider indirect network effects in structure-function relationships. This second result concords with results derived by Young et al. (2000), who used a similar model to investigate the consequences of lesions in a simple network.

SIMPLIFIED STRUCTURE-FUNCTION RELATIONSHIPS

Work on the superior colliculus (SC) and visual cortical areas of the cat has found that lesions can markedly *improve* behavioral performance (Sprague, 1966). Lomber and Payne (1996) found that reversible inactivation of the middle suprasylvian visual areas (MSS) makes the cat unresponsive to food items presented contralaterally to the cooled cortical site. However, cooling both left and right MSS simultaneously restores normal orienting to food items presented to either the left or right. In fact, unilateral lesions in either SC or MSS cause contralateral neglect of food items, whereas all combinations of bilateral lesions (SC and SC, SC and MSS, MSS and MSS) restore a high degree of function.

Hilgetag et al. (1999) have produced a synthetic model to try to account for these puzzling effects. Their model considers a subset of the anatomical pathways and structures that are thought to be involved in spatial orienting behavior, including the two retinae, the optic chiasm, cortical stations representing the middle suprasylvian cortices (Lomber and Payne, 1996), cell populations in the substantia nigra pars reticulata, and the two superior colliculi. Structures are linked by anatomically appropriate excitatory and inhibitory connections, including an inhibitory connection between the colliculi. To compute activity in the right and left colliculi in the model, Hilgetag et al., (1999) used a mathematical formalism similar to one frequently used in describing reaction rates in metabolic networks. Because the colliculi are linked by inhibitory connections, they compete directly to control orienting, and the model orients contralateral to the colliculus that is more active. The model made the correct orienting responses in its intact state and showed hemineglect with unilateral lesions to its SC and MSS analogus, paradoxical restoration of function with bilateral lesions, and a range of realistic biases to stimuli presented in the far periphery. A simplified version of the model is shown in Fig. 15-10 (Hilgetag et al., 1999).

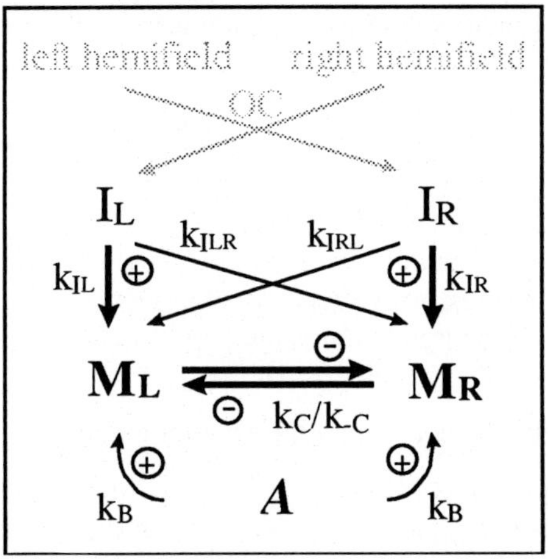

FIGURE 15-10. A minimal model for effects of intact, lesioned, and paradoxically restored spatial attention (Hilgetag et al., 1999; Young et al., 2000). This model contains only the midbrain activities M_L and M_R, representing left and right superior colliculi, as dynamical variables. Inputs related to stimuli in the left and right visual hemifields are relayed into the right and left midbrain structures, M_L and M_R, respectively, via middle suprasylvian cortex and other pathways, represented by I_L and I_R (OC, optic chiasm). A weaker, crossed connection to the contralateral side accounts for the representation of ipsilateral stimuli in the midbrain structures. The term A represents the average neural activity in the system. Plus and minus signs indicate excitatory and inhibitory interactions. Normal behavior: a food item is presented in the left visual hemifield activating I_R, which activates M_R, which inhibits M_L, causing orienting to the left. Unilateral cortical lesion: a food item is presented to the left hemifield; however, I_R is partially inactivated, so activity in M_R remains low, and the animal fails to orient to the left. Paradoxical restoration: a food item is presented to the left hemifield. Both I_L and I_R are partially inactivated, so visual stimulation induces higher in I_R than in I_L. Thus, M_R becomes more active than M_L, causing orienting to the left.

TOWARD QUANTITATIVE STRUCTURE-FUNCTION RELATIONSHIPS

An impressive approach to the structure-function problem has been taken by Vanduffel et al. (1997a), who have tried to relate qualitative data on the density of cortico-cortical projections to the projections' "functional-impact." Functional-impact was measured by combining metabolic mapping with radiolabeled 2-deoxyglucose (2-DG) with local cortical inactivation. Vanduffel et al. injected anterograde radioactive tracers into MSS cortex in several individuals and measured the density in transported label in a number of cortical and subcortical structures. In a second group of animals, 2DG was administered while MSS cortex was cooled with a chronically implanted cooling probe. By comparing the two hemispheres (cooled and uncooled), Vanduffel et al. were able to identify cortical regions where the metabolic rate had been suppressed by cooling. The degree of suppression provided a measure of functional-impact that was compared with the density of the corresponding cortico-cortical projections.

Vanduffel et al. (1997a) found good agreement between regions of the brain receiving projections from MSS cortex and regions where cooling had a substantial functional-impact. However, in quantitative terms, the correlation between functional-impact and anatomical density was rather low (the proportion of the variance in functional-impact explained by anatomical density was only $R^2 = 0.23$). Part of the discrepancy is due to the fact that projections to higher order structures have a greater functional impact than would be expected, whereas projections to lower order structures have a lower functional impact than would be expected on the basis of anatomical density.

COMMENTS ON STRUCTURE-FUNCTION STUDIES

We have briefly reviewed three studies that we think make substantial steps toward investigating structure-function relationships at the systems level. With regard to function, interpretable structure-function relationships are most likely to follow from studies that respect the evolutionary function of visual system. Kötter and Sommer (2000) considered the effects of an extremely unphysiological stimulus, strychnine, so their model is concerned with activity rather than function in the strict sense. In this respect, the model of Hilgetag et al. (1999) and the empirical studies on which it is based (Lomber and Payne, 1996) look promising. Orienting to food items is a naturalistic task with which to probe the visual system, and Hilgetag et al.'s model appears to perform well. However, the model fails to account for recent results from animals orienting to nonfood stimuli (Lomber and Payne, 2001; Lomber et al., 2001). In contrast to Hilgetag et al., Vanduffel et al.'s (1997a) experimental animals did not perform visually guided behavior during 2 DG administration. Furthermore, cooling MSS cortex induces behavioral changes, so the functional impact measure may reflect indirect consequences of changes in behavior or attention. Finally, in the three studies outlined, and other similar work (Young et al., 2000), the functional measure has been extremely simple—the amount of activity within a cortical area or network component representing an area. The cortex almost certainly interprets the activity of a neural population as a vector, and not a scalar, quantity (Georgopoulos et al., 1986; Lennie, 1998). Therefore, models should try to address the quality of neural activity and not just the quantity.

Concerning structure, the anatomical data suggest, and Young et al. (2000) and Kötter and Sommer (2000) have argued, that it is necessary to consider connection data in a systematic and comprehensive way because the function of individual cortical areas depends on their overall relationship with the network and because activity can spread via direct and indirect pathways. Kötter and Sommer (2000) did consider global aspects of cortico-cortical connectivity. However, their model was based on the low-level ordinal data collated by Scannell et al. (1995). Hilgetag et al. (1999) were also limited to low level nonquantitative connection data and used only a limited subset of connections. Vanduffel et al. (1997a) used quantitative connection data of a kind that is extremely laborious, but worthwhile, to collect. However, Vanduffel et al. did not consider wider aspects of the organization of the network, such as indirect pathways between areas.

None of the studies when considered in isolation can firmly specify systems level structure-function relationships. However, these articles show that the right elements are in place. If we look across studies we can see formal consideration of network properties, naturalistic behavioral protocols, and quantitative neuroanatomy. The challenge remains to bring these elements together within a single experimental program.

CONCLUSIONS

This chapter began by asking the questions: What is the function of primary visual cortex, and how does its function relate to the structure of network within which it is embedded? We argued that no satisfactory answer is available at present; thus we considered the framework within which an answer might develop. Before we summarize this framework, we consider two simpler systems in which considerable progress has been made in determining the relationship between structure and function. The first example concerns the relationship between structure and (mal)function in a set of faulty amplifiers that our electrophysiological laboratory inherited. The second example concerns the relationship between the structure and function of the beaks of the finch species that inhabit the Galapagos Islands.

Given an unknown electrical circuit, it is impossible to conclusively impute f_c or f_l to individual components by removing the components and investigating the change in the circuit's global input-output function f_g (Lewis, 1970). In fact, one cannot be certain about the function (f_c or f_l) of the components with anything short of complete system decomposition because interactions between the components can have complicated consequences (Lewis, 1970). Despite these problems, our engineer, Roger Mason, was able to quickly identify and replace the faulty components in a set of faulty amplifiers without resorting to complete system decomposition. The problem of imputing (mal)function to structure was overcome by examining local and global input-output relationships in conjunction with two additional sources of information. These were, first, knowledge of what the amplifiers were designed to do (the amplifiers' "evolutionary function") and, second, the amplifiers' wiring diagram (their "connection matrix"). The task of identifying the faulty component, imputing (mal)function to structure, was relatively straighforward because we knew the function and organization of the amplifiers.

The second example is more biological, but is also substantially simpler than the problem we face in the cerebral cortex. Darwin was impressed by the huge differences in beak morphology between finch species in the Galapagos Islands and suggested that they might represent adaptations to particular kinds of food (Darwin, 1859). However, for more than 100 years, the functional significance of the structural variability in beaks was controversial. Repeated observations showed that species of finch with very different beak morphologies had very similar diets (Winer, 1994). In other words, the beaks appeared to be general purpose food processors, whose function was independent of marked structural variability. The issue was not resolved until the 1970s, when the Galapagos Islands were visited by researchers equipped with calibrated pliers for measuring the hardness of the seeds eaten by the finches. They found a small poppy seed would crack at around 10 newtons, and a tough but nutritious cactus seed might yield at over 100 newtons. They combined their analysis of the statistical properties of natural foods with laborious surveys of the distribution and abundance of seeds, as well as the size and shape of the beaks of the finches that ate them. Remarkably, their initial studies also showed very little difference in the food items taken by finches with different beak structures. However, when the dry season came, the birds stopped being generalists and became specialists. When food was scarce and survival was critical, there developed clear within- and between-species correlations between beak morphology and food selection (Weiner, 1994). The relationship between the beak structure and beak function became apparent only via exhaustive and quantitative observation of the way the beaks were used during natural behavior. Only under a limited set of real-world conditions did the structure-function relationship make itself clear.

In conclusion, we return to the three themes have linked the ideas presented in this chapter. These themes might provide a framework for exploring the relationship between systems level structure and function in visual cortex;

1. The visual system is a specific biological adaptation and not a general purpose computer. Therefore, we must clarify the visual system's evolutionary function if we are to elucidate its structure-function relationships. Tasks and stimuli used in vision research should respect the visual system's evolutionary function.
2. Psychophysical, electrophysiological, and anatomical evidence suggests that the function in context, f_c, of visual cortex may be closer to inference than analysis. Tasks and stimuli used in vision research should respect the visual system's prior expectations about the visual world.
3. The function in context, f_c, of visual areas depends on global aspects of the organization of the network. Local and global processing are tightly coupled. Therefore, it is vital to have good anatomical data for as much of the network as possible and not just parts of the network traditionally thought of as interesting (e.g., the projection from LGN to area 17). In the long term, this can be best achieved if neuroanatomy takes a lead from genetics and aims to automate its methods for data acquisition.

ACKNOWLEDGMENTS

Supported by the Wellcome Trust. We thank Drs Simon Grant and Claus Hilgetag for access to unpublished material on the hierarchical arrangement of the cat visual system.

REFERENCES

Aarts, E. H., Korst, J. H., and Van Laarhoven, P. J. (1988). A quantitative analysis of the simulated annealing algorithm—a case study for the traveling salesman problem. *J. Statistical Physics* **50,** 187–206.

Abeles, M. (1990). *Corticonics: neural circuits of the cerebral cortex,* Cambridge, Cambridge University Press.

Agmonsnir, H., and Segev, I. (1993). Signal delay and input synchronization in passive dendritic structures. *J. Neurophys.* **70,** 2066–2085.

Aloimonos, Y., and Rosenfeld, A. (1991). Computer vision. *Science* **253,** 1249–1259.

Anderson, J. C., Binzegger, T., Martin, K. A. C., and Rockland, K. S. (1998). The connection from cortical area of V1 to V5: a light and electron microscopic study. *J. Neurosci.* **18(24),** 10525–10540.

Atick, J. J., Li, Z. P., and Redlich, A. N. (1992). Understanding retinal color coding from 1st principles, *Neural Comput.* **4,** 559–572.

Atick, J. J., and Redlich, A. N. (1992). What does the retina know about natural scenes? *Neural Comput.* **4,** 196–210.

Atick, J. J. (1992). Could information theory provide an ecological theory of sensory processing? *Network Comput. Neural Syst.* **2,** 213–251.

Azzopardi, P., and Cowey, A. (1993). Preferential representation of the fovea in primary visual cortex. *Nature* **361,** 719–721.

Azzopardi, P., and Cowey, A. (1996). The overrepresentation of the fovea and adjacent retina in the striate cortex and dorsal lateral geniculate nucleus of the macaque monkey. *Neuroscience* **72,** 627–639.

Baddeley, R. J., and Hancock, P. J. B. (1991). A statistical analysis of natural images matches psychophysically derived orientation tuning curves. *Proc. Soc. Lond. B.* **246,** 219–223.

Baddeley, R. (1996). An efficient code in V1? *Nature* **381,** 560–561.

Baddeley, R., Abbott, L. F., Booth, M. C. A., Sengpiel, F., Freeman, T., Wakeman, E. A., and Rolls, E. T. (1997). Responses of neurons in primary and inferior temporal visual cortices to natural scenes. *Proc. R. Soc. Lond. B* **264,** 1775–1783.

Barlow, H. B. (1953). Summation and inhibition in the frog's retina. *J. Physiol. Lond.* **119,** 69–88.

Barlow, H. B., Blakemore, C., and Pettigrew, J. D. (1967). The neural mechanisms of binocular depth discrimination. *J. Physiol. Lond.* **193,** 327–342.

Barton R. A. (1998). Visual specialization and brain evolution in primates. *Proc. R. Soc. Lond. B.* **265,** 1933–1937.

Barton R. A., and Dean P. (1993). Comparative evidence indicating neural specialization for predatory behaviour in mammals. *Proc. R. Soc. Lond. B.* **254,** 63–68.

Barrow, H. G., and Tenebaum, J. M. (1986). Computational approaches to vision. In: *Handbook of perception and human performance* (K. R. Boff, L. R. Kaufman, and Thomas, J. P. Eds.), pp. 38:1–38, 70, New York, Wiley.

Berson, D. M., and Graybiel, A. M. (1978). Parallel thalamic zones in the LP-Pulvinar complex of the cat identified by the afferent and efferent connections. *Brain Res.* **147,** 139–148.

Berson, D. M., and Graybiel, A. M. (1983). Organization of the striate-recipient zone on the cat's lateralis posterior-pulvinar complex and its relations with the geniculostriate system. *Neuroscience* **9,** 337–372.

Boyd, J. D., and Mastubara, J. A. (1996). Laminar and columnar patterns of geniculocortical projections in the cat: relationships to cytochrome oxidase. *J. Comp. Neurol.* **365,** 659–682.

Budd, J. M. (1998). Extrastriate feedback to primary visual cortex in primates: a quantitative analysis of connectivity. *Proc. R. Soc. Lond. B* **265,** 1037–1044.

Burns, G. A. P. C., and Young, M. P. (2000). Analysis of the connectional organisation of neural systems associated with the hippocampus in rats. *Phil. Trans. R. Soc. Lond. B* **355,** 55–70.

Burton, H., and Kopf, E. M. (1984). Connections between the thalamus and the somatosensory areas of the anterior ectosylvian gyrus in the cat. *J. Comp. Neurol.* **224,** 173–205.

Bush, P. C., and Sejnowski, T. J. (1994). Effects of inhibition and dendritic saturation in simulated neocortical pyramidal cells, *J. Neurophys.* **71,** 2183–2193.

Catania, K. C., and Kaas, J. H. (1997a). Somatosensory fovea in the star-nosed mole: behavioral use of the star in relation to innervation patterns and cortical representation. *J. Comp. Neurol.* **387,** 215–233.

Catania, K. C., and Kaas, J. H. (1997b). Organization of somatosensory cortex and distribution of corticospinal neurons in the eastern mole *(Scalopus aquaticus). J. Comp. Neurol.* **378,** 337–353.

Cavanagh, P. (1989). Multiple analyses of orientation in the visual system. In: *Neural mechanisms of visual perception* (D. M-K. Lam and C. Gilbert, Eds.), pp. 261–280, Woodlands, TX, Portfolio.

Chance, F. S., Nelson, S. B., and Abbot, L. F. (1998). Synaptic depression and the temporal response characteristics of V1 cells, *J. Neurosci.* **12,** 4785–4799.

Cherniak, C. C. (1990). The bounded brain: towards quantitative neuroanatomy. *J. Cog. Neurosci.* **2,** 58–68.

Cherniak, C. C. (1992). Local optimization of neuronal arbors. *Biol. Cyber.* **66,** 503–510.

Churchland P. S., Ramachandran V. S., and Sejnowski T. J. (1993). A critique of pure vision. In: *Large scale neuronal theories of the brain* (C. Koch, and J. Davis, Eds.), pp. 23–60 Cambridge, Mass, MIT Press.

Coppola, D. M., White, L. E., Fitzpatrick, D., and Purves, D. (1998a). Unequal representation of cardinal and oblique contours in the ferret visual cortex. *Proc. Natl. Acad. Sci. U.S.A.* **95,** 2621–2623.

Coppola, D. M., Purves, H. R., McCoy, A. N., and Purves, D. (1998b). The distribution of oriented contours in the real-world. *Proc. Natl. Acad. Sci. U.S.A.* **95,** 4002–4006.

Cowan, W. M., Gottlieb, D. I., Hendrickson, A. E., Price, J. L., and Woolsey, T. A. (1972). The autoradiographic demonstration of axonal connections in the central nervous system. *Brain Res.* **37,** 21–51.

Crick, F., and Koch, C. (1998). Constraints on cortical and thalamic projections: the no-strong loops hypothesis. *Nature* **391,** 245–250.

Cumming, B. G., and Parker, A. J. (1997). Responses of primary visual cortical neurons to binocular disparity without depth perception. *Nature* **389,** 280–283.

Dan, Y., Atick, J. J., and Reid, R. C. (1996). Efficient coding of natural scenes in the lateral geniculate nucleus: experimental test of a computational theory. *J. Neurosci.* **16,** 3351–3362.

Darwin, C. (1859). *The origin of species.* London, John Murray.

Dreher, B. (1986). Thalamocortical and corticocortical interconnections in the cat visual system: relation to the mechanisms of information processing. In: *Visual neuroscience* (J. D. Pettigrew, K. J. Sanderson, and W. R. Levick, Eds.), pp. 290–314, Cambridge, Cambridge University Press.

Felleman, D. J., and Van Essen, D. C. (1991). Distributed hierarchical processing in the primate cerebral cortex. *Cerebral Cortex* **1,** 1–47.

Field, D. J. (1987). Relations between the statistics of natural images and a response properties of cortical cells, *J. Opt. Soc. Am. A* **4,** 2379–2394.

Field, D. J. (1994). What is the goal of sensory coding? *Neural Comput.* **6,** 559–601.

Field, D. J., Hayes, A., and Hess, R. F. (1993). Contour integration by the human visual system: evidence for a local "association field." *Vis. Res.* **33,** 173–193.

Georgopoulos, A. P., Schwartz, A. B., and Kettner, R. E. (1986). Neuronal population coding of movement direction. *Science* **233,** 1416–1419.

Gerfen, C. R., and Sawchenko, P. E. (1984). An anterograde neuroanatomical tracing method that shows the detailed morphology of neurons, their axons, and terminals—immunohistochemical localization of an axonally transported plact lectin, Phaseolus-vulgaris leukoagglutinin PHA-L. *Brain Res.* **290,** 219–238.

Gilbert, C. D., and Wiesel, T. N. (1983). Clustered intrinsic connections in cat visual cortex. *J. Neurosci.* **2,** 1116–1133.

Gilbert, C. D., and Wiesel, T. N. (1989). Columnar specificity of intrinsic horizonttal and cortico-cortical connections in cat visual cortex. *J. Neurosci.* **9,** 2432–2442.

Gilbert, C. D., and Wiesel, T. N. (1990). The influence of contextual stimuli on the orientational selectivity of cells in primary visual cortex of the cat. *Vis. Res.* **30,** 1689–1701.

Goodhill, G. J., Simmen, M. W., and Willshaw, D. J. (1995). An evaluation of the use of multidimensional scaling for understanding brain connectivity. *Phil. Trans. R. Soc. B* **348,** 265–280.

Gower, J. C. (1971). Statistical methods of comparing different multivariate analyses of the same data. In: *Mathematics in the archeological and historical sciences* pp. 138–149, Edinburgh, Edinburgh University Press.

Graepel, T., and Obermayer, K. (1999). A stochastic self-organizing map for proximity data. *Neural Comput.* **11,** 139–155.

Graybiel, A. M., and Berson, D. M. (1980). Histochemical identification and afferent connections of subdivisions in the lateralis posterior pulvinar complex and related thalamic nuclei in the cat, *Neuroscience* **5,** 1175–1238.

Grosof, D. H., Shapley, R. M., and Hawken, M. J. (1993). Macaque V1 neurons can signal illusory contours, *Nature* **365,** 550–552.

Gulyas, B., Lagae, L., Eysel, U., and Orban, G. (1990). Corticofugal feedback influences the responses of geniculate neurons to moving stimuli. *Exp. Brain Res.* **79,** 441–446.

Gulyas, B., Orban, G. A., Duysens, J., and Maes, H. (1987). The suppressive influence of moving texture background on responses of cat striate neurons to moving bars. *J. Neurophys.* **57,** 1767–1791.

Heeger, D. J. (1992a). Normalization of cell responses in cat striate cortex. *Vis. Neurosci.* **9,** 181–197.

Heeger, D. J. (1992b). Half-squaring in responses of cat striate cells. *Vis. Neurosci.* **9,** 427–443.

Henry, G. H., Salin, P. A., and Bullier, J. (1991). Projections from area's 18 and 19 to cat striate cortex: divergence and laminar specificity. *Eur. J. Neurosci.* **3,** 186–200.

Hilgetag C-C., O'Neill M. A., and Young M. P. (1996). Indeterminate organization of the visual hierarchy. *Science* **271,** 776–777.

Hilgetag C-C., Burns, G. A. P. C., O'Neill M. A., Scannell, J. W., and Young M. P. (2000). Anatomical connectivity defines the organisation of clusters of cortical areas in macaque monkey and cat. *Phil. Trans. Roy. Soc. Lond. B.* **355,** 91–110.

Hilgetag C-C, Kötter R., and Young M. P. (1999). Paradoxical restoration of function: a mathematical model based on anatomical connectivity. *Prog. Brain Res.* **121,** 125–146.

Holmes, W. R., and Rall, W. (1992). Estimating the electrotonic structure of neurons with compartmental models. *J. Neurophys.* **68,** 1438–1452.

Hupé, J., Payne, B. R., Lomber, S. G., Girád, P., and Bullier, J. (1998). Cortical feedback improves discrimination between figure and background by V1, V2, and V3 neurons. *Nature* **394,** 784–787.

Jacobs, R. A., and Jordan, M. I. (1992). Computational consequences of a bias towards short connections. *J. Cog. Neurosci.* **4,** 323–336.

Johnstone, T. (1995). *Magic 3D,* London, Stanley Paul.

Jones, E. G. (1985). *The thalamus.* New York, Plenum.

Knill, D. C., Kersten, D., and Mamassian, P. (1996). Implications of a Bayesian formulation of visual information for processing in psychophysics. In: *Perception as bayesian inference* (D. C. Knill, & W. Richards, Eds.), pp. 239–285, Cambridge, Cambridge University Press.

Knill, D. C., and Richards, W. (1996). *Perception as bayesian inference.* Cambridge, Cambridge University Press.

Kötter, R., and Sommer F. T. (2000). Global relationship between anatomical connectivity and activity propagation in the cerebral cortex. *Phil. Trans. R. Soc. Lond. B* (in press).

Kristan, W. B., and Shaw, B. K. (1997). Population coding and behavioural choice, *Curr. Opin. Neurobiol.* **7,** 826–831.

Kristensson, K., Olsson, Y., and Sjostrand, J. (1971). Axonal uptake and retrograde transport of exogenous proteins in the hypoglossal nerve. *Brain Res.* **32,** 399–406.

Kruskal, J. B. (1964a). Multidimensional scaling by optimizing goodness of fit to a nonmetric hypothesis. *Psychometrika* **19,** 1–27.

Kruskal, J. B. (1964b). Non-metric multidimensional scaling: a numerical method. *Psychometrika* **29,** 115–129.

Land, M. F., and Barth, F. G. (1992). The quality of vision in the ctenid spider *Cupiennisu salei. J. Exp. Biol.* **164,** 227–242.

Laughlin, S. B., van Steveninck, R. R. D. and Anderson, J. C. (1998). The metabolic cost of neural information. *Nature Neurosci.* **1,** 36–41.

Le Gros Clark, W. E. (1932). The structure and connections of the thalamus. *Brain* **55,** 261–273.

Le Gros Clark, W. E. (1942). The visual centres of the brain and their connections. *Phys. Rev.* **22,** 205–232.

Lennie, P. (1998). Single units and visual cortical organization. *Perception* **27,** 889–935.

Lettvin, J. Y., Maturana, H. R., McCulloch, W. S., and Pitts, W. H. (1959). What the frog's eye tells the frog's brain. *Proceedings of the Institute of Radio Engineers* **47,** 1940–1951.

Lewis E. R. (1970). Neural subsystems: goals, concepts, and tools. In: *The neurosciences second study program* (F. O., Schmitt, Ed.), pp. 384–396, New York, Rockefeller University Press.

Lewis, J. E., and Kristan, W. B. (1998). A neuronal network for computing population vectors in the leech. *Nature* **391,** 76–79.

Lomber, S. G., and Payne, B. R. (1996). Removal of two halves restores the whole—reversal of visual hemineglect during bilateral cortical or collicular inactivation in the cat. *Vis. Neurosci.* **13,** 1143–1156.

Lomber, S. G., and Payne, B. R. (2001). Task specific reversal of visual hemineglect during reversible deactivation of posterior parietal cortex: a comparison with deactivation of superior colliculus. *Vis. Neurosci.*

Lomber, S. G., Payne, B. R., and Cornwell, P. (1996a). Learning and recall of form discriminations during reversible cooling deactivation of ventral-posterior suprasylvian cortex in the cat. *Proc. Natl. Acad. Sci. U.S.A.* **93,** 1654–1658.

Lomber, S. G., Payne, B. R., and Cornwell P. (2001). Role of superior colliculus in analyses of space: upper and intermediate layer contributions revealed by a battery of tasks and reversible deactivation. *Proc. Natl. Acad. Sci. U.S.A.*

Lomber, S. G., Payne, B. R., Cornwell, P., and Long, K. D. (1996b). Perceptual and cognitive visual functions of parietal and temporal cortices in the cat. *Cerebral Cortex* **6,** 673–695.

MacLean, P. D., and Pribram, K. H. (1953). Neuronographic analysis of medial and basal cerebral cortex. I. Cat. *J. Neurophys.* **16,** 312–323.

Marchi, V., and Algeri, G. (1895). Sulle degenerazioni discendenti consecutive a lesioni della corteccia cerebrale. *Nota Pre. Riv. Sper. Di Freniat.* **11,** 429.

Marr, D. (1982). *Vision.* San Francisco, W. H. Freeman and Co.

Martin, J. H., and Ghez, C. (1999). Pharmocalogical inactivation in the analysis of the central control of movement, *J. Neurosci. Methods* **86,** 145–159.

Maunsell, J. R., and Ferrera, V. P. (1995). Attentional mechanisms in visual cortex. In: *The cognitive neurosciences* (M. S. Gazzaniga, Ed.), pp. 451–462 Cambridge, MIT Press.

McClurkin, J. W., Optican, L. M., and Richmond, B. J. (1994). Cortical feedback increases visual information transmitted by monkey parvocellular lateral geniculate nucleus neurons. *Vis. Neurosci.* **11,** 601–617.

McCulloch, W. S. (1944). The functional organization of the cerebral cortex. *Physiol. Rev.* **24,** 390–407.

Michalski, A., Wimbourne, B. M., and Henry, G. H. (1993). The effect on reversible cooling of cat's primary visual cortex on the responses of area 21a neurons. *J. Physiol. Lond.* **466,** 133–156.

Michalski, A., Wimbourne, B. M., and Henry, G. H. (1994). The role of ipsilateral and contralateral inputs from primary cortex in responses of area 21a neurons in cats. *Vis. Neurosci.* **11,** 839–849.

Murphy, P. C., and Sillito, A. M. (1987). Corticofugal feedback influences the generation of length tuning in the visual pathway. *Nature* **329,** 727–729.

Musil, S. Y., and Olson, C. R. (1988a). Organization of the cortical and subcortical projections to the medial prefrontal cortex in the cat. *J. Comp. Neurol.* **272,** 219–241.

Musil, S. Y., and Olson, C. R. (1988b). Organization of the cortical and subcortical projections to the anterior cingulate cortex in the cat. *J. Comp. Neurol.* **272,** 203–218.

Musil, S. Y., and Olson, C. R. (1991). Cortical areas in the medial frontal-lobe of the cat delineated by quantitative-analysis of thalamic afferents. *J. Comp. Neurol.* **308,** 457–466.

Nauta, W. J. H., and Gygax, P. A. (1954). Silver impregnation for degenerating axons in the central nervous system: a modified technique. *Stain Technol.* **29,** 91–93.

Nicolelis, M. A. L., Tinone, G., Sameshima, K., Timo-Iaria, C., Hong, Y. C., and Van de Bilt, M. T. (1990). Connection, a microcomputer program for storing and analyzing structural properties of neural circuits. *Comput. Biomed. Res.* **23,** 64–81.

Nikara, T., Bishop, P. O., and Pettigrew, J. D. (1968). Analysis of retinal correspondence by studying receptive fields of binocular single units in cat striate cortex. *Exp. Brain Res.* **6,** 353–372.

Nowak, L., and Bullier, J. (1997). The timing of information transfer in the visual system. In: *Cerebral cortex,* Vol. 12 (K. S. Rockland, J. H. Kaas, and A. Peters, Eds.), pp. 205–241, New York, Plenum Press.

O'Carrol, D. C., Laughlin, S. D., Bidwell, H. J., Harris, R. A. (1997). Spatio-temporal properties of motion detectors matched to low image velocities in hovering insects, *Vis. Res.* **23,** 3427–3439.

O'Carrol, D. C., Bidwell, H. J., Laughlin, S. D., and Warrant, E. J. (1996). Insect motion detectors matched to visual ecology. *Nature* **382,** 63–66.

Olshausen, B. A., and Field, D. J. (1996). Emergence of simple-cell receptive field properties by learning a sparse code for natural images. *Nature* **381,** 607–609.

Olson, C. R., and Musil, S. Y. (1992). Topographic organization of cortical and subcortical projections to posterior cingulate cortex in the cat—evidence for somatic, ocular, and complex subregions. *J. Comp. Neurol.* **324,** 237–260.

Osorio, D., Marshall, N. J., and Cronin, T. W. (1997). Stomatopod photoreceptor spectral tuning as an adaptation for colour constancy in water. *Vis. Res.* **37,** 3299–3309.

Osorio, D., and Vorobyev, M. (1996). Colour vision as an adaptation to frugivory in primates. *Proc. R. Soc. Lond. B.* **263,** 593–599.

Pandya, D. N., and Yeterian, E. H. (1985). Architecture and connections of cortical association areas. In: *Cerebral Cortex,* Vol. 4 (A. Peters, and E. G. Jones, Eds.), pp. 3–61, New York, Plenum Press.

Pasternak, T., Horn, K. M., and Maunsell, J. H. R. (1989). Deficits in speed discrimination following lesions of the lateral suprasylvian cortex in the cat. *Vis. Neurosci.* **3,** 365–375.

Pasternak, T., and Maunsell, J. H. R. (1992). Spatiotemporal sensitivity following lesions of area 18 in the cat. *J. Neurosci.* **12,** 4521–4529.

Pasternak, T., Tompkins, J., and Maunsell, J. H. R. (1995). The role of striate cortex in visual function of the cat. *J. Neurosci.* **15,** 1940–1950.

Payne, B. R., Lomber, S. G., Geeraerts, S., Vandergucht E., and Vandenbussche, E. (1996a). Reversible visual hemineglect. *Proc. Natl. Acad. Sci. U.S.A.* **93,** 290–294.

Payne, B. R., Lomber, S. G., Villa, A. E., and Bullier, J. (1996b). Reversible deactivation of cerebral network components. *Trends Neurosci.* **19,** 535–542.

Peterhans, E., and von der Heydt, R. (1989). Mechanisms of contour perception in monkey visual cortex. I. Contours bridging gaps. *J. Neurosci.* **9,** 1749–1764.

Peters, A., and Payne, B. R. (1993). Numerical relationships between geniculocortical afferents and pyramidal cell modules in cat primary visual cortex. *Cerebral Cortex* **1,** 69–78.

Peters, A., Payne, B. R., and Budd, J. (1994). A numerical analysis of the geniculocortical input to striate cortex in the monkey. *Cerebral Cortex* **4,** 215–229.

Petersen, S. E., Fox, P. T., Posner, M. I., Mintun, M., and Raichle, M. E. (1988). Positron emission topographic studies of the cortical anatomy of single-word processing. *Nature* **331,** 585–589.

Poeppel, D. (1996). A critical review of PET studies of phonological processing. *Brain Language* **55,** 317–351.

Polat, U., Mizobe, K., Pettet, M. W., Kasamatsu, T., and Norcia, A. M. (1998a). Collinear stimuli regulate visual responses depending on cell's contrast threshold. *Nature* **391,** 580–584.

Polat, U., Mizobe, K., Pettet, M. W., Kasamatsu, T. and Norcia, A. M. (1998b). Collinear stimuli regulate visual responses depending on cells contrast threshold. *Nature* **391,** 580–584.

Polyak. S. (1927). An experimental study on the association, callosal, and projection fibres of the cerebral cortex of the cat. *J. Comp. Neurol.* **44,** 197–254.

Polyak, S. (1933). *The main afferent fiber systems in the cerebral cortex of primates* Vol. 2. Berkeley, University of California Publ. Anat.

Price, D. J., and Zumbroich, T. J. (1989). Postnatal development of cortico-cortical efferents from area 17 in the cats' visual cortex. *J. Neurosci.* **9,** 600–613.

Raczkowski, D., and Rosenquist, A. C. (1983). Connections of the multiple visual cortical areas with the lateral posterior-pulvinar complex and adjacent thalamic nuclei in the cat. *J. Neurosci.* **3,** 1912–1942.

Rao, R. P. N., and Ballard, D. H. (1999). Predictive coding in the visual cortex: a functional interpretation of some extra-classical receptive field effects, *Nature Neurosci.* **2,** 79–87.

Rieke, F., Warland, D., de Ruyter van Steveninck, R., and Bialek, W. (1997). *Spikes. Exploring the neural code,* Cambridge, MIT Press.

Rockland, K. S., and Pandya, D. N. (1979). Laminar origins and terminations of cortical connections of the occipital lobe in the rhesus monkey. *Brain Res.* **179,** 3–20.

Rose, J. E., and Woolsey, C. N. (1948). Structure and relations of limbic cortex and anterior thalamic nuclei in rabbit and cat. *J. Comp. Neurol.* **89,** 279–438.

Rosenquist, A. C. (1985). Connections of visual cortical areas in the cat. In: *Cerebral Cortex,* Vol. 3 (A. Peters, and E. G. Jones, Eds.), pp. 81–117, New York, Plenum Press.

Salzman, C. D., Murasugi, C. M., Britten, K. H., and Newsome W. T. (1992). Microstimulation in area MT: effects on direction discrimination performance. *J. Neurosci.* **12,** 2331–2355.

Savage, R. J. G. (1986). *Mammalian evolution: an ilustrated guide.* London, British Museum (Natural History).

Scannell, J. W., Sengpiel, F., Tovee, M. J., Benson, P. J., Blakemore, C., and Young, M. P. (1996). Visual-motion processing in the anterior ectosylvian sulcus of the cat. *J. Neurophys.* **76,** 895–907.

Scannell, J. W., Blakemore, C., and Young, M. P. (1995). Analysis of connectivity in the cat cerebral-cortex. *J. Neurosci.* **15,** 1463–1483.

Scannell, J. W. (1997). Determining cortical landscapes. *Nature* **386,** 452.

Scannell, J. W., Burns, G. A. P. C., Hilgetag, C. C., O'Neil, M. A., and Young, M. P. (1999). The connectional organization of the cortico-thalamic system of the cat. *Cerebral Cortex* **9,** 277–299.

Scannell, J. W., Grant, S., Payne, B., and Baddeley, R. (2000). On variability in the density of corticocortical and thalamo-cortical connections, *Phil. Trans. R. Soc. Lond. B* **355,** 21–35.

Scannell, J. W., and Young, M. P. (1999). Neuronal population activity and functional imaging. *Proc. R. Soc. Lond. B* **266,** 875–881.

Schonemann, P., and Carrol, R. M. (1970). Fitting one matrix to another under a choice of a similarity transformation and rigid motion. *Psychometrika* **35,** 245–255.

Schmolesky, M. T., Wang, Y. C., Hanes, D. P., Thompson, K. G., Leutgeb, S., Schall, J. D., and Leventhal, A. G. (1998). Signal timing across the macaque visual system. *J. Neurophys.* **79,** 3272–3278.

Schroeder, C. E., Mehta, A. D., and Givre, S. J. (1998). A spatiotemporal profile of visual system activation revealed by current source density analysis in the awake macaque. *Cerebral Cortex* **8,** 575–592.

Sherk, H. (1986). Location and connections of visual cortical areas in the cats' suprasylvian sulcus. *J. Comp. Neurol.* **247,** 1–31.

Simmen, M. W., Goodhill, G. J., and Willshaw, D. J. (1994). Scaling and brain connectivity, *Nature* **369,** 448–449.

Somers, D. C., Nelson, S. B., and Sur, M. (1995). An emergent model of orientation selectivity in cat visual cortical simple cells. *J. Neurosci.* **15,** 5448–5465.

Sommer, F., and Kötter, R. (1997). Simulating a network of cortical areas using anatomical connection data in the cat. In: *Computational neuroscience. Trends in research* (J. M. Bower, Ed.), pp. 511–517, New York, Plenum Press.

Sprague, J. M. (1966). Interaction of cortex and superior colliculus in mediation of visually guided behavior in the cat. *Science* **153,** 1544–1547.

Srinivasan, M. V., Laughlin, S. B., and Dubs, A. (1982). Predictive coding: a fresh view of inhibition in the retina. *Proc. R. Soc. Lond. B.* **216,** 427–459.

Steele, G. E., Weller, R. E., and Cusick, C. G. (1991). Cortical connections of the caudal subdivision of the dorsolateral area (V4) in monkeys. *J. Comp. Neurol.* **306,** 495–520.

Stephan, K. E., Hilgetag, C-C., Burns, G. A. P. C., O'Neill, M. A., Young, M. P., and Kötter R. (2000). Computational analysis of functional connectivity between areas of primate cerebral cortex, *Phil. Trans. Roy. Soc. Lond. B.* **355,** 111–126.

Stratford, K. J., Tarczy-Hornoch, K., Martin, K. A. C., Bannister, N. J., and Jack, J. J. B. (1996). Excitatory synaptic inputs to spiny stellate cells in cat visual cortex. *Nature* **382,** 258–261.

Symonds, L. L., and Rosenquist, A. C. (1984). Laminar origins of visual cortico-cortical connections in the cat. *J. Comp. Neurol.* **229,** 39–47.

Takane, Y., Young, F. W., and de Leeuw, J. (1977). Non-metric multidimensional scaling: an alternating least squares method with optimal scaling features. *Psychometrika* **42,** 7–67.

Tehovnik, E. J. (1996). Electrical stimulation of neural tissue to evoke behavioural responses. *J. Neurosci. Methods* **65,** 1–17.

Teuber, H. L. (1955). Physiological psychology. *Annu. Rev. Psychol.* **6,** 267–296.

Tooby J., and Cosmides L. (1995). Mapping the evolved functional organisation of mind and brain. In: *The Cognitive Neurosciences* (M. S. Gazzaniga, Ed.) pp. 1181–1185, Cambridge, MIT Press.

Treisman, A. M. (1986). Properties and parts of objects, In: *Handbook of perception and human performance* (K. R. Borf, L. R. Kaufman, and J. P. Thomas, Eds.), pp. 35:1–35:70, New York, John Wiley.

Treue, S., and Maunsell, J. H. R. (1996). Attentional modulation of visual motion processing in cortical areas MT and MST, *Nature* **382,** 539–541.

Tsotsos, J. K. (1987). A complexity level analysis of immediate vision. *Int. J. Comput. Vis.* **1,** 303–320.

Ungerleider, L. G., and Mishkin, M. (1982). Two cortical visual systems. In: *Analysis of visual behavior* (D. G. Ingle, M. A. Goodale, and R. J. Q. Mansfield, Eds.), pp. 549–586, Cambridge, MIT Press.

Updyke, B. V. (1981). Projections from visual areas of the middle suprasylvian sulcus onto the lateral posterior complex and adjacent thalamic nuclei in the cat. *J. Comp. Neurol.* **201,** 477–506.

Updyke, B. V. (1983). A re-evaluation of the functional organization and cytoarchitecture of the feline lateral posterior complex, with observations on adjoining cell groups. *J. Comp. Neurol.* **219,** 143–181.

van Hateren, J. H. (1992). Real and optimal neural images in early vision. *Nature* **360,** 68–70.

van Hateren, J. H. (1997). Processing of natural time series of intensities by the visual system of the blowfly. *Vis. Res.* **23,** 3407–3416.

van Hateren, J. H., and van der Schaaf, A. (1998). Independent component filters of natural images compared with simple cells in primary visual cortex. *Proc. R. Soc. Lond. B.* **265,** 359–366.

Vanduffel, W., VanDenbussche, E., Singer, W., and Orban, G. A. (1995). Metabolic mapping of visual areas in the behaving cat: a [14C]2-deoxyglucose study. *J. Comp. Neurol.* **354,** 161–180.

Vanduffel, W. Vandenbussche, E., Singer, W., and Orban, G. A. (1997b). A metabolic mapping study of orientation discrimination and detection tasks in the cat. *Eur. J. Neurosci.* **9,** 1314–1328.

Vanduffel, W., Payne, B. R., Lomber, S. G., and Orban, G. (1997a). Functional impact of cerebral connections. *Proc. Natl. Acad. Sci. U.S.A.* **94,** 7617–7620.

Winer, J. (1994). *The beak of the finch.* London, Jonathon Cape.

Williams, S. M., McCoy, A. N., and Purves, D. (1998a). The influence of depicted illumination on brightness. *Proc. Natl. Acad. Sci. U.S.A.* **95,** 13296–13300.

Williams, S. M., McCoy, A. N., and Purves, D. (1998b). An empirical explanation of brightness. *Proc. Natl. Acad. Sci. U.S.A.* **95,** 13301–13306.

Worgotter, F., Nelle, E., Li, B., and Funke, K. (1998). The influence of corticofugal feedback on the temporal structure of visual responses of cat thalamic relay cells. *J. Physiol. Lond.* **509,** 797–815.

Young, F. W., Takane, Y., and Lewyckyj, R. (1978). ALSCAL: a non-metric multidimensional scaling program with several differences options, *Behav. Res. Methods Instrument.* **10,** 451–453.

Young, M. P. (1992). Objective analysis of the topological organization of the primate cortical visual-system *Nature* **358,** 152–155.

Young, M. P. (1993). The organization of neural systems in the primate cerebral-cortex. *Proc. R. Soc. Lond. B* **252,** 13.

Young, M. P. (1994). Turn-on, tune-in, and drop-out. *Curr. Biol.* **4,** 51–53.

Young, M. P., Scannell, J. W., and Burns, G. A. P. C. (1995b). *The analysis of cortical connectivity.* Austin, R. G. Landes.

Young, M. P., Scannell, J. W., O'Neill, M. A., Hilgetag, C. C., Burns, G., and Blakemore, C. (1995a). Nonmetric multidimensional-scaling in the analysis of neuroanatomical connection data and the organization of the primate cortical visual system. *Phil. Trans. R. Soc. Series B* **348,** 281–308.
Young, M. P., Hilgetag, C. C., and Scannell, J. W. (2000). On imputing function to structure from the behavioural effects of brain lesions. *Phil. Trans. R. Soc. Lond. B.* **355,** 147–161.
Yuille, A. L., and Bulthoff, H. (1996). Bayesian decision theory and psychophysics. In: *Perception as bayesian inference* (R. Knill, and W. Richards, Eds.), pp. 123–161, Cambridge, Cambridge University Press.
Zeki, S. M. (1969). Representation of central visual fields in prestriate cortex of monkey. *Brain Res.* **14,** 271–291.
Zeki, S. M. (1971). Cortical projections from two prestriate areas in the monkey. *Brain Res.* **34,** 19–35.
Zeki, S. M. (1974). Functional organization of a visual area in the posterior bank of the superior temporal sulcus of the rhesus monkey. *J. Physiol. Lond.* **236,** 549–573.
Zeki, S., and Shipp, S. (1988). The functional logic of cortical connections. *Nature* **335,** 311–317.
Zeki, S., and Shipp, S. (1995). Segregation and convergence of specialised pathways in macaque monkey visual cortex. *J. Anat.* **187,** 547–562.

16

Behavioral Analyses of the Contributions of Cat Primary Visual Cortex to Vision

Donald E. Mitchell

Psychology Department, Dalhousie University, Halifax, Canada

INTRODUCTION

Study of the contributions that the primary visual cortex (area 17) makes to vision in the cat is necessarily constrained by prevailing knowledge of the anatomical organization of the visual system. As Zeki's (1) excellent and scholarly review of cerebral achromatopsia makes very clear, the same constraint applies equally well to study of the role of particular cortical areas in humans. The explosive increase in the level of understanding of both the anatomical organization of the visual system and the functional properties of neurons in the various cortical visual areas in the last 20 years have had a profound impact on the way in which the role of particular cortical areas in visual perception have been studied.

From a historical perspective, it is possible to identify three behaviorally based experimental approaches to study of the contributions that the primary visual cortex (area 17) makes to vision in the cat. The first and oldest of these approaches represents applications to the visual cortex of a generic method that has been used widely to ascertain the functional role of particular neural structures elsewhere in the central nervous system. This method attempts to define the function of area 17 by documentation of the behavioral deficits that follow when it is either ablated as a whole or when lesions are made at known locations with respect to the retinotopic map. The second approach is quite recent and to this point has been applied exclusively to examination of the contributions of area MT to vision in the macaque monkey (2,3). The method uses standard chronic recording techniques to permit weak (microstimulation) of a subset of neurons in a very small region (typically a functionally defined column) of a neural structure while the animal makes a perceptual decision concerning a visual stimulus presented within their receptive fields. Comparison of the decisions made during microstimulation with those made in the absence of such stimulation permits strong inferences to be drawn concerning the involvement of the population of stimulated neurons in a particular perceptual experience. The ability to simultaneously combine behavioral and electrophysiological measures in the same animal provides more compelling evidence of a link between particular neurons and certain perceptual experiences than can be obtained from electrophysiological study of the response characteristics of those neurons alone (4,5). In the latter situation, the evidence of a role that such neurons play in perception relies on the demonstration of a high degree of correlation between the changes in the visual response characteristics of neurons with alteration of stimulus parameters and those that occur in psychophysical performance of other animals or humans in response to the same stimulus manipulations. Finally, the third approach examines the consequences for vision that follow various forms of early selective visual deprivation that are known to produce well-documented and long-lasting changes to the visual response characteristics of neurons in area 17. As the focus of this chapter is on behavioral analyses, emphasis is placed on study of the consequences for vision of lesions and of early selected visual deprivation. Nevertheless, because most of the lesion studies conducted in the last 30 years have been guided by knowledge gleaned from electrophysiological studies of the visual cortex, it is not possible to discuss one without reference to the other.

THE BEHAVIORAL CONSEQUENCES OF LESIONS OF THE VISUAL CORTEX

Despite a long history of study of the effects on vision of lesions of the cat visual cortex, only limited conclusions could be drawn from many of the early studies concerning the role of the visual cortex in vision. In part this can be attributed to certain methodological considerations but, more important, by constraints set by prevailing knowledge of the anatomical and functional organization of the visual system. Following a discussion of these methodological and conceptual issues, the key conclusions that can be drawn from this rich corpus of studies are described beginning with those that were apparent even from the early studies.

METHODOLOGICAL ISSUES

The Intended and Reconstructed Extent of the Lesion

The most important and common methodological concern relates to the extent of the lesion. Areas 17 and 18 in the cat are contiguous, and at the border between them they share a representation of the vertical meridian of the visual field. It is thus difficult to ablate completely one of these two areas without impinging on the other. The presence of substantial variations in the sulcal patterns within the cat population (6) and indeed between the two hemispheres of individual cats (7,8), adds an added dimension to the difficult task of identification of the border between areas 17 and 18 on the basis of morphological criteria. Partly for this practical reason and also because of past lack of a conceptual framework that would make it necessary to examine the consequences of lesions restricted to just one of these areas, the majority of published studies on the effects on vision of lesions of visual cortex report the consequences of ablation of both areas 17 and 18. In most cases the authors' intended to restrict the lesion to these two areas, but in some cases the lesions have included parts of area 19 as well.

The ability to reconstruct the extent of the lesion with accuracy has improved over the years with increasingly detailed documentation of both the representation of the visual field on the different visual cortical areas and on the dorsal lateral geniculate nucleus (dLGN [see Chapter 1]). With respect to the latter, the careful documentation of the retinotopic representation of the visual field on the dLGN and the median interlaminar nucleus (MIN) of the LGN by Sanderson (9) has permitted reconstruction of the extent of lesions of areas 17 and 18 based on the extent of retrograde degeneration. Through exhaustive application of single-unit recording methods, the visual field topography in each of areas 17, 18, 19, 20, and 21 has been documented (10–13). These studies have revealed the presence of significant variability in the topographical organization of area 17, which makes it likely that similar variation in the retinotopic maps may exist for the other visual areas. A further refinement to reconstruction of lesions has been introduced recently by combining information gained from anatomical indicators with single unit recordings conducted in the cortex surrounding the lesion (14–16). In this way it is possible to determine whether islands of remaining tissue are functional by characterization of the visual response properties of single cells within any remaining tissue. In addition it is possible to determine the extent to which cells in areas adjacent to the lesion have altered in terms of their visual response properties. This group of investigators have also taken advantage of retinotopic maps of Tusa et al. in the different visual areas in order to plot the extent of the lesion on the visual fields for each of the relevant cortical areas (14–16). But since the visual field plots do not show the cortical magnification, they also display plots of the lesion on flat maps of the visual cortex and adjacent visual areas (15–17), modified from the maps published by Van Essen and Maunsell (18).

Partly because of the difficulty of achieving complete lesions of a particular cortical area and also to provide a within-animal control, an alternative approach that uses restricted lesions at known locations within a particular cortical area was developed for use with monkeys and subsequently has been applied for use with cats. On the basis of spatial maps based on electrophysiological recordings, injections of neurotoxins such as ibotenic acid or local anesthetics are placed in a particular cortical area to inactivate cortex on which a known region of the visual field is represented. Subsequently the animal is trained to fixate so that targets can be displayed in this particular region of the visual field. Thresholds for stimuli located in this region can then be compared with thresholds in neighboring regions or else for stimuli placed in the visual field on the opposite side of the midline. Although this approach has been used in monkeys for study of effects on vision of localized lesions (19–21), to date it has been exploited by only one group (22,23) on a small number of cats.

The Nature of the Visual Discrimination

Until the last two decades, the majority of studies of the consequences of lesions of visual cortex for vision examined the ability of animals to discriminate between suprathreshold stimuli that were intended to differ along a single stimulus dimension such as luminance or shape (form). The usual performance measure on such tasks was the number of trials required to achieve a defined criterion, which led to problems of interpretation of any observed deficits. Losses in performance following lesions of the visual cortex, whether reflected by an increased number of trials to reach criterion, or even a failure to perform the task, could as easily be attributed to deficits of visual learning as to a reduction in the ability to process and recognize the stimuli. Moreover, learning measures such as the number of trials to reach criterion performance are influenced highly by attentional and motivational fluctuations. The use of large suprathreshold stimuli raise a number of other issues that have been debated in the literature over the years. One prominent issue concerns the degree to which the stimuli are equated for luminance both globally and in terms of local regions within the stimuli that the animals could conceivably use to make their judgments. Even when luminance is equated, several studies have shown that lesioned cats can use cues relating to the total amount of contour or complexity in their judgments (24–26). Another issue that has received some attention is the possible use by the animal of different strategies postoperatively than those used preoperatively that may mask the presence of a functional deficit. With respect to the stimuli themselves, one possible strategy would be for the animal to attend postoperatively only to local regions of the stimuli, for example, the lower left corner, that may provide sufficient information to solve the discrimination in situations where it could no longer process the intended global aspects of the stimuli that it used before the lesion.

Studies based on the use of such shape discrimination tasks resulted in a number of key observations including the important finding that cats were able to relearn preoperatively learned visual discriminations following lesions of areas 17 and 18 but that as the lesion encroached progressively more onto area 19 the number of trials to reacquire the discrimination relative to the initial acquisition increased dramatically (24,26,27–36). However, over the years the rapid increase in knowledge of the visual response characteristics of neurons in different visual areas leed to a change in the nature of the visual stimuli used and an increased level of sophistication of the psychophysical probes of visual function. Together, these changes were introduced in an attempt to uncover specific functions that particular cortical areas may play in vision.

Arguably the study that can be considered the cornerstone in this respect was that of Berkley and Sprague (35), who articulated a new approach to the study of the role of neural structures that required measurements of psychophysical thresholds for particular stimuli that were chosen as the most likely candidates to be processed by the particular pathway or structure to be lesioned. In their words:

> An additional prerequisite in the analysis of individual pathways is the functional isolation of the pathway. This may be achieved by choosing appropriate stimuli which can only be processed (or best processed) by one pathway. The use of psychophysical threshold testing procedures, in which one stimulus dimension is systematically varied, fulfills the requirement of functional isolation. Thus, threshold testing permits specifying the stimulus dimension that the animal is using to make the discrimination, and requires the organism to use the neural system most sensitive to the dimension being varied. (35, p. 68).

To this statement can be added a point raised by Kruger et al. (37), that rather than testing the capacity to learn, detection threshold experiments begin from overtrained knowledge. The variability introduced by motivational and attentional fluctuations to conventional pattern discrimination tasks are largely removed as the number of trials can be increased to achieve the desired level of overtraining. The strategy of Berkley and Sprague is reminiscent of the approach pioneered by Stiles (38) to first isolate and then characterize individual cone mechanisms in human color vision. With increasing knowledge of the organization of the visual pathways and the different spatiotemporal sensitivities of its component neural streams such as the X-, Y- and W-pathways, it becomes possible to tailor psychophysical tasks to preferentially probe these individual streams.

In addition to the conceptual breakthrough that followed from measurements of thresholds, especially spatial thresholds, was an important practical benefit that negated, to some extent, concerns relating to the extent (completeness) of the lesion. Because spatial thresholds are best (i.e., lowest) in central vision and rise quite rapidly with eccentricity (39), cats must image the stimuli on their area centralis to obtain the best thresholds. Consequently lesions of the visual cortex that include the representation of the central visual field should produce the maximum effect on thresholds. Incomplete lesions that leave intact regions of cortex on which the peripheral visual field is represented would not support low thresholds. Moreover if the animals used such islands of intact tissue in areas 17 and 18 to mediate vision, it would be apparent as a change in the animal's head posture following the lesion as the animal rotated its head to locate the stimuli in those regions of the peripheral visual field that project to the remaining intact tissue in areas 17 and 18.

The use of threshold testing procedures has an added benefit in terms of the ability to interpret the results of lesion studies in situations where the deficits are small or nonexistent (40). When animals are required to discriminate between complex stimuli (such as shape discrimination tasks), it is possible for the animal to use any of a large number of possible features or dimensions contained within the stimuli to aid with the discrimination. As the animal can use a different set of such cues following the lesion than before, it is possible that an animal can compensate completely for a cue eliminated from use by the lesion. In threshold testing procedures, only one dimension of the stimulus is varied and if manipulation of this dimension alters behavior, then this is good evidence that the animal is attending to this cue. With such a procedure, shifting of cues between preoperative and postoperative testing is not possible. Moreover such a procedure contains within it a performance control for a nonspecific effect of the lesion, as only the latter would produce reduced discrimination performance of the animal for suprathreshold stimuli.

No matter whether the task is a pattern discrimination or measurement of a spatial threshold, it is important to have control over the observation distance. For a pattern discrimination task, such control is necessary to ensure that the whole figure falls within the visual field at the point at which the discrimination is made to avoid forcing the animal to use local cues. Measurements of spatial thresholds demand precise knowledge of the observation distance to calculate accurately the angular dimensions of the stimuli at threshold. A surprising number of studies that used two-choice discrimination boxes where the cat was required to push panels on which the stimuli were displayed at best could only provide a range of distances over which the animal could have made its choice. As a consequence, the specification of thresholds could not be precise and uncertainty is introduced as to the nature of the cue used in pattern discrimination tasks.

The Age of the Animal at the Time of the Lesion

There is a considerable body of evidence to show that lesions of areas 17 and 18 made on or around the day of birth (P1) have different anatomical (41,42) and functional (43–46) consequences than those observed following identical lesions made either in the second to fourth week or in adulthood. Following cortical lesions in the first week of postnatal life widespread retrograde degeneration of X- type retinal ganglion cells are observed (see Chapter 14), a result that in turn would be expected to impact on the subsequent ability of the animal to process fine spatial detail. Identical lesions in the second to fourth week result in widespread anatomical reorganization of the visual pathways (see reference 41 for review) that may explain the smaller functional deficits that have been observed in such animals, which are generally less than those observed after adult lesions and which in turn are less than those observed following the same lesions at birth (42,46).

CONCEPTUAL ISSUES

It is widely recognized that the deficits that follow ablation of particular cortical areas can be difficult to interpret (47). This is certainly true of the common situation in cats with lesions of areas 17 and 18 where the visual deficits can be subtle. But even when the deficits are so large as to represent a complete loss of a particular visual function, it does not necessarily follow that the neural processing for that function was primarily conducted in the specific brain area that was lesioned. Strictly speaking, there are a number of caveats surrounding attempts to link the site of a lesion with loss of a specific visual function. First, the lesioned area may represent just a part, albeit an important fraction of the cortical neural machinery required for processing the particular visual attribute. Second, the lesioned area may provide the main afferents to another brain area where that visual attribute is processed. Third, the afferents to the latter area may simply pass through the lesioned cortical area or lie in close proximity so that they are damaged by the lesion. Finally, the lesioned area may form part of a network or system such that the visual function represents a property of the system as a whole. Furthermore, if this distributed system was nonlinear, the visual function of concern may not be apparent at any of the individual components of the system, including the component located at the area lesioned. With these caveats in mind, Regan et al. (48) suggested that in situations where a lesion of cortical area *X* results in a loss of visual function *V*, rather than stating that area *X* mediates visual function *V*, it is more appropriate to state that this area is "necessary" for visual function *V*.

EARLY OBSERVATIONS: TESTS OF PATTERN DISCRIMINATION

For many years it has been apparent that the consequences for vision of ablation of area 17 alone or in combination with area 18 yield quite different results in cats from those observed following lesions of just area 17 in either humans or monkeys. Primates experience profound deficits in spatial vision and visually guided motor behavior following lesions of striate cortex, and only with lengthy training do they relearn a limited capacity to make pattern discrimination between luminance-equated figures (49–51). On the other hand, the impairments observed in cats with lesions of area 17 or even of both areas 17 and 18 are far more subtle. Casual observation of the visual behavior of cats with cortical lesions reveals no obvious visual deficits or overall changes in visually mediated behavior (35,52). In cats with lesions of visual cortex incurred as adults, visuomotor behavior appears relatively normal (35,53), although with rigorous testing certain deficits such as a restriction of the visual field (particularly in the nasal visual field) become apparent (54). As mentioned previously (see page 658), formal testing revealed that cats with lesioned areas 17 and 18 are capable of making discriminations between complex patterns and shapes (see also references 53 and 55). On the other hand, lesions of areas 17, 18, *and* 19 have much larger effects that include a substantial loss of preoperatively learned discriminations. Nevertheless they can, with long relearning periods, acquire the ability to discriminate between patterns on empty backgrounds (24,26,29,31,33,53,56–61). From this large body of work, it would not be unreasonable to conclude that the behavioral consequences for pattern vision of a combined lesion of areas 17, 18, and 19 that represent the primary destination of thalamocortical projections from the dLGN are similar to the deficits observed in monkeys following lesion of the primary projection from the dLGN in this species, namely area 17.

For many years, the majority of studies of the consequences of cortical lesions for vision continued to use pattern discrimination tasks to determine the cues that the animals used before and after the lesion and to define the functional roles of particular visual areas. By suitable modification of conventional pattern discrimination tasks, such as the addition of noise or masking patterns (59), some progress was made in defining further the contributions of the major visual cortical areas to vision. For example, in two studies Krüger et al. (37,62) added gaussian noise of different spatial frequencies to geometrical outline patterns (circles and triangles) and measured the threshold signal-to-noise ratios before and after ablation of areas 17 and 18 or areas 17, 18, and 19. Lesions of areas 17 and 18 reduced the performance at low signal-to-noise ratios and especially for small patterns, a result consistent with the X-type input to area 17. By contrast, a combined lesion of all three major visual cortical areas leads to a complete loss of the ability to distinguish the patterns in noise. With long (3 months) remedial training, the performance for small and large patterns returns to the levels observed for just small patterns after 17/18 lesions. This recovery was observed only for stationary patterns; unlike cats with the smaller lesions, cats with 17/18/19 lesions were unable to discriminate moving patterns, a result that suggests that area 19 may be necessary for the perception of motion. Similar results were obtained by Hoffmann and Von Seelen (63), who also found that the deficits for stationary masked patterns following lesions of just area 17 were no different from those exhibited after lesions of 17 and 18. In addition, they found that lesions of area 19 alone had a profound effect on recognition of patterns moved at slow velocities, but had no effect at all on the performance with stationary patterns

In an ingenious study, Hughes and Sprague (36) devised pairs of rectilinear dot patterns in which the cue to be discriminated was not based on analysis of local elements, but instead was contained only within the pattern's global structure (the spatial distribution of the dots). Cats with lesions of 17 and 18 showed small deficits that mimicked the performance of normal cats with the stimuli blurred. Following an additional second stage lesion in the lateral suprasylvian cortex (LSA), striking deficits were observed that were considerably larger than those observed with lesions of LSA alone. Lesions of LSA that included area 19 produced large deficits. These various results complement data with patterns masked with lines obtained by Cornwell et al. (59), who observed only mild deficits on such tasks with lesions of area 17 and 18 but an extremely severe impairment when the lesion included area 19. These results, taken together with the different effects of these lesions on visual acuity (see page 663), led to the suggestion of a double dissociation of deficits in acuity or global form perception with lesions of areas 17/18 or suprasylvian cortex, respectively.

With the application of tests of visual thresholds to cats with lesions of various visual cortical areas, substantial deficits that could be related to the location and the extent of the lesion became apparent. Because of their ability to define specific deficits that remained hidden by conventional tests of shape discriminations, this chapter focuses on application of threshold testing as a means of shedding light on the combined and separate contributions of areas 17 and 18 to vision in the cat.

THE WAY FORWARD: INSIGHTS GAINED FROM STUDY OF THE EFFECTS OF CORTICAL LESIONS ON VISUAL THRESHOLDS

Spatial Resolution

The ability of the visual system to partition space, its resolving capacity, is assessed in terms of measurements of visual acuity. Although it can be measured with a number of different targets, including letters (Snellen acuity) and small spots of light (the minimum separable), the most common stimulus used in laboratory settings is either a sinusoidal or more commonly a square-wave grating of maximum contrast (grating acuity). During the last three decades, increasing emphasis has been placed on measurements of contrast sensitivity functions (CSF), in part because they provide a succinct description of the visibility of objects of *all* sizes, not just the very smallest objects (64). This function is a plot, on logarithmic scales, of the contrast sensitivity (the reciprocal of the threshold contrast) for sinusoidal gratings as a function of their spatial frequency (the reciprocal of the period of the grating expressed in degrees of visual angle). Visual acuity, when measured with a grating of maximum contrast (Michelson contrast is defined as [$L_{max} - L_{min}/L_{max} + L_{min}$] where L_{max} and L_{min}, respectively, are the maximum and minimum luminance values across the grating pattern) represents a specific point on this curve, the cut-off spatial frequency, so called because it represents the highest spatial frequency that can be detected by the visual system. The cut-off spatial frequency is indicated by the arrows on Fig. 16-1 for the CSF of a monkey prior to and following ablation of area 17.

Berkley and Sprague (35) reported results for three of four animals with lesions intended to remove areas 17 and 18 on a two-alternative grating detection task between a grating and a blank field (actually a grating having a spatial frequency well beyond the resolution limit of a normal cat) of the same space average luminance. Reconstructions of the lesions revealed that lesions were essentially complete in two of the animals, Zelda and Scarlett. At spatial frequencies below 2 cyc/deg, postoperative performance was no different from that before the lesion. However, at higher spatial frequencies, the postoperative performance progressively worsened. Grating acuity, defined as the spatial frequency for which performance was estimated to be 56% correct, was reported (35) as being reduced by the cortical lesion by only 33% for Zelda and even less (20%) for Scarlett. Expressed in octaves (where an octave is a factor of 2), the reduction in grating acuity following the lesion was 0.5 and 0.3 octaves, respectively, for these two animals. The third animal had an incomplete lesion and was tested only postoperatively. Its postoperative grating acuity lay at the lower extreme of the range measured in normal animals by the same testing procedure and apparatus.

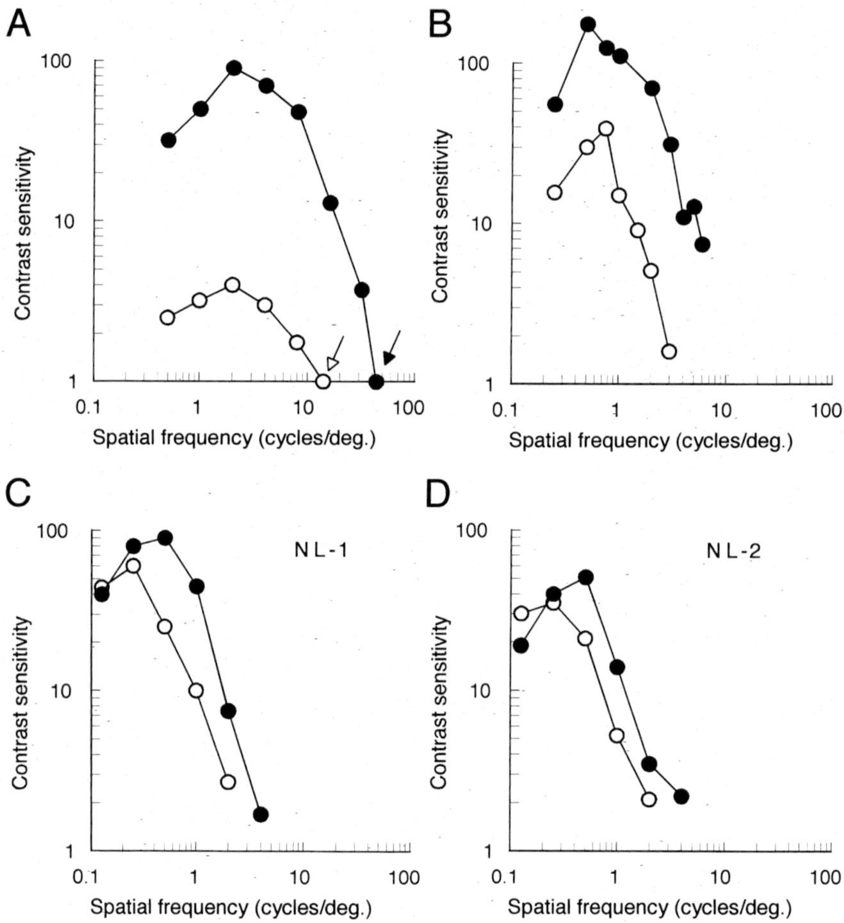

FIGURE 16-1. Contrast sensitivity functions before (filled symbols) and after (open symbols) cortical lesions. **(A)** Mean data from four monkeys after ablation of area 17. The arrows show the cut-off spatial frequencies **(B)** The CSF for a cat after ablation of areas 17 and 18. **(C and D)** CSFs for two cats following ablation of area 17 and part of area 18. (Data in A, C and D were redrawn from Miller et al. [66] and Lehmkuhle et al. [52], respectively. The data in **(B)** is unpublished data from my laboratory.)

Other groups have reported comparable mild effects of lesions of areas 17 and 18 on grating acuity. Although their grating acuities were not tested at all preoperatively, the two cats studied by Kaye et al. (65) had postoperative acuities of between 2.3 and 3.0 cyc/deg, values that were reduced from the lower limit of normal values (6.4–8.6 cyc/deg) as measured in the same manner by 1.47 and 1.10 octaves, respectively. An additional four animals studied subsequently in my laboratory had a very similar mean postoperative acuity of 3.03 cyc/deg (range 2.30–3.79 cyc/deg), a value that is lower than the mean of the normal range by 1.3

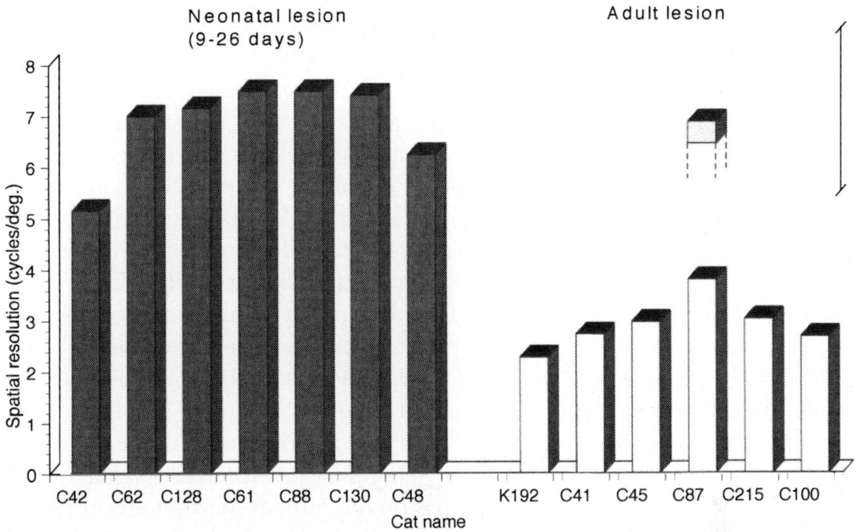

FIGURE 16-2. Histograms that depict the postoperative grating acuity of individual cats after ablation of cortical areas 17 and 18 in either early postnatal life (P9–P26) or as adults. The bracket on the right shows the range of acuities measured in normal adult cats under the same condition. (Data redrawn from Mitchell [46].)

octaves. All of these animals were tested on a detection task between a grating and a uniform field of the same average luminance (100 cd/m^2). The final postoperative grating acuities of the six animals for which areas 17 and 18 were ablated in adulthood are displayed by the histograms on the right of Fig. 16-2. The preoperative acuity of C87, the only animal for which measurements were made before the lesion, is indicated by the partial bar above the postoperative data. The bracket to the right displays the range of acuities among normal age-matched cats measured under similar conditions. Lehmkuhle et al. (52) measured contrast sensitivity functions on two animals with lesions of area 17 and that part of area 18 on which the vertical meridian was represented. The postoperative reduction in acuity for the two animals as estimated from extrapolation of the CSFs to the abscissae (Fig. 16-1C, D) was also about an octave. In comparing data from different laboratories, it may be relevant that the luminance of the stimuli and their overall dimensions were not the same and the testing procedures also varied. It is possible that the somewhat higher acuities encountered in my laboratory may reflect the high luminance and large dimensions of the stimuli (>20 deg dia) we used.

Because they probe the sensitivity of the animal to a complete range of spatial frequencies, contrast sensitivity functions have the potential to provide additional information on the nature of the visual impairments produced by lesions of the visual cortex. Unfortunately, such measurements have been made on only a very limited number of animals. Lehmkuhle et al. (52) measured the CSF of two cats

(NL-1 and NL-2) before and after a virtually complete lesion of area 17 and their data for gratings that were temporally modulated at a slow rate (1.5 Hz) have been replotted in Fig. 16-1 C and D. No deficits at all were apparent at the two lowest spatial frequencies but for spatial frequencies of 0.5 cyc/deg and above, the postoperative contrast sensitivities were worse. Interestingly, this data contrast with the effects of ablation of area 17 in the monkey where the contrast sensitivity loss has been characterized by Miller at al. (66) as a uniform 26 dB reduction of contrast sensitivity at all spatial frequencies. The mean data of the four monkeys tested both preoperatively and postoperatively by Miller et. al. (66) have been replotted in Fig. 16-1A. The loss of contrast sensitivity was considerably larger than that observed in cats following lesions of the same cortical area. Moreover, the loss in visual acuity, from 43 to 12 cyc/deg, a reduction of 1.94 octaves, was also considerably larger than that of cats with even larger lesions that include both area 17 and 18.

Figure 16-1B shows the results of both preoperative and postoperative measurements of CSF on an animal from my laboratory that received a virtually complete lesion of both areas 17 and 18. Possibly because of the significantly higher luminance of the display (100 cd/m^2) compared with that used by Lehmkuhle et al. (20 cd/m^2), the preoperative and postoperative contrast sensitivities were higher in this animal at all spatial frequencies than those of the two animals studied by Lehmkuhle et al. (52). In addition to this difference, the postoperative sensitivity of this animal was reduced at all spatial frequencies with respect to the corresponding preoperative values, a picture similar to that observed in monkeys following ablation of just area 17. As was originally pointed out by Lehmkuhle et al. (52), the contrast sensitivity deficits in their animals are entirely consistent with the lesion having eliminated the X-cell pathway. X-cells are commonly thought to be most sensitive to high spatial and low temporal frequencies, whereas Y-cells are preferentially sensitive to the opposite, namely low spatial and high temporal frequencies (67). Although contested (68,69), it is generally held (67) that X-cells project through the A laminae of the dLGN to area 17 almost exclusively whereas the Y- and W- cells project to this and other cortical areas (see Chapter 1). Thus a lesion of area 17 would be expected to effectively eliminate the X-cell pathway from the cortex leaving a significant amount of the Y- and W- pathways. The more extensive loss of contrast sensitivity evident at even low spatial frequencies in the animal with a lesion of both area 17 and 18 (Fig. 16-1B) may reflect the greater involvement of the Y- pathway with extension of the lesion to area 18.

Two elegant and painstaking studies by Pasternak's group (22,23) used circumscribed chemical lesions into electrophysiologically determined regions within either area 17 or area 18 of one hemisphere to investigate the roles of these two cortical areas for vision. The animals were trained to fixate on stimuli presented to the lesioned region within the visual field. To ensure that the lesions were well within the desired cortical area they were placed such that the affected area was located below the representation of the horizontal meridian and several degrees from the midline in the lesioned hemifield. The locations of the lesions were determined by measurements of contrast thresholds for small grating patches (dia. 6°). Eye position was monitored accurately with scleral search coils during psychophysical testing in both the lesioned region and at a control location in the corresponding portion of the normal hemifield of two cats used in the two studies. A variety of psychophysical thresholds were measured including contrast sensitivities for gratings at both low and high temporal frequencies, orientation discrimination thresholds, and tests of motion perception. Because the lesions had to be placed in the peripheral representation of the visual field, the measured sensitivities and resolution were lower than those characteristic of central vision. Moreover, the limited extent of the lesions (~ 8–10° dia.) restricted the number of cycles in gratings of low spatial frequency, which may have further reduced sensitivity. Despite these methodological considerations, the results of the two studies provide strong indications that the two areas are involved in the processing of both form and motion for stimuli that possess different but somewhat overlapping spatiotemporal properties.

Following a lesion within area 17 of two cats, measurements of contrast sensitivity in a detection task revealed a loss at middle spatial frequencies but, surprisingly, none at high spatial frequencies as observed in previous studies of combined 17 and 18 lesions. The losses were most evident at low temporal frequencies and were absent at the highest temporal frequency tested (18 Hz). Because area 17 represents the major projection of X-cells that are presumed to convey high spatial frequency information to the cortex, area 17 lesions would be expected to lead to a loss of contrast sensitivity at high spatial frequencies since Y-cells that project to area 18 do not respond in a linear fashion to such frequencies. By contrast, a very different result was obtained when the measurements were repeated as a discrimination task where the cat had to discriminate between horizontal and vertical gratings. Contrast sensitivity loss was substantial at both middle and high spatial frequencies with slightly more than a one octave reduction in the cut-off spatial frequency. The latter reduction was virtually identical to the loss reported in central vision in the studies mentioned earlier. A possible explanation for the discrepant results from detection versus discrimination tasks can be gleaned from the observation that the cut-off spatial frequency measured with the former task was twice as high as that obtained with the latter and exceeded the value that would be expected from the sampling density (Nyquist frequency) of either Y-cells or W-cells. This in turn suggests that in a detection task the cat may be responding to aliased (nonveridical) patterns (70,71) provided by undersampling of Y- and/or W-cells that project to areas (e.g., areas 18,19 and lateral suprasylvian cortex) and that mediate vision in the absence of area 17. A discrimination task eliminates the ability to use such aliased signals.

In an earlier study by Pasternak and Maunsell (22), a lesion located within area 18 of another two cats, resulted in a 4- to 10-fold loss of contrast sensitivity for gratings of low and medium spatial frequencies located within the lesioned visual field, but none at all at the highest spatial frequencies. Only a detection task was used for these measurements so that it is possible that the latter result may represent the detection of an aliased pattern. The contrast sensitivity losses extended across all temporal frequencies and were most pronounced at 4.5 Hz. The measurements of spatiotemporal sensitivity point to complementary roles for areas 17 and 18 in vision. Whereas the former area appears to play a major role in the perception of stimuli that contain high spatial frequencies and that move or flicker at low rates, area 18 is primarily involved in the perception of drifting stimuli of targets containing low or intermediate spatial frequencies (see Chapter 1). The complementary roles for the two areas in vision receive further support from measurements of the effects of lesions of these areas on the perception of motion (see page 683).

Effects of Lesions at Different Ages

There is evidence that lesions of areas 17 and 18 at different ages have very different consequences for spatial resolution. Preliminary data (Mitchell, in preparation) obtained from three kittens that had areas 17 and 18 ablated on postnatal day 3 indicate that a lesion at that early age has a far more drastic effect on contrast sensitivity functions than does ablation of the same cortical visual areas in adult cats. For example, the peak contrast sensitivity observed in the three animals that received the perinatal lesion was between 10 and 14 compared with a value of 40 in the animal (Fig. 16-1B) that received the same lesion as an adult. These peak values occurred with gratings having a spatial frequency of 0.75 cyc/deg in both groups. The reduced spatial resolution of the animals that received the early lesion may be a consequence of the loss of X-cells that have been reported after such early lesions (72; see also Chapter 14).

As a counterpoint to the profound loss of spatial resolution experienced by the animals that received a lesion on P3, other animals for which areas 17 and 18 were ablated in the second to fourth postnatal week (specifically, days 9–26), exhibited only mild or nonmeasureable deficits in grating acuity. This point is made evident in the individual data from the seven animals in this study, to date published only in preliminary form (70), that are displayed on the left of Fig. 16-2. The acuity of only one animal (C42) lay outside the range of values encountered among normal cats. As would be anticipated from the minimal effects on grating acuity of lesions in the second to fourth week, CSF appear virtually normal (46). The minimal effects of lesions at this age may reflect the dramatic rearrangement of the visual pathways that occurs following these lesions when they occur in this strict window of time (41).

Hyperacuity Tasks

The measures of spatial resolution discussed to this point are limited by the density or spacing of the relevant sampling elements whether they be photoreceptors, such as cones in the human fovea, or particular classes of retinal ganglion cells, as most likely is the case in the cat (71,73). A retinal image of a grating having a period finer than that limit (the Nyquist limit) dictated by the spacing of the sampling elements will not be processed in a veridical fashion and may be perceived as an alias of lower spatial frequency (i.e., a larger period) than that of the actual image (74). By contrast, another class of acuity tasks have in common the discrimination of a difference in location of one or more elements of the display. Collectively, these acuity tasks have been referred to as hyperacuities (75) in acknowledgment of the fact that in angular terms they have values much smaller than the angular spacing of sampling elements that provides a physical limit to conventional resolution measures such as grating acuity. Vernier acuity, which measures the minimum detectable displacement from colinearity between two or more elements, is perhaps the most widely known example of a hyperacuity. For certain stimuli such as two lines separated by a small gap, humans can detect a displacement of one line with respect to the other of only 5 seconds of arc (75) compared with foveal cone spacing of about 30 seconds (76). The detection of differences in location of the elements of the stimuli in hyperacuity tasks such as vernier acuity in principle requires accurate identification of the center of the retinal light distribution (i.e., the centroid) of the key feature elements. Unlike resolution of double stars as separate entities, or resolution of the bars of a grating, there is no physical restriction on identification of the *location* of the centroid of the retinal light distribution of stimulus features (75). The possibility that the mechanism that underlies the accurate identification of location of stimuli involves processing within one of both of area 17 and 18 receives strong support from studies of the consequences of lesions of these visual areas on vernier acuity.

Berkley and Sprague (35) were the first to measure vernier acuity in cats with lesions of various cortical visual areas. The same animals that displayed only an octave reduction in grating acuity showed massive losses in vernier acuity. Two cats (Streak and Zelda) were virtually unable to perform the vernier task at all postoperatively, and the postoperatively measured displacement threshold for the third animal (Francis) was about 1°, a value about 10 times larger than the average preoperative acuity (5–6 minutes). Measurements conducted in my laboratory (in preparation) on one animal with ablation of areas 17 and 18 revealed a 25-fold increase between the preoperatively measured threshold (1.3 minutes) and those measured after the lesion (33 minutes). In contrast to the massive postoperative reduction of vernier acuity reported in these two studies, a third group (77) obtained results that suggested that any postoperative deficits was small. However, as they themselves point out, their results must be interpreted with caution, as there was no control in their study of the observation distance; the animals could make their choice from a maximum distance of 57 cm (the door of the start chamber) to the stimulus panels themselves. Moreover, the ability to approach and even touch the stimuli meant that the overall angular subtense of the stimuli would vary widely, and as well the animals may have been able to exploit different cues at close observation distances than at 57 cm so that the character of the task could have been different for the two groups of animals. When the data are presented in terms of the minimum physical displacement that could be detected, there appeared to be no difference between the thresholds for the normal and lesioned animals. However, as the decisions of the two groups of animals could have been made at very different distances from the stimuli, the thresholds in angular terms could in fact differ by a factor of 10 or more.

With these caveats in mind concerning the data from the latter group, it appears that the weight of data indicates that vernier acuity is reduced substantially following lesions of areas 17 and 18. Indeed for both groups that found massive changes in postoperative vernier thresholds, the values were elevated well above their grating acuity. For example, the postoperative grating acuity of the cats of Berkely and Sprague (35) was about 3 cyc/deg (equivalent to a grating period of 20 minutes) compared with vernier acuities of 1° or worse. Postoperatively, vernier acuity was no longer a hyperacuity but was now *worse* than resolution acuity! On the basis of the studies of vernier acuity, it could be tentatively argued that areas 17 and/or 18 are necessary for stimulus features to be located with the exquisite accuracy that permit the hyperacuities such as vernier acuity to achieve values that are considerably better than resolution acuity.

Orientation Discrimination

Two of the most salient properties of cells in cat area 17 and 18 are their selectivity for the orientation of line stimuli and their distributed responsiveness to stimulation of the two eyes that, since Hubel and Wiesel's (78) introduction of the term, has been referred to as their ocular dominance. The latter characteristic implied that cortical cells received input from both eyes, the functional importance of which was revealed by later work that indicated that cortical cells were selective for retinal disparity (see Chapter 1). The strong selectivity of cortical cells for orientation and disparity has led to explorations of the effects of ablation of area 17 alone or in combination with area 18 on orientation discrimination thresholds and on stereoscopic vision (see page 674).

There have been a number of attempts to relate the psychophysical performance of normal cats on orientation discrimination tasks with the orientation selectivity of subclasses of cells in area 17 (80–83). Although certain findings suggest that sufficient information may be present in the responses of the most sensitive cortical cells to sustain behaviorally measured thresholds, such experiments cannot show how or whether this information is the actual basis on which the perceptual performance is based. Indeed, De Weerd et al. (83) argued on the basis of marked differences between behavioral performance and the responses of single cells to changes of stimulus contrast that orientation discrimination performance cannot be linked to the response strength of individual cells in area 17. This conclusion received strong support from an elegant series of studies (14,15,17,84,85) of the effects on orientation discrimination thresholds of lesions restricted to either area 17, areas 18 and part of 19 (Tier 1 lesions), or of a lesion of those areas (19,20a,21a,21b, and two divisions of the lateral suprasylvian cortex, namely AMLS and PMLS) that receive afferents originating in Tier 1 areas and which collectively were referred to as Tier 2 lesions.

From the perspective of the functional roles of areas 17 and 18, the most interesting observations generated from this exhaustive series of studies were the graded changes in orientation discrimination thresholds as the lesion expanded from area 17 to include progressively more of area 18. Measurements were made of orientation discrimination thresholds in a two-alternative spatial forced-choice task that determined the just noticeable difference (JND) in orientation between two elongated bars, a reference bar that was always either horizontal (0°) or at 135° and a test bar rotated clockwise from the reference. Most of the data were collected with a standard bar that was 12° long and 0.2° wide but measurements were also made with bars of different lengths, width, and contrast to document deficits in situations where they could not be assessed with the standard bar.

Perhaps the most surprising result to emerge from the first two of these studies (14,84) was the complete retention of the task and the lack of any change in the JND for a standard bar following ablation of area 17 alone. Small deficits were apparent only with very short bars (4° or less) or bars of low contrast. Animals with lesions of area 18 and part of area 19 also exhibited no retention deficits and no changes in their JNDs for standard bars. However, ablation of *both* areas 17 and 18 produced very large effects that began as a retention deficit that required extensive retraining on the task and large eventual changes in JNDs that were apparent with bars of all lengths. In one animal (cat 21), the JNDs increased from preoperative values of 1–3° to a median of 23°. A second animal (cat 27) showed a similarly large 15-fold increase in the JND following the lesion while the changes in the other two were somewhat less. Comparison of the deficits in orientation discrimination with the extent of the lesion revealed that the former increased proportionally as the lesion expanded from area 17 to encroach on progressively more of area 18. The latter finding led to the suggestion (14,84) that orientation discrimination is a distributed function within and across areas 17 and 18, with area 17 contributing more than area 18 for narrow and long bars.

Because lesions of areas 17 and 18 result in a loss of contrast sensitivity it could be argued that the deficits in orientation discrimination are nothing more than a consequence of the contrast sensitivity loss. Indeed, Vandenbussche et al. (14) reported that area 17/18 lesions did elevate the contrast threshold for detection of the bar stimuli they used. However, a number of arguments were advanced to refute this interpretation of the data. Not only did thresholds change very little as a function of suprathreshold contrast except when very close to detection threshold (14, Fig. 29), but short stimuli for which orientation discriminations could not be made were nonetheless easily detectable.

The negligible consequences of a lesion of area 17, as well as the proportional increase in orientation discrimination JNDs as the lesion encroached onto area 18, led to two important conclusions. First, since area 17 is the major destination of retinal X-cells, it must be concluded that the X-system is not necessary for orientation discrimination of long thin bars. Second, although cells in area 17 are more narrowly tuned for orientation than cells in area 18 (86), areas 17 and 18 apparently contribute about equally to orientation discrimination. This result is not consistent with the view that orientation discrimination can be linked to the tuning of individual cells (81,82) but instead suggests that perceptual performance may depend on an opponent process that computes the ratio of the outputs from several different cells. An additional important conclusion was drawn from the finding of subsequent studies (17,85) that Tier 2 lesions produced only minor effects on orientation discrimination thresholds. This leads to the conclusion that, at least for contours defined by luminance, the representation of bar orientation depends heavily on Tier 1 but little on Tier 2 visual cortical areas.

The boundaries of most objects that are encountered in everyday life are defined by a difference of luminance. However, boundaries defined by other stimulus attributes are still visible even in the absence of a luminance difference. One such boundary can be produced by suitable arrangement of line ends or of corners (Pacmen) to produce illusory contours (Fig. 16-3). De Weerd et al. (15) studied the effects of Tier 1 and Tier 2 lesions on the ability of cats to make orientation discriminations between the two types of illusory contours defined by opposing line ends and displayed in Fig 16-3. One such illusory contour (GIC) is defined by a gap (a luminance discontinuity), and the other (PSIC) by a displacement or phase shift between the lines. The mean orientation discrimination JNDs (n = 3 cats) for a luminance bar and the two illusory bars before and after Tier 1 lesions are displayed in Fig 16-3. The smaller elevation in JND for a standard luminance bar following the lesion (a factor < 3) than that reported earlier (14) was attributed to differences in testing procedures and the number and nature of preoperative discriminations that were learned. The moderate deficits observed with luminance-defined bars stand in contrast to the devastating effects of Tier 1 lesions on the ability to make orientation discriminations with illusory contours. Postoperatively, no animal was able to make such discriminations with either stimulus, a result that points to the important role of areas 17 and 18 in the processing of illu-

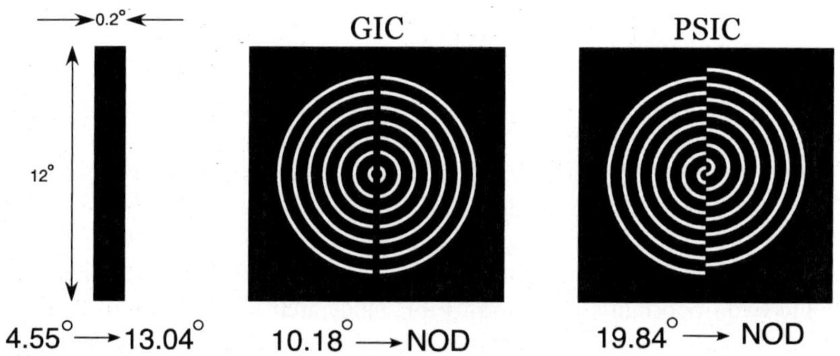

FIGURE 16-3. The stimuli used to test orientation discrimination before and after lesions of area 17 and 18 (15). The numbers below indicate the mean thresholds obtained by three animals before and after the lesion. The bar on the left was the standard luminance-defined bar used in a number of studies (14,15,17), and the other two represent illusory bars defined by gaps (GIC, middle) or by a phase shift (PSIC, right). After the lesion none of the animals were able to make orientation discriminations (NOD). (Redrawn from DeWeerd et al. [15].)

sory contours. However, unlike what is observed with luminance-defined bars, where Tier 2 lesions produce negligible effects, Tier 2 lesions resulted in substantial deficits with illusory contours that were greater for PSIC than GIC stimuli. Also in contrast to the permanent nature of the deficits of Tier 1 deficits, there was a gradual improvement in thresholds following Tier 2 lesions but even after many months a significant deficit (approximately 50%) remained. The fact that Tier 2 lesions produce substantial initial deficits and significant long-term changes of orientation discrimination with illusory contours suggest that the perception of such contours is not entirely based on early processing in area 17 and 18.

A more recent study (16) of visual texture segregation following Tier 1 and Tier 2 lesions provides more information on the different roles of Tier 1 and Tier 2 areas. The two texture discrimination tasks used in this study are displayed in two forms in Fig. 16-4. The column on the left shows a texture defined by line orientation, and that on the right is a texture defined by dots and annuli. The top row displays the textures in regular form where the features are arranged in regular rows and columns, and the bottom row illustrates textures in which the features are randomized in position. The lesions had different effects on the ability of cats to segregate the two types of textures. Ablation of areas 17 and 18 (Tier 1 lesions) produced severe deficits in the ability to segment textures based on line orientation but spared the ability to segment the textures based on dots and annuli. On the other hand, Tier 2 lesions destroyed the ability to make either texture discrim-

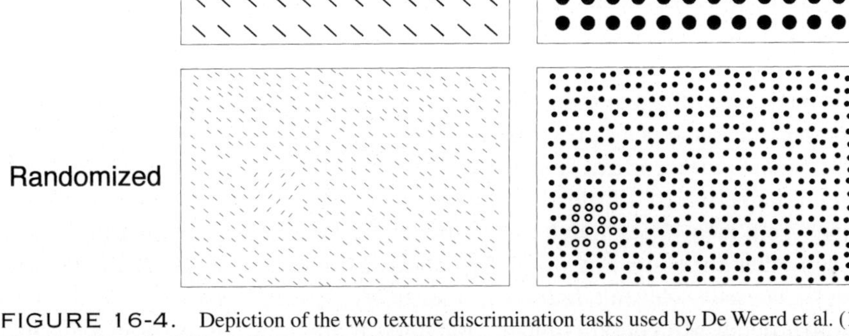

FIGURE 16-4. Depiction of the two texture discrimination tasks used by De Weerd et al. (16). The stimuli on the left are line textures, and the feature elements of the textures on the right are dot/annuli. The textures are displayed in either regular or irregular form depending on whether the features are arranged in regular rows and columns or else randomized in position. (Redrawn from DeWeerd et al. [16].)

ination. The authors argue that Tier 2 areas conduct a comparison (the search for different regions in the display) on the basis of filtered information computed in areas 17 and 18. To the extent that the features in the display match the filtering operations in areas 17 and 18, lesions of those areas will destroy the ability to segment. The important role of Tier 2 areas to texture segregation is a direct challenge to the widely held view that the latter operation is entirely an early process in vision.

Stereoscopic Vision

As with primates and many species of raptors, the two eyes of a cat are located in a frontal position in the head so that their visual fields overlap to a considerable extent. According to measurements of Hughes (87), the binocular visual field of the cat is 99°, a value close to the average value (114°) observed in humans (79). The presence of a large binocular field has long been regarded as one of the essential elements for attainment of stereopsis, the perception of relative depth based on the detection of retinal disparity cues (73,79,88,89). Tests for the presence of stereopsis in cats have been made with ever increasing levels of sophistication designed to eliminate extraneous depth or shape cues leaving only retinal disparity cues to depth. Early tests strongly suggested that cats possessed stereopsis (90–92), but arguably the definitive demonstration was provided by Ptito et al. (77) who used random-dot stereograms (93) as stimuli. In principle, such stereograms eliminate all monocular depth cues as well as monocular cues to the shape of the stimuli. They used anaglyphic presentation (79,93) of the stimuli whereby the stereograms for the two eyes were of different color that were viewed by the cat through narrow bandpass colored filters placed in the contact lenses worn by the cat.

A neural substrate for stereopsis in area 17 of the cat was discovered more than 30 years ago (94,95), and since then much of our knowledge of the physiological basis of stereoscopic vision has been derived from studies of the cat visual cortex. Even though there is as yet no consensus as to whether the basis for stereopsis is captured best in terms of the sensitivity of cortical cells to positional disparity or to their sensitivity to relative phase information, or on another level, whether receptive fields of individual cortical cells differ in terms of their relative positions on the two retinae or their profiles (phases) in right and left eyes, there is no doubt that cells in area 17 provide a rich source of information about retinal disparity. Since other cortical areas in which disparity specific cells have been reported, namely area 18 (96,97), 19 (98–101), and 21a (102,103), receive projections from area 17 (although they also receive parallel projections from thalamic and other structures), it would be anticipated that lesions of area 17 would seriously degrade or even eliminate stereopsis. A growing body of evidence suggests that depth perception is indeed impaired following lesions of area 17 and that stereopsis may be lost entirely.

The earliest studies of the effects of cortical lesions on depth perception used a modification of the visual cliff, an apparatus that was originally developed by Walk and Gibson (105) for testing the depth perception of human infants. The apparatus consists of a large box having a clear plastic or glass top. Two adjacent plastic transilluminated surfaces on which a black and white textured pattern such as a checkerboard are printed, are located at different distances below the transparent surface of the box (Fig. 16-5). One of the surfaces (the shallow side of the visual cliff) is typically placed directly below the top, and the other (the deep side) is placed as much as 1 meter below. For tests of the depth perception of animals, a narrow opaque surface (the starting platform) is placed on top of the glass plate above the junction of the two surfaces that are visible underneath. Depth perception is judged on the basis of the results of a number of trials that begin with the animal being placed on the starting platform and noting the side (shallow or deep) to which the animal first steps from the platform. The early lesion studies used just one (binocular) viewing condition, thereby making it difficult to determine the extent to which any deficits could be attributed to an impairment of stereopsis as opposed to effects on the many monocular depth cues that are available with this apparatus. Moreover, in none of these studies were the distances of the two surfaces below the glass plate

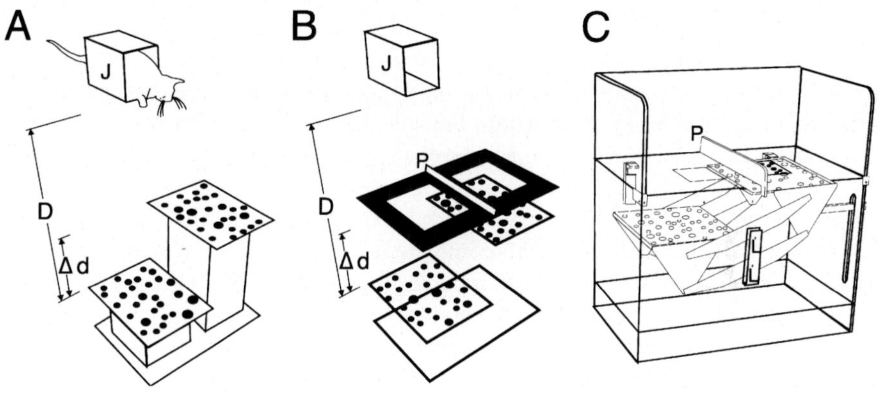

FIGURE 16-5. Schematic representations of the method used to assess depth perception in cats. **(A)** A real version of the task in which the cat is required to jump from the jumping platform (J) onto the nearer of two surfaces (the cat's left surface as depicted) placed directly below. The two surfaces are covered with circular dots of three different sizes arranged in a random fashion with 18% coverage and are supported by laboratory jacks (not shown) that permit the surfaces to be placed at variable heights. Because many depth cues are available, this apparatus was only used for initial training but in rare situations where depth perception was poor, quantitative estimates of depth perception could be obtained by determination of the minimum detectable difference in distance (ΔD) that permitted criterion levels of performance. Performance is defined in terms of the ratio ΔD/D where D is the distance to the distal stimulus. To reduce the ability to use many of the salient monocular depth cues, the vast majority of animals were tested in the situation depicted schematically in **(B)** where the animals are required to jump onto a glass surface through which the two stimuli are visible with their edges hidden by a mask (shown in black). The animal is required to jump toward the nearer of the two stimuli (on the cat's left as depicted). A vertical wooden divider separates the positive from the negative landing surface. **(C)** Implementation of the task as depicted schematically in **(B)** See text for

manipulated to provide quantification of the impairment of depth perception. The differences between the depths of the shallow and deep sides of the visual cliff used in these studies were typically very large (e.g, 79 cm), as in the study of Cornwell et al. (106), so that even if animals continued to choose the shallow side of the visual cliff after ablation of the visual cortex, it is possible that a substantial deficit in depth perception may still have been present.

The results of the five oldest studies that used a visual cliff to test the depth perception of cats before and after lesions of the visual cortex were mixed. Four studies (24,106–108) reported that animals continued to prefer the shallow side, and in one (106) the results were similar for tests conducted on both visual cliffs and situations where the glass plate was removed from the top of the deep side to produce a physical cliff. In contrast, the fifth of these investigations (26) reported that seven of their nine lesioned animals no longer showed a preference for the shallow side. The intended sizes of the lesions in these studies were quite different and the published reconstruction of their location and extent suggest that many were incomplete even in area 17. The possibility that the discrepant behavioral findings could be attributed to variations in the extent of the lesions was explored systematically by Cornwell et al. (106) from examination of performance on the visual cliff of three groups of animals that had lesions intended to be of different extents or else targeted at different cortical visual areas. In one group (Group MS) the lesion was intended to include all of area 17 and most of areas 18 and 19, whereas in a second group (M), a deliberate attempt was made to spare that part of area 17 deep within the splenial sulcus, thereby replicating the lesions of Dalby et al. (24) and of Meyer et al. (108). The lesion in the third group was placed lateral to the marginal gyrus and included many of the visual areas within the suprasylvian and ectosylvian gyri but was intended to spare areas 17, 18, and 19 (109). The animals with complete lesions of area 17 (Group MS) performed at chance on the visual cliff, whereas those in the other two groups chose the shallow side with the same high frequency as did normal control animals. On the basis of reconstruction of the lesions, as assessed by degeneration in the dLGN of the animals in the various groups, Cornwell et al. (106) concluded that preference for the shallow side on the visual cliff was retained in animals in which the lesion spared the lateral part of the dLGN (from 10–20% of the nucleus) on which the peripheral visual field was represented.

In addition to their efforts to delineate the extent of cortical lesion necessary to disrupt preference for the shallow side of the visual cliff, Cornwell et al. (106) also attempted to identify the depth cues the animals used on their apparatus. Experiments conducted on three groups (N = 14) of normal cats revealed that the group tested binocularly performed significantly better in terms of the proportion of choices for the shallow side than did the two groups that were tested monocularly. Nevertheless, even with monocular viewing, the cats continued to prefer the shallow side, suggesting that the depth cue provided by binocular disparity was not essential for successful performance. Although the lower proportion of shallow responses under monocular testing conditions could be explained in terms of the poorer depth perception that resulted from the absence of binocular disparity cues, the pattern of results suggested that the poorer performance in this situation may have resulted from difficulties (such as the reduced visual field) adjusting to the contact lens occluders used to restrict vision to one eye. The visual cliff is rich in monocular depth cues, with linear perspective, texture gradients, and motion parallax being the most salient to a human observer. The failure of Cornwell et al.'s (106) cats with complete lesions of area 17 to choose the shallow side suggests that their ability to utilize both binocular cues and these salient monocular cues may have been eliminated by the lesion. This view received support from the lack of any change in performance in such lesioned animals even when dowels were introduced between the glass top and the floor of the cliff to enhance the ability to use motion parallax cues.

Although it may seem attractive to conclude that failure to exhibit a preference on the visual cliff results from an inability to discriminate differences in depth between the deep and shallow sides, it is important to recognize that this is a test of preferences and that negative results do not necessarily constitute evidence for elimination of the ability to *discriminate* differences in depth. To make the latter claim, it is necessary to use differential reinforcement whereby choices to one side are rewarded and responses to the other side are not. Surprisingly, differential reinforcement with the visual cliff has been attempted in only one study (26) and with the display turned on its side. Interestingly, two of the three cats with lesions of the visual cortex that performed at chance on the visual cliff could nevertheless be trained to discriminate the deep from the shallow surface with differential reinforcement. Another manifestation that the visual cliff, as usually applied, is a test of preference, is the fact that performance tends to deteriorate over trials so that performance on the first trial can differ from the mean response over subsequent trials (106,107).

About 30 years ago, new methods were developed to assess depth perception in cats. These methods went beyond simple tests of preference to both quantify the accuracy of depth perception and ascertain the presence or absence of stereoscopic vision. The first of these new methods to be applied (90,91) were intended to be direct tests for the presence of stereoscopic vision and used anaglyphs (79,93) with stimuli presented by a shadow casting technique to provide only retinal disparity cues to depth. Later versions of this method used projected random dot stereograms as stimuli (77). Both versions of the tests that used anaglyphs required that the animals wore red/green filters in front of the eyes held in either helmets or on contact lenses so that only one of the two stimuli was seen by each eye. However, before the application to lesioned animals of methods that involved anaglyphic presentation of stimuli, a more simple and less direct approach was used to infer the presence or absence of stereopsis. It is a commonplace observation that humans are better at making depth judgments between real objects with two eyes than with one, a reflection of the presence of a depth cue, presumably stereopsis, that is uniquely available in the former situation. This observation formed the basis of a screening instrument for stereopsis developed by Howard

(110), which enjoyed widespread use in the 1940s. It required subjects to adjust two vertical rods (that were masked so as to minimize the ability to use monocular depth cues) to appear equidistant under both binocular and monocular viewing conditions. Superior performance (defined operationally as a lower variance in the equidistance settings) in the former viewing situation was taken as evidence of stereoscopic vision, whereas comparable performance in the two viewing situations implied that stereopsis, if present at all, was poor and at best provided no more information concerning relative depth than could be obtained from monocular depth cues.

As applied to cats, the method (92) exploited the natural tendency of kittens to jump from an elevated platform to the nearer of two adjacent surfaces beneath them and uses a modified version of the apparatus and jumping procedure that had been devised earlier for testing the spatial resolution of young kittens (111,112). Although a real version of such a task (Fig. 16-5A) in which the two surfaces were separated in depth (ΔD) by as much as 23 cm was constructed to assist with the initial training of some animals and to quantify the capabilities of others with very poor depth perception, the apparatus used for the vast majority of the tests used stimuli viewed through a glass plate onto which the animals jumped (Fig. 16-5B and C). Two transilluminated clear plastic plates, on which circular opaque circles of three sizes (9–20 mm diameter with 18% coverage) were randomly placed, served as the two discriminanda. As indicated by the diagram of Fig. 16-5C, the two plates were located at different distances from the glass surface of the jumping stand and were arranged on a seesaw arrangement that permitted the position of the two plates to be moved symmetrically about a central pivot point (L). The closer of the two stimuli (the positive) was always 3 cm below the glass top, and the other (negative) plate could be placed a variable distance below the other to a maximum separation of 23 cm. The stimuli were viewed by the cat through square apertures that were cut in a sheet of cardboard placed immediately below the glass landing surface. In this way the edges of the stimuli were masked from view so that to the cat, the stimuli appeared as two sheets of randomly placed dots that floated in two planes beneath the apertures. On each trial, jumps to the closer of the two stimuli were rewarded, and jumps to the other stimulus located ΔD below the other were not rewarded. As with the conventional jumping stand procedure, a modified method of limits was used whereby the separation of the stimuli (ΔD) was progressively reduced between blocks of trials until the animal could no longer perform the task. In a typical testing session, depth thresholds were first measured binocularly after which a contact lens occluder was placed in one eye to obtain a monocular threshold. With normal animals there was a striking change in the animal's behavior on switching from binocular to monocular viewing (92); the animal's jumps were more hesitant and threshold values for ΔD were about eight times greater than their binocular thresholds. It is likely that animals experience greater difficulty monocularly because the task is essentially different and has to be solved on the basis of depth cues other than retinal disparity such as texture density, size, or possibly motion parallax. The monocular depth cues that might be available can be appreciated from Fig. 16-6, which shows photographs of the jumping stand. In Fig. 16-6A, the stimuli are separated by 23 cm, so that differences in texture density are very apparent. On the other hand, Fig. 16-6B illustrates the monocular view of the stimuli when they are separated by only 2 cm, a typical binocular threshold for a normal animal.

A very different pattern of behavior was reported in a short published paper (65) that used this testing procedure on two animals following ablation of visual cortical areas 17 and 18 at 15 months of age. As illustrated in Fig. 16-7, both cats behaved as other normal animals before the lesion, but postoperatively they achieved only chance levels of performance on the jumping stand illustrated in Fig. 16-5B and C even with the largest separation (23 cm) of the two stimuli. Subsequent tests of monocular and binocular depth thresholds were made on the apparatus of Fig. 16-5A that provided a more diverse set of cues to depth. However, even with the rich variety of depth cues provided by this apparatus, binocular depth thresholds were still extremely poor and, moreover, no better than the animal's monocular thresholds, which were far worse than the monocular thresholds of normal cats. Subsequently, measurements of depth thresholds have been made (Mitchell, in preparation) in a similar manner on four other animals following lesions of areas 17 and 18 made in adulthood, and on six other animals that received similar lesions at between either 9 and 26 days of age or on postnatal day 3 (N = 3). Without exception, these animals performed no better binocularly than monocularly, a result that supports the conclusion drawn from the earlier results (65), that ablation of areas 17 and 18 eliminates stereopsis. Profound deficits of binocular depth perception on a similar jumping stand were reported by Hovda et al. (113) following bilateral ablation of areas 17 and 18.

It was argued earlier (65) that the poor performance on depth discriminations following cortical lesions made in adulthood cannot be attributed to a reduction of

FIGURE 16-6. **(A)** Photographs of the jumping stand viewed from above with the stimuli separated in depth (D) by a large interval (23 cm) to illustrate the monocular static depth cues (principally texture density and size) that are potentially available to the cat. **(B)** Photographs of the stimuli as viewed from 75 cm with D equal to 2 cm, a value close to the binocular threshold of a normal animal. Note that monocular depth cues are minimal in this situation.

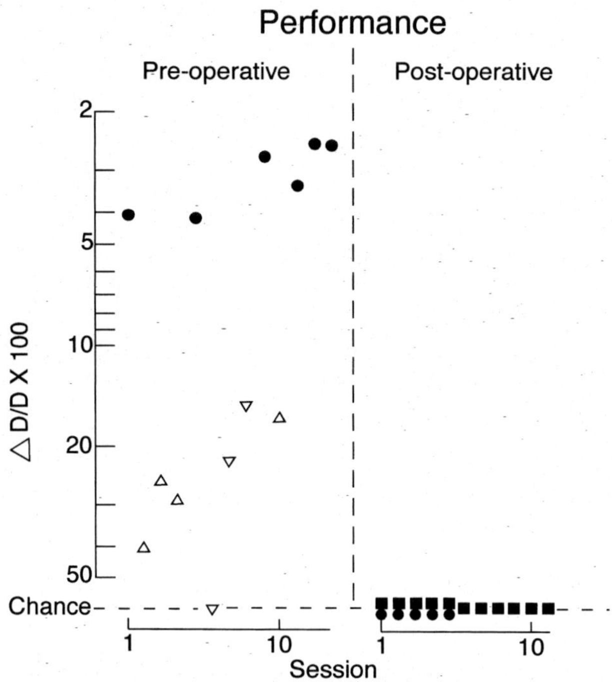

FIGURE 16-7. Binocular depth thresholds for one cat (filled circles) before and after ablation of areas 17 and 18. The monocular performance of this animal before the lesion is indicated by the open triangles. The postoperative data for the second animal of this study are indicated by squares. See text for details. (Data redrawn from Kaye et al. [65].)

visual acuity, as resolution acuity is reduced by no more than a factor of 2 or 3 (to approximately 3 cyc/deg). Even the smallest circular dots on the stimuli subtended angular dimensions at the usual viewing distances that were at least five times larger than the resolution limit. Moreover, the animals that received cortical lesions in the second to fourth week of life achieved virtually normal visual acuities but nevertheless showed no evidence of superior performance on depth judgments when tested binocularly.

The conclusion that ablation of cortical areas 17 and 18 eliminates stereopsis receives even stronger support from an elegant study (77) that evaluated the abilities of animals to perceive patterns generated by retinal disparity in random-dot stereograms. The tests were conducted on four animals both before and after lesions of areas 17 and 18 that were intended to include only those regions on which the central visual fields were represented. Reconstruction of the lesions on the basis of degeneration of the dLGN revealed a pattern of atrophy consistent with the intended ablation with a variable amount of sparing of the peripheral (upper) visual fields that was somewhat less than that intended. Before the lesion, the cats were trained in a conventional two-choice discrimination box on the series of discriminations that are shown in Fig. 16-8. The animal could make its choice at any point from its release at the start gate 57 cm from the stimuli. After training on a simple luminance discrimination, they were tested on a series of three discriminations between vertical and horizontal rectangles of exactly the same dimensions for all three tests. The three tests were designed as a progression from shape defined by luminance alone (Fig. 16-8A, black figures against a white background) to shape defined solely by retinal disparity where the stimuli were random-dot stereograms and the rectangles were defined by retinal disparity (Fig. 16-8C). The retinal disparity between the two stereopairs was 20 minutes of arc if the choice was made at the greatest distance possible (57 cm) but correspondingly

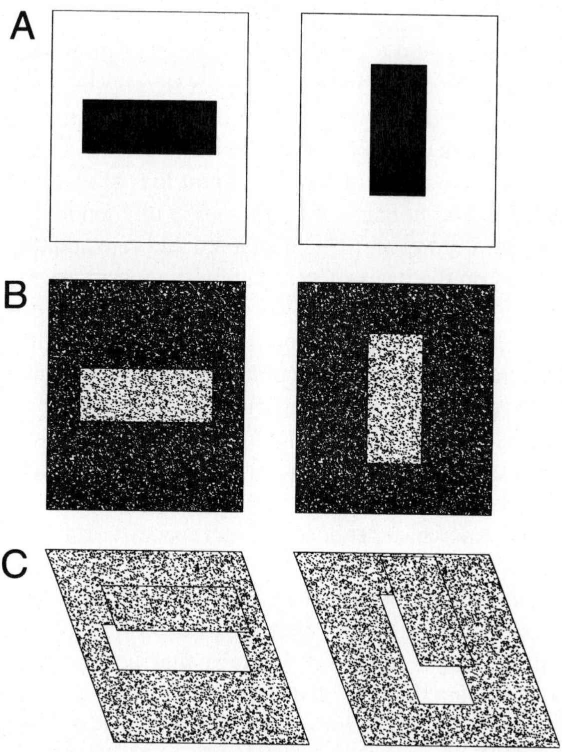

FIGURE 16-8. The sequence of three-form discrimination tasks used by Ptito et al. (77) **(A)** The stimuli (rectangles) are defined by luminance. **(B)** (Pseudostereoscopic task) The rectangles, although still defined by luminance, were drawn on a background of random dots identical to those used for the true stereoscopic task **(C)** where the rectangles are defined exclusively by retinal disparity. To a human, the rectangles appear to lie in a different depth plane than the background. (Redrawn from Ptito et al. [77].)

larger than that if the choice was made closer to the stimuli. Between the tests with stimuli defined by luminance or by disparity was a (pseudostereoscopic) test in which the rectangles were drawn on a background of random dots identical in size and density to those in the true stereoscopic discrimination, and with the cats wearing the same contact lenses and filters that were used in the true stereoscopic task (Fig. 16-8B). Before the lesion of area 17/18, all four cats learned the luminance and the first shape discrimination task very quickly. But as a possible consequence of having to wear the contact lenses and filters, they required five to eight times as many sessions (each of 40 trials) to attain a level of 90% correct with the pseudostereoscopic shape discrimination test. Even after learning this task, the cats took a similarly long time to attain the same criterion level of performance on the true stereoscopic task. After the lesions, the animals quickly relearned the luminance discrimination but began all the other tasks at chance levels. However, although they relearned the shape and pseudostereoscopic tasks quite rapidly compared with their preoperative performance, their performance on the stereoscopic task remained at chance levels over 50 sessions (1,000 trials) at which time testing was ceased as there was no sign of improvement.

On the basis of both direct and indirect tests for stereopsis, it would appear that lesions of areas 17 and 18 destroy the cat's ability to utilize retinal disparity cues for either depth judgments or to extract form information. Alternate explanations for the loss of stereopsis in terms of the concurrent loss of visual acuity were discussed and rejected in the context of studies (65) of monocular and binocular depth discriminations. Ptito et al. (77) considered and rejected the same explanation for their findings on the basis of vernier acuity measurements made on their animals that indicated that they could discriminate offsets smaller than those used in their stereograms. Moreover, they argued on the basis of this finding, as well as data of others, that the lesioned animals could discriminate the random-dots of the size they used. Although it is possible to refute explanations for the apparent loss of stereopsis in terms of loss of acuity, it could still be argued that the results follow from the loss of contrast sensitivity after lesions of area 17 alone or combination with area 18. The loss of contrast sensitivity may be small (Fig. 16-2), but, strictly speaking, the known dependence of stereoacuity in humans on stimulus contrast (114,115) requires that in the future an additional control experiment be conducted that examines the performance of normal cats on stereoscopic depth judgments as a function of stimulus contrast to explore the somewhat unlikely possibility that the contrast sensitivity deficits that accompany lesions of the visual cortex could explain the loss of stereoscopic vision.

Cells sensitive to retinal disparity or interocular phase differences have been reported in cat cortical areas other than areas 17 and 18. Disparity-sensitive cells have been reported in area 19 (98–101), but according to one group (98,99), the proportion of such cells is substantially smaller and their tuning broader than those observed in areas 17 and 18, which provides the major anatomical input to area 19. The latter area also receives substantial afferents from such subcortical regions as the C laminae of the dLGN, the lateral pulvinar complex, the medial interlaminar nucleus, and the geniculate wing (67). Collectively, the input to area 19 from these subcortical regions would be characterized as Y- and W- type (67). Moreover, recordings from cells in the LP complex reveal that they have large receptive fields and poor spatial selectivity (116,117), characteristics that are consistent with the coarse disparity selectivity of cells within area 19. Cells selective for retinal disparity and/or that show significant binocular phase interactions are also observed in area 21a, a visual cortical area situated on the posterior part of the middle suprasylvian gyrus (102–104). Like area 19, area 21a receives its major input from areas 17 and 18 and also from the lateral posterior nucleus, but unlike area 19, area 21a contains a much higher proportion of disparity-selective cells. Approximately two thirds of cells in area 21a are disparity selective and/or exhibit profound modulation to change in the relative phase of identical sinusoidal gratings presented to the two eyes (103,104). There is evidence (102) that the latter selectivity is a result of interactions occurring within area 21a itself rather than being a passive reflection of prior processing in area 17. That lesions of areas 17 and 18 abolished the ability of cats to detect shapes defined by retinal disparity suggests that the surviving subcortical inputs to areas 19 and 21a that follow elimination of the cortical inputs to these areas are unable, by themselves, to extract meaningful disparity information.

Effects of Lesions at Different Ages

Early studies of depth perception on the visual cliff revealed that cats as adults exhibited profound impairment in their ability to chose the shallow side of a visual cliff irrespective of the age at which cortical areas 17 and 18 had been removed (44,45,106,113). By contrast, a more recent and systematic study (54) on a larger groups of animals lesioned at different ages found "sizeable and statistically reliable differences" ([54] p. 638) in the proportion of choices for the shallow side between animals lesioned in adulthood as opposed to lesions incurred on days P3–6, even when comparisons were restricted to animals in each group with less than 10% sparing of layers A and A1 of the dLGN. Approximately two thirds of the cats with such complete neonatal lesions chose the shallow side of the visual cliff significantly greater than chance, whereas only one third of the cats with comparable lesions in adulthood achieved the same performance.

In contrast to tests of visual preference, depth thresholds measured on a jumping stand in my laboratory do not show evidence of superior performance in cats lesioned in early postnatal life as opposed to adulthood, suggesting that stereoscopic vision cannot develop following neonatal lesions. On the other hand, data obtained on the visual cliff suggest that such animals may be more proficient in the use of monocular depth cues than are animals lesioned as adults.

Motion Perception

It has been known for many years that neurons in the visual cortex, as well as other cortical visual areas, are selective for both the direction and speed of motion of stimuli moved through their receptive fields. In turn the various classes of retinal ganglion cells (the X-, Y-, and W-cells) that provide the inputs to these various cortical areas have different velocity preferences (67) (see Chapter 1). A small number of studies have been conducted on motion perception in the cat to compare behavioral performance with the properties of motion selective neurons in various cortical areas in this species (118,119). Studies of the consequences for motion perception provide another avenue by which it may be possible to ascertain the role of these selective neuronal responses in motion perception. However, until comparatively recently, there were few studies of the effects of lesions of the major cortical visual areas on the perception of motion. Kennedy (120,121) measured the minimum velocity of rotary motion that permitted cats to distinguish a moving from a stationary cross both before and after lesions that appeared to include areas 17,18, and most of area 19. One cat could not perform the task at all after the lesion, and the thresholds for the other two cats with complete lesions were elevated by a factor of between 5 and 20. The importance of areas 17/18/19 for detection of motion was reinforced by the observations (62) of the substantial effects of combined lesions of these areas on signal-to-noise thresholds for moving patterns. Without question, however, the most insightful work on the role of motion selective neurons in the visual cortex on motion perception is provided by the multifaceted approach of Pasternak's group to this issue. In addition to study of motion perception in normal cats (119,122), she has investigated the alterations of motion thresholds following cortical lesions as well as those that follow a form of selective visual deprivation, namely strobe rearing, that virtually eliminates directionally selective neurons in area 17 (see page 686).

The most comprehensive studies to date of the effects of cortical lesions for motion perception were the comparisons (22,23) of the various motion thresholds described next between the intact and lesioned hemifield of two cats with a lesion within either area 17 or area 18.

Comparison of the Detection Thresholds for Drifting versus Counterphase Modulated Gratings

Both cats (122) and normal humans (123) typically show higher contrast sensitivity for detecting drifting gratings than counterphase modulated gratings by approximately a factor two. Systematic tests conducted on just one cat with a lesion in area 17 revealed an almost identical pattern in both the lesioned and intact hemifield for a grating of low spatial frequency (23). Similar tests performed on two cats with a lesion within area 18 revealed a similar reduction of contrast sensitivity in the lesioned visual field for both drifting and counterphase modulated gratings so that the ratio of the two thresholds was preserved (22). Thus, lesions of neither area 17 nor area 18 appear to disrupt the ratio of the sensitivities to drifting and counterphase modulated gratings.

Direction Discrimination Task

Contrast thresholds were measured for the ability to discriminate between rightward and leftward motion of drifting gratings of three spatial frequencies as a function of temporal frequency. The cat with a lesion within area 17 showed no sensitivity loss at the lowest spatial frequency over a broad range of temporal frequencies within the lesioned hemifield (23). At higher spatial frequencies a substantial contrast sensitivity loss was evident but only at the lowest temporal frequency. This result implicates area 17 in the discrimination of direction of motion of stimuli of higher spatial frequency and moving at low speeds. By striking contrast, the two cats with a lesion in area 18 were completely unable to discriminate the direction of motion of stimuli even at the highest contrast within the lesioned hemifield (22). This was true despite repeated attempts over a broad range of spatial and temporal frequencies. Despite the known presence of directionally selective neurons in many cortical areas outside area 18, including area 17 and the lateral suprasylvian areas, apparently the directional signals from these neurons are not used by cats following a lesion of area 18. It is possible that directionally selective cells in areas other than area 18 may nonetheless depend on input from area 18 and thus lose their selectivity once area 18 is lesioned or inactivated.

Motion Perception with Random Dot Stimuli

Paradoxically, a lesion within area 17 appears to improve performance with such stimuli irrespective of the nature of the task. Thus performance in the lesioned field was superior by a factor of 2 in terms of the range of directions that can be integrated into a global motion percept and was superior in terms of coherence thresholds (the proportion of motion signal-to-random motion "noise" dots). Finally, the maximal spatial displacement of coherently moving dots that permitted detection of the direction of motion, D_{max}, was larger in the lesioned field. However, when the dots in the stimuli were spatially filtered to remove the higher spatial frequency content of the stimuli, the performance in the intact hemifield improved to the extent that the superior performance on all three measures vanished in the lesioned visual field. This result is consistent with other indications (123) that the inferior performance in the intact hemifield with the unfiltered dots was due to masking of the dominant low frequency content of the broadband dot stimuli by the high frequencies.

Lesions of the lateral suprasylvian area in the cat have been shown (124) to induce moderate deficits in the perception of the direction of motion of random dot stimuli, but over time substantial recovery can occur. That direction discrimination is still possible suggests that integration of motion information inherent in such tasks may occur at earlier stages of the visual pathway. Further, the observation that lesions of area 17 have no effect on motion integration rules out the possibility that neurons in area 17 are necessary for motion integration of random dot stimuli. However, the possible contribution of area 18 to this function has yet to be evaluated. On the other hand, the somewhat complementary nature of the data obtained from tests conducted with drifting gratings on the cats with lesions of area 17 or 18 are consistent with the view that directionally selective neurons in the latter area are necessary for processing coarse targets moving at relatively high speeds, whereas such neurons in the former area may be critical for the perception of motion of slow-moving gratings or other stimuli that contain a predominance of high spatial frequencies.

INSIGHTS GAINED FROM CATS REARED WITH SELECTED FORMS OF EARLY VISUAL DEPRIVATION

Manipulations of the early visual input of kittens can produce long-lasting if not permanent alterations of the visual response characteristics of neurons in area 17 and 18, as well as changes in visual perception (125–129). To the extent that the changes in neuronal response properties in response to selected visual deprivation are documented in these particular cortical areas, they can in principle provide information on the functional role of these cortical areas or of particular classes of cells. The effects of different forms of early selective deprivation vary widely in their effects on area 17 and possibly in terms of the number of cortical areas that are affected. At one extreme is monocular deprivation by eyelid suture, which can result in the nondeprived eye dominating most cells in extragranular layers within area 17 to the extent that the nondeprived eye becomes ineffective in its ability to excite cortical cells. However, some residual ability to modulate the response of cells can be revealed by their binocular response to interocular differences in phase between gratings drifted across corresponding regions in the two eyes (130). The traditional index of the severity of the effects of monocular deprivation is an extremely skewed distribution of ocular dominance among cortical cells. The behavioral consequences of prolonged early monocular deprivation are extreme; initially the animal appears blind when forced to use its deprived eye but some recovery occurs, the rate and extent of which are determined by factors such as the length of the deprivation, the age at which it was introduced, and the nature of the vision to the two eyes during the recovery period (i.e., whether both eyes are open or whether the formerly nondeprived eye is covered) (131). With periods of monocular deprivation that begin near birth and extend to 3 months of age, the vision recovered in the deprived eye is much worse than that which follows combined ablation of areas 17 and 18. Whereas the latter reduces grating acuity by only about an octave and has only a minor effect on the contrast sensitivity for gratings of low spatial frequency, the effects of monocular deprivation result in a much greater reduction of grating acuity and an equally profound impairment of contrast sensitivity for gratings of all spatial frequencies (52 Fig. 13; 131 Fig. 4). The clear implication of the vast difference in the consequences of these two manipulations is that the effects of monocular deprivation extend beyond areas 17 and 18 and must include the area(s) that mediate vision after ablation of areas 17 and 18.

To the extent that the effects of monocular deprivation are widespread and are not restricted to a particular cell class, they do not provide much insight into the roles of area 17 and/or 18 for vision. The ideal requirement for such an approach is a form of early selected visual deprivation that could be considered a "silver bullet" or "environmental knock-out" that eliminates one particular class of cell, and especially one restricted to a particular cortical area, so that the visual consequences of its absence can be documented with a view to ascertaining both the functional role of such neurons and, within a larger context, the role of the particular visual area in which the affected neurons are located. Although it remains to be seen whether such a restricted cell-class *and* area-specific "lesion" does in fact exist, there are indications that certain forms of early selected deprivation exert very specific effects that may be restricted to a particular class of cell. The most informative type of deprivation in this respect is to rear kittens in an environment that is illuminated stroboscopically by a brief (3 μsec) light flash at a rate of 8 Hz. To a human such intermittent illumination destroys the perception of smooth motion and, in strobe-reared cats, virtually eliminates directionally selective neurons in a number of visual cortical areas including area 17 (132), area 18 (133), and the lateral suprasylvian area (134). The consequences for vision of this form of deprivation have been studied in detail by Pasternak's group (135–137) and were found to be quite selective; whereas certain spatial and temporal thresholds of strobe-reared animals were normal or only slightly elevated, substantial deficits were evident in the perception of motion. Thus, while the contrast sensitivity for detection of drifting gratings was normal at low spatial frequencies and slightly elevated at higher spatial frequencies, the contrast thresholds for detecting the direction of motion of drifting gratings were elevated by a factor of 10 (135,136). However, despite a dramatic loss in the proportion of directionally specific cells in area 17, the deprived cats had no trouble discriminating the direction of motion of stimuli of high suprathreshold contrast from which it must be concluded that the small remaining proportion of directionally selective neurons can support normal directional discrimination if the targets are of high contrast (136). Profound deficits were also apparent in the ability of such deprived cats to perceive global motion in random dot displays (137,138).

CONCLUSIONS

As measurements of various spatial thresholds make abundantly clear, lesions of area 17 and 18 can produce selective deficits with only small changes on some measures such as grating acuity but profound impairment on others. Examples of the latter include a complete loss of stereoscopic vision and profound deficits of vernier acuity and in the ability to make orientation discriminations. Lesions designed to ascertain the separate roles of area 17 and 18 suggest that functions such as the processing of orientation, motion, and spatiotemporal thresholds are best described as being shared or distributed between the two areas with some specialization for stimuli with certain dimensions or temporal properties. Although it is easy to identify differences between the visual response characteristics of cells in these two areas, it is important to recognize that there are many similarities between them, such as in receptive field organization (139; see also Chapter 1). It is the similarities and not the differences between the two areas that appear to fit best with the behavioral findings from lesion studies. Finally, a more complete understanding of the role of areas 17 and 18 for vision requires better understanding of the functional organization of the two areas. For example, the recent discovery (140) of cytochrome blobs in area 17 of cats points to our lack of complete understanding of the functional architecture of area 17 (see Chapter 1).

ACKNOWLEDGMENTS

The preparation of this chapter and the author's studies on the effects of cortical lesions were supported by a grant (A7660) from the Natural Sciences and Engineering Research Council of Canada.

REFERENCES

1. Zeki, S. (1990). A century of cerebral achromatopsia. *Brain* **113,** 1721–1777.
2. Salzman, C. D., Murasagi, C. M., Britten, K. H., and Newsome, W. T. (1992). Microstimulation in visual area MT: effects on direction discrimination performance. *J. Neurosci.* **12,** 2331–2355.
3. DeAngelis, G. C., Cumming, B. G., and Newsome, W. T. (1998). Cortical area MT and the perception of stereoscopic depth. *Nature* **394,** 677–680.
4. Parker, A. J., and Newsome, W. T. (1998). Sense and the single neuron: probing the physiology of perception. *Annu. Rev. Neurosci.* **21,** 227–277.
5. Albright, T. D. (1991). Motion perception and the mind-body problem. *Curr. Biol.* **1,** 391–393.
6. Otsuka, R., and Hassler, R. (1962). Über Aufbau und Gliederung der corticalen Sehsphäre bei der Katze. *Arch. Psychiat. Z. Ges. Neurol.* **203,** 212–234.
7. Webster, W. G., and Webster, I. H. (1975). Anatomical asymmetry of the cerebral hemispheres of the cat brain. *Physiol. Behav.* **14,** 867–869.
8. Webster, W. G. (1981). Morphological asymmetries of the cat brain. *Brain Behav. Evol.* **18,** 72–79.
9. Sanderson, K. J. (1971). The projection of the visual field to the lateral geniculate and medial interlaminar nuclei in the cat. *J. Comp. Neurol.* **143,** 101–118.
10. Tusa, R. J., and Palmer, L. A. (1980). Retinotopic organization of areas 20 and 21 in the cat. *J. Comp. Neurol.* **193,** 147–164.
11. Tusa, R. J., Palmer, L. A., and Rosenquist, A. C. (1978). The retinotopic organization of area 17 (striate cortex) in the cat. *J. Comp. Neurol.* **177,** 213–236.
12. Tusa, R. J., Rosenquist, A. C., and Palmer, L. A. (1979). Retinotopic organization of areas 18 and 19 in the cat. *J. Comp. Neurol.* **185,** 657–678.
13. Tusa, R. J., Palmer, L. A., and Rosenquist, A. C. (1981). Multiple cortical visual areas. Visual field topography in the cat. In: *Cortical sensory organization.* Vol. 2. Multiple visual areas (C. N. Woolsey, Ed.), pp. 1–31. Clifton NJ, Humana Press.
14. Vandenbussche, E., Sprague, J. M., De Weerd, P., and Orban, G. A. (1991). Orientation discrimination in the cat: its cortical locus. I. Areas 17 and 18. *J. Comp. Neurol.* **305,** 632–658.
15. De Weerd, P., Sprague, J. M., Raiguel, S., Vandenbussche, E., and Orban, G. A. (1993). Effects of visual cortex lesions on orientation discrimination of illusory contours in the cat. *Eur. J. Neurosci.* **5,** 1695–1710.
16. De Weerd, P. D., Sprague, J. M., Vandenbussche, E., and Orban, G. A. (1994). Two stages in visual texture segregation: A lesion study in the cat. *J. Neurosci.* **14,** 929–948.
17. Sprague, J. M., De Weerd, P., Xiao, D., Vandenbussche, E., and Orban, G. A. (1996). Orientation discrimination in the cat: Its cortical locus II. Extrastriate cortical areas. *J. Comp. Neurol.* **364,** 32–50.
18. Van Essen, D. C., and Maunsell, J. H. R. (1980). Two-dimensional maps of the cerebral cortex. *J. Comp. Neurol.* **191,** 255–281.

19. Merigan, W. H., Nealey, T. A., and Maunsell, J. H. R. (1993). Visual effects of lesions of cortical area V2 in macaques. *J. Neurosci.* **13,** 3180–3191.
20. Merigan, W. H. (1996). Basic visual capacities and shape discrimination after lesions of extrastriate area V4 in macaques. *Vis. Neurosci.* **13,** 51–60.
21. Newsome, W. T., and Wurtz, R. H. (1988). Probing visual cortical function with discrete chemical lesions. *Trends Neurosci.* **11,** 394–400.
22. Pasternak, T., and Maunsell, J. H. R. (1992). Spatiotemporal sensitivity following lesions of area 18 in the cat. *J. Neurosci.* **12,** 4521–4529.
23. Pasternak, T., Tompkins, J., and Olson, C. R. (1995). The role of striate cortex in visual function of the cat. *J. Neurosci.* **15,** 1940–1950.
24. Dalby, D. A., Meyer, D. R., and Meyer, P. M. (1970). Effects of occipital neocortical lesions upon visual discrimination in the cat. *Physiol. Behav.* **5,** 727–734.
25. Ritchie, G. D., Meyer, P. M., and Meyer, D. R. (1976). Residual spatial vision of cats with lesions of the visual cortex. *Exp. Neurol.* **53,** 227–253.
26. Wetzel, A. B. (1969). Visual cortical lesions in the cat: a study of depth and pattern discrimination. *J. Comp. Physiol. Psychol.* **68,** 580–588.
27. Smith, K. U. (1938). Visual discrimination in the cat. VI. The relation between pattern vision and visual acuity and the optic projection centers in the nervous system. *J. Genet. Psychol.* **53,** 251–272.
28. Chow, K. L. (1968). Visual discriminations after extensive ablation of optic tract and visual cortex in cats. *Brain Res.* **9,** 363–366.
29. Spear, P. D., and Braun, J. J. (1969). Pattern discrimination following removal of visual neocortex in the cat. *Exp. Neurol.* **25,** 331–348.
30. Winans, S. S. (1971). Visual cues used by normal and visual-decorticate cats to discriminate figures of equal luminous flux. *J. Comp. Physiol. Psychol.* **74,** 167–178.
31. Cornwell, P., Warren, J. M., and Nonneman, A. J. (1976). Marginal and extramarginal cortical lesions and visual discrimination by cats. *J. Comp. Physiol. Psychol.* **90,** 986–995.
32. Cornwell, P., Overman, W., and Campbell, A. (1980). Subtotal lesions of the visual cortex impair discrimination of hidden figures by cats. *J. Comp. Physiol. Psychol.* **94,** 289–304.
33. Murphy, E. H., Mize, R. R., and Schechter, P. B. (1975). Visual discrimination following infant and adult ablation of cortical areas 17, 18, and 19 in the cat. *Exp. Neurol.* **49,** 386–405.
34. Sprague, J. M., Berlucchi, G., and Antonini, A. (1985). Immediate postoperative retention of visual discrimination following selective cortical lesions in the cat. *Behav. Brain Res.* **17,** 145–162.
35. Berkley, M. A., and Sprague, J. M. (1979). Striate cortex and visual acuity functions in the cat. *J. Comp. Neurol.* **187,** 679–702.
36. Hughes, H. C., and Sprague, J. M. (1986). Cortical mechanisms for local and global analysis of visual space in the cat. *Exp. Brain. Res.* **61,** 332–354.
37. Krüger, K., Heitländer-Fransa, H., Dinse, H., and Berlucchi, G. (1986). Detection performance of normal cats and those lacking areas 17 and 18: a behavioral approach to analyse pattern recognition deficits. *Exp. Brain. Res.* **63,** 233–247.
38. Stiles, W. S. (1959). Colour vision: the approach through increment threshold sensitivity. *Proc. Natl. Acad. Sci. U.S.A.* **75,** 100–114.
39. Pasternak, T., and Horn, K. (1991). Spatial vision of the cat: variation with eccentricity. *Vis. Neurosci.* **6,** 151–158.
40. Berkley, M. A., and Sprague, J. M. (1982). The role of the geniculocortical system in spatial vision. In: Analysis of visual behavior (J. D. Ingle, M. A. Goodale, and R. J. W. Mansfield, Eds.), pp. 525–547. Cambridge MA, MIT Press.
41. Payne, B. R., and Cornwell, P. (1994). System-wide repercussions of damage to the immature visual cortex. *Trends Neurosci.* **17,** 126–130.
42. Payne, B. R. (1999). Immature visual cortex lesions; global rewiring, neural adaptations and behavioral sparing. In: *The Changing nervous system: Neurobehavioral consequences of early*

brain disorders (S. Broman, and J. Fletcher, Eds.), pp. 114–148. New York, Oxford University Press.

43. Cornwell, P., Herbien, C., Corso, C., Eskew, R., Warren, J. M., and Payne, B. R. (1989). Selective sparing after lesions of visual cortex in newborn kittens. *Behav. Neurosci.* **103,** 1176–1190.
44. Cornwell, P., and Overman, W. (1981). Behavioral effects of early rearing conditions and neonatal lesions of the visual cortex in kittens. *J. Comp. Physiol. Psychol.* **95,** 848–862.
45. Cornwell, P., Overman, W., and Ross, C. (1978). Extent of recovery from neonatal damage to the cortical visual system in cats. *J. Comp. Physiol. Psychol.* **92,** 255–270.
46. Mitchell, D. E. (1990). Sensitive periods in visual development: insights gained from studies of recovery of visual function in cats following early monocular deprivation or cortical lesions. In: *Vision: Coding and efficiency* (C. Blakemore, Ed.), pp. 234–246. Cambridge, Cambridge University Press.
47. Glassman, R. B. (1978). The logic of the lesion experiment and its role in the neural sciences. In: *Recovery from brain damage* (S. Finger, Ed.), pp. 3–31. New York, Plenum Press.
48. Regan, D., Giaschi, D., Sharpe, J. A., and Hong, X. H. (1992). Visual processing of motion-defined form: selective failure in patients with parietotemporal lesions. *J. Neurosci.* **12,** 2198–2210.
49. Schilder, P., Pasik, P., and Pasik, T. (1972). Extrageniculostriate vision in the monkey. III. Circle vs. triangle and "red vs. green" discrimination. *Exp. Brain Res.* **14,** 436–448.
50. Dineen, J., and Keating, E. G. (1981). The primate visual system after bilateral removal of striate cortex. *Exp. Brain Res.* **41,** 338–345.
51. Keating, E. G., and Dineen, J. (1982). Visuomotor transforms of the primate tectum. In: *Analysis of visual behavior* (D. J. Ingle, M. A. Goodale, and R. J. W. Mansfield, Eds.), pp. 335–365. Cambridge MA, MIT Press.
52. Lehmkuhle, S., Kratz, K. E., and Sherman S. M. (1982). Spatial and temporal sensitivity of normal and amblyopic cats. *J. Neurophysiol.* **48,** 372–387.
53. Sprague, J. M., Levy, J., Di Berardino, A., and Berlucchi, G. (1977). Visual cortical areas mediating form discrimination in the cat. *J. Comp. Neurol.* **172,** 441–448.
54. Shupert, C., Cornwell, P., and Payne, B. (1993). Differential sparing of depth perception, orienting, and optokinetic nystagmus after neonatal versus adult lesions of cortical areas 17, 18, and 19 in the cat. *Behav. Neurosci.* **4,** 633–650.
55. Doty, R. W. (1971). Survival of pattern vision after removal of striate cortex in the adult cat. *J. Comp. Neurol.* **143,** 341–370.
56. Alder, S. W., and Meikle, T. H. (1975). Visual discrimination of flux-equated figures by cats with brain lesions. *Brain Res.* **90,** 23–43.
57. Doty, R. W. (1973). Ablation of visual areas in the central nervous system. In: *Handbook of sensory physiology* (R. Jung, Ed.), Vol. VII/3B, pp. 483–542. Berlin, Springer.
58. Baumann, T. P., and Spear, P. D. (1977). Evidence for recovery of spatial pattern vision by cats with visual cortex damage. *Exp. Neurol.* **57,** 603–612.
59. Cornwell, P., Overman, W., and Ross, C. (1980). Lesions of visual cortex impair discrimination of hidden figures by cats. *Physiol. Behav.* **24,** 533–540.
60. Spear, P. D., and Baumann, T. P. (1979). Neurophysiological mechanisms of recovery from visual cortex damage in cats: properties of lateral suprasylvian visual area neurons following behavioral recovery. *Exp. Brain Res.* **35,** 177–192.
61. Berlucchi, G., and Sprague, J. M. (1981). The cerebtal cortex in visual learning and memory, and in interhemispheric transfer in the cat. In: *The organization of the cerebral cortex* (F. O. Schmitt, F. G. Worden, G. Adelman, and S. G. Dennis, Eds.), pp. 415–440. Cambridge MA, MIT Press.
62. Krüger, K., Donicht, M., Muller-Kusdian, G., Kiefer, W., and Berlucchi, G. (1988). Lesion of areas 17/18/19: effects of the cat's performance in a binary detection task. *Exp. Brain Res.* **72,** 510–516.
63. Hoffmann, K. P., and Von Seelen, W. (1984). Performance in the cat's visual system: a behavioral and neurophysiological analysis. *Behav. Brain Res.* **12,** 101–120.
64. Oyster, C. W. (1999). *The human eye: Structure and function.* pp. 82–92. Sunderland MA, Sinauer.
65. Kaye, M., Mitchell, D. E., and Cynader, M. (1981). Selective loss of binocular depth perception after ablation of cat visual cortex. *Nature* **293,** 60–62.

66. Miller, M., Pasik, P., and Pasik, T. (1980). Extrageniculate vision in the monkey. VII. Contrast sensitivity functions. *J. Neurophysiol.* **43,** 1510–1526.
67. Stone, J. (1983). Parallel processing in the visual system. *The classification of retinal ganglion cells and its impact on the neurobiology of vision.* New York, Plenum.
68. Ferster, D. (1990). X- and Y- mediated synaptic potentials in neurons of areas 17 and 18 of cat visual cortex. *Vis. Neurosci.* **4,** 115–133.
69. Ferster, D. (1990). X- and Y- mediated current souces in areas 17 and 18 of cat visual cortex. *Vis. Neurosci.* **4,** 135–145.
70. Mitchell, D. E. (1989). Normal and abnormal visual development in kittens: insights into the mechanisms that underlie visual perceptual development in humans. *Can. J. Psychol.* **43,** 141–164.
71. Hall, S. E., and Mitchell, D. E. (1991). Grating acuity of cats measured with detection and discrimination tasks. *Behav. Brain Res.* **44,** 1–9.
72. Pearson, H. E., Labar, D. R., Payne, B. R., Cornwall, P., and Aggarwell, N. (1981). Transneuronal retrograde degeneration in the cat retina following neonatal ablation of visual cortex. *Brain Res.* **212,** 470–475.
73. Hughes, A. (1977). The topography of vision in mammals of contrasting life style: Comparative optics and retinal organization. In: *Handbook of sensory physiology.* Vol. VII/5. *The visual system in vertebrates.* (F. Crescitelli, Ed.), pp. 615–756. New York, Springer.
74. Wilson, H. R., Levi, D., Maffei, L., Rovamo, J., and DeValois, R. (1990). The perception of form. Retina to striate cortex. In: *Visual perception. The neurophysiological foundations* (L. Spillmann, and J. S. Werner, Eds.), pp. 231–272. San Diego, Academic Press.
75. Westheimer, G. (1979). The spatial sense of the eye. *Invest. Ophthalmol. Vis. Sci.* **18,** 893–912.
76. Curcio, C. A., Sloan, K. R., Packer, O., Hendrickson, A. E., and Kalina, R. (1987). Distribution of cones in human and monkey retina: individual variability and retinal asymmetry. *Science* **236,** 579–582.
77. Ptito, M., Lepore, F., and Guillemot, J. P. (1992). Loss of stereopsis following lesions of cortical areas 17–18 in the cat. *Exp. Brain Res.* **89,** 521–530.
78. Hubel, D. H., and Wiesel, T. N. (1962). Receptive fields, binocular interaction and functional architecture in the cat's visual cortex. *J. Physiol. (London)* **160,** 106–154.
79. Howard, I. P., and Rogers, B. J. (1995). *Binocular vision and stereopsis.* New York, Oxford University Press.
80. Orban, G. A., Vandenbussche, E., and Vogels, R. (1984). Meridional variations and other properties suggesting that acuity and orientation discrimination rely on different neuronal mechanisms *Ophthalmol. Physiol. Opt.* **4,** 89–93.
81. Bradley, A., Skottun, B. C., Ohzawa, I., Sclar, G., and Freeman, R. D. (1987). Visual orientation and spatial frequency discrimination: a comparison of single neurons and behavior. *J. Neurophysiol.* **57,** 755–772.
82. Scottun, B. C., Bradley, A., Sclar, G., Ohzawa, I., and Freeman, R. D. (1987). The effects of contrast on visual orientation and spatial frequency discrimination: a comparison of single cells and behavior. *J. Neurophysiol.* **57,** 773–786.
83. De Weerd, P., Vandenbussche, E., and Orban, G. A. (1990). Bar orientation discrimination in the cat. *Vis. Neurosci.* **4,** 257–268.
84. Orban, G. A., Vandenbussche, E., Sprague, J. M., and De Weerd, P. (1990). Orientation discrimination in the cat: a distributed function. *Proc. Natl. Acad. Sci. U.S.A.* **87,** 1134–1138.
85. Sprague, J. M., De Weerd, P., Vandenbussche, E., and Orban, G. A. (1993). Orientation discrimination in the cat and its cortical loci. In: *Progress in brain research* (T. P. Hicks, S. Molotchnikoff, and T. Ono, Eds.), Vol. 95, pp. 381–400. The Netherlands, Elsevier.
86. Hammond, P., and Andrews, D. P. (1978). Orientation tuning of cells in area 17 and 18 of the cat's visual cortex. *Exp. Brain Res.* **31,** 341–351.
87. Hughes, A. (1976). A supplement to the cat schematic eye. *Vis. Res.* **16,** 149–154.
88. Bishop, P. O. (1973). Neurophysiology of binocular single vision and stereopsis. In: *Handbook of sensory physiology* (R. Jung, Ed.), Vol. VII/3A, pp. 255–305. Berlin, Springer-Verlag.

89. Walls, G. L. (1942). *The vertebrate eye and its adaptive radiation.* New York, Hafner.
90. Fox, R., and Blake, R. R. (1971). Stereoscopic vision in the cat. *Nature* **233,** 55–56.
91. Packwood, J., and Gordon, B. (1975). Stereopsis in normal domestic cat, siamese cat and cat raised with alternating monocular occlusion. *J. Neurophysiol.* **38,** 1485–1499.
92. Mitchell, D. E., Kaye, M., and Timney, B. (1979). Assessment of depth perception in cats. *Perception* **8,** 389–396.
93. Julesz, B. (1971). *Foundations of cyclopean perception.* Chicago, University of Chicago Press.
94. Barlow, H. B., Blakemore, C., and Pettigrew, J. D. (1967). The neural mechanisms of binocular depth discrimination. *J. Physiol. (Lond.)* **193,** 327–342.
95. Nikara, T., Bishop, P. O., and Pettigrew, J. D. (1968). Analysis of retinal correspondence by studying receptive fields of binocular single units in cat striate cortex. *Exp. Brain Res.* **6,** 353–372.
96. Cynader, M., and Regan, D. (1978). Neurons in cat parastriate cortex sensitive to the direction of motion in three-dimensional space. *J. Physiol. (London)* **274,** 549–569.
97. Ferster, D. (1981). A comparison of binocular depth mechanisms in areas 17 and 18 of the cat visual cortex. *J. Physiol. (London)* **311,** 623–655.
98. Guillemot, J. P., Paradis, M. D., Samson, A., Ptito, M., Richer, L., and Lepore, F. (1993). Binocular interaction and disparity coding in area 19 of visual cortex in normal and split-chiasm cats. *Exp. Brain Res.* **94,** 405–417.
99. Guillemot, J. P., Richer, L., Ptito, M., and Lepore, F. (1993). Disparity coding in the cat: a comparison between areas 17–18 and area 19. In: *Progress in brain research* (T. P. Hicks, S. Molotchnikoff, and T. Ono, Eds.), Vol. 95, pp. 179–187. The Netherlands, Elsevier.
100. Bacon, B. A., Lepore, F., and Guillemot, J. P. (1998). Striate, extrastriate and collicular processing of spatial disparity cues. *Arch. Physiol. Biochem.* **106,** 236–244.
101. Pettigrew, J. D., and Dreher, B. (1987). Parallel processing of binocular disparity in the cat's retinogeniculate pathways. *Proc. R. Soc. Lond. Ser. B* **232,** 297–321.
102. Morley, J. W., and Vickery, R. M. (1999). Binocular interactions in area 21a of the cat. *Neuroreport* **10,** 2241–2244.
103. Vickery, R. M., and Morley, J. W. (1999). Binocular phase interactions in area 21a of the cat. *J. Physiol. (London)* **514,** 541–549.
104. Wang, C., and Dreher, B. (1996). Binocular interactions and disparity coding in area 21a of cat extrastriate visual cortex. *Exp. Brain Res.* **108,** 257–272.
105. Walk, R. D., and Gibson, E. J. A. (1961). A comparative and analytical study of visual depth perception. *Psychol. Monogr.* **75,** Whole No. 519.
106. Cornwell, P., Overman, W., Levitsky, C., Shipley, J., and Lezynski, B. (1976). Performance on the visual cliff by cats with marginal gyrus lesions. *J. Comp. Physiol. Psychol.* **90,** 996–1010.
107. Meyer, P. M. (1963). Analysis of visual behavior in cats with extensive neocortical ablations, *J. Comp. Physiol. Psychol.* **56,** 397–401.
108. Meyer, P. M., Anderson, R. A., and Braun, M. G. (1966). Visual cliff preferences following lesions of the visual cortex in cats and rats. *Psychon. Sci.* **4,** 269–270.
109. Cornwell, P., Warren, J. M., and Nonneman, A. J. (1976). Marginal and extramarginal cortical lesions and visual discrimination by cats. *J. Comp. Physiol. Psychol.* **90,** 986–995.
110. Howard, H. J. (1919). A test for the judgement of distance. *Am. J. Ophthalmol.* **2,** 656–675.
111. Mitchell, D. E., Giffin, F., Wilkinson, F., Anderson, P., and Smith, M. L. (1976). Visual resolution in young kittens. *Vis. Res.* **16,** 363–366.
112. Mitchell, D. E., Giffin, F., and Timney, B. (1977). A behavioral technique for the rapid assessment of the visual capabilities of kittens. *Perception* **6,** 181–193.
113. Hovda, D. A., Sutton, R. L., and Feeney, D. M. (1989). Amphetamine-induced recovery of visual cliff performance after bilateral visual cortex ablation in cats: measurements of depth perception thresholds. *Behav. Neurosci.* **103,** 574–584.
114. Halpern, L., and Blake, R. (1988). How contrast affects stereoacuity. *Perception* **17,** 483–495.

115. Westheimer, G., and Pettet, M. W. (1990). Contrast and exposure duration differentially effect vernier and stereoscopic acuity. *Proc. R. Soc. Lond. Ser. B* **241,** 42–46.
116. Casanova, C., Freeman, R. D., and Nordmann, J. D. (1989). Monocular and binocular response properties of cells in the striate-recipient zone of the cat's lateral posterior-pulvinar compex. *J. Neurophysiol.* **62,** 544–557.
117. Chalupa, L. M., and Abramson, B. P. (1989). Visual receptive fields in the striate-recipient zone of the lateral posterior-pulvinar complex. *J. Neurophysiol.* **62,** 347–357.
118. Berkley, M. A., Warmath, D. S., and Tunkl, J. E. (1978). Movement discrimination capacities in the cat. *J. Comp. Physiol. Psychol.* **92,** 463–473.
119. Pasternak, T., and Merigan, W. H. (1980). Movement detection by cats: invariance with direction and target configuration. *J. Comp. Physiol. Psychol.* **94,** 943–952.
120. Kennedy, J. L., and Smith, K. U. (1935). Visual thresholds of real movement in the cat. *J. Genet. Psychol.* **46,** 470–476.
121. Kennedy, J. L. (1939). The effects of complete and partial occipital lobectomy upon thresholds of visual real movement discrimination in the cat. *J. Genet. Psychol.* **54,** 119–149.
122. Pasternak, T. (1986). The role of cortical direction selectivity in detection of motion and flicker. *Vis. Res.* **26,** 1187–1194.
123. Cleary, R., and Braddick, O. J. (1990). Direction discrimination for band-pass filtered random dot kinematograms. *Vis. Res.* **30,** 303–316.
124. Rudolph, K. K., and Pasternak, T. (1996). Lesions in cat lateral suprasylvian cortex affect the perception of complex motion. *Cerebral Cortex* **6,** 814–822.
125. Mitchell, D. E., and Timney, B. (1984). Postnatal development of function in the mammalian visual system. In: *Handbook of physiology, section I: The nervous system,* Vol. 3, part 1. *Sensory processes* (I. Darian-Smith, Ed.), pp. 507–555. Bethesda, American Physiological Society.
126. Movshon, J. A., and Kiorpes, L. (1990). The role of experience in visual development. In: *Development of sensory systems in mammals* (J. R. Coleman, Ed.), pp. 155–202. New York, Wiley.
127. Kiorpes, L., and Movshon, J. A. (1990). Behavioral analysis of visual development. In: *Development of sensory systems in mammals* (J. R. Coleman, Ed.), pp. 125–154. New York, Wiley.
128. Rauschecker, J. P. (1991). Mechanisms of visual plasticity: Hebb synapses, NMDA receptors, and beyond. *Physiol. Rev.* **71,** 587–613.
129. Daw, N. W. (1995). *Visual development.* New York, Plenum.
130. Freeman, R. D., and Ohzawa, I. (1988). Monocularly deprived cats: binocular tests of cortical cells reveal functional connections from the deprived eye. *J. Neurosci.* **8,** 2491–2506.
131. Mitchell, D. E. (1988). The extent of visual recovery from early monocular or binocular visual deprivation in kittens. *J. Physiol. (London)* **395,** 639–660.
132. Cynader, M. S., and Chernenko, G. (1976). Abolition of directional selectivity in the visual cortex of the cat. *Science* **193,** 504–505.
133. Kennedy, H., and Orban, G. A. (1983). Response properties of visual cortical neurons in cats reared in stroboscopic illumination. *J. Neurophysiol.* **49,** 686–704.
134. Spear, P. D., Tong, L., McCall, M. A., and Pasternak, T. (1985). Developmentally induced loss of direction selective neurons in cat lateral suprasylvian visual cortex. *Dev. Brain Res.* **20,** 281–285.
135. Pasternak, T., Schumer, R. A., Gizzi, M. S., and Movshon, J. A. (1985). Abolition of visual cortical direction selectivity affects visual behavior in cats. *Exp. Brain Res.* **61,** 214–217.
136. Pasternak, T., and Leinen, L. J. (1986). Pattern and motion vision in cats with selective loss of cortical direction selectivity. *J. Neurosci.* **6,** 938–945.
137. Pasternak, T., Albano, J. E., and Harvitt, D. M. (1990). The role of directionally selective neurons in the perception of global motion. *J. Neurosci.* **10,** 3079–3086.
138. Rudolph, K. K., Ferrera, V. P., and Pasternak, T. (1994). A reduction in the number of directionally selective neurons extends the spatial limit for global motion perception. *Vis. Res.* **34,** 3241–3251.

139. Tretter, F., Cynader, M., and Singer, W. (1975). Cat parastriate cortex: a primary or secondary visual area?. *J. Neurophysiol.* **38,** 1099–1113.

140. Murphy, K. M., Jones, D. G., and van Sluyters, R. C. (1995). Cytochome-oxidase blobs in cat primary visual cortex. *J. Neurosci.* **15,** 4196–4208.

INDEX

Edwards Brothers Malloy
Thorofare, NJ USA
December 29, 2015